THE PHYSICS OF SEMICONDUCTORS

18th International Conference on

THE PHYSICS OF SEMICONDUCTORS

VOLUME 1

Stockholm, Sweden
August 11-15, 1986

EDITOR: **OLOF ENGSTRÖM**

Chalmers, Göteborg

World Scientific

Published by

World Scientific Publishing Co Pte Ltd.
P. O. Box 128, Farrer Road, Singapore 9128

18TH INTERNATIONAL CONFERENCE ON THE PHYSICS OF SEMICONDUCTORS

ISBN: 9971-50-197-X

Printed in Singapore by Kim Hup Lee Printing Co. Pte. Ltd.

PREFACE

The 18th International Conference on the Physics of Semiconductors was held in Stockholm from August 11 to 15 1986. More than 420 papers were presented to 850 delegates from 42 different countries. The Proceedings of the Conference, contained in these two volumes, include all but a few of the contributions.

The Conference was sponsored by the International Union of Pure and Applied Physics as well as other professional and industrial organizations, which are listed on the following pages.

Editing this influential scientific material is an exciting experience, not just because it represents a focus in an intense beam of current results, but also because it is a concrete reminder of the large variety of physical phenomena which belong to the field of semiconductors. Sorting the material in a book is a matter of minimizing the entropy of an ensemble of titles and abstracts to give the reader a text with surveyable order. To use strictly logical principles in this process, would probably give a distribution of titles which would appear to be irrelevant. Instead, the titles of the chapters and their subdivisions have been determined by dividing the material into groups of related papers with different partition principles for different areas, reflecting the pattern of Semiconductor Physics 1986. As an exception the plenary presentations have been placed together at the beginning of Volume I because of their general character and interest. I hope that the reader will find this method convenient.

I am grateful to the authors for their work and cooperation in creating these Proceedings and to the members of the Organizing, Program and Advisory Committees, whose names are found on the following pages. Also the Sponsoring Organizations and all other people involved are acknowledged for their contributions to making a successful Conference.

November 1986   Olof Engström

ICPS SPONSORS

Professional Organizing Sponsors

Swedish Ministry of Industry
Swedish Ministry of Education and Cultural Affairs
Swedish Board for Technical Development
Swedish Natural Science Research Council
International Union for Pure and Applied Physics
USARDSG
EOARD
ONR (USA)

Contributing Sponsors

ASEA
Ericsson
IBM
Volvo

Organizing Committee

Chairman	Hermann G. Grimmeiss, Lund University
Secretary	Lars Samuelson, Lund University
Treasurer	Christer Åsberg, Ericsson, Stockholm
Program	Bo Monemar, Linköping University
Proceedings	Olof Engström, Chalmers, Gothenburg
Satellites	Karl-Fredrik Berggren, Linköping University
Arrangements	Günter Grossmann, Lund University
	Pär Omling, Lund University

Program Committee

B. Monemar, Sweden *Chairman*

S. Alferov, USSR

A.Baldereschi, Switzerland

M.Balkanski, France

I. Balslev, Denmark

H.G.Grimmeiss, Sweden

T. Masumi, Japan

G.O.Müller, GDR

J.Mycielski, Poland

V.Narayanamurti, USA

S.T.Pantelides*, USA

H.J.Queisser, FRG

L. Samuelson, Sweden

F.W.Saris, The Netherlands

L.J.Sham, USA

W.E.Spear, UK

Y.Toyozawa, Japan

Xie Xide, China

*Chairman of sub-committee for "Science Underlying
Semiconductor Technology"

Advisory Committee

Opening Session

H.G. Grimmeiss

Department of Solid State Physics, University of Lund

Box 118, S-221 00 Lund, Sweden

Your Royal Highness, Your Excellencies, Ladies and Gentlemen :

On behalf of the organizing committee I am glad to welcome you to the 18th International Conference on the Physics of Semiconductors in Stockholm. It is a great honour for me to welcome in particular His Royal Highness, Prince Bertil, Duke of Halland, and the Swedish Minister of Education and Cultural Affairs, Lennard Bodström. During the 36 years of its existence, this is the first time the ICPS is being held in Scandinavia. It is also as near the Artic Circle as the Conference is ever likely to be, but I think you'll see that summer is quite beautiful up here in Stockholm - provided the weather does not do something unexpected, of course.

A conference like the ICPS can only be organized by the dedicated work of many people. Let me therefore express my gratitude to all of you who made this conference possible. In particular I would like to thank the organizing committee for its efficient and inventive work. Among other things Lars Samuelson suggested the logo and supervised the tedious work of printing matters. The program which I hope all of you will enjoy during the conference has been puzzled out by the international program committee chaired by Bo Monemar. We appreciate the valuable suggestions made by the advisory committees and all the help we obtained from the IUPAP Secretary General and Semiconductor Commission.

We are also grateful to all our sponsors. In particular I would like to thank the Swedish Board for Technical Developement through Mr. S. I. Ragnarsson, Ericsson through Mr. G. Lindberg and Asea through Dr. B. Kredell. Without their quick and unbureaucratic support the ICPS 1986 could never have been organized in Sweden. Last but not least we would like to thank all the participants of this conference for their support and confidence in our efforts .

The importance and significance of semiconductor physics has been commented on at previous conferences. I have nothing to add. Although technical applications of semiconductors have already been discussed at previous conferences we tried to further emphasize this aspect by particularly calling your attention to papers dealing with Science underlying Semiconductor Technology. We feel obliged to Sokrates Pantelides for his efforts in this regard. Nobody, I suppose, questions the role semiconductor physics played and still plays in the development of microelectronics and information technology. Most of the industrial and consumer products of the future will contain electronics. This is why the electronic industry holds a key position among the high-technology industries. It is not unreasonable to say that the competitiveness of many industrial products will, in the first instance, be determined by the competitiveness of the electronics contained in the product. This is one of the reasons for the increasing interest of governmental agencies in micro-electronics. Semiconductor physics is the basis of modern electronics. No invention or innovation has ever owed more to pure, abstract science. Governments are therefore well advised when promoting basic research with a free hand to the scientists in the area of semiconductor physics rather than supporting applied research with only narrowly defined immediate objectives relevant only to specific national industries. As Marvin Cohen said at our last meeting, there is no reason why we should be preoccupied with the possible death of semiconductor physics.

I want to thank IUPAP's Semiconductor Commission for giving us the honour of hosting the ICPS. This gave us the opportunity to introduce to you one of the most beautiful cities in Europe. I hope you will find enough time to visit the archipelago with its 24.000 islands, the Vasa ship, the old city with the Royal Castle and many other exciting places. You will probably get some insight into Swedish history and Swedish life if you take part in the Conference excursion on Wednesday and the Conference banquet on Thursday or the reception in the Town Hall this evening. The reception has been kindly arranged by the City of Stockholm, which is most appreciated.

May I add here that concurrently with the ICPS, Stockholm is also the venue of the International Security Conference. It is worth bringing to the attention of our guests that ICPS over the years has, apart from its scientific contribution, also acted as a harbringer of peace and fraternity by bringing together scientific communities from East and West, North and South, regardless of cast or creed. Let us therefore hope that this city of peace helps us in a small way to achieve that end through this conference.

As Minko Balkanski said four years ago in Montpellier, some innovations are expected at all conferences. As you have already noticed, we tried to adorn the opening session of this conference with a few pieces of Swedish Baroque Music performed by the Scandinavian Baroque Orchestra (and not by the Nordic Baroque Orchestra as announced in the Conference program) under the conductorship of Sören Hansen.

I hope you will enjoy your stay in Stockholm and I wish all of you great success with the scientific program.

May I now ask Your Royal Highness to open the conference.

Hermann G. Grimmeiss
Chairman

OPENING ADDRESS

H.R.H. Prince Bertil

Conference delegates, ladies and gentlemen,

It is my great pleasure to be here today to wish you all very welcome to the 18th International Conference on the Physics of Semiconductors.

For a small country like Sweden it is an honour to be given the opportunity to arrange an international conference as prestigious as this one. I think it is most gratifying to see that our country is able to play an important role in this advanced and highly competitive field of semiconductor physics out of which has grown the whole modern electronics industry.

It is my sincere hope that this conference will be fruitful and that you will remember with pleasure your days here in Stockholm - not only that the scientific sessions will be successful but also that you will have a chance to experience the charm and traditions of our beautiful capital.

With these words I declare the 18th International Conference on the Physics of Semiconductors open and wish you all the best of luck.

OPENING ADDRESS

Minister of Education, Lennart Bodström

It is a pleasure and an honour for us that the 1986 International conference on the Physics of Semiconductors of the International Union of Pure and Applied Physics is being held here in Sweden, and to have you, the participants, as our guests. We are especially happy because this is the first time in the thirty-six years of its existence that the ICPS meeting has come to Scandinavia.

As representatives of the world-wide scientific community you are most cordially welcome to Stockholm. The physics of Semiconductors, however, also has a special significance, since it is the basis of semi-conductor technology, on which the whole field of microelectronics - and hence modern information technology - depends. We all know how important information technology is for the industrial nations, and Sweden not least is certainly becoming increasingly dependent on it. Earlier, our prosperity was mainly based on such natural resources as iron ore and forests, but nowadays we have to look more to various fields of high technology - and here information technology is begin-ning to occupy a more and more central place. These technological de-velopments, however, are not just something we are compelled to take part in - they also fit in with the economic and technical structure of our country.

Sweden has a great tradition both in technology and in scientific research. Once a year it is a special focal point for the interest of scientists, when the prizes which we owe to the vision of the Swedish inventor and friend of science, Alfred Nobel, are awarded. The fact that last year's prize for physics went to semiconductor physics is a sign of the scientific stature of this branch of physics.

A small country like Sweden is inevitably more dependent on international contacts and international co-operation than other, larger nations. Therefore we particularly hope that this conference will help to foster new contacts between scientists in different countries, and naturally not least for those of the country in which it is being held. This is obviously something which our Government is glad to encourage.

Ladies and gentlemen,
It is estimated that Sweden accounts for only 1 - 2 % of the world's total resources used in research development. As a small country with a limited number of researchers, it is not possible for us to conduct research in all fields of importance. Research findings from other countries make up a large proportion of the total stock of knowledge in Sweden.

Sweden participates in international scientific co-operation in a variety of ways. At the government level, it is primarily the Ministry of Education that is responsible for the overall international involvement in research co-operation, but other Ministries are becoming involved to an increasing extent. There is now a growing need for co-operation, both in financial and organisational terms. The Swedish Government and Parliament regard international scientific co-operation as a very positive development. But their rule must be to define broad areas of support and to decide on the balance between different fields of research from the national perspective. It must be left to the researchers to decide on the more detailed definition of priorities and also on which projects within a certain area ought to be funded.

For some time we have been well aware of the national importance of microelectronics and recently we initiated a national microelectronic programme, backed by about a hundred million dollars over a five-year period. This programme is designed to foster research and deveopment in the field of microelectronics. Another way in which we endeavour to encourage development in this field is through higher education in computer science and microelectronics.

Even a layman like myself is aware of the fantastic advances that have been made in semiconductor science and technology in recent decades. They began with the first transistor - which was made of germanium, I understand - and went on to silicon and all the astonishing things that can be done with a tiny chip of this material - more and more complicated, and smaller and smaller all the time. Now there are these new materials which are made by putting atoms together in ways that nature never invented - I believe you physicists talk about "super-lattices", "quantum wells", and so on. And there is laser light coming from tiny semiconductor devices to carry messages many kilometres through tiny fibres - soon even across the oceans, they say. Yes - even for a layman these are fantastic advances, and it is obviously a matter of great public interest where they will lead in the future.

The Swedish Government is well aware that the technology of tomorrow will not just grow automatically, like a plant, out of the technology of today. General progress in technology and technological innovations are of course very important, but new things will also come from a deeper, scientific level of understanding and inventiveness. These things will emerge from basic scientific research, and what a govern-ment can do to help them on their way is to encourage basic research in this field of semiconductors. Our Government intends to provide such encouragement. Of course in this field, as in others, we are still a small country. But ideas come from individuals, and where individual achievements and the climate that encourages them are con-cerned, a small country need to always be so small.

And so, to conclude. On behalf of the Swedish Government I would like to welcome you all to Sweden and wish you both a pleasant stay here on a very successful conference. We hope that the conference will further the cause of international collaboration in your very important field of science.

ICPS OPENING ADDRESS

Hiroshi Kamimura

Department of Physics, Faculty of Science
University of Tokyo
Bunkyo-ku, Tokyo, Japan

Your Royal Highness, Mr. Chairman, Ladies and Gentlemen.

On behalf of the Semiconductor Commission, I am very happy to welcome you to Stockholm and to the 18th International Conference on the Physics of Semiconductors.

This conference has a long tradition. It moves from country to country every two years. As you know, this is the 18th in the series which started in 1950 in Reading in U.K. This conference is the first semiconductor physics conference that has been held in a Scandinavian country. From the marvelous arrangements, both scientific and social, we can already congratulate our Swedish host. Attendance here is approximately 850 delegates and 50 accompanying persons. I would like to express our satisfaction that so large an audience has come together here, and I extend our sincere gratitude to Professor Hermann Grimmeiss and all the members of the organizing committee for the wonderfully prepared conference.

Nowadays the field of semiconductor physics is prosperous. This conference is 36 years old since its birth in 1950, and this series has played an important role in the development of semiconductor physics and technology. The age of 36 is the age of maturity in human being. However, we are not mature, but still young as I mention below. According to the suggestion by Professor Marvin Cohen in his opening address of the 1984 Semiconductor Conference, first I would like to look back briefly to what was said in the early conferences. A good example of this kind was given by Professor Yasutada Uemura in his closing address in the 1980 Semiconductor Conference in Kyoto. He divided the activity in semiconductor physics into three decades: the dawn of modern semiconductor physics opened by the discovery of transistor action; extension and sophistications; and then transformations. These decades roughly correspond to the 1950s, 60s and 70s. How about in the 1980s? In the 1980s the semiconductor physics has been growing enormously, not only by the interaction between theory and experiment but also by the interactions among semiconductor physics, materials science and technology. For example, by these interactions the technology related to making artificial materials not found in nature such as superlattices has been playing an important role in the development of semiconductor physics. Thanks to this technology a number of new exotic phenomena such as normal and fractional quantum Hall effects have been discovered, new physical concepts have been created, and new devices have been invented. The birth and rapid growth of new physics and new phenomena and the invention of new devices and new experimental techniques such as scanning tunneling microscope have given tremendous impact not only to other fields of physics but also to other areas of science and engineering, and the outlook for the future of the field of semiconductor physics is promising. In this view I would like to propose the heading "interactions and development" to characterize the present decade.

As an evidence that my optimistic view is not too subjective, I would like to recall that the Nobel Physics Prize has been awarded twice in the 1970s, and one last year to the achievements in the field of semiconductors: Leo Esaki in 1973, Philip Anderson and Sir Nevill Mott in 1977, and in 1985 Klaus von Klitzing for the discovery of the Quantum Hall effect.

Well, I don´t want to keep you for very long. I just want to say "enjoy Stockholm", "enjoy this conference" as an occasion to exchange physical ideas and results, to meet old friends and to make new friendship. I am sure that this Conference will be as exciting as any of previous conferences. Good luck to the Conference.

Thank you very much.

OPENING ADDRESS

K. Siegbahn

Department of Physics, Uppsala University

Box 256, S-751 05 Uppsala, Sweden

Your Royal Highness, Mr. Minister, Mr. President of the Conference, Mr. Chairman, Ladies and Gentlemen:

As we all know, the physics of semiconductors has been one of the dominating fields in scientific research for a long time, and will no doubt continue to be so for an unforseeable time to come. What is particularly exciting with this field is its intimate contact and influence on daily human life. Rarely has a discovery in science had such a rapid and far-reaching effect on technology and therefore on society as the invention of the transistor by Shockley, Bardeen and Brattain at the Bell laboratories in 1948. This event marks a milestone not only in the field of basic solid state physics but also in communication technology with all the consequences in the further developments of radio, television, space technology, not to speak of the modern computerized society. As a matter of fact there are hardly any installations in factories, hospitals, offices, banks or households which do not contain transistors and other semiconductor devices as basic components.

The invention of the transistor is interesting from another point of view. Unlike many other practical inventions it was entirely a result of an interplay between very fundamental basic science and advanced technology, cleverly applied to electronic industrial demands.

One comes to think of the old philosophical paradoxical problem: Which
came first, the hen or the egg? Are discoveries in science nowadays
dependent and even consequences of technological improvements and
tools or the reverse? The right answer to this question is of course:
You can't tell, because they are both aspects of the same thing,
similar to the complementary principle in quantum physics. This situation has not been so obvious before. Galilei could make his fundamental discoveries in physics without access to technology. Newton was
of course dependent on previous experimental observations but could
nevertheless proceed with his fundamental scientific work without
technology. Einstein did not have much use of his background as a
patent engineer when he treated relativity or when he invented the
photon. The consequence of the latter achievement had to wait for half
a century until it could be explored in the field of surface science
and technology in the form of photo-electron spectroscopy. Equally
loose was the connection in time between his treatment of stimulated
photon-emission and its practical realisation in the form of the maser
and laser.

The connection between progress in science and in technology
becomes, however, more and more pronounced the nearer we approach our
own time. The time elapse between a discovery or an achievement in
basic research and its industrial exploration is getting correspondingly shorter. It took 400 years to explore the anticipation of the
airplane by Leonardo da Vinci, to take an extreme example (he did not
provide it with an engine!). The full technological consequences of
Maxwell's and others' research on the electro-magnetic field more than
a hundred years ago took also a long time to explore. Probably a unique
exception to this general rule from previous times is provided by
Röntgen's discovery of the X rays in 1895 which almost instantaneously
lead to manufacturing and general utilization in hospitals.

Even today there is a certain time needed between a scientific
discovery or a new conceptional idea and its actual exploration. The
importance of this difficult phase in the development should by no
means be underestimated, but the general trend is still that this time

period is decreasing due to the much intensified research and development activities nowadays performed in industrial laboratories. This is
of course particularly true in the field of semiconductors. The im-
pressing list of contributions to this conference from such laboratories bears strong witness of the important role many industrial
research laboratories play in fields which previously would have been
classified as basic science. In fact the borderlines between basic
university research and frontier technological research performed in
industry are indeed becoming more and more difficult to recognize. In
the near future decision makers concerned with research policy applied
to universities have to seriously consider this new situation. Already
we find new types of laboratories growing up around and in close contacts with university institutions. This is a valuable ingredient in
modern science. There are some specific problems associated with these
types of scientific activities, with bearing upon the freedom of
science and scientists and which should be continuously subject to
attention. The situation is similar for many other sciences with close
connections to applications. One of the main actual problems concerns
the balance between industry and university in regard to the recruit-
ing of qualified scientists. It is the duty of the university to edu-
cate scientists to be employed by industry and by other bodies in
society and also to be employed by the university itself so that it
can fulfill its mission to do university research and teaching at a
scientific level. It is therefore necessary and important to create
such opportunities also at the universities that the scientific
research there can expand and develop to the benefit of our society as
a whole, including industrial and other activities.

To achieve this and to stimulate the cooperation between scientists of various origins, international conferences like the present
one play a very important role. It is particularly gratifying to
notice that this conference on semiconductor physics here in Stockholm
has attracted so many leading scientists. Some of our participants
will report on new basic achievements and recent further developments
of previous discoveries. Other speakers are bringing us up to new
frontiers by the use of new techniques and methods. We are looking

forward to listening to you and to discussing with you during the
next few days.

Let me finally wish you all good luck in your further scientific
work either in the field of basic science, technology or in the
equally rewarding symbiotic field between the two.

CONTENTS

Volume I

Contents

Contents

I. PLENARY TALKS

SEMICONDUCTOR HETEROSTRUCTURES AND SUPERLATTICES AT THE FRONTIER OF SOLID STATE ELECTRONICS AND PHOTONICS

V. Narayanamurti
AT&T Bell Laboratories
Murray Hill, New Jersey 07974 USA

ABSTRACT

The ability to artificially structure new materials on an atomic scale, using advanced crystal growth methods such as Molecular Beam Epitaxy (MBE) and Metal-Organic Chemical Vapor Deposition (MOCVD), has led recently to the observation of unexpected new physical phenomena as well as the creation of entirely new classes of devices based on band-gap and wave function engineering. Recent progress in this area, in both the lattice matched $GaAs/Al_xGa_{1-x}As$ system as well as in strained-layer systems, such as Ge_xSi_{1-x} on Si, will be reviewed. Our ability to create new high speed transistors, optically bistable switches, modulators, low threshold lasers, and novel photodetectors, at the frontier of solid state electronics and photonics, will be shown to be limited only by the laws of quantum mechanics.

Introduction

Ever since the invention of the transistor we have witnessed a spectacular growth in silicon technology, leading to increasingly higher densities of devices and more complex functions. Almost as revolutionary as the invention of the transistor in 1947 was the invention of the laser a decade later. Thus, nearly concurrent with the electronics revolution, we have seen another technological revolution, the so-called photonics revolution: using beams of laser light for information transmission. The lasers that provide light for today's lightwave communication systems are made not from silicon but from compound semiconductors such as GaAs and InP.

The first continuously operating semiconductor lasers were made in 1970. They were grown by liquid phase epitaxy and were a direct consequence of the ability to build heterostructures of different but lattice matched semiconductors which enabled one to confine both electrons and light in an efficient manner. Subsequent advances in new crystal growth techniques[1,2] such as molecular beam epitaxy (MBE) and metal-organic chemical vapor deposition (MOCVD) has allowed the construction of virtually arbitrary, atomic scale, potential profiles for electrons and holes, including abrupt band discontinuities used to confine carriers in a two-dimensional (2-D) state. Because of their applications to novel and potentially useful devices, the physics of thin film semiconductor heterostructures has become[3] a subject of intense study throughout the world. In this article I will review some of the important advances in this area over the last few years and will try to illustrate the unique synergy between basic science, new materials and new technology.

High Carrier Mobility Structures: Besides the obvious technological importance of compound semiconductor lightwave devices, there has long been an interest in such materials for their electronic properties because of the higher electron mobilities in GaAs when compared with silicon. The principal building block of Si IC's is the MOSFET which is basically the inversion layer of 2-D electrons (or holes) at the interface between crystalline Si and its native oxide SiO_2. In the case of GaAs, the inversion layer can be formed at the interface of a $GaAs/Al_xGa_{1-x}As$ heterostructure. Until 1978, carrier

4 V. Narayanamurti

mobilities in such structures were extremely low. It was then that Stormer, Dingle, Gossard and Wiegmann introduced modulation doping[4] in which the band edge discontinuity in the heterostructure is utilized to separate the electrons (holes) from their parent donors (acceptors). This has led to record carrier mobilities for both electrons and holes in work done at many laboratories. See Figure 1.

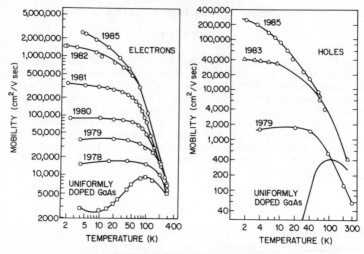

Figure 1 Highest reported (a) electron and (b) hole mobilities during the period 1978 to 1985 in modulation doped GaAs. Compiled by A. C. Gossard.

A key feature of modulation doping is the greatly enhanced conductivity in the plane of the layered structure as compared to that perpendicular to the layers. This led to the fabrication of modulation doped field effect transistors (MODFET's) at many industrial laboratories and universities. They are characterized by high switching speeds and low power consumption. At 77 K where the modulation doping advantage is already significant, a switching time as short as 5.8 ps has recently been observed[5]. Such transistors will probably form the basis of advanced high speed GaAs digital circuits. Already in the laboratory a 4 kbit static RAM and a 4x4 multiplier have been fabricated. The recent demonstration of high mobility holes has led to the first fabrication of low power, complementary transistor circuits in the GaAs system.

Perpendicular Transport and Tunneling Devices: In 1970 Esaki and Tsu[6] first proposed that perpendicular transport in semiconductor superlattices might show negative differential resistance (NDR) associated with electron transfer into the negative mass region of the minizone and Bloch oscillations. Even though they first observed resonant tunneling in 1974, it is only in the last two years that the full potential of such structures is being realized. This is in large part due to remarkable progress in engineering new materials by MBE. The physics of NDR for a Double Barrier Quantum Well Structure is illustrated in Fig. 2. Assuming conservation of lateral momentum during tunneling, only those emitter electrons whose momenta lie on a disk $k_z = k_o$ are resonant. At high bias voltages, V, resonant electrons no longer exist which results in a sharp drop in current. Resonant tunneling with very large (3:1) peak to valley ratios and microwave oscillations to frequencies up to 18 GHz has recently been observed by several groups[7].

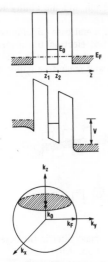

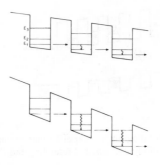

Figure 2   Schematic of the energy band diagram of a double barrier resonant tunneling diode in equilibrium and with an applied bias V. The bottom illustrates the Fermi surface for a degenerately doped emitter. After Luryi. Ref. 7.

Figure 3   Schematic of resonant tunneling of electrons through a superlattice for potential drops corresponding to the energy difference of the first excited state (top) and the second excited state (bottom). See Ref. 8.

Sequential Resonant Tunneling: In the case of an ideal semiconductor superlattice consisting of a large number of equally spaced identical quantum wells, one can expect a resonant transmission if the applied field is such that the potential difference, acquired by an electron over many periods of the superlattice is less than the width of the lowest miniband as first predicted by Kazarinov and Suris. See Figure 3. The effect has been observed recently in some beautiful experiments by Capasso et al[8] who placed the superlattice in the i region of a reverse biased p+-i-n+ junction and observed peaks in the photocurrent characteristics at their expected positions.

Effective Mass Filters: The large difference in effective masses (and hence tunneling rates) of electrons and holes has led Capasso et al[8] to develop a new class of quantum mechanical photoconductors. See Figure 4 which shows the band diagram of a superlattice photoconductor with applied bias. Photogenerated (heavy) holes remain relatively localized whilst photoelectrons are transported through the superlattice by phonon assisted tunneling or miniband conduction. This effective mass filtering effect produces a photoconductive gain given by the ratio of the electron lifetime to the electron transit time. The gain and the gain bandwidth product of these novel photoconductors can be tuned over a very wide range by varying the superlattice period and/or duty factor, a unique feature not available in conventional photoconductors.

"Ballistic" Transistors and Hot-Electron Spectroscopy: As the dimensions of semiconductor devices shrink, hot electron and mean free path effects become increasingly important. In the early 1960's, considerable effort was directed toward the development of metal base and "ballistic" electron transistors which in principle would have a very high speed. With the improvement in crystal perfection and the ability to tailor atomic scale structures by MBE, there has been renewed recent interest in such structures. There has been considerable activity[7] in the design of metal base (silicide on Si) transistors, permeable base transistors, planar doped barrier hot electron transistors and tunneling hot electron transistor amplifiers. Greatest progress has been made in basic experimental and theoretical understanding of hot electron effects and mean free paths in semiconductors such as GaAs and Si.

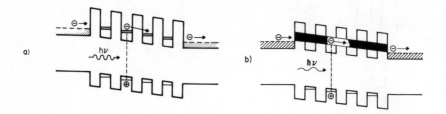

Figure 4   Band diagram with applied bias for superlattice effective mass filter in the case (a) of phonon assisted tunneling and (b) miniband conduction. After Ref. 8.

Hayes and Levi[9] have used a planar doped barrier structure (see Figure 5) for the study of hot electron transport across thin GaAs layers. Here one barrier serves as an injector of monoenergetic hot electrons and a second barrier is used as an energy analyzer for electrons traversing the layers. The sharply peaked "ballistic" part corresponding to the injector energy was found to disappear at thicknesses greater than about 1000Å and suggests mean free paths of ≈300 to 400Å. Detailed analysis of various scattering mechanisms suggests that a more favorable material for the fabrication of ballistic electron transistors might be InAs.

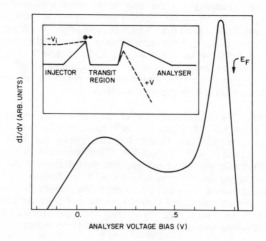

Figure 5   Inset shows structure for hot electron spectroscopy. The curve corresponds to the hot electron spectrum for a GaAs sample with a 650Å transit region. Peak near 0.1V results from nearly ballistic electrons. After Ref. 9.

Lateral Quantum Confinement: MBE up to now has been primarily used to provide quantum confined structures in the growth direction. Confinement in one and zero dimensions (quantum wires and dots) requires fabrication of laterally confined structures with a dimensional scale≈100Å . Using enhancement of interdiffusion in selected (by electron beam defined masks) areas of multilayer structures by ion bombardment, Cibert et al[10] have recently seen the effects of carrier confinement to one and zero degrees of freedom in quantum well wires and boxes fabricated in the GaAs-GaAlAs system. See Figure 6. Low temperature high resolution cathodoluminescence measurements show new luminescence lines attributed to transitions arising from ground and excited levels of electrons within these low dimensional structures. Exploitation of such structures, their electric field and transport properties is likely to lead to entirely new classes of devices and new physics.

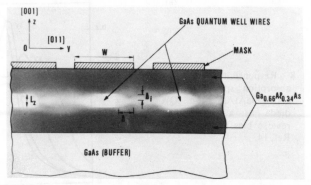

Figure 6   Schematic (not the scale) of GaAs quantum well wire structure after implantation and annealing. The mask position during implantation is indicated. The wires are actually 1μm apart. After Ref. 10.

Superlattice Avalanche Photodetectors (APD's): An excellent example of the modification of bulk properties through band gap engineering is the superlattice (APD). In bulk GaAs the ionization coefficients α for electrons and β for holes is almost equal. Capasso et al[11] first showed experimentally several years ago that it is possible to alter α and β dramatically over the values found in nature in a superlattice APD consisting of alternating layers of AlGaAs and GaAs.

Recently Capasso et al[12] have designed superlattice structures where hot carriers in the barrier layers can collide with carriers confined or dynamically stored in the wells and impact ionize them out, across the band edge discontinuity. See Figure 7. They have observed large ratios of multiplication factors for holes and electrons implying $\beta > \alpha$ in the case of superlattices of Al.48In.52As/Ga.47In.53As as well as of AlSb/GaSb. Both these materials systems are of interest for long wavelength lightwave applications.

Electro-Optic Effects in Quantum Wells: The 2-D confinement of excitons in quantum wells has been the subject of intense interest for many years. Recent perfection[13] of MBE GaAs/AlGaAs structures with low background doping has led to the observation of large electric field effects in the absorption spectrum at room temperature[14]. See Figure 8. The Quantum Confined Stark Effect has led to the fabrication of several novel devices, including high speed modulators and optically bistable devices with potential for application in many different areas of optical signal processing.

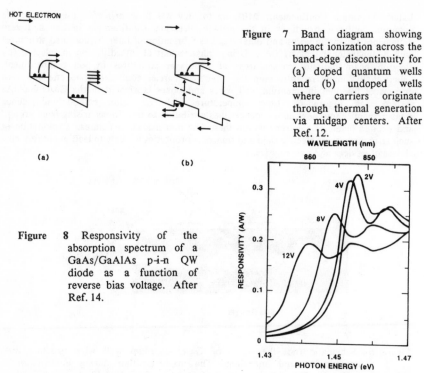

HOT ELECTRON

(a)

(b)

Figure 7 Band diagram showing impact ionization across the band-edge discontinuity for (a) doped quantum wells and (b) undoped wells where carriers originate through thermal generation via midgap centers. After Ref. 12.

Figure 8 Responsivity of the absorption spectrum of a GaAs/GaAlAs p-i-n QW diode as a function of reverse bias voltage. After Ref. 14.

Gas Source MBE and Band-Gap Engineering in GaInAsP: Recently Panish et al[15] have made considerable progress in handling phosphorous containing compounds in MBE using gaseous sources for the growth of InP and GaInAsP. This material system is of great importance for $1.3-1.5\mu$ lightwave communication. Single quantum wells and superlattices with near monolayer abruptness have been grown. See Figure 9. Also shown in this figure are the wavelength shifts in the photoluminescence spectra due to the quantum size effect. The perfection of these materials has led recently to the observation of room temperature excitons and their electric field effect, effective mass filters of the type previously grown only in the AlInAs/InGaAs system; superlattice APD's with high gain and low dark current; heterojunction bipolar transistors with properties superior to any previously reported bipolars.

Heteroepitaxy on Si: The use of silicon in semiconductor electronics and photonics is ultimately limited by its incompatibility with other semiconductor materials. In the last few years, there has been considerable progress in the growth of new materials on Si, especially using the concept of strained layer epitaxy[16]. In particular, Si/SiGe heterostructures and superlattices have been grown by Bean using MBE. The larger GeSi lattice is simply compressed in the growth plane to match the Si lattice spacing. A key observation was the growth of dislocation-free materials ten times thicker than predicted by equilibrium theories of dislocation formation. The band gap[17] of the strained Ge:Si films is substantially smaller than that of corresponding bulk alloys. See Figure 10. The growth of GeSi heterostructures on Si thus allows one to exploit "band gap engineering" techniques such as modulation doping, FETs, bipolar and novel photodetectors in silicon based technologies.

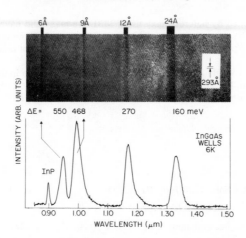

Figure 9 TEM photograph of GaInAs/InP single quantum wells of various thicknesses and associated photo-luminescence spectra. After ref. 15.

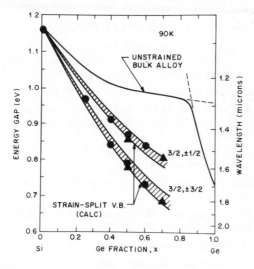

Figure 10 Energy gap as a function of Ge fraction or Ge_xSi_{1-x} alloys grown on Si. After Ref. 17.

MBE also allows the growth of compatible metals, such as silicides and insulators on Si. From these some very novel device structures are emerging. A buried heterostructure[18] of $Si/CoSi_2/Si$ has been grown which demonstrates transistor action with the metal layer as a base. Epitaxial fluorides[19] on Si have been used both for dielectric isolation and for fabrication of FETs.

The possibilities opened by heteroepitaxy using MBE are many and have only just begun to be explored. The low growth temperature and control permitted by MBE are unparalleled, and will permit much further development in the area of heteroepitaxy.

Acknowledgements: I would like to thank several colleagues for help in the preparation of this article: J. C. Bean, F. Capasso, D. S. Chemla, A. Y. Cho, A. C. Gossard, D. V. Lang, A. F. J. Levi, M. B. Panish, P. M. Petroff, S. S. Pei, R. People, N. Shah, H. L. Stormer, and H. Temkin.

REFERENCES

[1]   Cho, A. Y., Thin Solid Film *100*, 291 (1983).

[2]   Dupuis, R. D., Science *226*, 623 (1984).

[3]   For a review see Narayanamurti, V., Physics Today *37*, 24 (1984).

[4]   For a review see Stormer, H. L., Surf. Sci. *132*, 519 (1983).

[5]   Shah, N. J., Pei, S. S., Tu, C. W., and Tiberio, R. C., IEEE Trans. on Elec. Devices *ED-33*, 543 (1986).

[6]   Esaki, L., and Tsu, R., IBM J. Res. Dev. *14*, 61 (1970).

[7]   For a review see Luryi, S., in "Heterojunctions: A Modern View of Band Discontinuities and Device Applications", F. Capasso and G. Margaritondo Eds., North Hollow Publishing House.

[8]   Capasso, F., Mohammed, K., and Cho, A. Y., IEEE J. Q. Elec, Sept (1986).

[9]   Hayes, J. R., and Levi, A. F. J., IEEE J. Q. Elec., Sept (1986), see also Heiblum, N., Calleja, E., Anderson, I. M., Dumke, W. P., Knoedler, C. M., and Osterling, L., Phys. Rev. Lett. *56*, 2854 (1986).

[10] Cibert, J., Petroff, P. M., Dolan, G. J., Pearton, S. J., Gossard, A. C., and English, J. H., App. Phys. Lett. (to be published).

[11] Capasso, F., Tsang, W. T., and Williams, G., in IEEE Trans. on Electron Devices, *ED-30*, 381 (1983).

[12] Capasso, F., Allam, J., Cho, A. Y., Mohammed, K., Malik, R. J., Hutchinson, A. L., and Sivco, D., App. Phys. Lett. *48*, 1294 (1986).

[13] Gossard, A. C., in IEEE J. Q. Elec, Sept (1986).

[14] Chemla, D. S., in Proceedings of this Conference.

[15] For a review see Panish, M. B., in Proceedings of York MBE Conference, In J. Cryst. Growth (1986); also see Temkin, H., Panish, M. B., and Chu, S. N. G., App. Phys. Lett., in press.

[16] For a review see Bean, J. C., Science *230*, 127 (1985).

[17] People, R., Bean, J. C., and Lang, D. V., J. Vac. Sci. Technol. A*3*, 846 (1985).

[18] Hensel, J. C., Levi, A. F. J., Tung, R. T., and Gibson, J. M., App. Phys. Lett. *47*, 151 (1985).

[19] Smith III, T. P., Phillips, J. M., Augustyniak, W. M., and Stiles, P. J., App. Phys. Lett. *45*, 907 (1984).

SCIENTIFIC CHALLENGES IN MICROELECTRONICS

P. Chaudhari

IBM Thomas J. Watson Research Center
Yorktown Heights, NY 10598, USA

ABSTRACT

The level of complexity of integration in semiconductor devices has been increasing at a remaιкable pace. The principal technological challenges that nɔed to be overcome in
order to sustain this rate of progress will be described and
translated into the parlance of science. Several examples
will be used to illustrate the coupling between current
semiconductor science and technology.

Over the last two decades advances in semiconductor technology have
moved in parallel and, to a large extent, independent of advances in semiconductor physics. As the technology advances towards using semiconductor devices with line widths in the submicron raɪge the interdependencies
between science and technology are likely to sharply increase. In this note
we sketch in a very superficial manner some of the essential directions required for a semiconductor science program underlying future technology.

In the preceding paper Dr. Narayanmurti [1]) has discussed the advances we
can expect from deliberately structuring materials particularly GaAs and

their derivatives. Superlattices or strained superlattices ever since their inception [2] have been extensively studied by the semiconductor science community. Much has been written about their potential in technology and, in fact, their scientific properties are reported in a great many papers at this conference. I shall therefore not address these materials in this article. I shall confine myself entirely to silicon, where a material is discussed, and its role in the evolution of semiconductor technology.

There are two silicon devices that form the backbone of, say, the computer technology. These are the field-effect (FET) and the bi-polar transistors. A schematic cross-section of a FET is shown in figure 1. This is a C-MOSFET structure. It can be shown that the successful fabrication and operation of such a transistor or a bipolar transistor is directly related to controlling seven generic items. These are (1) selective deposition of atoms, (2) selective removal of atoms (or etching), (3) dopant introduction, (4) interaction of radiation with matter, (5) interfaces (6) materials, such as insulators, and (7) transport associated with both atomic and electronic motion. The rest of this article is concerned with addressing some of these items. Space does not permit addressing all or for that matter addressing anyone of these in any reasonable depth.

DEPOSITION AND REMOVAL OF ATOMS

Starting from silicon crystal growth the fabrication of a semiconductor device is a series of steps requiring the addition of selected atoms by evaporation or chemical means. One of the more interesting deposition steps involves the epitaxial growth of a thin film with the requisite dopants and dopant profiles. Although epitaxial silicon is routinely grown by chemical vapor deposition techniques we do not as yet understand the mechanism of growth or its rate controlling step. This understanding will become particularly important as the requirement for thinner epitaxial film increases. At the present time the epitaxial films are ion implanted to obtain the required dopant levels. The material is subsequently annealed to eliminate crystalline damage caused by implementation. This process changes the dopant profiles due to thermal diffusion. In order to obtain a sharp and defined profile it will be essential to go to lower epitaxial temperatures and introduce the dopant by evaporation or chemical means so as to avoid the annealing process associated with ion implantation. In order to accomplish this goal and predict properties before the experiment is completed we need to understand the atomistic mechanism of growth. This will involve not only understanding how atoms migrate on a surface and find appropriate attachment sites but also how atoms interact with the surface and in some cases the chemistry of the gas-phase and gas-surface interactions.

Based on what we know it is possible to devise reactors as, for example, recently reported by Myerson [3]) in which low pressure chemical vapor deposition is used to make device quality epitaxial films. In Fig. 2 we show the beneficial effect of low temperature deposition in retaining a sharp dopant profile when the concentration of Boron is changed.

In order to develop a detailed atomic scale knowledge of film deposition we need to develop new techniques to observe what is going on surfaces during reaction. Apart from the usual complement of surface science techniques there are two new methods that are particularly appealing. The first is a relatively low spatial resolution instrument that has the potential of extremely high temporal resolution. It is a laser based technique and therefore has the advantage of being used in an ambient environment. It relies on the absence of a center of symmetry at or near the surface to generate a second harmonic signal. [4]) Recent experimental observations on silicon growth by this technique yielded information about the activation energy associated with the growth of epitaxial films. This number provides phenomenological limits on epitaxial growth rates but does not provide unequivocal evidence about atomistic mechanism. For this we may have to use the scanning tunneling microscope developed by Binnig and Rohrer [5]) and discussed at some length in this conference [6,7]).

Epitaxial deposition is, of course, one special form of growing a material on a surface. Oxidation is another. In the case of silicon devices this oxide layer can become very important for it is essential as an insulator. The thickness of the insulators in future devices is expected to reduce and approach a number of the order of magnitude of 10 nanometers. Not only the dielectric strength, defects and materials properties become issues but also at these thicknesses tunneling is an increasingly important consideration. Gate oxides are currently grown at high temperatures which will not be compatible with a low temperature epitaxial processes. As oxidation is a diffusion controlled process one cannot drop the temperature very low without adverse impact on throughput. The approach then is to develop low temperature oxidation (or nitride formation) methods. A promising approach is to use external radiation-photons or particles to enhance kinetics. This opens up a vast area where radiation is used not only to enhance the reaction but also to localize it where desired. For example when silicon is exposed to ammonia [8]) it is observed that nitrogen goes below the surface but that the hydrogen bonds to surface atoms. This bonding prevents further nitrogen from forming. The hydrogen bond to silicon can be broken by heating to high temperature or by electron irradiation. In the latter approach thin films of nitride can be formed at 90 ° K!

Just as radiation is responsible for enhancing oxidation it can also be used to enhance removal or etching of surfaces. For example tetrafluorides can be used to selectively etch silicon devices in a reactive ion etching environment. There a number of papers on this subject including the invited talk by Winters [9]) at this conference and I shall therefore not develop this theme any further. It seems to me, however, that the general theme of gas-surface interaction in the presence of radiation is an important area for scientific investigations as well as rich in potential for application.

INTERFACES

There are a number of different interfaces in a semiconductor device. These include interfaces formed with insulators, semiconductors, polymers, and metals. None of these interfaces are really understood. For example, the semiconductor metal interface can form an an ohmic contact or a Schottky barrier. Both are required in device operation. After several decades of research we still do not have a widely accepted model for Schottky barrier heights [10]). The problem is enormously complicated by the presence of impurities and defects. I believe one of the few techniques that can provide unique information which may overcome many of the constraints imposed by other experimental approaches is to use the tunneling microscope. We can determine the topological and electronic properties of the surface and also measure the Schottky barrier properties with the tunneling tip. This would be a local measurement and not averaged over a large area with the attendant uncertainties associated with impurities and defects.

The tunneling microscope has already found a role in studying defects associated with silicon -silicon oxide interface. The jump of an electron back and forth from a trap in the insulator was monitored by watching the change in tunneling current [11]) This is shown in Fig. 3.

In this measurement the spatial location of the trap site and the associated binding energy could be determined.   One hopes that the process of oxidation or insulator formation can also be studied at this level of resolution. The properties of these interfaces are particularly important in future devices where the charge is carried in a shallower channel.

TRANSPORT

Atomic transport in silicon remains a controversial subject despite decades of research. Calculations [12] show that the formation energies for vacancies and interstitial is comparable so that many of the discussions as to a vacancy versus an interstitial mechanism should be viewed with this perspective. Adding to this is Pandey's [13] recent insight supported by calculations that one may not need a vacancy or an interstitial defect for diffusion. The open structure of silicon may allow atoms, to exchange locations. The role of defects, regardless of transport, will continue to be important in semiconductor technology. These defects range in size from point defects associated with, say, dopants to line or pipe defects (dislocations) and grain boundaries or more generally interfaces. Grain boundaries, for example, are important in determining the electrical properties of polycrystalline silicon which is providing increasing use in silicon devices as a metallic conductor. Defects associated with silicon crystal growth will remain an active subject of interest for the diameter of the crystal is expected to continue to increase with time.

Electronic transport in crystalline semiconductors was an active area of research over a decade ago. In the last decade and a half electronic transport in disordered systems has been center stage. In amorphous silicon this may have practical value for this material may be the basis of large area thin film transistor technology useful for displays. The study of transport in disordered systems also provided insight into the notion of localization and, in particular, the role of quantum interference in weakly localized systems. [14]) I believe these ideas will become increasingly important in very small devices when the elastic and inelastic mean free paths become comparable to or larger than the source-drain distance. The influence of hot elections are recognized and discussed but not fully understood in device design. This area of research will increase with time. When the inelastic scattering lengths become comparable to critical device dimensions the effects of quantum interference both in designing novel devices and also in possibly introducing conductance flutuations will become important. I see this area as one coupled to silicon devices in the tenth micron range at room temperature and larger dimensions at lower temperatures.

In preparing this talk I have used new material from work going on in IBM. I apologize for this parochial act and recognize that creative and elegant work is common to science around the world. I defend my action on the basis of expediency, which is really not defensible in science.

1.   Narayanmurti, V., proc. this conference.

2.   Esaki, L., IBM Jour. of Res. and Dev., 14, 61 (1970).

3.   Meyerson, B.

4.   Heinz, T., Loy, M., Thompson, W.A., Phys. Rev. Lett, 54, 63 (1985).

5.   Binnig, G., Rohrer, H., Gerber, C.L. and Weibel, E., Appl. Phys. Lett. 40, 178 (1982); Binnig, G. and Rohrer, H., Sci. Amer. 253, 50 (1985); see also special issue of the IBM Jour. of Res. and Dev., 30, (1986).

6.   Golevchenko, J., proc. this conference.

7.   Demuth, J.E., Hamers, R.J., Tromp, R.M., proc. this conference; Hamers, R.J., Tromp, R.M., and Demuth, J.E., Phys. Rev. Lett., 56, 1972 (1986).

8.   Boszo, F. and Avouris, P., Phys. Rev. Lett., 57, 1185 (1986).

9.   Winters, H., proc. this conference.

10. Ho, P., proc. this conference.

11. Koch, R.H., and Hamers, R.J., Surface Science, to be published.

12. Car, R., Kelly, P.J., Oshiyama, A., Pantelides, S.T., Phys. Rev. Lett., 52, 1814 (1984).

13. Pandey, K., proc. this conference.

14. Ando, T., Fowler, A.B., Stern, F., Rev. Mod. Phys. 54, 437 (1982).

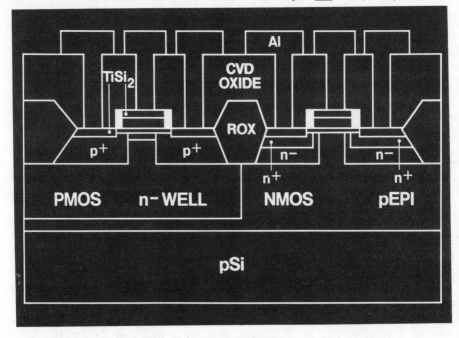

FIG. 1.   Schematic sketch of a typical C-MOSFET

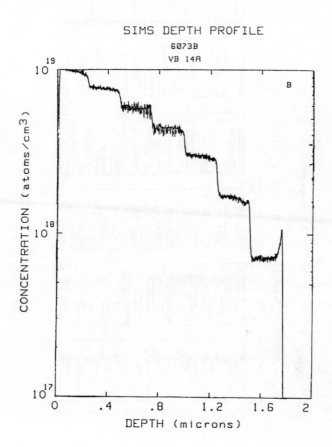

FIG. 2. Depth profile of boron concentration in an epitaxial film grown at tempeature below 700 degrees C.

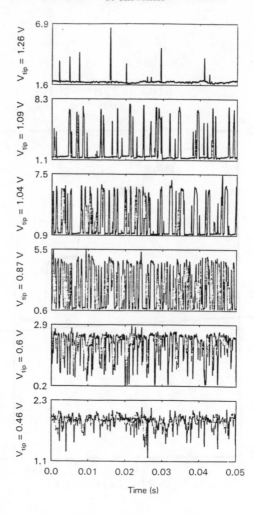

FIG. 3. Time traces of tunneling current for six
values of applied tip voltage, V_{tip}, when
the tip was positioned over a trap on the
surface. The average tunneling current was
set to be 2 nA, and the y-scale is in nA.

TRANSITION METAL IMPURITIES IN SEMICONDUCTORS

Alex Zunger

Solar Energy Research Institute, Golden, Colorado, 80401

Abstract

First-principles self-consistent Green's function methods are used to understand the apparent dichotomy between the atomically-localized and covalently-delocalized characteristics of 3d impurities in semiconductors, and to unravel intriguing chemical trends and universalities.

1. DUALITY OF 3d IMPURITIES: LOCALIZED AND EXTENDED CHARACTERISTICS

Despite extensive experimental studies over the past 30 years (reviewed recently in Ref. 1), the electronic structure of 3d impurities in semiconductors has received theoretical attention only rather recently. Rapid progress has, however, been made in this area, through use of a wide range of theoretical techniques, including cluster models[2], continued-fraction Extended Hückel methods[3], empirical tight-binding models[4], Anderson-type model Hamiltonians[5], and more recently, by first-principles, self-consistent Green's function models[6-10] utilizing the local density formalism. This system has attracted my attention not only because of its obvious importance in technological device issues (e.g., solar cells, semi-insulating substrates, LED's, lifetime-controlling dopants, etc.), but primarily due to an intriguing **duality** that the experimental data on these impurities seem to exhibit. The intellectual tools established since Van-Vleck and Bethe considered the structure of 3d atoms in non spherical chemical environment have prepared one to think about impurities as exhibiting **either** atomically-localized characteristics (as is the case for 3d atoms in ionic coordination compounds), **or** covalently-bonded, "extended" characteristics (as is the case for shallow, "hydrogenic" impurities). This dichotomy is rooted in

numerous text books and review articles and constitutes the working
paradigm in impurity physics. Examination of the experimental data
for 3d impurities in semiconductors[1] suggested to me, however, that
depending on which experimental probe is looked at, **both points of
view are valid at the same time.** This can be appreciated as
follows. Consider the observed free-ion and impurity ionization
energies (donors and acceptors) depicted in Fig. 1, the observed
hyperfine coupling constants shown in Fig. 2, the observed g-values of
Table I, and the absorption spectra of Fig. 3.

Arguing for the covalently-delocalized model of such impurities,
I note the following:

(A) The **rate** at which impurity ionization energies change with atomic
number (Fig. 1 b-f) is ~ 10 times smaller than in free-ions (Fig. 1a),
suggesting bond-formation and chemical interactions in the solid.
(B) The Mott-Hubbard Coulomb repulsion energy for impurities [i.e.,
the difference between two consecutive ionization energies such as
E(0/+) - E(0/-)] is ~ 50-100 times smaller than for free-ions, suggesting strong (and nonlinear) screening in the solid. (C) The **absolute** impurity ionization potential (e.g., the difference between the
host work function and the impurity energy of Fig. 1b-f taken with
respect to the valence band maximum] is about 5-7 eV in all systems
(Fig. 6 below), exactly as the redox potential (relative to vacuum
too) of such ions in **strongly polarizeable media,** (e.g., water), or
even for Fe in Heme proteins[1], suggesting again strong polarization
and interactions with the host crystal. (D) The hyperfine-coupling
constants in semiconductors (Fig. 2 and Table I) are reduced
dramatically relative to free-ions or to 3d impurities in ionic
compounds, suggesting strong covalency. (E) The angular momentum part
of the g-values (Table I) are quenched, and the spin-orbit
splittings[1] are reduced, both suggesting once more covalency.
(F) The Mössbauer Isomer shifts (e.g.[1], Fe in Si and III-V's) seems
not to depend much on the formal oxidation state, suggesting that the
host crystal resupplies the impurity with s electrons when ionized,
again a manifestation of impurity-host interactions.

Arguing for the opposite point of view, namely that 3d impurities
in semiconductors are "localized", I note the following:

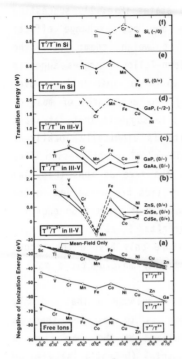

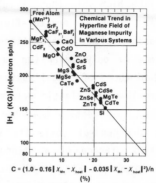

Fig. 2: Trends in the observed[1] hyperfine field vs the phenomenological electronegativity C, for Mn.

Fig. 1: Observed[1] donor and acceptor (b-f) and free-ion (a) ionization energies of 3d ions.

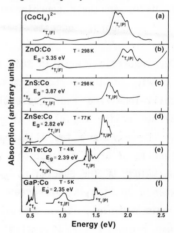

Fig. 3. Absorption spectra[1] of Co in different media. ⟶

Table I: Calculated and observed[1] g-values and hyperfine coupling constants A (10^{-4} cm^{-1}) for impurities in Si.

Impurity	g_{calc}	g_{exp}	A_{calc}	A_{exp}
Ti$^+$, J=3/2	1.9912	1.9986	+4.7	±5.2
Cr$^+$, J=5/2	1.9986	1.9978	+11.0	+10.7
Cr^0, J=2	3.0828	2.97	+14.0	±15.9
Fe^0, J=1	2.0430	2.0104	-5.2	±6.9
Fe$^+$, 3=1/2	3.5582	3.524	-3.9	±3.0

(I) the atomiclike multiplet structure (indicative of localized states) is largely retained in the solid (Fig. 3), with similar ranges of d→d* excitation energies as in free-ions; (II) the existence of

atomiclike high-spin (Hund's rule) ground states (Fig. 1.), suggesting
that as large as crystal-field effects may be, exchange effects
(increasing with localization) usually outweigh them; (III) the
existence of only small (or vanishing) Jahn-Teller energies suggests
the dominance of localization-induced exchange interactions over
symmetry-lowering elastic deformation; (IV) the appearance of most of
the ENDOR spin density near the impurity site suggests that, at least
inside the central cell, the spin density is atomically localized;
(V) the occurrence of a strong half-shell d^5/d^4 stabilization (see
local minimum in Fig. 1b-f) like that in the **free** 3d ions (Fig. 1a)
suggests that much of the atomic characteristics are preserved in the
solid; and, (VI) the great similarity between d→d* excitation energies
of 3d impurities in zincblende and wurtzite II-VI semiconductors
(which differ in their coordination structure only beyond the second
shell of neighbors) suggests that the impurity wavefunctions may be
localized.

Our theoretical work in this field had convinced me that the only
correct way to think about these systems is to describe them as
atomically-localized and covalently-delocalized at the same time; a
duality not unlike the classical <u>vs</u> the quantal (DeBroglie's) view on
light. I describe a few illustrations of this in the next section,
were results of Green's function calculations are given.

2. THE NATURE OF THE DUALITY

When a 3d atom is introduced into a wide-band-gap narrow-valence-
band **ionic material**, its d orbital energies fall naturally inside the
band gap region. The 3d system is then only weakly coupled to the
host crystal, in the sense that (external) perturbations exerted on
this ion (e.g., ionization, external fields, etc.) must be absorbed by
the ion itself. In contrast, we find that a 3d impurity in a **semiconductor** distributes most of its orbital character as broad hybridized resonances within the (wide) valence band, and only a faint
shadow of it is captured inside the (narrow) band gap region. This is
illustrated in Fig. 4, showing that most of the effective charge (Q)
and local magnetic moment (μ) is contributed by the valence band resonances (VBR). Symmetry, however, disallows any coupling between the
e-type (or Γ_{12}) 3d orbitals (which, in substitutional tetrahedral symmetry point to the **next** nearest neighbors) and the nearest ligand

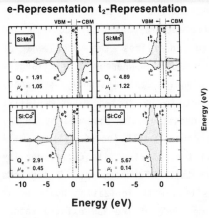

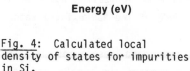

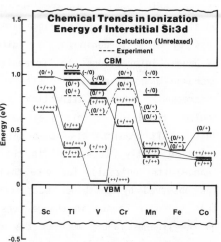

Fig. 4: Calculated local density of states for impurities in Si.

Fig. 5: Calculated and observed[1] impurity ionization energies in Si.

atoms. These e orbitals hence remain localized, whether they resonate in the valence band (heavy 3d impurities) or are in the band gap (light 3d impurities). This simple picture, emerging from our studies[6-9] on 3d impurities in Si, III-V's and II-VI's contains the essential physics of the coexistence between localized and delocalized characteristics. This situation has a number of significant physical implications. When a gap level is ionized in an absorption or thermal experiment, reducing thereby the charge around the impurity atom, the valence band resonance wavefunctions respond by becoming more local-ized around the impurity[6], returning thereby much of the charge lost by the gap level. The **effective** charge around the impurity atom in a semiconductor hence remains nearly constant in different ionization states, explaining the observed constancy of the Mössbauer isomer shift (point F). This "self regulating response"[6] (analogous to Homoestasis in biological systems) explains also why so many different charge states can exist in a narrow (band gap) energy range (Fig. 1b-f) as manifested by a small Mott-Hubbard Coulomb energy (point B). It is important, however, to note that only direct Coulomb interactions are screened effectively by this self-regulating response: whereas the band gap levels and the valence band levels have opposing (hence, compensating) contributions to the **impurity**

charge, both types of levels contribute in the same direction to the **local magnetic moment**; exchange interactions are hence affected far less by this screening. The different responses of Coulomb and exchange interactions to screening underlines the phenomena of exchange-induced negative "effective U"[8], whereby the (largely unscreened) exchange **attraction** overwhelms the (strongly screened) Coulomb **repulsion**. This idea is yet to be tested experimentally. This result predicts also that **spin**-densities (as observed by ENDOR) would be mostly "localized" in the central cell (point IV) and will change with ionization far more than the impurity site **charge** density (as observed by Mössbauer measurements). Furthermore, it suggests that the main change in charge densities with ionization would occur on the **ligands** (where much of the amplitude of the VBR exists) rather than on the the **impurity**.

The occurrence of hybridized valence band resonances is also largely responsible for the covalent quenching of the angular momentum part of the g-value (Table I), and the spin-orbit splitting (point E), the reduction in the hyperfine coupling constants (point D and Table I) the attenuation of the rate of change of ionization energies along the 3d series (point A), and for the overall reduction in the magnitude of the ionization energies relative to free-ions (point C). The substantial coupling of the impurity to its ligands through the valence band resonances is manifested also by atomic relaxation effects: direct calculations[7] of the force exerted by the impurity on its ligands has shown for interstitial impurities in Si that the first ligand shell (containing 4 atoms) moves **outwards**, whereas the second ligand shell (containing 6 atoms) moves **inwards**. This results in an effective **~10-fold coordinated 3d atom**, a rare situation in 3d coordination chemistry, but nevertheless one exhibited by numerous bulk 3d silicides.[7] This coupling is so pronounced that it would seem that the correct chemical picture of the system should involve a "complex" of the impurity with its 10 ligand neighbors, viewed as a "local silicide".

In contrast with the strong Coulomb screening and hybridization, the weak screening of exchange interactions is responsible for the survival of atomiclike high-spin (Hund's rule) ground states

(point II) and the characteristic break in the trends of the ionization energies at the d^5/d^4 "Hund's point" seen in Fig. 1 (Point D).
Combined with the symmetry and angular momentum imposed localization
of the e-type 3d orbitals, these two effects are responsible also for
the existence of atomiclike multiplet effects (point I) and for the
overall signature of localized 3d orbitals (points III and VI).

3. TRENDS IN IONIZATION ENERGIES

While the calcu-
lated[6-8] donor and acceptor energies agree rather
well with experiments
(e.g., see Fig. 5 for
interstitial impurities in
Si), they do give the im-
pression of an absence of
recognizable and simple
chemical trends. Caldas
et al.[19] have observed,
however, that if one

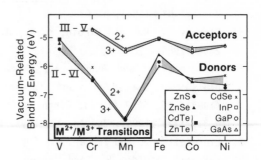

Fig. 6: Universality of VRBE of 3d
impurities in various hosts.

refers these energies to an intrinsic reference energy characteristic
of the host (e.g., the vacuum level), these "vacuum referred binding
energies" (VRBE) are approximately constant for the same impurity in
different host crystals (Fig. 6). This rule can therefore be used to
predict the binding energy of an impurity in a host AC if the value in
another crystal, BC, is known[11]. Conversely, I showed[9] that the
knowledge of the impurity level both in AC and in BC (or in their
alloy) can be used to deduce the intrinsic valence band offset between
the two semiconductors, thereby providing complementary information to
that deduced by photoemission from interfaces, (essential in understanding Schottky barriers too).

REFERENCES

(1) Zunger, A., in Solid States Physics, Edt. H. Ehrenreich and D.
     Turnbull (Academic Press, New York, 1986), Vol. 36, pp. 275-464.

(2) Roitsin, A. B., and Firshtein, L. A., Sov. Phys. Solid State 13, 50 (1971); Hemstreet, L. A., Phys. Rev. B 15, 834 (1977); Fazzio, A., and Leite, J. R., Phys. Rev. B 21, 4710 (1980); DeLeo, G. G., Watkins, G. D., and Fowler, W. B., Phys. Rev. B 25, 4972 (1982).

(3) Masterov, V. F., Sov. Phys. Semicond. 18, 1 (1984).

(4) Pecheur, P., and Toussaint, G., Physica 116B, 112, (1983); Vogl, P., and Baranowski, J., in 17th Int. Conf. Semicond. (Springer-Verlag, NY, 1985), p. 623.

(5) Fleurov, V. N., and Kikoin, K. A., J. Phys. C 9, 1673 (1976); 17, 2354 (1984).

(6) Lindefelt U., and Zunger, A., Phys. Rev. B 24, 5913 (1981); 26, 846 (1982); Zunger, A., and Lindefelt, U., Phys. Rev. B 27, 1191 (1983); 26, 5989 (1982); Singh, V. A., and Zunger, A., Phys. Rev. B 31, 3729 (1985); Zunger, A., Phys. Rev. Lett. 50, 1215 (1983).

(7) Lindefelt, U., and Zunger, A., Phys. Rev. B 30, 1102 (1984).

(8) Katayama-Yoshida, H., and Zunger, A., Phys. Rev. Lett 53, 1256 (1984); 59, 1618 (1985); Phys. Rev. B 32, 8317 (1985); 33, 2961 (1986).

(9) Zunger, A., Phys. Rev. Lett. 54, 848 (1985).

(10) Beeler, F., Anderson, O. K., and Scheffler, M., Phys. Rev. Lett. 55, 1498 (1985).

(11) Caldas, M., Fazzio, A., and Zunger, A., Appl. Phys. Lett. 45, 671 (1984).

SEMICONDUCTOR SURFACES AND INTERFACES- A LOOK AT THEIR ATOMIC STRUCTURE AND FORMATION

J.F. van der Veen and A.E.M.J. Fischer

FOM-institute for Atomic and Molecular Physics,
Kruislaan 407, 1098 SJ Amsterdam, The Netherlands

ABSTRACT

A progress report is given on the crystallography of semiconductor surfaces and interfaces. The use of high-resolution ion scattering spectroscopy as an in-depth probe of interface structures is illustrated.

1. INTRODUCTION

The geometric arrangement of atoms in a solid surface influences many of its essential properties, notably the chemical reactivity and the electronic structure. In order to understand these properties, it is important to know 'where the atoms are'. It is not surprising, therefore, that the determination of surface structures has become a research area of major interest. Surfaces of semiconductor crystals represent a true challenge, since they often exhibit very complex atomic rearrangements, leading to periodic structures having a unit cell size much larger than that of the truncated bulk lattice. An added complication is that the atomic displacements in such a reconstructed surface are not restricted to the top layers only; the associated bond angle strains are partially relieved by lattice distortions in deeper layers of the crystal.

Recently, substantial progress has been made with unravelling some of the most complex structures, among them the reconstructions appearing on the (111) surfaces of Si and diamond and on the polar surfaces of III-V compound semiconductors. These successes are largely due to the use of novel structure probes and the increasing reliability of theoretical predictions of surface structures.

Gratifying as it may be to know 'where the atoms are', it is perhaps more important to know 'to where they move', if the surface is brought in contact with a layer of different elemental composition. Here we will discuss the epitaxial $NiSi_2$ on $Si(111)$ and $CoSi_2$ on $Si(111)$ systems. These silicide contacts, with their well-defined and atomically abrupt interfaces, serve as ideal systems for testing theoretical models of Schottky barrier formation. The atomic structure of the interface has been determined for both systems.

2.   SURFACES

The reconstructions occurring on the (111) surfaces of silicon and diamond are intriguing. When a Si crystal is cleaved in vacuum along the $[\bar{1}\bar{1}2]$ direction, the exposed (111) surface exhibits a single-domain 2x1 LEED pattern, with the two-fold periodicity along the direction of cleavage. The now commonly accepted model for its structure is the π-bonded chain model (fig. 1b) proposed by Pandey[1]. It involves a drastic change in bonding topology. The dangling-bond orbitals on the top-layer atoms, which for a bulk-like surface (fig. 1a) would reside on next-nearest neighbors, now reside on nearest neighbors. The proximity of these bonds along the upper chains allows them to interact and to form a π-bond, causing a reduction of the energy by ~ 0.3 eV per atom.

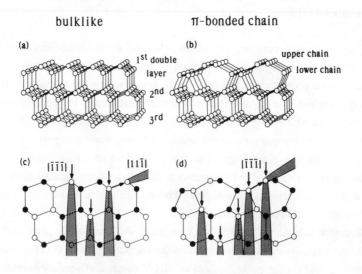

Fig.1. View of the diamond (111) surface, (a) assuming bulk-like atomic positions, (b) incorporating the π-bonded chain reconstruction proposed by Pandey [2]; (c) and (d) are projections on the (110) plane of the above two structures respectively, showing schematically the ion scattering experiment with the beam incident along $[\bar{1}\bar{1}\bar{1}]$. Shadow and blocking cones are indicated as hatched regions.

The π-bonded chain model also explains the 2x1 reconstruction of the clean, thermally annealed diamond (111) surface[2,3]. For diamond such a bonding configuration is quite plausible; π-bonding is commonly observed in organic carbon chemistry. The diamond surface can also be prepared such, that it is covered by a full layer of hydrogen[4]. A 1x1 LEED pattern is then observed, suggesting that this surface is bulk-like, with the adsorbed hydrogen atoms saturating the dangling bonds. Precisely this behavior has been observed in an ion scattering experiment by Derry et al.[5], where use was made of shadowing and blocking effects[6] in the first few surface layers of the diamond crystal. A beam of 100 keV protons was aligned with the [$\bar{1}\bar{1}\bar{1}$] shadowing direction and protons backscattered from the surface region were detected in a 20° range around the [11$\bar{1}$] blocking direction (fig. 1c and 1d). A plot of the backscattering intensity versus the detection angle shows that, on average, two extra monolayers (i.e., one double-layer) are 'visible' on the 2x1 surface (fig. 2). That is, two monolayers do *not* contribute to shadowing and blocking along the [$\bar{1}\bar{1}\bar{1}$] and [11$\bar{1}$] directions, respectively. This result is fully consistent with the π-bonded chain model, in which atoms in the top double-layer are displaced from bulk lattice sites (neglecting the strain in deeper layers).Monte Carlo computer simulations of the experiment for the π-bonded and bulk-like structures (solid curves in fig. 2) show good agreement with the measured backscattering yields from the clean and hydrogen-covered surface, respectively. Equally good fits have been obtained for the 2x1 surface of Si(111)[7,8], confirming Pandey's prediction that the nature of the

Fig.2. Backscattering intensity of ions versus exit angle with respect to the surface plane. Data taken in the scattering geometry of fig. 1 are shown for the (1x1)H and 2x1 reconstructed surfaces (squares). The solid curves are best-fit computer simulations for a bulk-like and a π-bonded surface, respectively. For further details on the simulations, see ref. 5.

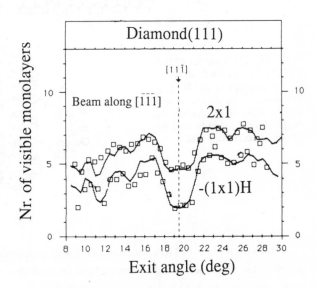

reconstructions on these surfaces is very similar. Minor differences, to which the ion scattering data are rather insensitive, may still exist. The upper chains on the Si(111) surface appear to be slightly tilted[8,9], while the chains on the diamond surface may be dimerized [2,5]. The latter feature, which implies the formation of double or even triple C-C bonds on the surface, needs further experimental verification.

The Si(111)-2x1 surface, obtained after cleavage, is metastable; upon heating, the 2x1 unit cell changes into a 7x7 unit cell. A major step towards solving the 7x7 structure was made by Binnig et al.[10] in a scanning tunneling microscope (STM) study. The tunneling images clearly showed 12 maxima in the 7x7 unit cell, indicating the presence of Si adatoms. More recently, Takayanagi et al.[11] studied the surface with transmission electron diffraction and derived a model which contains, apart from the 12 adatoms, stacking faults in one half of the unit cell, subsurface dimers, and holes in the corners (fig. 3). This dimer-adatom-stacking fault (DAS) model is consistent with ion scattering data[12] and has also been confirmed by glancing-incidence X-ray diffraction experiments by Robinson et al.[13]. Recent STM images taken by Tromp et al.[14] provide striking evidence for the correctness of Takayanagi's model and even the electron states associated with the various dangling bonds and back-bonds in the DAS unit cell have been identified and spatially resolved[15].

Fig.3. The DAS model of the Si(111)-7x7 surface, derived by Takayanagi et al.[11]. The 7x7 unit cell is shown in top and in side view. Atoms in deeper layers are indicated by circles of decreasing size.

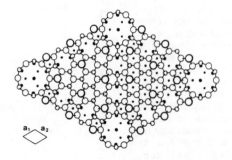

The origin of the 7x7 reconstruction is not well-understood . The notion that a bulk-like Si(111)-1x1 surface would be under a large compressive stress[16,17] may provide an important clue. An efficient way of reducing the compressive stress is to have a reconstruction into a large unit cell in which some first-layer atoms become adatoms (provided the *total* energy is lowered). Evidence for the influence of stress on the size of the reconstructed unit cell has recently been obtained by Nakagawa et al. [18] in a structural investigation of epitaxial Si_xGe_{1-x} layers on Ge and Si substrates, in which the sign and magnitude of the stress were varied.

The reconstructions occurring on the other low-index faces of Si will not be discussed here; the structure of the Si(100)-2x1 surface is well known [19] and the Si(110) surface has so far remained virtually unexplored. As to compound semiconductors, the (110)-1x1 surfaces of the III-V compounds all exhibit a well-documented relaxation involving a ~30° rotation of the atomic bonds out of the surface plane [20,21]. Crystallographic studies of the polar (111) and (111) surfaces of III-V materials have just begun to appear in the literature [22,23]. The variable stoïchiometry of these surfaces gives rise to a multitude of reconstruction phenomena, for which recently some structure models have been proposed [24-26].

3.   INTERFACES

If the surface of a semiconductor is overgrown with an epitaxial lattice-matched crystal of a different material, the reconstruction is usually lifted; the surface dangling bonds are 'used-up' and new bonds are formed across the interface. Molecular beam epitaxy (MBE) then allows for the continued growth of high-quality single-crystal layers with monolayer control over their thickness. In case the overlayer lattice does not quite match with that of the substrate, defect-free growth is still possible if the mismatch is accomodated by lattice strain. This concept of strained-layer epitaxy opens up many possibilities for growing novel device stuctures. Examples are (monolayer) superlattices of the GeSi/Si system [27-31] and 'buried' epitaxial layers of $CoSi_2$ and $NiSi_2$ silicides in Si/silicide/Si heterostructures [31,32]. The electrical properties of such systems are dependent not only on crystal quality and morphology of the films grown, but also on the atomic bonding arrangement at the interface. In particular, the Schottky barrier (SB) height of a metal-semiconductor junction is known to be a property of the very interface. However, it is not yet possible to unequivocally explain the magnitudes of the SB heights measured on the different junctions. An example is the nearly lattice-matched $NiSi_2$ on Si(111) system. Single-crystal $NiSi_2$ layers can be grown epitaxially with an orientation which is either aligned with the substrate (type A) or rotated 180° about the substrate normal (type B) [33]. Recently a controversy has arisen over the question whether the SB is intrinsically dependent on the type of epitaxy [34] (A or B) or correlated with structural defects at the interface[35].For a proper understanding of any such structure dependence, the atomic bonding arrangement at the interface has to be known accurately for both types of epitaxy, including possible relaxation effects.

Recently, van Loenen et al.[36] employed high-resolution ion scattering as a technique to determine the structure of the B-oriented interface. For a given orientation there are two obvious possibilities for the bonding arrangement: the dangling bonds on the Si(111) substrate are either attached to the Ni atoms in the silicide (fig. 4a) or to the Si atoms (fig. 4b), resulting in 5- or 7-fold coordinated Ni atoms at the interface, respectively. In order to distinguish between these two possibilities, use was made of an ion focusing effect in the channels of the overlayer crystal, as shown in fig. 4. Provided the overlayer is thin enough (~ 25 Å in that study) and a specific angle- of-incidence is chosen, the ion beam is focused onto the atoms in the layers labeled 4 and 5 at the substrate side of the interface. The ions backscattered from these atoms can reach the vacuum again through the [110] NiSi$_2$ crystal channels and emerge from these within a well-defined range of exit angles. For a 5-fold coordinated interface the location of atoms 4 and 5 is such that this angular range is very wide (fig. 4a), whereas for the 7-fold interface this range is quite narrow (fig. 4b). Angular-dependent measurements of the backscattering intensity (fig. 5b) show immediately that the interface is 7-fold coordinated; a sharp peak is

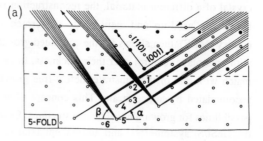

Fig. 4. Ion backscattering geometry for determination of the type-B NiSi$_2$-Si(111) and CoSi$_2$-Si(111) interface structures, showing the (1$\bar{1}$0) plane for the (a) five-fold and (b) seven-fold coordinated interfaces. Open circles denote Si atoms, filled circles denote Ni atoms.

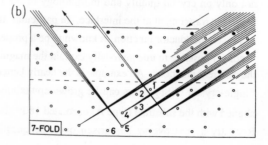

seen near the [110] exit angle, which fits exactly to Monte Carlo computer simulations of this yield for the 7-fold structure. This result has been confirmed in an X-ray standing wave (XSW) analysis by Vlieg et al.[37]. They measured also 7-fold coordination for type-A epitaxy and found in addition that the bond lengths across the interface are slightly contracted by 0.04±0.05 Å and 0.11±0.03 Å for type A and type B, respectively.

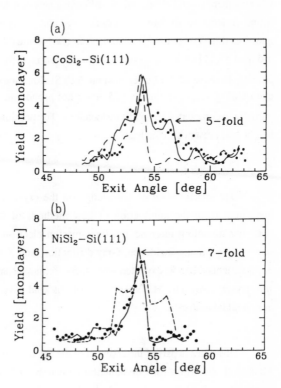

Fig. 5. Angular distributions of ions backscattered from atom layers 4 and 5 in the scatttering geometry of fig. 4. The measured distributions in panels (a) and (b) correspond with the ones schematically indicated in fig. 4(a) and (b) for the five- and seven-fold coordinated structures, respectively.

Surprisingly, the interface of the very similar $CoSi_2$ - Si(111) system is 5-fold coordinated[38]. This is evident from the broad angular distribution of backscattered ions shown in fig. 5a . The striking contrast with the result for $NiSi_2$ is puzzling since $CoSi_2$ and $NiSi_2$ have the same lattice structure and differ only in their lattice match to Si (1.2 % mismatch for $CoSi_2$ and 0.4 % for $NiSi_2$). It is interesting to note, that earlier TEM lattice imaging studies had already hinted at 7- and 5-fold coordination at the $NiSi_2$-Si(111) and $CoSi_2$-Si(111) interfaces[39,40].

Let us finally return to the controversial issue regarding an intrinsic dependence of the Schottky barrier height on the interface atomic structure. The A- and B-type interfaces differ in structure only in the positions of the third- and higher-nearest neighbors with respect to the atoms in the last Ni layer, but the 5- and 7-fold coordinated interfaces have a quite different bonding arrangement (fig. 4) and also different numbers of unpaired electrons at the Ni atoms nearest to the interface. Thus, if there is an intrinsic structure dependence at all, then one expects a difference between $CoSi_2$ and B-type $NiSi_2$, whereas there should almost be no difference between A- and B-type $NiSi_2$. Liehr et al.[35] found indeed no difference between A- and B-type $NiSi_2$, but these results were questioned by Tung et al.[41], who measured a 140 meV higher barrier for B-type $NiSi_2$ on n-doped Si(111).For $CoSi_2$ such a comparison is difficult: the $CoSi_2$ interface contains defects, because of the larger mismatch to Si. Ion blocking analyses by us indicate that even $CoSi_2$ layers as thin as ~25 Å are not pseudomorph, i.e., the mismatch with Si is partially accomodated by misfit dislocations. The possible effect of such defects on the SB height is at present still unclear.

4.   CONCLUSION

Our present state of knowledge on the crystallography of surfaces has greatly benefitted from the application of new methods of structure determination, such as scanning tunneling microscopy, transmission electron diffraction, high-resolution ion scattering, glancing-incidence X-ray diffraction, and X-ray standing wave analysis. New developments are to be expected when these techniques are combined with a facility for *in-situ* crystal growth by MBE. Activities of this kind have just started or are now underway in several laboratories.

This work is part of the research of the Stichting voor Fundamenteel Onderzoek der Materie (Foundation for Fundamental Research on Matter) and was made possible by financial support from the Nederlandse Organisatie voor Zuiver-Wetenschappelijk Onderzoek (Netherlands Organisation for the Advancement of Pure Research).

REFERENCES

1) Pandey, K.C., Phys. Rev. Lett. 47,1913 (1981).
2) Pandey, K.C., Phys. Rev. B25, 4338 (1982).
3) Vanderbilt, D., and Louie, S.G., Phys. Rev. B30, 6118 (1984).
4) see, e.g., Pate, B.B., Surface Sci. 165, 83 (1986).
5) Derry, T.E., Smit, L., and van der Veen, J.F., Surface Sci. 167, 502 (1986).
6) van der Veen, J.F., Surface. Sci. Rept. 5, 199 (1985).
7) Tromp, R.M., Smit, L., and van der Veen, J.F., Phys. Rev. Lett. 51, 1672 (1983).
8) Smit, L., Tromp, R.M., and van der Veen, J.F., Surface Sci. 163, 315 (1985).

9) Himpsel, F.J., Marcus, P.M., Tromp, R.M., Batra, I.P., Cook, M.R., Jona, F., and Liu, H., Phys. Rev. B30, 2257 (1984).
10) Binnig, G., Rohrer, H., Gerber, Ch., and Weibel, E., Phys. Rev. Lett. 50, 120 (1983).
11) Takayanagi, K., Tanishiro, Y., Takahashi, S., and Takahashi, M., Surface Sci. 164, 367 (1985).
12) Tromp, R.M., and van Loenen, E.J., Surface Sci. 155, 441 (1985).
13) Robinson, I.K., J. Vac. Sci. Technol. A4, 1309 (1986).
14) Tromp, R.M., Hamers, R.J., and Demuth, J.E., to be published.
15) Hamers, R.J., Tromp, R.M., and Demuth, J.E., Phys. Rev. Lett. 56, 1972 (1986).
16) Phillips, J.C., Phys. Rev. Lett. 45, 905 (1980).
17) Pearson, E., Halicioglu, T., and Tiller, W.A., Surface Sci. 168, 46 (1986).
18) Nakagawa, K., Marée, P.M.J., and van der Veen, J.F., these proceedings.
19) Tromp, R.M., Hamers, R.J., and Demuth, J.E., Phys. Rev. Lett. 55, 1303 (1985), and refs. therein.
20) Kahn, A., Surface Sci. Rept. 3, 193 (1983).
21) Smit, L., and van der Veen, J.F., Surface Sci. 166, 183 (1986).
22) Tong, S.Y., Xu, G., and Mei, W.N., Phys. Rev. Lett . 52, 1693 (1984).
23) Bohr, J., Feidenhans'l, R., Nielsen, M., Toney, M., Johnson, R.L., and Robinson , I.K., Phys. Rev. Lett. 54, 1275 (1985).
24) Kaxiras, E., Pandey, K.C., Bar-Yam, Y., and Joannopoulos, J.D., Phys. Rev. Lett . 56, 2819 (1986).
25) Chadi, D.J., Phys. Rev. Lett. 57, 102 (1986).
26) Kaxiras, E., Bar-Yam, Y., Joannopoulos,J.D., and Pandey, K.C., Phys. Rev. Lett. 57, 106 (1986).
27) Kasper, E., Herzog, H.J., and Kibbel, H., Appl. Phys. 8, 199 (1975).
28) Abstreiter, G., Brugger, H., Wolf, T., Jorke, H., and Herzog, H.J., Phys. Rev. Lett. 54, 2441 (1985).
29) People, R., Bean, J.C., and Lang, D.V., J. Vac. Sci. Technol. A3, 846 (1985).
30) Bevk, J., Mannaerts, J.P., Feldman, L.C., Davidson, B.A., and Ourmazd, Appl. Phys. Lett. 49, 286 (1986).
31) Saitoh, S., Ishiwara, H., and Furukawa, S., Appl. Phys. Lett 37, 203 (1980).
32) Rosencher, E., Badoz, P.A., Pfister, J.C., Arnaud d'Avitaya, F., Vincent, G., and Delage, S., Appl. Phys. Lett. 49, 271 (1986).
33) Tung, R.T., Gibson, J.M., and Poate, J.M., Phys. Rev. Lett. 50, 429 (1983).
34) Tung, R.T., Phys. Rev. Lett. 52, 461 (1984),
35) Liehr, M., Schmid, P.E., LeGoues, F.K., and Ho, P.S., Phys. Rev. Lett. 54, 2139 (1985), see also these proceedings.
36) van Loenen, E.J., Frenken, J.W.M., van der Veen, J.F., and Valeri, S., Phys. Rev. Lett. 54, 827 (1985).
37) Vlieg, E., Fischer, A.E.M.J., van der Veen, J.F., Dev, B.N., and Materlik, G., Surface Sci., in press.
38) Fischer, A.E.M.J., Gustafsson, T., Nakagawa, K., and van der Veen, J.F., to be published.
39) Cherns, D., Anstis, G.R. Hutchinson, J.L., and Spence, J.C.H., Phil. Mag. 46A, 849 (1982).
40) Gibson, J.M., Bean, J.C., Poate, J.M., and Tung, R.T., Appl. Phys. Lett. 41, 818 (1982).
41) Tung, R.T., Ng, K.K., Gibson, J.M., and Levi, A.F.J., Phys. Rev. B33, 7077 (1986).

II. SURFACES

PHOTOELECTRON SPECTROSCOPY STUDIES OF SEMICONDUCTOR SURFACE STATES

Göran V. Hansson and Per Mårtensson
Department of Physics and Measurement Technology
Linköping Institute of Technology
S-581 83 Linköping
SWEDEN

ABSTRACT

The electronic structure of Si, Ge, and Si/Ge surfaces has been studied with angle-resolved photoemission. Using highly n-doped crystals, the minima of normally empty surface-state bands have been investigated together with the filled surface-state bands for Si(111)2x1, Ge(111)2x1, and Si(100)2x1 surfaces. A description of the characteristics of the different surface states on the Si(111)7x7 surface will be given and a comparison will be made with results for the Ge-covered Si(111)5x5-Ge and Si(111)7x7-Ge surfaces.

1. INTRODUCTION

Angle-resolved photoelectron spectroscopy (ARPES) has been used extensively in studies of the electronic structure of semiconductor surfaces. Although the first photoelectron spectra showing surface states on the Si(111) surface were presented almost fifteen years ago, it is only during the last few years that the electronic structure of semiconductor surfaces has been determined in such detail as to be used in direct comparisons with theoretical surface band structure calculations.

The cleaved Si(111)2x1 and Ge(111)2x1 surfaces will be used as examples of surfaces for which the calculated band structures for a model geometry can explain the experimentally obtained dispersions. The Si(100)2x1 surface on the other hand has several main features in photoemission that cannot be accounted for in calculations made so far. Finally, the Si(111)7x7 surface is an example of a surface with a unit cell too large for a full calculation of the electronic structure. The experimental surface state studies can, however, still be used to draw conclusions about the geometry of the surface. Photoemission results for the Ge-covered Si(111)5x5-Ge and Si(111)7x7-Ge surfaces will also be shown, illustrating effects of long range order and local bonding.

2.   Si(111)2x1 AND Ge(111)2x1

The atomic and electronic structures of the cleaved silicon and germanium surfaces are quite similar. Both semiconductors form 2x1 reconstructed domains that can have three different orientations on the (111) surface. Using special cleavage procedures, it has been possible to prepare large single-domain surfaces and to determine the energy dispersion of the filled surface-state bands.

For both the Si(111)2x1 and the Ge(111)2x1 reconstructions, it is now established that there is a filled surface-state band that has a strong dispersion ($\sim$0.8 eV) along the outer half of the $\bar{\Gamma}$-$\bar{J}$ line in the 2x1 surface Brillouin zone (SBZ), with maximum energy at the SBZ boundary.[1-3] This characteristic dispersion is found in theoretical calculations for the π-bonded chain model of the surface, providing strong support for the type of reconstruction suggested in this model.[4,5]

A fundamental problem in semiconductor surface physics is the understanding of Fermi-level pinning, which often has been attributed to defect states in the gap. In recent ARPES studies using highly n-doped crystals, it has been possible to detect emission from the minima of almost empty surface-state bands. In Fig. 1, we show photoemission spectra measured from a highly n-doped ($1\cdot10^{18}$ cm^{-3}) Ge(111)2x1 surface.[3] The geometry is such that electrons are collected on the same side of the surface normal as the light is incident. With this geometry, the emission from the filled surface state band, peak A, is suppressed relative to emission from the almost empty band, peak B. The peak at ~1.5eV is a bulk contribution. This set of spectra shows the rapid dispersion of the filled band, having maximum energy at the J̄-point on the SBZ boundary. Emission from the almost empty band is observed only near the J̄-K̄ line at the SBZ boundary.

The narrow angular range of the emission from the almost empty band (4° full width half maximum (FWHM)) corresponds to a k̄-space localization of 0.06 Å^{-1} (FWHM). Using the uncertainty principle, one can estimate a lower limit of 45 Å for the real space extent of the surface states. Considering that the experimental angle-resolution was ±2°, it is clear that the true k̄-space localization is even stronger than estimated here. This rules out the existence of very localized states in flat polaron bands as was suggested for Si in Refs. 6 and 7.

The measured energy dispersion for the filled and almost empty surface-state bands are shown in Fig. 2, together with the calculated bands for the π-bonded chain model. Although the calculated bands have to be shifted to lower energies to coincide with the experimentally

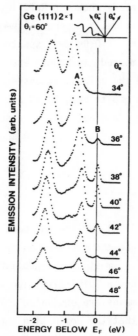

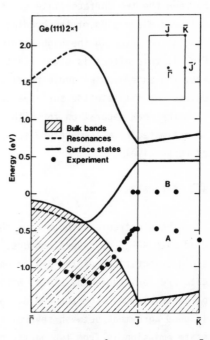

Fig.1. Photoelectron spectra[3] probing near the J̄ point ($\theta_e \sim 40°$) The photon energy used is 10.2 eV.

Fig.2. Measured[3] and calculated[5] dangling bond band dispersions for Ge(111)2x1.

obtained dispersions, there is very good agreement concerning the characteristic shape of the filled (bonding) and the position of the minimum of the almost empty (antibonding) band, which is pinning the Fermi
level at 0.1 eV above the valence band edge.

Measurements on the Si(111)2x1 surface give very similar results[8],
i.e., the Fermi level for highly n-doped samples is pinned by the minimum of the antibonding band at the $\bar{J}$-point in the SBZ. The agreement
with calculations[9] is better than for germanium concerning the absolute
energy positions of the surface state bands.

In a recent study[7] of the temperature dependence of transitions
across the surface-state band gap on Si(111)2x1, a detailed comparison
was made with the photoemission results. Since a high accuracy of the
band gap determination is needed for a discussion of, e.g., electron-
phonon effects, it is necessary to discuss the difference between the
surface-state band gap and what is measured as the energy difference
between the two surface-state peaks. As already shown in Ref. 10, the
measured dispersion of the bonding band critically depend on the $k_{//}$-
resolution. Even in the high resolution ($\Delta k_{//} < 0.06$ Å^{-1}) spectra in
Fig. 1, this effect is significant when compared to phonon energies.
From the experimental angle resolution and the calculated dispersion
curves, we estimate the surface-state band gap to be 0.03 eV smaller
than the peak separation for both Si(111)2x1 and Ge(111)2x1, resulting
in estimated room temperature surface-state band gaps of 0.40 and 0.47
eV respectively.

3.   Si(100)2x1

The electronic structure of the Si(100)2x1 surface has been studied with ARPES by several groups.[11-13] Since the surface has two possible orientations of the 2x1 reconstruction, most of the published
data are for studies of states with $k_{//}$ along a [010]-direction whereby contributions from the two different domains are equivalent. There
is good agreement between several studies that the dominating surface-
state emission is from two partly overlapping surface-state bands approximately 1 eV below the Fermi level. In Fig. 3 we show the dispersion
of these two filled surface-state bands A and B. Included is also the

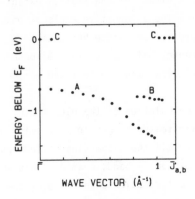

Fig.3. Dispersion of surface state bands on Si(100)2x1 measured along a [010]-direction using 10.2eV photon energy[13].

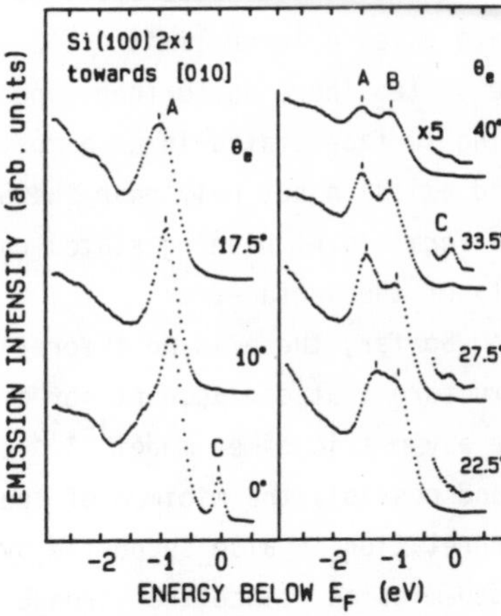

Fig.4. Photoelectron spectra from Si(100)2x1 showing emission from three surface states A,B and C.

peak position of a recently discovered surface state at the Fermi level (C).

In Fig. 4, there are some spectra showing the emission from the different surface states on a highly n-doped silicon crystal ($N_D \sim 1 \cdot 10^{19}$ cm^{-3}, Sb). The photon energy used is 21.2 eV. The metallic surface state (C) is observed around normal emission, but also around $\theta_e = 33°$, which corresponds to emission from a $\bar{J}'$ point in both 2x1 domains. Localized emission from the same high symmetry points has also been observed in photoemission experiments[13] using lower photon energies ($\hbar\omega$ = 10.2 eV) on samples with the same level of n-doping. The intensity of the metallic surface state emission near the normal direction was much lower in that experiment indicating a strong photon energy dependence like the one reported for a similar surface state on the Ge(100)2x1 surface.[14]

Goldman et al.[12] recently reported emission from the metallic surface state of the Si(100)2x1 surface. They found emission only near the surface normal, with much lower intensity than that shown in Fig. 4. This is reasonable since they used a sample with low p-doping, having few electrons in the states pinning the Fermi level. It has been suggested that the metallic surface states on Ge(100)2x1 and Si(100)2x1 are due to localized defects,[12,14] but as discussed in Ref. 13, we find

that the strong $\bar{k}$-space localization found for the germanium surface
state gives a lower limit of 50 surface atoms for the spatial extent of
the state. There is further support for the alternative explanation
using surface states in an almost empty surface-state band, since we
find emission not only near the SBZ-center but also at the SBZ-boundary.
The Fermi level pinning states on Si(100)2x1 thus reflect the periodicity of the surface.

So far, there is no theoretical calculation of the electronic
structure that can account for all main features seen in photoemission.
The asymmetric dimer model[15,16] can explain the dispersing filled band
A and possibly the minimum of the almost empty band, C at $\bar{J}'$. This interpretation is also supported by the polarization dependence of the
photoemission, since the strongest P_z-character of the surface states
is found for structure A and structure C near the $\bar{J}'$ point, where they
would correspond to dangling bond states on the upper and lower atoms
of the dimer. However, for calculations giving the correct bandwidth of
the filled band[15,16], the calculated surface-state band gap is $\sim$0.2 eV
while it is 0.70 eV in the experiment. Further, there is no explanation
for the second filled band B and the minimum of the almost empty band
at $\bar{\Gamma}$.

4. Si(111): THE CLEAN 7x7 AND Ge-COVERED 5x5 AND 7x7 SURFACES

The surface electronic structure of the clean Si(111)7x7 surface
is dominated by three easily separable surface states S_1, S_2 and S_3
that are seen over large solid angles. Some typical spectra measured
along the [01$\bar{1}$] azimuthal direction using p-polarized light with 21.2
eV photon energy ($\theta_i=45°$) are shown in Fig. 5a. S_1 is at $\sim$ -0.2 eV relative to the Fermi level (E_F) and it has its intensity maximum at 15°
emission angle. S_2 is at $\sim$ -0.8 eV in normal emission but has a small
(0.1 eV) upwards dispersion with maximum energy at $\sim$15°. Finally, S_3
is not seen in normal emission, but grows to maximum intensity at
$\theta_e \sim 25°$. It has a downwards dispersion of $\sim$0.35 eV with the minimum in
energy at the 1x1 SBZ boundary.

The polarization dependence of emission shows that all three surface states are suppressed if the light is polarized parallel to the

surface. This is evidence for significant P_z-character of all three surface states. The opposite polarization dependence for S_3 has been reported in an early study.[17] In experiments using 21.2 eV radiation, a structure is found in normal emission with similar energy as S_3 that is excited by light polarized parallel to the surface. From the photon energy and emission angle dependence of this structure, it can, however, be concluded that this is a bulk contribution from the uppermost two valence bands near the L-point. Similar bulk emission has also been reported from the Si(111)2x1 surface.[18]

During the last few years, there has been a large interest in Si/Ge alloys and superlattices. By annealing thin deposits of Ge on Si(111) surfaces, it has been possible to prepare Si(111)5x5-Ge and Si(111)7x7-Ge surfaces.[19] From the attenuation of the 92 eV Si(LVV) Auger transition the estimated Ge-content is $\sim$50% for the 5x5-Ge surface and $\sim$30% for the 7x7-Ge surface assuming a homogeneous alloy within the probe depth (λ_e=4.4 Å). In Figs. 5b and 5c, we show photoemission spectra measured from a 7x7-Ge and a 5x5-Ge surface, with the same experimental configuration as for the data of the clean 7x7 surface in Fig. 5a.

There is a gradual change in electronic structure going from the clean 7x7 surface to the 7x7-Ge and 5x5-Ge surfaces. The S_1 state is replaced by the A_1 and B_1 states respectively. They all have similar energies and emission patterns. The S_2 state seen strongly in normal emission corresponds to peaks A_2 and B_2 which are shifted by $\sim$0.15 eV to higher binding energies. Finally, the strong peak S_3 at $\theta_e \sim 25°$ is replaced by the structures A_3 and B_3 which are shifted to lower binding energies by $\sim$0.25 and $\sim$0.40 eV respectively. These shifts have a major effect on the shape of the spectra, since for the Ge-covered surfaces the two lower lying surface states are partly overlapping.

It is worth noting that the changes in the spectra seem to correlate more with the Ge-content on the surfaces than with the surface periodicity. The changes in the spectra when changing the reconstruction from 7x7-Ge to 5x5-Ge are not larger than the differences between the spectra of the clean and Ge-covered 7x7 surfaces. This is consistent with STM-studies[20] of the 5x5-Ge surface which showed strong similarities with the 7x7 reconstruction.

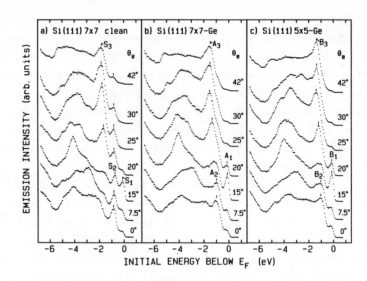

Fig.5. ARPES spectra[19] measured along a $[10\bar{1}]$-azimu-
thal direction with 21.2 eV photon energy. The light
is p-polarized, incident at θ_i=45.

REFERENCES

1. R.I.G. Uhrberg et al., Phys. Rev. Lett. 48, 1032 (1982).
2. F.J. Himpsel et al., Surf. Sci. 132, 22 (1983).
3. J.M. Nicholls et al., Phys. Rev. Lett. 54, 2363 (1985).
4. K.C. Pandey, Phys. Rev. Lett. 47, 1913 (1981).
5. J.E. Northrup and M.L. Cohen, Phys. Rev. B27, 6553 (1983).
6. C.D. Chen et al., Phys. Rev. B30, 7067 (1984).
7. F. Ciccacci et al., Phys. Rev. Lett. 56, 2411 (1986).
8. P. Mårtensson et al., Phys. Rev. B32, 6959 (1985).
9. J.E. Northrup and M.L. Cohen, Phys. Rev. Lett. 49, 1349 (1982).
10. J.M. Nicholls et al., Phys. Rev. Lett. 52, 1555 (1984).
11. F.J. Himpsel and D.E. Eastman, J. Vac. Sci. Technol. 16 (5), 1297
    (1979).
12. A. Goldmann et al., Surf. Sci. 169, 438 (1986).
13. P. Mårtensson et al., Phys. Rev. B33, 8855 (1986).
14. S.D. Kevan, Phys. Rev. B32, 2344 (1985).
15. J. Ihm et al., Phys. Rev. B21, 4592 (1980).
16. M. Schmeitz et al., Phys. Rev. B27, 5012 (1983).
17. D.E. Eastman et al., Proc. of the 14th Int. Semiconductor Conf.
    (Inst. of Phys. Conf. Series 43), 1059 (1979).
18. R.I.G. Uhrberg et al., Phys. Rev. B31, 3805 (1985).
19. P. Mårtensson et al., Phys. Rev. B34 (in press) and unpublished.
20. R.S. Becker et al., Phys. Rev. B32, 8455 (1985).

ATOMIC IMAGING OF SURFACE ELECTRONIC STATES ON Si†

J. E. Demuth, R. J. Hamers and R. M. Tromp

IBM Thomas J. Watson Research Center
Yorktown Heights, New York 10598
USA

ABSTRACT

A method for delineating geometric and electronic contributions to STM images is described. Atomically resolved images of the filled and empty surface states of Si(111)-7×7 and Si(100)-2×1 are obtained and directly related to the local atomic structure of the surface.

1. INTRODUCTION

The Scanning Tunneling Microscope has become an important method to directly examine semiconductor surface structure.[1-5] Tunneling is determined by the overlap of the wavefunctions between the tip and sample, and thereby depends upon both the magnitude of these wavefunctions and the location of the atoms to which they are tied. Recent theories of STM[6,7,8] quantify this notion and show that for small bias the constant current feedback condition used for scanning makes the tip follow a contour of constant charge density for the states at the Fermi level.[7,8] However, in most semiconductors the Fermi level is pinned in or near the gap by surface states, which can then dominate the STM image. This has in one case, for example, allowed the selective imaging of defect-derived states near the surface state gap of Si(111) 2×1.[4] For most semiconductors studied, however, STM images are difficult to obtain at low bias voltages and higher bias voltages must be used.[1,5,9] In this case tunneling occurs from the overlap between all filled (empty) states of the tip and empty (filled) states of the sample between their Fermi levels. Differences between these states have led to dramatic bias-dependent changes in STM images.[9] Such effects while problematical for atomic scale structure analysis provide a means to spatially probe surface state charge densities sampled at different bias voltages.

Here, we summarize a new method of control and measurement with the STM which allows us to separate electronic and geometric contributions so as to image the surface states of Si(111)-7×7 and Si(100)-2×1. All other attempts to do atomic scale spectroscopic imaging have utilized taking dI/dV in the constant current feedback mode,[10,11] and are limited by: feedback control derived divergence of dI/dV below ~1eV and complications in interpreting these images due to strong non-spectroscopic terms in this partial derivative.[11]

† Work partially supported by the U.S. Office of Naval Research

50 J.E. Demuth et al.

2. 2-D IMAGING OF SURFACE STATES

For small bias, V (in eV), and neglecting the momentum dependence of the tunneling, the tunneling at one x,y point of the surface is roughly approximated as:

$$I(x,y) \cong \int_{\varepsilon_F}^{\varepsilon_F+V} dE \rho_s(x,y;E-V) e^{-2s(x,y)\sqrt{2(W-E)+V}} \rho_t(E)$$

where s(x,y) represents the effective barrier thickness, W, the average barrier height and ρ_s or ρ_t, the density of states of the sample or tip respectively.[12] Maintaining a constant current feedback condition forces the barrier thickness, s, to vary at each position and for different bias voltages. If this barrier thickness is maintained constant for each position and bias voltage, and assuming ρ_t to be weakly varying, then I(x,y) gives a voltage weighted integral of $\rho_s(x,y)$. This measurement condition can be realized experimentally by using a bias voltage where the STM image follows a contour of constant atomic charge density. (Recall that the barrier is derived from the total effective potential determined by *all* the electrons.) If this barrier is further maintained constant for different bias voltages, then I(x,y) for each V can be compared to derive a 2-D map of the different surface states.

3. EXPERIMENTAL METHOD: CURRENT IMAGING TUNNELING SPECTROSCOPY

We have developed a method for controlling the feedback of the STM that allows us to operate the feedback independent of I-V measurements.[13] This is achieved by repetitively ramping the sample bias over some energy range at 2.2 kHz and sampling the tunneling current at particular bias voltages. We use one of the bias voltages sampled to control the feedback which is chosen from prior analysis to maintain a constant barrier height, and we measure a series of tunnel currents at the other voltages. The current responds rapidly (~200 kHz bandwidth), and we become limited by our A/D conversion rates and ability to manage the data stream as it is acquired. Our present system allows us to acquire 49 channels for each 12 bit pixel element of a 100×100 array in 5 minutes. One of these channels is used to simultaneously acquire the normal topographic image as it arises from the feedback loop. This provides a direct comparison between each pixel in the topograph and its corresponding I-V data, or with 2-D current images at any bias. The STM used for these experiments is described elsewhere.[14] We use an electropolished tungsten tip, n-type (5mΩ cm) Si samples and make no corrections for thermal drift in the images.

4. THE ELECTRONIC STRUCTURE OF Si(111)-7×7 and Si(100)-2×1

We have found that tunneling into the unoccupied states of Si(111)-7×7 at +2eV produces an image that closely follows a contour of constant atomic charge density calculated by atomic charge superposition (ACS) for the DAS model.[9] Using this bias to control the feedback we obtain the current images shown in Fig. 1 for a variety of voltages on the sample. We have previously reported I-V spectra for several points in the unit cell[13] which show onsets that correspond closely to the surface states observed in normal and inverse photoemission, PE.[15] At −.25eV we see the highest lying surface states located on adatoms on the faulted half of the unit cell. At −0.70eV we see a superposition of the ad-atom derived states and states located on atoms in between the ad-atoms, called "rest atoms". At −1.45eV the tunneling from rest atom states now dominate and at −2eV we see additional diffuse intensity near the corner holes and in between rest atoms. This diffuse intensity is located near the edges of atoms in the 7×7 structure and has been attributed to back bonds.[13]

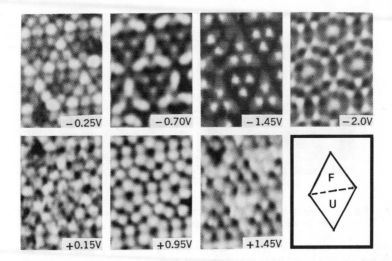

Fig. 1.   Current images of Si(111)-7×7 at selected negative and positive bias as indicated with the feedback stabilized at +2V. The orientation and location of the faulted (F) and unfaulted (U) halves of the unit cell are indicated.

For +.15eV bias, we see a similar pattern of empty states as for the filled states.   Starting above +.25 the ad-atoms states are nearly equally distributed but with a slight 3 fold symmetry about the corner hole rotated 180° to that for the occupied ad-atom state.   Above +1.3eV an overall increase in intensity occurs for the unfaulted side of the unit cell.   These changes in tunneling symmetry seen for both filled and empty states also correspond to the onsets we find in the I-V curves.

Analysis of our STM results for Si(100)-2×1 reveals a close correspondence between our STM images taken at −2V and ACS calculations for dimer structures.[16] Using this bias for the feedback control we have also examined the current images between ±2eV.   Fig. 2 shows the main surface state features found with current imaging at −1.2 and +1.2eV which correspond to tunneling out of (into) the filled (empty) surface states seen in UPS.[15]   The top right corner of each image shows buckled dimers whose charge density is consistent with that calculated for the asymmetric dimer.[17]   The energy of the states we see on the "symmetric" dimers do not differ from those on the asymmetric dimers and argue that the "symmetric" dimers we observe are time-averaged asymmetric dimers.   The irregular and sometimes enhanced charge transfer seen along the rows of "symmetric" dimers would argue that small local defects readily perturb the charge to disrupt this dynamic buckling.

5. SUMMARY AND CONCLUSIONS

We have atomically imaged a variety of surface states on Si(111) and (100) seen previously by PE and can directly relate the surface state orbitals to the geometric structure on an atomic scale.   For the Si(111)-7×7, the surface states are spatially very complex and appear to correspond to orbitals of a large surface compound.   One important aspect of our surface state tunneling images relative to PE is that these images may be 'distorted' for large bias and/or because of momentum dependent effects, i.e. tunneling is stronger for states at the zone center than at the

J.E. Demuth et al.

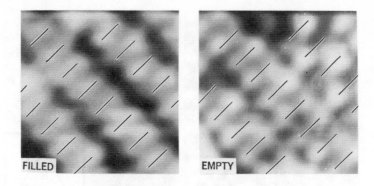

Fig. 2. Images of the filled and empty surface state orbitals of Si(100)-2×1. The dimers in each row are indicated by the bars and are 3.8Å apart.

zone edges. Finally, in other systems where we have no a priori knowledge of the detailed surface structure, such atomically resolved I-V spectra and surface state orbital locations provide important additional information for evaluating and testing geometric and electronic structure models consistent with one another.

6. REFERENCES

1. Binnig, G., Rohrer, H., Gerber, C., and Weibel, E., Phys. Rev. Lett. **50**, 120 (1983).
2. Becker, R. S., Golovchenko, J. A., and Swartzentruber, B. S., Phys. Rev. Lett. **54**, 2678 (1985).
3. Tromp, R.M., Hamers, R.J., and Demuth, J.E., Phys. Rev. Lett. **55**, 1303 (1985).
4. Feenstra, R.M., Thompson, W.A., and Fein, A.P., Phys. Rev. Lett. **56**, 608 (1986).
5. Feenstra, R. M. and Fein, A. P., Phys. Rev. B **32**, 1394 (1985).
6. Baratoff, A., Physica **127B**, 143 (1984).
7. Tersoff, J. and Hamann, D. R., Phys. Rev. B **31**, 805 (1985).
8. Lang, N.D., Phys. Rev. Lett. **56**, 1164 (1986).
9. Tromp, R. M., Hamers, R. J., and Demuth, J.E., Phys. Rev. B**34**, 1388 (1986).
10. Becker, R. S., Golovchenko, J. A., Hamann, D. R., and Swartzentruber, B. S., Phys. Rev. Lett. **55**, 2032 (1985).
11. Binnig, G. and Rohrer, H., IBM J. of Res. and Dev., **30**, 354 (1986).
12. Lang, N. D., Phys. Rev. B, to be published.
13. Hamers, R.J., Tromp, R.M., and Demuth, J.E., Phys. Rev. Lett. **56**, 1972 (1986).
14. Demuth, J. E., Hamers, R. J., Tromp, R. M., and Welland, M. E., J. Vac. Sci. Tech. A4, 1320 (1986).
15. Himpsel, F. J., Straub, D., and Fauster, Th., Proc. 17th Int'l Conf. Phys. Semi, J. D. Chadi and W. A. Harrison, Springer-Verlag, NY, p. 39 (1985).
16. Hamers, R.J., Tromp, R.M., and Demuth, J.D., Phys. Rev. B, to appear.
17. Ihm, J., Cohen, M. L., and Chadi, D. J., Phys. Rev. B **21**, 4592 (1980).

Photothermal Modulation of the Gap Distance in Scanning Tunneling Microscopy

Nabil M. Amer[a], Andrew Skumanich[b], and Dean Ripple[c]

[a]IBM Thomas. J. Watson Research Center, P.O. Box 218, Yorktown Heights, NY 10598.
[b]IBM Almaden Research Center, San Jose, CA 95120.
[c]Physics Department, Cornell University, Ithaca, NY 14853.

ABSTRACT

The photothermal effect has been employed to modulate the the gap distance in a tunneling microscope. In this approach, optical heating induces the expansion and buckling of laser-illuminated sample surface. The surface displacement can be modulated over a wide frequency range, and its height (typically $< 1\text{Å}$) can be varied by changing the illumination intensity and modulation frequency. This novel method provides an alternative means for performing tunneling spectroscopy and microscopy.

Scanning tunneling microscopy [1] has been shown to yield structural, and potentially spectroscopic information on an atomic scale. In both microscopy and spectroscopy, the modulation of the tunneling gap distance is an essential step in obtaining the desired information. Typically, the distance modulation is accomplished using piezodrives. In this paper, we describe a novel approach which employs the photothermal effect [2] to modulate the gap.

The physical mechanism underlying the photothermal effect is that when an intensity-modulated beam of electromagnetic radiation is incident on an absorbing medium, heating will ensue. A typical temperature rise associated with the process is on the order of 10^{-4} °C. This optical heating causes the expansion and buckling (typically $< 1\text{Å}$) of the illuminated surface [3] as schematically shown in Fig. (1).

The height of the surface displacement, h, is given by [3]

$$h = P \, (\beta/\rho C) \, (1-R) \, (1/2\pi af)$$

where P is the incident power, β the thermal expansion coefficient of the sample, ρ its density, C the heat capacity, R the reflectivity, a the area of the illuminated surface, and f the modulation frequency of the incident light.

Next, we show how the photothermal displacement effect can be quantitatively applied to tunneling microscopy. The tunneling current, J_T, is given by [4]

$$J_T = (V/x) \exp(-A \, \varphi^{1/2} \, x)$$

where V is the applied voltage, x is the tunneling distance, $A \approx 1.025$ $eV^{-1/2}\overset{\circ}{A}^{-1}$,
and φ is the average of the work functions of the tip and the sample. The modulated tunneling current can then be written as

$$dJ_T/dx = -x^2 \, V \exp (-A \, \varphi^{1/2} x) + x^{-1} \, V \, (-A \, \varphi^{1/2}) \exp (-A \, \varphi^{1/2}x)$$

and

$$([dJ_T/dx])/J_T = x^{-1} + A \, \varphi^{1/2}$$

Since dx = h, then

$$(dJ_T/J_T \,)(h^{-1}) = x^{-1} + A \, \varphi^{1/2}$$

Consider the following two limiting cases:

1) $x \gg (A\varphi^{1/2})^{-1}$, then

$$(dJ_T/J_T) (h^{-1}) = A \, \varphi^{1/2}$$

which provides a direct measure of the work function; and

2) $x \ll (A \varphi^{1/2})^{-1}$, then

$$(dJ_T/J_T)(h^{-1}) = x^{-1}$$

which yields the value of x.

The experimental arrangement is shown in Fig. (1). The design of the tunnel junction combined the use of high-resolution micrometers and piezoelectric tubes.[5] The results reported here were obtained by operating the microscope in air. Phase-sensitive detection was used to measure the AC component of the tunneling current produced by the the photothermal modulation of the gap.

We performed a series of experiments aimed at testing the validity of the theoretical treatment outlined above. Since $dJ_T \propto h$, then dJ_T should reflect the dependence of h on both the modulation frequency and the intensity of the illuminating beam. The results are shown in Figs. (2,3), indicating excellent agreement between theory and experiment. In addition, the displacement height shows the expected exponential decay away from the optically heated region (the case where the thermal diffusion length is smaller than the laser beam spot size) as seen in Fig. (4). The observed behavior clearly exhibits the known characteristics of photothermal displacement[3].

In summary, we have presented a new method for modulating the gap distance in tunneling junctions. It enables modulation over wide frquency and temperature ranges , and opens up the prospect of performing time-resolved studies with scanning tunneling microscopes/spectrometers.

REFERENCES

1. G. Binnig, H. Rohrer, Ch. Gerber, and E. Weibel, Phys. Rev. Lett. 49, 57 (1982).

2. N. M. Amer, J. Phys. (Paris) 44, C6-185 (1983).

3. N. M. Amer and M. A. Olmstead, Surf. Sci. 132, 68 (1983); M. A. Olmstead, N. M. Amer, S. Kohn. D. Fournier, and A. C. Boccara, Appl. Phys. A 32, 141 (1983).

4. R. H. Fowler and L. Nordheim, Proc. R. Soc. London A119, 173 (1928); J. Frenkel, Phys. Rev. 36, 1604 (1930).
5. D. Ripple, A. Skumanich, and N. M. Amer, to be published.

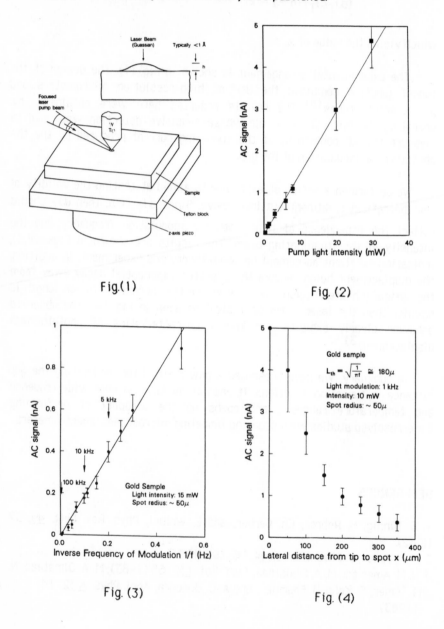

Fig.(1)

Fig. (2)

Fig. (3)

Fig. (4)

INVESTIGATION OF GRAPHITE SURFACE USING SCANNING TUNNELING MICROSCOPE AND SCF-ELECTRONIC STRUCTURE CALCULATIONS

Inder P. Batra*, S. Ciraci[†], N. Garcia[‡], H. Rohrer,

H. Salemink, and E. Stoll

IBM Zurich Research Laboratory, 8803 Rüschlikon, Switzerland

Graphite surface is proving to be an important system for Scanning Tunneling Microscopy (STM).[1] Here, we present an STM study of the (0001) surface of pyrolytic graphite, together with a theoretical analysis based on the self-consistent calculations for the charge density. The issues we address are: (i) can the STM distinguish between two inequivalent atoms in the graphite surface layer; (ii) why the images seen do not show the trigonal symmetry of the lattice, and (iii) what may be the cause of unexpectedly large corrugations?[1,2] Following Tersoff et al.[3] we assume implicitly that the STM images are essentially contours of local density states at the Fermi energy, $\rho(\vec{r}, E_f)$.

Experiments were performed with a "pocket size" instrument mounted inside an UHV chamber, at a pressure of 5×10^{-11} mbar. The scanning velocity was chosen between 45 and 150 Å/sec, and the tunnel voltages ranged from 10 to 700 mV with a fixed tunneling current of 1.0 nA. A typical set of line scans are shown in Fig. 1(a) demonstrating the high lateral resolution of less than 0.5 Å with the reproducibility of the peaks and valleys. Corrugations around 2-3 Å and sometimes even higher were recorded. Figure 1(b) shows contour plots of STM line scans. Note the incomplete threefold symmetry.

* Permanent address: IBM Almaden Research Center, K33/801, San Jose CA 95120, U.S.A.
† Permanent address: Department of Physics, Middle East Technical University Ankara, Turkey.
‡ Permanent address: Departmento de Fisica Fundemental, Universidad Autonoma de Madrid, Spain.

(a)   (b)

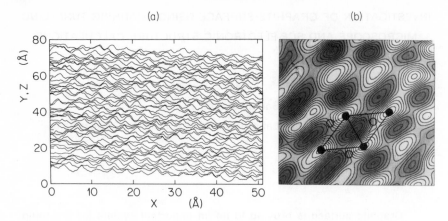

Fig. 1.(a) STM line scans, and (b) corrugation contours with a spacing corresponding to 0.4 Å vertical displacement of the tip. Full lines deliniate the unit cell, and open and full circles denote two types of carbon-atom positions.

Calculations are performed by using the standard SCF-pseudo-potential method for a monolayer, three-layer slab (in ideal and slipped configuration). In the ideal configuration, two types of surface atoms — one with a second-layer atom below (A-site), and the other without (B-site) — are identified. The center of the hexagon is denoted by H.

For graphite $\rho(\vec{r}, E_f)$ arises only from a few states at the corner of BZ, and consequently the STM image may be quite different from the total charge density $\rho_T(\vec{r})$.[4] To examine this, we carried out an extensive charge-density analysis. First, we found that $\rho_T(\vec{r})$ above the surface yields a small corrugation of $\sim$ 0.3 Å, but is able to resolve two different carbon-atom positions, the B-site having higher corrugation. The charge density $\rho(\vec{r}, \grave{E}_f)$ for the three-layer slab, relevant to STM, is calculated only approximately by summing slab states at the zone corner with a Fermi-Dirac thermal distribution. The contour plots shown in Fig. 2(a) reveal two important points. First, the corrugation is $\sim$ 1 Å between A- and H-sites. It is somewhat higher between B- and H-sites. STM has resolved this differential corrugation, but it has been incorrectly interpreted to suggest that the higher corrugation is between A- and H − sites.[1] Secondly, as far as the topology of the corrugation is

(a) $\rho(x,y,\ E_F)$ at z = 3.2 a.u.   (b)

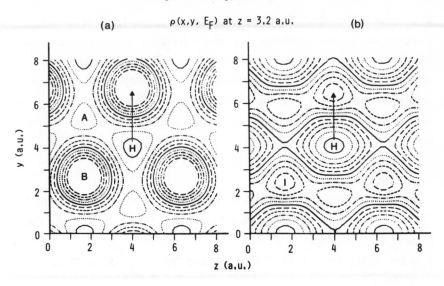

Fig. 2. Contour plots of $\rho(x,y, E_f)$ on a plane z = 3.2 a.u. above the surface; (a) ideal three-layer slab; (b) slipped surface.

concerned the total charge density and $\rho(\vec{r}, E_f)$ show the same trend. However, the dramatic difference lies in the value of the corrugation. The local density of states $\rho(\vec{r}, E_f)$ has a trigonal charge distribution with B-sites forming the vertices, and yields five times larger corrugation as compared to the total charge density. This result clearly contrasts graphite with other metal surfaces. Also, the huge corrugation in $\rho(\vec{r}, E_f)$ seen in our and earlier monolayer calculations[4] are substantially reduced in a realistic three-layer slab. These results are in agreement with Selloni et al.[5]

     Finally, to understand the origin of the missing trigonal symmetry we have investigated the three-layer slab with the top layer slipped by half the C-C bond distance. The A′-site atom lies in the bridge position relative to second-layer atoms, and thus continues to be different from the B′-site atom. In the new surface geometry, the total energy is ~ 7 mRy/atom higher than the ideal configuration, implying that the slipped surface may be easily generated. As seen in Fig. 2(b), $\rho(\vec{r}, E_f)$ does not show trigonal symmetry and H′-site minimum is oblate shaped.

In agreement with the STM corrugation contours, two types of atomic positions are readily identified. The total corrugation is somewhat higher than the ideal case, but is not sufficient to explain the corrugation observed greater than 2 Å. In summary, it can be stated that STM can resolve A- and B-sites, and the missing trigonal symmetry can be due to slipped surface layer. The large corrugation is difficult to understand in terms of charge-density analysis.

REFERENCES

1. Binnig, G., Fuchs, H., Gerber, Ch., Rohrer, H., Stoll, E., and Tosatti, E., Europhys. Lett. 1, 31 (1985); Park Sang-Il and Quate, C. F., Appl. Phys. Lett. 48, 112 (1986).

2. Soler, J. M., Baro, A. M., Garcia, N., and Rohrer, H., Phys. Rev. Lett. (1986).

3. Tersoff, J. and Hamann, D. R., Phys. Rev. Lett. 50, 1998 (1983); Garcia, N., Ocal, C. and Flores, F., Phys. Rev. Lett. 50, 2002 (1983); Stoll, E., Baratoff, A., Selloni, A., and Carnevali, P., J. Phys. C 17, 137 (1984).

4. Tersoff, J., Phys. Rev. Lett. (to be published).

5. Selloni, A., Carnevali, P., Tosatti, E. and Chen, C. D., Phys. Rev. B 31, 2602 (1985).

ADATOM ENERGETICS AND SURFACE STATES ON Si(111)7x7

John E. Northrup

Arizona State University, Tempe, AZ   85287 USA

ABSTRACT

First principles total energy calculations of the surface energies of four Si(111) adatom reconstructions are reported. An estimate of the surface energy for Takayanagi's DAS model for Si(111)7x7 is given. The structural origin of the surface states present on Si(111)7x7 is discussed.

1. ATOMIC STRUCTURE

The atomic structure of the Si(111)7x7 surface has recently been determined by a variety of experimental probes[1,2,3]. These studies indicate that the dimer–adatom–stacking fault (DAS) model proposed by Takayanagi et al.[1] is the correct 7x7 structure. The DAS model contains 12 adatoms, 7 rest atoms, 9 subsurface dimers, and a stacking fault in each unit cell. Rest atoms are threefold coordinated surface atoms not bonded to an adatom. In this paper we calculate the surface energy for several Si adatom structures and present an estimate of the surface energy of the DAS model.

Total energy calculations were performed for four different types of adatom geometries including two $\sqrt{3}$x$\sqrt{3}$ structures with the adatom in either the hollow site (H_3) or the filled site (T_4) and two geometries with 2x2 symmetry having the adatom in the T_4 or H_3 site[4]. These calculations were carried out using the first principles pseudopotential method employing the local density approximation. Force calculations were made for each of the adatom geometries and used to minimize the energy with respect to the atomic positions. The energies of these surfaces relative to the ideal unrelaxed surface are shown in Table I. From these calculations we conclude that adatoms are strongly bound to the relaxed 1x1 surface. The T_4–$\sqrt{3}$ surface energy is about 0.11 eV/(surface atom) lower than the relaxed 1x1 surface.

Table I.   Surface energy of adatom structures.
           Values given are in eV/(surface atom).

	Relative energy (eV)	Surface energy (eV)
Ideal 1 × 1	0	1.63
Relaxed 1 × 1	−0.17	1.46
H_3 − $\sqrt{3}$ × $\sqrt{3}$	−0.07	1.56
H_3 − 2 × 2	−0.24	1.39
T_4 − 2 × 2	−0.24	1.39
T_4 − $\sqrt{3}$ × $\sqrt{3}$	−0.28	1.35

It is interesting that the total dangling bond density in the DAS structure, 19/49, is intermediate between 1/3 for the $\sqrt{3}\times\sqrt{3}$ and 1/2 for the 2×2 structures. If minimization of the total dangling bond density were the only criterion, the T_4–$\sqrt{3}$ structure would be favored over the DAS model. However, this argument is oversimplified because the adatom dangling bonds (type A) and the rest atom dangling bonds (type R) are not equivalent in surface energy. One must consider the two types of broken bonds separately. The type (A,R) density is (12/49, 7/49) for the DAS model, (1/3, 0) for $\sqrt{3}\times\sqrt{3}$, and (1/4, 1/4) for the 2×2 surface. The DAS model has a lower type A density than either surface and a type R density lower than the 2×2 but higher than $\sqrt{3}\times\sqrt{3}$. The superior stability of the DAS model could result if rest atom dangling bonds were energetically less costly than adatom dangling bonds. We will now show that this is the case.

Assuming no interaction between the T_4–$\sqrt{3}$ adatom and rest atom building blocks of the T_4–2×2 geometry, we can write $E(2\times2)=(R+A)/4$, and $E(\sqrt{3})=A/3$, where R is the surface energy of a rest atom dangling bond and A is the surface energy of an adatom dangling bond. Then one obtains from Table I that A=4.05 eV and R=1.51 eV. Since R is very close to the calculated surface energy of the relaxed 1×1 surface (1.46 eV), the neglect of interaction between the adatoms and rest atoms is justified, and we can use this model to estimate the surface energy of the DAS model. We propose that $E(DAS) = (12A + 7R + 9S)/49$ where S is the energy cost of the strained dimer bonds in the substrate. The atoms in these substrate dimers are fourfold coordinated and have a local bonding topology similar to the

reconstructed 30° partial dislocation, which has bulklike bond
lengths and bond angles[5]. We estimate that 0.1 < S < 0.6 eV. Then
the estimated surface energy for the DAS model ranges from 1.22 to
1.32 eV/(surface atom). The T_4-√3 surface has a calculated surface
energy of 1.35 eV/(surface atom). The reason for the greater
stability of the DAS model is that it eliminates adatom dangling
bonds and replaces them with rest atom dangling bonds which are a
factor R/A (approximately 3/8) lower in surface energy. The Pandey
chain structure[6] has a calculated surface energy of 1.27 eV/(surface
atom). Thus S must be less than 0.35 eV for the DAS model to have
lower energy than the chain model.

2. ELECTRONIC STRUCTURE

We now discuss the electronic structure of the DAS model. A more
detailed discussion can be found in Ref. 4. From angle resolved
photoemission experiments[7,8] it is known that there are three
prominent types of surface states (S_1, S_2, and S_3) on the Si(111)7×7
surface. The experimental dispersion of these three states along the
ΓM azimuth of the 1×1 Brillouin zone is shown in Fig. 1. In the
calculations for the T_4-√3 structures, we find two types of surface
states. To compare with experiment we must map the bands from the
√3x√3 Brillouin zone into the 1×1 zone as discussed previously[4,9].
The mapped bands are shown in Fig. 1 as solid lines. The band
denoted Σ_1 is half filled and pins the Fermi level near mid gap. The
character of the Σ_1 band is that of a p_z orbital localized mainly on
the adatom, whereas Σ_3 is comprised of adatom p_x and p_y orbitals
which couple to p_z orbitals on the substrate. From the comparison
between theory and experiment it is apparent that Σ_1 corresponds to
S_1 and Σ_3 to S_3. However, there is no state in the T_4-√3 calculation
which corresponds to the S_2 band. This state is not present in the
√3x√3 calculation because there are no rest atoms in that geometry.
This state is present in calculations for the T_4-2×2 surface,
although at a slightly higher energy than the experimental S_2 band.
The presence of the rest atom in the 2×2 unit cell gives rise to the
additional surface state. These results are consistent with scanning
tunneling microscope studies by Hamers et al.[2], which probe the
spatial extent of the surface state wave functions in the unit cell.

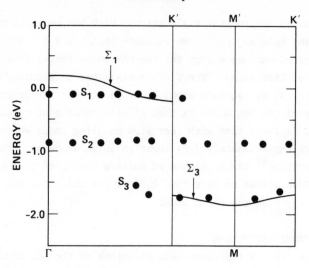

Figure 1. Experimental surface state dispersion from Ref. 8 for Si(111)7×7 along the ΓM direction of the surface Brillouin zone. Solid lines indicate the calculated dispersion for the Si(111)√3–T_4 model after mapping into the 1×1 zone.

REFERENCES

1. K. Takayanagi, Y. Tanishiro, M. Takahashi, and S. Takahashi, J. Vac. Sci. Technol. A3, 1502 (1985).

2. R.J. Hamers, R.M. Tromp and J.E. Demuth, Phys. Rev. Lett. 56, 1972 (1986).

3. I.K. Robinson, W.K. Waskiewicz, P.H. Fuoss, J.B. Stark and P.A. Bennett, Phys. Rev. B 33, 7013 (1986).

4. J.E. Northrup, Phys. Rev. Lett. 57, 154 (1986).

5. J.R. Chelikowsky, Phys. Rev. Lett. 49, 1569 (1982).

6. K.C. Pandey, Phys. Rev. Lett. 47, 1913 (1981).

7. F.J. Himpsel, D.E. Eastman, P. Heimann, B. Reihl, C.W. White and D.M. Zehner, Phys. Rev. B 24, 1120 (1981).

8. R.I.G. Uhrberg, G.V. Hansson, J.M. Nicholls, P.E.S. Persson and S.A. Flodstrom, Phys. Rev. B 31 3805 (1985).

9. J.E. Northrup, Phys. Rev. Lett. 53, 683 (1984).

ATOMIC STRUCTURE OF THE Si(111)7x7 SURFACE

W. S. Yang and R. G. Zhao
Department of Physics, Peking University
Beijing, China

The Si(111)7x7 surface is still the most challenging object to surface scientists, even if it has been studied extensively with virtually every surface analytical technique available.[1] The general situation is encouraging: many plausible models for the surface are emerging one after another.[2-7] However, none of them can pass or even has been tested with all of the major techniques. On the other hand, none of the techniques can make clear-cut discriminations among all models proposed.[1] If we note the very large size of its unit cell, this is natural, then. Consequently, our philosophy in studying the mysterious surface is listening to what every careful work says, even if it sounds inharmonic with the others, and modifying a model if it fails in explaining the experimental data of a careful work, even if it does well in explaining the others.

Following this philosophy, we have reached the model shown in Fig. 1. It consists of modified milkstool[6] or piramidal[5] buliding blocks arranged in a short-range 2x2 order,[8,7] and contains stacking faults,[4,9] dimers,[3] and an oscillatory multilayer relaxation.[9] Besides, there is an asymmetry between the two halves of the cell, i.e., in the left (right) half, the 1st-layer atoms shift downwards (upwards) by 0.1 $\overset{\circ}{A}$; the 2nd-layer atoms shift sideways by 0.15 (0.10) $\overset{\circ}{A}$; and the 3rd-layer atoms with dangling bond shift upwards by 0.45 (0.25) $\overset{\circ}{A}$. Clearly, the model is compatible with the scanning tunneling microscopy results.[2,10]

Fig. 2 shows that the model can pass the transmission electron diffraction (TED) tests, at least, not worse than the model proposed by Takayanagi et al.[7] can. Besides, it is quite clear that as long as a model can give a calculated TED pattern which is in agreement with the experiment, the Fourier synthesis of the kind in Ref. 7 always gives the same model, even if it is not correct.

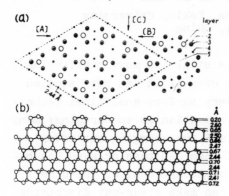

Fig. 1. Top view (top) and and side view (bottom) of the model proposed in this work. Note the __three__ stacking faults in both halves of the unit cell (outlined with dashed lines). For shifts of the atoms in the topmost three layers, see text.

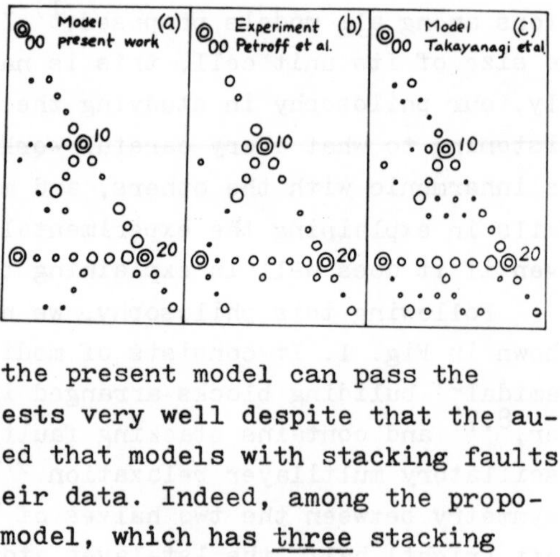

Fig. 2. Repeating triangle of the TED patterns. For the relevant details, see Ref. 3.

Fig. 3. shows that the present model can pass the x-ray standing wave[12] tests very well despite that the authors of Ref. 12 concluded that models with stacking faults are incompatible with their data. Indeed, among the proposed models[2-7] only our model, which has __three__ stacking

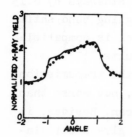

Fig. 3. X-ray standing wave germanium fluorecence angular yield for (220) Bragg diffraction from the Si(111)7x7 surface (closed circles, Ref. 12) and their calculated counterpart (curve) of the present model. In the calculations, the same assumptions as those in Ref. 12 were adopted. For details, see Ref. 12.

faults in both halves, can pass the tests. Of course, the details are also important to the good agreement.

The model can pass the LEED-CMTA (constant-momentum-transfer averaged low-energy electron diffraction)[4] tests

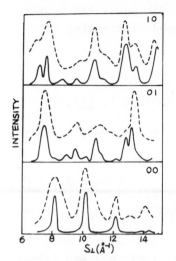

Fig. 4. LEED-CMTA spectra (dashed curves, Ref. 4) and their calculated counterparts of the present model. The calculations were done in a maner described in Ref. 9.

reasonably well as Fig. 4 shows. In fact, it was these LEED-CMTA curves that led Bennett et al. to the idea about stacking faults.[4] However, none of the existing models[2-7] could pass the tests since they do not have the right stacking sequense and layer spacings.

Calculations indicate (not shown) that the present model, comparing to the others, does quite well in passing the low-energy ion scattering[5] tests. The small asymmetry between the two halves of the cell is responsible for the slight difference between the azimuths A and B in Ref. 5, thus being an important ingredient of the model.

Our model is also compatible with medium-energy ion scattering,[1] as it contains a "triangle-dimer stacking fault" which is the "key ingredient"[1] that makes a model to be compatible with the technique.

Besides, as the buliding block geometry of the present model is very similar to that of the modified milkstool model proposed by Snyder,[6] the mechanism suggested by him for the building block is likely to be true. We discuss the short and long range order mechanisms of the surface together with the photoemission results[14] elsewhere.[15]

In addition, the present model is compatible with high-energy ion scattering,[16] high-resolusion infrared

spectroscopy,[17] He scattering,[18] and grazing-incidence
x-ray diffraction.[19] We discuss these elsewhere.[15]

In summary, we have proposed a new semi-quantitative
model for the clean Si(111)7x7 surface, which has been the
only model that can pass the tests of virtually every surface analytical technique without a need of changing its
parameters.

REFERENCES:

1. Tromp, R.M. and van Loenen, E.J., Surf. Sci. 155, 441
   (1985), and references therein.

2. Binnig, G., Rohrer, H., Gerber, C., and Weibel, E., Phys.
   Rev. Lett. 50, 120 (1983).

3. McRae, E.G., Phys. Rev. B28, 2305 (1983).

4. Bennett, P.A., Feldman, L.C., Kuk, Y., McRae, E.G., and
   Rowe, J.E., Phys. Rev. B28, 3656 (1983).

5. Aono, M., Souda, R., Oshima, C., and Ishizawa, Y., Phys.
   Rev. Lett. 51, 801 (1983).

6. Snyder, L.C., Surf. Sci. 140, 101 (1984).

7. Takayanagi, K., Tanishiro, Y., Takahashi, M., and Takahashi, S., J. Vac. Sci. Technol. A3, 1502 (1985).

8. Yang, W.S. and Jona, F., Solid State Commun. 48, 377
   (1983).

9. Yang, W.S. and Zhao, R.G., Phys. Rev. B30, 6016 (1984).

10. Becker, R.S., Golovchenko, J.A., McRae, E.G., and Swar-
    tzentruber, B.S., Phys. Rev. Lett. 55, 2028 (1985).

11. Petroff, P.M. and Wilson, R.J., Phys. Rev. Lett. 51,
    199 (1983).

12. Patel, J.R., Golovchenko, J.A., Bean, J.C., and Morris,
    R.J., Phys. Rev. B31, 6884 (1985).

13. Yamaguchi, T., Phys. Rev. B32, 2356 (1985).

14. Himpsel, F.J., Eastman, D.E., Heimann, P., and Reihl,
    B., Phys. Rev. B24, 1120 (1981).

15. Yang, W.S. and Zhao, R.G., to be published.

16. Culbertson, R.J., Feldman, L.C., and Silverman, P.J.,
    Phys. Rev. Lett. 45, 2043 (1980).

17. Chabal, Y.J., Phys. Rev. Lett. 50, 1850 (1983).

18. Cardillo, M.J., Phys. Rev. B23, 4279 (1981).

19. Robinson, I.K., Waskiewicz, W.K., Fuoss, P.H., Stark,
    J.B., and Bennett, P.A., to be published.

TEMPERATURE DEPENDENT SURFACE OPTICAL ABSORPTION IN Si(111)2x1

G.Chiarotti, F.Ciccacci, and S.Selci

Dipartimento di Fisica, Università di Roma "Tor Vergata"
via Orazio Raimondo, 00173 Roma
ITALY

P.Chiaradia, A.C.Felici, and C.Goletti

Istituto di Struttura della Materia del CNR
via Enrico Fermi 38, 00044 Frascati
ITALY

Measurements of Surface Differential Reflectivity as a function of temperature in Si(111)2x1 are presented. The results give evidence of a strong electron-phonon interaction affecting the optical transition across the surface gap. The dependence upon sample temperature of the sticking coefficient for oxygen on the same surface is also reported.

The temperature dependence of electronic transitions in solids is a probe of the electron-lattice interaction. It is well known that dangling bond states at the Si(111)2x1 surface cause an optical absorption peak at approximately 0.5 eV [1]. In order to investigate the electron-phonon coupling on this surface, we have studied the temperature dependence of such transition by Surface Differential Reflectivity (SDR) measurements.

Silicon samples with (111) orientation were cleaved at room temperature in Ultra High Vacuum (pressure in the 10^{-10} torr range). After cleavage the samples were brought to the required temperature and the reflectivity of the clean surface recorded as a function of wavelength. The surface was then fully oxidized, at that temperature, and a new reflectivity spectrum was taken. The ratio:

$$\frac{\Delta R}{R} = \frac{R_{clean} - R_{ox}}{R_{ox}} \tag{1}$$

proportional to the imaginary part of the surface dielectric function, is shown in Fig. 1 versus the energy of the photons for three different temperatures, all of them well below that of the (2x1) ⟶ (7x7) transition.

It is seen from the curves of Fig. 1 that the peaks of nearly gaussian shape display a red shift with increasing temperature, while their half-widths increase considerably. The shapes of the curves are slightly asymmetric, while their areas remain sensibly constant at the various temperatures.

Figure 2 shows the temperature dependence of the full width at half maximum W(T) of the curves of Fig. 1. The width increases roughly as the square root of the absolute temperature.

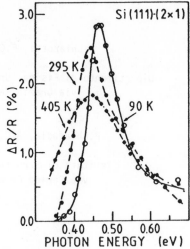

FIG. 1. Differential reflectivity vs photon energy in Si(111)2x1 at three different temperatures.

All the features of Figs. 1 and 2 are characteristic of localized excitations with strong electron-phonon interaction as in, for example, the F-center or the polaronic exciton in alkali halides[2]. In the theory originally developed by K.Huang and A.Rhys for the F-center[3], the optical absorption is described as a vertical transition between electronic states, whose energy minima are displaced, in the configurational coordinate scheme, by the electron-lattice interaction. The half-width is proportional to $S^{\frac{1}{2}}$, where S is the so called Huang-Rhys factor that measures the strength of the electron-phonon interaction. For an Einstein model with a phonon frequency ω_o, the number of phonons emitted in the transition corresponding to the peak of the absorption curve at zero

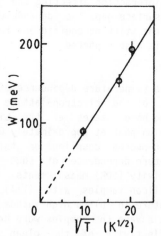

FIG.2. The FWHM of the curves of Fig. 1 vs the square root of the absolute temperature.

temperature is simply S, while the half-width is:

$$W(T) = W(0)[\coth(\hbar\omega_s/2kT)]^{\frac{1}{2}} \qquad (2)$$

which gives the T behaviour at high temperatures. Within such a model a red-shift of the peak is explained by a softening of the mode when the electron is raised to the excited state, as well as by a change in the lattice parameter.

From the slope of the straight line of Fig. 2, representing the high temperature expansion of W(T), we obtain [4] that the frequency of the surface phonon created in the transition is smaller than 15 meV, while S is larger than 7, i.e. at least 7 phonons are created during the transition, which confirms that we are dealing with the case of a strong interaction. This localized model seems to give a qualitatively correct picture of the surface optical transitions, even if the dangling bond bands show a large dispersion characteristic of a delocalized system. However, it has been recently suggested that both localized and delocalized features can coexist at semiconductor surfaces.

2.   OXYGEN ADSORPTION

From the variation of the reflectivity as a function of oxygen exposure, information on the chemisorption kinetics can be obtained: in particular the sticking coefficient can be measured.

Using this approach, a detailed study of the oxidation of the Si(111)2x1 exposed to molecular oxygen at various temperatures has been undertaken. Though this system has been extensively studied, a model for the chemisorption process in agreement with the experimental data is still missing. Fig. 3 shows the dependence of the sticking coefficient on the inverse of absolute temperature. It is seen that a linear relation between the logaritmic value of the sticking coefficient and 1/T exists.

From the slope of the straigth line in Fig. 3, an activation energy of approximately 16 meV can be evaluated. This relatively small value of the energy barrier seems to rule out a dissociative adsorption, whereas the presence of a weakly bonded precursor state, unable to saturate dangling bonds, before a molecular chemisorption, is suggested.

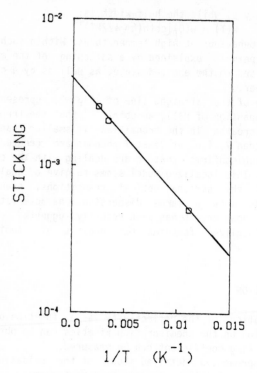

FIG. 3. The sticking coefficient of O_2 on Si(111)2x1 vs the substrate temperature.

1) Chiarotti, G., Nannarone, S., Pastore, R., and Chiaradia, P., Phys. Rev. B4, 3398 (1971).
2) Toyozawa, Y., "The Physic of Elementary Excitation", Springer Series in Solid State Sciences Vol. 12, Chap. 7 (Springer, Berlin 1980).
3) Huang, K. and Rhys, A., Proc. Roy. Soc. London, Ser. A 204, 403 (1950).
4) Ciccacci, F., Selci, S., Chiarotti, G., Chiaradia, P., Phys. Rev. Lett. 56, 2411 (1986).
5) Zunger, A., Proc. of the 18th International Conference on the Physics of Semiconductors, Stockolm 1986.

ROUGHNESS DEPENDENCE OF THE DANGLING
BOND STATES ON CLEAN Si(001)

I. Andriamanantenasoa, J.P. Lacharme, C.A. Sébenne and F. Proix

Laboratoire de Physique des Solides, associé au CNRS 154,

Université Pierre et Marie Curie 75252 PARIS Cedex 05 FRANCE

ABSTRACT

The shape and size of the dangling bond occupied state band is deduced from high resolution photoemission yield spectra of clean Si(001) surfaces. Their variations upon annealing or oxidation-desoxidation cycles of (001) or vicinal faces are correlated to the smoothing or roughening of the surface at the atomic scale.

Both structural and electronic properties of the (001) face of silicon are fairly well known and theoretically understood. The latest observations, using scanning tunneling microscopy [1], confirmed the surface dimers, buckled or not, as the basic reconstructed units, and the existence of several defective geometries correlated to them. The electronic structure has been deduced mostly from angle resolved UV photoemission spectroscopy [2,3] in accordance with calculations based on the buckled dimer model [3,4]. Summarizing the results in terms of density of occupied surface states in the dangling bond band, a two-peak structure has been found, the positions of the maxima being 0.8 and 1.3 eV below Fermi level respectively, plus or minus 0.1 eV. Moreover it is well known that surface defects highly influence structural and electronic properties of Si(001) but a comprehensive study remains to be done. In the present work, we take advantage of the high energy resolution and absolute values which are obtained using photoemission yield spectroscopy [5] to analyze the shape and size of the dangling bond band of a clean (001) Si surface under different conditions.

Optically polished Si wafers, 4x20x0.5 mm^3, of different types
and doping levels were used. The large area is either (001) or vicinal.
The latter makes an angle of 1.8° with the (001) plane along the [010]
direction, giving a double set of steps along [110] for one and [$\bar{1}$10]
for the other, with an average terrace width of 8 surface units
assuming a step height equal to the distance between consecutive (001)
planes. The sample was simply rinsed in acetone and mounted on the
sample holder. It was put under ultrahigh vacuum (1 to 2x10^{-10}torr)
where it was cleaned by Joule heating at 900°C, as checked with an
optical pyrometer. The surface was characterized by Auger electron
spectroscopy and LEED observation and was studied only when clean
enough and displaying a 2x1 reconstructed diagram. In some cases,
the presence of residual carbon was found which was not observed to
influence the results. Photoemission yield spectra taken between 4.4
and 6.6 eV photon energy were first used to characterize the cleaning
procedure: they have shown that heating beyond 900°C generates
electrically active defects below the surface, making the sample
irreversibly more and more p-type within the escape depth of
photoemitted electrons.

From the yield spectra, several properties can be obtained [5]:
the work function from the photon energy threshold, the ionization
energy from the comparison of spectra taken on differently doped n
and p-type samples and the density of filled surface states $N_s^*(E)$
by separating bulk and surface contributions to the yield Y(E). In
the case of Si(001) the procedure is tedious but straightforward since
the bulk contribution to the effective density of states takes the
form of a power law $(E-E_{vs})^{5/2}$ leading to :

$$N_s^*(E) \propto dY/dE - \alpha(E-E_{vs})^{5/2}$$

where E is the photon energy and E_{vs} the valence band edge with respect
to vacuum level, i.e., the ionization energy. α is a parameter adjusted
in the high energy part of the spectra: it depends only slightly on
the experimental adjustments and appears as a good index of the degree
of surface roughness. This point has been checked many times on cleaved
surfaces: the smoother and better ordered the surface, the higher
the absolute value of the photoemission yield.

The general results are the following. The work function of the

clean Si(001) surface is found at 4.8 eV whatever the doping level
of the sample below 10^{18} cm^{-3} ; it is decreased by about 50 meV for
the roughest surfaces. The ionization energy is found remarkably stable
at 5.3 eV. Then, there is a quasi-pinning of the Fermi level 0.5 eV
above E_{vs}. Not only the absolute value of Y, but the size and shape
of the effective density of surface states $N_s^*(E)$, the dangling bond
peak, are found roughness-dependent. Various sizes and shapes are
illustrated in fig.1 and 2. To render more visible the variations,
it is convenient to analyze the band as the sum of three gaussian
peaks I, II, and III respectively 0.2, 0.45 and 0.75 eV below E_{vs}
(that is 0.7, 0.95 and 1.25 eV at least below Fermi level) with a
constant width of 0.4 eV. Tentatively, we attribute I and III to the
perfect surface (we keep their intensity ratio constant) and II to
the roughness effect. Under such a scheme, fig.1 shows the evolution
of a (001) surface after cycles of oxygen exposure and thermal
cleaning: the evolution can be interpreted as an overall roughening

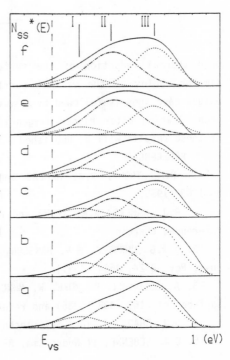

Fig. 1: Evolution of the dangling
bond surface state band on a
(001) Si surface a) after thermal
cleaning (TC), b) after 1 L of
O_2 and TC, c) after further TC,
d) after several cycles involving
successively 5, 10, 3, 30 L of
O_2 with TC in between, c) after
50 L of O_2, TC, 100 L of O_2 and
TC, f) after 50 L of O_2, TC,
300 L of O_2 and TC.

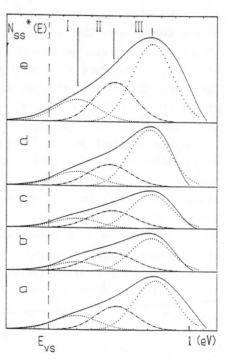

Fig. 2: Evolution of the dangling bond surface state band on a vicinal Si(001) face (see text) a) after thermal cleaning (TC), b) after 1 L of O_2 and TC, c) after four cycles involving each a few L of O_2 and TC, d) after 5 L of O_2, TC, 50 L of O_2 and TC, e) after 1 L of O_2, TC, 100 L of O_2, TC, 300 L of O_2 and TC.

upon thermal reduction of the surface. Fig. 2 shows the effects of similar cycles on the vicinal surface where a roughening after low oxygen coverages is followed by a smoothing and ordering of the surface upon cleaning after higher oxygen coverages. A more detailed account of these results will be published elsewhere.

REFERENCES

1. R.M. TROMP, R.J. HAMERS and J.E. DEMUTH, Phys. Rev. Lett. 55,1303 (1985).

R.J. HAMERS, R.M. TROMP and J.E. DEMUTH, to be published, and references therein.

2. R.I.G. UHRBERG, G.V. HANSSON, J.M. NICHOLLS and S.A. FLODSTRÖM, Phys. Rev. B 24, 4684 (1981).

3. A. GOLDMANN, P. KOKE, W. MÖNCH, G. WOLFGARTEN and J. POLLMANN, Surface Sci. 169, 438 (1986) and references therein.

4. D.J. CHADI, J. Vac. Sci. Technol. 16, 1290 (1979).

5. C.A. SEBENNE, Il Nuovo Cim. 39 B, 768 (1977).

PHOTOEMISSION STUDIES OF THE TEMPERATURE DEPENDENCE
OF SURFACE STATES ON n+-DOPED Si(100)2x1 SURFACES.

A. Cricenti, P. Mårtensson, L.S.O. Johansson, and G.V. Hansson

Department of Physics and Measurement Technology
Linköping Institute of Technology
S-581 83 Linköping
SWEDEN

ABSTRACT

The temperature dependence of the electronic and atomic structures of Si(100) surfaces of a heavily n-doped crystal have
been studied with angle-resolved photoelectron spectroscopy
(ARPES) and low-energy electron diffraction (LEED). When the
temperature of the sample is lowered to ~150 K the surface
structures observed in the ARPES spectra become sharper, the
surface-state band gap becomes ~40 meV larger than at room
temperature, and the LEED pattern shows diffuse streaks through
the half-order diffraction spots.

The temperature dependence of the electronic and atomic structures of Ge(100) surfaces have recently been investigated with ARPES
and LEED by Kevan and Stoffel[1]. It was shown that the reconstruction
of the Ge(100) surface changes from two-domain 2x1 at room temperature
to two-domain c-4x2 below 200 K. This change in the reconstruction is
accompanied by a metal-insulator transition. Very recent ARPES studies[2,3] have proven the existence at room temperature of a metallic
surface state also at Si(100)2x1 surfaces. Due to these similarities
between Si and Ge(100) surfaces at room temperature, we have investigated with ARPES the temperature dependence of the electronic and
atomic structures of Si(100) surfaces.

ARPES spectra were recorded in an ultrahigh-vacuum chamber at a pressure of $<2\times10^{-10}$ Torr. Unpolarized light from a resonance lamp was used ($h\nu$=21.2 eV). A more detailed description of the apparatus will be given elsewhere[4]. The angle of incidence θ_i and the emission angle θ_e are defined according to the inset of Fig.1. The sample was a heavily n-doped, mirror polished Si(100) single crystal (ρ~6 mΩcm, N_D~1$\times10^{19}$ Sb atoms/cm^3, Wacker-Chemitronic) with a size of 14x8x0.3 mm^3. The cleaning procedure for the sample is the same as described in Ref.3.

Figure 1 shows photoemission spectra recorded at room temperature for various angles of emission θ_e in the [010] azimuthal direction. Structures A, B and C are due to emission from previously reported surface states[2,3,5,6]. In Fig.2 photoemission spectra recorded around the $\bar{\Gamma}$ point in the surface Brillouin zone (SBZ) are shown for two

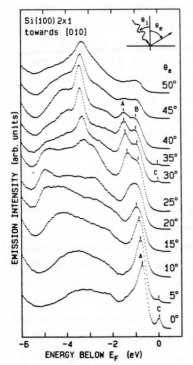

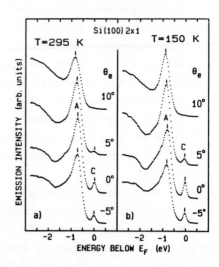

Fig. 2. Photoelectron spectra showing emission around the $\bar{\Gamma}$ point for two different sample temperatures.

Fig. 1. Photoemission spectra recorded with 21.2 eV photon energy along a [010]-direction. The angle of incidence is θ_i=60°.

different sample temperatures, room temperature for Fig. 2(a) and
150 K for Fig. 2(b). The emission from the state C at the Fermi level
remains after lowering the temperature of the sample to 150 K, i.e. no
metal-semiconductor transition is observed for Si(100) surfaces. The
LEED pattern at 150 K is showing a 2x1 reconstruction with diffuse
streaks between the half order diffraction spots. This LEED pattern is
very similar to the one described in Ref. 1 for the Ge(100)2x1 surface
at room temperature[1], well above its metal-semiconductor transition
temperature. This could indicate that a temperature lower than 150 K
is needed in order to observe a phase transition on Si(100) surfaces.
From the photoemission spectra of Fig.2 it can be observed that the
surface structures A and C become sharper and narrower when the temperature of the sample is lowered. Furthermore the energy distance
between the surface states A and C is increasing by ~40 meV, giving
rise to a larger surface-state band gap at 150 K for Si(100)2x1
surface. A similar effect, i.e. increasing of the surface-state band
gap when lowering the temperature, has been observed with optical
absorption experiments on Si(111)2x1 surfaces[7].

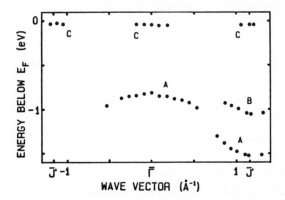

Fig. 3. Dispersion of surface state
bands measured along a [010]-direction.

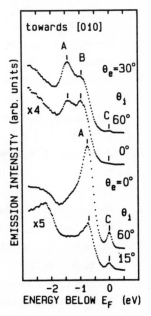

Fig. 4. Dependence of the surface
state emission intensities on the
angle of incidence.

Figure 3 shows the measured initial-energy dispersions $E_i(k_{//})$ for the three surface states A, B and C along the [010] azimuthal directions with the sample at low temperature. The plotted dispersions are in good agreement with a previously reported study using 10.2 eV photon energy[3].

The emission from the metallic state C was very contamination sensitive so that the emission intensity decreased by ~50% in one hour with the sample at room temperature. After lowering the temperature to 150 K no emission intensity decrease was observed for the peak C during the whole photoemission experiment (~6 hours). Although the contaminant involved is not known, this behaviour is similar to what is found on Si(111)2x1 surfaces where the sticking coefficient for oxygen decreases two orders of magnitude when the sample temperature is changed from room temperature to liquid-nitrogen temperature[8].

Figure 4 shows the dependence of the surface-state emission intensities on the angle of light incidence. The surface states A and C are affected more than the state B when the electric field component perpendicular to the surface is reduced. This means that the surface states A and C have stronger p_z-character than state B.

REFERENCES

1.   Kevan, S.D. and Stoffel, N.G., Phys. Rev. Lett. 53, 702 (1984); Kevan, S.D., Phys. Rev. B32, 2344 (1985).

2.   Goldmann, A., Koke, P., Mönch, W., Wolfgartner, G., Pollmann, J., Surf. Sci. 69, 438 (1986).

3.   Mårtensson, P., Cricenti, A., Hansson, G.V., Phys. Rev. B33, 8855 (1986).

4.   Johansson, L.S.O. et al., to be published.

5.   Himpsel, F.J. and Eastman, D.E., J. Vac. Sci. Technol. 16, 1297 (1979).

6.   Uhrberg, R.I.G., Hansson, G.V., Nicholls, J.M., Flodström, S.A., Phys. Rev. B24, 4684 (1981).

7.   Ciccacci, F., Selci, S., Chiarotti, G., Chiaradia, P., Phys. Rev. Lett. 56, 2411 (1986).

8.   Ciccacci, F., Selci, S., Chiarotti, G., Chiaradia, P., Habib, Z., Felici, A.C., Goletti, C., Vuoto XVI, 16 (1986).

SELFCONSISTENT ELECTRONIC STRUCTURE OF SEMIINFINITE Ge(001)-(2x1)

J. Pollmann, P. Krüger, A. Mazur, and G. Wolfgarten

Institut für Physik, Universität Dortmund, D-4600 Dortmund 50, Germany

We present the first selfconsistent electronic structure calculation of semiinfinite Ge(001) with a 2x1-reconstructed surface. The theoretical results are used to elucidate controversial interpretations of ARUPS data.

1. INTRODUCTION

The technologically most important low index faces of Si and GaAs as well as the Ge(111) surface have been investigated in detail by angle-resolved ultraviolet photoelectron spectroscopy (ARUPS) within the last decade. It was only in recent years, that Ge(001)-(2x1) has been studied employing ARUPS with synchrotron radiation. Using cleaved as well as MBE-grown samples, both Nelson et al.[1] and Hsieh et al.[2] found essentially the same normal emission spectra. Their interpretation in Refs. 1 and 2 seems controversial, however. While Nelson et al. claim to have identified two *surface states* at 0.6eV and at 1.3eV below valence band maximum, Hsieh et al. challenge this interpretation and associate the observed peaks at the Γ-point almost exclusively with *bulk states*. This situation calls for a surface electronic structure calculation using an approach that allows to unambiguously discern surface from bulk features within the projected bulk bands.

2. THEORY

To resolve the issue, we have applied our newly developed self-consistent scattering theoretical method[3]. It is a first-principles version of the empirical scattering theoretical method used previously for surface electronic structure calculations[4]. The sc STM is based on ionic pseudopotentials and the local density approximation and makes use of a Gaussian-orbital basis set. It describes the surface as a two-

dimensionally periodic perturbation of the bulk solid, which is highly
localized in the surface perpendicular direction. Solving the resulting
problem is most easily accomplished by referring to potential scattering theory and one-particle Green's functions. The Bloch-symmetry
parallel to the surface and the strong localization of the surface
creating perturbation perpendicular to the surface allow the numerical
solution of the Lippman-Schwinger and Dyson equations for the semi-
infinite solid. Formal details are given in Ref. 3.

The surface structure was determined by minimizing the total
energy of a ten layer Ge(001) slab using the semiempirical approach
suggested by Chadi[5]. A realistic second-nearest-neighbour Hamiltonian
was employed. The resulting structural changes turn out to be
essentially restricted to the first three layers. Asymmetric dimers are
found as the main building blocks of the reconstruction at the surface.
In our selfconsistent calculations we have used exactly this geometry
for the first three layers of the semiinfinite crystal.

3.   RESULTS

The total valence charge density near the surface in the dimer
plane is shown in Fig. 1. We see that the surface gives rise to a
strong *dimer bond* between the up-atom and the down-atom in the surface
dimer. We note that the charge density on the third layer is already
bulk-like.

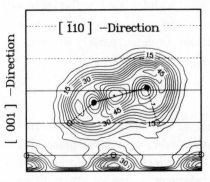

The selfconsistent surface
band structure of Ge(001)-(2x1) is
shown in Fig. 2 for the gap energy
region. Similar to Si(001)-(2x1)[4],
the Ge surface gives rise to four
pronounced surface bands D_i, D_{up},
D_{down} and D_i^*, which are related to
the dimer bond, the dangling bonds
at the up- and down-atom in the
surface dimer, respectively, and
to an antibonding dimer state. The
full lines represent bound surface

Fig. 1   Valence charge density
at Ge(001)-(2x1) in the
dimer plane.

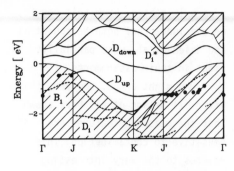

Fig. 2   Surface bandstructure of Ge(001)-(2x1).

states while the dashed lines show pronounced resonances. Experimental ARUPS data of Ref. 1 are included by the heavy dots whose thickness corresponds to the experimental error bars. Measurements have been carried out only from Γ to J and J' to Γ. The general agreement between the measured peak positions in the electronic density curves and the computed D_{up} band is good although the theory yields a somewhat larger bandwidth. The deviations could partially be related to difficulties in the determination of surface-derived features in the experimental spectra when they strongly overlap with bulk features.

This fact seems to be the origin of the above mentioned controversy in the interpretation of *normal emission spectra*, as well. In our

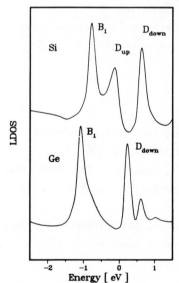

Fig. 3   Surface layer DOS for Si(001)-(2x1) and Ge(001)-(2x1) at the Γ-point.

theoretical results, the D_{up} resonance becomes very broad at ΓJ/2. This is in marked contrast to Si(001)-(2x1), where the resonance remains sharp up to the zone center[3]. In Fig. 3, a direct comparison of the surface layer density of states at Γ is shown for Si(001)-(2x1) and for Ge(001)-(2x1) in the gap energy region. Both graphs show a B_1 and a D_{down} peak. The B_1 feature is a predominantly p_x, p_y-like backbond resonance. While at Si(001)-(2x1) a clearly resolvable D_{up}- peak is found, the corresponding feature becomes a very broad resonance at Ge(001)-(2x1) giving rise to a shoulder in the LDOS only. The measured feature, therefore, is related to a very broad surface resonance that can hardly be distinguished

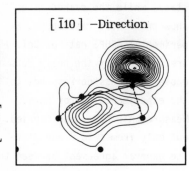

Fig. 4   Charge density at the
Γ-point for E=-0.8eV.

from bulk states. From this result, we conclude, that both Nelson et al. and Hsieh et al. were partially right in their assignments.

One might argue, however, that after all Nelson et al. have shown that the measured features are sensitive to contamination. This relates to the very interesting fact, that the Γ-point D_{up}- resonance at Ge(001)-(2x1) seems to reside on the borderline between surface and bulk states. Resonances may be broad in energy and yet their charge density *at the surface* may resemble a surface state. In Fig. 4, we show the charge density at 0.8eV below E_{VBM} (in the middle of the broad resonance) at the Γ-point. The dangling bond character of the surface charge density is clearly visible and indicates that the related spectral feature must react to contamination.

A detailed discussion of the complete electronic structure of Ge (001)-(2x1) in comparison with experiment and in relation to Si(001)-(2x1) will be given elsewhere.

4.   REFERENCES

1] Nelson J.G., Gignac W.J., Williams R.S., Robey S.W., Tobin G.G., and Shirley D.A., Phys. Rev. B27, 3924 (1983)

2] Hsieh T.C., Miller T., and Chiang T.-C., Phys. Rev. B30, 7005 (1984)

3] Krüger P., Mazur A., Pollmann J., Wolfgarten G., submitted to Physical Review Letters

4] see e.g., Pollmann J., Krüger P., Mazur A., and Wolfgarten G., Surf. Science 152/153, 977 (1985), and references therein.

5] Chadi D.J., Phys. Rev. Lett. 43, 43 (1979)

LOW-ENERGY ELECTRON-LOSS SPECTROSCOPY OF THIN Ge LAYERS ON Si(111)-7x7 SURFACE

C. Tatsuyama, H. Ueba and N. Ishimaru

Department of Electronics, Toyama University
Toyama, Toyama 930
JAPAN

Ge was deposited onto the Si(111)-7x7 surface maintained at RT, 350°C and 500°C from a tungsten spiral filament. At RT, low-energy electron-loss spectroscopy(LEELS) spectrum of Si evolves to that of amorphous Ge with increase in Ge thickness. Below one-monolayer(1 ML) coverage, the surface plasmon of Si splits into two peaks due to Ge induced transition. At high temperatures, the Si(111)-7x7 surface structure is replaced by a (5x5) one by the formation of Ge-Ge bonds after Ge deposition of 1.5-2 ML. However, the LEELS measurements reveal that the backbond surface states of Si remaines up to some critical thickness depending on the deposition and/or annealing temperature.

1. INTRODUCTION

Recently, interface formation of Ge on Si substrate has been extensively investigated from several point of views: One is surface superstructures induced by Ge deposition on Si(111)-7x7 surface[1-5]. The other is a heteroepitaxial growth of GaAs through Ge layers on Si substrate.[6,7] Furthermore, strained-layer superlattices consisting of Si and Ge_xSi_{1-x} on Si substrate are now a most exciting subject.[8-9]

In previous papers,[1,2] we have reported that the Si(111)-7x7 structure is replaced by a (5x5) one, when about 2-ML Ge grows heteroepitaxially on Si(111)-7x7 surface, and pointed out the similarity of the LEED patterns between (7x7) and (5x5) structures. The similar results have also been confirmed by McRae et al[3] and Ichikawa et al.[4] Furthermore, Gossmann et al have shown that Ge-Si alloy films grown by MBE on Si(111) also exibits a (5x5) structure.[5]

In the present paper, we report the systematic studies of initial stage of heteroepitaxy of Ge on Si(111)-7x7 surface observed by LEELS combined with low-energy electron diffraction(LEED) and Auger-electron spectroscopy(AES). The special effort has been devoted to elucidate the electronic structure of a Ge-induced (5x5) superstructure.

2.   EXPERIMENTAL

Si(111) single crystal used here was n-type with resistivity of 200 Ωcm. Ge was deposited onto the Si(111)-7x7 surface maintained at RT, 350 and 500°C from a tungsten spiral filament in an ultra-high vacuum chamber (base pressure ~1.5x10^{-10} Torr). A clean (7x7) surface was prepared by direct heating of substrate up to ~1200°C. LEED, AES and LEELS spectra were measured for each step of Ge deposition by using a four-grid retarding-field-type energy analyzer (PHI model 10-120). All measurements were carried out after the substrate temperature was cooled down to RT. The temperature of substrate was measured by a Pt-Pt(Rh) thermocouple and a pyrometer. Ge thickness on Si was estimated by measuring a decrease in the peak-to-peak intensity of Si(LVV)-Auger signal at 92 eV in the first-derivative spectra, and also monitored by a quartz oscillator (ULVAC model CRTM-1D). The second-derivative LEELS spectrum was measured at a primary-electron beam energy E_p=120 eV, and a beam current 0.7 µA.

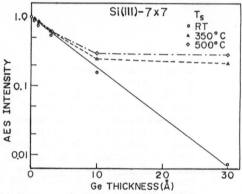

Fig.1. Intensity of Si(LVV)-92 eV $vs.$ Ge thickness.

3.   RESULTS AND DISCUSSION
3.1   Growth Mode

Figure 1 shows the decrease in intensity of Si(LVV) -92 eV AES signal with Ge deposition. At RT, the intensity decreases exponentially with Ge thickness with electron-escape depth of λ=5.7 Å, indicating a layer-by-layer growth. While at 350 and 500°C, it decreases in accordance with the Stranski-Krastanov-type growth mode with islands formation after Ge deposition of ~4 ML, where 1 ML is defined as the number of Ge atoms being equal to the surface atomic density of Si(111) surface. These results coincide with previous results.[1,2]

3.2   LEELS Measurements
3.2.1   RT deposition

Figure 2 represents the evolution of LEELS spectrum with Ge deposition at RT. Ge coverage in ML units is shown in parentheses in the figure. LEED pattern for each step is also shown. The spectrum of Si (111)-7x7 surface consists of well known bulk plasmon $\hbar\omega_p$, surface

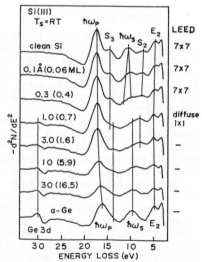

Fig.2. Ge deposition at RT.

plasmon $\hbar\omega_S$, bulk interband transition E_2, and back-bond surface-related transitions S_2 and S_3. The spectrum at 30 Å deposition is very similar to that of amorphous Ge (a-Ge) shown in the lowest spectrum in the figure.

A surface plasmon $\hbar\omega_S$ splits into two peaks with small deposition. This splitting may be explained by a simple model calculation in the framework of the dielectric scattering theory as shown for oxygen[10] and gold[11] depositions on Si surfaces. Figure 3 shows a comparison of experiment with calculation, where

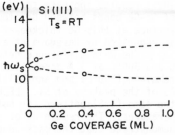

Fig.3. Splitting of $\hbar\omega_S$

the values of the oscillator-strength parameter, and transition energy of a Ge-induced transition are taken as A=0.1 ,and $\hbar\omega_0$=11 eV, respectively, in the model. The agreement is fairly good.

3.2.2 High-temperature deposition

Figures 4 and 5 show the evolution of LEELS spectrum with Ge deposition at 350 and 500°C, respectively. Ge coverage in ML units estimated by AES intensity of Si(LVV)-92 eV is shown in parentheses in the figures. The lowest spectrum in each figure is that of clean Ge(111)-(2x8) surface. The LEELS spectrum evolves in a similar way upon Ge deposition for both temperature. However, it does quite differently from RT deposition.

The (7x7) LEED pattern of Si(111) surface is replaced by a (5x5) one at 3 Å deposition for both temperature. At the same time, in LEELS spectrum, a new peak at ~9 eV appears. This new peak is a characteristic structure of Ge(111) surface as denoted S_2 in the lowest spectrum.

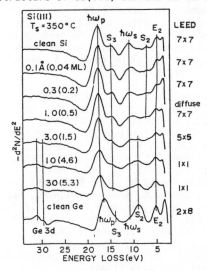

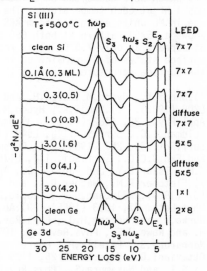

Fig.4. Ge deposition at 350°C.   Fig.5. Ge deposition at 500°C

These results indicate the formation of Ge-Ge bonds between Ge atoms on Si surface at this Ge thickness(~1.5 ML), as suggested by Hasegawa et al.[12]

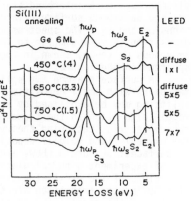

Fig.6. Annealing effect.

The main difference between 350° C and 500°C at 3 Å deposition is the remaining(500°C) and annihilation(350° C) of the peak S_2 of Si. LEED shows the same (5x5) pattern for both temperatures. The remaining of S_2 should demonstrate that the back-bond surface states of Si are still present. The similar result is also obtained in the annealing effect as shown in the next subsection.

Both bulk and surface plasmons keep about the same energy up to 30 Å deposition. We note here that Ge atoms form islands on a few monolayers (~4 ML) of Ge at these high depositions. Furthermore, the bulk plasmon observed at these high depositions should come from these thin Ge layers, not from Si under Ge layers, since the bulk-plasmon intensity does not decrease so much as Si(LVV)-AES signal at 92 eV. These results are explained by assuming that the electron density in these thin Ge layers is the same as in Si. This assumption is consistent with the pseudomorphic growth of thin Ge layers to Si substrate.

3.2.3 Annealing effect

Figure 6 shows the annealing effect on LEELS spectrum of Ge layers which were deposited on Si at RT up to 6 ML. The annealing time was 1 hr at each temperature. The number in parentheses is Ge coverage in ML units estimated from AES intensity after annealing. LEED shows a (5x5) pattern at both 650 and 750°C. However, the peak S_2 of Si appears at only 750°C. This result also shows that the back-bond surface states of Si remain up to some critical Ge thickness, even if the surface structure exibits a (5x5) LEED pattern.

1).Shoji,K., Hyodo,M., Ueba,H. and Tatsuyama,C., Jpn.J.Appl.Phys.22, L200 (1983). 2).Shoji,K., Hyodo,M., Ueba,H. and Tatsuyama,C., Jpn.J. Appl.Phys. 22, 1482 (1983). 3).McRae,E.G., Gossmann,H.-J. and Feldman,L.C., Surf.Sci.146, L540 (1984). 4).Ichikawa,T. and Ino,S., Surf.Sci.136, 267 (1984). 5).Gossmann,H.-J., Bean,J.C, Feldman,L.C. and Fan,J.C.C., Surf.Sci. 138, L175 (1984). 6).Windhorn,T.W., Metz, G.M., Tsaur,B-Y. and Fan,J.C.C., Appl.Phys.Lett. 45, 309 (1984). 7).Fletcher,R.M., Wagner,D.K. and Ballentyne,J.M., Appl.Phys.Lett. 44, 967 (1984). 8).Lang,D.V., People,R., Bean,J.C. and Sergent,A.M., Appl. Phys.Lett. 47, 1333 (1985). 9).Abstreiter,G., Brugger,H. and Wolf,T., Jorke,H. and Herzog,H.J., Phys.Rev.Lett. 54, 2441 (1985). 10).Ibach, H. and Rowe,J.E., Phys.Rev.B 9, 1951 (1974). 11).Perfetti,P., Nannarone,S., Patella,F., Quaresima,C., Capozi,M., Savoia,A. and Ottaviani,G., Phys.Rev.B 26, 1125 (1982). 12).Hasegawa,S., Iwasaki,H., Li,S.T. and Nakamura,S., Phys.Rev.Lett. 32, 6949 (1985).

ARSENIC PASSIVATION OF THE Si(111) SURFACE

R. I. G. Uhrberg*, R. D. Bringans, Marjorie A. Olmstead**, R. Z. Bachrach, and John E. Northrup***

Xerox Palo Alto Research Center
3333 Coyote Hill Road, Palo Alto, CA 94304
USA

ABSTRACT

The electronic structure and stability of As-terminated Si(111) have been investigated with angle-resolved photoemission. The surface was found to be extremely resistant to oxygen and air exposures.

1.   INTRODUCTION

The atomic and electronic structure of Si(111) 7x7 and Ge(111) c(2x8) have attracted much attention recently.   These reconstructions are very complicated compared to the simplest form of the (111) surface, which is a truncation of the bulk with all atoms in their bulk positions.   This ideal surface is, however, energetically very unfavorable because of the large number of dangling bonds (one per surface atom).   A close approximation to an ideal surface can be achieved by exposing the Si(111)[1] or Ge(111)[2] surfaces to As. The result is a surface where the As atoms have replaced the outermost Si (Ge) atoms.   Due to the extra valence electron provided by the As, the dangling-bond states are replaced by doubly occupied As lone-pair states.   Dangling-bond reduction is no longer a possible driving force for reconstruction and the resulting surface structure stays very close to that of an ideal, 1x1, surface.

An interesting consequence of the lone-pair states on the (111) surface is the extreme resistance of this surface to contamination.   This property was investigated for the Si(111):As 1x1 surface by exposing it to controlled amounts of oxygen, air and hydrogen.   The surface was almost unaffected, as monitored by the lone-pair surface state emission, after exposures as high as 10^7 L and 10^{11} L of oxygen and air, respectively.   This should be compared with the 15 L oxygen exposure which was sufficient to completely remove the dangling-bond state emission on the Si(111) 7x7 surface. Hydrogen on the other hand had a much stronger effect on the As terminated surface[1], causing disorder and a strong attenuation of the lone-pair emission at exposures of 1×10^4 L.

2. EXPERIMENTAL DETAILS

The silicon samples used in this experiment were cut from polished, p-type (Boron-doped, ~ 10 Ωcm), Si wafers. Clean Si(111) 7x7 surfaces were obtained by repeated cycles of Ar^+-ion sputtering (500 eV) and annealing (~850°C). Arsenic was added to the 7x7 surface in the form of As_4 molecules by evaporating elemental As from an MBE cell. The substrate temperature was brought to 800°C while the As pressure was 5×10^{-7} Torr, and then the sample was held in the As_4 flux (effective pressure ~10^{-3} Torr) for 10 seconds while being held at 350°C. The result is a well-ordered Si(111):As 1x1 surface.

3. RESULTS AND DISCUSSION

3.1 Surface Lone-Pair State Dispersion

The energy dispersion of surface states depends strongly on the local bonding geometry at the surface. By comparing experimental surface state dispersions and calculated surface band structures it is possible to make reliable statements about the atomic structure on the surface. The angle-resolved photoemission results for the initial energy of the lone-pair state are shown in Fig. 1 as a function of $\bar{k}_\parallel$. The strong downward dispersion (1.35 eV from $\bar{\Gamma}$ to $\bar{K}$) makes it very different from any of the dangling bond states observed on the Si(111)7x7 surface[3]. The downward dispersion is qualitatively similar to those calculated for dangling bond states on ideal 1x1 Si(111) surfaces[4].

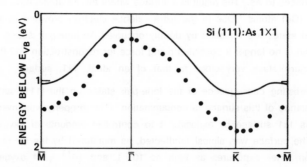

Figure 1.   Comparison between the experimental (dots) and the calculated (solid line) As lone-pair dispersion. $\bar{\Gamma}\rightarrow\bar{K}$ and $\bar{\Gamma}\rightarrow\bar{M}$ correspond to the major symmetry lines of the hexagonal 1x1 surface Brillouin zone.

The solid line in Fig. 1 shows the calculated dispersion of the As lone-pair state for a model in which the As atoms replace the outermost Si atoms. The calculations were performed using the first principles pseudopotential method and the local density functional approximation. The dispersion was calculated for the minimum-energy geometry and, as can be seen from the figure, is in very good agreement with the experimental data points. The calculated bandwidth (~1.1 eV) is somewhat smaller than the experimental value (1.35 eV), but part of this discrepancy is due to the different shape of the experimental and theoretical curves near $\bar{\Gamma}$. The rigid shift between the two curves appears to be a systematic failing of the density functional approximation used in the calculation[2].

The good agreement between the experimental and theoretical surface state dispersions is a strong evidence for the model of the Si(111):As 1x1 surface in which the As atoms are actually replacing the outermost Si layer.

3.2   **Arsenic Passivation of Si(111)**

The stability of the Si(111):As 1x1 surface, which is due to the existence of lone-pair states instead of dangling-bond states, was investigated by exposing the surface to oxygen, air and hydrogen. The sensitivities of the Si(111) 7x7 and Si(111):As 1x1 surfaces to oxygen exposure are compared in Fig. 2. Angle-resolved photoemission spectra obtained at $\Theta_e = 10°$ for both the clean 7x7 surface and for the same surface after 15 L of O_2 exposure are shown in Fig. 2a. The emission angle of 10° was chosen because all three surface states on the 7x7 surface are clearly observable. The surface states are totally removed after the 15 L exposure and a new peak corresponding to oxygen is observed at ~ 6 eV below the top of the valence band (E_{VB}). Noticeable changes in the surface state emission occur after exposures of less than 1 L of oxygen to the 7x7 surface.

In Fig. 2b spectra obtained at the $\bar{K}$ point for the Si(111):As 1x1 surface are shown before (bottom spectrum) and after oxygen and air exposures. The intensity of the surface state is almost unaffected by an exposure of 10^7 L of oxygen (80% of original peak height). Molecular oxygen thus seems to have only a marginal effect on the Si(111):As 1x1 surface. In particular we do not observe any evidence of an oxygen peak near 6 eV below E_{VB}. To further investigate the stability of the As terminated surface it was exposed to air for 5 minutes at atmospheric pressure. Except for a ~25% reduction of the lone-pair peak height the spectrum is essentially unaffected by the air exposure. Half of the lost intensity was recovered after a 1 minute anneal at 600°C. The Si(111):As 1x1 surface used for the air exposure had not previously been exposed to oxygen.

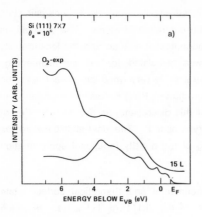 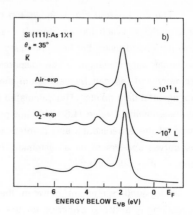

Figure 2a.   Angle-resolved photoemission spectra obtained with 21.2 eV photons before and after oxygen exposure. The three dangling bond surface states located within 2 eV from E_F are completely removed after the 15 L oxygen exposure.
2b. Arsenic lone-pair state probed at the $\bar{K}$ point with angle-resolved photoemission ($h\nu = 21.2$ eV). Both the 10^7 L oxygen and the 10^{11} L air exposure have only marginal effects on the lone-pair emission.

Hydrogen exposure (in the presence of a hot filament) has a stronger effect on the Si(111):As 1x1 surface, causing disorder and a strong attenuation of the lone-pair intensity for exposures of 1×10^4 L. The original surface could be returned, however, by a 2 min., 600°C anneal[1].

4.   REFERENCES

1.   Uhrberg, R.I.G., Bringans, R.D., Olmstead, M.A., Bachrach, R.Z. and Northrup, J.E., submitted to Phys. Rev. B.

2.   Bringans, R.D., Uhrberg, R.I.G., Bachrach, R.Z. and Northrup, J.E., Phys. Rev. Lett., 55, 533 (1985).

3.   Uhrberg, R.I.G., Hansson, G.V., Nicholls, J.M., Persson, P.E.S., and Flodström, S.A., Phys. Rev. B31, 3805 (1985).

4.   For example. Pandey, K.C. and Phillips, J.C., Phys. Rev. Lett., 32, 1433 (1974).

Present addresses:
*   Dept. of Physics and Measurement Technology, Linköping Institute of Technology, S-581 83 Linköping, Sweden.
**   Dept. of Physics, Univ. of California, Berkeley, CA 94720, USA.
***   Dept. of Physics, Arizona State Univ., Tempe, Az 85287, USA.

STRAIN-INDUCED SURFACE RECONSTRUCTION IN THE EPITAXIAL Si-Ge(111) SYSTEM

K. Nakagawa[*], P.M.J. Marée and J.F. van der Veen

FOM-Institute for Atomic and Molecular Physics, Kruislaan 407, 1098 SJ Amsterdam, the Netherlands

[*]permanent address: Central Research Laboratory, Hitachi Ltd., Kokubunji, Tokyo 185, Japan

ABSTRACT

The surface reconstruction of $Si_xGe_{1-x}(111)$ alloy films is shown to be related to lateral stress forces. An external strain field was applied by growing these films epitaxially on either a Si(111) or Ge(111) substrate. The observed 5×5 and 3×3 RHEED patterns are thought to be induced by compressive and tensile stress, respectively. The appearance of a 1×1 pattern is ascribed to a bulk-like surface structure.

Surface reconstructions have been often observed on almost every face of clean semiconductors. A typical example is the Si(111) - 7×7 surface reconstruction, which has been studied for a long time. According to the Takayanagi structure model[1] of the Si(111) - 7×7 reconstruction, which explains the observations most satisfactorily, there are vacancies, stacking faults and adatoms in the surface region, and surface atoms are displaced from their bulk sites. The driving force of the reconstruction is thought to be the compressive stress caused by inward relaxation of outer layers[2,3]. That is, the surface atomic layers of semiconductors are likely to be expanded laterally as a consequence of inward relaxation (enhanced back-bonding). As a result these layers feel compressive stress from the substrate in the lateral direction. When the strain energy exceeds the threshold energy for reconstruction (i.e., the energy for the formation of vacancies, stacking faults or adatoms), the surface will reconstruct[2].

We carried out Reflection High-Energy Electron Diffraction (RHEED) measurements and high-resolution Rutherford Backscattering Spectroscopy (RBS) measurements in conjunction with surface blocking[4] to study the influence of lateral stress on the surface reconstruction of the Si-Ge system. In addition to the internal strain field, an external strain field was applied, of which direction and magnitude were tunable by varying the composition x of Si_xGe_{1-x} alloy films on both Si(111) and Ge(111) substrates.

Samples were prepared in a standard MBE system (base pressure $\simeq 5 \times 10^{-9}$ Pa) equipped with RHEED, facilities for sample sputtering and heating, a Knudsen cell for evaporation of Ge and a Si e-gun evaporator. This system is coupled to a UHV RBS chamber. Si(111) and Ge(111) substrates were cleaned by ultrasonic rinsing in ethanol, followed by several cycles of heating and Ar^+ ion sputtering in the MBE chamber. After this treatment, clear 7×7 and $c2 \times 8$ RHEED patterns were observed on Si and Ge substrates, respectively (Fig.1(b) and 1(c)). Thin Si_xGe_{1-x} films of about 4 monolayers thickness (1 monolayer $\simeq 7.5 \times 10^{14}$ atoms/cm^2) were grown by co-deposition of Si and Ge at room temperature and crystallized by subsequent heating at 500°C. In order to determine the composition x and the film thickness, a substrate and a graphite plate were mounted on the same sample holder simultaneously and the amounts of deposited Si and Ge on the graphite plate were measured accurately by RBS.

Figure 1 shows a series of RHEED patterns as a function of the composition x. A 5×5 reconstruction was observed on Si_xGe_{1-x}/Si(111) ($x \leq 0.75$) (Fig.1(a)), as reported previously[5]. A sharp 1×1 pattern without superlattice reflections was observed on Si_xGe_{1-x}/Ge(111) ($x \simeq 0.5$) (Fig.1(d)) and a 3×3 pattern on Si_xGe_{1-x}Ge(111) ($x \geq .75$) (Fig.1(e)). The $c2 \times 8$ reconstruction on the Ge(111) substrate may be attributed to a lateral compressive stress in analogy with the Si(111) - 7×7 structure, as described above.

The 1×1 structure observed at $x \simeq 0.5$ on the Ge(111) substrate (Fig.1(d)) was thermally stable even during annealing at temperature up to 700°C. (Above this temperature, Ge atoms sublimate.) All our experiments were carried out under ultra-high vacuum conditions (pressure $\simeq 1 \times 10^{-8}$ Pa during deposition) and the surfaces were free of impurities. Therefore, this structure is different from those obtained by impurity-stabilization[6] and laser irradiation[7] which causes thermally unstable structures. We believe that the 1×1 pattern indicates the presence of a bulk-terminated semiconductor surface on which there is no reconstruction. The lateral lattice spacing of the mixed crystal, which is expected to be expanded by inward relaxation, appears to coincide with that of the Ge substrate. Consequently, there is no net stress on this film, which is believed to be the driving force for reconstruction. The 3×3 reconstruction, which is observed for the first time, can be ascribed to the presence of large tensile stress.

Gossmann et al., reported that alloying of Ge and Si was responsible for the formation of the 5×5 surface reconstruction even for pure Ge deposited on Si (111)[5]. They deduced alloying from a significant tailing of the Ge peak in a RBS energy spectrum.

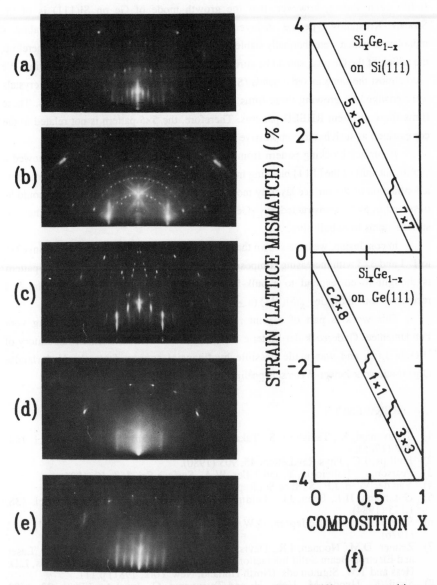

Fig.1.   Observed RHEED patterns for Si_xGe_{1-x} alloy films of different composition on both Si(111) and Ge(111) substrates, listed in order of decreasing compressive stress: (a) 5×5, for x $\sim$ 0.5 and Si substrate; (b) 7×7, for x=1 and Si substrate; (c) c2×8, for x=0 and Ge substrate; (d) 1×1, for x $\sim$ 0.5 and Ge substrate; (e) 3×3, for x $\sim$ 0.75 and Ge substrate.   (f) Corresponding phase diagram of superstructures observed on strained Si_xGe_{1-x} films. (+) indicates a compressive strain introduced in the film by growth on a Si substrate and (-) indicates a tensile strain introduced by growth on a Ge substrate.

It has been shown, however, that the growth mode of Ge on Si(111) is of the Stranski-Krastanov type, i.e., a layer-by-layer growth (in this case up to about 4 monolayers which were thermally stable) followed by island formation[8,9,10]. Therefore, the tailing of the Ge peak should be attributed to island formation instead of alloying. We used almost the same mixed crystals ($Si_{0.5}Ge_{0.5}$) but the stresses on the mixed crystals were changed by growing these films on both Si(111) and Ge(111) substrates. These films show different RHEED patterns. Therefore, the 5×5 pattern is not related to the composition as such but to compressive stress.

A surface blocking pattern from the 1×1 $Si_{0.5}Ge_{0.5}$/Ge(111) system showed a significant shift of the [1 1 $\overline{1}$] blocking minimum toward lower scattering angle, indicating a contraction of the surface layer by more than 4%. On the other hand, no shift could be observed in 5×5 reconstructed $Si_{0.5}Ge_{0.5}$/Si(111) because of a large displacement of surface atoms from bulk sites.

In conclusion, we have shown that the surface structure can be changed from c2×8 to 1×1 and 3×3 with increasing composition x of Si_xGe_{1-x} on Ge(111). The 1×1 pattern is thought to correspond to a bulk-like structure. It is also shown that the 5×5 reconstruction on Si_xGe_{1-x}/Si(111) (x $\simeq$ 0.5) is related to compressive stress.

This work is part of a joint research program between the Stichting voor Fundamenteel Onderzoek der Materie (FOM) and the Central Research Laboratory of Hitachi Ltd., and was made possible by financial support from the Nederlandse Organisatie voor Zuiver-Wetenschappelijk Onderzoek (ZWO).

REFERENCES

1) Takayanagi, K., Tanishiro, S., Takahashi, S. and Takahashi M., Surface Sci. **164**, 367 (1985).
2) Phillips, J.C., Phys.Rev.Letters **45**, 905 (1980).
3) Pearson, E., Halicioglu, T. and Tiller, W.A., Surface Sci. **168**, 46 (1986).
4) Van der Veen, J.F., Surface Sci.Rept. **5**, 199 (1985).
5) Gossmann, H.J., Bean, J.C., Feldman, L.C. and Gibson W.M., Surface Sci. **138**, L175 (1984).
6) Shih, H.D., Jona, F., Jepsen, D.W. and Marcus, P.M., Phys.Rev.Letters **37**, 1622 (1976).
7) Zehner, D.M., Noonan, J.R., Davis, H.L., White, C.W. and Ownby, G.W., "Laser and Electron-Beam Solid Interactions and Materials Processing", J.F. Gibbons, L.D. Hess and T.W. Sigmon eds. (North Holland, New York, 1981) p.111.
8) Shoji, K., Hyodo, M., Ueba, H. and Tatsuyama, C., Jpn.J.Appl.Phys. **22**, 1482 (1983).
9) Narusawa, T. and Gibson W.M., Phys.Rev.Letters **47**, 1459 (1981).
10) Marée, P.M.J., Nakagawa, K., Mulders, F.M. and Van der Veen, J.F., to be published.

PHOTOEMISSION SURFACE CORE-LEVEL STUDY OF SULFUR ADSORPTION ON Si(100)[+]

T. Weser, A. Bogen, B. Konrad, R. D. Schnell, C. A. Schug and
W. Steinmann

Sektion Physik der Universität München
Schellingstr. 4, D-8000 München 40
Fed. Rep. Germany

Chemisorption of elementary sulfur on Si(100) is studied using
LEED, AES and photoemission surface core-level spectroscopy.
Room temperature adsorption results in sulfur bonded on bridge
sites at the surface, at higher adsorption temperatures sulfur
penetrates into the crystal volume.

Stochiometric saturation of all surface valences of a semiconductor
surface may result in a (1x1) reconstruction of high order and chemical stability. Such ideally terminated surfaces as As/Si,Ge(111)(1x1)
and Cl/Ge(111)(1x1) attract current interest [1,2]. As will be reported
elsewhere we have recently shown that the Ge(100) surface can be ideally terminated by one monolayer of sulfur. Our attempts to obtain an
analogous result for the Si(100) surface are reported in this paper.

Some work has been done on sulfur adsorption on metal surfaces where
the standard preparation method uses H_2S. This technique cannot be
applied to semiconductor surfaces as it leads to coadsorption of H,
HS and S. We prepared our surfaces by exposition to elementary sulfur which was produced in situ by dissociation of Ag_2S in a solid-
state electrochemical cell in a separate chamber connected to the
UHV system. The photoemission experiments were carried out at the
dedicated storage ring BESSY in Berlin. The photoelectrons were
analyzed with an ellipsoidal mirror display spectrometer [3] operated
angle integrating in an acceptance cone of about 90°. The Si2p elec-
trons were excited in the photon-energy range $h\nu = 110 - 130$ eV, $h\nu = 130$ eV gave the best surface sensitivity.

[+]This work has been supported by the Bundesministerium für Forschung
und Technologie, T. W. received a scholarship of the Studienstiftung
des Deutschen Volkes.

We used Si samples with low p-type doping. The surfaces were prepared
with repeated cycles of mild sputtering (600 - 800 eV Ar^+) and heating
to $\sim$1000°C. This resulted in a (2x1) LEED pattern. All surfaces were
controlled before and after adsorption by LEED and AES.

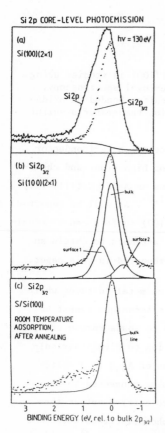

Si 2p CORE-LEVEL PHOTOEMISSION

(a) Si(100)(2×1)   hν = 130 eV

Si2p   Si2p$_{3/2}$

(b) Si2p$_{3/2}$ Si(100)(2×1)   bulk   surface 2   surface 1

(c) Si2p$_{3/2}$ S/Si(100)
ROOM TEMPERATURE
ADSORPTION,
AFTER ANNEALING   bulk line

3   2   1   0   -1
BINDING ENERGY (eV, rel. to bulk 2p$_{3/2}$)

Fig. 1

Fig. 1 shows the evaluation procedure and
the determination of the bulk line shape.
Fig. 1a demonstrates the deconvolution of
an experimental spectrum due to spin-orbit splitting in two equally shaped contributions ($2p_{1/2}$ and $2p_{3/2}$; splitting
$\Delta^{LS} = 0.61 \pm 0.01$ eV; $2p_{1/2}$ to $2p_{3/2}$
branching ratio B = 0.53 $\pm$ 0.02) and the
secondary electron background. After subtraction of the $2p_{1/2}$ contribution and
the background the spectra are fitted
using a convolution of a Gaussian and a
Lorentzian. In agreement with other
authors [4] fig. 1b shows two surface
contributions S1, S2 for the clean Si(100)
(2x1) surface (S1: binding energy shift
$\Delta E = 0.34$ eV, intensity ratio R = I_{S1}/I_{tot}
= 0.23; S2: $\Delta E = -0.43$ eV, R = 0.09). The
comparison with a low coverage S/Si(100)
spectrum (fig. 1c) confirms the determination of the bulk line parameters (Lorentzian width: 200 meV FWHM, Gaussian
width: 500 meV). The surface contributions of the clean surface disappear.

The adsorption behaviour of sulfur on Si(100) depends critically on the
crystal temperature. After room temperature deposition the (2x1) superstructure LEED spots are weakened, the background intensity is uniformly increased. Higher temperatures ($\sim$200°C) lead to enhanced adsorption,
the (2x1) reconstruction disappears, the LEED spots are broadened.

Si2p$_{3/2}$ CORE-LEVEL PHOTOEMISSION

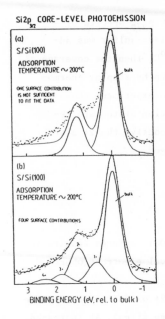

Fig. 2

Si2p$_{3/2}$ CORE-LEVEL PHOTOEMISSION

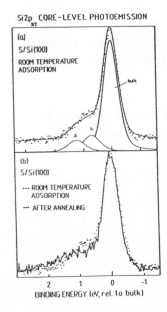

Fig. 3

Fig. 2 shows Si2p$_{3/2}$ core-level photo-emission data of a Si(100) surface after sulfur adsorption at about 200°C. The spectrum is characterized by a broad surface contribution with a maximum at ΔE = 1.23 eV. As shown in fig. 2a, this surface contribution cannot be explained with a single bulk-like line as one would expect for ideal surface termination. In fig. 2b the spectrum is fitted with four surface contributions (oxidation states) of equidistant binding energy shift of 0.62 ± 0.02 eV per S-Si bonding. The existence of 3+ and 4+ states indicates that sulfur penetrates into the Si crystal. The surface contributions show a broader Gaussian (600 meV) than the bulk, this is probably due to disorder.

Fig. 3 shows Si2p$_{3/2}$ core level spectra of a Si(100) surface after room temperature adsorption. The fit in fig. 3a shows that only the first two oxidation states are formed. From this we draw the following conclusions: (i) the sulfur is bridge-bonded (not on-top-bonded); (ii) no sulfur islands are formed; (iii) saturation was not reached; otherwise in all cases one would expect only a 2+ contribution. Fig. 3b shows that after annealing to ~200°C higher oxidation states appear at the cost of the lower ones (sulfur penetration into the volume).

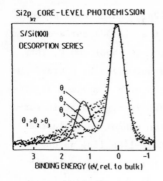

Fig. 4 shows Si2p$_{3/2}$ spectra of a desorption series: sulfur was adsorbed at $\sim$200°C, then the coverage was reduced step by step by thermodesorption. The experiment shows that all oxidation states are reduced simultaneously, the 2+ state is not privileged.

Fig. 4

SUMMARY

1. Chemisorption of elementary sulfur on Si(100) surfaces results in a binding energy shift of the Si2p core level electrons where ΔE = 0.62 eV per S—Si bonding.

2. Room temperature adsorption yields 1+ and 2+ oxidation states (the sulfur stays at the surface bonded on bridge sites). At increased temperatures ($\sim$200°C) additional higher oxidation states occur (penetration of sulfur into the volume).

3. There is no ideal (1x1) termination as in the case of S/Ge (100). During thermodesorption of an oversaturated surface the emission of all surface oxidation states is reduced simultaneously.

ACKNOWLEDGEMENTS

We would like to thank the staff of BESSY for support. We thank D. Rieger, F. J. Himpsel, K. Wandelt and W. Moritz for valuable discussions.

REFERENCES

1. R. D. Bringans, R. I. G. Uhrberg, R. Z. Bachrach, and John E. Northrup; Phys. Rev. Lett. 55, 533 (1985).

2. R. D. Schnell, F. J. Himpsel, A. Bogen, D. Rieger, and W. Steinmann; Phys. Rev. B 32, 8052 (1985).

3. D. Rieger, R. D. Schnell, W. Steinmann, and V. Saile; Nucl. Instrum. Methods 208, 777 (1983).

4. for example F. J. Himpsel, P. Heimann, T.-C. Chiang, and D. E. Eastman; Phys. Rev. Lett. 45, 1112 (1980).

Alkali Overlayer Bands and Their Collective Excitations on Si Surfaces

M. Tsukada, [*]H. Ishida, N. Shima and [*]K. Terakura

Department of Physics, University of Tokyo, Bunkyo-ku, Tokyo, JAPAN

*Institute for Solid State Physics, University of Tokyo,
Minato-ku, Tokyo, JAPAN

ABSTRACT

Overlayer band structures of various alkali monolayers
on Si(111)2×1 and Si(100)2×1 surfaces are calculated by
the first-principles LCAO-Xα-slab method. It is
elucidated how the surface layers are metallized. The
alkali interband overlayer plasmon on Si(100)2×1/K
surface shows anisotropic positive linear dispersion
characteristic to the chain structure.

1. INTRODUCTION

    Alkali overlayers on silicon surfaces provide a novel two-
dimensional (2D) electron gas system confined within a top few atomic
layers of surface. The 2D electron gas is expected to show a variety
of fascinating physical and chemical properties.
    Based on numerical calculations by the first-principles LCAO-Xα-
slab method, we discuss some remarkable features of the alkali
overlayer bands on Si surfaces. Peculiar properties of the overlayer
plasmons, which are the collective excitation of the 2D electron gas
in the overlayer bands are presented.

2. BAND STRUCTURES OF THE ALKALI OVERLAYER ON Si

2.1 Si(111)2×1/K,Cs

    The overlayer bands on Pandey's Π bonded chain model on
Si(111)2×1 surface are calculated for the slab model of Fig.1. The

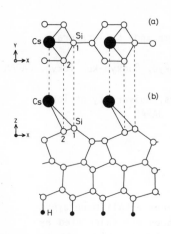

Fig.1 (a) top view of the slab model of Si(111)2×1/Cs. (b) side view seen from [1T0] direction.

adatom-top Si interatomic distance is chosen as the sum of the atomic radii. We do not use the repeated slab model, but the isolated slab model, to avoid the artificial surface potential from the adjacent slabs.

The obtained overlayer bands for Si(111)2×1/Cs are shown in Fig.2 and compared with the π, π^* dangling bond bands of the substrate. The alkali s band, which is located between π and π^* band at the Γ point, crosses with the π^* band along the Γ-J line. Except the crossover region, the alkali character is preserved in this band. The remarkable intactness of the alkali and the dangling bond band has been inferred by the UPS experiment.[1] The overlayer bands of the Si(111)2×1/K surface is almost the same as that of the Si(111)2×1/Cs surface.

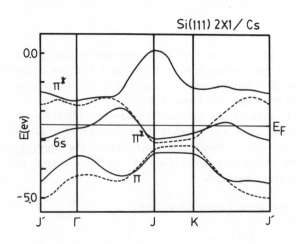

Si(111) 2X1 / Cs

Fig.2 The overlayer band structure of the Si(111)2×1/Cs (full lines). The π and π^* bands of the clean Si(111)2×1 surface are shown (dashed lines) for comparison, which are shifted by -3.0 ev.

2.2 Si(100)2×1/K, Cs

The overlayer bands of the K monolayer on the symmetric dimer surface are shown in Fig.3.[2] The adatom-Si distance is chosen as the sum of the atomic radii. There appear three overlayer bands, a,b,c, which are the antibonding state of K 4s-dimer σ , the bonding state of K $4p_x$-dimer π*, and the bonding state of K $4p_z$-dimer π around the Γ point, respectively. In the Γ -X axis region, the alkali state and the dangling bond are strongly admixed. This feature is different from that of Si(111)2×1/alkali surface.

The overlayer band structure of Si(100)2×1/Cs is similar to that for Si(100)2×1/K surface.

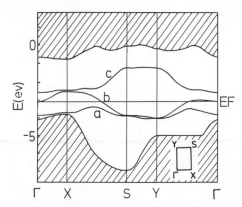

Fig.3 The overlayer band structure of the Si(100)2×1/K for the bridge site.

3. OVERLAYER PLASMONS OF Si(100)2×1/K

Microscopic RPA leads to the following dispersions of the s→p ($p=p_x,p_z$) like interband overlayer plasmon mode near the wave vector $Q \sim 0$:

$$\omega^2(Q) = \omega_{sp}^2 + 4\Pi N_s \omega_{sp} e^2$$

$$\times \{\frac{s}{2\Pi}\int dr_1 \int dr_2 \Phi^*(Q=0,s\to p,r_1)|r_1-r_2|^{-1}\Phi(Q=0,s\to p,r_2)$$

$$+\frac{2}{Q(\kappa+1)}|\int dr \phi_s^*(r)(Q_x X+Q_y Y)\phi_p(r)|^2$$

$$-\frac{2\kappa Q}{\kappa+1}|\int dr \phi_s^*(r) Z \phi_p(r)|^2\}. \qquad (1)$$

In the above $\Phi(Q,\nu,r)$, Ns, κ are the induced charge by the mode ν , the overlayer electron density and the substrate dielectric constant, respectively. ω_{sp} is the energy difference between the p-like and s-like band, with the wave function ϕ_p, ϕ_s, respectively. For the case of Si(100)2×1/K, s,p_x,p_z band correspond to the a,b,c band of Fig.3, respectively. From eq.(1) the dispersion is isotropic and negative linear for a→c mode, while anisotropic and positive linear for a→b mode. The latter is the observed feature for the main ELS peak.[3]

The first term in the bracket { } of eq.(1) does not vanish, even for
the a → b (s → p$_x$) mode.   This is due to the non-vanishing
depolarization field in the direction vertical to chain axis.   The
numerical Wannier functions used as the basis of the induced charge
penetrate into several substrate layers.   The calculated a → b mode
dispersion is shown in Fig.4.   The anisotropic positive linear
dispersion is excellently reproduced.   The agreement of the plasmon
energy is better for the bridge site than the hollow site geometry.

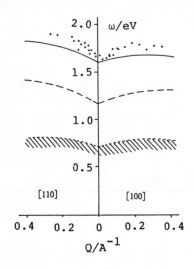

Fig.4 Dispersion relation of
the a→b (s→p$_x$) interband
overlayer plasmon for the
bridge site (full line) and
the hollow site (dashed line)
model. The dots are the
AREELS results.[3]

The observed overlayer plasmon energies are about 1.2eV, 1.6eV,
3.0eV for Cs, K and Na chains on Si(100)2×1.   Such dependence on the
alkali species can not be explained by the individual excitations,
since the change of the energies is much smaller.   Alkali dependence
should be ascribed to the difference in the depolarization shift, the
first term in { } of eq.(1), which is sensitive to the alkali orbital
form.

4. REFERENCES

1)   Tochihara, H., Kubota, M., Miyano, M. and Murata, Y., Jpn. J.
      Appl. Phys. 23, L271 (1984); Surf. Sci. 158, 497 (1985).
2)   Ishida, H., Shima, N. and Tsukada, M., Phys. Rev. B32, 6246
      (1985); Surf. Sci. 158, 438 (1985).
3)   Aruga, T., Tochihara, H. and Murata, Y., Phys. Rev. Letters 53,
      372 (1984).

BUILD-UP AND ELECTRONIC PROPERTIES
OF "ARTIFICIAL" III-V SURFACES ON BULK III-V COMPOUNDS

J.M.MOISON, M.VAN ROMPAY, C.GUILLE,
F.HOUZAY, F.BARTHE, and M.BENSOUSSAN

Laboratoire de Bagneux
Centre National d'Etudes des Télécommunications
196, Avenue Henri Ravera F-92220 BAGNEUX FRANCE

ABSTRACT

We report a study of the molecular-beam-epitaxy (MBE) build-up of III-V monolayer-sized overlayers on III-V substrates, InAs on InP and GaAs, and GaAs on GaP. We have shown previously that InAs overlayers on InP have the morphology of a surface and give InP the surface electronic properties of InAs [1]. Similar results obtained on the other systems suggest that overlayers behave like "artificial surfaces", whose electronic properties may be foreseen and hence tailored, since they belong to the overlayer material.

1. BUILD-UP AND MORPHOLOGY OF THE OVERLAYERS

Overlayers on InP and GaP are obtained through As/P exchanges by annealing under As flux, at temperatures where surface P atoms desorb and are replaced by incoming As atoms [2]. InAs overlayers on GaAs are grown by MBE to a thickness of one monolayer (1ML = one In and one As substrate-like (100) planes) calibrated by reflexion high-energy electron diffraction (RHEED) oscillations.

Though many studies of the reaction of group III and V materials with III-V compounds have been reported [3], only

a few of them have been performed in MBE conditions (high
temperature, (100) substrates and presence of As pressure).
Therefore, the morphology of our structures was checked
in-situ. The RHEED patterns of the overlayers are streaky,
with the same integer-streak spacing as the substrate.
This confirms the overlayer flatness and shows that their
parameter parallel to the surface is reduced to fit the
substrate (pseudomorphism). Thickness values deduced from
Auger and X-ray photoemission intensities [1] with probe
depths lying between 5 Å and 20 Å (table I) are in
agreement with each other and with the deposited 1-ML
thickness for InAs/GaAs, indicating that the overlayer is
pure InAs, and with a 2-ML thickness for InP or GaP-based
structures, showing that the As/P exchange is limited to
the two top P layers. Finally, the morphology of the
overlayers described here is found to be correct; this is
not true for other systems, as AlAs/GaAs or AlAs/InAs [4].

Table I : lattice mismatches and overlayer parameters

Overlayer/Substrate	InAs/GaAs	InAs/InP	GaAs/GaP
Lattice mismatch (%)	-6.7	-3.1	-3.6
Normal parameter (Å)			
- of strain-free overlayer	6.06	6.06	5.65
- of overlayer strained to fit the substrate	6.96	6.45	6.06
Thickness from AES/XPS (Å)	3.0±0.9	6.0±1.0	5.3±0.6

2.   ELECTRONIC PROPERTIES OF THE OVERLAYERS

The positions below the vacuum level of the band gap
(Ev, Ec), of the Fermi level Ef, of the valence band
features B1 and B2, of the In4d or Ga 3d 5/2 core levels,
and of the highest occupied surface state band OSS1 are
obtained by UV photoemission; the positions of the lowest

empty ESS and second-highest occupied OSS2 surface bands are obtained by electron energy loss spectroscopy (see [1,5,6] for details). The absolute incertitude involved in these data (Table II) is rather high (0.2 eV ?), but since the same experimental set-up and interpretation scheme are used in all cases, various materials can be compared on a much more accurate basis.

Table II : distance to the vacuum level of main electronic features for overlayers and bulk materials (cleaned-annealed InP and GaP, MBE-grown InAs and GaAs).

System	InP	InAs/InP	InAs	InAs/GaAs	GaAs	GaAs/GaP	GaP
Ef	4.65	4.75	4.75	4.70	4.80	4.85	4.6
Ev	5.65	5.65	5.40	5.55	5.60	5.60	5.65
Ec	4.30	4.30	5.05	4.15	4.20	3.35	3.4
B1	6.8	7.1	6.7	6.8	6.7	7.1	7.3
B2	11.4	11.6	11.3	12.2	12.1	12.6	12.6
In4d	22.15	22.25	22.25	22.10			
Ga3d				24.05	24.10	24.50	24.80
ESS	4.1	3.7	3.7	3.6	3.3	3.4	3.
OSS1	5.60	5.75	>5.5	>5.6	>5.7	>5.7	>5.
OSS2	6.6	6.5	6.7	6.3	6.7	6.7	7.3

From this table it appears that bulk features of InAs/InP are roughly those of InP and surface ones those of InAs: InAs/InP may be seen as a surface of bulk InAs with its dangling-bond and back-bond states, fitted on InP. Results on InAs/GaAs and GaAs/GaP confirm this wild framework, in which effects such as deformations are not considered. We may have here a very simple way to tailor the surface of covalent materials. For instance, "grafting" an artificial InAs surface on InP gives it the InAs surface states, i.e. moves them away from the band gap and decreases their density within it. This leads to a ten-fold decrease of the surface recombination velocity and a three-fold increase of the luminescence yield [7]. An InAs surface on GaAs has the inverse effect on the surface states and then decreases by half the luminescence yield.

3. CONCLUSION

With MBE techniques, artificial III-V surfaces can be "grafted" on III-V compounds: in spite of huge lattice mismatches, we have grown ML-sized overlayers (InAs on GaAs, InAs on InP, and GaAs on GaP) strained to fit the bulk substrate. Because of their thinness, they cannot be described as heterojunctions, but rather as the overlayer surfaces fitted on top of the bulk substrates. The electronic properties of these artificial surfaces are similar to those the overlayer has when fitted on its own bulk and may then be foreseen, which may open the way to genuine surface engineering.

[1] J.M.Moison, M.Bensoussan, and F.Houzay,
    Phys.Rev. B 34,2018(1986)
[2] K.Y.Cheng, A.Y.Cho, W.R.Wagner, and W.A.Bonner,
    J.Appl.Phys. 52,1015(1981)
[3] A.Kahn, Surf.Sci.Rep. 3,193(1983)
    L.J.Brillson, Surf.Sci.Rep. 2,123(1982)
[4] R.A.Stall, J.Zilko, V.Swaminathan, and N.Schumaker,
    J.Vac.Sci.Technol B3,524(1985)
    J.M.Moison, Rev.Phys.Appl., in press
[5] J.M.Moison and M.Bensoussan, Surf.Sci. 168,68(1986)
[6] J.VanLaar, A.Huijser, and T.L.VanRooy,
    J.Vac.Sci.Technol. 14,894(1977)
[7] J.M.Moison, M.Van Rompay and M.Bensoussan,
    Appl.Phys.Lett., 48,1362(1986)

THE NATURE OF SURFACE MIGRATION DURING MBE GROWTH OF III-V COMPOUNDS:

A STUDY VIA THE RHEED SPECULAR BEAM INTENSITY DYNAMICS

P. Chen,* A. Madhukar, J.Y. Kim and N.M. Cho

Department of Materials Science
University of Southern California
Los Angeles, CA 90089-0241, U.S.A.

The surface migration and interaction of the deposited species during MBE growth determine the surface morphology of III-V semiconductors, such as GaAs(100). Under the normally employed As-stabilized surface growth conditions, the generally held view of the growth process considers the surface migration of the individual Ga atoms as the predominant diffusion process controlling surface morphology[1]. This process is inhibited when Ga atom reacts with As atoms which constitute the next As layer (fig.1-a). Our recent measurements[2,3] of the RHEED specular beam intensity from static and growing GaAs(100) surfaces show the existence of a small temperature range below which the above noted view is consistent with findings. At higher temperatures, appearance of new migration processes is suggested by the data. The bonds formed between the depo-sited Ga and As are sufficiently thermally disturbed and the As atom association reaction and subsequent desorption sufficiently enhanced to give rise to significant Ga-As bond breaking and reformation at an appropriate As_4 pressure. This "binding-breaking" process enhances the opportunity for Ga migration and is stopped only when the Ga-As pair formed is at

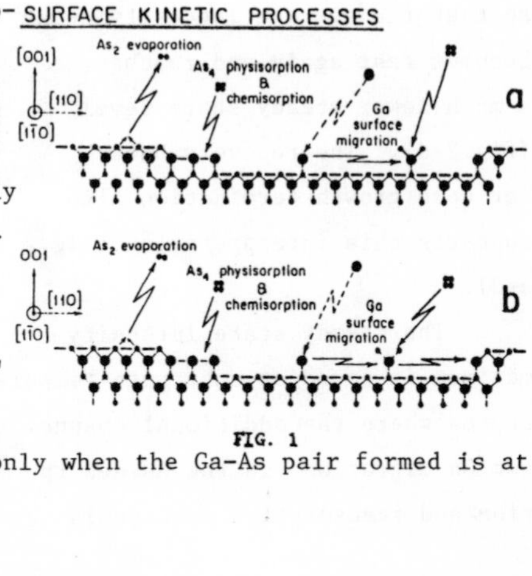

SURFACE KINETIC PROCESSES

FIG. 1

an incorporated epitaxial site (fig.1-b). As shall be seen in the following, the critical temperature range referred to above happens to be near the temperature of bulk GaAs congruent vaporisation. Even though under the mentionned conditions no congruent vaporisation is operative, the Ga migration processes might be similar in nature to that which may occur under congruent vaporisation conditions.

The measurements were done on the GaAs(100) surface along the 2-fold direction at the first off-Bragg condition $S_\perp d=\pi$, where $S_\perp$ is the momentem transfer normal to the surface and d is the step height between adjacent As layers[2,3]. Although the RHEED intensity dynamics cannot be interpreted simply by a single scattering process[4], the specular intensity measured under such condition is nevertheless inversely related to the surface step density[5].

The specular beam intensity variation near a "congruent" temperature shows quick stabilization of the average intensity of oscillation and fast damping of the oscillation amplitude (fig.2 a & b), which implies a rapid stabilization of the step distribution on the growing surface through the above noted new Ga migration mechanism. At lower temperatures (fig. 2-c), the slow decay of average intensity and amplitude are indicative of lesser effective Ga migration. Further decrease in temperature slows the usual Ga migration process so that the average intensity decay becomes fast again and reaches a much lower steady state level (fig.2-d). The recovery behavior upon growth termination also supports this interpretation (fig.2 a-d).

The steady state intensity maximum is found at substrate temperatures where the additional channel for Ga migration via the As desorption and readsorption process is

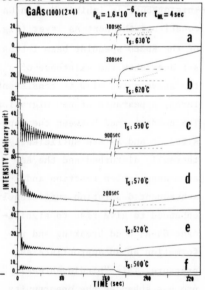

FIG. 2 Oscillation and recovery behavior of the RHEED specular beam intensity of GaAs(100)(2x4) at a constant P_{As} but different T_s. The arrows indicate the instant of growth termination.

stimulated (fig.3-b). Thus the smoothest growth fronts are formed
under this condition (fig.3-b) even though the corresponding static
intensity is not at its maximum (fig.3-a) since, in the absence of
arriving Ga, this stimulated situation perturbs the static surface
stability and tends to degrade surface smoothness. Smoothness of such
a surface cannot be restored by simply decreasing sample temperature[2].
Thus a metastable state of surface step distribution is found after
growth near the "congruent" temperatures. An irreversibility in the
the step distribution on growing surface also is found at lower
temperatures where only the usual Ga migration is significant[2].

When the As_4 pressure is so low that the As evaporation near
the "congruent" temperature cannot be adequately compensated, the
steady state intensity variation with the substrate temperature becomes almost flat (fig.3 4×10^{-7} torr), with a weak maximum at the
high T_s side of static intensity plateau, the oscillations being also
at a low average intensity level (fig.4) since the new Ga migration
mechanism is not stimulated. The consequences of the surface kinetic
processes have been examined via Monte Carlo simulations as a function
of the Ga hoping rate and As incorporation rate for a given Ga flux[1].
The results are consistent with the observations presented here and
in the related papers[2,3].

The effect of the surface
migration mechanism on the surface
or interface morphology has been
investigated simultaneously in our
group by other techniques, such as
photoluminescence study of the
bulk material or multilayer structures, etc.[5,6]. When growth of
multilayer structure is performed
following the usual practise of
no growth interruption[5,6,7] the
growth conditions should be
chosen to optimize group III
migration via the new "binding-
breaking" process. Because of

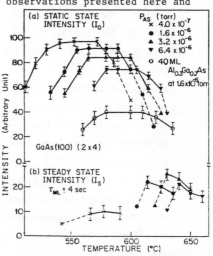

FIG. 3 Relative positions of (a) I_0 cap and
(b) I_s at different P_{As} values. The points
corresponding to reversible and irreversible
regions are connected by solid and broken
lines, respectively.

the existence of the metastable state
such a kinetic process can be fruit-
fully exploited also when the growth
interruption technique[7] is applied
in order to improve the interface
quality of multilayered structures[5,6]. A monitoring of the
metastability might be needed if
growth at very high temperatures
is desired since the life time of
the metastable state will become
short in this case. Details
related to this work will be
published elsewhere.

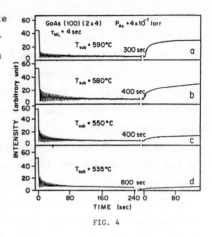

FIG. 4

This work was supported by AFOSR and JSEP.

REFERENCE

*   Permanent Address: Physics Dept., Fudan Univ., Shanghai, China.

1.   A general review can be found in: Madhukar, A. and Ghaisas, S.V.,
      CRC Reviews in Solid State and Materials Sciences (To appear).

2.   Chen, P., Madhukar, A., Kim, J.Y. and Lee, T.C., Appl. Phys.
      Lett. 48, 650 (1986).

3.   Chen, P., Kim, J.Y., Madhukar, A. and Cho, N.M., J. Vac. Sci.
      Tech. B4,

4.   Larson, P.K., Dobson, P.J., Neave, J.H., Joyce, B.A., Bolger, B.
      and Zhang, J., Surf. Sci. 169, 176 (1986).

5.   This was verified by the combined study of photoluminescence
      and RHEED specular beam intensity measurement. See: Voillot, F.,
      Madhukar, A., Kim, J.Y., Chen, P., Cho, N.M., Tang, W.C. and
      Newman, P.G., Appl. Phys. Lett. 48, 1009 (1986).

6.   Kim, J.Y., Voillot, F., Chen, P., Madhukar, A., Tang, W.C. and
      Cho, N.M., Paper presented at the Electronic Materials Confer-
      ence, June 24-26, 1986, Amherst, Mass. (U.S.A.), and to be
      published.

7.   Madhukar, A., Lee, T.C., Yen, M.Y., Chen, P., Kim, J.Y., Ghaisas,
      S.V. and Newman, P.G., Appl. Phys. Lett. 46, 1148 (1985).

VACANCY–COMPLEX STRUCTURE FOR GaAs (1̄1̄1̄)-(2x2) RECONSTRUCTED SURFACE

D. J. Chadi

Xerox Palo Alto Research Center
3333 Coyote Hill Road, Palo Alto, California 94304
USA

ABSTRACT

Among various structures for the As-stabilized (1̄1̄1̄) surface of GaAs the vacancy-complex model is found to provide the best agreement with experimental data on surface As coverage and surface electronic states. Total-energy calculations indicate that the model is energetically very favorable.

1. INTRODUCTION

Significant progress in the determination and understanding of the surface atomic structure of diamond and zincblende type semiconductors has been achieved in the last decade as a result of the development of new experimental and theoretical techniques. The major portion of this effort has been concentrated on the cleavage surfaces of group IV, III-V, and II-VI semiconductors, and on the annealed (111) surfaces of Si and Ge. The atomic structures of GaAs and ZnSe (110) surfaces, Si(111)-(2x1) and even Si(111)-(7x7) surfaces are now known with a high degree of confidence. Much less is known, however, about the nature of the polar (1̄1̄1̄) and (100) surfaces of III-V materials. An important consideration in the study of these surfaces is that, in contrast to the nonpolar (110) cleavage surfaces, they can have a variable surface stoichiometry. This leads to the occurrence of a large number of surface reconstructions, particularly on the (100) surface, and to a lesser extent on the (1̄1̄1̄) surface. An interesting exception is the polar (111) surface which seems to have only a single stable (2x2) reconstruction.

There is considerable evidence that the reconstruction of the Ga-rich GaAs (111)-(2x2) surface arises from the ordering of 1/4 monolayer of Ga vacancies.[1-4] Each 2x2 cell contains, therefore, a single Ga vacancy. For the As-stabilized (1̄1̄1̄)-(2x2) surface low-energy-electron-diffraction studies[5] indicate that a similar type of reconstruction involving 1/4 of monolayer of

As vacancies is not appropriate. This result is consistent with measurements of the surface As coverage on the same surface. The coverage as determined by mass spectrometry and other techniques is very nearly 0.5 monolayer for ($\bar{1}\bar{1}\bar{1}$) surfaces grown by molecular-beam-epitaxy[6] (MBE). The simple vacancy model, with a surface As coverage of 0.75 monolayer, is inconsistent with this result. In the following a model which is consistent with this key experimental data is presented.

2.   VACANCY-COMPLEX STRUCTURE

A top view of the proposed multiple vacancy structure[7] is shown in Fig. 1. Per (2x2) unit cell, the surface bilayer contains two As atoms (large dark circles) and three Ga atoms (circles with dots inside) in a lower plane. The surface bilayer contains, therefore, two As and one Ga vacancy as compared to the ideal (2x2) surface. There are six threefold coordinated atoms, equally divided among Ga and As, per cell. Each Ga atom in the surface bilayer is bonded to two surface As atoms and to one third layer as atom. Figure 1 shows clearly that the surface As atoms are not all in their normal '1x1' sites: One-half the As atoms occupy stacking fault sites. This arrangement of atoms does not lead to any strains as compared to the normal stacking sequence.

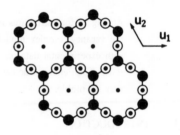

● **Surface As Atom**
● **Substrate As Atom**
⊙ **Surface Ga Atom**
  Above ●

Fig. 1

The main advantage of the stacking fault sequence is that it makes the surface Ga atoms coplanar with their three nearest-neighbor As atoms by construction. This is energetically very favorable for Ga because it leads to an optimal sp^2 rehybridization of its bonding orbitals. The relaxations of the surface As atoms which lead to this optimal configuration also improve the bond angles around the As atoms themselves. The surface energy is reduced when the angular distributions about threefold coordinated As atoms is decreased from $109.47°$ towards $90°$. The surface As atoms have to relax outward from the surface to achieve this. This relaxation also brings the angular distributions around the surface Ga atoms closer to the optimal $120°$. The tight-binding based energy-minimization calculations[8] predict an outward relaxation of 0.26 Å for the As atoms. The angular distributions about the surface As atoms are about $103°$. For the Ga atoms the corresponding values are approximately $115°$, $116°$, and $129°$. The stucture has threefold axes of symmetry going through the As atoms located in the hollow sites of the unit cell (indicated by small dark circles in Fig. 1). The relaxed atomic coordinates of the first two bilayers at the surface are shown in Table I.

An important test of the adequacy of the proposed vacancy-complex structure is its ability to properly describe the surface electronic structure. The surface is calculated to be nonmetallic as expected from simple electron counting in which each Ga [As] threefold coordinated atom is assigned 3/4 [5/4] of a "dangling-bond" electron, and two electrons are assigned to each full band. The calculated surface state bandwidths of 0.4-0.5 eV and band separations of 0 to 1 eV at different points of the Brillouin zone for the two most surface localized states near the valence band maximum are in very good agreement with the experimental results of Bringans and Bachrach[9] from angle-resolved photoemission spectroscopy. For the simple single vacancy model the bandwidths are found to be a factor of two smaller and some bands have the wrong dispersion, even though there is very good agreement with experiment for the Γ point of the surface Brillouin zone.

The multivacancy model is in much better agreement with experiment than the single vacancy model. This does not imply, however, that the latter model cannot occur at the surface. Since the surface As coverage depends on experimental conditions such as the Ga and As fluxes during growth, the substrate temperature, and the surface reaction kinetics, it may be possible to form this structure, and total-energy calculations[10] suggest that it should be a stable structure as well. The multivacancy model (Fig. 1) with a surface As coverage of 0.5 monolayer appears to be the only model that can satisfactorily explain the experimentally measured[6] surface As concentration. The relaxed atomic coordinates for this model are given below.

Table I.   The *reduced* atomic coordinates (a,b,c) in a hexagonal coordinate system   for the (2x2) ideal, and for the vacancy-complex (2x2) reconstructed ($\bar{1}\bar{1}\bar{1}$) surfaces of GaAs are shown. Atomic coordinates are determined from the relation $r = au_1 + bu_2 + cu_3$ where u_1 and u_2 are lattice vectors of the 1x1 lattice shown in Fig. 1 and u_3 is the outward normal to the surface with magnitude equal to the bulk bilayer spacing of $\simeq 0.816$ Å. The largest relaxations occur at the surface As atoms.

Table I

	(a, b, c)	(a, b, c)
Atom	Ideal 2x2 Surface	Reconstructed Surface
As	(0.667, 0.333, 0)	Vacancy
As	(1.667, 1.333, 0)	Vacancy
As	(0.667, 1.333, 0)	(0.667, 1.333, 0.323)
As	(1.667, 0.333, 0)	(1.333, 0.667, 0.323)
Ga	(0.000, 0.000,-1)	Vacancy
Ga	(1.000, 0.000,-1)	(1.000, 0.000, -1.014)
Ga	(0.000, 1.000,-1)	(0.000, 1.000, -1.014)
Ga	(1.000, 1.000,-1)	(1.000, 1.000, -1.014)
As	(0.000, 0.000,-4)	(0.000, 0.000, -3.648)
As	(0.000, 1.000,-4)	(0.005, 1.003, -4.029)
As	(1.000, 1.000,-4)	(0.997, 1.003, -4.029)
As	(1.000, 2.000,-4)	(0.997, 1.995, -4.029)
Ga	(0.333, 0.667,-5)	(0.319, 0.638, -4.989)
Ga	(0.333, 1.667,-5)	(0.319, 1.681,-4.989)
Ga	(1.333, 0.667,-5)	(1.333, 0.667,-4.981)
Ga	(1.333, 1.667,-5)	(1.362, 1.681,-4.989)

3.   CONCLUSIONS

In summary, the proposed vacancy-complex model for the GaAs($\bar{1}\bar{1}\bar{1}$)-(2x2) surface provides an energetically attractive structure which is consistent with experimental data on surface As coverage and surface electronic structure for MBE grown samples. Further experimental work is needed to test whether this structure is the correct one for the surface. It would also be interesting to examine the 2x2 surfaces prepared by simultaneous ion bombardment and annealling[11] to see if they have the same structure as the MBE grown samples.[12] It is presently unclear whether the two differently prepared surfaces even have the same chemical composition, particularly since the ratios of their Ga to As Auger peak intensities appear to be quite different.[11-13] It is conceivable that two (or more) different metastable 2x2-ordered structures occur at this surface as has been suggested by several researchers.[10-12]

This work is supported in part by the U. S. Office of Naval Research under Contract No. N00014-82-C-0244.

4.   REFERENCES

1.   S. Y. Tong, G. Xu, and W. N. Mei, Phys. Rev. Lett. **52**, 1693 (1984).

2.   D. J. Chadi, Phys. Rev. Lett. **52**, 1911 (1984).

3.   J. Bohr, R. Feidhans'l, M. Nielsen, M. Toney, R. L. Johnson, and I. K. Robinson, Phys. Rev. Lett. **54**, 1275 (1985).

4.   E. Kaxiras, K. C. Pandey, Y. Bar-Yam, and J. D. Joannopoulos, Phys. Rev. Lett., **56**, 2819 (1986).

5.   S. Y. Tong (unpublished).

6.   J. R. Arthur, Surf. Sci. **43**, 449 (1974).

7.   D. J. Chadi, Phys. Rev. Lett. **57**, 102 (1986).

8.   D. J. Chadi, Phys. Rev B **19**, 2074 (1979), and **29**, 785 (1984).

9.   R. D. Bringans and R. Z. Bachrach, Phys. Rev. Lett. **53**, 1954 (1984).

10.   E. Kaxiras, Y. Bar-Yam, J. D. Joannopoulos, and K. C. Pandey, Phys. Rev. Lett. **57**, 106 (1986).

11.   M. Alonso, F. Soria, and J. L. Sacedon, J. Vac. Sci. Technol. A **3**, 1598 (1985).

12.   H. H. Farrell, D. W. Niles, and M. H. Bakshi (unpublished).

13.   K. Jacobi, C. v. Muschwitz, and W. Ranke, Surf. Sci. **82**, 270 (1979).

VARIABLE STOICHIOMETRY SURFACE RECONSTRUCTION : NEW MODELS AND PHASE TRANSITIONS ON GaAs {111}2x2

E.Kaxiras, Y. Bar-Yam and J. D. Joannopoulos
Department of Physics, Massachusetts Institute of Technology
Cambridge, Massachusetts 02139, USA
and
K. C. Pandey
I B M Thomas J. Watson Research Center
Yorktown Heights, New York 10598, USA

The reconstructions of the (111) and ($\overline{1}\,\overline{1}\,\overline{1}$) polar surfaces of GaAs are examined using the results of total energy calculations and incorporating the effect of the relative chemical potential of Ga and As atoms. A phase transition is predicted between the Ga vacancy and As triangle (2x2) reconstructions for the (111) surface, and new (2x2) models are proposed for the As and Ga rich reconstructions of the ($\overline{1}\,\overline{1}\,\overline{1}$) surface.

1. INTRODUCTION

Polar surfaces of binary compound semiconductors exhibit a rich variety of reconstructions. Two reasons contribute to this diversity : first, the polarity of the ideal unreconstucted surface indicates that large atomic rearrangements are required to produce a stable configuration; in addition to this, the presence of two kinds of atoms allows the formation of surfaces with variable stoichiometry.

In this work we will concetrate on the (2x2) reconstructions of the (111) and ($\overline{1}\,\overline{1}\,\overline{1}$) surfaces of GaAs which are representative examples of polar surfaces. For each surface reconstruction we calculate the total energy, allowing for extensive relaxation. We chose to perform the calculations in the pseudopotential formalism for two reasons : i) It is an *ab initio*, parameter free formalism, which is important when dealing with systems involving large distortions of the bond lengths and bond angles; ii) It gives a consistent evaluation of the energy of free atoms with respect to the surface, which is necessary for comparisons of systems with different number of surface atoms. Details of the calculational procedure have been published elsewhere [1].

2. TOTAL ENERGIES OF (2x2) RECONSTRUCTIONS

The calculated energy (in eV per (2x2) surface unit cell) of each model is given in

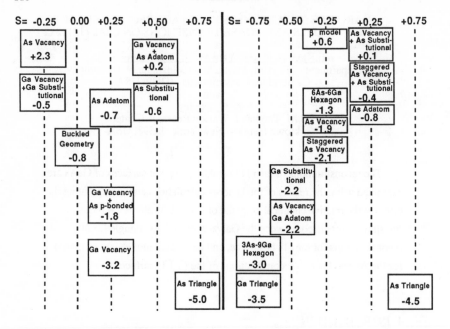

Figure 1. (111) Surface total energies.   Figure 2. (1̄1̄1̄) Surface total energies.

Figures 1 and 2 for the (111) and (1̄1̄1̄) surfaces respectively. The zero of energy is defined by the ideal unreconstructed surface in each case. The various models are divided in stoichiometric classes according to $S = (N_{As} - N_{Ga})/4$ where N_{As} (N_{Ga}) is the number of As (Ga) atoms in the surface bilayer of the (2x2) unit cell. The denominator 4 was chosen so that $|S| = 1$ indicates a difference by a whole monolayer from the ideal surface. In determining the total energy, equilibrium with bulk GaAs is assumed. For the reservoirs of Ga and As atoms we chose Ga bulk and As_2 gas respectively, as representative of experimental conditions. The binding energies of these reservoirs, which were used in the energy comparisons, are 6.8 eV per pair for bulk GaAs, 2.8 eV per atom for Ga bulk and 2.0 eV per atom for As_2 gas. The calculated total energy of each added (or removed) atom minus the appropriate binding energy was added to (or subtracted from) the surface total energy to obtain the final total energy values.

2.1   (111) 2x2 Surface

This surface in unreconstructed form consists of a plane of Ga atoms with one dangling bond on each surface atom. The lowest energy geometry is the As triangle [2] at $S = 0.75$ (Fig. 1), consisting of three As adatoms bonded in an equilateral triangle configuration to three surface Ga atoms. The lowest energy models in the other stoichiometric classes are: the As substitutional model [3] at S = 0.50, in which one As atom replaces one surface Ga atom per (2x2) unit cell; the Ga vacncy model [4],[5] at S = 0.25, with a Ga atom missing from each (2x2) cell; the buckled geometry [6] at S = 0, in which one surface Ga atom has moved outward from the surface and the other three surface Ga atoms have moved inward; and the Ga vacancy + substitutional geometry at S = -0.25, in which a surface Ga atom has replaced an As atom in the first bilayer.

2.2 ($\bar{1}\bar{1}\bar{1}$) 2x2 Surface

The unreconstructed configuration of this surface is similar to that of the (111) surface, with all surface atoms being As. The As triangle geometry at S = 0.75 is the lowest energy configuration for this surface as well (Fig. 2). The lowest energy geometries for the other stoichiometric classes are: the As adatom [7] at S = 0.25, consisting of one As atom bonded to three surface As atoms; the staggered As vacancy [8] at S = -0.25, in which one surface As atom is missing and the remaining surface As atoms are staggered by 60° with respect to their ideal positions; the As vacancy + Ga adatom and the Ga substitutional geometries (with equal energy) at S = -0.50; and the Ga triangle at S = -0.75, consisting of three Ga adatoms in a triangular configuration.

3. EFFECT OF CHEMICAL POTENTIAL

The assumption that each surface reconstruction is in equilibrium with its natural environment (e.g. an As-rich surface in an As-rich environment) does not necessarily hold in realistic situations. In particular, given the very low sticking coefficient of As [9], an As-rich environment might be very hard to realize experimentally. It is therefore important to include the effect of the relative chemical potential of As and Ga atoms. Experimentally this is controlled by adjusting the partial pressures or the temperature of the sample [9]. Theoretically, at zero temperature (where all our calculations are performed) the relative chemical potential can be taken into account as follows : given the binding energy of the Ga (2.8 eV) and As (2.0 eV) atoms and the cohesive energy of GaAs (6.8 eV per pair) we conclude that the range of variation of the relative chemical potential $\delta\mu$ is 2.0 eV. Since the zero of energy in this range is arbitrary, we define the interval of values for $\delta\mu$ to be [-1.0 eV, 1.0 eV] with $\delta\mu$ = 1.0 eV ($\delta\mu$ = -1.0 eV) corresponding to As-rich (Ga-rich) environment. The energy of the As-rich or Ga-rich surface configurations is then given by :

$$E_{As\text{-}rich}(\delta\mu) = E^{eq.}_{As\text{-}rich} + 4S(1.0\,eV - \delta\mu)$$

$$E_{Ga-rich}(\,\delta\mu\,) = E^{eq\cdot}{}_{Ga-rich} - 4S\,(1.0\,eV + \delta\mu\,)$$

where $E^{eq\cdot}{}_{As-rich}$ ($E^{eq\cdot}{}_{Ga-rich}$) is the total energy of the As-rich (Ga-rich) surface reconstruction in equilibrium with As-rich (Ga-rich) environment (given in Fig.1 and 2) and S is the stoichiometry as defined above.

Using the above analysis we find that for the (111) surface, in Ga-rich environment the lowest energy geometry is the Ga vacancy with an energy of -1.2 eV, whereas the As triangle in Ga-rich environment has an energy of +1.0 eV. This is consistent with experimental studies [4],[10], which observe the Ga vacancy under Ga-rich preparation conditions. In As-rich environment the lowest energy geometry is the As triangle. We therefore predict a phase transition between the As triangle and the Ga vacancy geometries at $\delta\mu \sim 0.1$ eV.

For the ($\overline{1}\,\overline{1}\,\overline{1}$) surface, by the same analysis, the lowest energy geometry in a Ga-rich environment is the Ga triangle. A phase transition is predicted between the Ga triangle and staggered As vacancy geometries at $\delta\mu \sim -0.4$ eV. However, experiment shows that in Ga-rich environment a ($\sqrt{19}$x$\sqrt{19}$) stable configuration appears on this surface [9]. Such a geometry, incorporating features of both the Ga triangle and staggered As vacancy geometries, with intermediate As coverage, has been proposed recently [8]. Finally the lowest energy geometry of the ($\overline{1}\,\overline{1}\,\overline{1}$) surface in As-rich environment is the As triangle, though this particular reconstruction might be hindered by kinematics, due to the many As-As bonds on the surface.

References

1) E. Kaxiras, Y. Bar-Yam, J. D. Joannopoulos and K. C. Pandey, Phys. Rev. B 33, 4406 (1986)

2) E. Kaxiras, K. C. Pandey, Y. Bar-Yam and J. D. Joannopoulos, Phys. Rev. Lett. 56, 2819 (1986).

3) A. U. MacRae and G. W. Gobeli, J. Appl. Phys. 35, 1692 (1964).

4) S. Y. Tong, G. Xu and W. N. Mei, Phys. Rev. Lett. 52, 1693 (1984).

5) D. J. Chadi, Phys. Rev. Lett. 52, 1911 (1984).

6) D. Haneman, Phys. Rev. 121, 1093 (1961).

7) W. A. Harrison, J. Vac. Sci. Technol. 16, 1792 (1979).

8) E. Kaxiras, Y. Bar-Yam, J. D. Joannopoulos and K. C. Pandey, Phys. Rev. Lett. 57, 106 (1986).

9) A. Y. Cho and I. Hayashi, Solid State Electron. 14, 125 (1971).

10) J. Bohr, R. Feidenhans'l, M. Nielsen, M. Toney, R. L. Johnson and I. K. Robinson, Phys. Rev. Lett. 54,1275 (1985).

ATOMIC GEOMETRY AND ELECTRONIC STRUCTURE OF THE (311)-(1x1) SURFACES OF GaAs

C.B. Duke, C. Mailhiot and A. Paton

Xerox Webster Research Center
Webster, NY 14580
USA

and

A. Kahn and K. Stiles

Department of Electrical Engineering, Princeton University
Princeton, NJ 08544
USA

ABSTRACT

The polar (311) surfaces of GaAs exhibit four inequivalent bulk (1x1) terminations each of which possesses a metallic electronic character associated with partly occupied surface states. Two inequivalent surfaces have been prepared experimentally by cutting and polishing followed by ion-bombardment and vacuum annealing. Both surfaces exhibit (1x1) structures for which the elastic low-energy electron diffraction (ELEED) intensities of fifteen beams have been measured at T=300K. Analysis of these beams using dynamical calculations of the ELEED intensities for several hundred model structures suggests that both inequivalent structures are associated with relaxed truncated-bulk atomic geometries. Detailed results are given for one set of ELEED intensity data. Energy minimization calculations predict the qualitative features of the geometries determined via ELEED intensity analysis. Calculations of the surface electronic structure indicate metallic-surface-state behavior for the relaxed truncated-bulk structures.

1.   INTRODUCTION

The (311) surfaces of GaAs are of particular interest for several reasons. First, two inequivalent (311) polar surfaces occur, each of which exhibits a stable (1x1) surface structure[1,2]. Second, these surfaces should lead to high-quality epitaxial growth[3]. Third, both inequivalent bulk terminations are predicted to exhibit metallic behavior. Consequently, surface reconstruction is expected[2]. Fourth, elastic low-energy electron diffraction (ELEED) intensity data are available for this material[2,4] so that a surface structure analysis can be performed. In this paper we report the results of such an analysis for both the Ga terminated ("A") and As-terminated ("B") variants of the bulk B' substrate.

2.   EXPERIMENTAL ELEED INTENSITIES

Experimentally prepared GaAs(311) surfaces display a (1x1) symmetry. Two inequivalent surfaces have been prepared experimentally by cutting and polishing followed by ion-bombardment and vacuum annealing[1]). The probable character of their termination was assessed by etching prior to the vacuum ion-bombardment-and-anneal cycle. On that basis the two inequivalent surfaces are labelled "A" and "B" indicating suggested Ga and As terminations, respectively. Further etching experiments indicated that these two inequivalent surfaces correspond to the A' and B' substrate orientations, respectively. Since a B surface structure corresponds to a Ga overlayer on an As terminated A' substrate, the "A" and "B" labels of the data do not necessarily imply Ga and As terminated surface geometries. Only the ELEED intensity analysis can establish the actual surface structure associated with each set of intensity data.

For each surface the elastic low-energy electron diffraction (ELEED) intensities at T = 300K were measured for fifteen diffracted beams resulting from normally incident electrons in the energy range 40 eV$\leq$E$\leq$ 240 eV. These beams are labelled by the indices (01), ($\overline{1}$0), (11) ($\overline{1}\overline{1}$), (02), ($\overline{2}$0), (12) ($\overline{2}\overline{1}$), (22), ($\overline{2}\overline{2}$), (23), ($\overline{3}\overline{2}$) (33) and ($\overline{3}\overline{3}$) defined relative to the surface unit mesh specified in Ref. 2.

Table I suggests that the experimental "B" surface most probably corresponds to a relaxed A atomic geometry. Analogous results have been obtained for the experimental "A" surface[4]. This class of surface geometries describes the experimentally measured intensities for GaAs(311)-(1x1) as well as corresponding calculations performed for the (110) surfaces of zincblende structure compound semiconductors[9,10]. Therefore the model descriptions of the measured ELEED intensities are now sufficiently refined that a search for the predicted metallic surface-state behavior, e.g., via photoemission spectroscopy, becomes useful in refining the structures further.

We conclude, therefore, that plausible structural models of GaAs(311)-(1x1) surfaces have been established, that most of these models predict metallic surface-state behavior, and that the observation of this behavior (or its absence) can provide an important clue either to the surface structures, or to the validity of a tight-binding model description for their electronic structure, or to both.

REFERENCES

1. Stiles, K., and Kahn, A., J. Vac. Sci. Technol. B3, 1089 (1985).

2. Duke, C.B., Mailhiot C., Paton, A., Kahn, A. and Stiles, K., J. Vac. Sci. Technol. A4, 947 (1986).

3. Sangster, R.C., in "Compound Semiconductors", ed. Williamson, R.K. and Goering, H.L. (Reinhold, London, 1962), Vol. 1, p. 241.

4. Mailhiot, C., Duke, C.B., Paton, A., Kahn, A., and Stiles, K., J. Vac. Sci. Technol. B4, in press.

5. Ford, W.K., Duke, C.B., and Paton, A., Surf. Sci. 115, 195 (1982).

6. Meyer, R.J., Duke, C.B., Paton, A., Kahn, A., So, E., Yeh, J.L. , and Mark, P., Phys. Rev. B19, 5194 (1979).

7. Duke, C.B., Richardson, S.L., Paton, A., and Kahn, A., Surf. Sci. 127, L135 (1983).

8. Chadi, D.J., J. Vac. Sci. Technol. B3, 1167 (1985).

9. Duke, C.B., Adv. Ceram. 6, 1 (1983).

10. Kahn, A., Surf. Sci. Repts. 3, 193 (1983).

C.B. Duke et al.

TABLE I: Best-fit atomic geometries associated with the "B" set of ELEED intensity data for GaAs(311). Layer numbers are indicated as subscripts and correspond to the unreconstructed bulk surface. The atomic displacements are given relative to the corresponding bulk structures. The threefold-Ga-on-B' geometry is specified relative to the unreconstructed A truncated-bulk structure.

Structure	Atom	Atomic Displacements			R Factors	
		d_x (A)	d_y (A)	d_z (A)	R_x	R_I
A (twofold Ga on B')	Ga_1	0	0	0.04	0.21	0.08
A: relaxed substrate	Ga_1	0	0	0.02		
	As_2	0	0	0	0.25	0.08
	Ga_3	0	0.20	0.45		
B'	As_1	0	0	-0.02	0.23	0.11
Threefold Ga on B' termination	Ga_1	0	1.03	1.08		
	As_2	0	0	0	0.29	0.06
	Ga_3	0	0	0		

THE ELECTRONIC STRUCTURE OF THE GaSb-(110)-SURFACE STUDIED BY COMBINED ANGULAR RESOLVED PHOTOEMISSION AND INVERSE PHOTOEMISSION

H. Carstensen, R. Manzke, I. Schäfer, and M. Skibowski

Institut für Experimentalphysik, Universität Kiel
2300 Kiel, FR Germany

ABSTRACT

Energy position and dispersion of the empty and occupied electron states parallel to a p-type GaSb-(110)-surface have been studied for the main symmetry lines $\overline{\Gamma X}'$ and $\overline{\Gamma X}$ of the surface Brillouin zone for energies up to 8 eV on both sides of the Fermi level. The results are compared with recent calculations for the relaxed (110)-surface in view of cation and anion derived surface states, surface resonances and bulk states.

1. INTRODUCTION

Angular resolved photoemission and inverse photoemission yield detailed information about the electronic band structure of both occupied and unoccupied states of solid surfaces. For the GaSb-(110)-surface there are only few experimental studies [1] concerning the dispersion of the occupied bands and none dealing with the unoccupied states. With regard to several band structure calculations [2-4] providing detailed but differing information on the relaxed GaSb-(110)-surface it is necessary to test these results experimentally, in particular the results for the unoccupied states, and to determine energy position and dispersion of the bands including the energy gaps between occupied and unoccupied states, especially the fundamental surface gap. Combined investigation of photoemission and inverse photoemission for the same sample with a common energy calibration is a powerful method for such studies.

2. EXPERIMENTAL

The experiments were performed on UHV-cleaved p-type GaSb-(110)-surfaces using an experimental set-up for combined angular resolved photoemission and inverse photoemission [5]. Photoelectrons were excited by He I radiation (21.2 eV). Their energy was measured by a cylindrical mirror analyzer ($\Delta E \approx 100$ meV). The electron induced bremsstrahlung was detected at $\hbar\omega = 9.8$ eV by an efficient band pass detector consisting of an open Cu-Be multiplier with a CaF_2 entrance window ($\Delta E \approx 500$ meV). A common energy scale for both spectroscopies was established by measuring the energy of the electrons from the inverse photoemission electron source (5-20 eV) with the photoemission energy analyzer. No explicit determination of the Fermi level was needed for comparing energy differences between occupied and unoccupied levels.

3. RESULTS AND DISCUSSION

Figure 1 displays some inverse photoemission spectra of the GaSb-(110)-surface in the $\bar{\Gamma}\bar{X}'$ direction (mirror plane) for various angles of electron incidence on both sides of the surface normal over an energy range of about 8 eV. Up to five different maxima labeled A-E can be observed in the spectra representing either a surface state (resonance) or a bulk feature. Contamination experiments have shown that especially the peaks A and C can be considered as closely related to the surface. A detailed analysis gives a weak but distinct dispersion with the wave vector parallel to the surface for all peaks. The spectra also exhibit remarkable intensity modulations with angle due to matrix elements. In particular peak A is very strong for positive angles of incidence, i.e. on the side of the unoccupied dangling bond. This behaviour, its energy minimum at $\bar{X}'$ ($\vartheta = 25°$) and its surface sensitivity identifies peak A as being associated with the cation related empty dangling bond state (C_3). It was predicted by all available surface band structure calculations which, however, give different energy locations with respect to the conduction band minimum or the edge of the projected bulk band structure in the $\bar{\Gamma}\bar{X}'$ direction. It is the counterpart of

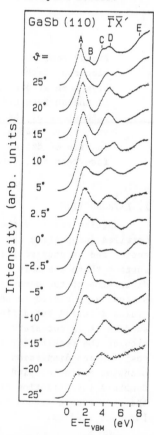

Fig. 1: Inverse photoemission spectra for $\bar{\Gamma}\bar{X}'$

the well known occupied anion derived
dangling bond state (A₅) which can be
studied by angle resolved photoemission.
The latter appears around X̄' as a prominent peak close to the valence band
maximum in the photoemission spectrum
(see Fig. 2, middle, left side) when the
photoelectron direction is opposite to
that of electron incidence for which the
inverse photoemission spectrum exhibits
the prominent empty dangling bond structure (Fig. 2, middle, right side). This
antisymmetry in intensity concerning the
electron directions in both processes
can be understood by taking matrix element effects into account. In plane wave
approximation for the high energy states
involved in the measuring processes the
intensity behaviour simply indicates
that the occupied and unoccupied dangling bonds extend to opposite sides of
the surface normal.

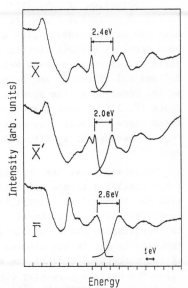

Fig. 2: Determination of
energy gaps for different
points in the surface BZ

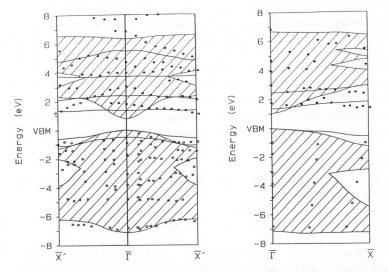

Fig. 3: Experimental band structure for Γ̄X̄' and Γ̄X̄ (dots) and
the band calculation of Bertoni et al. [2] including surface
bands (solid lines) and projected bulk bands (hatched area)

128   H. Carstensen et al.

    Combination of both types of spectra with a common energy scale as
in Fig. 2 using the procedure described in Section 2 allows to determine the surface energy gaps at $\bar{X}'$ 2.0 (± 0.05) eV and at $\bar{X}$ 2.4 eV with
good accuracy without relying on a determination of the Fermi level.
These values compare quite well with the theoretical ones lying between
2.0 and 2.1 eV for $\bar{X}'$ and between 2.1 and 2.4 eV for $\bar{X}$ [2-4]. The broad
maxima observed in the spectra for normal electron emission and incidence ($\bar{\Gamma}$, 2.6 eV apart) are considered to include bulk and surface features where the bulk component is not exactly associated with the bulk
Γ-point because of $k_\perp$ dispersion effects which are not studied in this
work.

    Figure 3 shows the experimental two dimensional band structure
$E(k_{\shortparallel})$ resulting from a detailed analysis of all spectra. It is compared
with the theoretical calculation of Bertoni et al. [2] because this
band structure is calculated up to about 7 eV above the valence band
maximum whereas other calculations [3,4] only show the low energy part.
Emphasis is put in our studies on the experimentally new investigation
of the unoccupied part of the electronic structure. It is seen that besides the surface gaps many others important features compare in general fairly well with the calculations, e.g. the two low lying experimental empty bands (C_3,C_4) separated by about 1 eV at $\bar{X}'$ and the band
located around 4 eV. The empty and occupied surface dangling bond
states at $\bar{X}'$ are removed from the fundamental bulk gap (0.72 eV).
Closer inspection of the experimental data and the calculations, however, shows apparent differences, also between different theoretical
results. So it is not quite clear yet in which region of the Brillouin
zone the empty surface bands behave more like surface resonances than
like surface states split off the projected bulk bands. Refined calculations including the correct surface geometry and more experimental
data are necessary to clarify these points.

    Work supported in part by the Bundesministerium für Forschung und
Technologie and the Deutsche Forschungsgemeinschaft of the FR Germany.

5. REFERENCES

1. Chiang, T.-C. and Eastman, D.E., Phys. Rev. B22, 2940 (1980)
   and references therein
2. Bertoni, C.M., Bisi, O., Calandra, C., and Manghi, F.,
   Inst. Phys. Conf. Ser. No. 43, 191 (1979)
3. Beres, R.P., Allen, R.E., and Dow, J.D.,
   Solid State Commun. 45, 13 (1983)
4. Mailhiot, C., Duke, C.B., and Chadi, D.J.,
   Phys. Rev. B31, 2213 (1985)
5. Babbe, N., Drube, W., Schäfer, I., and Skibowski, M., J. Phys. E18,
   158 (1985), improved version with KBr coated first dynode

THE INFLUENCE OF DOPING ON THE ETCHING OF Si(111)

Harold F. Winters and D. Haarer*

IBM Almaden Research Center
650 Harry Road
San Jose, California 95120-6099

Summary Abstract: It has been recognized for some time that the doping level in silicon influences etch rate in plasma environments.[1-8] We have now been able to reproduce and investigate these doping effects in a modulated-beam, mass spectrometer system described previously[9] using XeF$_2$ as the etchant gas. The phenomena which have been observed in plasma reactors containing fluorine atoms are also observed in our experiments. The data has led to a model which explains the major trends.

The influence of doping could be related to the electronic structure of the surface, the chemical influence of the dopant atoms, or some sort of impurity related phenomena. Using Auger spectroscopy, we have been unable to find a correlation between the surface concentration of dopant atoms or impurities and etching characteristics. Therefore, it is concluded that these doping phenomena are related to the electronic structure of the surface. Similar conclusions have been drawn previously.[1,2,4,7]

Present address: Universitaet Bayreuth, Lehrstuhl fuer Experimentalphysik IV, Postfach 3008, D-8580 Bayreuth, F.R.G.

It was previously suggested[10] that etching reactions were similar to the growth of thin oxide layers and could therefore be described by the ideas proposed by Mott and Cabrera,[11,12] i.e., that the presence of negative ions and/or an electric field at the surface strongly influences the reaction rate. This abstract summarizes data and ideas which will be described in more detail in ref. 12 and which show that there is a strong correlation between the estimated negative ion concentration at the surface and the etch rate.

The experimental results which are related to the influence of doping on etch rate are summarized in this paragraph. References 12 and 13 refer to the etching of silicon with XeF_2 while the other references deal with plasma assisted etching of silicon with fluorine and chlorine. In situations where the XeF_2 and plasma experiments overlap, they give similar results. The symbol n^+-Si(111) means degenerately doped, n-type Si(111).

1. n^+-Si(111) etches much faster than undoped, lightly doped or p^+-Si(111). (See refs. 1,2,3,4,6,5,8,12)

2. Lightly doped p-type silicon etches faster than p^+-Si(111). (See refs. 12,2,4,8)

3. Both lightly doped p-type and n-type silicon have similar etch rates. (See refs. 12,4,8.)

4. The etch rates for n^+-Si(111) and p^+-Si(111) are similar when the fluorosilyl layer on the silicon surface is thin but differ greatly when the layer is thicker. (See ref. 12.)

5. Ion bombardment reduces the influence of doping on etch rates. (See refs. 12,2.)

6. Exposure of silicon to high intensity light increases the etch rate. (See ref. 13.)

7. The pressure of alkali metals on the silicon surface increases the etch rate. (See ref. 7.)

The experimental trends described above can be explained by
a model which is based on three ingredients: (1) exposure of
Si(111) to XeF_2 leads to the growth of a fluorosilyl layer on
the surface; (2) the affinity level for F at the silicon
surface lies below the top of the valence band; and (3) that
of some of the ideas proposed by Mott and co-workers to explain
the growth of thin oxide layers are also applicable to the
silicon fluorine system.

The fact that a fluorosilyl layer grows on a silicon surface
exposed to fluorine (XeF_2) has been adequately demonstrated by
the work of McFeely[14] and others.[15] The layer is composed of
mostly SiF_3 and continues to become thicker even for extremely
large XeF_2 fluences. Under most experimental conditions, the
thickness of this layer probably lies somewhere between 10Å
and 30Å and is in a quasi steady state. Etching occurs from
the surface of this layer. Therefore, the transport of either
silicon to the fluorosilyl surface or fluorine to the
SiF_x-Silicon interface is necessary.

The second important ingredient to the model is related to
the presence of negative ions "on" and/or "in" the fluorosilyl
layer. Bagus[16] has published calculations which show that the
lowest energy state for fluorine adsorbed on a silicon surface
is F^-. This indicates that the affinity level for fluorine
lies beneath the valence band of silicon. Arguments indicating
the plausibility of this result are easy to make. The affinity
level for fluorine atoms is 3.45 eV below the vacuum level
at an infinite distance from the silicon surface. As the
fluorine atom approaches the surface, this value becomes larger
as a consequence of the image potential which is given ap-

proximately by 3.6/Z in eV where Z is the distance to the
surface in Å. (The image potential concept is reasonably
correct because of the large dielectric constant of silicon.)
For this reason, it is reasonable to expect that the fluorine
affinity level will be somewhere below the valence band of
silicon. Similar arguments could be made for other species
such as SiF_3 which also have a large electron affinity. The
distance between the fluorosilyl surface and the silicon is
so small (10Å-30Å) that it is reasonable to expect a negative
ion to form on the surface with an electron from the silicon
whenever this process is energetically favorable; e.g., by
tunneling. Therefore, a small concentration of negative ions
is expected on or in the fluorosilyl layer.

The consequences of this situation can be deduced on the
basis of concepts proposed by Mott and Cabrera.[11] The formation of negative ions on the surface generates an electric
field which causes the bands to bend upward. The band bending
is a consequence of a change in the electrostatic potential
and the slope is the strength of the electric field. The
concentration of negative ions and the strength of the electric
field is determined by the fact that the affinity level and
the Fermi level must be equal on the immediate surface of the
fluorosilyl layer. The affinity level never goes above the
Fermi level because it would then no longer be favorable to
transfer an electron from silicon to a surface fluorine atom.
The number of negative ions on the surface is dependent upon
both the thickness of the insulating (assumed) fluorosilyl
layer and the doping of the silicon. A capacitor analogy is
helpful. The number of negative ions decreases as the thickness of the fluorosilyl layer increases and increases with
dopant concentration for "n" type silicon. If one assumes,
as did Mott and Cabrera, that the negative ions are on the

surface, than both the electric field and the negative ion concentration are proportional to

$$\frac{[E_F - E_a]_{surf}}{\ell} = \frac{V}{\ell} \quad \text{where} \quad (1)$$

$[E_F - E_a]_{surf}$ is the difference between the Fermi level and the affinity level at the fluorosilyl-silicon interface and ℓ is the thickness of the insulating layer. It can be shown[12] that there is a strong correlation between the etch rate and V/ℓ; i.e., when V/ℓ is large, the etch rate is large and when V/ℓ is small, the etch rate is small. (This could be a consequence of the fact that the electric field enhances the movement of ions which in turn increases rate of chemical reaction at the surface or at the silicon interface.)

The addition of negative charge at the surface of the fluorosilyl layer induces a net positive charge on or in the silicon which causes the bands to bend upwards, i.e., the Fermi level moves closer to the top of the valence band. When it has reached the top of the valence band, then the addition of negative charge produces little further band bending. This behavior is a consequence of the large electron density in the valence band. Therefore, for the parameter space of interest, one can schematically picture the XeF_2 reaction as moving the Fermi level at the silicon surface toward the valence band until it reaches the top, at which time it becomes effectively locked in place, i.e., V/ℓ changes little as negative charge on the fluorosilyl surface increases.

Let's now consider how this model qualitatively explains the seven experimental observations (henceforth designated by the symbols #1 through #7) described above. Let us first consider the case where the dopant concentration is small (#3). The bands will bend so that the Fermi level is at the top of the valence band for both n- and p-type dopants, i.e., V is about the same for both samples. Of course, the band bending is greater for n-doping than for p-doping; nevertheless, the etch rates as indicated by "V" are similar.

Heavily doped p-type silicon etches faster than lightly doped p-type material (#2). This result occurs because the Fermi level is near the top of the valence band for light, p-doping, whereas it is contained in the valence band for degenerate doping. Therefore, V (heavy doping) < V (light doping). It should be noted that for degenerately doped p-type silicon, the bulk Fermi level is already in the valence band (see ref. 17) and that the XeF_2 reaction produces little further band bending.

At this time we consider why heavily n-doped silicon etches faster than other types (#1). For heavily doped n-type silicon it takes a large quantity of negative surface charge to produce band bending. Therefore, for an appropriate thickness of the fluorosilyl layer, the Fermi level is near the conduction band while for other types of silicon it is near the valence band. Hence, V (degenerate n-type) > V (other types).

Next we consider #4, i.e., the dependence of etch rate on "ℓ": It is easiest to first focus our attention on n^+-Si(111). For very thin layers the bands are bent so that the Fermi level is near the top of the valence band. The etch rates for all types of silicon are similar under this condition. However, as the thickness, ℓ, of the fluorosilyl layer increases, the quantity of negative charge which is needed to make $E_F=E_a$ at the fluorosilyl surface decreases, i.e. V/ℓ decreases and the etch rate becomes smaller. This happens for all silicon samples. For most samples the increase in ℓ does not change V significantly. However, for degenerately n-doped silicon, the need for less negative charge reduces the band bending and the Fermi level moves toward the conduction band. Both V and ℓ become larger and therefore the etch rate does not decrease as rapidly as it would for the more lightly doped material.

Exposure of silicon to high intensity light (#6) is known to produce a flat band situation which increases $[E_F - E_a]_{surf} = V$ and therefore increases the etch rate. The presence of alkali metals on surfaces (#7) almost universally decreases the work function which is equivalent to increasing V and therefore the etch rate.

Finally, ion bombardment (#5) decreases the thickness of the fluorosilyl layer so that the reaction probability is very large, often approaching unity. The Fermi level is near the valence band for these layers of decreased thickness under most situations and therefore doping has little influence.

Hence, in a physically reasonable manner one can understand all seven of the experimental observations. These arguments can be made plausible on a semiquantitative basis by solving the Poisson-Boltzmann equation to determine the position of E_F as a function of the concentration of negative charge on the surface of the fluorosilyl layer, Q, and doping level. Q can then be determined as a function of ℓ. A complete description of this work will be presented elsewhere.[12]

1. Mogab, C. J. and Levinstein, H. J., J. Vac. Sci. Technol. 17, 721 (1980).

2. Lee, Y. H., Chen, M. M. and Bright A. A., Appl. Phys. Lett. 46, 260 (1985).

3. Ikawa, E. and Kurogi, Y., Nucl. Inst. and Meth. B7/8, 820 (1985).

4. Baldi, L. and Beardo, D., J. Appl. Phys. 57, 2221 (1985).

5. Schwartz, G. C. and Schaible, P. M., J. Electrochem. Soc. 130, 1898 (1983).

6. Jinno, K., Kinoshita, H. and Matsumoto, Y., J. Electrochem. Soc. 125, 827 (1978).

7. Makino, T., Nakamura, H. and Asano, M., J. Electrochem. Soc. 128, 103 (1981)

8. Koike, A., Imai, K., Hosoda, S., Tomozawa, A. and Agasuma, T., Extended Abstracts - Electrochemical Society, Spring Meeting (1982), Abstract #213.

9. Winters, H. F., J. Vac. Sci. Technol. B3, 9 (1985).

10. Mott, N. F., Trans. Faraday Soc. 43, 429 (1940).

11. Cabrera, N. and Mott, N. F., Rep. Prog. Phys. 12, 163 (1949).

12. Winters, H. F. and Haarer, D., to be published.

13. Houle, F. A., J. Chem. Phys. 79, 4237 (1983).

14. McFeely, F. R., Morar, J. F. and Himpsel, F. J., Surf. Sci. 165, 277 (1986).

15. Winters, H. F., Coburn, J. W. and Chuang, T. J., J. Vac. Sci. Technol. B1, 469 (1983).

16. Bagus, P. S., Mat. Res. Symp. Proc., Vol. 38, 179 (1985).

17. Wolf, H. F., "Silicon Semiconductor Data", Pergamon Press, New York. Reprinted 1976. See. p. 45.

MECHANISMS OF THE ION ENHANCED ETCHING OF SILICON

F. R. McFeely and J. A. Yarmoff

IBM T. J. Watson Research Center
Yorktown Heights, NY 10598 U.S.A.

ABSTRACT

Interaction of Argon ion beams with Si surfaces previously reacted with XeF_2 was investigated. Ion induced chemical processes were found to be important.

One of the more interesting phenomena observed in the etching reactions of
semiconductor surfaces is that of ion-bombardment-induced reaction rate enhancement.
While it is not in itself surprising that simultaneously sputtering a surface and removing
material from it by means of a purely chemical reaction should be faster than the effect
of the reaction alone, the two processes have been found not to be simply additive[1].
Instead, the reaction and the bombardment process exhibit a synergistic effect upon one
another, with the total rate of substrate volatilization being significantly larger than
would be predicted by a simple summing of the reaction and sputtering rates. The
interaction which produces this synergism is not obvious, and many models have been
proposed to explain it in the various systems in which it occurs. The experiments reported in this paper have as their goal to elucidate the operative mechanism in the most
intensively studied and best understood model etching system, the etching of Si (111)
surfaces by XeF_2.

A useful classification of the various interaction possibilities has been given by
Winters[2], whose terminology we shall adopt. He describes four generic interaction
mechanisms, which are as follows. First is "chemically enhanced physical sputtering".
This is a process in which the partial chemical attack on the surface would produce
species which are less tightly bound to the surface than the unreacted substrate, and thus
have a higher sputtering cross section. In the case we are considering, it would presumably be silicon subfluorides, SiF, SiF_2, and SiF_3, which have the enhanced sputtering cross section. A second but related mechanism is detrapping. This mechanism
presupposes that significant amounts of inherently volatile reaction products, e.g. SiF_4
molecules, are trapped within the reacting surface matrix, and that the overall reaction
is speeded up by liberating these molecules via ion bombardment. A third mechanism
is "chemical sputtering". In this scenario, incident ions have the effect of stimulating a
specific step or set of steps in the chemical reaction pathway, which leads to the increased production of volatile species. In order to investigate which, if any, of these
mechanisms are operative in the XeF_2 - Si(111) system we have performed high resolution soft x-ray photoemission measurements on surfaces which have been reacted with

XeF_2 without the presence of ion flux, and then to analyze the changes in these surfaces resulting from small increments of bombardment by 500 eV Argon ions. The spectrometer and associated experimental procedures have been described in detail elsewhere.[3] The photons for the experiments were provided by the VUV ring of the National Synchrotron Light Source at Brookhaven, L.I.

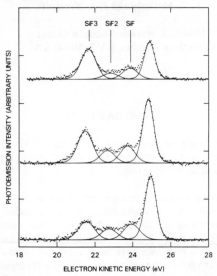

Fig. 1. Photoemission spectra of the Si $2p_{3/2}$ level after reaction with XeF_2 and before (a) and after (b and c) ion bombardment.

Figure 1a shows the Si (111) surface after exposure to approximately $1X10^{-3}$ torr of XeF_2 for 15 minutes. (The $2p_{1/2}$ contributions to these spectra have been numerically subtracted following standard methods[4].) In addition to the peak corresponding to unreacted silicon, this spectrum exhibits features characteristic of SiF, SiF_2, and SiF_3 reaction intermediates, and a small peak due to trapped SiF_4, the principal reaction product. The existence of the SiF_4 demonstrates that detrapping is at least a conceivable enhancement process, however, the trapped amount is in this case a truly negligable amount of the total SiF_4 produced in the reaction. The fluorine exposure which gave rise to this surface was sufficient to volatilize well in excess of 1000Å of the substrate, and the amount of trapped SiF_4 indicated by spectrum 1a corresponds to less than 1 monolayer of Si.

When the surface giving rise to spectrum 1a was subjected to a short bombardment to 500 eV Argon ions, the spectrum of Fig. 1b results. As expected some of the partially fluorinated layer has been removed. However, it is obvious that the chemical distribution of the intermediate species has also been changed. The primary effect of the ion beam has evidently been to depopulate the surface of SiF_3 groups. Additional ion bombardment furthers this process, as can be seen from the spectrum in

Fig. 1c. Furthermore, a careful analysis of the data reveal that not only has the SiF_3 intensity suffered a disproportionate decrease, but in fact the features arising from SiF_2 and SiF have actually increased in going from Fig. 1a to Fig. 1c.

Caution must be employed in interpreting the intensity increases in the SiF and SiF_2 peaks, as two entirely different physical effects could be responsible for this observation. First, the ion beam could be initiating chemical processes which result in actual increases in the concentrations of these species. On the other hand, the absolute numbers of these moieties need not increase at all in order for the photoemission intensities of their associated spectral peaks to increase. This is because in the sputtering process, a great deal of the SiF_3 groups are obviously removed from the surface. This serves to lessen the amount of inelastic scattering of the photoelectrons arising from the SiF and SiF_2 groups, thereby increasing the observed intensity. In order to estimate the relative importance of this second mechanism in the behavior of the SiF and SiF_2 peaks, we have calculated the evolution of the spectrum using a limiting case assumption that only SiF_3 was removed from the surface by the ion beam. The virtue of this model is that it defines the maximum possible increase in intensity the SiF and SiF_2 peaks can exhibit without any increase in the respective densities in the surface layer. The results of this calculation demonstrate that there is an actual increase in the number of SiF_2 groups present, but that this is probably not the case for SiF.

In summary, our data demonstrates the following. First, the sputtering process exhibits considerable chemical specificity, with SiF_3 moieties being the most efficiently removed from the surface. Second, SiF_2 species are synthesized in the bombardment process. To interpret these observations it is important to bear in mind the kinetic bias inherent in these experiments. Since we are examining surfaces after reaction, we are mostly observing those species which could not react. We therefore have a prejudice towards the slow processes. Thus the high concentration of SiF_3 units on the unbombarded surface strongly points to the further reaction of this moiety to form SiF_4 as the rate limiting process.

With the results presented above we are in a position to distinguish among the three proposed models for ion-induced rate enhancement. The mechanism of chemically enhanced physical sputtering clearly fails to describe the data satisfactorily. If the effect of the reaction is merely to render the entire surface more sputterable, then one would expect to see a uniform thinning of the fluoride layer, not the dramatic chemical specificity observed. It might be suggested that the species with dramatically enhanced sputterability is SiF_3, however, this strikes us as rather implausible. While SiF_3 groups are less tightly bound to the surface than unreacted atoms, they are still in all probability linked to the substrate by a reasonable chemical bond. In addition, if the sputterability were only a function of the remaining number of bonds to the surface, one would expect a monotonic increase in the sputterability with increasing fluorination, which is definitely not observed. Detrapping is also clearly ruled out as an enhancement mechanism for reaction conditions anything like those discussed here - there is simply not enough product trapped in the absence of ion bombardment to be of any significance. It is, however, possible that this mechanism plays some role at higher reaction pressures, where the amount of product incorporation can be larger[3]. By process of elimination, this brings us to the consideration of chemical sputtering as the operative mechanism. Certainly the chemical specificity of the bombardment and the observation of SiF_2

synthesis are consistent with this model. Furthermore it is kinetically significant that it is the SiF_3 groups which are most affected, since it is the reaction from SiF_3 to SiF_4 which is thought to be rate limiting in the neutral reaction. It is easy to imagine how the process might work. An Argon ion, incident on the surface, is most likely to encounter an SiF_3 group. A likely result of such an encounter would be the stripping off of an F atom. This would create two favorable entities for the furtherance of the reaction, an SiF_2 group with a dangling bond, which should be extremely reactive, and a very hot F atom, which would be able to overcome any activation barriers to fluorination, specifically that associated with the fluorination of SiF_3 to form SiF_4. This interpretation of our measurements implies that SiF_4 would be evolved as a consequence of the bombardment process. It would be interesting to attempt to observe this process directly in a post-bombardment experiment such as this one.

REFERENCES

1. Y. Y. Tu, T. J. Chuang, and H. F. Winters, Phys. Rev. B23, 823 (1981)

2. Harold F. Winters, J. Vac. Sci. Technol. A3, 700 (1985)

3. F. R. McFeely, J. F. Morar, and F. J. Himpsel, Surf. Sci. 165, 277 (1986)

4. F. R. McFeely, J. F. Morar, N. D. Shinn, G. Landgren, and F.J. Himpsel, Phys. Rev. B30, 764 (1984)

CATALYSIS OF SEMICONDUCTOR SURFACE REACTIONS

A. Franciosi, P. Philip, S. Chang,
A. Wall, A. Raisanen and N. Troullier
Department of Chemical Engineering and Materials Science
University of Minnesota, Minneapolis, MN 55455 USA

P. Soukiassian[*]
Synchrotron Radiation Center
University of Wisconsin-Madison, Stoughton, WI 53589 USA

ABSTRACT

The possibility of modulating semiconductor surface reactions through catalytic processes involving ultrathin metal overlayers has been demonstrated in a number of synchrotron radiation photoemission studies. We summarize here chemisorption results for Si(111) and GaAs(110) cleavage surfaces exposed to oxygen.

Recently a number of studies are providing a background indispensable for the understanding of catalytic effects involving semiconductor surfaces.[1-9] We have started systematic studies of the specific catalytic activity of metal overlayers in promoting semiconductor surface reactions with oxygen or water.[4,6,7] The goal is to clarify the relationship between catalytic activity of the overlayer, electronic parameters of the metal (electronegativity, density of states at the Fermi level, orbital character of the valence states), and character of the overlayer-substrate bonding (interface morphology, bonding ionicity, surface work function). In this paper we give a brief overview of some of our results.

All experiments were performed at room temperature on n-type single crystals cleaved in situ, following the methodology described in refs. 4, 6, 7 and 9. The photoemission measurements were performed at the Synchrotron Radiation Center of the University of Wisconsin-Madison using the radiation emitted by the 250 MeV storage ring Tantalus. A few selected results are summarized in angular integrated photoelectron energy distribution curves (EDC's) in Fig. 1. The results are shown after subtraction of a smooth secondary background and (approximate) normalization to the main emission feature in order to emphasize lineshape changes.

---

* Permanent address: SPAS, Commissariat a l'Energie Atomique, CEN Saday, France.

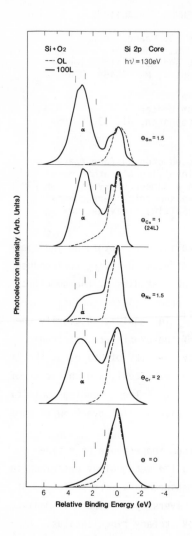

Si+O₂   Si 2p Core
--- OL   hυ =130eV
— 100L

$\Theta_{Sm}=1.5$

$\Theta_{Cs}=1$
(24L)

$\Theta_{Na}=1.5$

$\Theta_{Cr}=2$

$\Theta =0$

6   4   2   0   -2   -4
Relative Binding Energy (eV)

Photoelectron Intensity (Arb. Units)

Fig. 1. Si 2p core emission before (dashed line) and after
(solid line) line) exposure to
100 L (24 L for Cs) of oxygen.
Results for the free Si surface
(bottom-most EDC) are compared
with those for Cr, Na, Cs and Sm
overlayers on Si. The vertical
bars mark the position of characteristic Si 2p oxide features
(see text).

In Fig. 1 we show the effect
of oxygen exposure on the Si 2p
lineshape for a number of Si-metal
systems. The metal coverage on the
clean Si(111) cleavage surface is
given in monolayers (1 ML = 7.8x
10^{14} atoms/cm²). EDC's displaced
downward correspond to Sm (1.5 ML),
Cs (1 ML), Na (1.5 ML), and Cr (1
ML) overlayers on Si(111) before
(dashed line) and after (solid
line) exposure to 100 L of oxygen
(24 L for Cs). The bottom-most EDC
shows the evolution of the free Si
2p surface emission upon oxidation.
The vertical bars in Fig. 1 at 0.9,
1.8, 2.6 and 3.5 eV mark the position of the chemically shifted Si
2p contributions observed by
Hollinger and Himpsel[10] during
oxidation of the Si(111) surface,
and associated by these authors
with Si atoms bonded to 1, 2, 3
and 4 oxygen atoms, respectively.
Vertical bar α marks the position
of the dominant Si 2p oxide feature
observed by Riedel et al.[11] during
oxidation of amorphous silicon.

The overlayers included in
Fig. 1 are those which exhibit the
largest oxidation promotion activity observed to date[1-8] on Si. Fig.
1 shows that in the exposure range
explored the reaction products involve several non-equivalent oxidation states for silicon, with high

oxidation states being predominant, and that the nature of these products are consistent with what could be expected for a disordered, sub-
stoichiometric SiO_2 surface phase.[11] A rough estimate in the case of
Sm[7] suggests that 3-7 layers of silicon have reacted to form the surface oxide. For similar metal coverage and oxygen exposure Ag overlayers gives rise to minimal oxidation promotion, if any, on Si(111)
and no detectable oxidation promotion on GaAs(110).[4] Au overlayers,
however, do give rise to limited oxidation promotion on GaAs(110).[6]

The differences that can be observed in the Si 2p lineshapes of
Fig. 1 before oxygen exposure (dashed line) are related to the details
of the Si-metal interaction. The rather broad Si 2p lineshape observed
for Sm is the result of the formation of an interface silicide-like
phase and the emergence of the corresponding chemically shifted Si 2p
line.[12] Both Cr and Sm overlayers react with silicon for metal coverages above a critical threshold,[12,13] with the formation of silicide-
like phases. A striking result of our oxidation studies[4,6,7] is that
below this critical coverage for intermixing, both Sm and Cr overlayers
exhibit little or no catalytic activity for oxidation promotion.

The unoxidized Si 2p spectra for Cs and Na overlayers in Fig. 1
show, respectively, a broad high binding energy tail and a well defined
high binding energy feature some 2.7-2.8 eV below the main line. Such
features are associated with interface plasmon losses[9] rather than
with contaminants or other experimental artifacts. The valence band
spectra show no detectable oxide features. Their evolution with metal
coverage,[13] together with the coverage dependence of the Si 2p
emission,[9] suggests little interdiffusion or silicide formation for Cs
and Na on Si(111). Both alkali metals exhibit strongly polar chemisorption bonds and loss spectra characteristic of the "metallization" of
the Si surface.[9] We have proposed that an electrostatic overlayer-
oxygen interaction[9] mediated by the substrate might be responsible for
oxidation promotion.

Somewhat similar results have been observed for Cr,[6] Sm,[7] Cs[5]
and Na[14] on GaAs(110) surfaces. In particular, all of these overlayers give rise to substantial oxidation promotion effects, although
Cr and Sm overlayers exhibit a reactive, multi-stage interface formation process, while Cs and Na show little interdiffusion and a strongly

144   A. Franciosi et al.

polar chemisorption bond. The nature of the final reaction products is
more complex than in the Si case, and may involve inhomogeneous distributions of Ga and As oxide phases.[4,6)

     This work was supported by ONR under contract No. 0014-84-C-0545
and by the Minnesota Microelectronic and Information Sciences Center.
We thank the entire staff of the University of Wisconsin Synchrotron
Radiation Center, supported by NSF Grant No. DMR-80-20164.

REFERENCES

1. A. Cros, J. Derrien and F. Salvan, Surf. Sci. 110, 471 (1981); J.
    Derrien and F. Ringeisen, ibid. 124, L35 (1983); A. Cros, J. Phys.
    (Paris) 44, 707 (1983).
2. G. Rossi, L. Calliari, I. Abbati, L. Braicovich, I. Lindau, and
    W.E. Spicer, Surf. Sci. Lett. 116, L202 (1982).
3. I. Abbati, G. Rossi, L. Calliari, L. Braicovich, I. Lindau and
    W.E. Spicer, J. Vac. Sci. Technol. 21, 309 (1982).
4. A. Franciosi, S. Chang, P. Philip, C. Caprile and J. Joyce, J.
    Vac. Sci. Technol. A 3, 933 (1985).
5. L.Y. Su, P.W. Chye, P. Pianetta, I. Lindau and W.E. Spicer, Surf.
    Sci. 86, 894 (1979).
6. S. Chang, A. Rizzi, C. Caprile, P. Philip, A. Wall and A.
    Franciosi, J. Vac. Sci. Technol. A 4, 799 (1986).
7. S. Chang, P. Philip, A. Wall, A. Raïsanen, N. Troullier and A.
    Franciosi, Phys. Rev. B (in press).
8. A.D. Katnani, P. Perfetti, T.-X. Zhao and G. Margaritondo, Appl.
    Phys. Lett. 40, 619 (1982).
9. P. Philip, P. Soukiassian, S. Chang, A. Wall, A. Raisanen, N.
    Troullier and A. Franciosi, Phys. Rev. B (in press).
10. G. Hollinger and F.J. Himpsel, Phys. Rev. B 28, 3651 (1983) and J.
    Vac. Sci. Technol. A 1, 640 (1982).
11. R.A. Riedel, M. Turowski, G. Margaritono and C. Quaresima
    (unpublished).
12. A. Franciosi, P. Perfetti, A.D. Katnani, J.H. Weaver and G.
    Margaritondo, Phys. Rev. B 29, 5611 (1984).
13. A. Franciosi, D.J. Peterman, J.H. Weaver and V.L. Morussi, Phys.
    Rev. B 25, 4981 (1982).
14. S. Chang, P. Philip, A. Wall and A. Franciosi (unpublished).

CLEANING AND NITRIDATION OF III-V SEMICONDUCTOR SURFACES IN MULTIPOLAR PLASMAS

J.P. Landesman, P. Friedel, P. Ruterana, R. Strümpler[1]
Laboratoires d'Electronique et de Physique Appliquée*
3, avenue Descartes, 94451 Limeil-Brévannes Cedex (France)

[1] present address : II. Physikalisches Institut der Reinisch-
Westfälischen Technischen Hochschule Aachen,
D-5100 Aachen, Germany
* LEP : a member of the Philips Research Organization

ABSTRACT : In order to understand the mechanisms involved in the interaction of multipolar plasmas (using H_2 or N_2 as the gas), we have used X-ray photoemission spectroscopy with an analysis chamber coupled "in situ" (in ultra-high vacuum) with a plasma chamber. The interfaces obtained have then been studied by transmission electron microscopy.

The use of multipolar plasma treatments to de-oxidize and passivate the surface of III-V semiconductors before dielectric deposition should improve the interface properties of the MIS structures obtained[1].

In this study, GaAs (100) n- and p-type surfaces are analysed with X-ray Photoemission Spectroscopy (XPS) before and after exposure to H_2 or N_2 multipolar plasmas. "In situ" association of photoemission with a treatment chamber has yielded several interesting results in the domain of III-V semiconductor surfaces[2-6].

The samples are (100) wafers cut from bulk GaAs, with a mechano-chemical polishing. The n-type samples were Se-doped (N_D = $4x10^{17}$ cm^{-3}), the p-type samples were Cd-doped (N_A = 3 to $5x10^{18}$ cm^{-3}).

1. CHARACTERIZATION OF THE INITIAL SURFACE

Prior to the introduction in the system, the samples have been chemically cleaned in the following way :

.degreasing in a boiling solution of Ethanol-Trichlorethylen-Aceton

.de-oxidation in HCl/Ethanol (1:10)

.rinsing in boiling ethanol

The HCl/EtOH etching technique has been shown to minimize the carbon contamination and to keep the surface stoichiometry of GaAs[5].

Our samples, cleaned with the above procedure, are always covered with elemental As, as shown by Fig. 1. The chemical shift is 0.6 ± 0.1 eV, in good agreement with the values published[5]*.

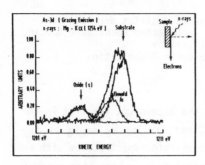

Fig. 1 : Initial surface (n-type sample) ; experimental spectrum (after background subtraction) and fitting spectra

Concerning the surface stoichiometry, we have checked that, when elemental As contribution is correctly taken into account, the measured As/Ga ratio is 1 ± 0.1.

2.   SURFACE CLEANING IN THE H_2 PLASMA

We have found that the best results (removal of the oxides and elemental As and preservation of the surface stoichiometry) are obtained for a filament bias of 75 V, a discharge current around 100 mA, a H_2 pressure around 4×10^{-4} torr and a temperature of 260° C (obtained by indirect heating from an IR lamp). The electron density is 10^{10} to 10^{11} cm^{-3}.

Under these conditions, and on the n-type as well as on the p-type samples, elemental As completely disappears from our spectra. The oxide and carbon contaminations are also completely removed.

---

*The method used to determine the positions in energy of the different contributions in the same core level spectrum consists in a fit (with a simplex procedure) using the measured spectrum on a very clean surface (where only the bulk contribution is observed) recorded with a good statistics.

The stoichiometry of the surface is maintained with the parameters set at the optimum values given above.

The important result concerning the Fermi level position after this treatment is that all the measured peaks (i.e. the bulk contributions on the As-2p3/2, Ga-2p3/2, As-3d and Ga-3d spectra) shift towards higher kinetic energy by 200 ± 50 meV. The same result is observed on n-type and on p-type samples (Fig. 2). Let us note that this phenomenon is probably different from a diffusion of hydrogen (which has been shown to lead to donor neutralization on n-type samples[7]) since a simple heating of the surface at 260° C, which reduces the quantity of elemental As and partly de-oxidizes the surface, produces the same movement of E_F on the n-type surface.

All our results about oxide and elemental As removal have been confirmed by in-situ kinetic ellipsometry measurements.

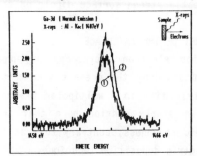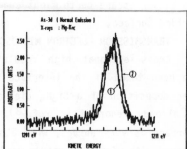

Fig. 2 : Ga and As-3d spectra before (①) and after (②) cleaning
in the H_2 multipolar plasma (n-type sample)

3.   NITRIDATION PLASMA

The N_2 plasma exposure was performed on an n-type sample only, with a N_2 pressure around 5×10^{-3} torr, a filament bias of 75 V and a discharge current of 100 mA. The temperature was 230°C.

As shown on Fig. 3, a shifted component appears both on the As and on the Ga spectra, which confirms the results on the MBE surface (2). At the same time, the Ga-3d and As-3d bulk components remain at their positions for the clean surface but, on the Ga-2p3/2 and As-2p3/2 spectra, the bulk contributions shift towards lower kinetic energy by 500 ± 50 meV ! On the MBE surface, the measurements indicated that the Fermi level is going up by 600 meV during nitridation.

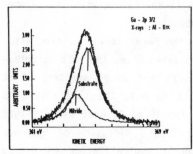 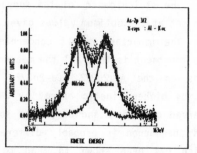

Fig. 3 : Ga and As-2p3/2 spectra (normal emission) after nitridation

From our observations, it can be concluded that N binds to both Ga and As ; moreover, the different behaviour of the core level spectra with small escape depth (10 Å for the 2p spectra) from those with larger escape depth might suggest a strong surface dipole on the nitrided surface.

4. TRANSMISSION ELECTRON MICROSCOPY OF THE INTERFACE

Cross-sectional high resolution electron microscopy pictures have been taken of the interface obtained by the present treatment, after deposition of a Si_3N_4 dielectric film in a multipolar plasma, and already reported in ref.[8]. We simply mention here that the interface is very abrupt, with atomic roughness, contrary to what is obtained with high temperature dielectric deposition techniques.

ACKNOWLEDGEMENTS
Thanks are due to A. Barrois for technical assistance.

REFERENCES
1   P. Friedel et al., Surf. Sci. 168, 635 (1986) and ref. therein.
2   S. Gourrier et al., J. Appl. Phys. 54 (7), 3993 (1983)
3   S. Gourrier et al., Surf. Sci. 152/153, 1147 (1985)
4   F. Proix et al., Sol. State Com. 57 (2), 133 (1986)
5   R.P. Vasquez et al., J. Vac. Sci. Technol. B1 (3), 791 (1983)
6   J. Massies et al., J. Appl. Phys. 58 (2), 806 (1985)
7   J. Chevallier et al., Appl. Phys. Lett. 47 (2), 108 (1985)
8   P. Ruterana et al., Appl. Phys. Lett., Sept. 15, 1986

EXCEPTIONALLY HIGH REACTIVITY OF ORGANIC MOLECULES
ON CLEAVED SILICON

M. N. Piancastelli

Department of Chemistry, University of Rome "La Sapienza",
00100 Rome, ITALY

M. K. Kelly and G. Margaritondo

Department of Physics and Synchrotron Radiation Center,
University of Wisconsin, Madison, WI 53706, USA

and

D. J. Frankel and G. J. Lapeyre

Department of Physics, Montana State University,
Bozeman, MT 59717, USA

ABSTRACT

The exposure of low-temperature cleaved Si to
benzene or pyridine vapors produces weakly-bound
physisorption states rather than the stable
chemisorption states obtained by room-temperature
exposure. Annealing to relatively low
temperatures unexpectedly converts the
physisorption states into stable chemisorption
states. In the case of thiophene adsorption, we
observe instead the annealing-induced
fragmentation of the molecules. These results
emphasize the unusually high reactivity of
organic molecules on cleaved silicon.

Recent experiments led to the discovery of an unexpected, stable
chemisorption state for benzene on cleaved Si(111) kept at room
temperature.[1],[2] A similar state was found for pyridine on the same
surface.[3],[4] No stable chemisorption state was found for aromatic

molecules on other semiconductor surfaces, including Si(111)7x7.[1] The stable chemisorption states on cleaved silicon exhibit rather unusual features. For example, benzene on cleaved Si forms σ-bonds, while on metals it chemisorbs through the donation of π-electrons to metal orbitals.

Other unexpected features were revealed by extending our experiments to low-temperature substrates. The main experimental technique was high resolution electron energy loss spectroscopy (HREELS), accompanied by photoemission spectroscopy with conventional and synchrotron radiation sources. C_6H_6 and C_5H_5N form weakly-bound physisorption states on low-temperature Si(111)2x1. Three different physisorption states were observed for increasing exposure to benzene, molecular benzene with the aromatic ring plane parallel to the substrate for 0.5-25 Langmuir exposures, benzene clusters for 30-75 Langmuirs and benzene multilayers for 100-500 Langmuirs. These states correspond to the HREELS spectra (b), (c) and (d) in the figure, taken in a specular mode with primary energy E_o = 6.5 eV. Curve (a) corresponds to the stable chemisorption state of C_6H_6 on room-temperature Si(111)2x1, while curve (e) is the off-specular counterpart of curve (d), taken at an angle of 7.5°.

Upon mild thermal annealing to 135°K, we unexpectedly observed the conversion of the last two states into stable chemisorption states, rather than simple thermal desorption. Specifically, the HREELS curves (c) and (d) were converted into spectra indistinguishable from curve (a).

For pyridine, only one adsorption state was observed on Si(111)2x1 at 85°K, probably consisting of clusters. Upon annealing to 275°K, this state converts to the stable chemisorption state. Ample evidence exists[1-4] that the C_6H_6 and C_5H_5N chemisorption states - obtained by room-temperature exposure or by low-temperature exposure and annealing - are established by breaking one of the C-H bonds and forming a C-Si bond. There is also evidence that this process is stimulated by cleavage steps.[2]

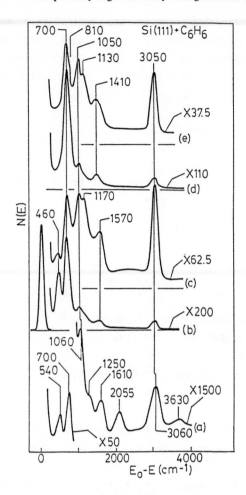

The behavior of benzene and pyridine is markedly different from that of thiophene, C_4H_4S. Upon exposure of $85°K$ Si(111)2x1 to 1-20 Langmuir of C_4H_4S, one obtains two different and simultaneous adsorption states. One is a chemisorption state, obtained by removal of a H atom in α-position and σ-bonding between Si and α-carbon. The second is a π-bonded state with the aromatic ring parallel or almost parallel to the substrate.

A partially ordered multilayer is formed for higher thiophene exposures. Upon annealing to ~270°K, the C_4H_4S decomposes. There is no evidence that the C_4H_4 chain breaks down, and we conclude that the chain is adsorbed as such on the substrate. Thus, the process is a desulfurization process. The experimental evidence for these conclusions is provided by HREELS data and by preliminary synchrotron-radiation photoemission results.[5] The latter reveal three different spectra, corresponding to the two low-temperature states and to the third state obtained by annealing.

In general, the above findings confirm the surprisingly high reactivity of aromatic molecules on cleaved silicon, in sharp contrast with all other semiconductor substrates. This unexpected activity stimulates further investigations of the chemisorption properties of chemical processes involving organic molecules on this surface.

ACKNOWLEDGMENTS

This research was supported in part by the NSF, Grants DMR-84-21212 and DMR-82-05581 and DMR-83-09460.

1. M. N. Piancastelli, F. Cerrina, G. Margaritondo, A. Franciosi and J. H. Weaver, Appl. Phys. Letters 42, 990 (1983).

2. M. N. Piancastelli, G. Margaritondo, J. Anderson, D. J. Frankel and G. J. Lapeyre, Phys. Rev. B30, 1945 (1984).

3. M. N. Piancastelli, G. Margaritondo and J. E. Rowe, Solid State Commun. 45, 219 (1983).

4. M. N. Piancastelli, M. K. Kelly, G. Margaritondo, J. Anderson, D. J. Frankel and G. J. Lapeyre, Phys. Rev. B32, 2351 (1985).

5. M. N. Piancastelli,   R. Zanoni,   M. K. Kelly,   D. Kilday, G. Margaritondo, M. Capozi, C. Quaresima and P. Perfetti, unpublished.

III. INTERFACES

FAILURE OF THE COMMON ANION RULE
FOR LATTICE-MATCHED HETEROJUNCTIONS

J. Tersoff

IBM Thomas J. Watson Research Center

Yorktown Heights, N.Y. 10598

ABSTRACT

Band-edge discontinuities have been calculated for several III-V and II-VI "common anion" heterojunctions. Contrary to the widely accepted common anion rule, large valence-band discontinuities are found in most cases, including all lattice-matched systems studied. Recent experiments for AlAs-GaAs, AlSb-GaSb, and HgTe-CdTe appear to confirm these predictions.

At a semiconductor heterojunction, the band lineup specifies the relative energies of the band edges in the respective semiconductors at the interface. In the past, because of the lack of direct data for many interfaces of interest, substantial reliance has been placed upon the so-called "common anion rule" in attempting to interpret experimental results. This rule[1,2] states that the valence-band discontinuity ΔE_V at the interface will be "very small" for semiconductors with the same anion. However, for the III-V and II-VI "common anion" interfaces treated here, the common anion rule is predicted to fail in most cases, including *all lattice-matched* heterojunctions.

Until recently,[3,4] all general theories of heterojunction band lineups were based on the idea that there is no dipole induced when two different semiconductors are joined to form a heterojunction.[5] The most successful[5] of these theories is the tight-binding approach[6] of Harrison. That theory provides a quantitative basis for the common anion rule, demonstrating its approximate validity (especially for lattice-matched heterojunctions) *if* interface dipoles may be neglected. In view of the large violations predicted here for lattice-matched common-anion heterojunctions, the success or failure of the common anion rule may provide and important test of the role of interface dipoles in determining band lineups.

The basic idea behind the present theory[3,4] is that the semiconductor heterojunction is closely analogous to a metal-metal junction. For that well-understood case, any difference in the workfunctions (i.e. in the electronegativities) of the two metals results in charge transfer, and hence a dipole, which may be viewed as growing until it is large enough to arrest further charge transfer. This equilibrium occurs when the Fermi levels in the two metals line up. In effect, the discontinuity in the electronegativity is screened by the (infinite) dielectric constant of the metal.

It was argued elsewhere[3,7,8] that there is an energy associated with each semiconductor, which plays a role analogous to E_F in a metal. That energy was called E_B in analogy to the branch point in the complex bandstructure of a one-dimensional semiconductor, and represents the energy (usually deep in the gap) where the states are non-bonding on the average. Any discontinuity in E_B at the interface amounts to a discontinuity in the electronegativity, which will give rise to a charge-transfer dipole. The net result is that the discontinuity in E_B is screened by the (large) dielectric constant of the semiconductor.[3] In the limit of large dielectric constant, $\varepsilon_\infty \to \infty$, the self-consistent band lineup should then be given by aligning E_B in the respective semiconductors at the interface, just as one aligns E_F at a metal-metal junction.

If E_B is given relative to E_V, then continuity of E_B across the interface implies

$$\Delta E_V = - \Delta E_B , \tag{1}$$

where Δ refers to differences across the interface. Similarly, at a metal-semiconductor interface, to obtain the (p-type) Schottky barrier ϕ_{bp} one aligns E_B with E_F in the metal, giving[7]

$$\phi_{bp} = E_B . \tag{2}$$

A simple method for calculating E_B directly from the bulk semiconductor bandstructure was described elsewhere.[7,9] This approach, in conjunction with (1-2), appears to give rather accurate predictions of band lineups and Schottky barriers,[8,10,11] at least for elemental and III-V semiconductors.

The results are summarized in Table I, which gives numerical results for a number of III-V and II-VI semiconductors. A more detailed discussion of the results, and of the common anion rule in general, has been presented elsewhere.[12] Valence-band offsets are given by (1); E_B for alloys may be estimated by linear interpolation. The most striking feature is that the predicted valence band offsets are quite large, of order 0.4 eV or more, for the majority of interfaces, including all telluride pairs except ZnTe-CdTe, and all III-V pairs except those where the two cations are Ga and In respectively.

The experimental situation for the interfaces of interest here is changing rapidly at the present time. Until recently, experimental data[13,14,15] for AlAs-GaAs, AlSb-GaSb, and HgTe-CdTe appeared to support the common anion rule. All those studies involved optical measurements, which give only indirect information on the lineup. However recent more direct

TABLE I

Semiconductor "midgap" energy E_B, and measured Fermi-level positions at metal-semiconductor interfaces, relative to valence maxima (eV).

	E_B	$E_F(Au)$	$E_F(Al)$
Si	0.36	0.32	0.40
Ge	0.18	0.07	0.18
AlP	1.27		
GaP	0.81	0.94	1.17
InP	0.76	0.77	
AlAs	1.05	0.96	
GaAs	0.50	0.52	0.62
InAs	0.50	0.47	
AlSb	0.45	0.55	
GaSb	0.07	0.07	
InSb	0.01	0.00	
ZnSe	1.70	1.34	1.94
MnTe	1.6		
ZnTe	0.84		
CdTe	0.85	0.73	0.68
HgTe	0.34		

measurements[13,16,17] have given large valence band discontinuities, of order 0.4 eV, for these three systems, in good agreement with the predictions here. It is worth mentioning that in all three cases, the theory predicted a violation of the common anion rule before it was observed experimentally.

In conclusion, the predicted failure of the common anion rule for lattice-matched heterojunctions appears to be supported by recent experiments. However the contradictory experimental results suggest the need for a re-examination of the relative reliability of the various available techniques for measuring band lineups.

REFERENCES

1. J. O. McCaldin, T. C. McGill and C. A. Mead, Phys. Rev. Lett. *36*, 56 (1976).

2. W. R. Frensley and H. Kroemer, Phys. Rev. B *16*, 2642 (1977); see also Ref. 5. However these authors noted that the interface dipole derived equally from the anion and cation, in agreement with the present conclusions.

3. J. Tersoff, Phys. Rev. B *30*, 4874 (1984).

4. C. Tejedor and F. Flores, J. Phys. C *11*, L19 (1978); F. Flores and C. Tejedor, J. Phys. C *12*, 731 (1979).

5. For a review see H. Kroemer, in *Proc. NATO Advances Study Institute on Molecular Beam Epitaxy and Heterostructures*, Erice, Sicily, 1983. Edited by L. L. Chang and K. Ploog (Martinus Nijhoff, The Netherlands, 1984).

6. W. A. Harrison, J. Vac. Sci. Technol. *14*, 1016 (1977). See however W. A. Harrison and J. Tersoff, J. Vac. Sci. Technol. (in press) for a recent reevaluation of the role of interface dipoles.

7. J. Tersoff, Phys. Rev. Lett. *52*, 465 (1984).

8. J. Tersoff, J. Vac. Sci. Technol. B 3 1157 (1985).

9. J. Tersoff, Surf. Sci. *168*, 275 (1986).

10. G. Margaritondo, Phys. Rev. B 31, 2526 (1985); A. D. Katnani and G. Margaritondo, Phys. Rev. B *28* 1944 (1983).

11. G. Margaritondo, Surf. Sci. *168*, 439 (1986).

12. J. Tersoff, Phys. Rev. Lett. *56*, 2755 (1986).

13. For a review of AlAs-GaAs measurements see G. Duggan, J. Vac. Sci. Technol. B *3*, 1224 (1985).

14. P. Voisin, C. Delalande, G. Bastard, M. Voos, L. L. Chang, and L. Esaki, J. Vac. Sci. Technol. B *1*, 152 (1983); C. Tejedor, J. M. Calleja, F. Meseguer, E. E. Mendez, A. Chang, and L. Esaki, Phys. Rev. B *32*, 5303 (1985).

15. Y. Guldner, G. Bastard, J. P. Vieren, M. Voos, J. P. Faurie and A. Million, Phys. Rev. Lett. *51*, 907 (1983).

16. G. J. Gualtieri, G. P. Schwartz, R. G. Nuzzo, and W. A. Sunder, (unpublished).

17. S. P. Kowalczyk, J. T. Cheung, E. A. Kraut, and R. W. Grant, Phys. Rev. Lett. *56*, 1605 (1986).

A SIMPLE MODEL FOR INTRINSIC BAND OFFSETS AT SEMICONDUCTOR HETEROJUNCTIONS.*

Chris G. Van de Walle

Stanford Electronics Laboratories, Stanford, CA 94305, and Xerox Palo Alto Research Center

and

Richard M. Martin

Xerox Palo Alto Research Center, 3333 Coyote Hill Road, Palo Alto, CA 94304, U.S.A.

*Work supported in part by ONR Contract No. N00014-82-C0244.

ABSTRACT

We present a new and simple approach to derive band discontinuities at semiconductor interfaces. We determine a reference potential for each material by considering a "model solid", which consists of a superposition of neutral atomic charge densities. The bulk band structures of the two materials are then aligned according to the average potentials in the model solids. The results are close to those obtained from full self-consistent interface calculations, and to reported experimental values.

1. INTRODUCTION

The discontinuities in valence and conduction bands at the interface between two semiconductors are the most important parameters which determine the device properties of the heterojunction. Accurate knowledge of these band offsets is essential for the design of novel semiconductor devices, such as high-mobility modulation-doped field-effect transistors, or superlattice photodetectors. Measurement of the discontinuities is complicated by a number of experimental difficulties, as was illustrated by the recent controversy over the GaAs/AlAs system.[1] Several theories have been developed to predict the band offsets.[2-4] Most of them rely on certain assumptions about the mechanism of the lineups, which are hard to test. The only way of determining the applicability of such theories is to compare the results with reliable experimental values, which are scarce.

We have undertaken a first-principles approach to the heterojunction problem. First, we have carried out self-consistent density functional calculations[5,6] on a large number of heterojunctions, using *ab initio* pseudopotentials.[7] In these calculations, the electrons are allowed to fully adjust to the specific environment created by the interface, and self-consistent charge densities and potentials are generated. From these, we can derive the lineup of electrostatic potentials, which will determine the relative position of the bulk band structures. Our systematic approach has allowed us to draw a number of general conclusions about the nature of the band lineups. We found that for a large class of ideal interfaces they are independent of orientation, and they obey the transitivity rule, which implies linearity. These conclusions were also found experimentally.[8,9] All this indicates that the lineups can be derived as a *difference* between quantities which are *intrinsic* to the bulk materials, an assumption that was made before by all of the linear theories,[2-4] but that now has been substantiated by our calculations. In the next section, we will describe how we obtain the appropriate intrinsic reference levels.

2. DEFINITION OF A REFERENCE MODEL SOLID

Because of the long range of the Coulomb interaction, the average potential in an infinite crystal is undefined[10]; it depends on the specific boundary conditions that are chosen to terminate the solid. This corresponds to the choice of a specific *reference surface*; the charge distribution near this surface can set up dipoles which will determine the position of the average potential with respect to the vacuum level. Since we want to use this potential to determine the lineup at an interface, our reference surfaces should be chosen such that at a junction between two semi-infinite solids the charge distribution is a good representation of the situation at a real interface. The model we propose is based on a superposition of neutral atomic charge densities. The potential outside a free atom goes exponentially to (an absolute) zero, which corresponds to the vacuum level. Since the electrostatic potential is linear in the charge density, superposition of atoms will lead to an average potential in the solid which corresponds directly to the average atomic potential.

When we use such neutral objects to construct a semi-infinite solid, the presence of a surface will not induce any shift in the average potential, since no dipole layers can be set up. That means that we can create a well-defined reference surface; the associated average electrostatic potential is an *intrinsic* property of the model solid. Full information about the atomic potential can be obtained by performing an atomic calculation in the local density approximation, using *ab initio* pseudopotentials.[7] Atomic configurations were obtained from tight-binding theory.[11] The average electrostatic potential in the model solid is proportional to the average atomic potential, and to the inverse of the volume of the unit cell. The *total* potential, of course, also includes an exchange and correlation contribution. This quantity is not linear in the charge density, and can therefore not be obtained from the superposition of atomic charge densities. However, it is local in nature, and can thus be regarded as a bulk property which can be added in afterwards.

3. RESULTS AND DISCUSSION

Once we know the reference potentials for the two semiconductors on either side of the interface, we can line up their band structures; these are obtained from standard bulk calculations. Table I contains an overview of results for valence band discontinuities for lattice-matched (110) interfaces. We explicitly rule out the application of the model to certain types of polar interfaces, where mixing at the interface can set up dipoles which significantly alter the band lineups.[12] The agreement between the model solid results and the full self-consistent values is typically better than 0.1 eV, and always within 0.25 eV. One should be careful in comparing with reported experimental values, since many of these are not well established yet. For the cases which we consider reliable (GaAs/AlAs, InAs/GaSb, and AlSb/GaSb) the model solid result is very close to experiment. In Table I, we also list values from Harrison's and Tersoff's theories. We have also applied the model to strained-layer interfaces,[13] where the dependence on strain turned out to be accurately taken into account.

TABLE I. Heterojunction band lineups for lattice-matched (110) interfaces, obtained by self-consistent interface calculations (SCIC), and by the model solid approach. Results from other theories and reported experimental values are listed for comparison.

Heterojunction		ΔE_v (eV)			
	SCIC	model solid	Harrison[4]	Tersoff[3]	Experiment
AlAs/Ge	1.05	1.19	0.78	0.87	0.95 [a]
GaAs/Ge	0.63	0.59	0.66	0.32	0.56 [b]
AlAs/GaAs	0.37	0.60	0.12	0.55	0.55 [c]
GaP/Si	0.61	0.45	0.69	0.45	0.80 [d]
ZnSe/GaAs	1.59	1.48	1.35	1.20	1.10 [e]
ZnSe/Ge	2.17	2.07	2.01	1.52	1.52 [e]
InAs/GaSb	0.38	0.58	0.42	0.43	0.51 [f]
AlSb/GaSb	0.38	0.49	0.18	0.38	0.45 [g]

[a] M. K. Kelly, D. W. Niles, E. Colavita, G. Margaritondo, and M. Henzler (unpublished).
[b] J. R. Waldrop, E. A. Kraut, S. P. Kowalczyk and R. W. Grant, Surface Sci. 132, 513 (1983).
[c] J. Batey and S. L. Wright, J. Appl. Phys. 59, 1200 (1986).
[d] P. Perfetti, F. Patella, F. Sette, C. Quaresima, C. Capasso, A. Savoia, and G. Margaritondo, Phys. Rev. B 30, 4533 (1984).
[e] S. P. Kowalczyk, E. A. Kraut, J. R. Waldrop, and R. W. Grant, J. Vac. Sci. Technol. 21, 482 (1982).
[f] J. Sakaki, L. L. Chang, R. Ludeke, C.-A. Chang, G. A. Sai-Halasz, and L. Esaki, Appl. Phys. Lett. 31, 211 (1977); L. L. Chang and L. Esaki, Surf. Science 98, 70 (1980).
[g] J. Menéndez and A. Pinczuk (private communication).

The fact that the model solid results are so close to those obtained from self-consistent calculations indicates that we have chosen a good *ansatz*. *Additional* dipoles (measured with respect to the model solid reference) due to deviations from the "neutral atom" picture are expected to be small; Frensley and Kroemer[2] constructed a model based on a superposition of charged ions, and found the dipole shift to be small. Therefore, it is better to maintain a model based on neutral atoms, since this uniquely defines a reference level, irrespective of boundary conditions. Another source of dipoles may be due to screening effects, of the type that Tersoff[3] (and more recently, Harrison[4]) consider to be the dominant effect that determines the lineup of "neutrality levels". It is essential to point out, in this context, that the concept of "dipole" at an interface is not uniquely defined – its magnitude depends on the choice of "reference surfaces" that are brought together to create an interface. It is therefore possible for different models to obtain good results, while claiming to deal with "dipoles" of very different magnitude. The charge distribution associated with our model solids can actually incorporate much of the dipole that Tersoff considers to be the dominant effect.

In conclusion, the main advantages of the present model are that it provides a well defined reference surface, that the numerical work is straightforward, and that, even though it should only be considered as an *ansatz*, the results are close to those obtained from self-consistent interface calculations.

REFERENCES

[1]G. Duggan, J. Vac. Sci. Technol. B $\underline{3}$, 1224 (1985).
[2]W. R. Frensley and H. Kroemer, Phys. Rev. B $\underline{16}$, 2642 (1977).
[3]J. Tersoff, Phys. Rev. B $\underline{30}$, 4874 (1984); Phys. Rev. B $\underline{32}$, 6968 (1985).
[4]W. A. Harrison and J. Tersoff, J. Vac. Sci. Technol. B, July-August 1986 (in press).
[5]C. G. Van de Walle and R. M. Martin, in *Computer-Based Microscopic Description of the Structure and Properties of Materials* (Materials Research Society Symposia Proceedings, Volume 63, M.R.S., Pittsburgh, 1986).
[6]C. G. Van de Walle and R. M. Martin, J. Vac. Sci. Technol. B $\underline{3}$, 1256 (1985); Phys. Rev. B (to be published).
[7]G. B. Bachelet, D. R. Hamann and M. Schlüter, Phys. Rev. B $\underline{26}$, 4199 (1982).
[8]W. I. Wang, T. S. Kuan, E. E. Mendez, and L. Esaki, Phys. Rev. B $\underline{31}$, 6890 (1985).
[9]A. D. Katnani and G. Margaritondo, Phys. Rev. B $\underline{28}$, 1944 (1983); A. D. Katnani and R. S. Bauer, Phys. Rev. B $\underline{33}$, 1106 (1986).
[10]L. Kleinman, Phys. Rev. B $\underline{24}$, 7412 (1981).
[11]D. J. Chadi (private communication); the tight-binding method is described in: Phys. Rev. B $\underline{19}$, 2074 (1979).
[12]K. Kunc and R. M. Martin, Phys. Rev. B $\underline{24}$, 3445 (1981).
[13]C. G. Van de Walle and R. M. Martin, J. Vac. Sci. Technol. B, July-August 1986 (in press).

LINEAR DECOMPOSITION OF THE BAND OFFSET

E. O. Kane

Bell Communications Research
Murray Hill, New Jersey 07974

ABSTRACT

We study the heterojunction between Si and Ge using the Pandey-Phillips tight-binding model for the band structure. We find that the difference in the pseudopotential of Si and Ge can be treated linearly to a fair approximation. We then decompose the band offset into a sum of bilayer contributions. The bilayers far from the junction lead to an "intrinsic" contribution to the offset which is much larger than the contribution from junction sensitive bilayers. This result explains the success of theories which treat the offset as an "intrinsic" property of the junction materials.

I. INTRODUCTION

The problem of heterojunction band offsets is an important topic for semiconductor superlattices.[1] A feature of many models of the "offset" is that it can be regarded as resulting from "intrinsic" properties of the component materials independent of the junction boundary conditions.[2]

In this paper we study a simple model of a Si-Ge heterojunction and show that the difference of the Si and Ge potentials can be treated linearly. This enables us to define an "intrinsic" contribution to the offset which justifies the success of the empirical models.

II. HETEROJUNCTION MODEL

We study a heterojunction based on the Pandey-Phillips[3] tight-binding model of silicon and germanium. This model has 4 states per atom, 1s and 3p, which interact up to second neighbor distances. The interactions are given by 7 empirically determined parameters. We study 001 junctions. The heterojunction is assumed to have perfect symmetry in the junction plane and is modelled by a single k-parallel point = $(1,0,0)\,\pi/a$ determined by the method of "special points" due to Cunningham.[4]

The junction is assumed to occur halfway between adjacent planes with silicon on the left and germanium on the right. The matrix elements across the junction are linearly interpolated.

In addition to the Pandey-Phillips couplings, we add a lattice site dependent diagonal potential resulting from the Coulomb potential.

$$<\phi_i(\mathbf{R}_1)\,|V|\,\phi_j(\mathbf{R}_2)> \,=\, V(\mathbf{R}_1 \cdot \hat{n})\,\delta_{ij}\delta_{\mathbf{R}_1,\mathbf{R}_2} \tag{1}$$

In Eq. 1 $\phi_i(\mathbf{R}_1)$ $(\phi_j(\mathbf{R}_2))$ refer to atomic orbitals on lattice sites $\mathbf{R}_1(\mathbf{R}_2)$. $\hat{n}$ is the junction normal. The Coulomb potential of the perfect crystal is implicit in the Pandey-Phillips[3] parameters. The V in Eq. (1) results from the excess (deficit) of charge $(\rho - 4\,|e|)$ which is obtained from solving the self-consistent Schroedinger equation.

We take the field and the potential equal to zero at $\mathbf{R}_n = \mathbf{R}_1$ and integrate to the right to obtain the potential as

$$V(\mathbf{R}_{n1}\cdot\hat{n}) = -\frac{4\pi}{|\mathbf{a}_1 \times \mathbf{a}_2|} \sum_{n2=1}^{n1} \tag{2}$$

$$(\mathbf{R}_{n1}-\mathbf{R}_{n2})\cdot\hat{n}\,\delta\rho(\mathbf{R}_{n2}) \; ; n2 \leqslant n1$$

In Eq. (2), $\mathbf{a}_1$ and $\mathbf{a}_2$ are primitive Bravais lattice vectors for the junction plane selected. The number of atoms per unit area per plane is $1/|\mathbf{a}_1 \times \mathbf{a}_2|$.

III. LINEAR DECOMPOSITION

We have found that the difference of the Pandey-Phillips[3] pseudopotentials for silicon and germanium is small enough to be treated linearly (approximately). We decompose the potential into a sum of bilayer contributions and calculate the self-consistent contribution for each bilayer separately. The total junction potential is the linear superposition of all the bilayer contributions.

We define the bilayer Hamiltonian, H_T^μ, by

$$H_T^\mu = \delta H^\mu + H_0^{Si} + V^\mu \tag{3}$$

where

$$\delta H = H_0 - H_0^{Si} = \sum_{\mu \text{ odd}} \delta H^\mu \tag{4}$$

V^μ is the self-consistent potential induced by the perturbation δH^μ. μ is the distance from the junction to the center of the bilayer in interplanar units.

For $\mu \leqslant -3$, $\delta H^\mu = 0$. $\mu = -1, 1$ are sensitive to the junction while $\mu \geqslant 3$ are translationally equivalent. For the translationally equivalent bilayers the charge is symmetric about the center of the bilayer. The V^μ are then also symmetric and, in particular, $V^\mu(\pm \infty) = 0$ for $\mu \geqslant 3$. The inequivalent bilayers need not be symmetric and may give finite contributions to $V^\mu(\infty)$. ($V^\mu(-\infty) = 0$ by construction).

The valence band offset V_B is defined to be the self-consistent potential at $+\infty$. The non-interacting junction is taken to have zero offset.

IV. CALCULATIONAL RESULTS

The junction sensitive contribution to the offset is .016 eV while the intrinsic contribution is .108 eV, very much larger. The total linear offset, $V^L = .124$ eV is increased to $V^T = .138$ eV when non-linear effects are included.

The bilayer potentials for $\mu = -1, 1, 3$ are shown in the figure. The terms $\mu = 5, 7, 9...$ are found by translating $\mu = 3$ by $2, 4, 6$ ... units to the right. The superposition of all the μ gives V^L which is corrected to V^T by non-linear effects. The linear approximation is seen to be quite good. The dominance of the "intrinsic" contribution justifies the assumptions made by Tersoff and others.[2]

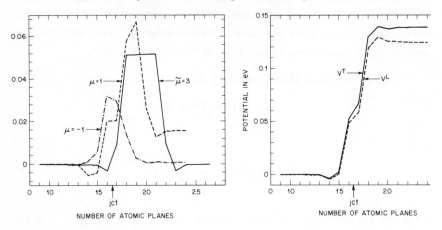

Acknowledgments

I am indebted to J. Tersoff for the stimulation of his work on band offsets and for a number of private conversations.

REFERENCES

1.   R. C. Miller, A. C. Gossard, D. A. Kleinman and O. Munteanu Phys. Rev. *B29*, 3740 (1984).

2.   J. Tersoff Phys. Rev. Lett. *52*, 465 (1984); Phys. Rev. *B30*, 4874 (1984).

3.   K. C. Pandey and J. C. Phillips Phys. Rev. B*13*, 750 (1976).

4.   S. L. Cunningham Phys. Rev. *B10*, 4988, (1974).

ROLE OF THE INTERFACIAL DIPOLES IN THE HETEROJUNCTION BAND LINEUP

C.Quaresima, P.Perfetti, M.Capozi

ISM-CNR, Via E. Fermi, 38 - 00044 FRASCATI

C. Coluzza

Ist. di Fisica Università di Roma I P.le A. Moro, 5 - Roma

D. G. Kilday and G. Margaritondo

Department of Physics University of Wisconsin - Madison

The effects of Metals (Cs , Al) and hydrogen intralayers on the valence band discontinuity (ΔE_v) of several heterojunction have been investigated using photoelectron spectroscopy . Metals, generally, induce an increase of ΔE_v and hydrogen couses a reduction. A simple model, based on electronegativity considerations, is proposed to explain these effects.

The possibility of tailoring the band discontinuities by a controlled procedure, could greatly improve a wide variety of heterojunction devices. Such possibility is suggested by our present data.

We discuss photoemission results on Si/SiO_2 Ge/CdS and Si/CdS heterostructure with or without thin intralayers of H, Cs and Al, i.e., elements with very different electronegativity values. As shown in Table I, a hydrogen intralayer decreases the Si/SiO_2 ΔE_v by 0.5 eV, while a cesium intralayer increases it by 0.3 eV. Intralayer effects in other heterostructures are reported in the same table. The data were obtained using the synchrotron radiation photoemission facility at the storage ring, ADONE, of the Frascati L.N.F. and the storage ring TANTALUS of the University of Wisconsin-Madison. The experimental setup and interface preparation procedure are reported in Ref's 1) and

4).

In Figure 1 we report the photoemission spectrum of an Si/SiO_2 structure compared to that of SiO_2 (dashed line). Comparison of this curve with those obtained with cesium and hydrogen intralayers reveals the ΔE_v modifications. Similar effects are shown by Figure 2 for Si/CdS with a hydrogen intralayer.

We propose a possible explanation of the intralayers effects, based on modifications of the interfacial dipole. In developping such model, we neglected all contributions to the interfacial dipole except that due to the charge tranfer between chemical species at the interface [2]. We simulated the effects of the charge tranfer by using plane charge distributions. To evaluate the charge transfers and the corresponding dipole voltage drop, V_d, caused by the plane distribution of charges, we applied the Sanderson criterion [3].

Since at least one of the two semiconductors is amorphous, we estimated the surface densities of the plane charge distributions from the bulk densities reported in the literature. The difference between the two semiconductor surface densities corresponds to unsaturated bonds on the site of the semiconductor that is more dense. Such bonds do not contribute to V_d, but are available to bind intralayer atoms. The charge transfer of the corresponding bonds causes changes in V_d, and therefore in ΔE_v. Furter details on this model will be given elsewhere [4]. Here, we emphasize two important consequences of our approach:

1) The site and magnitude of the modification of ΔE_v depends on the difference in electronegativity among the intralayer and the heterojunction components.

2) a linear relation exists between the number of intralayer atoms per surface unit and the modifications of ΔE_v in the submonolayer regime.

Point 1 was verified for all the heterojunctions tested in the experiments. Point 2 was verified only for Ge/Al/CdS system, while for the Si/H/SiO$_2$ system only a trend could be deducible from the data, due to the difficulty in estimating the number of the intralayer hydrogen atoms.

REFERENCES
1)   Niles, D.W. and Margaritondo, G., Perfetti, P., Quaresima, C. and Capozi, M., Appl. Phys. Lett. 47, 1092 (1985).
2)   Frenley, W.R. and Kroemer, H., Phys. Rev. B 16, 2642 (1977).
3)   Carver, J.C., Gray, R.C. and Hercules, D.M., J. Ann. Chem. Soc. 96, 6851 (1974).
4)   Perfetti, P., Quaresima, C., Capozi, M., Coluzza, C., Fortunato, G. and Margaritondo, G., to be published.

TABLE I.   Measured ΔE_v values and their modifications due to intralayers (columns 4 and 5). Also listed are the calculated dipole voltage drops, V_d, and their modifications according to our model (columns 6 and 7). Column 3 specifies the intralayer thickness (saturation values are given for hydrogen). Notice that the modifications of ΔE_v and V_d have opposite signs as discussed in Ref's 2) and 4).

Heterojunction	Intr.	Thick. Å	ΔE_v	$\delta \Delta E_v$	V_d	δV_d
Si/SiO$_2$	H	Sat.	4.4	-0.5	2.6	0.3
Si/SiO$_2$	Cs	0.5	5.2	0.3	2.1	-0.2
Si/CdS	H	Sat.	1.1	-0.5	1.6	0.5
Ge/CdS [1]	Al	0.5	1.9	0.1	-0.7	-0.3
Ge/CdS [1]	Al	1.0	2.0	0.2	-1.0	-0.6

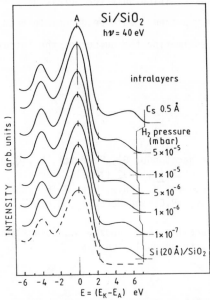

Fig. 1. Photoemission spectra of Si/SiO$_2$ valence band showing the
evolution of ΔE_V for increasing intralayer hydrogen densities.
Specified on the right-hand side of each curve is the hydrogen gas
pressure during 100 eV ion bombardment. The vertical lines emphasize
the changes in ΔE_V referred to the Si/SiO$_2$ dischontinuity without
intralayer, 4.9 eV. Also shown (top spectrum) is the modification
induced by a 0.5 A intralayer of Cs. All the spectra were corrected
for the secondary electron background and aligned to have peak A in
the same position.

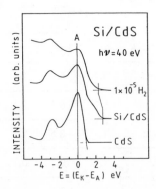

Fig.2 Photoemission spectra of the Si/CdS valence bands. The top
curve shows the reduction of ΔE_V due to hydrogen ion bombardment.

PREDICTION OF BAND OFFSETS AT SEMICONDUCTOR HETEROJUNCTIONS FROM A CHARGE DEPENDENT TIGHT BINDING APPROXIMATION

by

B. Haussy, C. Priester, G. Allan and M. Lannoo
Laboratoire de Physique des Solides,
ISEN, 41, boulevard Vauban, 59046 Lille Cedex
FRANCE

ABSTRACT

A charge dependent tight binding calculation of band offsets is presented for seven semiconductor heterojunctions and for the (110) interface. The relative accuracy of different numerical techniques is critically compared. Good average correlation with experiment is obtained (practically exact values are predicted in 4 cases). Our results are also compared with those of recent LDA calculations and conclusions concerning the ideality of some interfaces are drawn.

Available theories of band offsets at semiconductor heterojunctions are not yet capable of predicting them to better than 0.1 eV. Even the experimental values are not yet completely reliable [1]. At present the most successful predictions have been obtained from simple semiempirical theories but there have also been first principle calculations, based on a local density plus pseudopotential scheme (see [2] and [3] and references therein). We present here a calculation of band offsets based on the empirical tight binding approximation plus a local charge neutrality condition to obtain a simple and efficient determination of the dipole potential barrier at the interface.

As any quantitative theory should be capable of reproducing trends we have examined the following lattice-matched (110) interfaces: Ge-GaAs, Ge-ZnSe, GaAs-AlAs, Ge-AlAs, GaP-Si, InAs-GaSb, ZnSe-GaAs.

Throughout this work we shall systematically use the sp^3s* model of ref. [4], i.e., we take an atomic basis set built from two "s" states (s and s*) and three p states. Only nearest neighbors' interactions are taken into account in a two-center approximation. For each bulk material the tight binding parameters are determined from a fit to the known band structure. Such a fit allows to determine all

elements of the tight binding matrix except for the intraatomic terms
which are all determined only to an unknown additive constant within
each separate material. This is of no importance for bulk properties
but is essential in the interface problem. We choose as central
unknown ΔE_v the discontinuity between the top of the valence band of
the two bulk materials. The knowledge of ΔE_v completely fixes the
relative position of the whole set of bands of one material with
respect to the other one. The perturbation due to interface formation
leads to some charge transfer across the interface plane. This one is
responsible for the dipole potential barrier that is in principle
essential for determining ΔE_v. The charge disturbance will be limited
only to a few planes on each side of the interface. This is due to the
fact that, in all materials considered here, the typical screening
length of the perturbation potential due to neutral defects is of
order 2 Å [5]. Such charge transfer induces a change in the Coulomb
potential and the problem has to be solved in a selfconsistent way. As
is usual in charge dependent tight binding calculations we consider
that this effect only affects the diagonal matrix elements of the
Hamiltonian. This means that on each plane i,n parallel to the
interface (where i = 1 or 2 represents the material and n = 1 to ∞
numbers the plane position away from the interface) the diagonal terms
will be shifted by a quantity U_{in} from their value in the
corresponding bulk material. The unknowns of the problem are thus ΔE_v
and all U_{in} (with the boundary conditions $U_{in} \rightarrow 0$ when n $\rightarrow \infty$). As they
originate from Coulomb potentials they can be expressed linearly in
terms of the excess charge q_{in} of each plane. In turn these charges
can be calculated as functions of ΔE_v and the U_{in} and the whole
calculation has to be iterated to selfconsistency. In fact screening
in such systems with high dielectric constant is very efficient and we
have shown in earlier work [3] that a very good approximation to the
full selfconsistent solution is provided by the assumption of local
charge neutrality. This means that one can determine ΔE_v and the U_{in}
by imposing that all q_{in} be zero. In practice we have obtained a
converged solution by considering three planes on each side of the
interface (i.e., n = 1 to 3) leaving us six neutrality conditions q_{in}
= 0. This allows to determine six unknowns, i.e., 5 U_{in} and ΔE_v (in
such a case one has to arbitrarily choose 3 U_{in} on one side and 2 on
the other one but we have checked that the result is independent of
the initial choice).

A central issue of this problem is that if the desired precision
on ΔE_v is 0.1 eV then one has to calculate the q_{in} to better than
0.01e. We have checked this criterion numerically but it is also
understandable simply from the fact that the susceptibility (i.e the
ratio $\Delta q/\Delta U$) for bonds across the interplane is of order 0.1 Volt^{-1}.
For this reason we had to critically compare different numerical
techniques. The first method we have investigated is the Green's
function recursion technique. However, in that case, the nearest

neighbors' sps* model of ref. [4] causes convergence problems because of the narrow s* band which is almost impossible to describe in this method. We have then been obliged to drop the s* state in calculating the charges, which is permissible since it couples only weakly to the valence band states. We have applied this technique to (110) Ge-GaAs and obtained $\Delta E_v \simeq 0.62$ eV. However the expected numerical accuracy of the method might not be better than 0.1 eV. The second numerical procedure for calculating the charges is again based on Green's functions but their evaluation is made by the decimation method of ref. [6]. This is some sort of iterative procedure where, at each step, one suppresses one plane over two, replacing the direct interactions by renormalized energy dependent interactions. After p such operations each plane is only connected to its 2^P-th neighboring planes. If p is large enough the Green's functions connecting any two such planes become negligibly small and one directly obtains the expression for the intraplane Green's functions. The procedure converges fairly well needing about 9 iterations. To be completely sure of the numerical accuracy of our charges we have also applied a supercell method which is the one used in LDA calculations of band offsets. This is done by building a unit cell consisting of 2n planes (n for each material) and repeating this unit cell periodically. The advantage is that one can use Bloch's theorem in all three directions, the disadvantage is that one remains limited to small unit cells, i.e., to n = 6 or 7 in practice. It is thus not obvious that the error due to boundary effects can be neglected. Here we have found that this method with n = 6 gives, in all cases, exactly the same results as the decimation technique and this, within 0.01 eV for ΔE_v. We have also checked that the results obtained for n = 4 are not fully converged and can introduce an error of order 0.1 eV on ΔE_v.

Our predicted results for band offsets are compared to experiment on Figure 1. If one accepts the existing experimental values very good agreement is found in four cases : AsGa-Ge, AlAs-Ge, ZnSe-GaAs and ZnSe-Ge, for which the r.m.s. deviation is as low as 0.065 eV. This is not so for the other heterojunctions where discrepancies larger than 0.1 eV are found. On the seven lattice-matched systems considered here the r.m.s. deviation with experiment is found to be 0.175 eV. One can also notice on Fig. 1 that our values are consistently lower than those of ref. [2] calculated using local density theory and ab initio non local pseudopotentials which lead to substantial error ($\sim$ 0.6 to 0.7 eV) for ZnSe-GaAs and ZnSe-Ge. As in ref. [2] we find that our predictions, corresponding to ideal interfaces, strictly obey the transitivity rule. It is not so experimentally for GaAs-AlAs where the observed value (0.45 eV) is not equal to the difference in band offsets between AlAs-Ge and GaAs-Ge (0.27 eV). This tends to indicate that experimental interfaces are not ideal and that there can be deviations in stoechiometry for instance which certainly will modify the band line up.

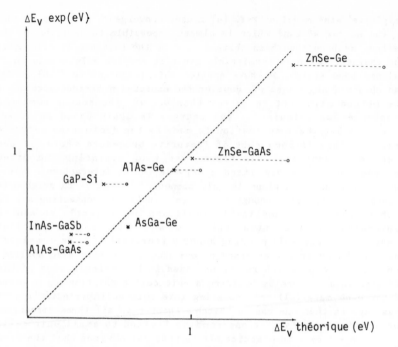

Fig. 1 - Comparison between predicted and observed values of band offsets ΔE_V - x : this work, o : values of ref.[2]

Acknowledgements : This work has been supported by a CNET contract N°84 6B 039. We also would like to thank J. Sanchez-Dehesa for a copy of his decimation program.

REFERENCES

1. Duggan, G., J. Vac. Sci. Technol. B3, 1224 (1985).
2. Van de Walle, C.G., and Martin, R.M., to be published in the M.R.S. Proceedings 1985, Fall Meeting.
3. Priester, C., Allan, G., and Lannoo, M., Phys. Rev. B 33, 7386 (1986)
4. Vogl, P., Hjalmarson, H.P., and Dow, J.D., J. Phys. Chem. Sol., 44, 365 (1983).
5. Lannoo, M., and Allan, G., Sol. State Comm., 33, 293 (1980)
6. Guinea, F., Sanchez-Dehesa, J., and Flores, F., J. Phys. C 16, 6499 (1983) and Guinea, F., Tejedor, C., Flores, F., and Louis, E., Phys. Rev. B 28, 4397 (1983).

TRANSITION METAL IMPURITY LEVELS, BAND OFFSETS IN HETEROJUNCTIONS, AND BEYOND

Jerzy M. Langer

Institute of Physics, Polish Academy of Sciences,
02-668 Warszawa, Al. Lotnikow 32/46, POLAND

Helmut Heinrich

Institut für Experimentalphysik, Johannes Kepler Universität,
A-4040 Linz, AUSTRIA

ABSTRACT

We postulate to connect the transition metal energy levels with the band-edge offset (BEO) in hetero-junctions (HJ). The case of GaAlAs/GaAs, CdHgTe/CdTe and GaInAsP/InP is treated. A possible extension to Schottky-barrier-height estimation is discussed.

So far purely theoretical approaches reached only a modest level of accuracy in predicting band-edge offsets (BEO) in heterojunctions (HJ) or barriers in metal-semiconductor (MS) junctions. Very recently we[1] and independently Zunger[2] have proposed to identify the difference between the binding energies of transition-metal (TM) impurities in two isovalent semiconductors with the BEO in the HJ. This novel approach results from the observation that a sequence of all available TM energy levels differs from the same sequence in another isovalent semiconductor only by a constant energy shift, yielding the relative valence-band (vb) energies summarized in Table I.

1.   VERIFICATION OF THE TM-ALIGNMENT PROCEDURE

For **GaAlAs/GaAs** the valence BEO[1,3] changes linearly with x (slope ≈ 0.45 eV/x; 36% of ΔE for x < 0.4). Exactly the same slope was found for Fe^{2+} and Cu^{2+} energy levels in $Ga_{1-x}Al_xAs$.

Table I. The relative vb edge energies (in eV) as inferred either from our TM hypothesis[1] or Fermi-level alignment of the barrier heights in Au based MS junctions (most of the MS data are taken from Ref. 4).

$A^{III}B^{V}$	"TM"	"MS"	$A^{II}B^{VI}$	"TM"	"MS"
InAs	0.05	−0.05	CdTe	0.80	0.47
GaAs	0	0	CdSe	0.09	0.12
InP	−0.16	−0.28	ZnSe	0	0
GaP	−0.33	−0.43	CdS	−0.37	−0.39
AlAs	−0.45	−0.48 (−0.60[5])	ZnS	−0.56	−0.34

At present no consensus exist on the BEO in HgCdTe/CdTe. The earliest[6] data suggest that the BEO between HgTe and CdTe is about 50% of the CdTe energy gap. In contrast, a more recent spectroscopic study of the multiquantum wells (MQW) indicates an almost perfect vb alignment[7]. Very recent external photoemission[8] experiments increase the MQW result by 0.3 eV, yielding the HgTe/CdTe valence BEO of $E_{vb}(HgTe) - E_{vb}(CdTe) = 0.35 \pm 0.06$ eV. Since all the impurity states in HgTe are resonant, one must be very cautious with regarding the TM energy level positions. There is, however, quite strong evidence[9] that the Fe^{2+} localized level lies about 0.2 eV below the top of the vb of HgTe. Since the same level is[10] about 0.15 eV above the vb of CdTe, a valence BEO of about 0.3 - 0.4 eV is expected, in good agreement with the most recent photoemission findings.

Similarly to the HgTe/CdTe case, a consensus on BEO in GaInAsP/InP has not yet been reached. For the end member of this HJ's family, i.e. the $Ga_{0.47}In_{0.53}As/InP$ HJ the estimated values of the valence BEO $E_{vb}(GaInAs) - E_{vb}(InP)$ vary from about 0.2 eV[11] to 0.4 eV[12,13] (40 to 60% of the energy gap difference). The straightforward approach of our method is not possible since the only data on TM's in GaInAsP are those on shallow Mn^{2+} acceptors, for which a compositional dependence of the ionization energy is given[14] by $E_{Mn} - E_{vb} = (0.23 - 0.185)$ eV in $Ga_xIn_{1-x}As_yP_y$ lattice matched to InP (y ≈ 2.2 x). Since only for small values of y Mn^{2+} can be considered to be a true deep acceptor, the value of 0.185 eV for the valence BEO is the lowest bound for this quantity, in agreement with all available BEO measurements.

2. IS THE TM-HJ RELATIONSHIP EXTENDABLE TO THE MS JUNCTIONS?

After the initial excitement raised by the Spicer et al.[15] proposal that the interface defects are the key factor causing such a behaviour, an older suggestion of Heine has recently been renewed that metal-induced gap states (MIGS) cause the Fermi-level pinning. In particular, Tersoff has suggested that the level which may be treated as a reference in Fermi-level alignment in MS junctions[16] and HJ's[17] is the neutrality level at which the MIGS change their character from conduction- to valence-band like (this level has been identified subsequently by Harrison and Tersoff[18] with the average sp^3 hybrids used in tight-binding-band-structure theory). Since this demarcation level is a property of a given semiconductor, the MS barrier heights and BEO's in HJ's should be directly related, namely, $\Delta E_{vb}(AB) = \phi_p(A) - \phi_p(B)$, where ϕ_p is the Schottky barrier height measured on an MS junction made on a p-type semiconductor. The immediate consequence of the Tersoff model and our proposal is the threefold connection[19] between the TM's energy levels, BEO's in HJ's and the barriers in MS junctions. It should be stressed that such a comparison like in Table I is more meaningful than comparing experimental HJ BEO's, since for the latter the lattice-mismatch induced stress may strongly affect the data. A comparison between the MS and TM data is favourable, especially for the III-V compounds. This procedure is hardly justifiable, however, for the II-VI compounds, since it assumes perfect Fermi-level pinning, which is not the case for most of them[4]. The critical test does not seem to lie in the III-V compound family, but rather in the group of weakly ionic II-VI semiconductors, such as CdSe or CdTe, for which Fermi-level pinning similar to III-V compounds has been found. For this pair, the TM valence BEO prediction is 0.71 eV, while the MS data yield only 0.35 eV. Since these data are reasonably accurate, a 100% disagreement makes the threefold connection hardly acceptable at present.

In conclusion we want to point out that all the currently available trustworthy experimental data on HJ BEO's agree quite well with our hypothesis. Recent model calculations by Harrison and Tersoff[20] also give some justification for our procedure. A consequence of our hypothesis is that pressure measurements of the TM energy levels should yield, in conjunction with the energy gap pressure dependence, the individual pressure coefficients for both the valence and the conduction bands. One of us (H. Heinrich) wants to appreciate support by the Austrian "Fonds zur Förderung der wissenschaftlichen Forschung".

3. REFERENCES

1) Langer, J.M. and Heinrich, H., Phys. Rev. Lett. 55, 1414 (1985) and Physica 143B, 444 (1985).

2) Zunger, A., Solid State Physics (to be published).

3) Heinrich, H. and Langer, J.M., in *Festkörperprobleme*, Vol. 26 (1986) and *Proc. 4th Int. Winterschool on Heterojunctions* (Solid State Sciences, Vol. 67, Springer, Berlin, 1986).

4) Sze, S.M., *Physics of Semiconductor Devices* (Wiley, N.Y. 1982).

5) Best, J.S., Appl. Phys. Lett. 34, 522 (1979).

6) Kuech, T.F. and McCaldin, J.O., J. Appl. Phys. 53, 3121 (1982).

7) Guldner, A. et al., Phys. Rev. Lett. 51, 907 (1983).

8) Kowalczyk, S. et al., Phys. Rev. Lett. 56, 1605 (1986).

9) Dobrowolski, W. et al., this Proceedings.

10) Lischka, K., et al., J. of Crystal Growth 72, 355 (1985).

11) Brunemeier, P.E. et al., Appl. Phys. Lett. 46, 755 (1985).

12) Forrest, S.R. et al., Appl. Phys. Lett. 45, 1199 (1984).

13) Skolnick, M.S. et al., Semicond. Sci. Technol. 1, 29 (1986).

14) Smith, A.W. et al., J. Phys. D16, 679 (1973).

15) Spicer, W.E. et al., Phys. Rev. Lett. 44, 420 (1980).

16) Tersoff, J., Phys. Rev. Lett. 52, 465 (1984).

17) Tersoff, J., Phys. Rev. B30, 4974 (1984).

18) Harrison, W. and Tersoff, J., J. Vac. Sci. Technol., (Aug. 1986).

19) Tersoff, J., Phys. Rev. Lett. 56, 675 (1986).

20) Harrison, W., in *Proc. 4th Int. Winterschool on Heterojunctions* (Solid State Sciences, Vol. 67, Springer, Berlin 1986).

VALENCE BAND OFFSETS IN HETEROJUNCTIONS: TRANSITIVITY, COMMUTATIVITY, AND A ROLE OF INTERFACE STRAINS[*]

P. Bogusławski

Institute of Physics, Polish Academy of Sciences
Al. Lotników 32/46, 02-668 Warsaw
POLAND

ABSTRACT

Ab initio pseudopotential calculations of valence
band offsets are reported. For lattice-matched heterojunctions good agreement with experimental data
is obtained. Important role of interface strains
in lattice-mismatched systems is demonstrated.

One of the principal characteristics of heterojunctions
(HJs) and superlattices is the valence band offset (VBO)
at the interface between two semiconductors A and B, defined
by $\Delta(A/B) = E_V(A) - E_V(B)$, where E_V is the position of the valence band top. For lattice-matched HJs, VBOs have been found
experimentally to be independent of interface orientation[1],
commutative[2] (independent of the growth sequence), and tran-
sitive[2,3]. Although these features have also been reported
to fail in a number of cases (see, e.g., Ref. 3), most of
the theoretical and experimental approaches to HJ physics
consider the VBO as a bulk band structure property of constituent semiconductors rather than an interface property,
consistently with the above findings. While the effect of
interface dipoles[4,5,6] has been discussed quite extensively recently, the effect of built-in strains at interface
in the lattice-mismatched HJs has commonly been neglected.

In order to verify these assumptions, VBOs were evaluated by comparing positions of the valence band tops of two
semiconductors at their respective theoretical lattice con-

stants[7,8)], neglecting thus all interfacial effects. The
calculations were based on ab initio atomic pseudopotentials
and the exchange-correlation potential of Ref. 9. The kinetic energy cutoff of 14 Ry for the plane-wave basis set was
used.

The zero of energy for band structure calculations is
a troublesome concept, because in an infinite crystal there
is no boundary conditions for the solution of the Poisson´s
equation at infinity. We shall fix the zero of energy for
bulk calculations by making the following anzatz. For reciprocal lattice vector G→0 the electron density may formally
be expanded as

$$n(G) = n(0) + \beta G^2 + O(G^4) + \ldots ,\qquad (1)$$

and the resulting Hartree potential (in atomic units) is

$$V_H(G) = \frac{8\pi\, n(0)}{G^2} + 8\pi \beta + O(G^2) + \ldots .\qquad (2)$$

The first term of V_H is cancelled out by the Coulomb part
of the ionic pseudopotential. The second term is undefined
in an infinite crystal[10)]. Here, we assume that $\beta = 0$ (or any
arbitrary constant). The total average electrostatic potential is therefore equal to the G=0 component of the short-
range part of the local ionic potential α_1 (for definition
and discussion see Ref. 11), what fixes the zero of energy.

For the lattice-matched systems one finds

$\Delta(Si/GaP)=1.11$ (0.8) eV, $\Delta(Ge/GaAs)=0.63$ (0.4) eV,
$\Delta(Si/AlP)=1.16$ eV, $\Delta(Ge/AlAs)=0.85$ (0.85) eV,
                          $\Delta(GaAs/AlAs)=0.22$ (0.45) eV.

The values for Ge, GaAs, and AlAs are corrected for the experimental spin-orbit interaction energy. In parentheses,
the (tentatively averaged) experimental data are given,
their uncertainty is typically about 0.2 eV. The overall
agreement with experiment confirms the essentially bulk character of the VBO, and justifies a posteriori our $\beta = 0$ anzatz.

For the lattice-mismatched systems, the built-in strains
at the interface modify drastically the energy band structures, and thus the VBOs. We evidence this effect for the
Si/Ge HJ. By comparing the positions of the valence band
tops of Si and Ge at equilibrium we find the "nominal" Δ(Si/Ge)
=0.48 eV. Actually, however, an overlayer grown on a thick
substrate adjusts its lattice constant parallel to the interface to the equilibrium lattice constant of the substrate.
The resulting uniaxial stress of the overlayer is computed
by minimizing the elastic energy, using the macroscopic elastic constants. The strain-induced changes of the band structure are evaluated according to Ref. 12. The calculated
hydrostatic deformation potentials are used (a(Si)=-10.3 eV,
a(Ge)=-10.5 eV), together with the experimental uniaxial deformation potentials, b(Si)=-2.1 eV, and b(Ge)=-2.5 eV.

Considering an overlayer of Ge on a (100) Si surface we
find that the hydrostatic component of the stress shifts upwards the Γ_8^+ Ge valence band top by 0.54 eV. The uniaxial
component splits the band, which farther shifts its top by
0.18 eV. The net shift of 0.72 eV changes the VBO from its
"nominal" value of 0.48 eV to -0.24 eV, in agreement with
experimental -0.2 eV[3]. However, if we consider the inverse
situation, i.e. a Si overlayer on Ge, we find the downwards
shift of the Γ_{25}' Si valence band top by 0.53 eV due to the
hydrostatic component of the stress, and the upwards shift
of 0.3 eV caused by the band splitting. The resulting offset
of 0.25 eV differs from the previously obtained by as much
as 0.49 eV.

In both considered cases, the hydrostatic stress compo-
nent results in a $\sim$|0.5| eV shift of the overlayer valence
band (with no change of symmetry), giving Δ(Si/Ge)$\cong$ Δ(Ge/Si)$\cong$
0 eV. The calculated pronounced non-commutativity is mainly
due to the uniaxial component which splits the valence bands
(with obvious consequences for transport and optical properties). Comparable results were obtained by Van de Walle

and Martin[13]. Similarly, the strain-induced shifts and splittings of the multi-valley conduction bands in Si/Ge HJ are expected to occur[14]. Finally we observe that in the lattice-mismatched HJs the offsets of the valence and of the conduction bands do not add up to the difference of the band gaps of two semiconductors, in contrast to what theory predicts for the lattice-matched HJs. This effect is due to the strain-induced change of the overlayer band gap.

It is a pleasure to thank S. Baroni, R. Car, I. Gorczyca, and J.M. Langer for helpful discussions.

REFERENCES

[*]Work partially supported by International School for Advanced Studies, Trieste.
1. Wang W.I., Kuan T.S.,Mendez E.E., Esaki L., Phys. Rev. B 31, 6890 (1985).
2. Katnani A.D., Bauer R.S., Phys. Rev. B 33,1106 (1986).
3. Katnani A.D., Margaritondo G., Phys. Rev. B28, 1944 (1983).
4. Frensley W.R., Kroemer H., Phys. Rev. B16, 2642 (1977).
5. Pickett W.E., Louie S.G., Cohen M.L., Phys. Rev. B17, 815 (1978).
6. Tersoff J., Phys. Rev. B30, 4874 (1984).
7. Bogusławski P., Acta Phys. Polonica, in print. For details, see Bogusławski P., Gorczyca I., to be published.
8. This approach is thus similar in its spirit to that of Harrison W.A., J. Vac. Sc. Technol. 14, 1016 (1977).
9. Bachelet G.B., Hamann D.R., Schlüter M., Phys. Rev. B26, 4199 (1982).
10. Kleinman L., Phys. Rev. B24, 7412 (1981), and the references therein.
11. Ihm J., Zunger A., Cohen M.L., J. Phys. C12, 4409 (1979).
12. Bir G.E., Pikus G.E., Symmetry and Deformation Effects in Semiconductors, Nauka, Moscow 1972.
13. Van de Walle C.G., Martin R.M., J. Vac. Sci. Technol. B3, 1256 (1985).
14. People R., Phys. Rev. B32, 1405 (1985).

INITIAL STAGES OF GaAs - Ge (100) and (110) INTERFACE FORMATION

A. Muñoz, J. Sánchez-Dehesa and F. Flores
Departamento de Física del Estado Sólido
Universidad Autónoma,Cantoblanco,28049 Madrid
SPAIN

ABSTRACT

We have analysed the initial stages of a GaAs-Ge (100) and (110) interface formation by using a consistent tight-binding method that introduces charge neutrality conditions at the interface.the results of the present calculation show a good agreement with experimental measurements and provide a strong support to the models based on the concept of charge neutrality level.

1. INTRODUCTION

The concept of charge neutrality level as introduced by Flores and Tejedor[1] in order to align the band offsets of semiconductor – semiconductor interface has received a strong support from recent theoretical[2,3] and experimental[4] works. Although the calculated offsets as compared with the experimental data show a remarkable agreement , several questions remain to be answered within the model. First of all, why do the band edge discontinuitiesand the Fermi level position seem to be completely formed with the deposition of one or two layers[4]? . Moreover, why does the Fermi level appear located at an energy wich seems to coincide with the Fermi energy at a metal – semiconductor junction[4]?.

Here we discuss the semiconductor – semiconductor interface formation by considering the deposition of 1, 2, 3, and 4 layers of Ge on GaAs (110) and (100). This system is analysed by using a tight – binding model where appropiate conditions are introduced to take into account the neutrality of charge associated with the interface dipole creation.

2. THE MODEL

We have assumed the different layers(from 1 to 4) of Ge to be deposited on the position of the ideal continued GaAs (110) or (100) structure. For the (110) interface, an abrupt planar interface is considered, but for the polar interface (100) we have assumed to have two transition planes following Harrison et $al.$[5] The electronic structures of both semiconductors are described by the sp^3s^* tight-binding model of Vogl et $al.$ [6]. Our method of calculation is similar to the procedure described in Ref.7, and uses the Green function method combined with a decimation technique to project the bulk structure into a few layers around the interface.

The consistency in our problem is achieved by means of diagonal perturbations, V_i, in the different layers of Ge and, in the last two subtrate semiconductor layers at most; we have checked that the effect of the interface perturbation penetrates only one layer in the GaAs (110), but two layers in the case of the homopolar (100)face[8]. We assume that for a zero induced electrostatic potential at the interface the mean sp^3 levels of both crystals are aligned, this has been checked in other cases to give good results and similar to the ones obtained by using the affinity rule. In a second step we allow changes in the charges and potentials of the layers around the interface. The consistent equations between charges and potentials are given elsewere[3,7].

3.   RESULTS AND CONCLUSIONS

Tables I and II gives the calculated V_i, in the different layers of Ge as a function of coverage for the GaAs-Ge (110) and (100), respectively (for the (100) case V_0, the perturbation in GaAs, is also shown). The local density of states in the different layers of Ge and GaAs are plotted in Figures 1, and 2, respectively, for the particular case of four layers of Ge deposited.

Table I.- Different values of the diagonal perturbation, and the Fermi energy as a function of the number of Ge - layers deposited on GaAs (110)

	V_1	V_2	V_3	V_4	E_f	ΔE_v
1 monolayer	0.73	–	–	–	0.80	0.16
2 layers	0.62	0.87	–	–	0.87	0.23
3 layers	0.50	0.61	0.89	–	0.89	0.30
4 layers	0.50	0.51	0.65	0.90	0.90	0.32

Table II.- Different values of the diagonal perturbation, and the Fermi energy as a function of the number of Ge - layers deposited on GaAs (100)

	V_0	V_1	V_2	V_3	V_4	E_f	ΔE_v
1 monolayer	0.56	0.17	-	-	-	0.68	-
2 layers	0.33	0.92	0.27	-	-	0.83	0.30
3 layers	0.33	0.66	0.87	0.27	-	1.03	0.40
4 layers	0.37	0.70	0.58	0.91	0.31	0.99	0.45

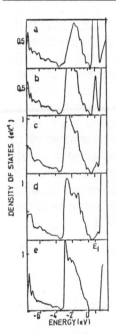

Figure 1.- Local density of states in different layers for the case of four layers of Ge deposited on GaAs (110);(a) Ge-fourth layer (surface),(b) Ge-third layer (c) Ge-second layer, (d) Ge-first layer (interface), and (e) bulk of GaAs.

Figure 2 (right).- Local density of states in different layers for the case of four layers of Ge deposited on GaAs(100); (A) Ge-surface layer, (B) Ge-third layer, (C) Ge-second layer, (D) Ge-interface, (E) Ga-interface, and (F) As-layer.

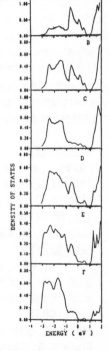

In the same tables, we give the calculated values ΔE_v for different layers. For the (110)-case, the Fermi level is pinned by the surface states of Ge; however, for the (100)-case we have assumed the Fermi level to be located at the mid energy between the two surface bands (one empty, another filled) of the Ge-(100) face. In this case, this gap is around 0.3 eV, and the Fermi

186   A. Muñoz et al.

level is expected to change by this value as a function of the different interface preparations[4].

The discontinuity between the valence band top of both semiconductors is practically formed for one or two layers (for a (100)-face, we cannot define ΔE_v for the case of a monolayer due to the Harrison's model used in the calculation). We have calculated $\Delta E_v = 0.5$ eV for the (110)-interface(see also Ref.9) and $\Delta E_v = 0.6$ eV for the (100)-interface, if two semiinfinite crystals are assumed. The slight differences between these values and the ones given in Tables I and II are a size effect, tending to zero with Ge deposition.

In conclusion, the important point coming out of our calculation is the following : i) the semiconductor – semiconductor interface is practically formed with one or two layers of deposited Ge. ii) The Fermi level at the interface is determined by the center of gravity of the density of states associated with the Ge– dangling bonds, this level practically coincides with the charge neutrality level of Ge. Since the energy of semiconductors electronic levels are refered to each other by aligning their charge neutrality levels, it is clear that the Fermi level, as determined by the surface states, coincides with the one wich would be obtained in a metal semiconductor interface. All these results explains the recent data obtained by Chiaradia et a l^A and present a beautiful example in which these alignements of the charge neutrality level apply at the same time for a heterojunction and a metal – semiconductor interface.

REFERENCES
1.   Flores F., and Tejedor C., J. Phys. C12, 731 (1979)
2.   Tersoff, J., Phys. Rev B30, 4874 (1984)
3.   Platero, G., Sánchez-Dehesa, J., Tejedor C., and Flores F., Surf. Sci.
     168, 553 (1986)
4.   Chiaradia, P., Katnani,A. D., Sang, H. W. Jr., and Bauer, R.S., Phys. Rev.
     Lett. 52, 1246 (1984)
5.   Harrison, W. A., Kraut, E. A. , Waldrop, J. R., and Graut, R. W., Phys. Rev.
     B18, 4402 (1978)
6.   Vogl, P. Hjalmarson, H. P., and Dow, J. D., J.Phys.Chem. Solids 44,
     365 (1983)
7.   Guinea, F., Sánchez-Dehesa J., and Flores, F., J. Phys. C16, 6499 (1983)
8.   Platero, G., Sánchez-Dehesa, J., Tejedor, C., Flores, F., and Muñoz, M.
     Surf. Sci. 172, 47 (1986)
9.   Priester, C., Allan, G., and Lanoo, M., Phys. Rev. B33, 7386 (1986).

A FINAL NAIL IN THE COFFIN OF THE ELECTRON AFFINITY RULE[*]

D. W. Niles and G. Margaritondo
Department of Physics and Synchrotron Radiation Center,
University of Wisconsin, Madison, WI 53706, USA

ABSTRACT

We performed a synchrotron-radiation photoemission test of the electron affinity rule on the prototypical interface ZnSe-Ge. Using directly measured quantities, we derived from the rule a valence band discontinuity ΔE_v = 2.21 eV. The measured discontinuity was 1.44 eV. The importance of this breakdown is enhanced by the recent discovery of a correlation between Schottky barrier heights and heterojunction band discontinuities.

The band lineup at the interface between two different semiconductors is one of the fundamental open questions in solid-state physics. For many years, the theoretical scene in this area was dominated by the electron affinity rule.[1] The rule simply states that the conduction band discontinuity, ΔE_c, equals the difference between the electron affinities of the two semiconductors. The conceptual limitations of this approach prompted the development of several alternate models. The electron affinity rule, however, is still very widely used in applied heterojunction research. Recent development demanded a stringent test of the rule, whose negative results we present here.

Among the new band lineup theories, Tersoff's midgap-energy model appears particularly successful in estimating band discontinuities with good accuracy.[2] Tersoff's model is conceptually linked to the fundamental Heine's work[3] on the tailing of metal wave functions at a metal-semiconductor interface, and to the charge-neutrality-point model of Flores and his collaborators.[4] The conceptual validity of this approach could be verified by testing its prediction of a linear relation between each heterojunction band discontinuity and the difference of the Schottky barrier heights between the two semiconductors and a given metal. Such prediction was positively tested in two cases, for semiconductor-gold interfaces[5] and for interfaces involving $Ga_{1-x}Al_xAs$.[6]

The problem with the above test is that the prediction of a link between heterojunction band discontinuities and Schottky barrier heights is not unique to the Heine-Tersoff-Flores theories. For example, it can also be obtained by combining the Schottky model for metal-semiconductor interfaces and the electron affinity rule -- or the most sophisticated version of these theories, the effective work function model.[7] This uncertainty suggested a stringent and final test of the electron affinity rule. Most previous tests were practically meaningless due to the experimental uncertainty in the electron affinity data. Photoemission spectroscopy with synchrotron radiation eliminates this difficulty. The essential elements of the test are illustrated in the figure. D, the distance between the two edges of the photoemission spectrum of a clean semiconductor surface, is clearly related to the electron affinity. The electron affinity rule can be rewritten as $\Delta E_v = D_1 - D_2$, where D_1 and D_2 are the D-values for the two semiconductors. ΔE_v is also be measured with photoemission techniques, by depositing the second semiconductor on top of the first. Thus, a single series of photoemission experiments provides all necessary elements for the test.

Negative results were obtained[8] by applying this test to GaAs-Ge. The results, however, left some uncertainty. The valence band discontinuity was indirectly estimated from core-level measurements, and this approach could be affected by systematic errors. We performed the test on the lattice-matched interface ZnSe-Ge. For this system, the heterojunction band discontinuity was directly visible as a double edge in the valence-band photoemission spectra. Thus, the test is direct and straightforward. The details of the experiments will be described in a forthcoming article. Much attention was dedicated to charging problems and to possible spurious low-energy signals in the photoemission spectra. We performed a careful analysis of all causes of experimental uncertainty, keeping the quantitative estimates on the pessimistic side. Our results for Ge deposited on cleaved ZnSe were: $D_1 - D_2 = 2.21 \pm 0.46$ eV and $\Delta E_v = 1.44\ ^{+0.08}_{-0.15}$ eV. Thus, there is a clear discrepancy between the electron affinity rule and the experiments. The experimental procedure was such that the test not only dealt with the original version of the electron affinity rule, but also with the effective work function theory.[7]

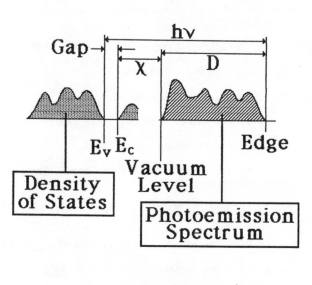

At present, therefore, the midgap-energy and related models would appear as the only band-lineup theories in good standing. This statement must be taken with prudence. The above test does not necessarily affect other band lineup models based on the concept of an absolute energy scale for the two band structures of the two semiconductors -- but not relying on the electron affinity to establish such scale. Furthermore, its successes notwithstanding, Tersoff's is not a perfect theory. It is affected, as all other current band lineup theories, by an underlying accuracy limit[9] due to their neglect of the details of the microscopic interface charge distribution which affect the band lineup. The importance of these factors is demonstrated by recent successful attempts to modify the band lineup by controlled interface contamination.[10] Their realistic treatment remains a formidable challenge to solid-state theory.

*   Research supported by the NSF, Grant DMR-84-21212, and performed at the University of Wisconsin Synchrotron Radiation Center, an NSF-supported facility.

1.   R. L. Anderson, Sol. State Electronics 5, 341 (1962).

2.   J. Tersoff, Phys. Rev. B30, 4875·(1984).

3.   V. Heine, Phys. Rev. A138, 1689 (1965).

4.   C. Tejedor and F. Flores, J. Phys. C11, L19 (1978).

5.   J. Tersoff and G. Margaritondo, unpublished

6.   M. Eizenberg, M. Heiblum, M. I. Nathan and N. Braslau, private communication and to be published.

7.   J. L. Freeouf and J. M. Woodall, Appl. Phys. Lett. 39, 727 (1981); Surface Science (in press).

8.   P. Zurcher and R. S. Bauer, J. Vac. Sci. Technol. A1, 695 (1983).

9.   A. D. Katnani and G. Margaritondo, Phys. Rev. B28, 1944 (1983).

10. D. W. Niles, G. Margaritondo, P. Perfetti, C. Quaresima and M. Capozi, J. Vac. Sci. Technol. (in press).

CORE LEVEL STUDY OF BONDING AT THE
Gaas - ON - Si INTERFACE

R. D. Bringans, M. A. Olmstead*, R. I. G. Uhrberg** and R. Z. Bachrach

Xerox Palo Alto Research Center

3333 Coyote Hill Road, Palo Alto, CA 94304

USA

ABSTRACT

The bonding of As, Ga and GaAs with Si(111), Si(100) and Ge(111) surfaces has been studied with photoemission core level spectroscopy. For thin GaAs layers, it was found that the substrate bonds predominantly to As.

1. INTRODUCTION

Many important scientific questions about the general features of compound growth on elemental semiconductors remain unanswered. Epitaxial growth of GaAs on Si substrates is a system for which the bulk materials are well understood, allowing the interface properties to be studied in detail. Because the interface chemistry is qualitatively different from that in the bulk, simple estimates of the atomic arrangement at the interface are not reliable. It is not clear, for example, how As and/or Ga atoms are bonded to the Si substrate. In order to address this issue, we have made a comparative study of the energetics of the interface bonding for As, Ga and GaAs with Si(111), Si(100) and Ge(111) surfaces. Use of these different substrates has allowed us to vary the geometry at the interface as well as the magnitude of the lattice mismatch. Several surface sensitive experiments were carried out and for thicknesses of a few monolayers we have found that the substrate bonds predominantly to As.

2. EXPERIMENTAL DETAILS

Thin layers of GaAs were grown in situ by Molecular Beam Epitaxy (MBE) on the clean Si and Ge substrates which were held at temperatures in the range 550 to 580°C. A comparison was made with results for isolated As and Ga layers on the same surface. The As-terminated surfaces were produced as described elsewhere[1-3] and the ~1 monolayer Ga layer was deposited at room temperature and then annealed at 570°C.

Surface sensitive core level spectroscopy measurements were carried out at the Stanford Synchrotron Radiation Laboratory.

3.　RESULTS AND DISCUSSION

3.1　As - Terminated　Surfaces

Arsenic 3d core level spectra are shown in Fig 1 for the substrates discussed. These spectra are fitted with a single spin-orbit pair with the statistical intensity ratio of 3:2 and a spin-orbit splitting of 0.69eV in each case.　The absence of any shifted components indicates that there is a single site for As on each surface.　In all cases, exposure of the surfaces to additional As did not increase the As 3d level intensity significantly, suggesting that the As coverage was complete.

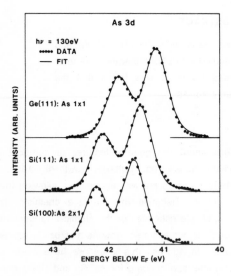

Fig 1.　Arsenic 3d core levels for As-terminated　Ge(111),　Si(111)　and Si(100) surfaces. A fit to a single spin-orbit pair is shown in each case.

The Si 2p (Ge3d) core levels shown in Fig 2 have 2 components in each case, corresponding to bulk Si (Ge) and to Si (Ge) bonded to As. The combination of these results is solid evidence in favor of the model[1] of the As-terminated (111) surface in which As replaces Si (Ge) atoms in the outer half of the surface double layer.　For the Si(100):As 2x1 surface, the results are consistent with the model proposed earlier[2] on the basis of the comparison between experiment and calculated surface state dispersions.　The structure in this case consists of the addition of symmetric As-As dimers to the Si(100) surface.

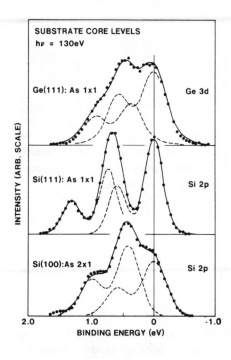

Fig 2. Si 2p and Ge 3d core levels for As-terminated Ge(111), Si(111) and Si(100) surfaces. The dashed lines show the results of deconvolution into bulk Si and chemically shifted components and the full line is the sum of the components. Intensities are shown on an arbitrary scale and the plots are aligned in energy at the bulk level.

3.2   Thin   GaAs   Layers

By examining the chemical shifts for the substrate core levels it is possible to determine the species in the GaAs overlayer which is bonded to the substrate. The Si 2p core level for Si(111):GaAs, for example, shows a single component shifted to higher binding energy by 0.62eV with respect to the bulk Si 2p energy. By comparison, the Si 2p level for Si(111):As has a shift in the same direction of 0.75eV whereas Si(111):Ga has a shift in the opposite direction of 0.30eV. This can be seen clearly in Fig 3 which shows only the 3/2 component of the Si 2p spectra. Deconvolution into 3/2 and 1/2 components was carried out using a spin-orbit splitting of 0.600eV and the statistical intensity ratio of 2:1 . These results for around one monolayer provide strong evidence that the surface Si atoms are predominantly bonded to As rather than Ga. We also found that these results did not change significantly if Ga was deposited on a room temperature substrate and then As was added at a raised substrate temperature. This indicates that the As atoms were able to move underneath the Ga atoms which had been bonded to the substrate atoms. Similar results showing a preferential bonding to As at the interface were found for Si(100):GaAs and Ge(111):GaAs.

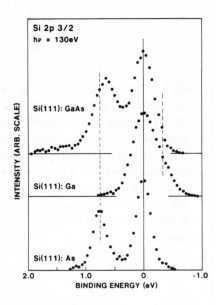

Fig 3. Si 2p 3/2 core level components for a Si(111) substrate with approximately 1 monolayer of GaAs, Ga and As.

At greater thicknesses than those considered here, a bonding arrangement in which the substrate atoms all bond to As atoms is expected to give rise to a prohibitively large dipole at the surface[4]. Consequently we expect a transition to another geometry, accompanied by defect formation as the lattice mismatch can no longer be accommodated by straining the overlayer. Similar results were found for Si(100):GaAs and Ge(111):GaAs.

Acknowledgements: Many useful discussions with Drs J. E. Northrup, D. J. Chadi and R. M. Martin are gratefully acknowledged. Part of this work was performed at SSRL which is supported by the Department of Energy, Office of Basic Energy Sciences.

4. REFERENCES

*   Present Address: Dept. of Physics, University of California, Berkeley, CA 94720

**   Present address: Dept. of Physics and Measurement Technology, Linköping Institute of Technology, S-581 83 Linköping, Sweden.

1.   Bringans, R. D., Uhrberg, R. I. G., Bachrach, R. Z. and Northrup, J. E., Phys. Rev. Lett., 55, 533 (1985).

2.   Uhrberg, R. I. G., Bringans, R. D., Bachrach, R. Z. and Northrup, J. E., Phys. Rev. Lett. 56, 520 (1986).

3.   Olmstead, M. A., Bringans, R. D., Uhrberg, R. I. G., Bachrach, R. Z., Phys. Rev. B, in press.

4.   Harrison, W. A., Kraut, E. A., Waldrop, J. R. and Grant, R. W., Phys. Rev. B, 18, 4402 (1978).

BAND OFFSETS IN PbTe-Pb$_{0.90}$Eu$_{0.10}$Se$_{0.096}$Te$_{0.904}$ HETEROJUNCTIONS

J. Heremans[*] and D. L. Partin
Physics Department, General Motors Research Laboratories
Warren, Michigan 48090-9055, U.S.A.

M. Shayegan[*]
Department of Electrical Engineering, Princeton University
Princeton, New Jersey 80544, U.S.A.

H. D. Drew[*]
Department of Physics and Astronomy, University of Maryland
College Park, Maryland 20742, U.S.A.

[*]Visiting scientists at the Francis Bitter National Magnet Laboratory, Massachusetts Institute of Technology, Cambridge, Massachusetts 02139, U.S.A.

ABSTRACT

We present results on the Shubnikov-de Haas (SdH) effect on 70A wide PbTe wells in the ⟨111⟩ plane grown lattice-matched between layers of the larger-gap semiconductor Pb$_{0.90}$Eu$_{0.10}$Se$_{0.096}$Te$_{0.904}$. Carrier densities from Hall measurement, Shubnikov-de Haas periods and cyclotron resonance data are fitted to a model we develop for the size and field quantized energy levels in PbTe quantum wells.

Magnetoresistance and Hall effect have been measured between 0.5 and 300K in fields up to 22 Tesla on single 70Å wide PbTe (E$_{gap}$ = 0.19 eV) quantum wells which were grown by molecular beam epitaxy. The wells are modulation-doped p-type. The barrier material was lattice-matched Pb$_{0.90}$Eu$_{0.10}$Se$_{0.096}$Te$_{0.904}$ which has a 0.5 eV gap at low temperatures. The wells were in the ⟨111⟩ plane; the magnetic field along the ⟨111⟩ axis. Sample preparation and experimental technique have been described previously.[1] The data are shown in Fig. 1 and reveal the presence of two carrier types, (a) and (b).[1] We have fitted the nonoscillatory part using four adjustable parameters for each temperature (P$_a$, μ_a, P$_b$, μ_b) to reproduce the field dependence of both ρ_{xx} and ρ_{xy} (Fig. 1); the values of the parameters are listed in Table I. In the oscillatory part of our data at 0.5K a Fourier analysis reveals mainly two frequencies Δ(1/H): 8.4 Tesla and 19 Tesla. From the extremal areas A of the pockets to which these

frequencies correspond one can estimate two-dimensional carrier densities of 4.1 and 9.2 10^{15} m^{-2}, respectively.   The fit of the 8.4T
period with the Hall areal density P_a (Table I) is quite convincing.
The Hall density of the b-pockets is much larger, and we have not yet
identified the origin of these low mobility holes.

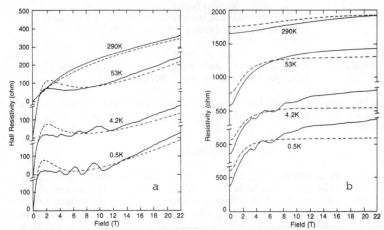

Fig. 1. (a) Areal Hall resistivity and (b) magnetoresistivity of a
70A PbTe quantum well at different temperatures. The full lines are
the experiment, the dashed line a fit to the non-oscillating part.

Table I: Areal carrier densities (P_a and P_b) and mobilities (μ_a and μ_b) deduced from resistivity and Hall data.

T	P_a	μ_a	P_b	μ_b
(K)	$(10^{15}m^{-2})$	$(m^2V^{-1}s^{-1})$	$(10^{17}m^{-2})$	$m^2V^{-1}s^{-1})$
0.5	4.1	1.2	3.9	0.014
4.2	5.7	0.91	4.7	0.012
53	4.6	0.77	3.8	0.012
290	8.5	0.078	6.4	0.005

We calculate in-plane dispersion relations and fan-charts of the
size-quantized subbands of the pocket located along the ⟨111⟩ axis
perpendicular to the well, an elongated ellipsoid with m_L = 0.26 m_o
mass along ⟨111⟩ and m_T = 0.023 m_o mass perpendicularly to it. The
dispersion relation for the well and boundary materials is assumed to
be given by a symmetric two-band model:

$$\left(E - V_s\right)^2 = \left(\frac{E_{gs}}{2}\right)^2 + \beta_{sT}^2 \, k_{sT}^2 + \beta_{sL}^2 \, k_{sL}^2 \tag{1}$$

where the indexes L and T stand for longitudinal and transverse and s
identifies either the boundary material (s = B) or the well material
(s = W). If we put the zero of energy at the midgap point of the well
material, V_W = 0 and the valence band offset is $(E_{gB} - E_{gW})/2 + V_B$.
The secular equation for the bound states in a well of thickness "a"
is:[2]

$$\left[r\, \beta_{WL} k_{WL} \tan \frac{k_{WL}a}{2} - i\, \beta_{BL} k_{BL} \right] \left[r\, \beta_{WL} k_{WL} \cot \frac{k_{WL}a}{2} + i \beta_{BL} k_{BL} \right]$$

$$+ k_T^2 \left(r\, \beta_{WT} - \beta_{BT} \right)^2 = 0 \quad \text{where } r = \frac{E + (E_{gB}/2) - V_B}{E + (E_{gW}/2)} \quad (2)$$

We further assume that the ratio m^*/E_g is the same in well and
barrier, so that $\beta_W = \beta_B = E_g\, \hbar^2/(2\,m^*)$. This allows us to plot for

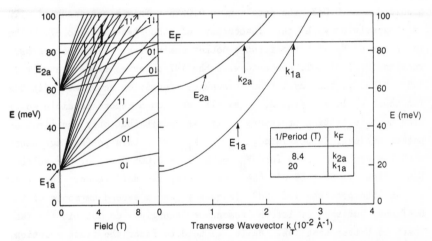

Fig. 2. Hole energy versus field, or "fan-chart", in the left frame,
and versus transverse momentum in the right frame. This was calculated assuming V_B = 0.1 eV. E_{1a} and E_{2a} are the first size-quantized
sublevels of the "a"-pocket along the $\langle 111 \rangle$ axis. The experimental
data on this figure are: magnetooptical transition energies (Ref. 3)
given as dark vertical lines on the fan-chart, and the frequencies of
the Shubnikov-de Haas oscillations (insert) which correspond to Fermi
wavevectors k_{1a} and k_{2a}. These data, as well as a "a-pocket" cyclotron mass of 0.0385 eV and Hall areal density of 4.1 10^{15} m^{-2}, are all
consistent with the indicated location of the Fermi level E_F.

each value of V_B the transverse dispersion relations for this pocket, which is our (a)-pocket (Fig. 2).   In a quantizing magnetic field we replace the transverse momentum terms $\beta_T k_T$ in Eq. (1) and (2) by $E_g \hbar w_c$ (n + 0.5 (1 $\pm$ γ)) where w_c is the cyclotron frequency, n the Landau level index and γ = 0.55 the spin-splitting factor, and thus plot the fan-charts (Fig. 2), again treating V_B as an adjustable para-meter.

To fit our data to this model, we proceed as illustrated in Fig. 2. The SdH period of 8.4T corresponds to a Fermi wavevector k_F of 1.6 10^{-2} $\overset{\circ}{A}{}^{-1}$ and a calculated carrier density of 4.1 10^{11} cm^{-2} which is well confirmed by the Hall data. We admit that $k_F = k_{2a}$ (see Fig. 2). If the value of V_B is now taken between 0.1 and 0.15 eV, leading to a valence band offset of 90 $\pm$ 10% of the total difference in energy gaps, we have the best fit to the cyclotron effective mass: the model predicts $m_c^* = 0.043$, the data[3] $m_c^* = 0.0385$. This discrepancy of ~12% is compatible with the inaccuracy of the two-band model.   The intercept of E_F with the first a-pocket sublevel E_{1a} leads to a cross-section (k_{1a}) of 20T, compared with the 19T observed experimentally in SdH. Finally, the cyclotron resonance transition energies[3] fit the fan chart well, see Fig. 2.   We also mention a second possible solu-tion to this fitting procedure: if we admit that only the first a-pocket sublevel is populated, $k_F = k_{1a}$, we obtain perfect agreement with the cyclotron mass for V_B = 0.01 $\pm$ 0.005 eV, i.e. the valence band offset is 55 $\pm$ 5% of the total energy gap difference. This time we can interpret the 19T SdH frequency as a second harmonic of the 8.4T one, but the cyclotron resonance energies do not fit the fan-chart to better than 15%. While we tend to favor the first solution, clearly more experiments are needed to distinguish between the two options.

REFERENCES

1.   Heremans, J., Partin, D. L., Dresselhaus, P. D., Shayegan, M., and Drew, H. D., Appl. Phys. Lett. 48, 928 (1986).

2.   This has been derived (R. E. Doezema and H. D. Drew, to be pub-lished) following G. Bastard: Phys. Rev. B25, 7584 (1982).

3.   Kim, L. S., Drew, H. D., Doezema, R. E., Heremans, J., and Partin, D. L., to be published.

DETERMINATION OF BAND DISCONTINUITY IN GaAs/AlGaAs
HETEROJUNCTIONS FROM C-V MEASUREMENTS: ROLE OF DEEP DONORS

S.Subramanian, R.Rajalakshmi, B.M.Arora and A.S.Vengurlekar

Tata Institute of Fundamental Research, Homi Bhabha Road,
Bombay 400005, India

1. INTRODUCTION

In the last few years there has been considerable interest in the
determination of band discontinuities at heterojunction (HJ) interfaces from C-V measurements by Kroemer's method [1]. In this paper,
we present a novel variation of this method by which the band discontinuity is determined from the intercepts of conventional C^{-2} vs
V plots. In the second part of the paper, we discuss the role of deep
donors on the determination of discontinuities from C-V measurements.

2. GRAPHICAL METHOD

Our graphical analysis is based on the fact that some segments of
the C^{-2} vs V plot of a HJ structure can be described by analytical
expressions derived under depletion approximation. Fig.1 shows typical simulated plots of a GaAs/AlGaAs structure for $\Delta E_c=300$ meV. One
can distinguish three distinct regions on these plots, consisting of
an S shaped region sandwiched between two linear regions. The linear
regions correspond to the reverse bias ranges in which the edge of the
Schottky barrier depletion region lies well inside either of the two
semiconductors, but far away from the HJ interface. The S shaped
region corresponds to the intermediate bias range in which the
Schottky barrier and HJ depletion regions overlap. In fact, Kroemer's
method essentially uses the C-V information in this range, whereas our
method uses the complimentary information in the two linear regions.

The C-V relationships in the two linear regions are easily
derived under depletion approximation to be:

$$c^{-2} = 2(\phi_b - \delta_1 - V)/ q\varepsilon N_1 \qquad ..(1)$$

in the small reverse bias region, and

$$c^{-2} = d^2/\varepsilon^2 + 2/\varepsilon q N_2 (\phi_b - \delta_2 - V + \Delta E_c/q - q N_1 d^2/2\varepsilon) \qquad ..(2)$$

in the large reverse bias region, where N_1 and N_2 are the doping
concentrations (assumed to be shallow donors), δ_1 and δ_2 are the

200   S. Subramanian et al.

Fermi energies (with respect to E_c) in the two materials, d the thickness of GaAs, and ε the dielectric permitivity of the two materials (assumed to be nearly same for simplicity).

If $N_1=N_2$, then the two linear regions are parallel (Curve A), but they are horizontally separated by an amount equal to the electrostatic potential difference across the heterojunction at zero bias given by

$$\Delta\Phi = \Delta E_c/q + (\delta_1 - \delta_2) \quad ..(3)$$

and from this shift ΔE_c is easily determined.

The procedure to determine

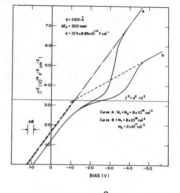

Fig.1: Simulated C^{-2} vs V plots for a GaAs/AlGaAs heterostructure at 300K, for ΔE_c=300 meV.

ΔE_c when $N_1 \neq N_2$ is as follows. We first note that the first inflection point at the minimum of slope $(d^2C^{-2}/dV^2=0)$ on the C^{-2} vs V plot (Curve B) corresponds to the accumulation maximum which occurs fairly close to the HJ interface as pointed out by Kroemer et al [1]. Thus, the value of C^{-2} at this point is d^2/ε^2. A horizontal line is first drawn through this point. Then, the large reverse bias linear region is extrapolated to intersect this horizontal line at

$$V_I = \phi_b - \delta_2 + \Delta E_c/q - qN_1 d^2/2\varepsilon \quad ..(4)$$

A third straight line is now drawn through this point and parallel to the first linear region, and this line is given by

$$\bar{C}^{-2} = \frac{d^2}{\varepsilon^2}+2/q\varepsilon N_1(\phi_b-\delta_2-V+\Delta E_c/q-qN_1d^2/2\varepsilon)=2/q\varepsilon N_1(\phi_b-\delta_2-V+\Delta E_c/q) \quad ..(5)$$

It is easily seen that the difference in intercepts made by the two parallel lines given by eqs. (1) and (5) is once again $\Delta\Phi$. The values of ΔE_c obtained from these intercepts ($\sim$320 meV) matches well with the value assumed for the simulation (300 meV), and also with the value obtained by Kroemer's method (310 meV).

3.   ROLE OF DEEP DONORS

When one of the HJ components contains a large concentration of deep donors (as e.g. DX centers in AlGaAs), both the graphical method described in section 2, and the Kroemer's method fail to give correct values of ΔE_c, since the deep donors (traps) introduce another source of distortion of free carrier profiles [2,3] in addition to the HJ effects. This distortion is caused by the traps in the transition region (see inset of Fig.2) in the AlGaAs, which respond to the frequency of the oscillator (ω_{osc}) and the variation of bias (ω_b) during C-V measurement depending upon their emission rate at any given temperature [3]. Only when $\omega_{osc} \geq \omega_b > e_n$, which is realized in practice by Copeland method [4], the distortion due to traps is absent. When $\omega_{osc} > e_n > \omega_b$, which is the case most frequently encountered in a normal C-V (1 MHz) measurement, the distortion caused by the traps is also

quite significant, as illustrated
by our simulated profiles shown in
Fig.2. For a given total donor
concentration N_s+N_t, as the deep
donor fraction $(K=N_t/(N_s+N_t))$
increases, the apparent profiles
(solid curves) shift away from the
HJ interface on the AlGaAs side by
a distance proportional to the
length of the λ region given by

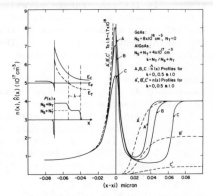

$$\lambda=\sqrt{2\varepsilon(E_F-E_t)/\{(N_s+N_t)(1-Kf_T)\}} \quad ..(6)$$

where f_T is the Fermi occupation
function for the deep donors. In
Kroemer's analysis, this additional
distortion results in a large
apparent negative interface
charge (σ_i) and the ΔE_c to be
overestimated. In Fig.2, while
profile $A(N_t=0)$ gives very small
value for σ_i ($\sim 1\times10^9 cm^{-2}$), and
the correct value of ΔE_c(300 meV)

Fig.2: Simulated apparent (Solid
curves) and true (dashed curves)
carrier profiles for a GaAs/
AlGaAs structure at 300K for $\Delta E_c=$
300 meV for different values of K

as used in the simulation, profiles B and C give much larger values
for ΔE_c (460 and 800 meV) as well as for σ_i (-4.1×10^{11} and -9.72×10^{11}
cm^{-2}, respectively).

4.   EXPERIMENTAL RESULTS AND DISCUSSION

The model given in section 3 was applied to the experimental results obtained from an MBE grown GaAs/AlGaAs structure in the temperature range 225-375K. C-V, DLTS and TSCAP measurements on the bulk
AlGaAs obtained from the same sample revealed that the shallow donor
concentration is AlGaAs is very small ($\sim 1\times10^{15}cm^{-3}$), as compared to
the total donor concentration ($4\times10^{17}cm^{-3}$). This corresponds to the
case K=1 in our model. The Al mole fraction in AlGaAs was found to
be $x\sim 0.36$ from the band-gap determined from photoluminescence measurements. Typical plots of $\hat{n}(x)$ profiles of the heterojunction sample at
a few typical temperature are shown in Fig.3. As the temperature
decreases, λ increases and once again the $\hat{n}(x)$ profiles on the AlGaAs
shift to the right . This in turn results in larger and larger values
of ΔE_c, and increasingly more negative apparent σ_i with decreasing
temperature in Kroemer's analysis as seen from Table 1. The value of

Table 1: Results from Kroemer's analysis

T(K)	375	348	303	277	257	226
$\sigma_i(10^{11}cm^{-2})$	0.57	0.62	2.85	5.45	8.0	12.9
ΔE_c (meV)	292	295	383	573	638	940

ΔE_c as obtained from Kroemer's analysis reaches a constant (~ 295 meV)
above 350 K, and σ_i also reaches a minimum ($\sim 6.0\times10^{10}cm^{-2}$). This is

202   S. Subramanian et al.

presumably the temperature at which
λ region nearly goes to zero, and
the influence of deep donors on the
ΔE_c measurement is negligible. The
value of ΔE_c at this temperature
($\sim$295 meV) is about 0.65 ΔE_g. The
value obtained at 300K ($\sim$380 meV)
is about 0.85 ΔE_g, and we feel this
is slightly overestimated due to a
finite λ region at this temperature.
Of course at lower temperatures, the
λ region is so large that the value
of ΔE_c obtained from Kroemer's analysis becomes even larger than ΔE_g!

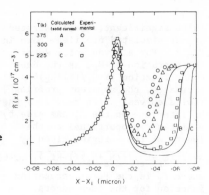

With ΔE_c set equal to 295 meV,
even though the calculated $\hat{n}(x)$
profiles (Solid curves) give
excellent fit with the experimental profiles on the GaAs side,
they depart considerably on the
AlGaAs side at all temperatures,
as seen in Fig.3, when we use the

Fig.3: Experimental and simulated
apparent carrier profiles ($\hat{n}(x)$)
for an MBE grown GaAs/Al$_x$Ga$_{1-x}$As
(x=0.36) structure at different
temperatures.

conventional density of states (N_c) for AlGaAs and E_t=0.1 eV. The
shift of the calculated profiles much farther away from the experimental profiles implies that the parameters assumed for the simulation give much larger values of λ than the experimental values. Good
fit of the calculated profiles on the AlGaAs side (that would give
matching values of λ with the experiment) could be obtained only by
assuming a much larger value of N_c for AlGaAs ($5 \times 10^{19} \text{cm}^{-3}$). Such a
large value of N_c might be caused by alloy disorder, or the deep
donor level being associated with L or X conduction band minima or
the temperature dependence of E_t as suggested by Schubert and Ploog
[5] to explain the Hall electron concentration in AlGaAs.

ACKNOWLEDGMENT

The authors are grateful to Prof.J.R.Arthur, Oregon State
University, Corvallis, USA for providing the samples.

REFERENCES

1. H.Kroemer, W.Y.Chin, J.S.Harris, Jr., and D.D.Edwall,
   Appl.Phys.Lett. 36, 295(1980).

2. G.L.Miller, IEEE Trans.Electron Devices, Vol.ED-19, 1103(1972).

3. L.C.Kimerling, J.Appl.Phys. 45, 1839 (1974).

4. J.A.Copeland, IEEE Trans.Electron Devices, Vol.ED-16, 445 (1969).

5. E.F.Schubert and K.Ploog, Phys.Rev. B30, 7021 (1984).

RADIATIVE AND NONRADIATIVE RECOMBINATION PROCESSES IN InGaAsP/InP and InGaAsP/GaAs   DOUBLE HETEROSTRUCTURES

Zh.I.Alferov and D.Z.Garbuzov

A.F.Ioffe Physico-Technical Institute, USSR Academy
of Sciences, 194021 Leningrad, USSR

The present paper is primarily concerned with the results of an investigation of recombination processes in
InGaAsP-quantum well double heterostructures (QW DHs)
grown by a modified liquid phase epitaxy (LPE) method. QW
DHs produced by LPE are well suited for the determination
of the recombination processes that limit the excess carrier lifetimes in the two-dimensional (2d) case, as these
processes are now fully understood for LPE grown DHs with
3d active regions.

The results of previous investigations of recombination processes in the DHs with 3d-active regions should be
briefly summarized here. The three types of DHs used in
these investigations are shown schematically in Fig.1.
The following results have been obtained on the basis of
the photoluminescence (PL) studies of these structures:[1-9]

(1) For a wide temperature range, the excess carrier
lifetimes (τ) in moderately doped direct-bandgap A^3B^5
compounds with the bandgap (E_g) between 0.8 and 1.9 eV
are determined by intrinsic radiative transitions ($\tau = \tau_r$)
(refs.1,3,8,9).

(2) The experimental data[1,4,6-9] obtained for excess
carrier lifetimes in the above compounds are in good qualitative and quantitative agreement with the calculations
based on a simple theory for band-to-band transitions,
which employs Kane's model and takes the momentum selection rules into account. Thus, in nondegenerate GaAs[4,6-8]
and $In_{0.7}Ga_{0.3}As_{0.65}P_{0.35}$[9-10], the measured and calculated values of lifetime are in accord, as are their tempera-

ture dependences ($\tau \sim T^{3/2}$) (refs.4,6-8). In addition,
as predicted by the theory, with the onset of degeneracy
τ_r ceases to decrease linearly with the concentration of
the opposite type carriers[9,10], and the radiative coefficient B=($\tau_r n$)$^{-1}$ decreases.

(3) At still higher levels of doping or pumping
($> 10^{18}$ cm^{-3}) and temperatures around 300K, nonradiative
processes begin to limit the PL efficiency and to affect
the excess carrier lifetimes[5,7,9-11]. It has been established that for InGaAsP compounds with moderately narrow
bandgaps (E_g=0.95 eV) the limiting factor is an Auger process involving a transition of hole to a split-off valence
band (the CHSH-process). The rate constant (R) for this
process has been estimated to be (1-2)×10^{-29} cm^{6}s^{-1}
(refs.9-11). Figure 2 displays PL efficiency reduction for
AlGaAs/GaAs DHs with active region being heavily doped by
Ge acceptors[5,7]. If this effect is to be interpreted in
terms of Auger-recombination, one must suppose that the
coefficient R is also about 10^{-29} cm^{6}s^{-1} for GaAs:Ge.

Now we turn to the recent results on quarternary DHs
with thicknesses of the active regions (d_a) varying over
a wide range. The structures of the type presented in
Fig.1b,c were grown by the LPE method on GaAs ⟨111⟩ and
InP ⟨100⟩ substrates at temperatures of 750 and 635°C respectively. All of the InGaAsP/InP DHs had the active region of the same composition ($In_{0.7}Ga_{0.3}As_{0.65}P_{0.35}$)
with E_g=0.95 eV (λ =1.3 µm). The composition of all
InGaAsP/GaAs DH active regions ($In_{0.13}Ga_{0.87}As_{0.74}P_{0.26}$) corresponded to E_g=1.58 eV (λ =0.78 µm). Several sets of DHs
with the active regions of varied thickness were grown.
They differed in the composition of the layers adjacent
to the active region (Fig.1b,c). The samples were prepared by a modified sliding technique[12] in which the active
region and the other layers thinner than 10^3 Å are grown
while the substrate is being moved[13] under a special, nar-

row growth cell. By varying the speed of sliding from
0.03 to $3\,m\,s^{-1}$, it is possible to grow layers ranging in
thickness within at least an order of magnitude.

The composition profiles of the DHs grown by the
above technique were determined by the recently developed
method of X-ray induced photoelectron emission[14]. The data obtained by this method for the content distributions
of Ga and As in the vicinity of the active region of an
InGaAsP/InP DH (type c-2) are shown in Fig.3. Interface
abruptness no worse than 20 Å were also observed for all
types of the InGaAsP/GaAs DHs investigated.

The results to be considered were obtained in the PL
studies of undoped DHs with residual donor concentrations
of about 10^{17} cm^{-3}. For simplicity, the level of optical
excitation is given in the units of equivalent current
density.

The changes of PL spectra occurring with decrease of
well thicknesses for b-1, b-2 and c-2 types of DHs, are
due to the quantum size effect, as is the case for the
thoroughly studied AlGaAs/GaAs QW DHs. The marked narrowing of the main emission peak for DHs with d_a between 200
and 70Å (Fig.4a,b), and the appearance of additional short
wavelength peaks in the PL spectra for such structures at
high pumping levels (Fig.5a,b) clearly demonstrate the
formation of the steplike density of states in their active regions.

A recent detailed comparison[15] of the positions of
the main and additional peaks in the PL spectra for QWs
of different sizes, with the calculated energetic positions of subbands has revealed that the conduction band discontinuities for both InGaAsP/InP and InGaAsP/GaAs DHs
should have the value of at least a half of the full difference between the E_g of the active region and that of
the cladding layer. The rectangular shape of the potential well for the DHs investigated was also confirmed by

the results of this comparison.

Fig.6a,b shows the external PL efficiencies (η_e) for
the DHs of all types (see Fig.1b,c) measured at 77 and
300K as a function of well thicknesses. The measurements
were performed while photoluminescence was excited by
short wavelength radiation with the equivalent density of
$10-20$ A/cm^2. A striking feature of the data in Fig.6 is
that the value of η_e remains as high as 2% throughout the
thickness range down to 100 Å. The estimates taking account of the total internal reflection and other output
processes$^{2,16)}$ have yielded the lower limit of internal
efficiency (η_i) of 80% for InGaAsP/GaAs DHs whose GaAs
substrates are fully absorbing, and of 70% for InGaAsP/InP
structures which have thick, Te-doped and hence also stron-
gly absorbing InP substrates. The reduction in η_e at $d_a \lesssim 100$Å
can be seen distinctly only at 300 K (Fig.6a,b). The fact
that for $d_a \lesssim 100$Å the efficiencies remain high at 77K, as
well as the similar character of temperature dependences
of efficiencies for thin quantum wells being pumped either
with infrared or short wave radiation (Fig.6c) seems to be
good evidence that the decrease observed in η_e at 300K is un-
related to the fall in capture efficiency with decreasing
well thickness.

Our results thus indicate that nearly all of the photoexcited carriers are captured by the wells of the thicknesses considered here; the carriers captured by the wells
with $d_a \gtrsim 100$Å have the radiative lifetimes (τ_r) that are
considerably shorter than the nonradiative lifetimes (τ_n),
like in the case of 3d active regions. It is unlikely that
the small value of τ_r/τ_n for the QW considered are attributable to decreasing of τ_r with decreasing d_a. Exciton enhancement effect$^{17)}$ which has lately been often advanced as a
likely cause of decreasing τ_r, is scarcely possible within
the range of high temperatures and concentrations with
which we are dealing here. As for the subband-to-subband

radiative transitions, which appear to be the dominant radiative process for the QWs in question, the corresponding
lifetimes are no shorter than those for the band-to-band
radiative transitions in the appropriate 3d conditions.[18]

Therefore, the high values which η_e and η_i retain down
to d_a=100Å imply that τ_n is independent of the well thicknesses between 10^2 and 10^3Å, and thus even for QWs with
d_a=100Å the nonradiative lifetimes due to interface recombination (τ_n^i) should be yet longer (by at least a factor
of 3) than τ_n in bulk material.

Using the conventional relations: $S = d_a (2\tau_n^i)^{-1}$ and
$\tau_n = \tau_r (\eta_i^{-1}-1)^{-1}$ and taking 10^{-8}s as the lower limit for τ_r, it
is easy to find that for all types of the DHs studied the
value of S is less than 5cm s^{-1}. It should be pointed out
that the minimum value of S so far determined for the interface recombination in MBE grown AlGaAs/GaAs QWs is an
order of magnitude greater than this value[19].

The distinctive features of the recombination processes in InGaAsP/InP QWs become apparent when the level of
pumping is increased. Examination of Fig.7 will show that
for the InGaAsP/InP QW with d_a=150Å, an excitation level
of 200 A/cm^2 is enough for the PL efficiency to be reduced by half. It is also evident that the pumping level at
which PL quenching occurred is proportional to the well
thickness, and the efficiency dependence for DHs with d_a
of both 150 and 900Å agree well with the result of a calculation (dashed curves) in which a "three-dimensional"
value of Auger coefficient (R=2×10^{-29}cm^{6}s^{-1}) was used[9-11].

In the case of the InGaAsP/GaAs QW (curve 3 in Fig.7),
the quenching of the PL efficiency occurs only at a pumping level 20 times higher than that for the InGaAsP/InP QW
which has nearly the same d_a. We are not certain that at
the corresponding range of very high excitation the Auger
process alone is responsible for the reduction of PL efficiency. However, assuming that to be the case, the value

of R, at which the calculated curve (broken line in Fig.7) fits with the experimental data ($R=0.7\times10^{-29}cm^6s^{-1}$), appears to be near to the theoretically predicted value of R for Auger recombination in GaAs[20]. This value is twice as small as R for Ge-doped GaAs (see Fig.2). In order to facilitate comparison with the data of Fig.2, curve 3 of Fig.7 has been rescaled in terms of concentration and plotted as such in Fig.2 (the broken line). Despite a smaller R, it can be seen that the broken curve is slightly shifted towards the lower concentrations. The shift can be accounted for by the radiative coefficient reduction occurring in highly pumped active region with increase of excess electron concentration[9,10].

1) Alferov Zh.I., et al., FTP 8, 561(1974); Alferov Zh.I., et al., FTP 9, 462 (1975).
2) Alferov Zh.I. et al., FTP 210, 1497 (1976).
3) Abdullaev A., et al., FTP 11, 481 (1977).
4) Garbuzov D.Z., et al., FTP 12, 1368 (1978).
5) Abdullaev A., et al., FTP 11, 272 (1977).
6) Garbuzov D.Z. in: Semicond.Optoelectr. Proc. of Second Intern.School, Cetniewo, 1978 ed. by M.A.Herman, pp.305-343, PWN-PSP, Warszawa. John Wiley and Sons. Chichester-New York.
7) Nelson R.J. and Sobers R.G., J.Appl.Phys. 49, 6103 (1978).
8) Garbuzov D.Z., Czechoslovak J.Phys. B30, 326 (1980).
9) Garbuzov D.Z. et al. FTP 17, 1557 (1983); Garbuzov D.Z. Czechoslovak J.Phys. B34, 365 (1984).
10) Su C.B. et al. Electron.Lett., 18, 595 (1982); Su C.B. et al., Appl.Phys.Lett., 44, 732 (1984).
11) Sermage B. et al. Appl.Phys.Lett. 42, 259 (1983).
12) Alferov Zh.I. et al. FTP 19, 1108 (1985).
13) Rezek E.A. et al. Appl.Phys.Lett. 31, 288 (1977); Rezek E.A. et al. J.Electron.Mater. 10, 255 (1981).
14) Konnikov S.G. et al. to be published in FTP 20 (1986); Arsentyev I.N. et al.to be published in FTP 20 (1986).
15) Garbuzov D.Z. et al. to be published in FTP 20 (1986); Arsentyev I.N. et al.to be published in FTP 20 (1986).
16) Garbuzov D.Z. et al. FTP 18, 1069 (1984).
17) Christen J. et al. Appl.Phys.Lett. 44, 84 (1984).
18) Khalfin V.B. et al. to be published in FTP 20 (1986).
19) Dowson P. et al. Appl.Phys.Lett. 45, 1227 (1984).
20) Takishima M. Phys.Rev. B 25, 5390 (1982).
21) Garbuzov D.Z. et al. FTP 19, 449 (1985).

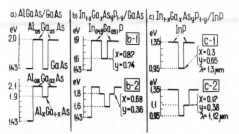

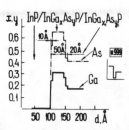

Fig.1 The DH types under discussion.

Fig.3 The distribution of Ga and As content in the vicinity of the active region, obtained from the photoemission analysis of the InGaAsP/InP QW.

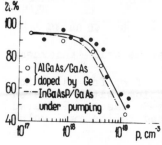

Fig.2 Internal PL efficiency at low level excitation for the active regions of AlGaAs/GaAs DHs as a function of hole concentration:●-our data(ref.5),O-Nelson's data(ref.7).The broken curve is the same as curve 3 from Fig.7 except that it is replotted on the concentration scale and normalized to the maximum η_i value.

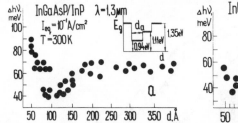

Fig.4 Half-widths of the emission peaks versus well thickness for quarternary DHs.

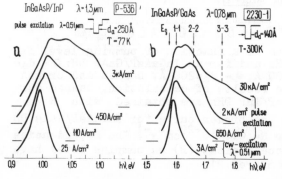

Fig.5 The evolution of PL spectra with increasing level of pumping.

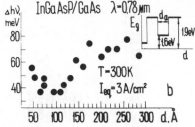

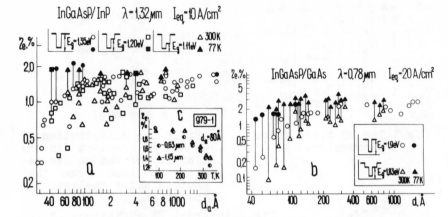

Fig.6 (a,b) External efficiency η_e versus well thickness
for quarternary DHs;
(c)   Temperature dependences of η_e for a thin
InGaAsP/InP QW under pumping with infrared
and red radiation from He-Ne laser. Infrared
pumping data are normalized to the real value
of η_e.

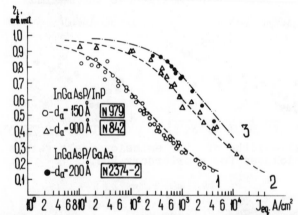

Fig.7 PL efficiencies as a function of pumping level for
InGaAsP/InP DHs and for an InGaAsP/GaAs DH: O - d_a=
150 Å, △ -d_a=900 Å, ● - d_a=200 Å. Dashed and broken
curves are the results of calculation. Dashed curves for InGaAsP/InP DHs are obtained with R =
$2 \times 10^{-29} cm^6 s^{-1}$. The broken curve for InGaAsP/GaAs DH
corresponds to R= $0.7 \times 10^{-29} cm^6 s^{-1}$. The dependencies
of radiative coefficient (B) on pumping level have
been taken from ref.21.

CARRIER HEATING IN GaAs BY NONRADIATIVE RECOMBINATION

W.W. Rühle, H.-J. Polland, and K. Leo

Max-Planck-Institut für Festkörperforschung, 7000 Stuttgart 80
Federal Republic of Germany

Cooling of a photoexcited hot eh-plasma in semiinsulating
GaAs is experimentally investigated by picosecond luminescence
spectroscopy. Due to the short lifetime of 100ps, recombination
heating strongly retards cooling. Additional heating processes
due to strong nonradiative recombination are experimentally
detected by resonantly generating a cool plasma and studying
the heating process.

1. INTRODUCTION

Cooling of an eh-plasma is discussed controversially: An apparent
reduction of the emission of LO phonons (i.e. the Fröhlich interaction) is observed at high densities and is explained by nonthermal
phonon populations[1] and/or screening[2]. Here we want to demonstrate
additional processes, which slow down cooling and which are related to
strong nonradiative recombination.

2. EXPERIMENTAL

Photoluminescence (PL) is excited by a synchronously pumped dye
laser (80MHz repetition rate, 4.5ps pulse width) with tunable wavelength (690 to 850 nm). Maximum excitation density is 6×10^{14} pho-
tons/cm^2. The temporal variation of PL at 10K is monitored at different wavelengths (1/4m spectrometer) using a synchroscan streak camera
with a time resolution of 20ps. The samples are liquid encapsulated

W.W. Rühle et al.

Czochralski grown semiinsulating GaAs (Cr-doping: 0 to $5 \times 10^{16} cm^{-3}$).

3. RESULTS AND DISCUSSION

Carrier lifetimes in semiinsulating GaAs are shorter than 300ps and determined by nonradiative transitions. Surface recombination can be neglected, since lifetimes are independent on surface treatment (polished, freshly etched). The transient spectra in Fig.1 are calculated from time-averaged spectra and streaks at different wavelengths. From the high energy tail the carrier temperature is deduced.

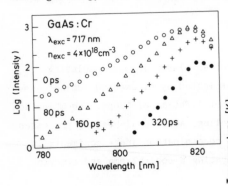

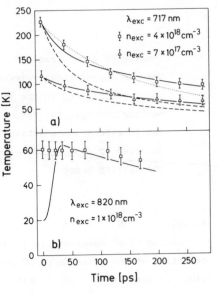

Fig.1: Transient spectra of the band to band transition

Fig.2: Evolution of carrier temperature with time. Curves are calculated with FD (solid lines) and MB statistic (dashed and dotted lines)

3.1 Recombination Heating

For short lifetimes, i.e. for rapidly changing Fermi level, "recombination heating"[1,3] becomes pronounced, even, if the transition probabilities are independent on carrier energy. The recombination heating is a natural consequence of Fermi-Dirac (FD) statistic: With decreasing Fermi level (recombination) the temperature of a Fermi gas must be increased in order not to change the mean energy of the Fermi

particles. Recombination heating is absent for Maxwell-Boltzmann (MB)
statistic where the energy per particle is 3/2 $k_B T$, i.e., independent
on carrier concentration.

With Figure 2 we want to demonstrate the difference using FD instead of MB statistic. Part a) depicts the decrease of carrier temperature for the sample with the highest Cr-doping (shortest lifetime).
The experimental values are obtained with an excitation photon energy
$h\nu$ which is larger than the band gap E_g. The following theoretical
curves are shown: i) For the solid lines we use FD-statistic adjusting
the polar optical (p.o.) scattering time to get a good fit with the
experimental points. ii) The dashed lines are calculated with exactly
the same parameters as in i) using MB statistic. iii) For the dotted
line we also use MB statistic, however, p.o. scattering is totally
neglected. Our calculations demonstrate that it is essential to use FD
statistic, i.e. to take into account recombination heating. MB statistic leads to completely wrong estimations, e.g., to the neglection of
the p.o. scattering as demonstrated by the similarity between the
dotted and upper solid line. Additionally, the elevated temperatures
at longer times cannot be understood using MB statistic.

Figure 2b shows the importance of recombination heating for reso-
nant excitation ($h\nu \approx E_g$) and high excitation density ($\approx 1 \times 10^{18} cm^{-3}$; the
high density is possible even for resonant excitation due to band gap
renormalization). Although cold carriers are generated and lattice
temperature is at 10K the carrier temperature raises immediately to
60K. The increase can only be explained by recombination heating as
shown by the solid theoretical fit. The heating process could not be
explained by MB statistic, since then recombination heating is absent.

The following scattering processes are included in the calculations: i) p.o. scattering; the scattering times $\tau_{p.o.}$, we use in Fi-
gure 2a, are 7.1 and 5.1 ps for $n_{exc} = 4 \times 10^{18} cm^{-3}$ and $7 \times 10^{17} cm^{-3}$, respectively and $\tau_{p.o.} = 0.2 ps$ for Figure 2b. This increase of $\tau_{p.o.}$ with
carrier density is generally observed[4] and reflects the influences of
hot phonons and/or screening. The values are consistent with results
on undoped GaAs.[5] ii) Deformation potential scattering; although it
is often neglected, it must be included, since energy loss by p.o.

scattering is reduced. We use a mean hole deformation of $4.8eV^{6)}$. iii)
Piezo-electric scattering can be neglected.

3.2 Heating by other nonradiative processes

Our experiments show that for resonant excitation and
$n_{exc}=2\times10^{17}cm^{-3}$ the carriers still achieve 54K, although recombination
heating for these lower densities is negligible. Even for low excitation into band tails (825nm) the carriers have 54K. Only at very low
densities $1\times10^{17}cm^{-3}$ the temperature drops to 40K. The additional
heating processes are less pronounced for the less Cr-doped samples.

Let us discuss three nonradiative heating mechanisms: i) Heating
by thermalization into band tails is improbable, since, even for excitation below band gap, carriers get and remain hot. ii) Heating via
absorption of LO phonons emitted by multiphonon nonradiative transitions seems unlikely: For the configuration coordinate model the capture cross section and therefore lifetime should vary strongly (exponentially) with temperature. Experimentally we observe, however, only
a slight decrease of lifetime with temperature (150K: 59ps, 290K:
35ps). iii) Impurity related Auger processes are in our opinion the
most probable heating process: Auger processes decrease with decreasing excitation density and saturate at the highest densities.

In conclusion, we have demonstrated that nonradiative recombination leads to strong recombination heating, via shortening lifetime.
Additionally, other heating processes are present, which can be
checked experimentally using resonant excitation.

We want to acknowledge valuable discussions with P. Kocevar.

References

1) Kocevar, P., Physica 134B, 155 (1985).

2) Zarrabi, H.J. and Alfano, R.R., Phys.Rev.B32, 3947 (1985)

3) Bimberg, D. and Mycielski, J., J. Phys. C19, 2363 (1986).

4) Kash, K., Shah, J., Block, D., Gossard, A.C., and Wiegmann, W.,
   Physica 134B, 189 (1985)

5) Rühle, W., Polland, H.-J., to be published.

6) Landolt-Börnstein, New Series, Vol.17 Semiconductors (Springer
   Verlag Berlin, Heidelberg, New York 1982)

LUMINESCENCE TRANSITIONS FROM EXCITED SUBBAND STATES
IN n-i-p-i HETEROSTRUCTURES

R. A. Street, G. H. Dohler,† J. N. Miller,† R. D. Burnham and P. P. Ruden*

Xerox Palo Alto Research Center, Palo Alto, California 94304 U.S.A.
†Hewlett-Packard Labs, Palo Alto, California 94304 U.S.A.
*Honeywell Physical Sciences Center, Bloomington, Minnesota 55420 U.S.A.

ABSTRACT

Luminescence of n-i-p-i heterostructures provide a new way of studying the subband structure of quantum wells. Excited state transitions originate when the excitation is sufficiently intense that the Fermi energy crosses into the first electron subband, when the hole subbands are populated thermally, and by non-equilibrium population of the higher levels

1. INTRODUCTION

The quantized subbands of GaAs/GaAlAs heterostructures are observed in luminescence.[1] Recently we have reported luminescence from a different type of heterostructure that is part of a n-i-p-i superlattice.[2] The electrons and holes are confined in quantum levels in the GaAs layer, but the well is approximately triangular rather than rectangular.

The well structure is shown in Fig. 1. Samples have been grown by both MBE and MOCVD with similar results. The alloy layers have the composition $Al_{0.3}Ga_{0.7}As$, and are alternately doped n-type and p-type. In addition there are undoped spacer layers to keep the dopants away from the GaAs. The undoped GaAs thickness is 150 Å for the data presented. The wafers were not rotated during MBE deposition so that the doping profile varied across the wafer depending on the position of the sources. The effects of small changes in the net doping level within a single growth can therefore be explored.

These n-i-p-i heterostructures have several different recombination properties compared to the usual rectangular well superlattice. The internal field causes the electrons and holes to be spatially separated at opposite sides of the GaAs. The internal field is modified by the population of electrons and holes so that the luminescence peak position is excitation intensity dependent, as observed for all n-i-p-i structures. The charge separation also causes long

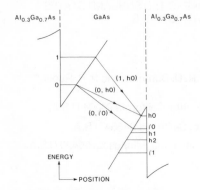

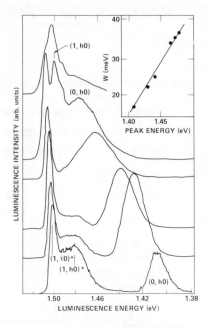

FIG. 1. Schematic diagram of the n–i–p–i heterostructure showing the tilted quantum wells, the subband structure and some possible transitions.

FIG. 2. Luminescence spectra (1.7K) for increasing excitation intensity. The identification of the peaks is discussed in the text. The insert shows the (0,h0) line width.

recombination lifetimes, due to the weak electronic overlap. The carrier population of the quantum wells can therefore be made very large, and for the structures studied here, reaches 10^{12} cm^{-2}.

The selection rules are also different in the two types of quantum well. The recombination is excitonic in a rectangular well so that in the absence of valence band mixing, only transitions from the same order subband are allowed.[3] Furthermore, each subband transition has roughly the same oscillation strength, so that the luminescence is always dominated by the lowest energy transitions. In the triangular wells of the n–i–p–i heterostructure, the recombination is not excitonic because of the spatial separation of the electrons and holes, and therefore transitions between different order subbands are allowed. Furthermore, the oscillator strength increases for higher subband transitions because the wavefunctions are more extended and the overlap increases.

2. EXPERIMENTAL RESULTS AND DISCUSSION

Fig. 2 shows the luminescence spectra of a n–i–p–i heterostructure for a range of excitation intensities. The lowest energy transition moves to higher energy as the intensity is increased and so is easily identified as the ground state n–i–p–i transition, denoted (0,h0) using the subband notation shown in Fig. 1. The peak shift agrees well with the values calculated for the structure. The

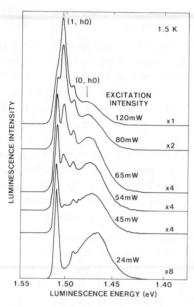

FIG. 3. Luminescence spectra (1.7K) at high excitation intensities, showing the onset of the (1,h0) line.

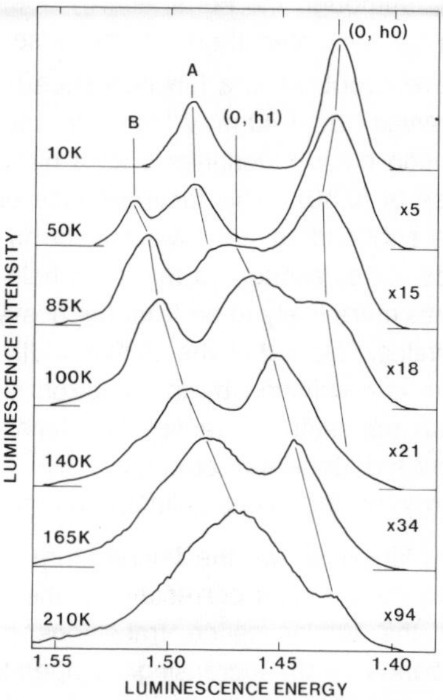

FIG. 4. Temperature dependence of the luminescence spectra, showing the evolution of the different transitions.

luminescence band width increases with excitation intensity as shown in the insert to Fig. 2. The width is due to the occupancy of the electron subband which has a much lower density of states than the heavy hole subband.[2]

Several other transitions are present in Fig. 2 beside the (0,h0) recombination. At the highest excitation intensities the luminescence is dominated by a peak at 1.50 eV, shown in more detail in Fig. 3. Below an excitation power of 45 mW this peak is absent, but grows rapidly in intensity at higher excitation intensity. We interpret this luminescence as a transition from the first excited electron subband to the lowest hole subband, denoted (1,h0). The threshold occurs when the Fermi energy just crosses into the next subband, at a density of about 10^{12} cm^{-2}. The (1,h0) transition is observed when the high energy side of the (0,h0) band just reaches the same energy, as expected if the Fermi energy extends up to the higher subband. Clearly the strength of the (1,h0) transition reflects the increased oscillation strength because the population of the two subbands where they overlap will be equal. We therefore estimate that the oscillator strength is larger by about a factor 5–10 compared to the (0,h0) transition. As further confirmation of the identity of the (1,h0) transition, its intensity is greatest for those samples cut from the more n–type side of the wafer, and the transition was not observed in the most p–type samples. Evidently a small excess electron concentration makes it easier to fill the subband.

Although the Fermi energy does not reach the higher hole subbands, they can be populated thermally because of the small heavy hole subband splitting. Above about 50 K, a luminescence band with about the same width as (0,h0) is observed about 40 meV higher in energy. The intensity dependence of the peak position clearly identifies it as a n–i–p–i transition, with shifts that are similar to those of (0,h0). This luminescence originates from recombination from a higher hole subband to the lowest electron band, and may be either (0,h1) or (o,l1), since calculations find that both hole bands are at a similar energy. However, it seems more likely to be from the heavy hole bands because of their larger density of states. Note that the similar width of (0,h0) and (0,h1) is expected since the width is dominated by the electron band. As the temperature is increased, the (0,h0) transition decreases in intensity relative to (0,h1), again because of the increased oscillator strength. The relative intensities are activated with the energy of the peak splitting, as expected for a thermalizing pair of lines.

Fig. 4 shows the temperature dependence of the luminescence extending up to 210 K. After correcting for the temperature shift of the band gap, it is clear that the luminescence transitions progressively move into the high energy subbands with increasing temperature, because of their larger oscillator strengths. The detailed temperature dependence is quite complex. In addition to the transfer of intensity from (0,h0) into (0,h1) near 85 K, a new transition at 1.52 eV appears above 30 K and eventually dominates the spectra. As the temperature increases, the (0,h1) line becomes narrower while the 1.52 eV line gets very broad. We tentatively interpret the high temperature luminescence as the unresolved transitions between many high energy subbands. As the subband splitting decreases, the total density of states will increase rapidly with energy. The transitions from these higher subbands dominate the luminescence because they have the largest oscillator strengths

Finally we discuss the transitions near 1.50 eV in the low temperature and low excitation intensity spectra in Figs. 1 and 4. The spectra depend on the net doping and also shift to higher energy with increasing excitation intensity (see Fig. 1) which confirms their origin as from within the n–i–p–i structure, rather than from the substrate. We propose that these transitions arise from the non–equilibrium population of excited subbands as the electrons and holes thermalize.[2]

4.   REFERENCES

1)   Petrou, A., Warnock, J., Ralston, J. and Wicks, G., Solid State Commun., 58, 581 (1986).
2)   Street, R. A., Dohler, G. H., Miller, J. N. and Ruden, P. P., Phys. Rev. B33, 7043 (1986).
3)   Miller, R. C., Gossard, A. C., Sanders, G. D., Chang, Y – C. and Schulman, J. N., Phys. Rev. B32, 8452 (1985).

EXCITON PHOTOLUMINESCENCE PROBE OF INTERFACE QUALITY IN MULTI-QUANTUM WELL STRUCTURES

Kop'ev P.S., Ledentsov N.N., Meltser B.Ya.,
Uraltsev I.N., Efros Al.L., Yakovlev D.R.

A.F.Ioffe Physico-Technical Institute, USSR Academy
of Sciences, 194021 Leningrad, USSR

Photoluminescence(PL) techniques have been used extensively to investigate the interface properties in MQW structures[1,2]. The present investigation focuses on the interpretation of the steady state PL and excitation spectra taken at extremely low excitation densities, when the strong competition between exciton and impurity related PL and nonradiative recombination occurs.

We have studied at 1.6 K a set of MBE grown GaAs MQW undoped samples with well width ranged from 13 to 7 nm and $Al_{0.35}Ga_{0.65}As$ barriers, 10 nm thick.

Dramatical changes of the exciton PL observed with an increase of incident photon energy are assumed to be governed by extrinsic properties associated with the interfaces in MQW.

The excitation spectrum of MQW samples is found to be sensitive to the detection wavelength at low excitation levels. Step-like density of states with two well resolved excitonic peaks corresponding to excitons involving heavy and light holes is exhibited in excitation spectrum when PL is detected at electron-neutral acceptor transition energy (Fig.1a, solid line). When detection is set on the heavy hole exciton peak (Fig.1a, dashed line) the exciton PL is practically absent at selective photoexcitation of

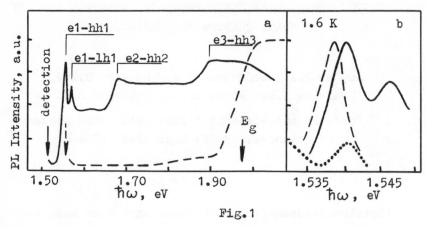

Fig.1

QWs, i.e. at $\hbar\omega < E_g$ (E_g is a bandgap of AlGaAs barriers)
and enhances strongly at incident photon energy $\hbar\omega > E_g$.
The energy position of the enhanced PL line (Fig.1b, dashed line) is Stokes shifted by 2 meV from excitonic peak
seen in excitation spectrum (Fig.1b, solid line). The
2 meV blue shift of the PL line is found both with temperature increasing from 15 to 20 K and when photoexcitation density becomes higher than 10 W/cm^2 (Fig.2 a,b).
These experiments have clearly demonstrated that Stokes
shift of the enhanced exciton line occurs due to exciton
trapping by interface fluctuations. The line width and
Stokes shift analysis[3] of the PL and excitation spectra
evidences for an island-like structure of the QW interface with 1 or 2 monolayer height. Furthermore, as we
shall see, the lateral extent of the islands can be determined by PL experiments.

    Excitons thermalize into localized states under the

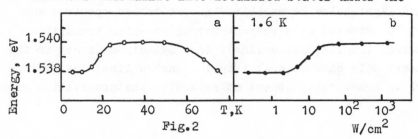

Fig.2

photoexcitation condition $\hbar\omega > E_g$. Localization by interface fluctuations prevents the excitons intralayer migration to impurities. We think, it is a reason for the exciton PL peak intensity being higher than that of acceptor
related one even at extremely low excitation level. Extrinsic PL dominates if effect of the exciton localization
is not important, as it does at $\hbar\omega < E_g$, when the complete quenching of the localized exciton PL occurs. The weak
shoulder of PL, as shown in Fig.1b by dotted line is due
to nonthermalized excitons. There are two experimental
findings in favour of this assumption: (i) the shoulder
energy position coincides with that of the peak in excitation spectra, i.e. this PL attributes to an average width
of the well; (ii) in contrast to the localized exciton PL
line, the shoulder does not show Zeeman patterns in Faraday configuration. The quenching of the localized exciton
PL at $\hbar\omega < E_g$ is believed to originate from electrostatic
potential of charged impurities spaced into barrier domains, closely to interface. When photoexcitation is absorbed into barrier domains, this effect reduces strongly
and the PL repartition due to localization occurs. To support this interpretation, the effect of an additional pulse
illumination of mcs duration on the steady state PL has
been studied. Recovery of the localized exciton PL, as
shown in Fig.3 by solid line is found if the PL at $\hbar\omega < E_g$
(dashed line) is illuminated by extremely low intensity
light with $\hbar\omega > E_g$. Evaluated time decay values of the
effect are a few tens of mcs as expected for tunnel relaxation time from deep impurities spaced in barrier domains.

We have measured Stokes shift of exciton line in the
presence of magnetic field applied normal to the QW layers. Fig.4 shows the energy position of the PL exciton
line at low (open circles) and high (full circles) excitation levels as a function of square magnetic field B^2.
When the PL line is related to delocalized excitons, the

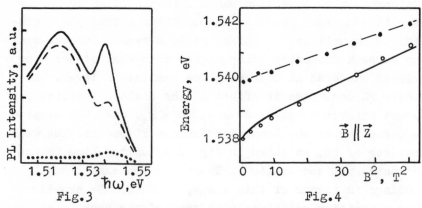

Fig.3   Fig.4

Fig.3. Illumination-induced change of PL spectrum,
       dotted line - PL under pulse illumination only.

linear law of B^2 is observed due to diamagnetic shift
(dashed curve). The localized exciton PL line exhibits additional shift, since the exciton wave function shrinks in
QW plane and   the exciton can be trapped by the islands
of lower lateral extent. Magnetic field induced diminish-
ing of Stokes shift has been calculated[4] using exciton
translation mass M=0.2 m_o (Fig.4, solid line). Measured
for the first time lateral extent of the islands trapping
effectively excitons turned out to be 350 Å in the MQW
samples studied.

REFERENCES

1) Weisbush C., Dingle R., Gossard A.C. and Wiegman W.,
   Solid State Commun. 38, 709 (1981).
2) Deveaud B., Emery I.Y., Chomette A., Lambert B.,
   Baudet M., Superlattices and Microstructures 1, 205
   (1985).
3) Bastard G., Journal of Luminescence 30, 488 (1985).
4) Kop'ev P.S., Meltser B.Ya., Uraltsev I.N., Efros Al.L.
   and Yakovlev D.R. JETP Lett. 42, 402 (1985).

p-ZnSe/n-GaAs HETEROJUNCTIONS FOR BLUE ELECTROLUMINESCENT CELLS

N. Stücheli, G.G. Baumann* and E. Bucher

Universität Konstanz, Fakultät für Physik, Postfach 5560,
7750 Konstanz, Fed. Rep. of Germany

Thin films of ZnSe were deposited on several types of
GaAs-substrates using the chemical vapor deposition (CVD)
method in an open system. Iodine and hydrogen were used as
transport agents. The layers were characterized by the
four-point resistivity and Hall techniques as well as by
photoluminescence measurements. The samples exhibit a low
p-type resistivity. Electroluminescent metal/p-ZnSe/n-GaAs
heterojunctions were made. A blue peak at 2.68 eV
dominates the emitted spectra.

1. EXPERIMENTAL

Zinc selenide (ZnSe) is one of the most promising materials for blue
electroluminescent devices. For such applications the growth of high-
quality epitaxial ZnSe layers is required. Since no single crystal
substrates of sufficient quality are available, our efforts have been
focused on the heteroepitaxial growth. Due to the small lattice
mismatch of 0.27 %, GaAs is one of the most favourable substrate
materials.
The ZnSe layers were prepared by chemical vapor deposition (CVD) in an
open system using hydrogen and iodine as transport agents. ZnSe single
crystal layers of about 1-5 μm thickness and of excellent quality
could be grown within the temperature range of 550-650 °C. The ZnSe

* Present address: Siemens AG, 8000 München, Fed. Rep. of Germany

powder source was kept at a temperature of 800 °C. The growth rate was
about 0.5 µm/h depending on the iodine concentration. The thickness of
the films were estimated from the weight increase. The layers were
deposited on differently orientated GaAs-substrates containing
selenium (n-type) or zinc (p-type) as dopants. For the Hall and
resistivity measurements semi-insulating (10^7 Ωcm) GaAs was used as
substrate material. The grown ZnSe layers showed smooth and
mirror-like behavior. For the electrical measurements the back side of
the substrates was etched to remove the back grown ZnSe films. Ohmic
contact to the GaAs was made by evaporating tin onto the n-type and
aluminum onto the p-type material. Evaporated gold was used for the
ohmic contacts to the ZnSe layers. Additionally a few experiments were
carried out by evaporating some metals with a lower work function such
as silver and aluminum on the epitaxial ZnSe.

2. CHARACTERISATION

2.1 Electrical Transport Properties

The electrical properties were determined by the four-point resitivity
and Hall measurements. Seebeck coefficient and Hall-voltages of the
ZnSe layers grown on semi-insulating GaAs (Cr as dopant) indicate a
p-type conduction of these films. The samples exhibit a low
resistivity of ρ = 0.5-1 Ωcm and a Hall mobility of μ_p = 20-40 cm²/Vs
at room temperature resulting in a net carrier concentration of
10^{17}-10^{18} cm^{-3}. The diode characteristics of the ZnSe heterojunctions
based on low ohmic n-GaAs substrates confirm the assumed p-type conductivity of ZnSe.

2.2 Photoluminescence Properties

The photoluminescence (PL) spectra of the ZnSe films were taken at
1.8 K and at room temperature. The samples were excited by a 20 mW
krypton ion laser at 406.75 nm.
The excitonic PL-spectra at 1.8 K of ZnSe layers grown on different
substrates are shown in Fig. 1. The emission peak at 2.8031 eV is due

to the radiative recombination of free excitons in the ground state [2]. The peaks at 2.7968 eV and 2.7941 eV originate from excitons bound to the neutral donors ($I_2(D^\circ,X)$) such as Al or Ga and to the neutral acceptors ($I_1(A^\circ,X)$) e.g. Li or Na, respectively [1-3]. The origin of the emission lines at 2.7991 eV, 2.7887 eV (I_b) and 2.7797 eV (I_c) are not elucidated. The I_b and I_c line may be caused by the radiative recombination of bound excitons at neutral acceptors judging from its energy position. The band emission at 2.695 eV results from the recombination of the donor-acceptor pair band Qo [3]. The weak lines at 2.7483 eV and 2.7167 eV are LO-phonon replicas with a phonon energy of 31.6 meV. At 1.8 K the green, yellow and red emission of the spectra are two order of magnitude weaker than the blue one. By increasing the thickness of the ZnSe layers the band at 2.25 eV is reduced in the room temperature spectrum. This dependence is shown in Fig. 2. This reduction is basically nescessary for good blue electroluminescent cells.

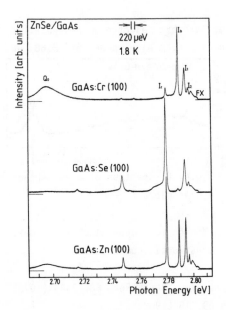

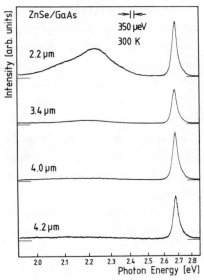

Fig.1. PL-spectra of ZnSe layers grown on different GaAs-substrates

Fig.2. Dependence of the PL-spectra on the layer thickness

3. DIODE CHARACTERISTICS

Metal/p–ZnSe/n–GaAs heterojunctions exhibit rectification with forward conduction occuring, when the ZnSe is positively biased. For these samples typical I–V curves are represented in Fig. 3. Three different kind of metals were eva-
porated on the ZnSe layers: gold, aluminum and silver. The fact that the reverse bias is independent on the metal confirms the assump- tion that the ZnSe is p-type. The diode quality factor is n = 2.5 for the Au/p–ZnSe/n–GaAs junctions. For the silver and aluminum diodes ohmic losses lead to an increased n of 18. Biased

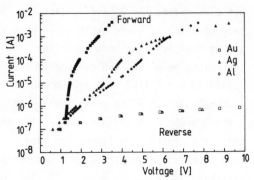

Fig.3. I–V curves of the metal/p–ZnSe/ n–GaAs heterojunctions

in forward direction this diodes exhibit electroluminescence. A typical spectrum is demonstrated in Fig. 4. A blue peak at 2.68 eV dominates the emitted spectrum. The light is emitted by single spots under the metal film. The intensity is relatively weak. In summary, we have shown the preparation of high quality p–ZnSe layers and their application as electroluminescent devices.

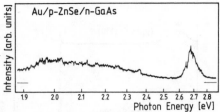

Fig.4. EL-spectrum of a Au/p–ZnSe/ n–GaAs diode

We would like to thank Siemens AG in Munich for supporting this work, in particular Dr.C.Weyrich for many constructiv comments.

[1] Merz, J.L., Nassau, K. and Shiever, J.W., Phys. Rev. B8, 1444 (1973)
[2] Bhargava, R.N., J. Crystal Growth 59, 15 (1982)
[3] Bhargava, R.N., Seymour, R.J., Fitzpatrick, B.J. and Herko, S.P., Phys. Rev. B20, 2407 (1979)

Indirect Resonant Tunneling in GaAs–GaAlAs

E. Calleja*, L. L. Chang, L. Esaki, C. E. T. Gonçalves da Silva**,

E. E. Mendez, and W. I. Wang

IBM T. J. Watson Research Center, Yorktown Heights, NY 10598, USA

ABSTRACT

The current-voltage characteristics of GaAlAs-GaAs-GaAlAs double barriers show features in addition to those corresponding to bound states of the GaAs quantum well. The new structures are explained in terms of resonant tunneling through confined states in GaAlAs, at the X point of the Brillouin zone.

The observation of resonant tunneling through double-barrier heterostructures provided the first proof of the formation of discrete states in a quantum well [1]. Since then, advances in materials preparation have made possible the use of this effect as a spectroscopic tool, for the study, e.g., of light- and heavy-hole states [2], and of Landau levels [3]. However, our knowledge, has been limited to band-edge states. In this paper we extend this technique to the study of confined states to other points of the Brillouin zone, namely the X point.

The experiments were done on (100) $Ga_{1-x}Al_xAs$-GaAs-$Ga_{1-x}Al_xAs$ heterostructures, grown by molecular-beam epitaxy, with an Al mole fraction of either 0.4 or 1. Thick n^+-GaAs regions on both sides of the devices served as electrodes, between which a voltage was applied. Under resonant conditions, the tunneling current increases dramatically, as the electron energy matches that of the states in the quantum well. This situation is illustrated in Fig.1, where we show the current-voltage (I-V) characteristics, at 4K, of a heterostructure with 100Å $Ga_{0.6}Al_{0.4}As$ barriers and a 60Å well. Two negative-resistance features at 0.11V and 0.45V correspond to resonant tunneling via the two bound states of the quantum well. Their energies in the absence of any electric field are 0.063

and 0.241eV, according to an envelope-function calculation [4], in agreement with the experimental voltage drop between the electrode and the well. At high bias, when the highest bound state is below the conduction-band edge of the emitter electrode, current is expected to flow to the collector via Fowler-Nordheim tunneling, which in the device of Fig.1 should occur for voltages ≥ 0.6V. However, at ~ 0.9V a well-defined feature is observed, even at 77K, for both bias polarities. Other samples with the same barrier widths but different well thickness show a similar structure, although its position differs slightly.

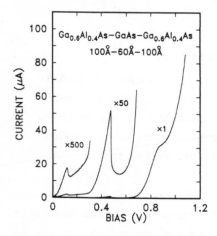
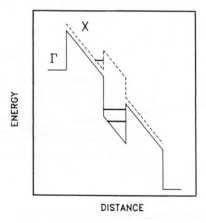

FIG 1. I-V characteristics, at 4K, of a heterostructure with 100Å $Ga_{0.6}Al_{0.4}As$ barriers and a 60Å GaAs well. The feature at ~ 0.9V corresponds to energies above the Γ-point barrier.

FIG.2. Γ-point and X-point potential profiles for the device of Fig.1. At Γ GaAs serves as a quantum well, while at X acts as a barrier.

We interpret the new structure as due to a confined state in $Ga_{1-x}Al_xAs$, that acts as a quantum well in a potential profile derived from the X-point of the Brillouin zone, while GaAs behaves as a barrier. Figure 2 sketches the energies of the Γ and X points along the heterostructure. The former, being lower in GaAs than in $Ga_{1-x}Al_xAs$, creates two bound states in the 60Å well, that lead to negative resistance, as mentioned above. In contrast, the X-point energy is lower in $Ga_{1-x}Al_xAs$ than in GaAs, and, therefore, confined states are formed in the alloy. It is then possible for electrons with wavevectors near the Γ valley of GaAs to tunnel through X-point barriers, via propagating states near the X valley of

$Ga_{1-x}Al_xAs$. Conservation of momentum perpendicular to the tunneling direction limits to one the number of X-minima involved, in (100) heterostructures.

The transmission probabilities of the $\Gamma \to \Gamma$ and $\Gamma \to X$ tunneling paths as a function of total bias have been calculated by numerical integration of Schrödinger's equation. The continuous line of Fig.3 corresponds to the probability for the $\Gamma \to \Gamma$ channel (intra-valley tunneling), calculated using a barrier height of 0.296eV and effective masses of $0.066m_0$ and $0.101m_0$, for GaAs and $Ga_{0.6}Al_{0.4}As$, respectively. The transmission probability for the $\Gamma \to X$ channel (inter-valley tunneling) is shown in Fig.3 as a dotted line, for the $\Gamma \to X \to X \to \Gamma \to \Gamma$ path, where the value $0.85\ m_0$ has been used for the effective mass at X, for both GaAs and $Ga_{0.6}Al_{0.4}As$. The arrows in Fig.3, showing experimental results, illustrate the agreement with the calculations. Other possible paths (e.g. $\Gamma \to X \to X \to X \to \Gamma$) differ in detail, but the existence of a strong transmission peak at ~0.9V remains valid. In principle, it would be possible to explain the observed high-bias structure in terms of virtual states in the GaAs quantum well, that is, resonant states above the confining barrier at Γ. Indeed, a recent optical experiment shows evidence of those states [5], which appear in Fig.3 as small modulations in the tunneling probability. However, neither their position nor their number can account for our results.

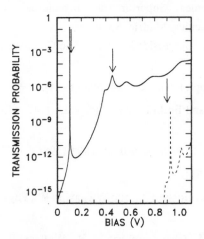

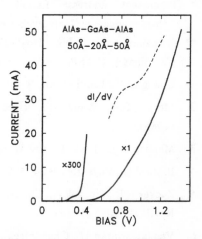

FIG 3. Transmission probability vs applied voltage for the structure of Fig.1. The arrows indicate the experimental positions of conductance minima.

FIG. 4. Current-voltage characteristics of an AlAs(50Å)- GaAs(20Å)-AlAs(50Å) double barrier, taken at 4K. The features observable in either I-V or dI/dV-V were absent at 77K.

The confined X-state on which we base our interpretation, has been observed in high-pressure luminescence experiments [6], in $GaAs-Ga_{1-x}Al_xAs$ multilayers with $x \leq 0.45$. It should be observable also at ambient pressure when $x \geq 0.45$, since, with suitable layer widths, it can become the ground state. Figure 4 shows the I-V characteristics of an AlAs-GaAs-AlAs double barrier, in which the thickness of GaAs is 20Å, while that of AlAs is 50Å. The lowest-energy Γ-state is, in the absence of field, at ~0.33eV, while the X-confined level is at ~0.17eV. Resonant tunneling through the latter is evident in the figure, at bias between 0.3V and 0.4V. For higher voltages, non-resonant tunneling through the X barriers dominates the current. An extremely weak feature at ~1V, enhanced in the conductance dI/dV (dotted line), corresponds to tunneling via the Γ-state. This reversal of the position of both states relative to the structure of Fig.1 provides clear proof of resonant tunneling via X-point confined states.

This work has been sponsored in part by the Army Research Office. One of us (CETGdaS) acknowledges partial support by the Conselho Nacional de Desenvolvimento Cientifico e Tecnologico (CNPq), Brasil.

* Permanent address: Escuela Tecnica Superior de Ingenieros de Telecomunicacion, Universidad Politecnica, Madrid, Spain.
** Permanent address: Instituto de Fisica, Universidade Estadual de Campinas, Campinas 13100 SP, Brasil.

1. Chang L. L., Esaki L., and Tsu R., Appl. Phys. Lett. **24**, 593 (1974).

2. Mendez E. E., Wang W. I., Ricco B., and Esaki L., Appl. Phys. Lett. **47**, 415 (1985).

3. Mendez E. E., Esaki L., and Wang W. I., Phys. Rev.B **33**, 2893 (1986).

4. Bastard G., Phys. Rev.B **24**, 5693 (1981).

5. Zucker J. E., Pinczuk A., Chemla D. S., Gossard A., and Wiegmann W., Phys. Rev.B **29**, 7065 (1984).

6. Venkateswaran U., Chandrasekhar M., Chandrasekhar H. R., Wolfram T., Fischer R., Masselink W. T., and Morkoç H., Phys. Rev. B, **31**, 4106 (1985).

RESONANT TUNNELING IN
DOUBLE BARRIER HETEROSTRUCTURES

Mark A. Reed and Jhang W. Lee

Central Research Laboratory
Texas Instruments Incorporated
Dallas, TX 75265
USA

ABSTRACT

Resonant tunneling in a variety of MBE-grown double
barrier heterostructures is investigated. We have
observed multiple negative differential resistance (NDR)
peaks in wide quantum well structures, resonant
tunneling in structures where the $Al_xGa_{1-x}As$ alloy
barriers have been emulated by thin, short period
superlattices, and resonant tunneling in a structure
where the center GaAs quantum well is replaced by a
different material, specifically InGaAs.

1. INTRODUCTION

Resonant tunneling heterostructures[1,2] have recently undergone a
renaissance[3-5] due to improved GaAs/AlGaAs molecular beam epitaxy
(MBE) techniques. In these structures, the essential carrier transport
mechanism is electron (or hole[6]) tunneling, typically through ultrathin
(~ 50Å) $Al_xGa_{1-x}As$ tunnel barriers and a GaAs quantum well. We present
in this paper the investigation of a variety of resonant tunneling structures
which are significantly different from previously investigated resonant
tunneling structures.

2. EXPERIMENTAL

The samples used in this study were grown in a Riber MBE-2300 on a
(100) GaAs substrate doped n-type with Si at 2-$3x10^{18}$ cm^{-3}. Following a

highly doped (n-type, Si @ 2×10^{18} cm^{-3}) buffer layer, the active undoped
resonant tunneling structure region was then grown, followed by a similar
top contact. Mesa diodes with diameters ranging from 2 to 225 microns
were fabricated by conventional photolithography and fabrication
techniques. Fig. 1 shows scanning electron micrographs and a schematic
cross-sectional diagram of the mesa device structures.

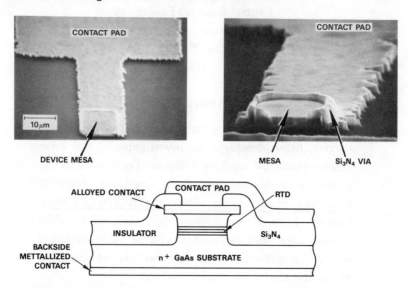

Fig. 1. SEMs and schematic cross-sectional diagram of the
device mesa structure used in this study for studying the
resonant tunneling diodes (RTDs).

3.   RESULTS

The observation of NDR in a double barrier heterostructure is due to
lowering the bound quantum well state through the Fermi level of the
doped contact. If there exist multiple bound states in the quantum well,
multiple NDR peaks should be observable. Fig. 2 shows the current-voltage
(I-V) characteristics of a 100 Å undoped GaAs quantum well confined by
50 Å undoped Al$_{.3}$Ga$_{.7}$As barriers. In terms of device bias voltage, the
resonant positions are 0.35 V for the first excited state (Fig. 2(a)) and 83 mV
for the ground state (Fig. 2(b)). The experimental results are in excellent
agreement with theory when the series parasitic resistance of 45 Ω is taken
into account. This resistance also explains the apparent "hysteresis" of the

excited state peak, which is due to the resonant tunneling diode - parasitic resistor series combination.

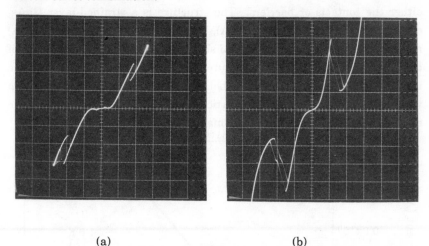

(a)   (b)

Fig. 2. I-V characteristics of the 100 Å GaAs quantum well sample. Mesa area = 5 μm x 5 μm, T = 77°K. The scale is (a) 500 mV / horiz. div., 5 mA /vert. div.; (b) 100 mV / horiz. div., 100 μA / vert. div.

We have also investigated a structure where the $Al_xGa_{1-x}As$ alloy barriers have been replaced by thin, short period superlattices. The heterostructure consisted of a 45 Å GaAs quantum well bounded by superlattices consisting of 3 components 7 Å AlAs / 2 components 7 Å GaAs. The structure does not exhibit the asymmetry in the electrical characteristics around zero bias normally seen in double barrier heterostructures. This observation suggests that the inverted GaAs surface is responsible for the observed asymmetry.[5] To determine the energy of the quantum well state independent of the parasitic resistance, thermal activation[7] through the state was investigated which directly determines the energy of the bound state relative to the Fermi level. This method was applied to a conventional 50 Å GaAs quantum well / 50 Å $Al_xGa_{1-x}As$ alloy barrier structure which gave the correct quantum well state. The superlattice barrier structure exhibited a quantum well state of ~120 meV, in disagreement with 677 meV calculated from an envelope function approximation.[8] Thus, these thin superlattice barriers exhibit an anomolously low barrier height.

Finally, we report the first observation of resonant tunneling NDR in a system where the center quantum well is of a different material than either the contacts or barriers. The structure is a strained-layer GaAs-contact, $Al_{.15}Ga_{.85}As$-barrier (40Å), $In_{.15}Ga_{.85}As$-quantum well (60Å) resonant tunneling structure. Fig. 3 shows the temperature dependence of the I-V characteristics of this structure. These types of structures have the intriguing possibility of "hidden" (below conduction band edge) quantum well states that are invisible to vertical carrier transport. The structure exhibits a complex temperature and magnetic field behavior.

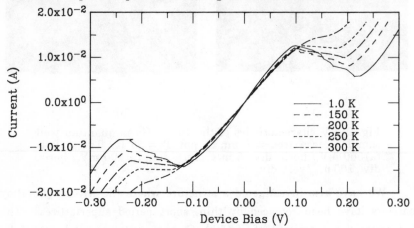

Fig. 3. Temperature dependence of the InGaAs quantum well structure I-V characteristics. Mesa area = 10µm x 10µm.

REFERENCES

1)   R. Tsu and L. Esaki, *Appl. Phys. Lett.* <u>22</u>, 562 (1973).

2)   L. L. Chang, L. Esaki, and R. Tsu, *Appl. Phys. Lett.* <u>24</u>, 593 (1974).

3)   T. C. L. G. Sollner, W. D. Goodhue, P. E. Tannenwald, C. D. Parker, and D. D. Peck, *Appl. Phys. Lett.* <u>43</u>, 588 (1983).

4)   A. R. Bonnefoi, R. T. Collins, T. C. McGill, R. D. Burnham, and F. A. Ponce, *Appl. Phys. Lett.* <u>46</u>, 285 (1985).

5)   T. J. Shewchuk, P. C. Chapin, P. D. Coleman, W. Kopp, R. Fisher, and H. Morkoç, *Appl. Phys. Lett.* <u>46</u>, 508 (1985).

6)   E. E. Menez, W. I. Wand, B. Ricco, and L. Esaki, *Appl. Phys. Lett.* <u>47</u>, 415 (1985).

7)   A. C. Gossard, W. Brown, C. L. Allyn, and W. Weigmann, *J. Vac. Sci. Technol.* <u>20</u>, 694 (1982).

ELECTRON AND HOLE MOBILITIES PERPENDICULAR TO INTERFACES IN AlGaAs SUPERLATTICES

J.F. Palmier, H.Le Person, C.Minot, A.Sibille, and F. Alexandre
Centre National d'Etudes des Telecommunications
EPM/PMM 196 avenue Henri Ravera 92220 BAGNEUX (FRANCE)

Abstract - Electrical transport perpendicular to interfaces in AlGaAs/GaAs superlattices is studied by means of nin photoconducting stuctures. The dark current data are interpreted with a deep acceptor center model. The picosecond time of flight data are interpreted in terms of Bloch conduction in electron and light hole minibands.

INTRODUCTION. New transport properties of semiconductor superlattices (SL) have been proposed and demonstrated by ESAKI's pionneering work [1]. A first approach of carrier mobility perpendicular to interfaces in very regular and pure type I superlattices is the Bloch transport theory [2]. A second approach is the phonon assisted tunneling [3], which does not require any Bloch function coherence over several superlattice periods. The present work reports on the mobility measurements in n+/SL/n+ structures, the SL being undoped. It is shown that dark current is often dominated by deep compensation centers with a concentration in the range $10^{15}- 10^{16}$ cm^{-3} giving space charge limited currents or diode exponential characteristics. More detailed results are obtained in picosecond photoconductive measurements. The time variation of the excess charges at different mean electric fields is a semi-direct approach of the carrier time of flight. As the electric field is not homogeneous in the photoconductor structures , the interpretation is assisted by a self consistent numerical solution of drift diffusion and Poisson equations. It is valid as long as local equilibrium quasi-Fermi levels can be defined and if the superlattice can be considered as an effective medium of specific band parameters, mobilities and lifetimes.

EXPERIMENTAL RESULTS. Samples are photoconducting four layers stuctures prepared by Molecular Beam Epitaxy (fig. (1)). The superlattice period consists in a GaAs well (here 45 A^O width for all samples) and variable width (B) GaAlAs barrier with an Al molar fraction of the order of 0.3 (see table I for B values).

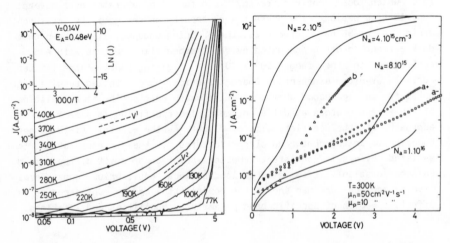

figure 1 - Cross section of processed samples. Typical lateral width=100 μm.

E=1-1.5 μm

D=0.2-0.3 μm

sam-ple	B (A^O)	Na cm^{-3}	μe dark	$'\mu e$ gain	μe tve	μh tvh
a	22.5	8.10^{15}	500			20
b	45	4.10^{15}	5			10
c	27	2.10^{15}	50	50	30	14

Table I . Summary of mobility results and barrier width B.

The structure is grown on n+ GaAs substrate ($10^{18}cm^{-3}$). It is terminated by a Ga Al As window layer (Nd < 3.10^{17} cm^{-3} and x<0.3), and a thick GaAs n+ contact layer. The current-voltage curves at different temperatures for sample (a) are plotted in fig(2).

figure 2 - J(V) versus T for sample a (log-log). Insert :log(J) vs 10 /T

figure 3 - Comparison of J(V) at 300K with the model (versus Na).

Three regimes can be distinguished: (i), a linear regime at low V
with an activation energy of 0.48   eV (insert of fig.(2)), (ii) a diode
exponential regime ($I \sim \exp(qV/\eta\, kT)$ with $\eta \sim$ 15-20 which is also T-
dependent, and,(iii), a V^2 regime at high voltages and low T. The dark
current cannot always be understood. It is dominated by deep acceptors,
and so very sensitive to MBE growth conditions. Applying the model, the
J(V) curves can be fitted in a restricted V range with 'only 3   parameters: the electron mobility in the SL, the acceptor concentration Na,
and the position of these acceptors from the SL conduction band Ea,which
of the order of 0.4 - 0.5 eV, as verified both by DLTS measurements [4]
and results of fig(2). Assuming μe = 50 $cm^2.V^{-1}.sec^{-1}$, Na is varied from
10^{15} cm^{-3} to 10^{16} cm^{-3} in fig(3) . This shows that the parameter η is
controlled by Na, as a rough space charge limited current estimate also
shows. The final adjustment is made with ue which is a simple curve
translation on the semi-log scale of fig(3). That procedure leads to the
μe and Na values of table I .

STATIC PHOTOCURRENT. If there is no trappping effect, the structures are readily usable to measure the photocurrent gain G = τ/tve,in
which tve is the electron time of flight across the SL and τ the hole
effective lifetime. The window layer acts as a barrier for holes on the
top electrode. So the gain is of the order of 1 for positive voltages,
i.e. when holes can be swept out towards the substrate , and possibly >1
for negative voltages. Sample (c) which has less electron traps presents
G values compatible with an electron time of flight of the order of 100
picosec.

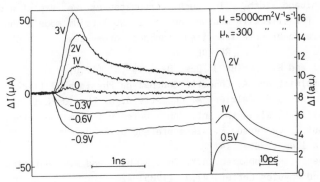

figure 4 - a) Photocurrent response versus V (sample c) ; b) Numerical
simulation results showing the hole sweepout effect (V>0).

TIME DEPENDENT PHOTOCONDUCTIVITY. A time of flight technique has
been developed to measure electron and hole mobilities(see the principle

in fig(1) and ref [5]). The phenomena occurring after the photon pulse
arrival are : (i), photovoltaic effects near the top electrode, the photovoltaic signal being present at V=0, (ii), recombination effects which
are supposed to be weakly field dependent, and, (iii), the time of
flight contribution. If the absorption is moderate (here 15000 cm^{-1}) the
electrode injection lasts a time close to the electron time of flight
(tve) ; during this the current increases, as a simple electrostatic
model shows. Then the slower carriers (holes) are either blocked(V<0)
or swept out. We have seen both effects in sample (c) (fig. (4)), but,up
to now, only the hole time of flight is reproduced by a numerical simulation. Time scale and hole mobility are different for figs. 4a and 4b.

DISCUSSION. A summary of the data is shown in table I. The electron
mobilities obtained here are compatible with earlier calculations [2]
which were made with the 85-15 offset hypothese. However we see the
discrepancy is high with a 60-40 hypothese. The variation of μe with B
is still steep, as also shown by recent Capasso's data on the GaAlInAs
system [6]. A transition from Bloch to hopping mobility has also been
observed in superlattice bipolar transistor structures[7]. All present
hole data give a value of the order of 10 $cm^2 v^{-1} s^{-1}$. At room temperature if we assume a local equilibrium the heavy to light hole ratio is a
few % depending on B value.The measured mobility is therefore light hole
contribution to the drift velocity, the heavy hole contribution being
negligible . Present data are a further confirmation of electron and
light hole minibands in such superlattices[8]. However at large applied.
fields (>10^4 V.cm) the potential variation between two wells is higher
than the miniband width. In such circumstances we expect a drastic
failure of the effective medium and local equilibrium models and a possible appearance of more localized transport processes.

REFERENCES
[1] L.Esaki and R. Tsu IBM J. Res. Dev. 14 ,61 (1970)

[2] J. F. Palmier and A. Chomette Journal de Physique 43 (1982) 381

[3] D. Calecki ,J.F. Palmier, A. Chomette J. of Physics C 17 5017(1984)

[4] A. Sibille et al 14 th Int. Conf. Defects in semiconductors (1986)

[5] C. Minot H. Le Person F. Alexandre and J. F. Palmier
    Physica 134B (1985) 514

[6] F. Capasso, K. Mohammed, A.Y. Cho, R. Hull and A.L. Hutchinson
    Applied Physics Letters 47,4 (1985) 420

[7] to appear in Appl. Phys. Lett.

[8] B.Deveaud, A.Chomette, B.Lambert, and A.Regreny
    Solid State Comm. 57 (1986),885

Vertical Transport Mass
in
GaAs/(Al,Ga)As Superlattices

T. Duffield[*], R. Bhat, M. Koza,
M.C. Tamargo, J.P. Harbison, F. DeRosa,
D.M. Hwang, P. Grabbe, K.M. Rush and S.J. Allen, Jr.[*]

Bell Communications Research, Inc.
Redbank, New Jersey 07701-7020

ABSTRACT

We have measured the transport mass along the
growth direction of GaAs/(Al,Ga)As superlattices. The mass that we observe can be
satisfactorily fit to a model of the
mini-band structure. The electrons are shown
to tunnel coherently through many
superlattice barriers.

1. INTRODUCTION

One of the more remarkable achievements in the field
of materials science has been the growth of semiconductor
single crystals atomic layer by atomic layer. Although the
scientific understanding and technological development of
these materials has been unusually rapid, the early
vision[1] that focused on the engineering of bulk,
3-dimensional, bandstructures has been largely forsaken in
favor of properties that emerge due to confinement of the
electron states to the two dimensions of a single layer.

Here we address the problem of vertical transport by
performing cyclotron resonance with the magnetic field
oriented perpendicular to the growth direction. In this
geometry a resonance condition can be achieved only if the
electrons are able to tunnel coherently through many
barriers. As a result the probe is sensitive and specific
to a number of crucial issues in vertical transport.

The recent theoretical and experimental work of Maan
et al.[2,3] approaches the issues in vertical transport from
much the same perspective. They observed
photo-luminescence in magnetic fields perpendicular to the
growth direction and showed that inter-band Landau
transitions disappearred for energies that exceeded the
mini-bandwidth.

2.   EXPERIMENT

The experiments were performed on superlattices grown
by organo-mettalic chemical vapor deposition (OMCVD) and
molecular beam epitaxy (MBE). Approximately 5-6 microns of
superlattice were grown on semi-insulating substrates. The
doping was kept in the 10^{15} cm^{-3} range to minimize
scattering and to diminish the effect of the plasma
resonance on the spectra. The resonance measurements were
made at liquid nitrogen temperature to avoid carrier freeze
out. Typical results are published elsewhere.[4]

The mass along the growth direction as well as in the
plane of the wells and barriers is shown in Fig. 1. The
latter is relatively insensitive to the Al fraction or
barrier height and the small departure from the bulk GaAs
mass at the band edge can be explained by band
non-parabolicity.

The mass along the growth direction increases
dramatically as the Al fraction is increased. There is
substantial scatter in the data that lies outside the
estimated uncertainties in the Al fraction or the measured
mass. The most glaring variations are seen when comparing
the MBE material to the OMCVD material. This indicates
that there are material parameters outside our control that
influence tunneling through the barriers.

3.   DISCUSSION

We calculate the transport mass along the growth
direction by assuming a Kronig-Penney model for the
superlattice band structure. We include the modifications
suggested by Bastard[5] that incorporate the change in
bandstructure mass in the barrier. After obtaining the
bandstructure we then approximate the cyclotron mass by
performing a classical calculation of the cyclotron mass
for an electron with energy $k_b T$ which corresponds to the
relatively high temperature at which these measurements
were made.. The results for the effective mass along the
growth direction are shown in Fig. 1.

At low Al fraction it appears that the model that
incorporates the Bastard corrections provides a more
convincing fit, although the scatter apparent at Al
fraction above .35 indicates that this conclusion can be
made firm only by continued experimental study and a better
understanding of the material parameters.

We have been unable as yet to successfully measure the
transport mass along the growth direction in superlattices
with barriers with large Al fraction. Work is continuing
to measure the tunneling rate in this regime.

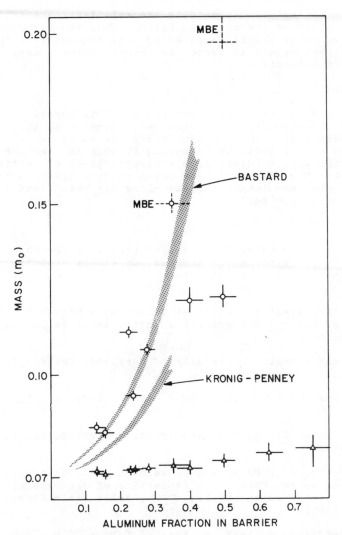

Fig. 1. Electrom mass in superlattices with 10nm period and 2nm barriers but with varying Al fraction in the barrier. Triangles denote the mass in the plane of the sample while circles are the mass in the growth direction. The shaded curves are the calculated mass using the Kronig-Penney potential with and without the prescription of Bastard.

We note that the linewidths are relatively insensitive
to orientation of the magnetic field. From the linewidth
we may deduce an elastic scattering time and conclude that
the electrons appear to tunnel coherently through many
superlattice barriers.

4. SUMMARY

We have measured cyclotron resonance in superlattices
with the magnetic field oriented perpendicular to the
growth direction. These measurements are shown to be
sensitive and specific to tunneling through the barrier and
as a result are an important microscopic probe of vertical
transport. The transport mass has been directly measured
and the coherence lenght for tunneling has been found to be
many superlattice periods.

ACKNOWLEDGEMENT

We wish to thank Larry Rubin and Bruce Brandt for their
support of the experiments carried out at the Francis
Bitter National Magnet Laboratory.

* Visiting guest scientist at the Francis Bitter National
Magnet Laboratory, Massachusetts Institute of Technology,
Cambridge Mass.

** Present Address: University of Maryland, College Park,
MD 20742.

1. L. Esaki and R. Tsu, IBM J. Res. Develop. $\underline{14}$, 61
(1971).

2. J.C. Maan, Springer Series in Solid State Sciences, $\underline{53}$
183 (1984).

3. G. Belle, J.C. Maan, and G. Weimann, VIth International
Conference on the Electronic Properties of Two-Dimensional
Systems, Kyoto, Japan (1985); —— —, Solid State Commun.
$\underline{56}$, 65 (1985).

4. T. Duffield, R. Bhat, M. Koza, F.DeRosa, D.M. Hwang, P.
Grabbe and S.J. Allen, Jr., Phys. Rev. Lett., $\underline{56}$, 2724
(1986).

5. G. Bastard, Phys. Rev., $\underline{B24}$, 5693 (1981)

ELECTRON SPIN RESONANCE AS A PROBE OF THE SILICON-METAL INTERFACE

D. C. Vier and S. Schultz

University of California, San Diego
Department of Physics, B-019
La Jolla, California 92093
USA

We introduce the use of electron spin resonance (ESR)
as a novel means of investigating the electron transport properties at the interface between phosphorus-
doped silicon (Si:P) and a metal, under intrinsic (i.e.,
zero bias) conditions. Our initial system is Si:P-Ni$_2$Si
formed by thermal annealing of a Ni layer evaporated
onto Si:P. From an analysis of the Si:P ESR linewidth
we are able to deduce the probability of an electron
crossing the Si:P-Ni$_2$Si interface.

Electron spin resonance (ESR) of donors in bulk silicon has been
extensively studied.[1-4] If the donor concentration is greater than
$3.74 \times 10^{18}/cm^3$, the silicon undergoes a transition from insulator to
metal,[5] with the donor electrons being mobile at all temperatures.
The ESR linewidth, ΔH, reflects the sum of all spin relaxation processes of the donor electrons. If the donor electrons are able to
tunnel from the silicon into a suitably prepared metal overlayer, the
additional spin relaxation channel will cause ΔH to increase. An
analysis of this increase in ΔH can provide a measure of the probability, p, of an electron crossing the potential barrier at a silicon-
metal interface. We have made computer simulation studies of ΔH as a
function of p for a phosphorus doped silicon (Si:P) layer in contact
with a suitable metal layer. The results indicate that p may be determined in the range from 10^{-4} to 1 by the ESR technique (see Fig.(1)).

Our initial experiments have focused on the determination of p
for electrons crossing the interface between Si:P and Ni$_2$Si. Thin
Si:P layers were made by diffusing phosphorus into <111> oriented,
n-type, polished Si wafers of 145 Ω-cm resistivity. Metal-Si:P interfaces were prepared by thermally evaporating Ni metal onto the surface
of the Si:P layers in a standard diffusion pumped vacuum system having

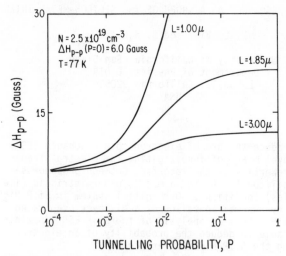

TUNNELLING PROBABILITY, P

Figure 1. Computer simulation of the peak-to-peak ESR linewidth, ΔH, in Si:P as a function of the probability, p, of an electron tunneling into a thick Ni_2Si metal layer. Results are shown for three different Si:P thicknesses, L.

a base pressure of $\sim 10^{-6}$ Torr. Cleaning of the silicon surfaces was accomplished by Ar ion bombardment at 5 kV, immediately prior to the metal deposition.

After deposition of the Ni, the samples were cooled to 77 K to perform ESR measurements. This was followed by an anneal at 270°C to promote Ni_2Si growth. The cooling/annealing process was repeated to give an ESR linewidth history as a function of annealing time.

The results of such experiments for three Si:P samples with 600 Å of Ni deposited are shown in Fig.(2). Sample A was doped to a nominal peak phosphorus concentration, n_p, of 5 x 10^{19}/cm^3 and a depth of about 2 µm. Sample B was identical to sample A except that a second more heavily doped layer was additionally diffused in at the surface, with n_p ~6 x 10^{20}/cm^3 and ~500 Å deep. The motivation for preparing sample B was to create a narrower potential barrier at the metal-Si:P interface for the electrons to tunnel through. Sample C was similar to A except that n_p was only 6 x 10^{18}/cm^3 (to give a wider metal-Si:P potential barrier than A).

Immediately after Ni deposition, the Si:P ESR linewidths were always found to be unchanged from the values before metal depositon,

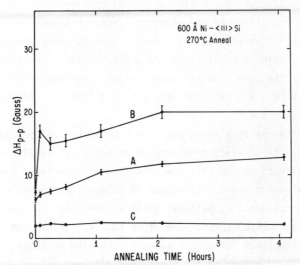

Figure 2. Peak-to-peak Si:P ESR linewidth, after depositing a 600 Å Ni layer on the surface, as a function of annealing time at 270°C. The data labeled "A", "B", or "C" correspond to Ni_2Si growth on Si:P samples A, B, or C (as described in the text). The lines connecting the data points are a guide to the eye.

which indicated that there was a small probability of electron transport across the "as evaporated" silicon-metal interface ($p < 10^{-4}$). This may be due to the poor vacuum conditions of our sample preparation method, as well as to the ion bombardment damaged silicon surface.

As can be seen from Fig.(2), for both samples A and B a brief anneal resulted in an increase of ΔH, which indicates an increase in the probability of an electron crossing the interface. Further annealing resulted in a continued slow increase of ΔH, which appeared to saturate. A likely explanation for the observed behavior is that, as the silicide layer grows deeper into the silicon, any contamination and/or defects which were present at the initial interface are left behind, improving the transport properties of the new interface. It is also possible that phosphorus snowplowing[6] takes place as the silicide grows, which would narrow the potential barrier through which the electrons must tunnel, and hence increase electron transport. We note that sample C demonstrated no measurable increase in linewidth over the entire annealing period, due to the low electron tunneling probability through the thicker potential barrier.

By a comparison of the observed linewidth broadening with numerical simulation studies, we can determine p, the probability of an electron crossing the interface (from the Si side). As expected, p is greater for the more heavily doped samples. For anneal times greater than 1 hour, $p \cong 5 \times 10^{-3}$ for sample A, 3×10^{-2} for sample B, and $<10^{-4}$ for sample C.

Now that we have shown that ESR studies can be used to determine the probability of electron transport across a metal-silicon interface, we plan to extend this new technique to interfaces prepared in ultra high vacuum and to the regime of submonolayer metal coverages.[7]

Work supported by National Science Foundation Condensed Matter Physics Grant DMR-83-12450. One of us (D.C.V.) thanks IBM for providing postdoctoral fellowship support. We wish to thank Pat Strehle for her help in preparing the phosphorus diffused wafers, Professor S. S. Lau and Charlie Hewett for RBS measurements, and Roy Troutt and the Burroughs Corporation for spreading resistance measurements. We also thank Professor Lu Sham for his helpful interest in our work.

REFERENCES

1. Pifer, J. H., Phys. Rev. B. 12, 4391 (1975).
2. Quirt, J. D. and Marko, J. R., Phys. Rev. B. 5, 1716 (1972), and Quirt, J. D. and Marko, J. R., Phys. Rev. B. 7, 3842 (1973).
3. Kodera, Hiroshi, J. Phys. Soc. Japan, 21, 1040 (1966).
4. Murakami, K., Masuda, K., Gamo, K., and Namba, S., Appl. Phys. Lett. 30, 300 (1977).
5. Rosenbaum, T. F., Milligan, R. F., Paalanen, M. A., Thomas, G. A. Bhatt, R. N., and Lin, W., Phys. Rev. B. 27, 7509 (1983).
6. Wittmer, M., and Tu, K. N., Phys. Rev. B. 29, 2010 (1984).
7. These proposed experiments are analogous to those reported for the determination of spin relaxation cross sections of Xe and Kr physisorbed on Li. Eigler, D. M. and Schultz, S., Phys. Rev. Lett. 54, 1185 (1985).

DIRECT AND INDIRECT GAP TUNNELLING IN A GRADED-PARAMETER
GaAs/AlAs SUPERLATTICE

N R Couch D G Parker M J Kelly and T M Kerr

GEC Research Ltd Long Range Research Laboratory
East Lane Wembley Middlesex HA9 7AP
UNITED KINGDOM

ABSTRACT

Evidence has been found of electron tunnelling through
multiple thin AlAs barriers via both direct and
indirect gap states. The strong resonant tunnelling
current is found to be controlled by the indirect
Γ(GaAs) - X(AlAs) band offset, while the fast
photoresponse exhibits a long time tail consistent
with the direct Γ(GaAs) - Γ(AlAs) band offset.
Initial magnetic field studies are reported.

1   INTRODUCTION

Modern crystal growth techniques have achieved atomic-scale control over
the composition of semiconductor multilayers.   One can now design structures
to investigate particular physical effects, and here we report initial studies
on the nature of electron states and transport in graded-parameter
superlattices.

2   EXPERIMENTAL PROCEDURES

The structure is a GaAs p-i-n diode where the i region is an asymmetric
GaAs/AlAs superlattice.   The well and barrier thicknesses vary so that the
average composition and its first moment are equivalent to that of a linearly
graded composition from 0 to 30% over the same distance.

The structure was grown by molecular beam epitaxy on an n+ (100) GaAs
substrate.   A 0.5 μm thick buffer layer of GaAs n-doped to 10^{18} cm^{-3} was grown
first.   The sequence of growth was then a further 0.5 μm of 10^{16} cm^{-3} Si-doped
GaAs, an undoped GaAs/AlAs superlattice, 0.5 μm of 10^{16} cm^{-3} Be-doped GaAs and
finally a top contact layer of 0.5 μm of 10^{18} cm^{-3} Be-doped GaAs.   The
GaAs/AlAs superlattice consists of five 10 nm periods, where the

AlAs barrier thickness starts at 3 nm and reduces by 0.6 nm at each period. The structure is shown schematically in Figure 1 (inset). The thicknesses of the layers in the superlattice have been verified by transmission electron microscopy (TEM)[1].

The DC I-V characteristics have been measured as a function of temperature, and at 4.2 K as a function of magnetic field. The response of the device to a short optical pulse has also been measured at room temperature using a mode-locked synchronously-pumped dye laser operating at a wavelength of 740 nm. The pulse length was 400 fs and the repetition rate was 84 MHz. The response of the device was recorded on a Tektronic sampling oscilloscope using an S4 sampling head with a ~25 ps response time. A large area device (180 µm square) was used and biased to -3V, the point of maximum sensitivity.

3   RESULTS

The DC I-V characteristics show excellent diode behaviour as a p-n junction (e.g. a reverse bias current density of $<6 \times 10^{-8}$ A/cm^2 at 3 V and 300 K), but with superposed a resonant tunnelling feature exhibiting strong negative differential resistance (n.d.r.) at low temperatures. (see Figure 1). Near the flat-band condition of the diode, the bound states in the progressively narrower wells are aligned with the Fermi energy in the n region to produce the resonant tunnelling. Our simulations provide quantitative agreement for both the position and magnitude of the maximum current density only if the barrier height is taken to be the difference between the Γ band minimum in GaAs and the X band minimum in AlAs, i.e. ~0.32 eV (see Figure 1). The current density, in a simple Kronig-Penney model is too small by three orders of magnitude if the Γ band in AlAs is invoked (a barrier of 1.2 eV). The hole current density is also too small to explain the data. The small amount of hystersis observed at turn-on is attributed to trapped charge.

The temperature (T) and magnetic field (B) dependences of the resonant tunnelling n.d.r. have been measured. The principle effect of increasing T or B ($\perp$I) is a linear decrease in the peak-to valley ratio (A/B), as shown in Figure 2 where we have plotted this dependence on an energy scale $-k_B T$ for temperature and $eB\hbar/m^*$ for magnetic field. The minimum current after the resonant tunnelling peak varies by as much as 30% between mesas, for reasons discussed below. For B (//I) the voltage range over which n.d.r. may be seen broadens, but the peak-to-valley ratio is unaffected. We also note that the decrease in band gap of GaAs with temperature is reflected in the reduction of the voltage required to create the n.d.r.

Under illumination our structure produces a photovoltage and an associated photocurrent[2]. The response of large area mesas to short, high intensity light pulses has been measured. Two distinct exponential decay rates may be seen. The rise time of 80 ps and the initial decay constant of 225 ps are essentially due to the RC time constant of the comparitively large

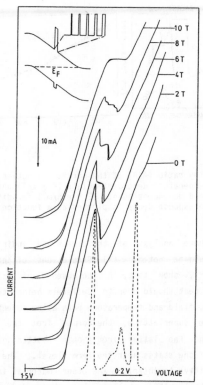

Figure 1:   I vs V at 4.2 K for magnetic fields perpendicular to I. The
broken line is the theoretical calculation. The inset (top) shows a
schematic diagram of the conduction and valence bands of the
structure. The dotted line is a simulation of the tunnel current in
zero magnetic field.

mesa[3].   A slower decay constant of 487 ps is observed at longer times.   The
difference in the decay times (487-225=262 ps) is taken to be the additional
time taken to tunnel through the thickest barrier.   (Shorter tunnelling times
should exist for the thinner barriers, but only the slowest tail will be
resolveable).   From Luryi's[4] consideration of the RC time constant of the
barrier, we obtain the tunnelling time as:

$$\tau = \varepsilon \, \alpha^{-1} \quad (\lambda/c) \, \exp (4\pi d/\lambda) \qquad (1)$$

where ε is the dielectric constant, α the fine structure constant, c the speed
of light, d the effective barrier thickness and λ the de Broglie wavelength.
From λ, the effective barrier height may be simply calculated. We obtain a
value of 1.06 V.   This is comparable to the Γ to Γ band gap offset between
GaAs and AlAs.   The results from tunnelling time calculations are very model
dependent, however this simple, well documented model gives a good explanation
of the observed tail.

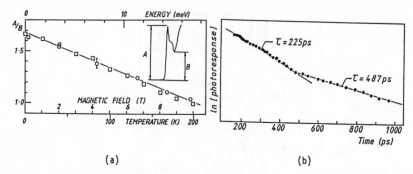

(a)   (b)

Figure 2: a) Peak to valley ratio (A/B) of the ndr as a function of energy, E.
∇ are the temperature dependent points (E = k_BT) and □ are the
magnetic field dependent points (E = eB$\hbar$/m*) for different mesas
b) Log of photoresponse decay is plotted as a function of time.

4 DISCUSSION

In addition to the above analysis of the direct and indirect tunnelling
two further points should be noted: (i) Simulations of this, and other
superlatttice structures[5], show that, in the absense of scattering, the
current after the n.d.r. peak should drop to low levels before the turn on of
the diode. The magnetic field and temperature both allow momentum conversion
from $k_\perp$ to $k_{//}$ in the superlattice, the former from the semiclassical
equations of motion, and the latter from phonon scattering and thermal
smearing of occupation of the states near the Fermi level. The $k_{//}$ component
of current is more sensitive to interface and other material inhomogeneities
in the superlattice. While this explains the observed decrease in the
peak-to-valley ratio, the coincidence of scaling with the relevant energies
k_BT and $\hbar\omega_c$ is not understood. (ii) The novel physics associated with this
structure is quite different from that anticipated for linearly graded-gap
structure that it was originally designed to simulate.

5 ACKNOWLEDGEMENTS

This work was supported in part by Directorate of Components, Valves and
Devices of the Procurement Executive of the UK Ministry of Defence and by
ESPRIT under programme number 514.

REFERENCES

1 Couch N R, Kelly M J, Kerr T M, Britton E and Stobbs W M, submitted for
   publication.
2 Couch N R, Parker D G, Kelly M J and Kerr T M, Elec. Lett. 22, 637 (1986).
3 Parker D G, Couch N R, Kelly M J and Kerr T M, To be published in Appl.
   Phys. Lett.
4 Luryi, S. Appl. Phys. Lett. 45, 490 (1985).
5 Davies R A, Kelly M J and Kerr T M Elec. Lett. 22, 131 (1986).

CaF_2/Si(111): ELECTRONIC STRUCTURE OF THE INTERFACE

U.O. Karlsson, D.Rieger, J.A. Yarmoff,
F.R. McFeely, J.F. Morar, and F.J. Himpsel.

IBM T.J. Watson Research Center
P.O. Box 218
Yorktown Heights, NY 10598

ABSTRACT

The core level spectra for epitaxial CaF_2 on Si(111) show that CaF_2 grows in a layer-by-layer mode. The electronic structure of the interface is mainly characterized by a charge transfer to Ca resulting in a Ca^{1+} at the interface. This is inferred from the interface core level shifts and from the Ca 2p NEXAFS spectra. The charge transfer induces a shift of the Fermi level to a new pinning position at the Si

A number of recent studies have shown that CaF_2 can be grown by molecular beam epitaxy (MBE) onto Si(111).[1,2] CaF_2 is a simple insulator, which has a cubic crystal structure similar to the diamond structure of Si and the lattice mismatch is only 0.6 % at room temperature. Another of CaF_2's more appealing features is that it sublimes as a molecule thereby solving the stoichiometry problem for films grown by MBE. For all MBE work it is of great importance to control and understand the electronic states at the interface. This is especially true when the device dimensions shrink and become comparable to characteristic lengths such as diffusion length or the Debye screening length. The interaction between the ionic CaF_2 and the covalent Si is an intriguing question which is the subject of this work.

The experiments were performed at the National Synchrotron Light Source at Brookhaven using a 6/10 m toroidal grating monochromator coupled to a display electron spectrometer. CaF_2 films were grown by MBE onto clean Si(111) substrates. Cleanliness and surface order of the Si(111) surfaces were monitored by photoelectron spectroscopy of surface states and surface core levels. The thickness of the CaF_2 films was determined from the intensities of the Si and CaF_2 core levels[2].

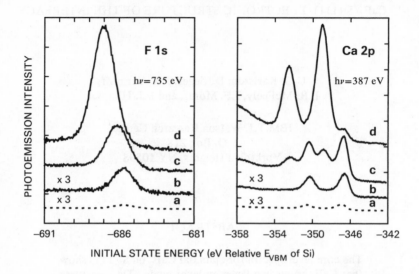

Fig 1. Photoemission spectra from the F 1s and Ca $3p_{3/2,1/2}$ core levels. Spectra a and b are for submonolayer coverages of CaF_2 on Si(111), spectrum c for 1.5 layers and spectrum d for 4 layers.

Figure 1 shows the Ca 2p and F 1s core levels for different CaF_2 coverages of the Si(111) substrate. For a CaF_2 coverage of less than one layer (3.15 Å) the Ca 2p core level shows a single spin-orbit split pair. When the coverage is increased a new set of Ca 2p peaks emerges on the higher binding energy side of the existing peaks. The photoemission intensity from the new peaks increases with coverage, since they are associated with photoemission from bulk CaF_2. The F 1s level exhibits only one peak for all CaF_2 coverages, however, it successively shifts to higher binding energy for increasing CaF_2 coverages between one and two layers. This shows that more than two CaF_2 layers are needed to establish a true interface. The intensity ratio between the Ca 2p and the F 1s emission is found to be approximately the same for all coverages from submonolayer up to more than ten layers. This strongly indicates that the Ca to F composition does not change and that the adsorption is mainly molecular. The behavior of the Ca 2p and the F 1s core levels as function of CaF_2 coverage is similar to the behavior reported earlier for the Ca 3p, Ca 3s, and F 2s respectively[2].

Ca 2p Absorption Edge   Band Line Up

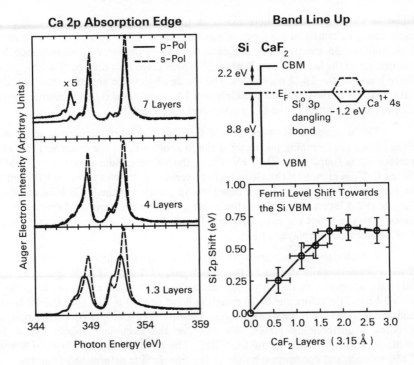

Fig. 2. The absorption edge fine structure of Ca2p measured via the yield of Ca Auger ectrons for different thicknesses of CaF$_2$ E_{KIN}=285 eV.

Fig. 3 Band line-up diagram of the CaF$_2$/Si(111) interface and proposed bonding model between Ca 4s and the Si dangling bonds.

The technique of polarization dependent near edge x-ray absorption fine structure (NEXAFS) extracts information about the bonding geometries and oxidation states of the atoms at the interface[2]. To distinguish the interface from the bulk contributions one can take advantage of the fact that the interface atoms exhibit shifted core levels as well as shifed Auger electron energies. In Fig. 2 the absorption spectra from 1.3, 4, and 7 layers of CaF$_2$ are shown for s- and p- polarized light. Two very prominent transitions from the $2p_{1/2}$ and $2p_{3/2}$ states into empty 3d states are observed. A number of extra lines are also observed. The intensities of these smaller peaks are attenuated with increasing film thickness suggesting an interpretation as interface transitions. This is also supported by the pronounced polarization dependence. The oxidation state of Ca can be deduced from the absorption spectra. Neutral Ca gives rise to a triplet[3] with two strong lines and a 50 times weaker line. For Ca^{2+} in CaF$_2$ we expect a similar mulitplet structure since it differs from neutral Ca only by a filled 4s shell. For Ca^{1+} there exists an unpaired 4s electron which creates a more complex multiplet via Coulomb and exchange interaction with the 2p core hole and the 3d electron. An estimate for this multiplet can be obtained by comparing to the isoelectronic K atom,[3] which gives four lines for the $2p_{3/2}$ and three lines for the $2p_{1/2}$ core holes. The Ca 2p absorption spectra strongly indicate, that Ca at the interface is mainly in a 1^+ oxidation state, in agreement with earlier interpretation of the Si 2p core level spectrum[2].

The interaction between Si and CaF_2 at the interface results in a new Fermi level (E_F) pinning position for the Si substrate. At the clean Si(111) 7x7 surface E_F is found to be 0.63 eV above the valence band maximum (VBM). When CaF_2 is grown on the clean Si(111) surface E_F moves towards VBM with increasing CaF_2 coverage until a new pinning position at the VBM is established for about 2.5 layers. This is also the coverage where the core level spectra indicates that the interface is completed. The Fermi level pinning as function of CaF_2 coverage is displayed in Fig. 3.

The electronic structure at the interface can be deduced from these data. This is schematically drawn in the upper part of Fig. 3. The Si atoms at the interface are nearly neutral if we juge from the relatively small Si core level shift of +0.4 eV. Therefore the Si 3p dangling bond orbital is still mainly half filled and can interact with the 4s electron of the interface Ca^{1+} by forming a pair of bonding - antibonding states. We observe the occupied bonding state in photoemission at 1.2 eV below the VBM. The band line-up between the Si and CaF_2 valence and conduction bands at the interface is determined from the position of the F2p valence band of the CaF_2. The upper edge of the F2p band is found $\simeq$ 8.8 eV below the Si VBM. Since the band gaps of CaF_2 and Si are 12.1 eV and 1.1 eV respectively, we find the conduction band minimum of CaF_2 to be $\simeq$ 2.2 eV above the conduction band minimum of Si.

REFERENCES

1. R.F.C. Farrow, P.W. Sullivan, G.M. Williams, G.R. Jones, and C. Cameron, J. Vac. Sci. Technol. **19**, 415 (1981); L.J. Schowalter, R.W. Fathauer, J. Vac. Sci. Technol. A4, 1026 (1986) .

2. F.J. Himpsel, F.U. Hillebrecht, G. Hughes, J.L. Jordan, U.O. Karlsson, F.R. McFeely, J.F. Morar, and D. Rieger, Appl. Phys. Lett. **48**, 596 (1984); F.J. Himpsel, U.O. Karlsson, J.F. Morar, D. Rieger, and J.A. Yarmoff, Phys. Rev. Lett. **56**, 1497 (1986).

3. M.W.D. Mansfield, Proc. Roy. Soc. London, Ser. A **348**, 143 (1976), and **346**, 555 (1975).

THE INTERFACE BETWEEN A COVALENT SEMICONDUCTOR
AND AN IONIC INSULATOR: CaF$_2$ ON Si(111)

M. A. Olmstead[†], R. I. G. Uhrberg[‡], R. D. Bringans, and R. Z. Bachrach

Xerox Palo Alto Research Center
3333 Coyote Hill Road, Palo Alto, CA 94304
USA

ABSTRACT

Initial stages of interface formation between calcium fluoride and silicon have been probed using photoemission spectroscopy. The Si(111)/CaF$_2$ interface contains Si – Ca bonds, and may be depleted of fluorine due to the dissociation of the CaF$_2$ molecule.

1. INTRODUCTION

The epitaxial interface between CaF$_2$ and Si(111) is a prototype for the study of interface formation between a polar insulator and a non-polar semiconductor. An important question is the nature of the bonding at the interface: whether it is of ionic or covalent character, or intermediate between these extremes. It is also important to address whether the silicon atoms bond to Ca or to F atoms, and whether the stoichiometry is modified at the interface.

The simplest model for the Si(111)/CaF$_2$ interface which contains Si – Ca bonds involves placing the linear CaF$_2$ molecules such that the Ca atoms are directly above the Si dangling bonds of a bulk-terminated lattice. The "type B" epitaxy (CaF$_2$ lattice axes are rotated by 180° from the Si substrate) places one F atom in each "hollow" site above the fourth layer Si atoms, and the other F atom about 4 Å above the second layer Si atoms. In this structure there are an odd number of electrons per interface unit cell, so it would be expected to be metallic. However, while there are some interface states in as-grown layers, there are fewer than the above model would suggest, and a rapid thermal anneal (1100°C, 20 sec) reduces the interface state density to less than 10^{-3}/interface Si.[1] This implies the absence of a partially-filled band, so that the interface either is reconstructed parallel to the interface or is missing one F atom per unit cell. There is no evidence for a parallel reconstruction in electron diffraction or microscopy measurements,[2,3] which suggests F atoms may be missing at the interface.

We have used core-level photoemission spectroscopy to study the bonding of thin films of CaF_2 grown on Si(111) substrates by molecular beam epitaxy. The core-level binding energies of the atoms at the Si(111)/CaF_2 interface are compared to those for submonolayer coverage of Ca and F at the surface. The results indicate that the primary bonding at the interface occurs between Si and Ca, and that the CaF_2 molecule may dissociate at the interface, leading to a depletion of F atoms.

2. RESULTS AND DISCUSSION

The photo-induced electron emission from the Ca 3p, F 2s, F 2p and Si valence band states is shown in Fig. 1. Spectrum 1(a) is characteristic of bulk CaF_2, and shows the F 2p states broadened into the CaF_2 valence band. The binding energy of the Ca 3p states corresponds to Ca atoms surrounded by eight F atoms, and would decrease if F atoms are missing or replaced by Si atoms at the interface. This is seen to be the case for a thin film [see Fig. 1(b)], where a Ca 3p component due to interface Ca atoms is shifted by 2.3 eV to lower binding energy. The F 2p states are still broadened into the CaF_2 valence band, even though the film is only ~3 molecular layers thick. The offset between the Si and CaF_2 valence bands is ~8 eV. This is ~1 eV larger than the offset for films grown at lower temperatures which have more disorder at the interface.[4]

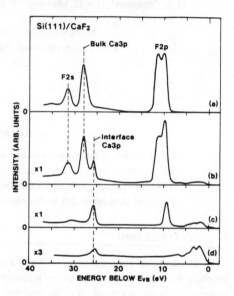

Figure 1. Shallow core levels for (a) >50 Å film of CaF_2 on Si(111) (hν = 130 eV, T_{dep} = 500°C); (b) 11 Å film (hν = 135 eV, T_{dep} = 700°C); (c) submonolayer film (5x1) obtained upon annealing (b) (hν = 135 eV).; (d) Si(111):Ca 3x1 surface obtained upon annealing (b) (hν = 135 eV).

Annealing these thin films of CaF_2 at 750–800°C leads to a re-evaporation of the film. Below a coverage of one monolayer, a series of reconstructions is observed, 1x1→2x1→5x1→3x1→7x7, with progressively smaller F : Ca ratios. Spectrum 1(c) is taken from a 5x1 surface, and the bottom spectrum from a 3x1 surface. The 3x1 structure can be seen to contain no fluorine from the absence of F 2p or F 2s emission. The ratio of the F 2p to Ca 3p emission in the 5x1 structure [1(c)] is about half that for the thin film [1(b)], indicating approximately equal numbers of Ca and F atoms. The F 2p state is a narrow, atomic-like level, as opposed to the broad band seen in the upper two spectra.

The binding energy of the Ca 3p on the 5x1 surface (relative to the Si valence band maximum) is the same as that at the Si(111)/CaF_2 interface. The Ca 3p state for the Si(111):Ca 3x1 surface is also close to this energy, with a slightly lower binding energy. The 5x1 structure likely involves bonding of the form Si – Ca – F, among others, while the Si(111):Ca surface contains only Si – Ca or Si – Ca – Si bonds. The similarity in binding energy between the interface Ca and that on the 5x1 surface suggests that the bonding environment is similar in both cases. If the CaF_2 molecule is intact at the interface, then the model described above would change the Ca environment from bulk CaF_2 only by interchanging one of eight fluorine atoms with a Si atom. It is unlikely that this would lead to a 2.3 eV shift in the Ca 3p binding energy. If each F atom in the layer between the Ca and Si atoms is removed, however, the interface Ca are in an equivalent environment to Si – Ca – F (four F neighbors above the Ca, each pulling ¼ electron). The Ca 3p binding energy on the Si(111):Ca surface is much closer to that of the interface Ca than is the bulk Ca 3p binding energy. This shows the Si – Ca bond to be largely covalent at the interface.

Further information can be obtained from the Si 2p core level, shown in Fig. 2. In addition to the bulk Si contribution, there are components shifted to both higher and lower binding energy. The primary interface component (29% of total intensity) is shifted by 0.36 eV to lower binding energy, the direction expected for Si – Ca bonding. There are also two smaller components, shifted by ~0.45 eV (6%) and ~0.85 eV (~3%) to higher binding energy. The larger value might be expected for Si – F bonding at the interface, with the F atoms directly above the Si dangling bonds, as it is about 80% of the shift found for F atoms alone on Si(111).[5] An interface with this type of alignment is seen over a small fraction of the interface with transmission electron microscopy, while most of the interface has the lattice aligned so that Ca atoms are directly above the Si.[3] The 0.45 eV shift to higher binding energy is close to that calculated for F atoms in the hollow site above the fourth layer.[6]

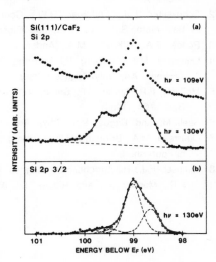

Figure 2. Si 2p core level for the same film as for Fig. 1(b). a) Top: Bulk sensitive spectrum. Bottom: Surface sensitive spectrum. Solid line is a least squares fit which is deconvolved in (c). b) Spin-orbit deconvolution of surface sensitive spectrum using a spin-orbit splitting of 0.600 eV and the statistical ratio of 2:1. The dashed curves are the components (constant width) of a least squares fit to the data.

Calibrating from the interface-to-bulk ratio for the Si(111):As 1x1 surface with the same photon energy and sample geometry,[7] we expect a single interface layer to comprise 33% of the total emission. An explanation consistent with the above results is that at least 90% of the interface consists of Ca atoms directly above Si dangling bonds, and that most of the fluorine layer between the Ca and Si layers is removed. A rapid thermal anneal might be expected to remove the rest of these F atoms. The F atoms which remain may sit in the hollow site or between the first and second silicon layers. Removal from the interface may also lead to interstitial F in the Si bulk. The Si 2p components shifted to higher binding energy could be due to any (or all) of these geometries. Another possible source for the component shifted to higher binding energy is a small portion of the interface (<10%) with direct Si – F bonds. Varying amounts of either this second type of interface or the amount of F remaining near the interface may account for the differences between this work and that recently reported by Himpsel et al.[8]

Acknowledgements: We are grateful to L.-E. Swartz for the assistance and to F. Ponce for useful discussions and the for communication of unpublished material. Part of this work was performed at the Stanford Synchrotron Radiation Laboratory which is supported by the U.S. Department of Energy, Office of Basic Energy Sciences.

3.   REFERENCES

† Present address: Dept. of Physics, Univ. of California, Berkeley, CA 94720 USA.
‡ Present address: Dept. of Physics Measurement Technology, Linköping Institute of Technology, S-581 83 Linköping Sweden.
1. Phillips, J.M., Materials Research Society Symposium Proceedings 71, to be published.
2. Schowalter, L.J., Fathauer, R.W., Goehner, R.P., Turner, L.G., DeBlois, R.W., Hashimoto, S., Peng, J.-L., and Gibson, W.M., J. Appl. Phys. 58, 302 (1985).
3. Ponce, F.A., O'Keefe, M.A., Anderson, G.B., and Schowalter, L.J., unpublished.
4. Olmstead, M.A., Uhrberg, R.I.G., Bringans, R.D., and Bachrach, R.Z., J. Vac. Sci. Technol. A4, July/August (1986).
5. McFeely, F.R., Morar, J.F., Shinn, N.D., Landgren, G., and Himpsel, F.J., Phys. Rev. B30, 764 (1984).
6. Seel, M. and Bagus, P.S., Phys. Rev. B28, 2023 (1983).
7. Olmstead, M.A., Bringans, R.D., Uhrberg, R.I.G., Bachrach, R.Z., Phys. Rev. B34, to be published.
8. Himpsel, F.J., Hillebrecht, F.U., Hughes, G., Jordan, J.L., Karlsson, U.O., McFeely, F.R., Morar, J.F., and Rieger, D., Appl. Phys. Lett. 48, 596 (1986).

ROUGHNESS AND ELECTRICAL CONDUCTION OF SiO$_2$/POLYCRYSTALLINE SILICON INTERFACES

G. Harbeke and A.E. Widmer

Laboratories RCA Ltd., Badenerstrasse 569
8048 Zurich, Switzerland

and

L. Faraone

RCA Laboratories, Princeton, NJ 08540, USA

ABSTRACT

The oxidation-induced roughening of amorphously deposited polycrystalline silicon films is quantitatively related to the effective barrier height for Fowler-Nordheim tunneling through the oxide.

1. INTRODUCTION

Previous studies have shown that thermally-grown SiO$_2$ films on amorphous-deposited polysilicon[1, 2] exhibit a higher dielectric breakdown strength and reduced Fowler-Nordheim leakage current when compared to SiO$_2$ formed from polycrystalline deposited silicon[3]. The purpose of the present paper is to correlate the optically determined interface roughness and the measured electrical conduction properties of polysilicon/SiO$_2$/polysilicon structures.

2. EXPERIMENTAL TECHNIQUES

All test structures (see inset of Fig. 1) were fabricated with 0.5 μm thick silicon films that were deposited in the amorphous phase by Low Pressure Chemical Vapor Deposition at 560°C onto thermally oxidized bulk silicon wafers[1, 2]. The films were n$^+$-doped by POCl$_3$ diffusion at 950°C for 15 minutes, and all thermal oxidations of these structures were carried out at 800°C in an ambient of 10% pyrogenic steam/90% dry O$_2$.

The oxide conduction properties were measured from ramped I-V curves generated with a constant field ramp-rate of 1 MVcm^{-1} sec^{-1}. For the optical measurements first the upper polysilicon layer was

removed by wet-etching. A silver film of 70 to 100 nm thickness was
then evaporated onto the exposed oxide. The reflectance of the sil-
vered surface at the wavelength of surface plasmon excitation, i.e. at
λ = 350 nm, is a direct measure of the r.m.s. surface roughness, σ [4].
After measuring the upper interface the interlevel polyoxide was dis-
solved in buffered HF acid and the roughness of the lower polysilicon
surface was determined by the same technique.

3.   RESULTS AND DISCUSSION

The high-field conduction properties of our SiO_2 films could be
described by the Fowler-Nordheim equation

$$J = A \left(E^2 / \phi_{B_{eff}} \right) \exp \left(B \cdot \phi_{B_{eff}}^{2/3} / E \right) \tag{1}$$

where A and B are constants, E is the electric field at the injecting
electrode, and $\phi_{B_{eff}}$ is the effective barrier height at the injecting
electrode.

Figure 1 shows the measured vertical surface roughness of both
interfaces as a function of oxide thickness, d_{ox}. The increase in
roughness with d_{ox} suggests that if the original surface is comparatively smooth (as in the present case), then the process of thermally
oxidizing phosphorus-doped polysilicon contributes considerably to the
final interface roughness. The upper interface was consistently found
to be between 0.1 nm and 0.4 nm smoother than the lower one. The
above observation is in agreement with the results that the oxide is
far more "conductive" across the lower oxide/polyoxide interface due
to higher roughness-induced local electric fields at this interface.

Figure 2 shows the effective barrier height for Fowler-Nordheim
tunneling, $\phi_{B_{eff}}$, as a function of measured roughness of the injecting
interface. It can be seen that with increasing σ there is a rapid
decrease in $\phi_{B_{eff}}$ and that this relation is essentially independent of
the voltage polarity used for electrical measurements, i.e. $\phi_{B_{eff}}$
appears to be uniquely determined by the interface roughness regardless of the fact that for any given sample-type the I-V curves were
highly asymmetric.

In contrast to oxides on polycrystalline-deposited silicon, where

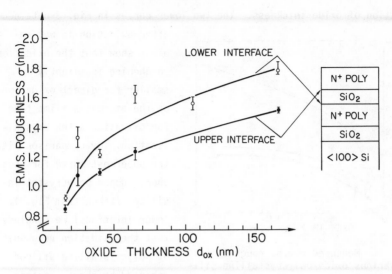

Fig. 1. Measured r.m.s. roughness of both oxide/polysilicon interfaces for thermally grown oxides.

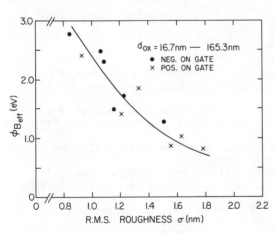

Fig. 2. Effective barrier height for Fowler-Nordheim tunneling under both positive and negative gate-bias polarities vs. interface roughness.

most of the interface texture is derived from the initial roughness of the polysilicon surface prior to oxidation[5] we have started with an extremely smooth polysilicon surface. Thus, the observed roughness is predominantly caused by differential oxidation of grains with varying surface-orientations as well as preferential oxidation along grain boundaries[6, 7].

We also determined quantitatively the oxidation-induced roughness of SiO_2/<u>undoped</u> amorphous (polycrystalline) silicon interfaces as a

262   G. Harbeke et al.

function of oxide thickness.   The two lower curves in Fig. 3 obtained

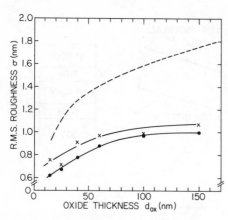

after oxidation in wet ambient show that the oxidation-
roughening is slightly
smaller for direct oxidation
of the amorphous films than
for oxidation after crystallization.   By comparison with
the uppermost curve for phosphorus doped polycrystalline
silicon (taken from Fig. 1,
lower interface) we conclude
that the oxidation roughening
of polycrystalline silicon is
enhanced by about a factor
two due to the presence of
phosphorus in the present concentration range of 10^{20} to
10^{21} cm^{-3}.

Fig. 3.   Measured r.m.s. roughness
of various oxide/polycrystalline silicon interfaces.   ●: undoped amorphous,
x: undoped polycrystalline, dashed
line: phosphorus doped polycrystalline.

4.   REFERENCES

1)   Harbeke, G., Krausbauer, L., Steigmeier, E.F., Widmer, A.E.,
     Kappert, H.F., and Neugebauer, G., Appl. Phys. Lett. 42, 249
     (1983).
2)   Harbeke, G., Krausbauer, L., Steigmeier, E.F., Widmer, A.E.,
     Kappert, H.F., and Neugebauer, G., J. Electrochem. Soc. 131,
     675 (1984).
3)   Faraone, L., Vibronek, R.D., and McGinn, J.T., IEEE Trans.
     Electron Dev. ED-32, 577 (1985).
4)   Harbeke, G., in "Polycrystalline Semiconductors - Physical
     Properties and Applications", G. Harbeke, ed., p. 156 (Springer,
     1985).
5)   Marcus, R.B., Sheng, T.T., and Lin, P., J. Electrochem. Soc.
     129, 1282 (1982).
6)   Irene, E.A., Tierney, E., and Wong, D.W., J. Electrochem. Soc.
     127, 705 (1980).
7)   Saraswat, K.C., and Singh, H., J. Electrochem. Soc. 129, 2321
     (1982).

LOW TEMPERATURE OXIDATION OF CRYSTALLINE SILICON

G. Lucovsky, M.J. Manitini, J.K. Srivastava* and E.A. Irene*

Department of Physics, North Carolina State University
Raleigh, North Carolina 27695-8202, USA
*Department of Chemistry, University of North Carolina
Chapel Hill, North Carolina 27514, USA

ABSTRACT

For SiO_2 films grown at temperatures (T_G) between 550 and 1000°C, the index of refraction (632.8 nm) increases and the frequency of the ir-active bond-stretching vibration decreases as T_G is decreased below 1000°C. This is explained in terms of a model in which the low temperature oxides become increasingly densified as T_G decreases.

1. INTRODUCTION

The trend in dielectric thin film research for Si VLSI applications is towards thinner films processed at low temperatures. For many applications it would be desirable to use procesing temperatures < 800°C and maintain the oxide and interface quality characteristic of higher processing temperatures. However, studies have shown that film and interface quality are degraded as processing temperatures are reduced below 1000°C[1,2]. In addition, the density[3] and intrinsic stress[4] increase substantially in films grown with T_G below 1000°C. We report on systematic changes in the refreactive index, n, and the frequency of the ir stretching vibration, v, in low temperature oxides. These are interpreted in terms of changes in the local atomic structure.

2. EXPERIMENTAL PROCEDURES AND RESULTS

Oxidation of ir-transmitting Si substrates (50-100 ohm-cm, n-type) was accomplished by conventional proceedures. Oxides were grown to approximately half an ellipsometric period (about 1000 A) to obtain reliable values for n and the sample thickness. Ir absorption was studied with emphasis on the dominant ir bond-stretching vibration (at about 1075 cm^{-1})[5].

Fig. 1 gives the temperature dependence of n and v for films grown at one atm. of dry oxygen. For the higher oxidation pressures (to about 300 atm.), n and v do not vary systematically with either T_G or pressure. However, we find correlations between n and v, shown in Fig. 2, for all of the films independent of T_G or the oxygen pressure.

3. CHEMICAL BONDING IN THE OXIDE FILMS

The results shown in Fig. 2 are interpreted in terms of a chemical bonding model. We consider two mechanisms that can produce changes in v and n. These are: (I) as T_G is reduced the films deviate from SiO_2 stoichiometry and the molar volume decreases; and (II) the films remain stoichiometric, but the density (p) increases with decreasing T_G. The increases in n, based on a Lorenz-Lorentz (L-L) model, scale with p and do not discriminate between the two mechanisms. Initially we felt that only (I) could explain the decrease in v. This was based on ir studies of suboxides, SiO_x 2<x, where v decreased with decreasing x[6]. However, the same studies showed that the ir line-width also increased signficantly as x decreased. We observed an essentially constant line-width and concluded that our films were not suboxides, ruling out (I). A problem with (II) is that decreases in v with increasing p are at odds with experience regarding stretching vibrations in densified or compressed three dimensional network structures[7]. However, there is experimental evidence that this does not apply for bulk silica where v for the stretching vibration has been reported to decrease with increased densification[8-10].

If 0 is the Si-0-Si bond angle, then v scales as $[(sin(0/2))^{1/2}]$[5]. Simon[8] has shown that in densified silica the Si-0 bond length, r_o is unchanged, but the Si-Si distance, d_{Si-Si}, is decreased. d_{Si-Si} is related to 0 by $d_{Si-Si} = 2r_o sin(0/2)$ and the density, p, is expected to scale as d_{Si-Si}^{-3}. Since n scales inversely as the molar volume, we can then expect a relationship between n and v. This turns out to be linear and decribes the data in Fig. 2. A second result of this model is that the density increase in the low temperature oxides is about 4 % as T_G is decreased from 1000 to 700°C.

4. COMPARISONS WITH BULK SILICA

We compare our results with densification studies on bulk silica and discuss two quantities, $dn/(dp/p)$ and $dv/(dp/p)$ There are three types of densified SiO_2 to consider: (i) thin films grown at low T_G; and bulk silica either (ii) densified below 800°C by static pressure, shock waves or radiation; or (iii) densified above 1000°C by thermal soaking. The value of $dn/(dp/p)$ is about the same for three densification mechanisms implying the L-L relationship holds independent of the densification process. On the other hand the values of $dv/(dp/p)$ are very different in films densified at low temperatures (i) and (ii), and high temperatures (iii). For (i) and (ii), $dv/(dp/p) = - 350$ cm^{-1}, whereas for (iii) it is more than 15 times larger.

We distinguish two regimes of densification; a low temperature regime extending to about 1000°C and a high temperature regime above 1000°C. In the low temperature regime increases in n and decreases in v can be understood in terms of the same atomic model, i.e., through changes in the Si–O–Si bond angle, O. Since $dv/(dp/p)$ is different in the high temperature densified glasses a different microscopic model must apply. Studies of the shear modulus, G, have shown that G increases between room temperature and about 900°C and then drops abruptly in the neighborhood of 1000°C[11]. The decrease in G has been interpreted in terms of bond rupture and a restructuring of the ring structure of the network above 1000°C.

It has been reported that the stress in low temperature oxides has two components, one from the mismatch of thermal expansion coefficients at the interface and the second an intrinsic stress that increases between 1000 and 700°C[4]. The densification we estimate from the model, 4 % for films grown at 700°C, is greater by more than a factor of two than the densification calculated from the total film stress. It is clear that additional studies of the local atomic structure by X-ray, EXAFS, etc are necessary before one can understand the differences between the local atomic structure in the densified low T_G oxides and bulk silica densified by high temperature annealing.

266   G. Lucovsky et al.

ACKNOWLEDGEMENTS

    This reserach has been supported by ONR contract N00014-C-79-0133
at NCSU and N00014-C-83-0571 at UNC-CH.

REFERENCES

1] Irene, E.A Semiconductor Int'l., June 1985, p.92
2] Deal, B.E., Sklar, M., Grove, A.S. and Snow, E.H., J. Electrochem
Soc. 114, 266 (1967).
3] Irene, E.A., Dong, D.W. and Zeto, R.J., J. Electrochem Soc. 127, 396
(1980).
4] Kobeda, E. and Irene, E.A., J. Vac. Sci. Tech. B4, 722 (1986).
5] Galeener, F.L. and Sen, P.N., Phys. Rev. B17, 1928 (1978).
6] Pai, P.G., Chao, S.S., Takagi, Y. and Lucovsky, G., J. Vac. Sci.
Tech. A4, 689 (1986).
7] Zallen, R. and Slade, M.L., Phys. Rev. B18, 5775 (1978).
8] Simon, I., in Modern Aspects of the Vitreous State, ed by J.D.
MacKenzie (Butterworths, London, 1960), p. 120.
9] Gaskell, P.H. and Grove, F.J., in Proc. 7th Int'l. Conf. (Int'L.
Comm. of Glass, Brussels, 1965), paper 363.
10] Geissberger, A.E. and Galeener, F.L., Phys. Rev. B28, 3266 (1983).
11] Sosman R.B., Properties of Silica (The Chemical Catalogue Co., New
York, 1927), p. 447.

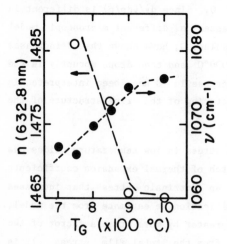

Fig. 1. n and v versus T_G
for films grown at one atm.
of dry oxygen.

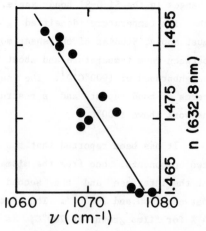

Fig. 2. n versus v for all
low temperature oxides. The
solid line is the model
calculation (mechanism II)

SEXAFS of Thin Thermal SiO$_2$ Films

Michael H. Hecht and F.J. Grunthaner

Jet Propulsion Laboratory 189-100, California Institute of Technology

Pasadena, CA 91109

P. Pianetta

Stanford Synchrotron Radiation Laboratory, Stanford, CA 94305

Abstract

We present SEXAFS measurements which address the distribution of Si-Si second nearest neighbor distances in the near-interface region of thermal SiO$_2$ grown on crystalline silicon. The data is consistent with a foreshortening of the Si-Si separation in this region as would be expected from lattice matching considerations.

The bond strain gradient model of the interface between thermal SiO$_2$ and Si suggests the presence of a structurally distinct region in the oxide approximately 30Å thick, characterized by reduced Si-O-Si bond angles.[1] We have examined this hypothesis using surface extended x-ray absorption fine structure (SEXAFS) spectroscopy to determine the distribution of Si-Si second neighbor distances in thin and thick oxides.

Examination of the structure of SiO$_2$ polymorphs indicates that the Si-O separation remains largely unchanged at 1.60$\pm$0.2Å.[2] The principal degree of freedom in the system is the angle of the Si-O-Si bond which joins the SiO$_4$ tetrahedra together. This angle can vary from 120° to 180°, resulting in corresponding variation in the Si-Si separation. SiO$_4$ tetrahedra tend to form rings of from 4 to 8 members. As a result, bond angle distributions tend to cluster around discrete values, with 144° corresponding to the average 6-membered ring angle (α-quartz structure) and 120° corresponding to the smaller 4-membered ring. While the larger ring is favored in bulk silica, the smaller rings might be expected to form a transition region to the higher density substrate.

Samples were prepared by thermal oxidation at 900°C of Si (100) wafers. Wafers were oxidized to 750Å and thinned to the desired oxide thickness using wet chemical techniques as described in reference 1. Oxide thicknesses of 20Å and 750Å were compared to bare Si surfaces. Peak heights in the photon-induced KLL Auger spectrum as well as total electron yield were measured as a function of photon energy using the JUMBO double crystal monochromator at the Stanford Synchrotron Radiation Laboratory. The photon energy was scanned repeatedly from 1800 eV to 2440 eV. Since the KLL lines from Si and SiO_2 are separated by approximately 7 volts, the contribution to the SEXAFS signal from the Si substrate and the oxide overlayer could be measured independently.

The distribution of atomic separations have been extracted from these spectra by examination of the Fourier transform. The spectra are first background-corrected using a polynomial spline through the nodes of the SEXAFS oscillations. Phase and amplitude functions were approximated using the parametric algorithm of reference 3. We note that since we are comparing scattering processes between the same types of atoms in two similar materials, the comparison should be relatively insensitive to the form of these functions. Prior to Fourier transform the momentum space interval is extended to resolve the peaks in the low frequency terms. The extended spectrum is tapered using a Gaussian window function, which minimizes Gibbs oscillations[4]. The inner potential E_0 is treated as an adjustable parameter. It is clear from the momentum space spectrum that the thin oxide has a more complicated frequency distribution than the bulk, which is consistent with the expected broader distribution of bond angles.

The results of the transform are shown in figure 1. The largest peak in each of the spectra in figure 1 corresponds to the Si-O bond. The two rightmost arrows indicate where the Si-Si peaks would be expected to fall for Si-O-Si bond angles of 120° and 144°. It can be seen instead that the Si-O peak falls about 0.1Å too high, and broad peaks are seen at approximately 2.5 and 3.3Å instead of the expected 2.8 and 3.0Å structures.

We attribute the anomalies in the spectra to the beating pattern between closely spaced frequencies, which dominates the spectra due to the shortness (and discreteness) of the interval being transformed. This phenomenon is fortuitous, as it exaggerates what would otherwise be subtle spectral differences. We have modeled this phenomenon by generating EXAFS spectra corresponding to an assumed distribution of Si-Si distances and repeating the analysis. We assumed four Si-O bonds with a bond length of 1.6Å, and a total of four Si-Si neighbors with a mixture of 2.77Å and 3.22Å separations, corresponding to Si-O-Si bond angles of 144° and 120°. We have not included Debye-Waller factors or the distribution of actual bond angles which would be expected to broaden the radial distribution function significantly.

The results of this modelling are shown in figure 2 for a family of evenly spaced 120°:144° ratios. The patterns observed are similar to those in figure 1. The 20Å and 750Å oxide spectra correspond to 120:144° intensity ratios of approximately 1.3 and 0.8 respectively. This result is clearly consistent with an increasing proportion of smaller rings in the near-interface region, as expected from the bond strain gradient model. The rigid shifts of the structures between the model and the data are presumably due to errors in the calculated phase. Approximately .05Å of the shift of the Si-O line is due to the use of the Si-Si phase factor. The continuous distribution of actual Si-Si separations as well as the Debye-Waller factor accounts for the reduced amplitude of the Si-Si structures compared to the mode.

In conclusion, we have used SEXAFS spectroscopy to compare thin and thick thermal SiO$_2$. By comparison of the result with a model, we were able to clearly distinguish different bond length distributions, and confirm a densification near the Si interface. This method should be equally applicable to other amorphous systems.

The research described here was performed by the Jet Propulsion Laboratory, California Institute of Technology, which is supported by the U.S. National Aeronautics and Space Administration. The work was performed at the Stanford Synchrotron Radiation Laboratory, which is supported by the U.S. Department of Energy and the U.S. National Science Foundation. The research was supported by the Director's Discretionary Fund at JPL and the Caltech President's fund.

References

1. F.J. Grunthaner, P.J. Grunthaner, R.P. Vasquez, B.F. Lewis, J. Maserjian, A. Madhukar, Phys. Rev. Lett. 43, 1683 (1979). Also, F.J. Grunthaner, P.J. Grunthaner, R.P. Vasquez, B.F. Lewis, J. Maserjian, A. Madhukar, J. Vac. Sci. Technol. 16, 1443 (1979).

2. For example, G.V. Gibbs, E.P. Meagher, M.D. Newton, D.K. Swanson, Structure and Bonding in Crystals 1, 195 (1981).

3. (a) B.K. Teo, P.A. Lee, A.L. Sims, P. Eisenberger, B.M. Kincaid, J. Am. Chem. Soc. 99, 3854 (1977); (b) P.A. Lee, B.K. Teo, A.L. Simons, *ibid.*, 99, 3856 (1977).

4. John Barton and D. A. Shirley, "Fourier Analysis of Extended Fine Structure with Auto-Regressive Prediction", to be published.

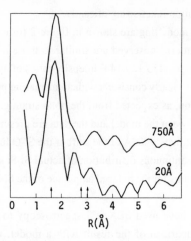

Figure 1: Approximate radial distribution function as determined using Fourier transform technique for thin (bottom) and thick (top) oxide. Arrows, from left, indicate expected Si-O distance, Si-Si distance for 120° bond angle, and Si-Si distance for 144° bond angle. Much of structure is artifactual (see text).

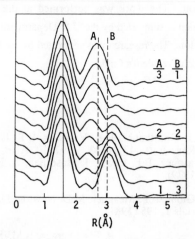

Figure 2: Simple model of functions in figure 1, incorporating 4 Si-O neighbors at 1.6Å and 4 Si-Si neighbors at distances of 2.77Å (A) and 3.04Å (B). Values of A and B are varied in increments of 0.25

PROPERTIES OF METAL-OXIDE-SEMICONDUCTOR STRUCTURES WITH THIN METAL GATES

M. Armgarth, T. Fare, A. Spetz, and I. Lundström
Laboratory of Applied Physics, Linköping Institute of Technology
S-581 83 Linköping
SWEDEN

ABSTRACT

The behaviour of metal-SiO_2-Si MOS structures with thin Pt gates in atmospheres containing low concentrations of NH_3 and H_2 is described. The action on NH_3 and H_2 is quite different as revealed by a study of the Pt film thickness dependence of their influence on high frequency and quasi-static C(V) curves. Whereas H_2 gives rise to a dipole layer at the metal/insulator interface, NH_3 appears to give a dipole layer on the metal surface, which is capacitively coupled through the voids in the Pt film. In both cases, the flatband voltage of the MOS structures is decreased.

1. INTRODUCTION

MOS capacitors with catalytic metal gates are valuable tools for the study of the physics of their different components: the metal, the metal/insulator interface etc. They enable us to study, in real time, chemical reactions in the devices. The hydrogen sensitive Pd-MOS structure is probably the most well known example.[1]

The NH_3 sensitive "thin Pt"-MOS capacitor lends itself to a more detailed study. It was first believed that NH_3 molecules were dissociated on the Pt film[2], releasing hydrogen atoms which diffuse through the Pt film to the Pt/SiO_2 interface. The hydrogen atoms would form a dipole layer changing the effective work function of the gate, and thus the flatband voltage (V_{FB}) as in the hydrogen sensitive Pd-MOS structure.[1] That only thin enough Pt films were sensitive to NH_3 was

ascribed to a promoting effect of the adsorption sites at the borders
of the voids in the Pt film on the catalytic dissociation of NH_3. Recent findings, however, do not support this model. This communication
deals with the results from high frequency (HF) and quasistatic (QS)
C(V) measurements on Pt films with a thickness ranging from 1.5 to 100
nm. We also describe a model for V_{FB} changes based on the capacitive
coupling of the surface potential change of the outer metal surface to
the semiconductor through the voids in the Pt film.[3]

2.   EXPERIMENTAL DETAILS

The samples were of p-type (100) Si, 7-13 ohmcm, with thermally
grown SiO_2 (100 nm). To ensure reliable electrical contacts, Al or Pd
dots, 200 nm thick and 1 mm diameter, were deposited through shadow
masks from resistively heated tungsten boats. Concentric dots of 1.5 to
100 nm Pt were deposited by electron gun evaporation. A schematic of
the sample is shown in Fig. 1. HF C(V) curves were taken with a Boonton
72B capacitance meter. V_{FB} changes were measured with a regulating cir-
cuit, maintaining a constant capacitance. QS curves were measured with
a Keithley 610C electrometer applying a voltage ramp of 100 mV/s over
the capacitor. All measurements were carried out in a background flow
of 20% oxygen in argon, into which different concentrations of NH_3 or
H_2 could be mixed. The samples were mounted on a temperature controlled
sample stage, typically at $150^{\circ}C$.

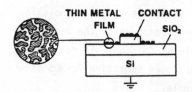

Fig.1: Schematic drawing of a test
structure. A schematic illustration of
the structure on the thin metal film
(as seen from above) is also given.

Fig.2: Dependence of the ammonia and hydrogen sensitivity on
the platinum film thickness.
Unfilled symbols: Structures
with Al-contacts. Filled symbols: Structures with Pd-con-
tacts. Squares: H_2. Circles:
NH_3, 500 ppm of H_2 or NH_3 in
synthetic air.

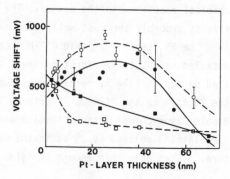

3.   RESULTS AND DISCUSSION

H_2 and NH_3 exposure of thin Pt gate MOS capacitors decrease V_{FB}, shifting the C(V) curve towards more negative voltages. A lower V_{FB} is interpreted as a lowering of the effective work function of the metal gate. Fig. 2 shows the shift of the HF C(V) curves for different Pt-MOS capacitors due to NH_3 and H_2 exposures. The NH_3 sensitivity shows a broad maximum between 10 and 40 nm of Pt. The H_2 sensitivity is fairly constant except for the thinnest films (<3nm). The larger response to H_2 for structures with Pd contacts is most likely due to the sensitivity of Pd itself to hydrogen.[1]

The surface resistivity is large for the very thin Pt-films. This means that the ac-signal does not spread out over the whole metal gate. The effective area of the Pt part of the gate decreases which causes a loss of ammonia sensitivity.[4] In case of hydrogen, there is an apparent increase of the sensitivity for films thinner than about 3 nm. C(V) measurements indicate that during H_2 exposure, the surface resistivity increases causing a further decrease of the active area, which in the constant capacitance mode shows up as a large voltage shift. Fig. 3 shows some results of QS C(V) measurements. In this case, it is observed that the Pt part of the capacitance "disappears" upon H_2 exposure whereas the minimum related to the contact (Al) remains. A similar effect was observed with Pt-MOS capacitors with Pd contacts. However, in that case, the minima coincide, due to the small difference in work function of Pd and Pt.

A careful examination of the HF C(V) curves shows that NH_3 exposures give the same V_{FB} shift on the Pt part of the capacitor irrespective of the choice of contact metal.[4] This implies that there are no hydrogen atoms involved in the detection mechanism. If "free" hydrogen atoms were produced from the dissociation of NH_3, they would rapidly diffuse through the contact[5] and, in case of Pd, cause a voltage shift also at the contact, which is not observed experimentally.[4]

We conclude that the H_2 and NH_3 sensitivity of the thin Pt films must have different origins. Preliminary Kelvin probe measurements show that the work function of both 10 nm and 100 nm Pt films is lowered by about .5V when exposed to 100 ppm NH_3. This observation supports the

274   M. Armgarth et al.

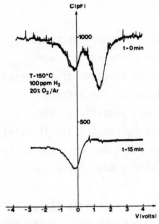

Fig.3: Quasistatic C(V) curve
of a 1.5 nm Pt film (with an
Al-contact) before and during
H$_2$ exposure.

idea that changes of the outer surface
potential (ϕ) may influence the electric field at the Si(SiO$_2$ interface.
A necessary condition is that the film
is porous. (ϕ) is then capacitively
divided over the stray capacitance C$_3$,
between the outer metal surface and
the SiO$_2$ surface. The V$_{FB}$ change becomes:

$$\Delta V = \phi/(\gamma \varepsilon_r C_i/C_s + 1/(1-\gamma)) \qquad (1)$$

where γ is the metal coverage, ε_r the
relative dielectric constant for SiO$_2$
and C$_i$ the oxide capacitance with $\gamma=1$.
Model calculations of C$_s$ indicate a
similar decrease in sensitivity with
increasing metal coverage or film thickness as shown in Fig. 2.

It is at present not clear if ϕ originates from adsorbed NH$_3$ molecules or from reaction intermediates with large dipole moments. The O$_2$
and temperature dependence of the observed V$_{FB}$ changes suggest that
chemical reactions take place on the metal surface.

REFERENCES

1.   I. Lundström and C. Svensson, in Solid State Chemical Sensors,
     Eds Janata and Huber, Academic Press, New York, 1985.

2.   F. Winquist, A. Spetz, M. Armgarth, C. Nylander, and I. Lundström,
     Appl. Phys. Lett., 43, 839-841, 1983.

3.   I. Lundström, M. Armgarth, A. Spetz, and F. Winquist, Proc. 2nd
     Int. Meeting on Chemical Sensors, Bordeaux, 1986, pp 387-390.

4.   A. Spetz, M. Armgarth and I. Lundström, Sensors & Actuators, in
     press.

5.   F. Enquist, M. Armgarth and I. Lundström, to be published.

STRUCTURAL PROPERTIES OF ULTRA THIN THERMAL SILICON OXIDES

M. Bader[*], W. Braun[xx], E. Holub-Krappe[*] and H. Kuhlenbeck[***]

[*]Fritz-Haber-Institut der Max-Planck-Gesellschaft
Faradayweg 4-6, D-1000 Berlin 33, Germany
[xx]Berliner Elektronenspeicherring-Gesellschaft für
Synchrotronstrahlung mbH (BESSY)
Lentzeallee 100, D-1000 Berlin 33, Germany
[***]Fachbereich Physik, Universität Osnabrück
Barbarastr. 7, D-4500 Osnabrück, FRG

ABSTRACT

The geometrical and electronic structures of ultra thin thermal silicon oxides have been investigated by SEXAFS and surface sensitive core level photoelectron spectroscopy. From the SEXAFS measurements we deduced Si-O bond lengths of ~ 0.163 nm. The analysis of the photoelectron spectra leads to the conclusion that the oxidation states are not homogeneously distributed across the interface and that the transition from crystalline Si to amorphous SiO_2 occurs within 0.5 nm.

1. INTRODUCTION

In modern device technology the Si - SiO_2 interface is at present the most important one. Its geometrical and electrical properties determine the performance of MOS devices and are crucial for the achievement of optimal metal contacts. The goal in the immediate future is the development of four-Megabit chips where the thickness of thermally grown gate oxides has to be of the order of 10 nm. This requires a very high perfection of the Si - SiO_2 interface. An understanding of the properties of the transition region of thermally grown silicon oxides is therefore of great importance.

2. EXPERIMENTAL

The experiments were performed at BESSY using synchrotron radiation from the HE-TGM (200 eV $\leq \hbar\omega \leq$ 1100 eV) and the TGM 3 (6.5 eV $\leq \hbar\omega \leq$ 200 eV). For the geometrical structure determination SEXAFS spectra were recorded above the oxygen K-absorption edge (~ 530 eV) in the partial electron yield mode using a two grid high pass filter combined with a channeltron electron multiplier as a detector. The angle of incidence of the light was 0^0 and 70^0 with respect to the surface normal. The SEXAFS signals of the oxidized samples were divided by the spectra of the in situ thermally cleaned Si (100)-wafer. High resolution core level spectra of the Si 2p levels were measured at photon energies of $\hbar\omega$ = 110 eV, 120 eV and 130 eV using a hemispherical electron energy analyser (ADES 400, Vacuum Generators). In this experiment the angle of incidence of the light was 20^0. The photoelectrons have been detected in normal emission. The overall resolution was 300 meV. The Si (100)-wafer was oxidized by exposing the clean sample to ultra pure oxygen at different temperatures and doses (see Table I).

3. RESULTS AND DISCUSSION

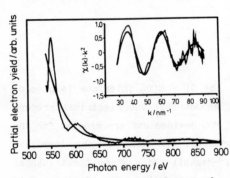

Fig.1: SEXAFS spectrum of thermal silicon oxide and comparison of back transform χ^1 of the main peak with the original χ (inset)

Figure 1 shows the oxygen K-edge SEXAFS spectrum (the main line is not shown) recorded at normal incidence from a sample which was oxidized at 650^0 C with an oxygen dose of 1 Pa.sec (10^{-3} Pa, 1000 sec). For background subtraction the sum of three polynomials of second and third order were fitted through the oscillatory structure. After background subtraction the SEXAFS spectra were Fourier transformed. The Fourier transform obtained for the oscillating term χ in the absorption coefficient is dominated by a single peak. The Fourier transform of χ does not peak at the nearest neighbour distance but is shifted due to a k-dependent phase shift. However, according to Citrin et al.[1], reliable bond length information can be obtained by using experimental phase shifts[2]. In Table I we summarize the bond lengths obtained for the differently prepared thermal oxides. These bond lengths are slightly larger (~ 0.002 nm) than that for α-quartz, but very close to that of β-quartz (0.163 nm).

Table I

preparation temperature/^{0}C	Dose/Pa sec	Thickness/nm	Si-0 bond length/nm
450 (± 30)	10^{-1}	0.68 (21)	0.164 (3)
450 (± 30)	1	0.82 (24)	0.164 (3)
650 (± 30)	10^{-1}	0.71 (22)	0.163 (3)
650 (± 30)	1	0.88 (26)	0.163 (3)

Back-transforming the main peak of the Fourier transform using a suitable window function yields the SEXAFS of the corresponding neighboring shell. This back-transform is compared with the original χ in the inset of fig. 1. As the Fourier transform is dominated by a single peak, this back-transform agrees well with the experimental χ.

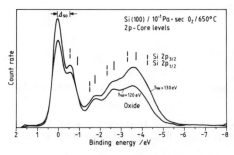

Fig.2: Si2p core level spectra of an oxidized Si (100)-wafer. The tic marks indicate the different pairs of the spin-orbit-split (Δ_{so}) Si2p dublets.

Figure 2 shows the Si2p core level photoelectron energy distribution spectra after a parabolic background subtraction. The sample has been oxidized at 650^0C with an oxygen dose of 0.1 Pa.sec (10^{-4}Pa, 1000 sec). The excitation energies were $\hbar\omega$ = 120 eV and 130 eV. From these spectra the ratio of the mean free paths of the photoelectrons could be determined to $\Lambda(130$ eV$)/\Lambda(120$ eV$) = 0.8$ and thus the spectra can be normalized with respect to each other. The energetic position of the chemically shifted Si 2p spin-orbit doublets were determined by a deconvolution procedure to 1.0, 1.8, 2.7 and 3.7 eV (Si^+, Si^{2+}, Si^{3+}, Si^{4+}) in agreement with the values given by Hollinger et al.[3] for a thermal oxide prepared under different conditions. The spectra shown in fig. 2 clearly demonstrate that the different oxidation states are not homogeniously distributed across the interface. At a shorter escape depth of the photoelectrons ($\hbar\omega$ = 130 eV) the highest oxidation state of silicon (Si^{4+}) dominates the energy distribution of the photoelectrons emitted from the oxide layer. The thickness of the oxide layers could also be estimated by using a mean free path of Λ (120 eV) = 0.54 nm[4]. The values obtained are listed in table I.

From the data shown here and a more elaborate evaluation of all our data it can be concluded that the transition from crystalline Si to SiO_2 occurs within two oxide layers and that there is a layer by layer growth of the SiO_x interface with x increasing for larger distances to the substrate.

Acknowledgement   This work was financially supported in part by the Bundesministerium für Forschung und Technologie under grant numbers 05 390 FX B/390 FX 4 and 05 230 NUP/I.

REFERENCES

1) Citrin, P.H., Eisenberger, P. and Kincaid, B.M., Phys. Rev. Lett. 36, 1346 (1976)

2) Stöhr, J., in: Principles, Techniques and Applications of EXAFS, SEXAFS and XANES   ed. by R. Prins and D.C. Konnigsberger, Wiley, New York (1985)

3) Hollinger, G. and Himpsel, F.J., Appl. Phys. Lett. 44, 93 (1984)

4) Himpsel, F.J., Heimann, P., Chiang, I.-C. and Eastman, D.E., Phys. Rev. Lett. 45, 1112 (1982)

BREAK OF Si-Si BONDS BY ATOMIC OXYGEN: FIRST STAGES OF THE SILICON OXIDATION

J. Plans, G. Díaz, E. Martínez and Félix Ynduráin

Departameto de Física Materia Condensada, C-III

Universidad Autónoma de Madrid,

Cantoblanco, 28049-Madrid, Spain.

ABSTRACT

We present total energy calculation of finite clusters of silicon atoms in the presence of atomic oxygen. We allow atomic relaxation to determine the equilibrium local configuration of atoms. The results show how oxygen breaks the covalent Si-Si bond. This breaking mechanism is essential to understand the oxygen uptake in the process of the oxidation of silicon.

The interaction of atomic oxygen with silicon plays an important role in both the bulk and surface governed electronic properties of this semiconductor. On one hand, interstitial oxygen is present in crystalline silicon forming a puckered bond[1,2] between two nearest neighbor silicon atoms. On the other hand, the formation of SiO_2 at the silicon surface is governed by the incorporation of oxygen in the silicon lattice[3-6]. In both cases the presence of oxygen can only be understood if it accommodates itself in the silicon network by breaking the silicon silicon bonds.

We have performed an ab initio calculation of the total energy of an oxygen atom in the vicinity of the silicon covalent network to find the equilibrium atomic configuration. Atomic oxygen is incorporated to a cluster formed by eight silicon atoms forming seven Si-Si bonds. The valence electrons associated to silicon and oxygen atoms are included in the calculations whereas the core ones are neglected and their effect is simulated by pseudopotentials[7].

The atomic valence electrons are described by contracted gaussians orbitals of single-ζ type. The calculation of the total electronic energy for each atomic configuration is performed in the restricted Hartree-Fock approximation.

We assume that the silicon silicon distance in the absence of oxygen is the nominal distance 2.35 Å of crystalline silicon and the Si-O distance is 1.61 Å as in α-quart (this introduces an error of less than 5% in the atomic distances[8-10]). Results of the calculated total energy versus the angle formed by the Si-Si and the Si-O bonds are shown in Figure 1.

Figure 1.- Total energy of the cluster Si_8H_{18} in the presence of atomic oxygen versus the angle formed by the central silicon-silicon bond and one of the Si-O bonds. The dots represent the atomic configurations where the calculations were performed. The solid line represents an interpolated polynomial.

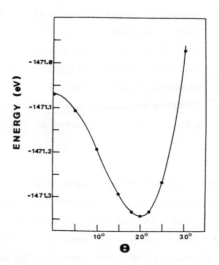

To describe how the electronic structure is perturbed by the presence of oxygen we have plotted the charge distribution in Figure 2. In Figure 2(a) we show the electronic distribution corresponding to the Si_8H_{18} cluster. In Figure 2(b) we show the charge distribution when oxygen is present at the equilibrium configuration. We observe how oxygen pushes apart the nearest neighbor silicon atoms forming the central bond of the original clusters. Also the back Si-Si bond are compressed by the presence of oxygen (the outer Si and H atoms of the original cluster are assumed to be fixed). To study how the original Si-Si bonds are perturbed by the presence of oxygen we have plotted in Figure 2(c) the charge density corresponding to the stable configuration of the oxidized cluster but excluding the oxygen contribution to the charge density. One observes the dramatic disappearance of the original Si-Si bond which is indeed replaced by two Si-O bonds.

The Si-O bond we have obtained is similar to the one in crystalline[11] SiO_2 with a very high ionic character, the charge transfer being $0.43e^-$ per Si-O bond. The electronic charge distribution is also similar to that of SiO_2.

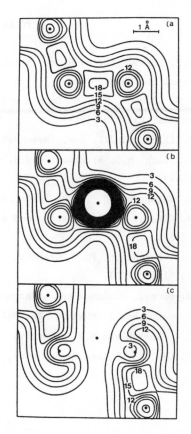

Figure 2.- Electronic charge distribution (in units of the crystalline silicon unit cell) of the clusters considered in this work. The dots represent the atomic positions. (a) Charge distribution of the bare Si_8H_{18} cluster. (b) Charge distribution of the oxidized cluster at the equilibrium atomic configuration. The constant charge contours larger than 135 are not represented. (c) Charge distribution of the oxidized cluster including only the silicon contribution to the charge (see text).

In symmary, we believe that this cluster calculation gives a tremendous amount of information concerning the important problem of the break of Si-Si bonds by atomic oxygen and the accompanied rebonding of the atoms. The oxygen incorporation mechanism is essential to understand the first stages of the silicon oxidation as well as the role played by oxygen as interstitial impurity in silicon.

Acknowledgements. This work has been supported in part by CAICyT.

REFERENCES.

1.- D.R. Bosomworth, W. Hayes, A.R.L. Spray and G.D. Watkins, Proc. Royal Soc. London, a. $\underline{317}$, 133 (1970).

2.- J.W. Corbett, R.S. Mc Donald and G.D. Watkins, j. Phys. Chem. Solids, $\underline{25}$, 873 (1964).

3.- H. Ibach and J.E. Row, Phys. Rev. $\underline{B10}$, 710 (1974).

4.- H. Ibach, H.D. Bruchmann and H. Wagner, Appl. Phys. $\underline{A29}$, 113 (1983)

5.- H. Wagner and H. Ibach, Festkorperprobleme, XXIII, 165 (1983).

6.- S. Ciraci, S. Ellialtioglu and S. Erkoc, Phys. Rev. $\underline{B26}$, 5716 (1982).

7.- J.C. Barthelat, PH. Durand and A. Serafini, Molecular Physics, $\underline{33}$, 159 (1977).

8.- R. Dovesi, M. Causa and G. Angonoa, Phys. Rev. $\underline{B24}$, 4177 (1981).

9.- I. Ohkoschi, J. Phys. C, $\underline{18}$, 5415 (1985)

10.- A.C. Kenton and M.W. Ribarsky, Phys. Rev. $\underline{B23}$, 2897 (1981).

11.- J. Chelikowsky and M. Schluter, Phys. Rev. $\underline{B15}$, 4020 (1977).

INTERFACIAL STRUCTURE OF ANODICALLY OXIDIZED $Hg_{1-x}Cd_xTe$

M. Seelmann-Eggebert and H.J. Richter

Fraunhofer-Institut für Angewandte Festkörperphysik

D-7800 Freiburg, Federal Republic of Germany

ABSTRACT

The formation of the compositional structure at the interface between $Hg_{1-x}Cd_xTe$ (MCT) and its common anodic oxide was investigated by electrochemical and XPS analyses. It was found that a bi-modal oxidation behaviour of MCT causes each of both interface regions (oxide and MCT) to deviate in its composition from the respective bulk.

1. INTRODUCTION

Anodically grown native oxides are known to be adequate surface passivants for n-type $Hg_{1-x}Cd_xTe$ (MCT). In particular the anodic oxide grown in a KOH electrolyte (pH13 EG: 0.1 M KOH in 90% ethylene glycol, 10% H_2O) at a current density of a few $100\mu A/cm^2$ is widely used for the formation of passivating films (1000 Å) on device quality MCT (P.C. Catagnus, C.T. Baker, U.S. Patent No. 3 977 018 (1976)). Since interfacial phenomena set a limit to the applicability of the present anodization techniques we attempted to establish a basis for a correlation of the physical and chemical properties which arise from the formation of the interface between MCT and its anodic oxides.

2. SAMPLE PREPARATION AND EXPERIMENTAL

MCT wafers of device qualtiy with x = 0.2 and Bridgman-grown single-crystalline HgTe and CdTe wafers were used . The MCT surfaces were etch-polished with 3% Br_2 in methanol and rinsed in pure methanol .After rinsing the MCT surfaces were found to be of stoichiometric composition, however, they were generally covered with a residue of elemental Te.

The prepared samples were electrochemically analyzed by voltammetry. To identify the nature of all electrochemical reactions which gave rise to peaks in the recorded current-voltage plots we examined by X-ray photoelectron spectroscopy (XPS) the surface composition of the samples before and after the occurence of each reaction.

For this purpose the scan was stopped at the desired voltage. The samples were immediately removed from the electrolyte, rinsed in pure methanol and examined by XPS. The XPS probing depth was varied by angular resolution of the photoemitted electrons to obtain depth-compositional profiles of a few 10 Å. The quantitative evaluation procedures have been described in detail elsewhere (M. Seelmann-Eggebert, Dissertation, Universität Tübingen (1986)).

3.   RESULTS AND DISCUSSION

All electrochemical reactions (oxidations A and reductions C) observed for MCT electrodes in the above electrolyte (pH13 EG) are listed in Tab. 1.

Table 1. Electrochemical reactions consistenly observed for electrodes of Te, Hg, Cd, HgTe, CdTe and MCT. Peak potential U_s refers to Hg/HgO standard and 5 mV/sec.

$$U_s \, [V]$$

$$A1 : HgTe + 6\ OH^- \longrightarrow HgTeO_3 + 3\ H_2O + 6\ e; \quad 0.4$$
$$C1 : HgTeO_3 + 2\ e \longrightarrow Hg + TeO_3^{2-} \quad ; -0.45$$
$$A2 : Te + 6\ OH^- \longrightarrow TeO_3^{2-} + 3\ H_2O + 4\ e; -0.1$$
$$C2 : TeO_3^{2-} + 3\ H_2O + 4\ e \longrightarrow Te + 6\ OH^- \quad ; -0.9$$
$$A3 : Hg + 2\ OH^- \longrightarrow Hg(OH)_2 + 2\ e; \quad 0.05$$
$$C3 : Hg(OH)_2 + 2\ e \longrightarrow Hg + 2\ OH^- \quad ; -0.1$$
$$A4 : Hg(ad) + 2\ OH^- \longrightarrow HgO + H_2O + 2\ e; \quad 0.2$$
$$C4 : HgO + H_2O + 2\ e \longrightarrow Hg(ad) + 2\ OH^- \quad ; -0.15$$
$$A5 : Te_2^{2-} \longrightarrow 2\ Te + 2\ e; -0.95$$
$$C5 : 2\ Te + 2\ e \longrightarrow Te_2^{2-} \quad ; -1.1$$
$$C6 : HgTe + 2\ e \longrightarrow Te^{2-} + Hg \quad ; -1.25$$
$$A7 : CdTe + 6\ OH^- \longrightarrow CdTeO_3 + 3\ H_2O + 6\ e; \quad 0$$
$$C7 : CdTeO_3 + 3\ H_2O + 6\ e \longrightarrow CdTe + 6\ OH^- \quad ; -1.3$$
$$A8 : Cd + 2\ OH^- \longrightarrow Cd(OH)_2 + 2\ e; -0.85$$
$$C8 : Cd(OH)_2 + 2\ e \longrightarrow Cd + 2\ OH^- \quad ; -1.05$$
$$C9 : CdTe + 2\ e \longrightarrow Te^{2-} + Cd \quad ; -1.7$$

Please note that there is only one oxidation potential for each of the binary compounds HgTe and CdTe, i.e. in either case the metal and the Te component are simultaneously oxidized. The HgTe and the CdTe component of MCT exhibit the same behaviour and the same oxidation potentials, i.e. HgTe and CdTe maintain individual modes of oxidation in their function as components of the alloy MCT (bi-modal oxidation).

Fig. 1 shows the anodic scans obtained for two MCT samples (x = 0.2), which were initially covered with different amounts of residual Te. The oxidation of this Te to TeO_3^{2-} is associated with a voltammetric signal A2 and hence A2

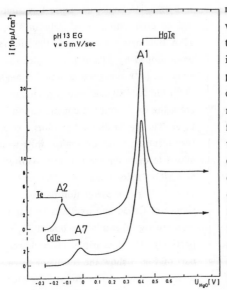

Fig. 1. Voltammetric analyses of MCT surfaces illustrating the onset of an anodic oxidation and the effect of residual Te.

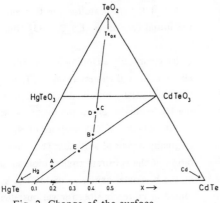

Fig. 2. Change of the surface composition with increasing electrode potential U as analyzed by XPS ($\vartheta = 45°$) (U < U_s of A1 for A, B, E; U > U_s of A1 for C, D)

may mask the more significant peak A7, which is connected with the oxidation of the CdTe component to $CdTeO_3$. Due to its lower oxidation potential the CdTe is preferentially oxidized, while the HgTe component is accumulated at the interface region of the MCT. The resulting barrier for the CdTe component is weakened when the HgTe component reaches its oxidation potential and initiates the dissolution phase of the dissolution-precipitation mechanism of the $HgTeO_3$ formation. The plateau following peak A1 is typical for the dynamic equilibrium of oxide growth characterized by a continuous formation of $HgTeO_3$ and $CdTeO_3$ at the electrolytic interface of the anodic film. The data triples of Fig. 2 represent the combined relative element concentration of Hg, Cd, and oxidized (tetravalent) Te analyzed after voltammetric scans (A-E) to different final potentials. During the onset of oxidation the data triples follow a straight line from a MCT phase point with x < 0.2 to the $CdTeO_3$ phase point indicating the presence of a substrate interface which is altered in its composition and covered with an oxide film of mainly $CdTeO_3$. The films resulting from scans beyond the peak A1 have thicknesses larger than the XPS probing depth, and their data triples indicate the presence of an oxide layer composed of $HgTeO_3$ and $CdTeO_3$. The initial data points of the investigated MCT surfaces are also shown. To obtain the set of XPS survey scans in the upper part of Fig. 3 the sample of run B in Fig. 2 was analysed at different elec-

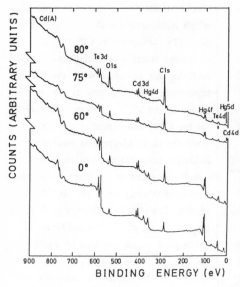

BINDING ENERGY (eV)

tron emission angles (ϑ). Since this sample was covered with an oxide film of about 17 Å thickness the oxide and the substrate were probed by XPS at normal electron emission (0°). For grazing emission angles (80°) the XPS signals mainly originated from the oxide and the contamination layer. The Te 3d doublet exhibits a structure resulting from a chemical shift which allows to differentiate between oxidized and MCT lattice bound Te. The set of scans indicates a correlation between the signals of Cd 3d and oxidized Te , and between Hg 4f and lattice bound Te, respectively. The data triples in the lower part of Fig. 3 show the results of three different signal combinations. The purpose of combining different signals was to assess the influence of systematic errors on the data evaluation procedure. Obviously, also the analysis of the depth - compositional profile indicates the alteration of the substrate composition and the initial formation of a $CdTeO_3$ film.

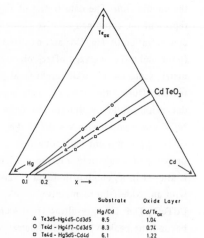

	Substrate	Oxide Layer
	Hg/Cd	Cd/Te$_{ox}$
△ Te3d5-Hg4d5-Cd3d5	8.5	1.04
○ Te4d - Hg4f7-Cd3d5	8.3	0.74
□ Te4d - Hg5d5-Cd4d	6.1	1.22

Fig. 3. Angle-resolved XPS analysis of sample B (Fig. 2) for depth-compositional profiling. The results obtained with three sets of different data triples are shown in the lower part.

After chemically stripping anodic films of about 1000 Å thickness from MCT samples the initially observed accumulation of the HgTe component at the interface was still detectable by XPS analyses of the originally anodized surfaces. This alteration of the substrate composition caused by the bi-modal oxidation of MCT is assumed to be responsible for various electronic properties which are peculiar to anodized MCT and which can be viewed in physical terms as caused by a reduced band gap at the interface.

AN IMPURITY MODEL FOR THE SCHOTTKY BARRIER: TRANSITION METALS ON III-V COMPOUND SEMICONDUCTORS.

R. Ludeke, G. Hughes*, W. Drube, F. Schäffler, F.J. Himpsel,
D. Rieger, D. Straub+ and G. Landgren†.

IBM T.J. Watson Research Center, P.O. Box 218, Yorktown Heights, N.Y.
10598, U.S.A.

ABSTRACT

We report the observation of filled and empty interface states derived from the interaction of transition metals (TMs) with GaAs. For Ti, V and Pd these states, attributed to impurity levels, determine the position of the interface Fermi level.

1. INTRODUCTION

The origin of the interface states responsible for the stabilization of the Fermi level at the metal-semiconductor interface has long been a controversial issue. Although a variety of sources have been proposed,[1-3] spectroscopic evidence in support of any model has been lacking up to now. Evidence for interface states has been obtained previously by spectroscopic[4] and electrical[5] measurements, however no correlations with a specific mechanism, structural defect or chemical species could be made. In this work we report the observation of filled and empty interface states derived from the interaction of transition metals (TMs) with the GaAs surface, and suggest a possible origin - a substitutional TM impurity at the interface - based on spectroscopic evidence and changes in the interface chemistry.

2. EXPERIMENTAL DETAILS

The TMs were sublimed from wires or foils onto freshly cleaved GaAs(110) surfaces in vacuums of 10^{-10} Torr. Both n and p-type substrates, doped respectively with Si and Zn to $\sim 10^{18}$ cm^{-3} were used. The thickness was monitored with a quartz microbalance. Details of the experimental procedures, including the line shape analysis of the core level spectra are given elsewhere.[6] Inverse photoemission was used to probe the conduction band states and empty interface states. The technique consists of directing a monochromatic electron beam onto a surface and spectrally analyzing the resulting photon flux.[7]

3. EXPERIMENTAL RESULTS

It has now been established that the TMs react strongly with GaAs and other semiconductor substrates.[7,8] This behavior is readily observed in the core level photoemission spectra of the semiconductor. Fig. 1 depicts the evolution of the Ga and As 3d core level spectra as a function of Mn coverage. The experimental spectra (dots) have been decomposed into chemically shifted individual components by computer fitting routines.[7,9] The spectra of the clean surface are made up of a bulk peak (large component) and a surface peak. The latter decreases with coverage of the TM. For coverages as low as 0.02 Å an additional chemically shifted peak can be observed in the Ga 3d core level spectrum; this structure is obvious in Fig. 1 for a coverage of 0.4 Å. This peak is attributed to elemental Ga, which diffuses to the surface of the growing Mn film. This behavior is strong evidence that an exchange reaction has occurred on the surface, with the TM (here Mn) replacing the Ga. The situation is less clear for the As since the emergence of a chemically shifted As component, obvious for a coverage of 3 Å, is disguised at the lower coverages by the surface peak. The general lack of experimental evidence for elemental As (except for Pd[7]) suggests that Ga is the predominantly replaced surface component, and that the As, both at the interface and as a diffused species, is predominantly bonded to the TM.

The spectral decomposition is necessary as well to accurately follow the shifts in kinetic energy with coverage of the bulk component, so that the resulting band bending and final positions of the Fermi level E_F can be ascertained.[9] The positions in the band gap of E_F as a function of coverage for several TMs on GaAs(110) are shown in Fig.2. We also indicate the final position of E_F for several other metals on the right hand ordinate. Three important conclusions can be drawn from the data: a) the final value of E_F (and the Schottky barrier) may vary by as much as 0.3 eV for different metals; b) the same final position of E_F is obtained independent of the doping type; and c) the band bending is essentially complete, except for Ag,[9] near a coverage of 0.1 Å, which is below the coverage necessary to establish a metallic behavior of the overlayer. These observations are generally inconsistent with present Schottky barrier models, and we must look for a different origin of the interface states which determine E_F at the interface. The strong reactions with the TMs suggest states induced by a new chemical environment, i.e. of extrinsic origin, which are readily observed in the valence band spectra.

Fig. 3 depicts valence band spectra for clean GaAs and for the indicated coverages of V, Mn and and Y. Strong spectral features are readily observed which originate from the d-electrons of the TMs. Difference spectra, obtained by subtracting the clean spectrum, are shown on the right of Fig.3. The sharp spectral features are already discernible at coverages of ~ 0.01 Å (not shown) and suggest a unique chemical environment of the TM, at least at low coverages. Based on the core level spectra, the Ga substitutional site is a likely site for the TM, but requires additional confirmation. We can, however, draw the conclusion that the TMs induce filled interface states which overlap the valence band of GaAs, and in the case of Ti, V, Pd and to a lesser extend Mn, extend into the band gap, where the emission edge coincides with the position of E_F. This is more clearly shown in Fig. 4, which is a composite of difference spectra (solid lines) of the valence bands (VB), obtained by phototemission, and the conduction bands (CB), obtained by inverse

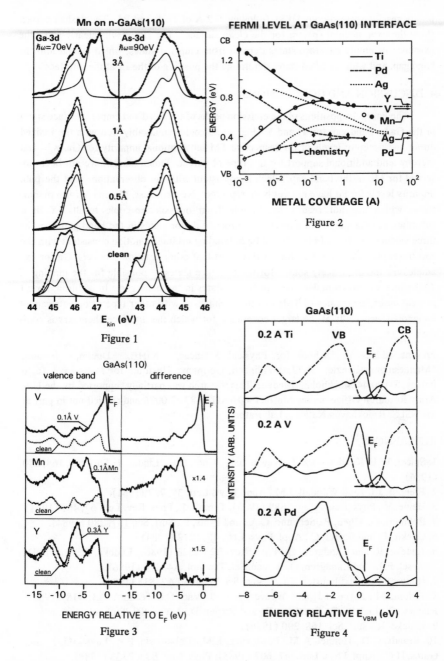

Mn on n-GaAs(110)

Ga-3d
$\hbar\omega = 70eV$

As-3d
$\hbar\omega = 90eV$

3Å

1Å

0.5Å

clean

44 45 46 47   43 44 45 46
E_{kin} (eV)

Figure 1

FERMI LEVEL AT GaAs(110) INTERFACE

CB

1.2

ENERGY (eV)

0.8

0.4

VB

—— Ti
--- Pd
···· Ag

Y
V
Mn

Ag
Pd

chemistry

10^{-3} 10^{-2} 10^{-1} 1 10 10^{2}

METAL COVERAGE (A)

Figure 2

GaAs(110)

valence band   difference

V E_F E_F
0.1Å V
clean

Mn
0.1ÅMn   x1.4
clean

Y
0.3Å Y   x1.5
clean

-15 -10 -5   0   -15 -10 -5   0

ENERGY RELATIVE TO E_F (eV)

Figure 3

GaAs(110)

0.2 A Ti   VB   CB
                   E_F

INTENSITY (ARB. UNITS)

0.2 A V
              E_F

0.2 A Pd
              E_F

-8 -6 -4 -2 0 2 4

ENERGY RELATIVE E_{VBM} (eV)

Figure 4

290   R. Ludeke et al.

photoemission spectroscopy (IPS), following 0.2 Å of Ti, V and Pd. The dashed curves show the experimental spectra for the clean surface. The IPS results clearly show the existence of empty interface states derived from empty d-states, which extend into the band gap, and with the filled states determine the position of the E_F at the interface.

4. DISCUSSION AND CONCLUSIONS

The energetic positions of the emission peaks of the filled and empty interface states in the GaAs band gap for Ti and V (Fig. 4) coincide remarkably well with the ionized donor and acceptor levels of the respective TM substitutional impurity for GaAs,[10] and thereby lend additional support for this type of defect. Donor states have not been observed for Mn in the band gap, which is consistent with our observations, and the bulk impurity levels for Pd have not yet been reported. Nevertheless, Mn and Y form pinning states which are not readily identifiable from our spectroscopic results. Y in a substitutional Ga site is not expected to produce donor states in the band gap, since its three valence electrons ($4d^1 5s^2$) will be in bonding orbitals, a notion consistent with the spectroscopic data. We speculate at this point that other reaction products, Ga for example, are the most likely source for the interface states responsible for the pinning of TMs which do not form d-derived interface states in the band gap. In addition, for all present cases, aggregates of TMs or other complexes with other type of defects cannot be ignored, especially at the larger coverages, for which the interface chemistry is complex.

Present addresses: *School for Physical Sciences, NIHE, Dublin, Ireland; +Mannesmann-Kienzle, D-7730 Villingen, Germany; †Institute for Microwave Technology, S-100 44 Stockholm, Sweden. This research was partially supported by the U.S. Army Research Office, under contract #DAAG29-83-C-0026 and carried out in part at the NSLS, Brookhaven National Laboratory.

REFERENCES

1. Spicer, W.E, Kendelewicz, T., Newman, N., Chin, K.K., Lindau, I., Surf. Sci. 168, 240 (1986)
2. Freeouf, J.L., and Woodall, J.M., Appl. Phys. Lett. 39, 727 (1981).
3. Heine, V., Phys. Rev. A138, 1689 (1965); Tersoff, J., Phys. Rev. Lett. 57, 465 (1984).
4. Bolmont, D., Chen, P., Sebenne, C.A., and Proix, F., Surf. Sci. 137, 280 (1984).
5. Chekir, F. and Barret, C., Appl. Phys. Lett. 45, 1212 (1984).
6. Ludeke, R. and Landgren, G., Phys. Rev. B33 5526 (1986); Ludeke, R., Straub, D., Himpsel, F.J., and Landgren, G. J. Vac. Sci. Technol. A4, 874 (1986).
7. Himpsel, F.J., and Fauster, Th., J. Vac. Sci. Technol. A2, 815 (1984).
8. Grioni, M., Joyce, J.J. and Weaver, J.H., J. Vac. Sci. Technol. A3, 918 (1985); Hughes, G., Ludeke, R., Schäffler, F., and Rieger, D., J. Vac. Sci. Technol. in press.
9. Ludeke, R., Surf. Sci. 168, 290 (1986).
10. Brandt, C.D., Hennel, A.M., Pawlowicz, L.M., Dabkowski, F.P., Lagowski, J. and Gatos, H.C., Appl. Phys. Lett. 47, 607 (1985); Phys. Rev. B33, 7353 (1986).

THEORY OF SCHOTTKY BARRIER FORMATION BASED ON UNIFIED DISORDER INDUCED GAP STATE MODEL

Hideki Hasegawa

Department of Electrical Engineering, Faculty of Engineering
Hokkaido University, Sapporo, 060 JAPAN

Schottky barrier formation is explained by a disorder induced gap state (DIGS) continuum produced by metal deposition whose charge neutrality point is given by the hybrid orbital energy E_{HO}. Deviation of Fermi level from E_{HO} results in a dipole which screens the metal electronegativity. A linear theory for barrier height is developed, and chemical trends are discussed.

1. INTRODUCTION

In spite of intensive efforts paid over past several decades, physics of Schottky barrier formation still remains unclarified. On the basis of a newly found strong universal correlation between metal-semiconductor (M-S) and insulator-semiconductor (I-S) interfaces which neither of the unified defect model[1] and the MIGS model[2] can explain, the unified disorder induced gap state (DIGS) model for both interfaces has recently been proposed[3]. This paper presents a theory of Schottky barrier formation based on the DIGS model. Chemical trends of barriers are discussed in a unified manner.

2. DIGS MODEL

Deposition of metal or insulator disturbs the crystalline perfection of semiconductor and forms a thin disordered semi-

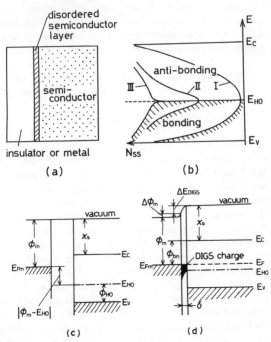

Fig.1 Unified DIGS model and Schottky barreir formation

conductor layer with a thickness δ as shown in Fig.1(a) characterized by DIGS continuum as shown in Fig.1(b). As compared with the carefully prepared I-S interface (I,II in Fig.1(b))where similar localized bonds meet, a highly disordered region is expected at M-S interface (III in Fig.1(b)) where localized bonds encounter nonlocalized bonds, being frequently combined with stoichiometry disrupting chemical reactions. From a tight-binding picture, the neutrality point of the DIGS spectrum is given by the hybrid orbital energy[3]:

$$E_{HO} = ((\varepsilon_s + 3\varepsilon_p))_{av.}/4 \tag{1}$$

where ε_s and ε_p are atomic term energies and av. denotes average over anions and cations. The position of E_{HO} is independent of details of disorder, and lies approximately at 5.0 eV from vacuum for major semiconductors[3]. Within a disordered layer, deviation of the Fermi level from E_{HO} results in appearance of space charge, which, in turn, screens electronegativity of metal as schematically shown in Fig.1(c) and (d). An approximate analysis ignoring the electric field in the ordered semiconductor region, gives the following:

$$\phi_{bn} = (S/A)(\phi_m + \Delta\phi_m - \chi_s) + (1 - S/A)(E_c - E_{HO}), \quad \phi_{bp} = E_g - \phi_{bn} \tag{2}$$

where ϕ_m : metal workfunction, ϕ_s : semiconductor electron affinity, S :index of interface defined by S $=d\phi_{bn}/dX_m$ (X_m = electronegativity of metal)[4] and A: a constant which relates the electronegativity scale to workfunction (= 2.86)[5]. $\Delta\phi_m$ represents possible spatial variation of potential on the metal side due to charge redistribution. S is derived for clean surfaces as:

$$S = A\ \mathrm{sech}(\delta/\lambda) \quad \text{with} \quad \lambda = \sqrt{\varepsilon/q^2 N_{DG}} \ \text{(DIGS screening length)}$$
$$N_{DG} = \text{DIGS density} \tag{3}$$

3. BARDEEN LIMIT(strongly pinned limit)

When N_{DG} is greater than 10^{19} cm^{-3}eV^{-1}, a few monolayers are sufficient to completely screen x_m, giving rise to the Bardeen limit where the Fermi level is firmly pinned at E_{HO} and the barrier height does not depend on the metal workfunction. The theoretical locations of E_{HO} may be estimated either directly from Eq.(1) or indirectly from the published band data. The theoretical barrier heights ϕ_{bp} for holes in the Bardeen limit are compared in Fig.2(a)-(c) with the experimental barrier heights for Au Schottky systems as complied by Sze[6]. In Fig.2(a), the universal parameter sp^3 tight-binding theory[7] was used, and the sp^{3}s* tight binding theory[8] was used in Fig.2(b). In Fig.2(c), the location of E_{HO} was deduced from the pseudopotential band calculation[9], using the eigenvalues at Γ point only under the nearest neighbor approximation. Excellent linear correlation with slope of unity is seen between theory and experiment in Fig.2(b) and (c) with correlation coefficients of larger than 0.95 and the rms deviation of about 90 meV. Much larger scatter of data is seen in Fig.2(a), most probably due to crudeness of parameters chosen to achieve universality. The theoretical values of ϕ_{bp} in Fig.2(a)-(c) which are based on the nearest neighbor tight binding interpretation of bands give consistently energy shifts of about 200 meV. This can be attributed to a self-energy correction for near neighbor interactions and presence of higher bands. The reason why Au Schottky barriers give the barrier height close to the Bardeen limit can be explained by the fact that the work-function of Au (5.1 eV) is close to E_{HO} (approximately 5.0 eV from vacuum) so that only a small degree of disorder causes pinning at E_{HO}.

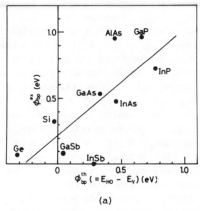

(a)

4. CHEMICAL TRENDS OF BARRIER HEIGHTS

As to the relation between barrier heights and metal workfunctions, there exists vast amount of scattered and conflicting data. On the statistical average, however, a linear relationship as in Eq.(2) has been recognized[5] with S describing the sensitivity of barrier height against metal electronegativity. Previously, S has been interpreted in terms of covalent-ionic transition[4], heat of compound formation[10] and dielectric screening[2].

Since the degree of disorder which determines the strength of pinning in the present model, depends obviously on the specific details of processing, it can explain the large scatter of data in the literature as well as the presence of a linear relationship of Eq.(2) on average, if one interprets S to be a characteristic quantity which is a measure of resistance against introduction of surface disorder during processing. The amount of loss of the cohesive energy as a result of a high degree of surface disorder, depends to a first approximation on the energy separation between E_{HO} and E_v provided that the bonding portion of DIGS is brought into the gap from the neighborhood of the top of the valence band as in bulk amorphous semiconductors. One would then expect a

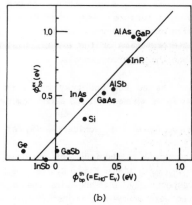

(b)

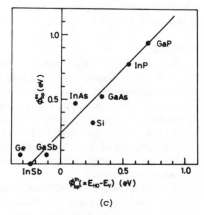

(c)

Fig.2   Theoretical(th) and experimental(ex) barrier heights

294   H. Hasegawa

correlation between S and the separation
$(E_{HO} - E_v)$. Experimental values of S^5
are plotted in Fig.3 vs. empirical values[11] of $(E_{HO} - E_v)$, which clearly
demonstrates existence of a strong correlation, suggesting an empirical exponential relationship between log S and
$(E_{HO} -_v)$. From Eq.(3), this means that
the DIGS density itself is roughly
proportional to an energy factor,
$\exp(-\alpha(E_{HO}- E_v))$ which   may   be
interpreted as a Boltzmann energy factor
corresponding to the process of
canceling the bonding-antibonding
splitting through disorder. Although a
further refinement is necessary, the
present picture seems to provide a much
simpler and more consistent overall
interpretation of the experimental data
than the previous models.

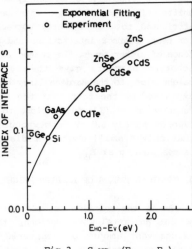

Fig.3   S vs. $(E_{HO} - E_v)$

As to the Schottky barriers formed on oxide covered semiconductor
surfaces where the MIGS model completely fails, the present theory gives

$$\phi_{bn}{}^{(bare)} - \phi_{bn}{}^{(oxide)} \doteqdot -(S_i /A)(\phi_m - E_{HO}) \qquad (4)$$

Here,S_i/A is the index of oxidized interface given approximately by 1/(1 +
$q^2 N_{DG}\delta t/ \varepsilon_i$)where t and ε_i are the thickness and permittivity of the oxide,
respectively. This explains the experimental result by Turner and Rhoderick[12]
who were annoyed by unjustifiable appearance of constant energy.

ACKNOWLEDGMENT: The present work is supported by a Grant-in-Aid for a Special
Promotion Research from Ministry of Education, Science and Culture.

REFERENCES
1. W.E. Spicer et al,J. Vac. Sci. Technol,**16**,1422(1979)
2. J. Tersoff, Phys. Rev. B,**32**,6968(1985)
3. H. Hasegawa and H.Ohno, presented at 13th PCSI(1986) to be published in
    J. Vac. Sci. Technol. July/Aug. issue(1986)
4. S. Kurtin et al, Phys. Rev. Lett.,**22**,1433(1969)
5. M. Schulter, Phys. Rev. B,**17**,5044(1978)
6. S.M. Sze:"Physics of Semiconductor Devices" 2nd.ed.(Wiley, New York(1981))
7. W.A. Harrison:"Electronic Structure and the Properties of Solids(Freeman,
    New York(1980)).
8. P. Vogl et al, J. Phys. Chem. Solids **44**,365(1983)
9. J.R. Chelikowsky and M.L. Cohen, Phys. Rev. B**14**,556(1976)
10. L.J. Brillson, Phys. Rev. Lett. **40**,260(1978))
11. H.Hasegawa, H. Ohno and T. Sawada, Jpn. J. Appl. Phys.**25**,L265(1986)
12. M.J.Turner and E.H.Rhoderick, Solid State Electron.,**11**,291(1968)

BARRIER FORMATION AND ELECTRONIC STATES

AT SILICIDE-SILICON INTERFACES†

M. Liehr, P. E. Schmid*, F. K. LeGoues, J. C. Chen, and P. S. Ho

IBM Thomas J. Watson Research Center
Yorktown Heights, New York 10598 USA

E. S. Yang, H. L. Evans, and X. Wu

Department of Electrical Engineering
Columbia University, New York, New York 10027 USA

ABSTRACT

This paper reviews the recent studies on Schottky barrier and interface states at silicide-silicon interfaces, with emphasis placed on the results obtained from the epitaxial Ni silicides. Measurements on the barrier heights of type A and type B epitaxial Ni silicides show that these interfaces can be formed with high degrees of perfection to yield a barrier 0.75 ± 0.05 eV. Similar interfaces formed under less ideal conditions or with impurity incorporation decrease the barrier to about 0.65-0.70 eV. The density and distribution of the interface states measured by a capacitance spectroscopy method correlate well with the structural perfection of the single and mixed-phase interfaces.

1. INTRODUCTION

For the past several years, extensive effort on the study of Schottky barrier formation on silicon has been focused on the silicide-silicon system. This system is highly reactive and the reaction which is responsible for silicide formation can be controlled to yield interfaces with distinctive characteristics, such as stoichiometry and microstructure. This provides a class of interfaces with a range of chemical and structural properties for basic studies of the formation of Schottky barriers.

Early studies have addressed the fundamental question concerning the effects of silicide chemistry and microstructure on the barrier height. Several phenomenological models have been proposed to correlate the barrier height to some thermodynamic parameters of the bulk silicide phase, e.g., the heat of formation[1] and the eutectic temperature of the silicide.[2] Results from studies addressing this question were reviewed in 1983.[3] The overall experimental evidence showed that variations in the material characteristics of the bulk silicide phase, e.g. stoichiometry and microstructure, do not influence the barrier height. In addition, synchrotron measurements on band bending[4] demonstrated that the barrier height can be established with a few monolayers of metal coverage. These observations led to the conclusion that the barrier at the silicide-silicon interface is a true interfacial property and the effect of silicide formation is to change the nature of the interface from one loaded with extrinsic imperfections, e.g. contaminations and process-induced defects, to one with intrinsic metal-silicon bonds.

It was emphasized that the central issue of understanding the barrier formation, namely the nature of the interface states, is not yet clear and has not even been measured properly. The difficulty in such measurements was attributed to the lack of sensitivity and energy resolution in the spectroscopy and electrical techniques employed for detecting the interface states.

Since then, new results have been obtained regarding the chemical trend,[5] the barrier formation at epitaxial interfaces as well as on the observation of interface states. In spite of a disagreement in the barrier height of certain nickel silicide interfaces, all these results can be assimilated to yield a reasonable and consistent picture for the barrier formation and interface states of the silicide-silicon system. This will be presented in this paper.

This paper first reviews the chemical trend of the barrier height. Then the results on the barrier heights and interface state measured for the epitaxial Ni silicides are discussed. There is a current controversy regarding the barrier heights of the type A and type B $NiSi_2$-Si(111) interfaces.[6,7] We will summarize the results of our studies including several recent experiments designed to examine the cause of the disagreement.

2.   CHEMICAL TREND OF THE BARRIER HEIGHT

In a recent paper,[5] the chemical trend in the variation of the barrier height of silicide-forming transition metals was analyzed. A correlation was proposed between the Schottky barrier height and the Miedema electronegativity of the transition metal. Among the various electronegativity or work function scales, the Miedema scale was used since it gives the most satisfactory results as well as expresses the chemical properties of the metal atom in solid state reactions.

The correlation of the barrier height to the Miedema electronegativity of the metal, without regard to the silicide stoichiometry was interpreted to reflect that the barrier height is defined by interfacial properties rather than by bulk properties of the two solids. It is proposed that the barrier height is dictated by the energy of the top-most occupied electron states at the interface, which may well be states bonding the metal to the semiconductor. Thus the electronegativity correlation expresses the energy dependence of the interface bond on the chemical nature of the metal atom of the silicide.

The correlation observed shows a systematic deviation from a straight line: there is a range of intermediate values of the electronegativity where the barrier height remains almost constant whereas it varies rapidly when the metal electronegativity is either very high or very low. It was proposed that this behavior could be explained by the presence in the semiconductor, close to the interface, of a half-filled, narrow band of electron states. These states are characterized by a rather small density (no more than 0.1 state per interface silicon atom), and independent of the metal in the silicide. As an alternative to a half-filled band, a density of localized states with a negative electron correlation energy, as proposed by Mailhiot and Duke,[8] would also explain the observed behavior.

The origin of such interface states is not well understood although capacitance spectroscopy measurements reveal electronics states of similar character at the $NiSi_2$-Si interfaces.[9] This will be discussed in a later section.

3.   BARRIER HEIGHTS OF EPITAXIAL Ni SILICIDES

To investigate barrier formation of Ni silicides, we have measured the barrier heights of type A, type B and mixed AB $NiSi_2$-Si(111) interfaces. To extend the range of structure studied, we have also studied a third epitaxial interface of NiSi (type C).[10] All measurements were made on clean interfaces formed under UHV conditions and to obtain meaningful statistics, a large number (more than 15 for each type except type C) of runs have been carried out. Since processing control and surface contamination can influence the barrier height, two series of experiments were performed to examine the effects of the silicide formation temperature and the incorporation of oxygen and carbon impurities. Finally, a series of experiments were made to check the effect due to the formation of a p layer during heat cleaning of the Si substrate.

Experiments were carried out in an UHV chamber with base pressure 10^{-8} Pa. Samples were prepared from n-type Si(111) wafers (phosphorus-doped and 8-10 Ω-cm) and an ohmic Ta back contact was provided for electrical measurements. Before metal evaporation, the sample surface was cleaned by several heating steps including a flash to 1050°C for about 10 seconds. Ni was e-beam evaporated from a separately pumped chamber in order to avoid contaminations. The Si surface and the Ni film prepared by this procedure showed no trace of C and O as detected by UPS and AES.

The epitaxial interfaces were prepared by first forming template structures.[11] For type A and type B interfaces, the template structures were formed by depositing 17 to 20Å and 5Å respectively, then annealed in a temperature range of 450 to 550°C for 3 to 5 minutes. The type C interface was formed by a similar procedure with template Ni thickness of 1.3 to 2Å and 250°C annealing.[10] The silicide phase and microstructure of the interface before and after reaction were identified by TEM and ion backscattering techniques.

In each run, Ni was evaporated through a metal mask defining three rows of diodes of 0.6, 1.3 and 2.5 mm in diameter and a blanket area. This was designed so that electrical measurement and structural analysis could be carried out on the same sample. Diode characteristics were measured by current-voltage (I-V) and photoresponse measurements. The values obtained from these two measurements coincided within 0.02 eV.

Barrier heights together with the ideality index are summarized in Fig. 1 for samples prepared using the procedures for type A, type B and mixed AB template structures. The structures of about 1/3 of these runs have been verified by cross-sectional TEM. The structure thus determined was found to correspond to the intended structure. For both A and B types of diodes with the high barrier height and good ideality, the interfaces exhibited atomically flat epitaxial structure.

Samples with a mixed orientation showed a less flat interface where interphase boundaries were clearly observable. The substrate alignment was found to influence only the ideality index but not the barrier height. The former clusters in the range of 1.00 to 1.08 for 0.25° misalignment and 1.05 to 1.3 for 1-2° misalignment.

The results in Fig. 1 show that a majority of the single-type diodes have ideality index between 1.0 to 1.2 and the scattering in the value of the barrier height is about 0.1 eV. Also, edge effects are often responsible for a high value of the ideality index. The variations in the data can be attributed to the difficulty in processing control for obtaining a highly perfect interfacial structure. An obvious evidence is the problem of substrate misalignment; the ideality index can be improved to the range of 1.00 to 1.08 by reducing the substrate misalignment from 1-2° to 0.25°. This is further supported by two series of experiments, one as a function of silicide formation temperature and the other by incorporating C and O impurities at the interface.[12] In the first series, the barrier height was found to be consistently high between 450° and 500°C and TEM observations for such samples revealed the most perfect, atomically flat interface. Above 500°C, the barrier height started to decrease and the silicide film began to aggregate and the perfection of the interfacial structure deteriorated. In the impurity experiments, carbon contamination of less than a monolayer coverage on the substrate surface was found to introduce atomic step defects at the interface and to reduce the barrier height to about 0.66 eV. The effect of oxygen is more complicated. Oxygen at submonolayer coverage does not lower the barrier height of the single-type interfaces. However, at higher coverages, oxygen can degrade the single-phase structure to mixed AB type or even polycrystalline, depending on the amount of coverage.

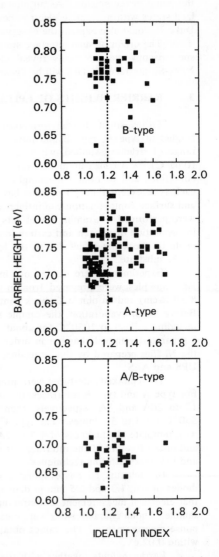

Figure 1. Summary of the barrier height and ideality index measured for samples prepared with type B, type A and mixed AB template methods.

In spite of these difficulties, it is clear from the results in Fig. 1 that type A and type B interfaces can be made reproducibly to have a barrier height of 0.75 ± 0.05 eV which is about 0.08 to 0.10 eV higher than the mixed type. The finding that the barrier height may not be determined by the specific epitaxy is further supported by the results from the type C interface. Although the data are more limited, the NiSi-Si(111) interface has been produced to yield barrier heights in the range of 0.73 to 0.78 eV. For this interface, the narrow range of template thickness (1.3 to 2Å) and the low formation temperature (about 250°C) make it more difficult to produce a structure with a high degree of perfection to yield the high barrier height.

Our results differ from other groups[6,13] who found a barrier height of 0.62 to 0.65 eV for the type A interface. The discrepancy may be related to different surface preparations. Tung and co-workers[6] used samples with a heavier doping and applied a cleaning procedure at a lower temperature, 850°C for several minutes, instead of 1050°C for 10 seconds used in our experiment. In the other study,[13] the Si surfaces were sputter cleaned and annealed at 850°C. We have found that the barrier height of both interfaces can be reduced by sputter cleaning even with subsequent 1100°C anneal. It has been suggested[14] that a p$^+$ surface layer can be induced on a moderately n-doped substrate by the heat cleaning used to prepare our samples and that such a p layer will raise the apparent barrier thus responsible for the high barrier observed for our type A samples. We have carried out experiments to check this possibility. Results are summarized in Fig. 2 where the acceptor density measured from spreading resistance and the expected barrier height increase calculated from the acceptor density are plotted as a function of heating time at 1035 and 1150°C. The thickness of the acceptor-accumulated layer was found to extend to about 100Å or less. Separate electrical measurements have been carried out to measure actual barrier changes. We found that a heating of 10 min. at 1100°C or 10 sec. at 1150°C is required to produce a measurable change of about 0.05 eV with an increase of n to about 1.4. This observation of high ideality index with increase of acceptor concentration indicates that when a p-layer develops after a high temperature anneal, the layer contains a large concentration of recombination centers. Such a layer would introduce certain changes in the IV and photoresponse characteristics, which were not observed in our measurements. In contrast, we have observed a significant number of A, B, and C interfaces exhibiting both a high barrier and a low value of the ideality index. From direct tests of heat cleaning effects on the Schottky barrier height, it is clear that the p layer created by our heat-clean procedure has a small effect. Based on the results in Fig. 2, we estimate the effect to be at most 0.01 eV, so it cannot account for the discrepancy. Another difficulty of this explanation is that since the same heating step was used to prepare all our samples, the p$^+$ layer should raise the barrier for all types of interfaces. The increase observed for type A barrier would have increased the barrier of the mixed AB interface to about 0.78 eV and the type B interfaces to above 0.9 eV; such high barriers have not been observed.

4.   INTERFACE STATES

The nature of the high and low barrier Ni silicide interfaces was investigated by measuring the interface states using an accurate-phase capacitance spectroscopy technique.[15] This method measures the junction capacitance under forward-bias

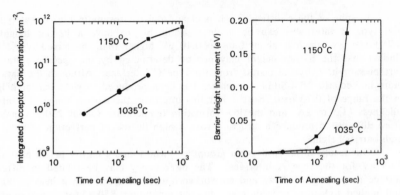

Figure 2.   Results of the measured integrated acceptor concentration and the calculated barrier change as a function of heating time for surface cleaning.

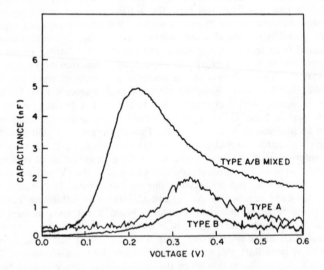

Figure 3.   Capacitance spectra observed at 100 Hz and R.T. for the type A, type B and mixed AB epitaxial $NiSi_2$-Si(111) interface.

voltages as a function of frequency and temperature. The capacitance is related to the capture and emission of electrons and holes by interface charge traps, thus capable of probing the density, and distribution of the interface states. A set of capacitance spectra measured at 100 Hz and room temperature are shown in Fig. 3 revealing characteristics similar for the two single-type interfaces. The overall features of the three spectra are similar except that their peak positions are shifted accordingly to the difference in barrier heights, indicating the same group of states is detected although the density is significantly higher at the mixed interface. Using a simple charge-capture model to calculate the density of states, we find that the density of the mixed interface is about $10\times$ that of the type B.

Although the role of these interface states in pinning the Fermi level has yet to be clarified, these measurements show that the single phases and mixed phase have a similar characteristic distribution of interface states. This reflects a correlation between the number of defect states and the degree of perfection of the interface.

5.   DISCUSSION

The epitaxial Ni silicides provide a unique system with interfacial structures which can be controlled to study the formation of Schottky barriers. Our experiments demonstrate that both A and B interfaces can be prepared to yield a high barrier height. Evidences collected from TEM observations, the type C interface, impurity incorporation and variation in silicide formation temperature all show that a high degree of perfection is of prime importance in obtaining the high barrier height. In spite of additional efforts, we are not able to resolve the discrepancy in the barrier height of the type A interface.

The observation of the interface states showed that the characteristics of the interface states are similar for the single and mixed-phase interfaces except that their density can vary by an order of magnitude according to the structure perfection of the interface. This suggests that different types of interfaces have the same group of interface states and their barrier heights are determined by the number of states available to pin the Fermi level. Thus the barrier height is determined primarily by the degree of perfection of the interface instead of the specific epitaxy.

In a related study of interface states at the Pd_2Si-Si(111) interface,[16] the temperature and frequency variations of the capacitance spectra can be explained by postulating an effective single-level state lying opposite the metal Fermi level with a density of about $1\times10^{12}/cm^2$. Since this is difficult to distinguish experimentally from a narrow group of states, this provides a direct evidence to support the previous postulate of existence of such states.[5] Additional measurements on Ti silicide-Si interfaces reveal interface state capacitance with characteristics similar to that of Ni and Pd silicides. Summarizing these results, a self-consistent picture seems to have emerged to correlate the observed interface states to the chemical trend of the barrier height at the silicide-silicon interface.

The origin of the interface states is not well understood although it seems that the results from the silicide-silicon system have advanced the study of Schottky barrier formation to a point where the central question on the microscopic origin of the interface states can now be investigated. Although Fermi level pinning by such defect centers has not been directly observed, this picture is consistent with the shape of one correlation observed with the Miedema electronegativity representing the work functions of the metal atom.

REFERENCES

† Work supported in part by the Office of Naval Research.

* Ecole Polytechnique de Lausanne, CH 1015 Lausanne, Switzerland.

1.   Andrews, J. M., and Phillips, J. C., *Phys. Rev. Lett.* **35**, 56 (1975).
2.   Ottaviani, G., Tu, K. N., and Mayer, J. W., *Phys. Rev. Lett.* **44**, 284 (1980).
3.   Ho, P. S., *J. Vac. Sci. Technol.* **A1**, 745 (1983).
4.   Purtell, R. J., Hollinger, G., Rubloff, G. W., and Ho, P. S., *J. Vac. Science Technol.* **A1**, 566 (1983); Purtell, R. J., Ho, P. S., Rubloff, G. W., and Schmid, P. E., *Physica* **117B**, 834 (1983).
5.   Schmid, P. E., *Helvetica Physica Acta* **58**, 371 (1985).
6.   Tung, R. T., *Phys. Rev. Lett.* **52**, 461 (1984).
7.   Liehr, M., Schmid, P. E., LeGoues, F. K., and Ho, P. S., *Phys. Rev. Lett.* **54**, 2139 (1985).
8.   Duke, C. B., and Mailhiot, C., *J. Vac. Sci. Technol.* **B3**, 1170 (1985).
9.   Ho, P. S., Yang, E. S., Evans, H. L., and Wu, X., *Phys. Rev. Lett.* **56**, 177 (1986).
10. Clabes, J., Liehr, M., LeGoues, F. K., and Wittmer, M., to be published.
11. Tung, R. J., Gibson, J. M., and Poate, J. M., *Phys. Rev. Lett.* **50**, 429 (1983).
12. Liehr, M., Schmid, P. E., LeGoues, F. K., and Ho, P. S., *J. Vac. Sci. Technol.* **B3**, 1190 (1985).
13. Hauenstein, R. J., Schlesinger, T. E., McGill, T. C., Hunt, B. D., and Schowalter, L. J., *Appl. Phys. Lett.* **47**, 853 (1985).
14. Tung, R. T., Levi, A. F. J., Gibson, J. M., Ng, K. K., and Chantre, A., *Proc. of MRS Symposium,* Vol. 54, 457 (1986).
15. Evans, H. L., Yang, E. S., and Ho, P. S., *Appl. Phys. Lett.* **46**, 486 (1985).
16. Evans, H. L., Wu, X., Yang, E. S., and Ho, P. S., to be published in J. Appl. Phys. November (1986).

STUDIES ON THE EVOLUTION FROM Ni/Si(111) AND Ni/Si(100) TO THE FORMATION OF NICKEL DISILICIDE/SILICON INTERFACE

Xie Xide,Zhang Kaiming,Tian Zengju and Ye Ling

Institute of Modern Physics,Fudan University

Shanghai, People's Republic of China

ABSTRACT

The cluster model and the DV-Xα methods are used to determine the configurations of Ni/Si(111) and Ni/Si(100) during the initial stage of deposition. The possible precursor states for the formation of nickel disililcides on silicon for both cases are dicussed.

1. Introduction

In recent years studies on transition metal silicides have attracted broad interest from both theoretical and experimental point of view. It has been shown by numerous experimental results[1-3] that transition metals are highly reactive, silicides with low resistivities can be formed at relatively low temperature. It was found by Tung et al.[4] that the epitaxial $NiSi_2$ has the same orientation as the Si(111) substrate (type A) if the initial deposit of nickel is large and that it can rotate $180°$ with respect to the substrate (type B) if the initial deposit of nickel is small. It is hard to tell about the orientation of the epitaxial layer if the coverage of Ni is extremely low. As for Ni on Si(100), results from both LEED and UPS showed that Ni can go into the largest interstitial hollow site[5,6] . Due to the complexity of the Ni/Si interfaces, although much studies have been carried out, there still exist quite a number of unsettled questions regarding the geometrical configuration of Ni on Si and its nature of interactions during the early stage of depositions. As for the mechanism of formation, it was suggested by Bisi et al.[7] that during the early stage of growth, Ni atoms could go into the tetrahedral interstitials below the Si

surface. However, from the results of SEXAFS, Comin et al.[8] suggested
that after the deposition of Ni on Si, rearrangement of surface bond
might be induced, incomplete sixfold or sevenfold coordination similar
to that in CaF_2 might be formed. Using the LDF and the LAPW method
Hamann et al.[9] supported the SEXAFS model. In order to make further
studies of the mechanism of interface formation of Ni/Si, in the
present work the problem was investigated using the cluster model and
the DV-Xα method. The possible adsorbed sites of Ni/Si(111) and
Ni/Si(100) were studied. Details of possible bonding under low
coverage were discussed and possible relationship between the early
stage of deposition to the formation of $NiSi_2$ explored.

2.   Methods of Calculation

In the present work the Discrete Variational Xα(DV-Xα) method[10]
and the frozen core approximation are used. Conventional cluster
models are adopted to study the possible chemisorbed sites.

Fig.1(a)(b)are configurations
for the top site and the threefold
hollow site adsorptions on Si(111).
Fig. 1 (c) (d) are the fourfold
hollow site and the bridge site
adsorption configurations on the
Si(100) surface. H atoms which
are used to saturate the dangling
bonds for simulating the semi-
infinite surfaces are not shown
in the figure. Bulk interatomic

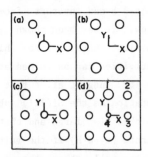

distances are used through out the
calculations. The z direction is

Fig.1 Various Adsorbed Configurations of Ni/Si(111)
and Ni/Si(100)

perpendicular to the surface and the origin is taken at the surface
of the substrate. In all configurations, the adsorbed sites of Ni
atoms vary along the z direction . The symmetry of configurations in
Fig.1 (a) (b) belong to the point group C_{3v} , whereas that of
configurations in Fig.1(c)(d) belong to point C_{2v}.The binding energy
E_b of the system is given by

$$E_b = E_t(A) - E_t(B)$$

where $E_t(A), E_t(B)$ are total energies of the cluster and that of the free atoms respectively.

3.   Results

For adsorption of Ni atoms on silicon surfaces, the optimum adsorbed site is determined by the minimum of the curve for total energy of the system versus $d_\perp$, where $d_\perp$ is the distance between the Ni atom and the surface. The energy levels of the cluster and the density of states (DOS) are calculated for the most stable adsorbed sites. The density of states are obtained through the Gaussian expansion of the energy levels, in the present paper the width of the Gaussian expansion is chosen as 0.2 ev.

3.1. Adsorption on An Ideal Si(111) Surface

Fig.2 (a)(b) give the curves of total energy for systems corresponding to the top site and the threefold hollow site adsorption on an ideal unrelaxed Si(111) surface. The stable adsorbed positions for the top adsorbed site and the threefold hollow site are at $d_\perp$ =2.27A and 1.69A respectively. The energy of adsorption for the latter is lower by 0.6ev. Therefore the threefold hollow site is more stable.

Fig.3(a)(b) give the density of states for the top site and the threefold hollow site adsorption configurations respectively, the dashed line represents the density of states for the clean surface. The half filled dangling bond peak lying at the center of the energy gap of the clean surface (zero energy) is contributed by the $3p_z$ orbital and 3s orbital of the first layer silicon atom. For the top site adsorption, the dangling bond is fully saturated by the adsorption and is represented by peak B in Fig.3(a). For the threefold hollow site the dangling bond is only partially saturated (Fig.3(b)). Peak A is formed by the interaction of the $3p_z$ states of the substrate and the Ni 3d orbitals. The strong adsorbed peak B at -1.1ev is the bonding state of the Ni 4s-3d electrons and the Si $3p_z$

orbitals, corresponding to the $5a_1$ and $6e$ states. At higher energies lie
the non-bonding states of Ni, corresponding to the $7e$ and $6a_1$
states of Fig.3(b). Due to the interaction with nickel, the 3e state
in the main peak of the $3p_z$ electron of the clean surface in the
valence band (Peak C in Fig.3(d)) moves towards the lower energy side
lying at peak C of Fig.3(b).

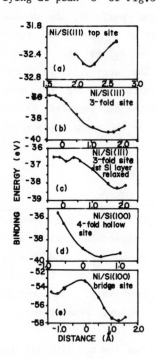

Fig.2 Various Curves of Energy
Corresponding to Fig.1

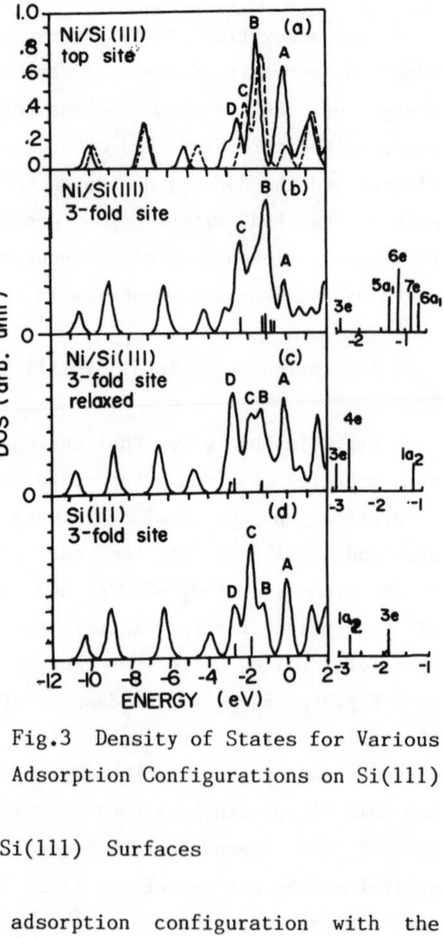

Fig.3 Density of States for Various
Adsorption Configurations on Si(111)

3.2. Adsorption on The Relaxed Si(111) Surfaces

A threefold hollow site adsorption configuration with the
outward relaxation of the first layer silicon atoms is used for
studying the microscopic mechanism concerning the possibility of
Ni atoms going into Si surfaces. Results show that if the Si atoms of
the first layer relax outward for 0.53 Å, it is possible for Ni atoms
to reach a position of energy minimum between the first two layers.

The curve of total energy versus $d_\perp$ is calculated for this configuration and the result is given in Fig.2(c). It can be seen that after overcoming the potential barrier caused by the first layer silicon atoms, nickel atoms can reach a stable position with the lowest potential energy at $d_\perp = 0.16$ A. It can be seen from Fig.3(c) that the peak A at E_f still keeps the characteristic of a dangling bond of the clean surface. After adsorption, the peak B near the top of the valence band has been enhanced significantly, this might be due to the fact that after the penetration of the Ni atom into the surface, the covalent bonds between Si atoms can be disrupted and the la_2 level in the peak D (Fig.3 (d)) moves towards the higher energy side. The density of states between peak A and B of Fig.3(c) increases due to the contribution of a small portion of the Ni 3d components. After the adsorption of Ni the characteristic peak C corresponding to the covalent bond of the clean surface splits, with one part moves towards the lower energy side forming a more stable sp-d hybridized peak D (the 3e and 4e states in Fig.3(c)) at -2.7ev.

3.3. Adsorpton of Ni on Si(100)

Fig.1(c) and Fig.1(d) represent the fourfold hollow site and the bridge site adsorption configurations respectively. The curves of total energy versus $d_\perp$ obtained are shown in Fig.2(d) and Fig.2(e). For the fourfold hollow site, the most stable adsorbed site determined by the energy minimum is at $d_\perp = 0.53$ A, whereas for the bridge site adsorption, the most stable positions are at $d_\perp = 1.33$A and at $d_\perp = -1.06$A respectively. The latter is nearer to the largest hollow site inside the diamond structure of Si. Except for one silicon atom missing, the Ni atom and the Si neighbors form the so-called adamantine structure. This result seems to agree with that given by the experiment and also with the previous theoretical analysis. The energy of adsorption at the bridge site is lower than that at the fourfold hollow site by 1.22 ev, this shows that the bridge site might be a more stable one for Si(100) surface.

Fig.4 gives the density of states for various configurations on Si(100). The solid line in Fig.4(a) corresponds to the case when Ni

atom lies at $d_\perp$ = 0.53A above the fourfold hollow site, whereas
the dashed line represents the density of states of the clean surface.
It can be seen that there exists states (peak A) consisting of
dangling bond states. States of Si 3s and $3p_z$ are occupied, whereas
those from $3p_x$ remain empty (energy level $5b_2$ and $7a_1$ in Fig.4(a)).
After the absorption of Ni atom, an enhanced peak shows up at the
energy region between -0.5ev and -1.5 ev, the center of the peak is
at B. This peak is due to the interaction of the Ni 3d orbital
and the $3p_z$ and part of 3s of the
Si atoms. As a result of interaction of the Ni 3d states and the
dangling bond states, the density
of states extends continuously
from the dangling bond peak A to
lower energies.The unfilled states
of $3p_x$ near the Fermi level ($5b_2$)
still exist, this might explain
the fact that after the adsorption
of Ni, the surface has metallic
property. Comparing to that of
the clean surface, the peak C at
-1.9 ev is somewhat enhanced due
to the interaction· of the Ni 3d
and the Si $3p_y$ orbitals. The
adsorbed peak D at -3.3 ev is the
bonding peak between the Si 3s and
the Ni 4s state.

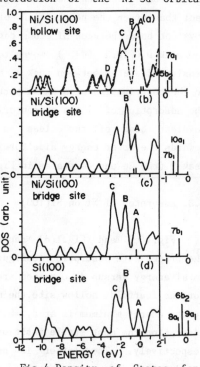

Fig.4 Density of States for
Various Adsorption Configurations on Si(100)

The density of states of
the clean surface for the bridge
site adsorption is given by
Fig.4(d). With the adsorption of Ni atom on the surface at a distance
of $d_\perp$ =1.33A above the bridge site, the density of states is given
in Fig.4(b). It can be seen that there exists strong interaction
between the Ni atom and the dangling bond of Si, forming a bonding
peak A at a position lower than peak A of the clean surface by
an energy of 0.5ev. There also exist some contributions from the Ni

non-bonding orbital of $3d(7b_1$ and $10a_1)$ in the peak A. The peak B at −1.4ev, the density of which is somewhat increased comparing with that before the adsorption,is mainly due to the interaction of the Si $3p_x$, $3p_y$ and the Ni 3d states, the density of states below −2.0ev remains unchanged. When Ni atom goes underneath the surface to the position at $d_\perp$ =−1.06A, the density of states is given by Fig.4 (c). The peak A near to E_f is enhanced comparing to that before the adsorption, this is due to the contribution of the Ni non-bonding states($7b_1$ states). The peak C due to $3p_z$ states changes significantly and forms a strong adsorbed peak C at −2.8ev which is characteristic for the adsorption at the interstitial site. The peak C is due to the bonding of the Ni atom with the neighboring Si atoms. The peak at −4.0ev is mainly due to the contribution of the Si 3s orbitals.

4.   Discussions

(1) From the results of UPS and LEED given by Chang et al.[5,6],Ni atom can sit in the tetrahedral interstitial site during the low coverage of deposition. It was discovered by the authors that there exist two strong peaks at −2.78ev and −1.90ev, and the peak at −2.78ev had significant d characteristic, the −1.90ev peak is due to surface adsorption. It was also discovered that with the increase of the Ni coverage, a strong peak appeared at −1.4ev. It is quite apparent that the −2.78ev might be the same as the peak C at −2.8ev for the interstitial adsorption configuration of the present study. The −1.9ev peak from the experiment might be the peak B at −1.4ev. The deviation might be due to the fact that the 2x1 reconstruction of Si(100) has not been taken into account.

(2) It was known that the basic characteristic of the Ni silicide is that the Ni d orbitals will hybridize with the Si 3p orbitals forming the bonding,non−bonding and the anti−bonding states. From the present work the bonding states of p−d also lies at lower energy side from the non−bonding d states. The density of states is not as large as that of the silicides since less Ni content is involved. With the penetration of Ni atoms into the surface, the bonding states move towards the lower energy side. For example, the peak B at −1.1ev

(Fig.3(b)) for the threefold hollow site adsorption above the surface
moves to D at $-2.7ev$ (Fig.3(c)) when Ni atoms move inside the surface,
the peak B at $-1.2ev$ (Fig.4(b)) for the bridge site adsorption above
the Si(100) surface moves to peak C at $-2.8ev$ (Fig.4(c)) when Ni
atoms go into the interstitial sites, the peak is getting near to the
bonding peak (about $-5.0ev$ below E_f) for $NiSi_2$. Therefore it is
quite likely that the diffusion of Ni into Si might be considered
as a precursor state for the formation of Ni silicide.

REFERENCES

1)   Miller,J.N., Schwarz,S.A., Lindau,I., Spicer,W.E., Demichelis,B.,
     Abatti,I.,and Braicovich,L., J.Vac.Sci.Technol.,17(5) 920 (1980).
2)   Kobayashi,K.L., Sugaki,S., Ishizaka,A., Shiraki,Y., Daimon,H.,
     and Murata,Y., Phys.Rev.,B25, 1377 (1982).
3)   Schmid,P.E., Ho,P.S., Foll,H., and Tan,T.Y ., Phys.Rev.B28 4593
     (1983).
4)   Tung,R.T., Gibson,J.M. and Poate,J.M., Phys.Rev.Lett., 50, 429
     (1983).
5)   Chang,Y.J., Erskine,J.L., Phys.Rev.B26,4766(1982). Phys. Rev.B26,
     7031(1982).
6)   Chang,Y.J.,and Erskine,J.L., Phys.Rev.B28,5766(1983).
7)   Bisi,O., Chiao,L.W.and Tu,K.N., Phys.Rev.B30 4664(1984).
8)   Comin,F.,Rowe,J.E.and Citrin,P.H., Phys.Rev.Lett.,51,2402(1983).
9)   Hamann,D.R.,Mattheiss,L.F., Phys.Rev.Lett.,54,2517(1985).
10) Ellis,D.E.,Painter,G.S., Phys.Rev.B2, 2887(1970).

ELECTRON TRANSPORT THROUGH
EPITAXIAL METAL/SEMICONDUCTOR HETEROSTRUCTURES

R. T. Tung, A. F. J. Levi, J. L. Batstone,
J. M. Gibson and M. Anzlowar
AT&T Bell Laboratories, Murray Hill, N.J. 07974
U. S. A.

ABSTRACT

Electron transport through epitaxial $NiSi_2$–Si interfaces have been studied and the Schottky barrier height (SBH) is found to depend on the epitaxial orientation. On n-type Si, SBH's of 0.49eV, 0.65eV and 0.79eV were obtained for $NiSi_2$/Si(100), type A $NiSi_2$/Si(111) and type B $NiSi_2$/Si(111) interfaces, respectively. Care must be exercised in substrate preparation to allow meaningful SBH determination as surface contaminants can give rise to p-n junction formation.

Advances in the technology of silicide fabrication have allowed the formation of epitaxial, single crystal, silicide films on silicon.[1] In particular, the system consisting of $NiSi_2$ on Si(111) and (100) has been shown to be nearly perfect in epitaxial structure,[1] and the atomic structure of the interface has been modelled by high resolution transmission electron microscopy (TEM).[2] [3] Recent interest in these epitaxial interfaces from the stand point of Schottky barrier (SB) formation[4] [5] [6] has stemmed from the hope that their unique, near perfect, structure can lend much-needed clues in solving the SB problem. For the first time, a direct correlation between the observed SBH and the physical parameters such as crystal orientation is possible because of the homogeneity of the interface structure.

The growth of epitaxial $NiSi_2$ layers was by the template technique[1] and experimental procedures regarding TEM analyses, device processing, current-voltage (I-V), capacitance-voltage (C-V), electron beam induced current (EBIC) have been detailed elsewhere.[3][7] It has been shown that the interface Fermi level position for type A is higher in the bandgap than that of type B $NiSi_2$.[4][6] This is manifested by a higher type B SBH on n-type Si over the type A SBH (0.79 eV vs. 0.65 eV), and a corresponding lower type B SBH on p-type Si than the type A SBH (0.34 eV vs. 0.47 eV). Richardson's plots for a type A and a type B $NiSi_2$ layer on p-type Si(111) are shown in Fig.1 as an example. Near ideal diode characteristics were observed, namely long and straight linear relationship for log I, and $1/C^2$ plots from I-V and C-V measurements and good ideality factors for I-V traces. SBH's determined from I-V, C-V, and activation energy studies are all in good agreement with each other. In addition, they were found to be independent of Si doping concentration and small Si wafer misorientation.[4] The dependence of SBH on orientation has recently been observed as contrast in EBIC images of mixed type A + B diodes.[8]

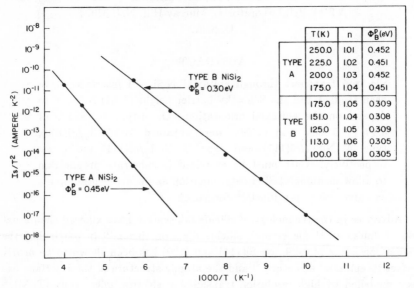

Fig. 1. Richardson's plots of a type A and a type B NiSi$_2$ layer on p-type Si(111).

On n-type Si(100), the SBH of uniform epitaxial NiSi$_2$ has been deduced to be 0.49 ± 0.03 eV. Layers grown from reacting ∼ 11−15 Å of deposited nickel fall into this category. Shown in Fig. 2 is a Richardson's plot of the extrapolated zero-bias forward currents of such a diode. along with the SBH's and the ideality factors determined at each individual temperatures. As may be seen the SBH's determined from I-V measurements are in good agreement from those deduced by the activation energy method (see Fig. 2). Moreover, the SBH deduced from C-V measurements is in good agreement with that obtained from I-V characteristics and the capacitance is independent of the testing frequency. A value of 0.60 ±0.3 eV has been deduced as the SBH of uniform single crystal NiSi$_2$ on p-type Si(100). Shown in Fig.3 is the C-V plot for a p-type Schottky diode.

TEM analyses of NiSi$_2$ layers under different conditions on Si(100) has revealed a large variation in the interface morphology. The associated SBH showed some dependence on interface parameters such as facet distribution and step density. A detailed discussion of this correlation is contained in a separate publication.[9]

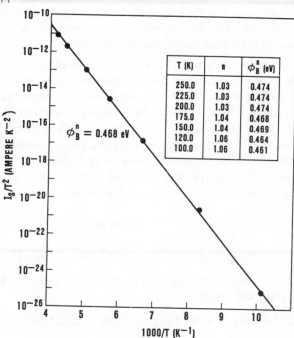

$$\phi_B^n = 0.468 \text{ eV}$$

T (K)	n	ϕ_B^n (eV)
250.0	1.03	0.474
225.0	1.03	0.474
200.0	1.03	0.474
175.0	1.04	0.468
150.0	1.04	0.469
120.0	1.06	0.464
100.0	1.06	0.461

Fig. 2. Richardson's plot of a NiSi$_2$ diode on n-type Si(100).

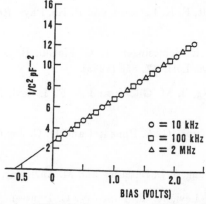

Fig. 3. C-V plot of a NiSi$_2$ diode on p-type Si(100), measured at 77 K.

Passivation of defect states at the interface through hydrogenation does not have a first order effect on the SBH of single crystal silicides[10] and DLTS studies of the silicide silicon interfaces show no evidence for the existence of defects. Although many effects may influence the SBH, it is believed that the microscopic SB mechanism for single crystal systems is not different from that for polycrystalline M-S systems. However, the systematic variation of the SBH is most clearly revealed in single crystal systems because they do not suffer from inhomogeneity.

Recently, Liehr et al[5] reported type A and type B $NiSi_2$ have the same SBH (0.78 eV) on n-type Si(111). However, because their substrate preparation technique has repeatedly been shown to induce shallow p-n junction on the surface,[7] it is not clear whether these experimentally obtained "apparent SBH" bear any relation to the true interface Fermi level position. The influence of a p-n junction on capacitance measurements may also be quite significant.[11] Under careful experimental conditions where no doping compensation occurs, the dependence of SBH on epitaxial orientation has been firmly established by us[4][7] and independent research groups.[6]

REFERENCES

1. R. T. Tung, J. M. Gibson and J. M. Poate, Phys. Rev. Lett. **50**, 429 (1983).

2. D. Cherns, G. R. Anstis, J. L. Hutchison and J. C. H. Spence, Philos. Mag. **A 46**, 849 (1982).

3. J. M. Gibson, R. T. Tung and J. M. Poate, MRS Symp. Proc. **14**, 395 (1983).

4. R. T. Tung, Phys. Rev. Lett. **52**, 461 (1984); J. Vac. Sci. Technol. **B 2**, 465 (1984).

5. M. Liehr, P.E. Schmidt, F. K. LeGoues and P. Ho, Phys. Rev. Lett. **54**, 2139 (1985).

6. R. J. Hauenstein, T. E. Schlesinger, T. C. McGill, B. D. Hunt and L. J. Schowalter, Appl. Phys. Lett. **47**, 853 (1985).

7. R. T. Tung, K. K. Ng, J. M. Gibson and A. F. J. Levi, Phys. Rev. **B 33**, 7077(1986).

8. J. M. Gibson, R. T. Tung, C. A. Pimentel and D. C. Joy, Inst. Phys. Conf. Ser. **76**, 173 (1985).

9. A. F. J. Levi, R. T. Tung, J. L. Batstone and M. Anzlowar, to be published.

10. A. Chantre, A. F. J. Levi, R. T. Tung, W. C. Dautremont-Smith and M. Anzlowar, Phys. Rev. **B**, in press.

11. P. S. Ho, E. S. Yang H. L. Evans and X. Wu, Phys. Rev. Lett. **56**, 177 (1986).

INITIAL STAGES OF LIGHT-INDUCED INTERFACE REACTION OF Ni FILMS WITH SILICON

A.C. Rastogi*, P.K. John, M. Dembinski+, B.Y. Tong, X.W. Wu and S.K. Wong‡

Department of Physics and Centre for Chemical Physics
University of Western Ontario, London, Canada N6A 3K7

* Permanent address: National Physical Laboratory, New Delhi
+ On leave from Polish Academy of Sciences, Warsaw
‡ Department of Chemistry

ABSTRACT

Changes in the Ni-Si interface when irradiated with intense ~60 μs light pulses have been studied by SEM, Auger, X-ray and Schottky analysis. The annealing results in diffusion of Ni into Si which introduces structural disorder in the Si. When the Ni concentration in Si reaches some finite value silicide formation takes place. NiSi is the first phase to grow after a total of ~ 1ms of light irradiation.

Most studies of silicides have been done under conditions of prolonged annealing (1-3). With considerable utilization of interface devices in VLSI, fast pulse annealing has become attractive. Annealing with successive short pulses would provide deeper insight into the evolution of the interface processes and their dynamics. We have investigated a proto-type interface between Ni film of a constant thickness ~60 monolayers and (100) silicon. The interface reactions were induced by intense pulses of white incoherent light irradiation of ~60 μs duration.

The samples were properly polished and cleaned Si (100) wafers on to which ~300ºA of Ni film is deposited by e-beam evaporation. The samples were mounted in a chamber and irradiated from the Ni side by incoherent white light from a pulsed plasma arc described elsewhere (4). The duration of the light pulse was ~60 μs and the incident energy density of each pulse, ~45 J/cm². Structural, electrical and

chemical changes at the interface were studied as a function of time
by examining the sample after successive light pulse irradiation.
Auger depth profile, Auger line shapes, SEM, glancing angle X-ray
diffraction and Schottky barrier measurements were used in the
analysis.

RESULTS AND DISCUSSION

Light Absorption: The incident light is absorbed within the Ni film
and the interface is directly heated by diffusion of the heat. The
thermal diffusion length in our case is ~100 μm, and thus one can
assume the sample is rather uniformally heated for the duration of
the pulse.   The rise in temperature ΔT of the interface depends on
incident light energy (E), its duration τ, reflectivity R of Ni,
thermal diffusivity k and conductivity K of Si, and is given by

$$\Delta T = 2E \ (1-R) \ k^{1/2} \Big/ K(\pi\tau)^{1/2} \text{ and is } \sim 1000 \text{ K for one pulse.}$$

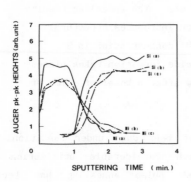
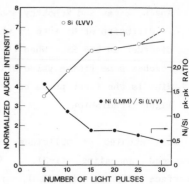

Fig.1 Auger Ni and Si Composi-   **Fig.2** Si(LVV) intensity at inter-
tion changes across interface.   face after light irradiation.

Compositional effects at interface: Auger depth profiles of the
sample after successive irradiations are shown in Fig. 1.   Initially
diffusion of Ni takes place into Si and continues for about the first
1 msec i.e. up to 15-20 pulses.   On further irradiation, a shift of
the Si profile towards the Ni side suggests that in addition to the
Ni diffusion, Si atoms tend to migrate toward the Ni. Figure 2 shows
first derivative normalized $Si(L_{23}VV)$ Auger peak intensities measured
directly at the interface for various light irradiation times, after
selectively etching away the bare Ni.   The $Si(L_{23}VV)$ Auger intensity
in the early stages is about 30% of that for Si before irradiation.

A progressive increase in its intensity with increasing pulse
irradiation indicates that Ni migrates deeper and/or silicon atomic
concentration in the intermix layers increases.   The Si Auger LVV
growth curve shows two linear regions; one until ~15 pulses and the
other between 15 and 27 pulses.   This break suggests a sudden change
in the structure of Ni-Si intermixed over-layers.   A comparison of
characteristic Si LVV line shapes at 92 eV in the two regions shown
in Fig. 3 with that of unannealed Si indicates that in the early
stages intermixing is purely physical in origin.   Subsequently a
splitting of the line shows that Si is now in an environment
chemically different from that of free Si suggesting transformation
of intermixed region into a (Ni- Si) chemical compound.   An estimate
of the stoichiometry of this compound from Si(LVV)/Ni(LMM) peak to
peak ratios suggest it to be either mixture of NiSi and $NiSi_2$ or
2NiSi and the remaining unreacted Si.   X-ray diffraction however
confirms it to be NiSi of preferred (200) crystalline orientation.

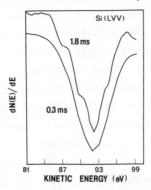

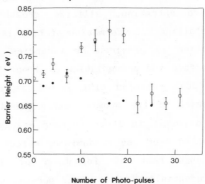

Fig.3 Si(LVV) line shapes   **Fig.4** Schottky barrier heights
before and after silicide growth.   as a function of light pulses.

Electrical behaviour of interface: Electrical characterization of
interface through an analysis of barrier heights after successive
light pulses is shown in Fig. 4.   Initial rise in barrier heights for
light irradiation is probably due to a shift of the Fermi level.
This could arise because of the formation of doubly charged Si
defects which makes the Si surface amorphous due to Si vacancies and
newly formed metallic bonds with Ni thus creating localized states in
the band gap.   A rapid decrease in barrier heights to a value usually

observed for NiSi/Si Schottky diodes indicates silicide formation. After the silicide formation has occurred additional pulse annealing does not change barrier height.   It seems that growth of silicide once nucleated occurs by simple atomic diffusion and reaction process and not through an amorphous-to-crystalline.   We have however observed evidence of the amorphous layer by X-ray on additional light annealing.   This is likely due to a melting process at the newly created interface.

Silicide nucleation model: The compositional and electrical data presented here suggest the following model of interaction of Ni atoms with Si surface.   Initially Ni diffuses into the Si.   As the diffusion of Ni continues, it results in the weakening of Si-Si covalent bonds, and formation of Si vacancies.   Thus an amorphous surface is created (5).   The Si surface after a short term annealing (~1 ms) thus supports an intermixed disordered Ni-Si overlayers.   On further annealing with continued migration of Ni, at some threshold Ni concentration, silicide growth occurs through amorphous-to-crystalline transformation of this layer.   The driving force for this reaction is provided by vacancies in the Ni lattice adjacent to interface and accumulation of free 'Si' due to out diffusion of Si. An <200> oriented NiSi growth is predominant which suggests that kinetics of the atomic rearrangement led to local ordering.   This would result in lower surface energy.   Accordingly this process is favoured over the randomizing effect of the atomic diffusion of Ni. Although heat of formation of Ni_2Si is higher, NiSi grows preferentially in our case, because of the heat of formation of NiSi is maximised due to presence of excess Si for the reaction (1).

References:

1. S.P. Murarka, Silicides for VLSI Applications, Academic Press,1983.

2. N.W. Cheung, P.J. Grunthaner, F.J. Grunthaner, J.W. Mayer and B.M. Ullrich, J. Vac. Sci. Technol. 18, 917, 1981.

3. K.N. Tu and J.W. Mayer, in Thin Films - Interdiffusion and Reactions, Eds. Poate, Tu and Mayer, John Wiley, New York, 1978.

4. P.K. John, S. Gecim, Y. Suda, B.Y. Tong and S.K. Wong, Can. J. Phys. 63, 876, 1985.

5. A.C. Rastogi, P.K. John, B.Y. Tong and S.K. Wong, to be published.

FORMATION OF EPITAXIAL SILICIDES:
IN SITU ELLIPSOMETRIC STUDIES

S. M. Kelso, R. J. Nemanich,* C. M. Doland, and F. A. Ponce

Xerox Palo Alto Research Center
3333 Coyote Hill Road, Palo Alto, CA 94304
USA

ABSTRACT

We have studied the formation of epitaxial Pd_2Si films on clean Si⟨111⟩ surfaces using *in situ* spectroscopic ellipsometry. Pd films of 5 to 50 Å thickness were studied both as deposited and after annealing at 300°C. In the as–deposited films the fraction of silicide increased with film thickness, up to about 40%, while the annealed films were completely reacted.

1. INTRODUCTION

Metal–semiconductor interfaces form an important class of interfaces between dissimilar materials. The reaction between certain metals and silicon results in an epitaxial silicide–silicon structure which is a building block for future multilayer electronic devices with unique properties.

The palladium–silicon system has been studied by Raman spectroscopy[1], Auger and photoemission spectroscopies[2,3], and transmission electron microscopy[3]. Pd reacts at room temperature to form crystalline Pd_2Si, and no other phases are expected for temperatures less than 700°C. Because of the lattice matching condition to Si⟨111⟩, epitaxial interfaces can be formed.

In this paper we study the formation of epitaxial Pd_2Si films on Si⟨111⟩ surfaces using *in situ* spectroscopic ellipsometry. This technique provides

- - - - - - - - - - - - - - - -

*Present address: Department of Physics, North Carolina State University, Raleigh, NC 27695 USA.

important new information about metal–silicon interactions: the amounts of silicide and unreacted metal are determined quantitatively, and interface widths and compositions can be probed.

2.   EXPERIMENTAL

The experiments were carried out in an ultra–high vacuum (UHV) chamber with strain–free quartz window assemblies. The rotating polarizer ellipsometer has a spectral range of 1.5–6 eV and an angle of incidence of 66.54°. Samples were cleaned in UHV by standard sputter–and–anneal or Shiraki cleaning procedures. Films were prepared by successive depositions of Pd, from 5 Å up to a total of 50 Å. Dielectric function spectra were measured *in situ* for as–deposited films and after annealing at 300°C for 10 min to complete the reaction to silicide. Surface conditions were monitored after each step by Auger electron spectroscopy. Further details are given elsewhere.[4]

Complex pseudodielectric function spectra $\langle \varepsilon \rangle = \langle \varepsilon_1 \rangle + i \langle \varepsilon_2 \rangle$ were recorded assuming a two–phase (substrate–ambient) sample configuration. Corresponding model spectra were calculated using various multi–layer models and the Bruggeman effective medium approximation and were fit to the data to determine the amounts of silicide and unreacted metal.

3.   RESULTS AND DISCUSSION

The condition of the UHV–cleaned Si surface was studied by Auger spectroscopy and ellipsometry. The Auger spectra typically showed less than 2% of a monolayer of C and O on the surface. The annealing temperatures were expected to result in a 7x7 reconstruction, which is usually interpreted to indicate a high degree of crystallization. The $\langle \varepsilon_2 \rangle$ spectrum of a UHV–cleaned surface is shown as the curve labelled 0 Å in Figure 1. The reduction in the peak value near 4.25 eV, as compared to chemically

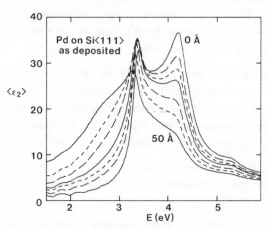

Figure 1.   Imaginary part $\langle \varepsilon_2 \rangle$ of the pseudodielectric function for Pd on Si.

cleaned crystalline Si,[5] indicates either a surface layer of disordered and/or density-deficient material much thicker than has been calculated for any reconstruction, or more extended subsurface damage of a lesser degree.

A series of $\langle \varepsilon_2 \rangle$ spectra is shown in Fig. 1 for as-deposited Pd films with thicknesses 0–50 Å. The spectra for annealed films look qualitatively similar, so modeling is necessary to obtain a quantitative description. Several models were considered, consisting of one or two overlayers and the starting spectrum of the series as the substrate. The models were (1) Pd only, (2) Pd_2Si only, (3) Pd on top of Pd_2Si, and (4) a mixture of Pd and Pd_2Si. Each model was fit simultaneously to both the $\langle \varepsilon_1 \rangle$ and $\langle \varepsilon_2 \rangle$ spectra measured for each film.

From the annealing conditions, it was expected that the annealed films had completely reacted to form crystalline, epitaxial Pd_2Si. Figure 2 shows a cross-sectional transmission electron micrograph of an annealed film after ellipsometry measurements. All areas of the film showed excellent epitaxy, and this area had very abrupt interfaces on both sides of the film. The dielectric function fits confirmed that the reaction was complete: models 3 and 4 gave negative Pd thicknesses and fractions, and the fitting residual was significantly better for model 2 than for model 1.

Figure 2.   Transmission electron micrograph of a Pd film annealed to form Pd_2Si on Si⟨111⟩.

The silicide thickness obtained from the dielectric function fits is shown as a function of the amount of deposited Pd in Fig. 3, where the Pd thickness is given by a thickness monitor in the UHV chamber. The line represents a fit from the origin through the data for thicknesses $\geq$ 15 Å. The slope of the line, 1.40, should be the ratio of molar volumes of Pd_2Si and Si, which is 1.6. Since the silicide thickness for this film is the same in Figs. 2 (71 Å) and 3 (68 Å), the discrepancy in slope could be explained if the thickness monitor slightly overestimated the amount of deposited Pd.

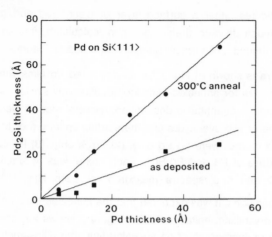

Figure 3.   Pd$_2$Si thickness as a function of deposited Pd thickness.

The as–deposited films were more complex. Models 3 and 4 gave better fits than models 1 or 2, indicating the presence of both Pd$_2$Si and unreacted Pd. Thus spontaneous reaction has occurred at room temperature, in agreement with previous experiments[1]. The fits did not distinguish between models 3 and 4, i.e., determine the location of the unreacted Pd relative to the Pd$_2$Si. This was due to only moderate optical contrast between Pd$_2$Si and Pd[4]. For the thicker films the amount of silicide, relative to the annealed films, was ~40%. For < 10 Å of Pd the silicide fraction is smaller, and could be zero within experimental uncertainty. This is consistent with a recent suggestion of a critical nucleation thickness of ~10 Å for crystalline Pd$_2$Si[1].

It may be noted that the dielectric function fitting residuals were somewhat larger than expected, both for as–deposited and annealed films. This is due at least in part to extra oscillator strength below 3 eV: note the broad extra peak in the 50 Å curve of Fig. 1. More complex models, including an amorphous Si layer or microscopic roughness at either interface, did not improve the fits. This will be the subject of further study.

4. REFERENCES

1)   Nemanich, R. J. and Doland, C. M., J. Vac. Sci. Technol. B3, 1142 (1985).

2)   Ho, P. S., Rubloff, G. W., Lewis, J. E., Moruzzi, V. L., and Williams, A. R., Phys. Rev. B22, 4784 (1980); Rubloff, G. W., Ho, P.S., Freeouf, J. F., and Lewis, J. E., Phys. Rev. B23, 4183 (1981).

3)   Ho, P. S., Schmid, P. E., and Föhl, H., Phys. Rev. Lett. 46, 782 (1981).

4)   Kelso, S. M., Nemanich, R. J., and Doland, C. M., Mat. Res. Soc. Symp. Proc. 54, 23 (1986).

5)   Aspnes, D. E. and Studna, A. A., Phys. Rev. B27, 985 (1983).

IR-ABSORPTION OF PtSi-Si INTERFACE STATES

Thomas Flohr, Max Schulz

Institut für angewandte Physik, Universität Erlangen
D-8520 Erlangen, Glückstr. 9
F.R.GERMANY

ABSTRACT

Photoacoustic absorption and transmission measurements
are performed to detect the contribution of interface
states at the PtSi-Si interface to the absorption of
the metal film. PtSi films in the thickness range
27Å-420Å on silicon substrates are employed. The
results are interpreted by a multilayer optical model.
Absorption in an interface layer representing interface
states is about 10%.

In this paper, we present direct evidence for interface states
at the PtSi-Si interface from optical absorption measurements in the
near infrared using photoacoustic spectroscopy (PAS). Absorption
and transmission measurements are performed with broad band illumination
in the wavelength range from 1.5µm to 2.7µm. The samples are PtSi
films with variable thickness on Si substrates. In the range from
1.5µm to 2.7 µm the high resistance (25 Ohm cm) Si-substrate is transparent. The films were fabricated by evaporation of platinum followed
by annealing at elevated temperatures (400-500°C) to form the platinum
silicide. These films show a reproducible Schottky barrier of 257 ± 5meV
(1). The thickness range of the PtSi-layers used in our measurements
is 27Å-420Å.

The sample structure is shown in Fig. 1a. The configuration of

the measurement chamber is shown in Figs. 1b-d. In PAS experiments,
heat generated in the sample by optical absorption is determined from
the pressure change in the gas ambient surrounding the sample in the
closed chamber (2). Usually, the incident light beam is chopped and
the pressure change then is an acoustic wave which can be detected
by a microphone (PAS-signal). The chopping frequency used in our
experiment is 44Hz.

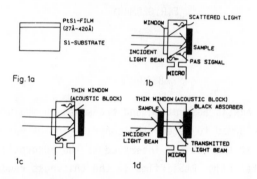

Fig. 1: Configuration of the PAS cell for measuring optical absorption
and transmission: a) sample structure, b) sample configuration for
absorption measurement, c) sample configuration for the measurement
of photoacoustic stray signals, d) sample configuration for transmission
measurement. Optical stray signals are indicated by straight lines.
The photoacoustic (PAS) signals are indicated by undulated arrows.

Photoacoustic stray signals arising from absorption of light at
the cell-walls limit the minimum detectable signal. In order to resolve
the small contribution of interface states to the absorption, we have
independently measured the stray signals by placing a thin (0.5mm)
transparent silica plate between the sample and the cell (Figs. 1b,c).
In this configuration, the optical properties of the system (e.g. reflectivity of the sample) are negligibly altered. Because of the small
thermal diffusion length of glass (50μm at 44 Hz), the PAS signal from
the sample cannot reach the cell and only stray light contributes to
the PAS signal. Transmission was measured by placing a porous black
absorber into the PAS cell as a detector for light transmitted through
the sample (Fig. 1d). The experimental results of absorption and
transmission measurements are presented in Figs. 2 and 3.

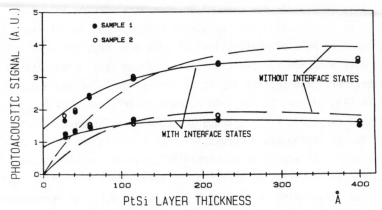

Fig. 2: Photoacoustic signal (as a measure of absorbance) of PtSi films on silicon as a function of the layer thickness for two samples. The upper points are for light incidence on the Si substrate. The lower points are for incidence on the PtSi film. Lines are calculated.

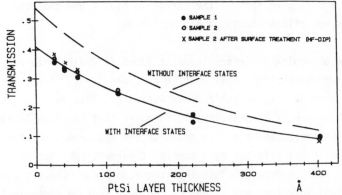

Fig. 3: Transmission of PtSi films on silicon as a function of the layer thickness for two samples. Lines are calculated.

For an evaluation of the experimental data, we treat the samples as multilayer optical systems. The interface states are assumed as an absorbing interfacial-layer of d_{int}=5Å thickness between the optically thick Si substrate and the thin PtSi-film. Each layer is characterized by its refractive index n and its absorption coefficient K. The values for PtSi (n=1.85, K=3.8 10^{-3}/Å in the range 1500nm - 2700 nm) are taken from the literature (3). The data for Si are n=3.4 and K=0. The refractive index n_{int} of the interface layer is considered equal to that of the Si-substrate (n=3.4). The absorption coefficient K_{int} of the interface layer is the parameter to be determined. The value

K_{int}=0 then represents the case of absence of absorbing interface
states. For the calculations, we assume normal incidence of the incoming
light beam and multiple reflections at each boundary. The details
of the matrix method for the calculation of the total transmission T,
the total reflection R, and the total absorption 1-T-R may e.g. be
found in (4). Without taking into account interface-states (K_{int}=0),
the dashed curves in Figs. 2 and 3 are computed. The measurement
points do not follow these curves. Even by changing the known optical
constants of PtSi and Si in an unrealistic manner, no consistent fit
of the three curves can be obtained. By taking into account an absorbing
interfacial layer with K_{int}=0.02/Å (K_{int} d_{int}= 0.1) the measurement
results can be consistently fitted. Since PAS measurements cannot
give absolute values for the absorption, the calculated curves in
Fig. 2 are multiplied by a scaling factor. It is noted that the same
scaling factor is used for all the curves in Fig. 2 and that the
factor does not affect the shape of the curves.

If the absorbing interface layer is identified with interface
states with an absorption cross section of typically 10^{-16}cm², an
interface state density of 10^{15}/cm² is estimated. This is a realistic
value of the order of a monolayer. It is noted that there is little
scatter in the measurement points and that the measured absorption and
transmission is not affected by a surface treatment (HF dip: crosses
in Fig. 3). We therefore conclude that the absorption observed occurs
at the PtSi-Si interface and is not caused by a surface contamination.
A surface roughness or a structural change of the layer with film
thickness are not expected to yield a smooth fit of the calculation.
We therefore attribute the observed absorption to interface states.

REFERENCES

1.   Cabanski, W., and Schulz, M., Int.Symp. SPIE, Innsbruck 1986
2.   Rosencwaig, A., Photoacoustics and Photoacoustic Spectroscopy
      (Wiley, New York 1971)
3.   Pimbley, J.M. and Katz, W., Appl.Phys.Lett. _42_, 984 (1983)
4.   Vasicek, A., Optics of Thin Films (North-Holland, Amsterdam,
      1960), p. 159

INTERFACIAL STRUCTURE OF
TITANIUM SILICIDE FILMS ON SILICON

A. Catana, M. Heintze, F. Lévy, P.E. Schmid and P. Stadelmann

Institut de Physique Appliquée
Ecole Polytechnique Fédérale de Lausanne
CH-1015 Lausanne, Switzerland

Thin films of $TiSi_2$ have been prepared on Si substrates by rapid thermal processing of a sputtered Ti layer. By high resolution microscopy, we establish a number of orientation relationships between the Si substrate and titanium silicide. In spite of the mismatch of the crystal structures, the interfaces are remarkably abrupt and planar.

Titanium silicide is a metallization material considered for high density and high speed MOS integrated circuits, mostly because it has the lowest electrical resistivity among all silicides. Research related with the properties of $TiSi_2$ contacts and interfaces are currently of acute interest.

Besides its technological interest, the $TiSi_2/Si$ interface raises questions of a more fundamental nature: The process of titanium silicidation is particularly complex[1], and the crystal structures of $TiSi_2$ are so different from that of silicon that one may fear that this silicon-silicide interface is very ill-behaved. We will show that this is not really the case.

For the most part, the samples used in this study have been fabricated using techniques close to production techniques: 400 Å of titanium were deposited by sputtering on Si (001) wafers doped with $10^{15} cm^{-3} P$. Subsequently, the wafers were submitted to rapid thermal processing (RTP) in an Ar atmosphere. In the following, we will refer to the three different RTP procedures used in this study as A, B and C. In procedure A, the wafers were treated at $600^\circ C$ for $60 s$. In procedure B, the temperature was $800^\circ C$. In procedure C, the wafers were treated at $600^\circ C$ for $60 s$, chemically etched to remove unreacted titanium, and finally heated to $800^\circ C$ for $60 s$ to optimize the electrical characteristics. Samples were also prepared using titanium evaporation under UHV conditions, UHV annealing and conventional furnace annealing in $Ar/N_2/H_2$ mixtures.

Figure 1: HRTEM of C49 $TiSi_2$ on Si [001]; $800^\circ C$ anneal

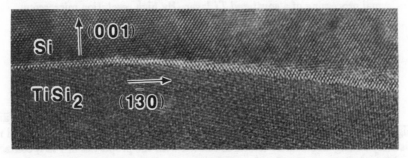

Figure 2: HRTEM of C49 $TiSi_2$ on Si [001], other orientation; $800^\circ C$ anneal

The investigation techniques included X-ray diffraction, X-ray electron spectroscopy, Auger etch profiling and high resolution transmission electron microscopy (HRTEM). A Phillips 430 ST microscope with a point to point resolution of 0.20 nm was used on both top-view (along the [001] direction of the Si substrate) and cross-sectional samples (along the [110] direction).

The crystal structure of bulk $TiSi_2$ is normally the C54 structure. However it has been shown that, at low temperature, the silicidation of titanium yields $TiSi_2$ with the metastable C49 structure[2]. Our observations by X-ray and electron diffraction confirm this conclusion. Regions with the C54 structure were detected in RTP samples which had been heated to $800^\circ C$. However, such samples also displayed areas with the C49 structure, as shown in Figs. 1 and 2. They show HRTEM cross-sectional views of samples obtained with procedure C. The $TiSi_2/Si$ interface is remarkably sharp and there is no evidence for a glassy interfacial layer, in agreement with another study[3]. On the contrary, the transition

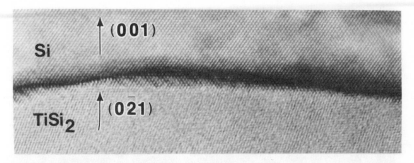

Figure 3: HRTEM of a C49 $TiSi_2$ / Si [001] interface with corrugation; $800^\circ C$ anneal

from Si to $TiSi_2$ occurs within a distance of the order of one interatomic spacing. The contrast modulation which is often visible along the interface is attributed to the fact that the interface is not perfectly planar and parallel to the viewing direction, as in Fig. 3, and to variations in sample thickness.

Optical diffraction on the high resolution images was used to assign the crystal structure of the silicide in the vicinity of the interface. In Fig. 1, the (131) planes of $TiSi_2$ are aligned with the ($\bar{1}$10) planes of Si, in agreement with a previous report[3].

We have observed several other epitaxial relationships. In Fig. 2, the (1$\bar{3}$0) planes of $TiSi_2$ are parallel to the ($\bar{1}$10) planes of Si and the [312] direction of $TiSi_2$ is parallel to the [110] direction of Si. The grain diameter was typically 250 nm. Fig. 4 shows the interface of a sample obtained with procedure A. Again, $TiSi_2$ has the C49, metastable structure. The 2.26 Å spacing is characteristic of the (060) planes and the 3-plane periodicity of 6.76 Å belongs to the (020) planes. A nearly parallel relation is observed between the (020) $TiSi_2$ planes and the ($\bar{1}$11) Si planes. Typical grain diameter was 250 nm for a thickness of the order of 40 nm.

In other samples, prepared with procedure C, the following orientation relationships have also been observed between C49 $TiSi_2$ and Si(001) :
(131)$TiSi_2$ // ($\bar{1}$1$\bar{1}$) Si, (1$\bar{3}$0) $TiSi_2$ // (001) Si and (0$\bar{2}$1) $TiSi_2$ // ($\bar{1}$1$\bar{1}$)Si. Other orientations have been reported previously[4].

With procedure B, we have observed (202) planes of C54 $TiSi_2$ parallel to (001) planes of Si. In this case, Ti silicidation was complete and grain thickness reached 100nm.

By HRTEM we have established a number of orientation relationships between Si and $TiSi_2$, both with the C49 and the C54 structures. The wide mismatch

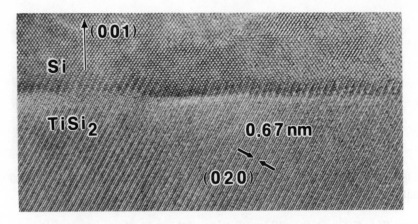

Figure 4: HRTEM of C49 $TiSi_2$ on Si [001]; $600^\circ C$ anneal

between the crystal structures of Si and of Ti silicide is responsible for the occurrence of several orientation relationships. However, it is not an obstacle to the formation of sharp epitaxial interfaces.

Acknowledgements:

We thank P. Weiss who prepared several of the wafers used in this study.

References:

[1] G. W. Rubloff, MRS Symposia Proceedings **54**, 3 (1986)

[2] F. M. d'Heurle, P. Gas, I. Engström, S. Nygren, M. Ostling and C. S. Petersson, IBM Research Report **RC 11151** (1985)

[3] R. Byers and R. Sinclair, J. Appl. Phys. **57**, 5240 (1985)

[4] M. S. Fung, H. C. Cheng and L. J. Chen, Appl. Phys. Lett. **47**, 1312 (1985)

TRANSPORT PROPERTIES AND ELECTRICAL CHARACTERIZATION OF Si/CoSi$_2$/Si HETEROSTRUCTURES

P.A. Badoz, E. Rosencher, S. Delage[a], G. Vincent and
F. Arnaud d'Avitaya
Centre National d'Etudes des Télécommunications
Chemin du vieux chêne – B.P. 98
38243 Meylan Cedex – FRANCE
[a]present address : IBM, Thomas J. Watson Research Center
Yorktown Heights, N.Y., 10598 (U.S.A.)

Thanks to advances in Molecular Beam Epitaxy (MBE) technology, it has recently become possible to realize the epitaxial Si/CoSi$_2$/Si heterostructure [1]. This Semiconductor – Metal – Semiconductor (SMS) structure is a promising ultra – low base resistance device for millimeter wave applications. In this communication we discuss the main features of the transport mechanisms along with the basic phenomena that set limits to the performance of the SMS. The question arose recently whether the perpendicular transport in the structure is controlled by Si pinholes in the metal base or by ballistic transport through the metal film [2,3]. A transconductance technique described in references 2 and 4, allows to detect pinholes in the base by electrical measurements. This technique, based on the screening of the collector potential by the metallic CoSi$_2$ film when no pinholes are present, ensures that in "pinhole – free" SMS, electron transport occurs almost entirely in the metal base.

The principle of functioning of the SMS, a device made of two back to back Schottky diodes, is the following : electrons are first injected in the metal by thermionic emission from the forward biased emitter – base Schottky diode. A fraction of the electrons crosses the metal base by ballistic transport and is eventually collected by the reversely biased base – collector Schottky diode. The ballistic transport is described by a mean free path l_b so that the emitter – collector transfer ratio of electrons is expected to be :

$$\alpha = \alpha_0(T)\exp(-W/l_b(T)) \tag{1}$$

where α_0 is the current gain extrapolated to zero metal base thickness, W is the base thickness and T the measuring temperature. The departure of α_0 from unity is then due to collector as well as emitter losses : indeed, α_0 is determined by :

$$\alpha_0 = \alpha_c \alpha_q \alpha_e \tag{2}$$

where α_c is the current gain upper limit associated with phonon scattering in the Si collector, α_q the quantum mechanical transmission of the base collector potential barrier and α_e the emitter efficiency coefficient. The phonon scattering contribution is expected to follow :

$$\alpha_c = \exp(-x_m/l_{ph}) \tag{3}$$

where x_m is the position of the maximum of the collector barrier potential in the image force approximation, and l_{ph}, the electron – phonon mean free path in the Si collector.

We shall now compare in the following theory with experiment. Figure 1 shows the transfer ratios obtained on samples with various values of CoSi$_2$ base thickness, at room temperature and at 77 K. Error bars correspond to the dispersion of values for different devices fabricated on the same wafer. The thicknesses are first determined in the MBE chamber by a quartz microbalance, calibrated by Rutherford backscattering measurements. They are then checked against values deduced from ex – situ resistance measurements, with which they agree within 5% for CoSi$_2$ thicknesses above 150 Å. For thinner films, surface scattering has to be taken into account, e.g. using the Fuchs – Sondheimer theory[5]. The results clearly show that Eq. (1) is verified, with values of l_b of 80±15 Å at 300 K and 350±50 Å at 77 K. This agreement is thus in strong favor of the ballistic theory. Moreover, the l_b values are very close to the mean free path l_r deduced from resistivity measurements[6], i.e. l_r is 85±15 Å at 300 K and 325±25 Å at 77 K. This indicates that the same scattering mechanisms control both the electron transport close to the Fermi level and the hot electron relaxation for energies in the 0.7 eV range above E_F. The variation of α_0 with temperature calls for a more detailed investigation : $\alpha(T)$ curves are recorded in differents SMS devices while Deep Level Transient Spectroscopy measurements are performed in the Si$_{epi}$/CoSi$_2$ junction of the same devices. The close similarity observed in the two sets of curves suggests that the detailed structure of the emitter – base space charge region has a major influence on the injection efficiency. The large influence of deep levels in

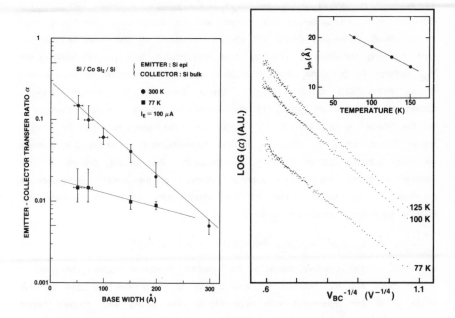

Fig. 1. Transfer ratio α versus CoSi$_2$ base thickness measured at 77 and 300 K with an emitter current of 100 µA.

Fig. 2. Transfer ratio α versus base – collector bias at different temperatures The mean free path l_{ph} deduced is shown in the inset as a function of temperature.

the emitter – base junction is also consistent with the observed systematic difference in metal – semiconductor barrier heights[4]. As a mater of fact, this difference provides another element supporting the ballistic transport hypothesis : experimental data confirm that when the higher Si$_{epi}$/CoSi$_2$ barrier ($\phi_{ms} \sim 0.69$ eV) is used as an emitter, the electrons are able to overpass the lower Si$_{bulk}$/CoSi$_2$ barrier ($\phi_{ms} \sim 0.63$ eV) while no electrons are transfered in the opposite direction.

In order to study the effect of phonon scattering on the electron transport, transfer ratio α is recorded at constant temperature and base thickness. Figure 2 shows α versus $V_{bc}^{-1/4}$ curves, taken at different temperatures, where V_{bc} is the base – collector bias. Equation (3) clearly holds, as $V_{bc}^{-1/4}$ is directly proportional to x_m. Furthermore, a mean free path l_{ph} is extracted from the slope of the curves and its temperature variation is shown in the inset. However, theory of phonon limitation of the effective mean free path involving variations of the quantum transmission and barrier lowering with V_{bc} remains to be done. Another problem requiring explanation is the low value of α_0 extrapolated to zero potential barrier thickness. The answer is clearly related to the quantum nature of the electron : indeed, the electron energy E_1 in the metal is in the 5 eV range, i.e. the Fermi energy plus the Schottky barrier height, while its value E_2 in the semiconductor is in the 50 meV range, i.e. the Schottky barrier lowering. Consequently, the abrupt change in the electron wavelength leads to a quantum reflection at the metal – semiconductor interface. If the crude model of the abrupt – step barrier is assumed, the quantum transmission coefficient α_q is

$$\alpha_q = 1 - [\,((m_2^* E_1)^{1/2} - (m_1^* E_2)^{1/2})/((m_2^* E_1)^{1/2} + (m_1^* E_2)^{1/2})\,]^2 \tag{4}$$

where m_i^* is the effective mass of the electron in the ith medium. Taking $m^* = 1$ in the metal and m^* between 0.2 and 0.4 in Si, one obtains from Eq.(4) $\alpha_q \sim 0.6$, which is in fair agreement with experimental data taking into account phonon reflection. More sophisticated calculations have been performed, by solving numerically the Schrödinger equation for image force type barrier, using various potential continuity procedure. These calculations lead to slightly lower transmission coefficient α_q. All our results, as well as those obtained earlier by Crowell and Sze[7] show that the quantum reflection is indeed a limiting factor for an almost symmetrical structure, i.e. in which the two barrier heigths are less than 0.1 eV apart. These results suggest, in order to reduce the reflection, the use of highly assymmetrical SMS structure, for instance by the use of two different semiconductors and/or metals.

REFERENCES

1) E. Rosencher, S. Delage, Y. Campidelli and F. Arnaud d'Avitaya, Electron. Lett. 20, 762 (1984).
2) R.T. Tung, A.F.J. Levy and J.M. Gibson, Appl. Phys. Lett. 48, 635 (1986).
3) E. Rosencher, S. Delage, F. Arnaud d'Avitaya, C. d'Anterroches, K. Belhaddad and J.C. Pfister, Physica B 134, 106 (1985).
4) S. Delage, P.A. Badoz, E. Rosencher and F. Arnaud d'Avitaya, Electron. Lett. 22, 207 (1986).
5) C.R. Tellier and A.J. Tosser, Size effect in thin films (Elsevier, Amsterdam, 1982)
6) E. Rosencher, P.A. Badoz, J.C. Pfister, F. Arnaud d'Avitaya, G. Vincent and S. Delage, Appl. Phys. Lett. (August 1986).
7) C.R. Crowell and S.M. Sze, J. Appl. Phys. 37, 2683 (1966).

INTERMIXING OF Cu AND Si IN THE EARLY STAGES OF THE Cu-Si(111) INTERFACE FORMATION

P. MATHIEZ, H. DALLAPORTA, G. MATHIEU, A. CROS and F. SALVAN

FACULTE DES SCIENCES DE LUMINY
U.A.CNRS 783 - Département de Physique - Case 901
13288 Marseille Cedex9, France

Despite numerous studies in the recent years a precise understanding of Schottky barrier formation at noble metals-silicon junction has not been achieved. Besides the theoretical problems, there is still the need for improved experimental characterization, particularly at the very first stages (metal coverage < 2-3 monolayers) of the junction formation. These first stages are of great importance since they govern the band bending and the Schottky barrier height.

The growth mode of Cu on the Si(111)7x7 surface has been previously studied /1,2/. There is a general agreement that Cu deposited on a Si surface maintained at room temperature (RT) grows in a layer by layer mode. When the deposition is made on a hot substrate (HT), one observes the formation of crystallites on top of a quasi 5x5 surstructure .

We present here a detailed study of the Cu-Si(111) interface for very thin metal coverages. The originality of our work is two-fold. Firstly, we have combined several in situ techniques to get a more complete understanding of the interface formation : (LEED), (AES) and (UPS) to check the hybridized bonds which form between Si and Cu valence states. Secondly, we have systematically used differential techniques for both AES and UPS to isolate the contribution of the true interface.The differential curves /3,4/ allow to overcome the limitations inherent to surface techniques.

The experimental set-up has been described in detail previously /1,4/. For the UPS, we use a He lamp ($h\nu$=21.2eV) ; the measurements were taken under angle integrated condition with a 45° incident photon beam. The AES and UPS analysis were performed with a CMA. For AES, the output of the lock in amplifier in the EdN/dE mode, and for UPS the direct EN(E) spectra are accumulated in a desk computer used as a multichannel detector. The differences between spectra for both AES and UPS are

336   P. Mathiez et al.

obtained numerically with the computer. The Si(111) 7x7 surface(B doped, 3 Ωcm) was prepared by standard procedures.The Cu coverage θ   was measured in Si(111) monolayer units(ML).

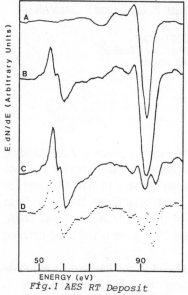

Fig.1 AES RT Deposit

Fig. 1 shows the evolution of AES spectra. For the clean Si surface (curve A) we observe the well-known Si(LVV) transition at 92 eV. This line is broadened by the presence of one metal ML (curve B). The structure at 60 eV is associated to the Cu (MVV) Auger transition. At higher Cu coverage (θ = 6, curve C), the Si(LVV) line shape changes drastically. It splits into two peaks, 90 and 94 eV. This splitting corresponds to the hybridization of Cu 3d and Si 3s 3p level in Cu rich alloy (typical stoichiometry : Cu_3Si) /5,6/. This hybridization produces bonding and antibonding levels which create the Si(LVV) splitting.This one is a signature of a metal rich alloy.

In this paper, we concentrate on the first stages of the interface formation (curve B). In this case an important contribution comes from the substrate. To isolate the signal from the interface, we have substracted from curve B the contribution from the substrate (curve A) corrected by an escape depth factor. The difference spectrum (curve D) is practically identical to curve C. The Si(LVV) lines are mostly sensitive to the p electrons /7/ and from this Auger analysis, we suggest that for these p states the nature of the bonding between Si and Cu atoms at the very beginning of interface formation is similar to that found in Cu rich Cu-Si alloys.

An interesting point concerns the long range order of the surface. For the Cu deposition on Si maintained at RT, one observes by LEED the disappearance of the Si 7x7 surface reconstruction and the absence of any long range order for the Cu ML.On the contrary, depositing Cu directly on a hot substrate (T 600°C) produces a quasi 5x5 LEED pattern /1,8/. The Auger spectrum corresponding to this reconstructed surface is identical to curve B. The difference spectrum (curve D) is the same and this implies that the hybridization responsible of the Si splitting is the same for both the amorphous and the ordered Cu ML at least for the Si p states seen in the Si(LVV) Auger transition.

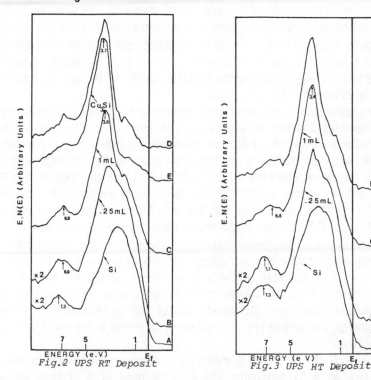

Fig.2 UPS RT Deposit   Fig.3 UPS HT Deposit

Fig. 2 shows the evolution of the UPS spectra. Curve A refers to the clean Si surface while B and C correspond respectively to $\theta=.25$ and $\theta=1ML$ of Cu deposited at RT.One observes on curve A two characteristic structures, a first one at 2.65eV below E_F corresponding to bulk p state, with a shoulder at 1eV due to the emission of dangling bonds ; the other one related to surface resonance with sp character occurs at 7.3 eV /9/. Curve B is dominated by the appearance of Cu 3d band emission and a correlative shift of 300 meV towards E_F of the surface resonance around 7 eV. This resonance stays at the same energy in curve C whereas the Cu d band emission is maximum at 3.6 eV. Curve E is the photoemission spectrum of a Cu_3Si compound obtained by annealing (200°C, 120 minutes) a thick Cu deposit on a Si substrate. In order to isolate the contribution of the interface from the substrate one and hence better compare the electronic structure of R.T. and H.T. deposits, we have substracted curve A from curve C and obtained curve D which is very similar to the Cu_3Si spectrum.

Fig. 3 concerns deposition on substrate maintained at HT (600°C) which gives the "5x5" reconstructed surface. From curve A (θ = 0) to curve C (θ = 1), one observes a rather important and progressive shift of the sp resonance (700 meV at θ = 1) whereas Cu d band emission has a maximum around 3.4 eV in curve C. Difference curve C - A gives curve D characteristic of the photoemission spectrum of the HT ML interface which is slightly different of curve D, Fig. 2.

The comparison of all these experimental data allows to stress both the similarities and the differences in the electronic structure of the disordered and ordered ML. It is clear that there is hybridization (i.e. bond formation) between the Si s-p and the Cu d states at the very beginning of interface formation. From the splitting of the Si (LVV) line, sensitive to Si p state, and the energy position of the Cu 3d band, we conclude that the bonding in the ML is similar to that found in Cu rich Cu-Si compounds of the Cu_3Si type. This is true for the ordered and disordered surfaces (the slight shift of the Cu 3d maximum upon annealing can hardly be associated to a strong modification of the bonding). Difference between the RT and HT deposit concerns essentially the s-p resonance structure. A possible interpretation is that, as suggested by Chambers et al. /10/, Cu slightly penetrates into the substrate which shifts the sp resonance to lower binding energy, because of rehybridization of Si atoms in the surface layer from sp^3 to sp^2.

This study clarify a long standing controversy on the growth mode of metal atom on a Si substrate. There is no need of a critical metal thickness to induce interface reaction and compound formation. In this way, Cu behaves very similarly to the nearest transition metals Ni, Pd, Pt where a metal rich phase grows at low metal coverages. However the Cu 5x5 ordered structure still presents some intriguing problems. Further studies should clarify the geometrical position of the Cu atoms as well as its peculiar behavior with respect to the oxidation /11/.

1   E. Daugy, P. Mathiez, F. Salvan and J.M. Layet, Surf. Sci. 154, 267 (1985).
2   A. Taleb-Ibrahimi, V. Mercier, C.A. Sébenne, D. Bolmont and P. Chen, Surf. Sci. 152/153 1228 (1985)
3   G. Rossi and I. Lindau, Phys. Rev. B 28,6, 3597 (1983).
4   H. Dallaporta and A. Cros, to be published in Surf.Sc. Proceedings ECOSS 8 (1986).
5   F. Salvan, A.Cros and J.Derrien,J. de Physique-Lettres 41, 337 (1980).
6   A. Hiraki, Surface Science Reports 3, 357 (1984) and references therein.
7   P.J. Felbelman, E.J. McGuire and K.C. Pandey, Phys. Rev. B 15, 2202 (1977).
8   J.T. Grant and T.W.Haas,Surf.Sci.19,347(1970)
9   J.E.Rowe and H.Ibach, Phys Rev. lett. 32,421(1974)
10   S.A. Chambers, S.B. Anderson and J.H. Weaver, Phys. Rev. B 32,2, 581 (1985)
11   E.Daugy ,P.Mathiez,F.Salvan,J.M.Layet and J.Derrien,Surf.Sci.152/153,1239(1985)

PHOTOEMISSION STUDY OF THE MECHANISM OF ROOM TEMPERATURE ALLOYED INTERFACE FORMATION AT AU-SI(111)2x1

Motohiro Iwami, Toshiyuki Terada*, Hiroshi Tochihara[+],
Masakazu Kubota[+] and Yoshitada Murata[+]

Research Laboratory for Surface Science, Faculty of
Science, Okayama University, Okayama 700, JAPAN
*Department of Electrical Engineering, Osaka University,
Suita, Osaka 565, JAPAN
[+]Institute for Solid State Physics, The University of
Tokyo, Minatoku, Tokyo 106, JAPAN

ABSTRACT

In Au-Si(111)2x1 system, binding energies (or
energy separations) of Si(2p), Au(4f) and Au(5d)
showed clear changes at Au coverage of $\theta \simeq 1$ (1ML$\equiv$
7.8×10^{14} atoms/cm^2). Also a resonant photoemission
study indicated a formation of a covalent-like bond
between Au and Si atoms. From these results, 'chemical bonding model' is discussed in relation to existing models.

1. INTRODUCTION

Many systems of metal(M)-semiconductor(S) contact show
intermixing reaction at M-S interface at room temperature.
So far, two different models as the origin of the intermixed
(alloyed) interface formation have been proposed; (a)
'screening model'[1] and (b) 'interstitial model'[2,3].

In the present study of Au-Si(111)2x1, the initial
stage ($\theta \lesssim 2$-3) was carefully examined. For this purpose,
the utilization of photoemission due to irradiation of photons with different energies(30-160eV) was very powerful.
A new model, 'chemical bonding model', is discussed in con-

trast with the two models described above.

2. EXPERIMENTS

Present experiment was carried out in the BL-2 system of SOR-Ring at the Universith of Tokyo. Analyzing and sample preparation chambers were constructed by stainless steel UHV systems.

3. RESULTS AND DISCUSSIONS

Au-coverage dependence of binding energies of Si(2p) and Au(4f) are shown in Fig.1. Several characteristics that can be deduced from results of E_b's above, in combination with those of $\Delta E[Si(2p)-Au(4f)]$ and $\Delta E[Au(5d_{5/2}-5d_{3/2})]$[4] are; (1) a clear change of these values at $\theta \simeq 1$, (2) at $\theta \gtrsim 1$ a gradual increase in $E_b[Si(2p)]$ and $\Delta E[Au(5d)]$, or decrease in $E_b[Au(4f)]$, (3) saturation of $E_b[Au(4f)]$ and $\Delta E[Au(5d)]$ above $\theta \simeq 20$ and (4) the existence of differences in the saturation values of $E_b[Au(4f)]$ and $\Delta E[Au(5d)]$ from those of pure Au. An Au-Si alloy formation must be the origin of the characteristic (3), because an Au deposition with enough amount induces Au-Si alloy formation on top of Si substrate at room temperature[1,3]. The fact (2) indicates a gradual change of Au-Si bond nature with θ. The feature (4) shows a difference of the electronic structure of Au in Au-Si alloy from that in pure Au. The most interesting feature among four shown above is the observation (1). This fact indicates a difference of the chemical bond nature of Au -Si bond between that at $\theta < 1$ and the one at $\theta > 1$.

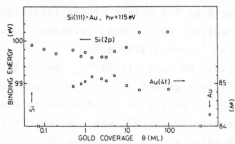

Fig.1 θ (Au) dependence of $E_b[Si(2p)]$ and $E_b[Au(4f)]$.

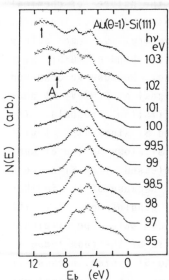

Fig.2 Photoemission
of Au(θ ≈1)/Si(111).

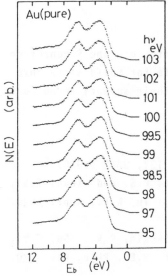

Fig.3 Photoemission
of pure gold.

At θ ≲ 1, we obtained an evidence of chemical bond formation between Au and Si atoms . Valence band photoemission spectra of a cleaved Si(111) surface with Au(θ ≈1) are given in Fig.2 in the photon energy region around the Si(2p)-core excitation threshold. Below the threshold, i.e. hν < 99eV, spectra show no special change. With increasing photon energy above the threshold, i.e. hν ≥ 99.5eV, the spectra show an intensity enhancement at E_b ≈ 7eV. At hν ≥ 101eV, another structure indicated by an arrow 'A' due to Si(LVV) Auger transition appears. By subtracting the contribution from substrate silicon, including the above Si(LVV) Auger signal, a clear enhancement of Au(5d) signal at E_b≈7eV can be deduced. This kind of enhancement of Au(5d) signal at E_b≈7eV can not be observed in pure gold(Fig.3). Therefore, the enhancement of Au(5d) signal at E_b≈7eV observed in the Au(θ ≈ 1)/Si(111) system must be due to a special situation in this system different from either pure Au and Si. Namely, the enhancement clearly indicates that Au atoms make chemical bond with Si atoms on the Si(111) surface. From this consideration, the possibility of physisorption of Au atoms on Si(111)2x1 surface can safely be eliminated. Therefore, the clear change in several characteristics at θ(Au)≈1 shown

above must be due to a difference in the nature of chemical
bond between Au and Si atoms. As mentioned above, the Au-Si
bond is metallic at $\theta > 1$. Therefore, we can safely
conclude that the nature of the chemical bond of Au-Si at θ
<1 is non-metallic, i.e. covalent-like, because we observed
the resonant enhancement of Au(5d) signal at $E_b \simeq 7eV$ at photon energies of around the Si(2p) excitation threshold. The
covalent bond formation between Au and Si atoms at $\theta < 1$ will
induce a weakening of back bonds of Si atoms on top of substrate Si(111). Si atoms will be released from Si-network
due to the weakening of Si-Si bonds because of the further
deposition of Au atoms: the 'chemical bonding model'[4]. The
model (b), the 'interstitial model', can safely be eliminated as a candidate to explain the origin of the room temperature alloyed interface formation in Au-Si system, because it
can not explain the feature (1) observed in the present experiment. The model (a), the 'screening model', could be a
candidate, although it is not necessary for the model that
Au and Si atoms are making covalent bonding at $\theta < 1$.

ACKNOWLEDGEMENTS

The authors are thankful to members of the Synchrotron
Radiation Laboratory of the Institute for Solid State Physics at the University of Tokyo for their support throughout
the work.

REFERENCES

1. Hiraki, A., J. Electrochem. Soc. 127, 2662 (1980).
2. For example, Freeouf, J.L., Ho, P.S. and Kuan, T.S.,
   Phys. Rev. Lett. 43, 1836 (1979), Tu, K.N., Appl. Phys.
   Lett. 27, 221 (1975).
3. Braicovich, L., Garner, C.M., Skeath, P.R., Su, C.Y.,
   Chye, P.W., Lindau, L. and Spicer, W.E., Phys. Rev. B20,
   5131 (1979).
4. Iwami, M., Terada, T., Tochihara, H., Kubota, M. and
   Murata, Y., Surf. Sci., submitted.

RARE EARTH METAL / SEMICONDUCTOR INTERFACE FORMATION

J. Nogami and I. Lindau,

Stanford Electronics Laboratories, Stanford University, Stanford USA

Interfaces formed at room temperature by deposition of rare earth (RE) metals onto semiconductor substrates have been studied with surface sensitive soft X-ray photoelectron spectroscopy. Trends in core level lineshape and intensity with increasing metal coverage have been used to deduce an outline of the evolution and the final morphology of the interfacial region on a microscopic scale. Measurements were taken of Yb,[1] Gd,[2] and Eu[3] on Si, and in addition Yb on Ge[4] and GaAs.[5] The Yb/Si interface work was supported by comparable measurements of bulk Yb silicide samples of known composition and crystal structure.[6]

Here we outline the general properties of this class of interfaces and point out some differences between the reactive behaviour of different RE's, demonstrating that both the valence and the size of the RE species have an effect on the interface formation.

In a general sense, the behaviour of all the systems studied is similar. At very low metal coverages, the metal atoms chemisorb and are weakly bonded to the substrate. Chemical interaction strengthens with coverage, culminating in the formation of a strongly reacted phase at between 1 and 3 monolayers (ML). This phase is stable and remains at the interface; further metal deposition results in the growth of a metal-rich overlayer that converges to virtually pure metal at coverages typically less than 10 ML. The strong reaction is limited to a narrow region at room temperature.

Figure 1 shows representative Ge $3d$ core level spectra from the Yb/Ge interface as a function of Yb deposition.[4] The substrate signal is rapidly attenuated with metal coverage Θ reflecting the relative narrowness of the interface formed. The broadening of the lineshape is due to the appearance of shifted emission from reacted Ge. The reacted component is clearly apparent at 2 ML, and dominates the total emission above this coverage.

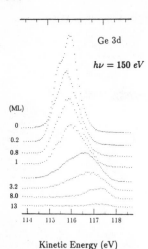

Fig1: Ge 3*d* core level emission from the Yb/Ge interface as a function of metal coverage.

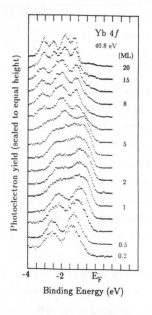

Fig 2: Yb 4*f* emission from the same interface. All spectra have been scaled to equal height to show changes in lineshape.

Figure 2 shows the Yb 4*f* emission from the interface. The spectra clearly demonstrate that there are several discrete stages in the interface formation. For $\Theta \leq$ 1 ML, the 4*f* emission has an "atomic-like" narrow doublet lineshape indicating that the metal atoms are semi-isolated and that there is little interaction with the substrate. The shift towards the Fermi level with Θ is due to increasing metal-metal atom coordination and changes in screening. The broadened and shifted lineshape at 2 ML is from a reacted phase covering the surface. This phase is stable with further metal deposition, as is apparent in these spectra by the continuing presence of 4*f* emission in the energy position seen at 2 ML as a shoulder in the higher coverage spectra. For $\Theta \geq 5$ ML, the spectra are consistent with the growth of an Yb-rich overlayer on top of the reacted phase at the interface. The lineshape is well described by two separate doublets originating from the "bulk" and the surface of the overlayer. The binding energy difference between the bulk and surface emission, the surface shift, decreases with increasing Yb concentration, just as is seen in Yb intermetallics of varied composition. It should be noted that the 4*f* core levels are a feature unique to RE interfaces; they provide information about the behaviour of the metal complementary to that deduced from the substrate core levels.

Figure 3 shows a comparision of Si 2*p* spectra from the Yb/Si interface and from bulk samples of the three stable silicide phases in the Yb/Si system. The silicide spectra show the core level binding energy decreasing with increasing Yb content, a trend we have also observed in Yb/Si reaction products of varying composition produced by annealing.[7,8] The spectrum from the interface is well described by a superposition of the spectra from clean Si and Yb_5Si_3, identifying the reacted phase at the interface as being metal-rich.

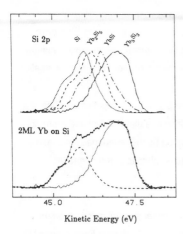

2ML Yb on Si

Kinetic Energy (eV)

Fig 3 (left): a comparision of Si 2p spectra from the Yb/Si interface and 3 bulk Yb silicide samples. Upper panel: Si 2p lineshapes of the clean Si surface, and the silicides of the indicated compositions. Lower panel: Si 2p spectrum for 2 ML Yb on Si. Superimposed are the spectra from clean Si (dashes) and the most metal-rich silicide (small dashes). The energy positions of the spectra have been shifted only slightly to account for band-bending effects.

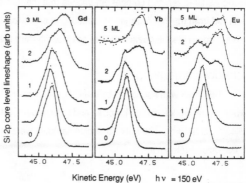

Fig 4 (right): Si 2p core levels for Gd, Yb, and Eu on Si. Metal coverage increased from bottom to top. The spectra have been scaled so as to emphasize changes in lineshape.

Kinetic Energy (eV)   hν = 150 eV

Figure 4 shows Si 2p core level spectra as a function of metal coverage at the Gd/Si, Yb/Si and Eu/Si interfaces. Several differences are apparent in comparing Eu (right) to Gd (left). The lineshape for $\Theta \leq 1$ ML is less broad and more similar to clean Si, indicating a lack of strong chemical interaction at the low coverages for Eu in comparsion to Gd. Once a strong reaction is apparent ($\Theta \approx 2$ ML) the Si 2p lineshape for Eu/Si has clearly separate peaks from unreacted (unshifted) and reacted (shifted) silicon. This is consistent with the formation of a single reacted phase at the interface. From the large size of the binding energy shift, we deduce that this reacted phase is metal-rich (Fig 3). For Gd/Si, the overall Si 2p lineshape is quite broad and cannot be described simply as a superposition of only two Si 2p peaks.

The behaviour of Yb on Si is much more similar to Eu than to Gd as might be expected from its valence. Divalent Eu and Yb are expected to have heats of adsorption and silicide formation different from those of trivalent rare earths (RE) such as Gd.[9] However, there is also a trend with the size of the RE atoms. Fig 4 is arranged so that the ionic radii of the RE species increases from left to right. In both the weak reaction at low coverage and the strong reaction at $\Theta \geq 2$ ML the behaviour of Yb is intermediate to that of Gd and Eu, as is its ionic radius. *Grioni et al* [10] show that Ce on Si behaves as Yb and Eu do, despite the fact that Ce is trivalent. Ce has an ionic radius between the divalent ionic radii of Yb and Eu. Thus, the size of the RE atom influences the reaction behaviour at the room temperature interface at least as much as the valence of the metal atom.

The formation of RE silicides is characterized by a strong volume contraction of the RE atom[11] which can in turn be correlated with a high heat of compound formation.[12] Atomic volumes for Gd, Yb, and Eu in silicides of several different stoichiometries show that for larger RE atoms (Eu) the volume contraction, ΔV, is higher for metal-rich phases, whereas for smaller RE atoms (Gd), ΔV is less dependent on stoichiometry.

This is consistent with the Si $2p$ data in Figure 4. For Eu on Si the rapid emergence of a strongly shifted reacted component implies that the formation of a metal-rich phase is strongly favoured from the initial stages of reaction, whereas for Gd on Si, the broadened lineshape at 2~3 ML suggests the coexistence of several different reacted phases at the interface.

ACKNOWLEDGEMENTS:

This work was supported by the U.S. Office of Naval Research. Collaborators on the experimental work included I. Abbati and L. Braicovich (Milano), C. Carbone (Jülich), and M.D. Williams (Stanford). The bulk Yb silicide samples were provided by A. Iandelli, G.L. Olcese and A. Palenzona (Genova).

REFERENCES:

1)   G. Rossi, J. Nogami, I. Lindau, L. Braicovich, I. Abbati, U. del Pennino
        and S. Nannarone, J. Vac. Sci. Technol. A1(2), 781 (1983)

2)   C. Carbone, J. Nogami, and I. Lindau, J. Vac Sci. Technol A3(3), 972 (1985)

3)   J. Nogami, C. Carbone, J.J. Yeh, I. Lindau, and S. Nannarone, in Proc. 17th ICPS,
        D.J. Chadi and W.A. Harrison, eds. p.201, (Springer, New York 1985)

4)   J. Nogami, C. Carbone, D.J. Friedman, and I. Lindau, Phys. Rev. B 33, 864 (1986)

5)   J. Nogami, M.D. Williams, T. Kendelewicz, I. Lindau, and W.E. Spicer,
        J. Vac. Sci. Technol. A 4(3), 808 (1986)

6)   I. Abbati, L. Braicovich, U. del Pennino, C. Carbone, J. Nogami, J.J. Yeh, I. Lindau,
        A. Iandelli, G. L. Olcese, and A. Palenzona, Phys. Rev. B (in press)

7)   G. Rossi, J. Nogami, J.J. Yeh, and I. Lindau, J. Vac. Sci. Technol. B1(3), 530 (1983)

8)   L. Braicovich, I. Abbati, C. Carbone, J. Nogami, and I. Lindau,
        Surf. Sci. 168, 193 (1986)

9)   A. Fujimori, M. Grioni, J.J. Joyce, and J.H. Weaver, Phys. Rev B 31, 8291 (1985)

10)   M. Grioni, J. Joyce, S.A. Chambers, D.G. O'Neill, M. del Giudice, and J.H. Weaver,
        Phys. Rev. Lett. 53, 2331 (1984)

11)   J. Nogami, PhD Thesis, Stanford University (1986)

12)   A. Iandelli, A. Palenzona, and G.L. Olcese, J. Less Common Metals 64, 213 (1979)

STUDY OF INTERMIXING OF NOBLE METALS/Si(100) AND Si(111) INTERFACES AT LNT, RT AND HT

M. Hanbücken[§], and G. Le Lay[§§]

CRMC2-CNRS, Campus de Luminy, Case 913,

13288 Marseille Cedex 09, FRANCE

and V. Vlassov

Institute of Crystallography, Academy of Sciences of the USSR,

Moscow B-333, Leninsky Prospekt 59, USSR

ABSTRACT

Noble metal/Si(100) and Si(111) interfaces, grown at LNT, RT and HT, have been studied using Auger lineshape spectroscopy. For both Si orientations the well known differences between non-reative (Ag) and reactive (Cu, Au) interfaces were confirmed. The onset of intermixing was detected for Au and Cu deposits at around one monolayer.

I.   INTRODUCTION

The noble metal/Silicon interfaces have been the subject of wide investigations in the last years looking at both the electronic states of the forming interfaces as well as their structural and morphological properties. Is is now agreed that Ag forms unreactive, sharp junctions in contrast to Au and Cu interfaces which were found to be strongly intermixed even at liquid nitrogen temperature (LNT) as was reported for Au/Si(111)[1] and Cu/Si(111)[2]. There is still disagreement whether a critical thickness of the metal deposit is required to trigger the intermixing reaction. Typically 4 ML of Au at RT are reported[3].

§ Also UA. CNRS 783, Département de Physique, Case 901, Faculté des Sciences de Luminy, F-13288 Marseille Cedex 09, France.

§§ Also UER de Physique, Université de Provence, Marseille, France.

Recently Dallaporta et al.[4] studied the initial stages of the formation of Au, Cu/Si(111) interfaces at RT, using a differential Auger
spectra analysis. In the derivative mode $dN(E)/dE$ the Si-LVV Auger line (92eV) splitts into two components at 90 and 94 eV when the metal/
Si interfaces is intermixed. As this characteristic splitting occured
in the difference curves at ~ 1 ML Au, Cu coverage these authors concluded that interdiffusion starts at the very beginning of the metal
deposition.

In the present work we studied directly the modification of the
Si-LVV lineshape in the integral $N(E)$ mode. Dependencies on the material (Ag, Au, Cu) and surface orientation Si(100) and (111), were analyzed in a wide temperature range from 100 to several hundred Kelvin.
All cases studied with Ag did not show any change in the lineshape of
Si, as is expected for non reactive interfaces. In the contrary, Au
and Cu/Si interfaces show clear intermixing beyond about 1 ML metal
deposit. In this paper we will focus on these later results.

2.   EXPERIMENTALS

All experiments were performed in an UHV chamber. A CMA with mi-
cron-size primary electron gun beam, operated at 2,5 kV and 200 nA was
used. The Si wafers were cleaned by flash heating to 1200 K. Cu and Au
were evaporated from W-wires heated with a controlled power supply. The
amount of deposit was calibrated with a quartz crystal mirobalance. The
coverage is expressed in substrate units : 1 ML corresponding to 7.8 x
10^{14} atoms cm^{-2} on Si(111) and 6.8 x 10^{14} atoms cm^{-2} on Si (100).

3.   RESULTS AND DISCUSSION

The Si-LVV transition reflects mostly the Si p density of states[5] : on the Si line a shoulder appears at 94 eV when certain metals
are deposited ; this is due to a transition involving Si-p and metal-d
states. Beyond 1 ML Au and Cu coverage we notice this shoulder on the
lineshapes of Si(111) and (100) at RT and LNT. In fig. 1 results of
the Au/Si(100) interface are presented. Surprisingly we observe for
Cu/Si(100) interfaces an increase of this new transition when Cu is deposited at LNT. This is even more pronounced at high coverages as
shown in fig. 2 for $\theta = 5$. Although not mentioned by the authors, the

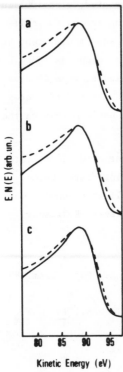

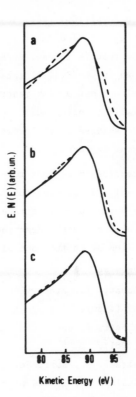

Fig. 1. Si-L$_{2,3}$VV Auger lineshape for 1,5 ML of Au on Si(100) at : a) 100 K ; b) RT ; c) 570 K.

Fig. 2. Si-L$_{2,3}$VV Auger lineshape for 5 ML of Cu on Si(100) at : a) 100 K ; b) RT ; c) 520 K.

same increase is apparent in fig. 3 and 4 by Rossi et al.[6], showing Auger curves in the derivative mode. We interpret this behaviour as due to a balance between two simultaneously occuring processes : metal indiffusion into the substrate and Si outdiffusion through the metal overlayer. Both diffusion channels must be thermally activated but the rate may be very different. By marker experiments[7] and polar angle resolved AES[8] a notable indiffusion of Au but non of Cu was established at RT, whereas it is known that Cu diffuses to a large extent into Si at high temperatures[9]. On the other hand, diffusion of Si through the metal already occurs at LNT[2] especially for Cu. Annealing the interface or depositing the metal at several hundred Kelvin produces a weakly reacted Au/Si(100) (fig. 1c) but a totaly unreacted Cu/Si(100) interface (fig. 2c). The later is probably due to the strong Cu indiffusion. These results for Au, Cu/Si(100) interfaces show a different behaviour than Au, Cu/Si(111) interfaces. Here no reaction was found between Au islands but a weak reaction between Cu islands, when studied with spatially resolved AES by Calliari et al.[10]. These differences reflect the changes in the growth morphologies of noble metals on the two different Si faces[11].

In conclusion we want to point out that clear evidence of intermixing at LNT was observed for Au and Cu/Si(100) interfaces beyond θ = 1. The build up of interfaces at RT and HT follows a complex mechanism, which cannot be described by a simple Stranski-Krastanov growth fashion as in the case of Ag.

4.   REFERENCES

1) Abbati, I., Braicovich, L.and Franciosi, A., Phys. Lett. 80A, 69 (1980).

2) Abbati, I. and Grioni, M., J. Vac. Sci. Technol. 19, 631 (1981).

3) Hiraki, A., Surf. Sci. Rep. 3, 357 (1984).

4) Dallaporta, H. and Cros, A, Surf. Sci. in press.

5) Calandra, C., Bisi, O. and Ottaviani, G., Surf. Sci. Rep. 4, 271 (1985).

6) Rossi, G., Kendelewics, T., Lindau, I. and Spicer, W.E., J. Vac. Sci. Technol. A1, 977 (1983).

7) Brillson, L.J., Katnami, A.D., Kelly, M. and Margaritondo, G., J. Vac. Sci. Technol. A2, 551 (1984).

8) Del Guidice, M., Grioni, M., Joyce, J.J., Ruckman, M.W. Chambers, S.A. and Weaver, J.H., Surf. Sci. 168, 309 (1986).

9) Hanbücken M., Métois, J.J., Mathiez, P. and Salvan, F., Surf. Sci. 162, 622 (1985).

10) Calliari, L., Sancrotti, M. and Braicovich, L., Phys. Rev. B30 4885 (1984) ; Calliari, L., Marchetti, F. and Sancrotti, M., Phys. Rev. in press.

11) Hanbücken, M. and Le Lay, G., Surf. Sci. 168, 122 (1986).

Electronic Structure and Interface States of the (100)Fe/Ge Interface

W.E. Pickett and D.A. Papaconstantopoulos

Naval Research Laboratory
Washington, DC   20375-5000
USA

ABSTRACT

We present calculations of the electronic states of the (110)Fe/Ge interface using tight-binding parameters derived by fitting to spin-polarized Fe bands, pseudopotential Ge bands and CsCl-structure FeGe bands. We utilize a generalized Slater-Koster scheme suitable for complicated geometries and capable of handling a large number of atoms per unit cell with realistic computational effort. Our results show that the Fe minority spin projected band structure has interface bands that lie in the fundamental gap of Ge; and that the majority spin has a large number of interface states just above the Fermi level strongly localized on the Ge atoms. These results are consistent with the reactive nature of this Fe/Ge interface shown by experiments.

1.   INTRODUCTION

Metal-semiconductor interfaces are of considerable technological interest and therefore have stimulated several experimental and theoretical studies. In this work we report calculations of the electronic structure of the (110)Fe/Ge interface. These calculations were motivated by the work of Prinz et al.[1-3] who, using molecular beam epitaxy, have grown single crystal (110)Fe films on GaAs but found that Fe films grown on Ge are heavily intermixed with Ge to a thickness of several tens of layers. In our study we determine the electronic states of the non-relaxed (110)Fe/Ge interface using a supercell geometry. We find minority interface bands which lie in the Ge energy gap, and a large density of majority spin interface resonances near the Fermi level E_F.

2. METHOD OF CALCULATION

The (110) interface requires 4 Fe and 2 Ge atoms per respective layer. In order to isolate the interface bands it was necessary to use 7 layers of Fe and 11 layers of Ge for a total of 50 atoms. Present day computer capabilities make such a calculation practically impossible from first principles. For this reason our calculations were done with the parametrized tight-binding method of Slater and Koster,[4] using a general computer code[5] that allows an arbitrary unit cell which can be periodically repeated in zero, one, two, or three dimensions. The tight-binding parameters needed in this approach were found by fitting to the energy bands of the host elements Fe and Ge, with a three neighbor orthogonal basis in the two-center approximation.[6] In addition we determined the Fe-Ge interaction parameters by fitting to the energy bands of a hypothetical CsCl-structure FeGe compound. Adjustment of the energy scale of this calculation was made via the on-site tight-binding parameters, to place E_F in the middle of the Ge gap (as suggested by experiments[7] on Ge-metal systems) and to eliminate unphysical charge transfer between atoms near the interface.

3. RESULTS AND DISCUSSIONS

In Fig. 1 the minority spin local densities of states (LDOS) are shown for several planes. The shaded lines indicate areas where the LDOS exceeds the bulk DOS. This excess LDOS appears mainly near E_F with very little elsewhere.

The majority spin LDOS is shown in Fig. 2. As is indicated by the shaded areas large excess LDOS is found on Ge in its fundamental gap. These states are metal-induced gap states like those observed at many other metal-insulator interfaces.

The excess LDOS can be understood qualitatively if we consider that an Fe atom at the interface has only half of its first two bcc shells occupied by Fe atoms. It is only when the Fe atoms are surrounded by all first- and second-neighbors that the bonding-antibonding separation characteristic of the bcc structure occurs. In the same way in Ge the energy gap forms only if several neighboring shells are all occupied by Ge. Thus independently of the specific geometrical arrangement of atoms, a large LDOS at the interface should occur near the gap/Fermi level.

Our results for the ideal geometry show a rich spectrum of minority interface states in the mutual gap, and large majority LDOS on the Ge atoms at the interface. The large interface DOS at E_F suggests that the system may gain electronic energy by intermixing[3] to form a disordered interface or an interfacial compound. This notion seems to be consistent with the experimental evidence for intermixing at the interface when Fe is deposited on the Ge surface.[3] A more detailed account of this work will be given elsewhere.[8]

This work was supported by Office of Naval Research Contract No. N0014-83-WR-30007.

4. REFERENCES

1. Prinz, G.A. and Krebs, J.J., Appl. Phys. Lett. <u>39</u>, 397 (1981).

2. Hathaway, K.B. and Prinz, G.A., Phys. Rev. Lett. <u>47</u>, 1761 (1981).

3. Prinz, G.A. and Krebs, J.J., unpublished.

4. Slater, J.C. and Koster, G.F., Phys. Rev. <u>94</u>, 1498 (1954).

5. Pickett, W.E., unpublished.

6. Papaconstantopoulos, D.A., "Handbook of the Band Structure of Elemental Solids", Plenum, New York 1986.

7. Sze, S.M., <u>Physics of Semiconductor Devices</u> (J. Wiley & Sons, New York, 1981).

8. Pickett, W.E. and Papaconstantopoulos, D.A., Phys. Rev. to be published.

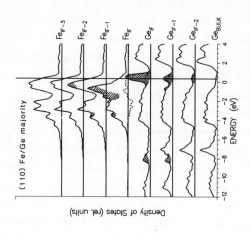

Figure 2. As in Figure 1, but for the majority spin.

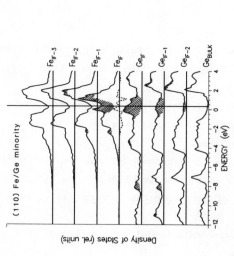

Figure 1. The minority local density of states (LDOS) per atom on several Fe and Ge layers near the Fe/Ge (110) interface. The hashed region indicates interface states, i.e. where the LDOS exceeds the bulk DOS. "Ge_{BULK}" indicates an average over the interior five layers of Ge. The dotted line on the Fe_{IF} layer indicates the difference between the LDOS on the Fe_{IF}^{nb} and Fe_{IF} sites. Note: the LDOS for the Ge layers have been multiplied by two for easier comparison.

PHOTOEMISSION STUDY OF METAL-INDUCED CORE LEVEL SHIFTS IN Ar-IMPLANTED GaAs

Janusz Kanski, Thorwald G. Andersson and Johan Westin

Department of Physics, Chalmers University of Technology
S-412 96 Göteborg, SWEDEN

ABSTRACT

By investigating photoemission spectra from Ar-implanted GaAs with Au overlayers, we show that chemical shifts of Ga core levels due to Au-Ga interaction are very small (well below 0.1 eV). It is argued that this result should also apply to MBE-grown GaAs(001), confirming that previously reported enhancement of core level shifts in the immediate vicinity of the Au-GaAs interface is not of chemical nature.

1. INTRODUCTION

Photoelectron spectroscopy is an extensively employed technique for investigations of metal-semiconductor interfaces. Due to the complexity of the interface problem it is realised that the detailed development of the electronic structure has to be studied on more or less ideal systems, i.e. systems which form abrupt and nonreacted interfaces. However, as the chemical interaction between the metal and the semiconductor is minimised, the overlayer tends to form three dimensional islands. This occurs for instance in the case of Ag on GaAs(110) [1]. The lateral inhomogeneity is of course disturbing, since it leads to simultaneous emission from clean and metallised surface regions and masks the appearance of metal induced spectral changes. On the other hand, if one investigates a system exhibiting two dimensional overlayer growth, the question of chemical effects inevitably turns up. This problem was recently encountered in a study [2] of Schottky barrier formation between Au and GaAs(001). In the present work we have analysed photoemission spectra of Ar-implanted GaAs in order to estimate the size of chemical shifts induced by an Au overlayer. The results are significant for our previous conclusions regarding metal induced screening at the Au/GaAs(001) interface.

2.   EXPERIMENT

The experiment was performed in a VG ADES 400 electron spectrometer system with He(I) and He(II) radiation at energy resolutions 0.15 and 0.30 eV respectively. The samples were n- and p-type GaAs wafers, normally used as substrates for MBE layers. Argon was implanted during conventional surface cleaning, i.e. bombardment with 500 eV ions under static Ar pressure of 5×10^{-5} torr. Au overlayers were deposited at room temperature from a shuttered W-coil evaporation source and the overlayer thickness was estimated by comparison with previous well calibrated photoemission data.

3.   RESULTS

The He(I) valence band spectrum of Ar-bombarded n-type GaAs is shown in figure 1. For comparison the figure also includes a normal emission spectrum of MBE-grown n-type GaAs(001)-c(2x8). Not unexpectedly the ion bombardment results in a rather smeared spectrum, since the surface disorder has an averaging effect. Another consequence of the Ar

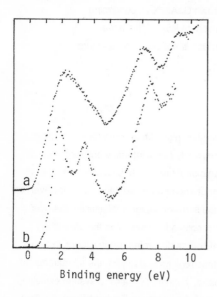

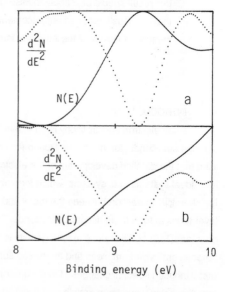

Figure 1 Valence band spectra of Ar-bombarded (a) and MBE-grown (b) GaAs

Figure 2 Ar(3p) emission before (a) and after (b) deposition of an Au overlayer

treatment is bond breaking and creation of surface states. This is seen via a shift of the whole spectrum towards higher energies so the valence band maximum is pinned 0.3 eV below the Fermi level for n- as well as for p-type samples. For the present purpose the most important effect of the Argon is the peak found just above 9 eV binding energy. This peak, identified as emission from Ar(3p) states appears in an energy range where the clean GaAs spectrum is structureless, and vanishes rapidly upon annealing at ~250 °C. The Ar(3p) peak is shown separately on an expanded scale in figure 2, before and after deposition of ~5 Å Au. Rather fortuitously, the emission of Au is also very smooth in the energy range of the Ar(3p) peak, which of course simplifies the extraction of the Ar structure. Using second derivative spectra (obtained by numerical differentiation), the Au induced shift of the Ar(3p) level is thus determined to be 0.15 eV. Since Au and Ar certainly do not interact chemically, this shift represents a change in the effective electrostatic potential at the surface and all other core levels in the surface region must be shifted the same way. Deviations from this behaviour can be directly taken as measures of local "chemical" shifts.

Let us then examine the Ga(3d) level. In figure 3a we show the He(II) excited Ga(3d) emission from the same Ar-implanted sample. For comparison the figure 3b shows the corresponding result on MBE-grown GaAs(001). The energy resolution in these spectra is just sufficient to confirm the presence of the spin-orbit components. It is immediately clear that the initial conditions are quite different in the two cases: due to the Ar bombardment the Ga(3d) levels are moved towards the Fermi level (just as the whole valence band) and some tailing is obtained on the low binding energy side, indicating contribution from atoms at various local environments on a disordered surface. Mainly as an effect of these initial

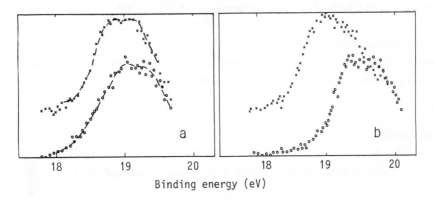

Binding energy (eV)

Figure 3 Ga(3d) emission from Ar-bombarded (a) and MBE-grown (b) GaAs before (oooo) and after (xxxx) deposition of Au overlayers

differences between the two samples, the Au induced shifts of the Ga(3d) levels are different: 0.16 eV for the Ar implanted- and 0.40 eV for the MBE grown surface. Quite independently of the initial conditions however, we obtain our main result: the Au induced shifts of the Ga(3d) and Ar(3p) levels are very similar, showing that in this case the effects of chemical interaction between Au and Ga are very small ($\leq$0.05 eV).

4.   CONCLUSION

The main purpose of the present study was to get information regarding the possible size of Au-induced chemical shifts of Ga(3d) levels on MBE-grown GaAs. Our interest in this goes back to an investigation[2] of the Schottky barrier formation between Au and GaAs(001), where we discovered that simultaneously with the formation of a metallic Au overlayer the Ga(3d) levels were shifted by different amounts in He(II)- and MgK$_\alpha$-excited spectra (~0.3 eV and ~0.1 eV respectively). While this difference could be directly attributed to different probing depths in the two cases, the underlying physical reason was not obvious. The present results show that in the case of Ar-implanted GaAs the chemical effect is indeed very small. Considering that the Ar-bombarded surface must contain a relatively large density of dangling bonds and other defects, it is reasonable to expect that it should be more reactive than other, better ordered surfaces. Therefore we believe that the result obtained here also applies to the interaction between Au and MBE-grown GaAs(001), i.e. the difference in the core level shifts observed at different photon energies is not due to chemical interaction between Au and Ga at the interface. This conclusion is in good agreement with other observations, e.g. the fact that the core level shifts occur at a critical Au coverage simultaneously with the formation of a metallic surface[2]. Our previous interpretation of the core level shifts in terms of metal induced screening is thus supported.

ACKNOWLEDGEMENTS
We wish to thank the Swedish Board for Technical Development and the Swedish Natural Science Research Council for financial support of this project.

REFERENCES
1. Ludeke, R., Chiang, T.-C. and Miller,T., J. Vac. Sci. Technol. B1 , 581 (1983)
2. Kanski, J., Svensson, S.P., Andersson, T.G. and Le Lay, G., to be published

Kinetics of E_F pinning at Au-GaAs(110) interfaces

K. Stiles and A. Kahn
Department of Electrical Engineering
Princeton University, Princeton, NJ 08544

D.G. Kilday, N. Tache and G. Margaritondo
Department of Physics and Synchrotron Radiation Center
University of Wisconsin, Madison, WI 53706

Abstract

The pinning of the Fermi level by Au on low-temperature GaAs(110) is investigated via soft X-ray photoemission spectroscopy. The kinetics of Schottky barrier formation on n- and p-substrates are drastically different, and suggest the interplay of several pinning mechanisms. Implications on the validity of current Schottky barrier models are discussed.

This study investigates the role of the substrate temperature in the Fermi level pinning on n- and p-GaAs. The Au-GaAs interface is particularly interesting because the pinning position of E_F is anomalously low, as compared to the position found with most other metal-GaAs interfaces. This was perceived by some authors as an indication of the failure of the defect models to predict barrier heights. Proponents of this model, however, argued that the reaction taking place between Au and GaAs, i.e. decomposition of the substrate and formation of AuGa and $AuGa_2$ compounds, might explain the formation of pinning defects with energy levels lower in the gap than those due to other adatoms[1]. It was also argued that the large electronegativity of Au induces an interface charge transfer sufficient to partially empty the pinning donor states[2] and pull E_F deeper toward the valence band maximum (VBM).

Studies of metals deposited on low temperature (LT) substrates were undertaken to investigate the specific roles of surface diffusion, cluster formation and chemical reactions in the pinning process. All three phenomena can presumably be altered by varying the substrate temperature. The present work was prompted by recent results obtained on the Al-GaAs(110) interface[3], which demonstrate dramatic differences in the rate of E_F pinning on n- and p-GaAs at LT. Al-GaAs(110) exhibits a dual structural complexity, i.e. metal clusterization and Al-Ga exchange reaction, during the initial stages of interface growth. Three years ago, Daniels et al. proposed the energy released by the aggregation of Al atoms could trigger the Al-Ga exchange reaction, and cause the formation of pinning defects[4]. The study of Al evaporated on LT GaAs(110) partly confirmed this hypothesis, but also suggested a more complex situation[3]. It was shown that the kinetics of formation of Al-GaAs Schottky barrier is strongly affected by the temperature of the substrate during deposition, as well as by its doping. The pinning of E_F by Al on LT n-GaAs is considerably slowed down with respect to room temperature (RT), whereas pinning on p-GaAs is as fast at LT than at RT. This asymmetry, which is observable only at low temperature, suggests the interplay of at least two distinct mechanisms responsible for the formation of pinning interface states. The n-type results remain consistent with the defect picture, and reflect a slower rate of formation of pinning defects due to reduced clustering and chemical reaction at LT. The p-type results, however, are inconsistent with reduced clustering and chemical activity. Other mechanisms must be invoked. It was suggested that interface structural modifications, which are more extensive and homogeneous across the semiconductor surface at LT, could play a role in this process[3].

We present here PES measurements for Au deposited on RT and LT GaAs(110). The soft X-ray photoemission (SXPS) work was done at the University of Wisconsin Synchrotron Radiation Center. The measurements were performed on Laser Diode n-type (Si-doped, $5 \times 10^{17} cm^{-3}$ and Te-doped, $4\text{-}7 \times 10^{17} cm^{-3}$) and p-type (Zn-doped, $1 \times 10^{18} cm^{-3}$) GaAs bars cleaved in ultra-high vacuum (base pressure $= 8 \times 10^{-11}$torr). Throughout the low temperature experiments, GaAs was kept below $80\,^{\circ}$K. Thicknesses of Au ranging between $0.25\overset{\circ}{A}$ and $12\overset{\circ}{A}$ were evaporated from a resistively heated W filament. At each coverage, core level and valence band spectra (Ga-3d, As-3d, Au-4f and VB) were recorded using photon energies corresponding to a minimum escape depth of the photoemitted electrons.

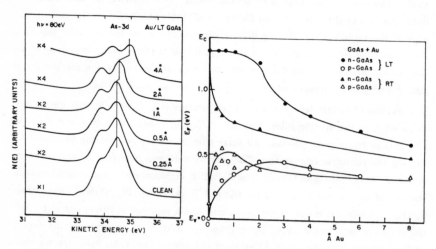

Figure 1 -- n-GaAs(110) band bending caused by Au deposition on a LT substrate as shown by the shift of the As-3d core level.

Figure 2 -- Interface position of the Fermi level in the gap of GaAs as a function of Au coverage on RT and LT n- and p-substrates.

The Au-5d doublet peak which appears at 4-6eV below the GaAs VBM distorts the VB spectrum and prevents measurements of band bending from the shift of the VB peaks. The movement of the Fermi level was determined from the energy shift of the Ga-3d and As-3d core levels as a function of Au coverage (figure 1). We considered the possibility that chemical shifts, due to the movement of Ga and As through the Au layer and the formation of intermetallic compounds, might invalidate the procedure. Our experimental results, however, always show parallel shifts of the As-3d and Ga-3d core levels in the low coverage regime where interdiffusion is limited. Thus, the main conclusion of our work, which concerns the kinetics of barrier formation during initial metal deposition (< 1-$2\overset{\circ}{A}$), is not affected by chemical shifts.

The deposition of Au on RT n- and p-GaAs(110) produced the E_F movement shown in figure 2. The fast initial movement of E_F on n- and p-GaAs for coverages below $0.5\overset{\circ}{A}$ has been attributed to the formation of near mid-gap donor and acceptor levels which account for E_F pinning with many other adatoms. The change of direction in the movement of E_F on p-GaAs at a coverage of $0.5\overset{\circ}{A}$ and the anomalously low final position of E_F on both n and p substrates at high coverage have been justified in terms of a kind of defects peculiar to this interface introduced by the specific Au-GaAs interface chemistry, and of the large adatom electronegativity.

The movement of E_F for Au deposited on LT GaAs(110) reveals a different kinetics of Schottky barrier formation. The movement of E_F on n-GaAs is *considerably slower* at LT than at RT (figures 1 and 2). E_F moves less than 100meV up to a coverage of $1.5\overset{\circ}{A}$, whereas pinning is almost complete at this coverage on the RT sample. Above $2\overset{\circ}{A}$, the pinning rate increases and E_F tends to its RT position, although a 0.15eV difference between the RT and LT positions remains up to fairly high coverages. On p-GaAs, the pinning rate is also somewhat slower than at RT, but remains *considerably larger* than on n-GaAs (figure 2). E_F moves up 300 - 400meV at a coverage of $1.5\overset{\circ}{A}$ and pinning is almost complete at $2\overset{\circ}{A}$. After reaching a maximum position in the gap at 2 - $3\overset{\circ}{A}$, E_F moves down and tends toward the RT position.

The results obtained for LT n-GaAs can be divided in two coverage regimes. Below $2\overset{\circ}{A}$, the very slow pinning implies that the mechanism which introduces acceptor states at RT is dramatically influenced by the substrate temperature. This agrees with the results obtained with Al[3] and In[5] on LT GaAs, and is consistent with models where acceptor states are created via energy released by the formation of metal clusters, and/or interface reaction, both mechanisms presumably being inhibited at LT. It also suggests that the heat of condensation of the metal on GaAs, which is not affected by temperature, is not the direct cause of pinning. Above $2\overset{\circ}{A}$, the pinning accelerates substantially for reasons which are still unclear. The surface mobility of Au and the overlayer morphology on RT and LT GaAs(110) are not as well characterized as in the case of Al. Preliminary LEED and AES measurements[6] indicate that Au deposited on LT GaAs does not form as smooth a layer as Al does, nor does it appear to unrelax the substrate surface structure as Al does[7]. The increase in pinning rate could therefore correspond to the onset of cluster formation and/or interdiffusion which might occur with higher concentrations of metal on the surface.

On LT p-GaAs, the rate of E_F pinning is considerably larger than on n-GaAs, and slower than on RT p-GaAs. The first result confirms the trend already observed with Al on GaAs[3]. This is extremely interesting, as it suggests again an interplay between two or more mechanisms. The initial pinning on p-GaAs cannot be attributed solely to the mechanism responsible for pinning on n-GaAs, e.g. clustering or interface reaction, as this mechanism is greatly inhibited at LT. The second result points out a departure from the Al case where pinning at LT was as fast as at RT. This difference might reflect a difference between the morphology and structure of the two interfaces at LT. We believe that a possible additional pinning mechanism in the case of Al on LT p-GaAs was the change in the substrate relaxation[7] accompanied by a movement of the substrate dangling bond states back into the gap. As mentioned above, the structural behavior of Au on GaAs is more complex, and an incomplete reversal of the GaAs relaxation might produce a slower pinning rate than with Al.

In summary, the study of Au on LT GaAs(110) confirms the principal conclusions reached with Al on the same substrate, namely that more than one mechanism must be considered to explain initial pinning on n- and p-GaAs. The n-GaAs data are consistent with pinning by cluster- or reaction-induced defects, whereas the p-GaAs data require a mechanism less dependent on substrate temperature.

Acknowledgement

Support of this work by the ONR, the NSF (Grant DMR-84-06820), General Electric and Xerox Corporation is gratefully acknowledged.

References

1. W. Mönch, Surf. Sci. *132* 92 (1983).
2. W.E. Spicer, S. Pan, D. Mo, N. Newman, P. Mahowald, T. Kendelewicz and S. Eglash, J. Vac. Sci. Technol. *B2* 476 (1984).
3. M.K. Kelly, A. Kahn, N. Tache, E. Colavita and G. Margaritondo, Sol. State Communications *58* 429 (1986).
4. R.R. Daniels, A.D. Katnani, T.X. Zhao, G. Margaritondo and A. Zunger, Phys. Rev. Lett. *49* 895 (1982).
5. K.K. Chin, T. Kendelewicz, C. McCant, R. Cao, K. Miyano, I. Lindau and W.E. Spicer, J. Vac. Sci. Technol. *A4* 969 (1986).
6. K. Stiles and A. Kahn (to be published).
7. C.R. Bonapace, K. Li and A. Kahn, Journal de Physique *4* C5-409 (1981).

FORMATION AND THERMAL STABILITY OF LaB$_6$–GaAs(001) SCHOTTKY BARRIER

Hisao Nakashima, Tatsuo Yokotsuka, Yoko Uchida and Tadashi Narusawa

Optoelectronics Joint Research Laboratory
1333 Kamikodanaka, Nakahara-ku, Kawasaki 211, Japan

Formation and thermal stability of LaB$_6$–GaAs(001) interfaces have been studied by XPS, RBS and I-V measurements. The uniform LaB$_6$ films are formed on GaAs surfaces without nucleation. Any appreciable interface reaction is not detected even after high temperature ($\sim$ 800°C) annealing.

We have proposed LaB$_6$ as a self-aligned gate material for GaAs MESFET's.[1] Because of excellent properties of the thermal stability, high electrical conductibity and easiness in film formation, LaB$_6$ is expected as a very promising gate material. The study of this stable LaB$_6$ and GaAs interfaces will provide important information to the understanding of Schottky barrier formation as well as to the practical use of LaB$_6$. In this paper, we describe the formation and thermal stability of LaB$_6$–GaAs(001) interfaces studied by XPS, RBS and I-V measurements.

Samples used here were made by evaporating LaB$_6$ on MBE-prepared c(4x4) and chemically etched GaAs(001) surfaces. XPS measurements with AlKα excitation were carried out using VG-ADES 400 combined with MBE. RBS channeling spectra were taken in a separate vacuum system using 3 MeV He^{++} ion beam. I-V characteristics of Schottky diodes which were made on MBE-prepared and chemically etched surfaces were measured after being annealing at 300 ~ 850°C.

Core photoemission spectral intensities normalized to the clean surface value as a function of LaB_6 coverage are shown in Fig. 1. LaB_6 films were deposited at a rate of 2∼ 3 Å/min on the c(4x4) GaAs(001) surfaces at room temperature. Exponential decays of Ga3d, Ga2p, As3d and As2p intensities with the LaB_6 coverage are observed. This result indicates that the LaB_6 films were continuously and uniformly formed on GaAs surfaces without nucleation. This is very consistent with the RHEED observation which showed that the LaB_6 films deposited at room temperature were amorphous.

Usually, the Ga2p and As2p peak positions did not change with the LaB_6 coverage (case 1). The barrier height obtained from I-V characteristics of the LaB_6– c(4x4) GaAs(001) Schottky diodes was 0.7 eV, which coincides with

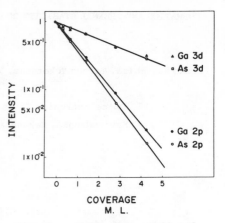

Fig. 1   Ga3d, Ga2p, As3d and As2p XPS intensities as a function of LaB_6 coverage.

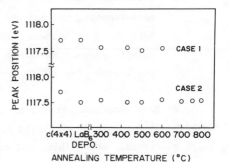

Fig. 2   Peak position of Ga2p spectrum as a function of annealing temperature.

the Fermi level pinning position of the c(4x4) GaAs(001) surface.[2] These results suggest that the LaB_6 deposition does not introduce much defects which cause further band bending. However, with high deposition rate or slight substrate temperature rise, these peak positions shifted by 0.2 eV toward lower binding energy side (case 2).

The Ga2p peak position as a function of UHV-annealing temperature is shown in Fig. 2 for both cases. LaB_6 film thickness is about 7 Å. For case 1, the extra band bending of 0.2 eV is induced by annealing at above 400°C. It has been reported that excess As(Ga) at the metal-MBE prepared GaAs interface gives a lower (higher) barrier height.[2-4]

Then, we think that this induced band bending is mainly due to the release of As atoms from the LaB_6-As stabilized c(4x4) GaAs interfaces by annealing.

No change in As3d and Ga3d spectral shapes which are very sensitive to chemical reaction was observed even after annealing at 800°C. This result indicates that no interface reaction occurred at high temperature. RBS channeling measurements also support this result. The disorder-sensitive surface peak (SP) intensities[5] which were calibrated using As implanted Si wafer were measured as a function of annealing temperatue, as shown in Fig. 3. The SP intensity of $\sim$ 5.0 atoms/row shown as a dashed line in Fig. 3 is the calculated value assuming a bulk-like structure with enhanced surface thermal vibration. This value agrees fairly well with an experimental result for (1x1) GaAs(001) surface[5]. Before LaB_6 deposition, both SP intensities are $\sim$ 0.7 atoms/row higher than that of the (1x1) GaAs(001) surface due to relaxed Ga and As atoms for the c(4x4) surface and native oxide for the chemically etched surface. The SP intensities decrease by $\sim$ 0.5 atoms/row due to LaB_6 deposition (chemically etched surface) or annealing (c(4x4) surface) and remain nearly constant after higher temperature annealing. These results show that the number of displaced Ga and As atoms after annealing are not more than $\sim$ 1 monolayer ($\frac{1}{4}$ atoms/row), i.e. 5×10^{14} cm^{-2}. Further, it shows that the interface disorder does not proceed by higher temperature annealing. Then, we conclude that any appreciable reaction does not take place between LaB_6 and GaAs.

Figure 4 shows the barrier height ϕ_B and ideality factor n of Schottky diodes as a function of annealing temperature. For the

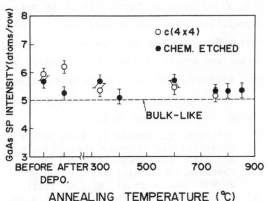

Fig. 3   Annealing temperature dependence of GaAs SP intensities for LaB_6-GaAs(001). The incident beam is 3 MeV, He^{++} and aligned 001 direction.

LaB$_6$-c(4x4) GaAs(001) Schottky
diode, ϕ_B gradually increases
with increasing annealing
temperature. This annealing
behavior of ϕ_B is quantitatively
different from that obtained
from XPS measurements. Thick
LaB$_6$ films ($\sim$ 2000 Å) of
Schottky diodes are considered
to prevent the release of As
from the interface. Annealing
behavior (400 $\sim$ 850°C) of ϕ_B
of the LaB$_6$-chemically etched
GaAs(001) Schottky diodes can
not be explained at the moment.
There may be some reaction
between LaB$_6$ and native oxide.

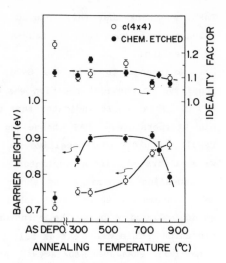

Fig. 4   The barrier height and
ideality factor of LaB$_6$-GaAs
Schottky diodes as a function of
annealing temperature.

In any case, the values of n and ϕ_B do not change much by high
temperature annealing and are satisfactory for an application to GaAs
MESFET's.

In summary, we have studied the formation and thermal stability
of LaB$_6$-GaAs(001) interfaces by XPS, RBS and I-V measurements. The
LaB$_6$ films were uniformly formed on GaAs surfaces by electron beam
evaporation. Any appreciable reaction between LaB$_6$ and GaAs was not
observed even after high temperature annealing. Since the ideality
factor and barrier height do not change much by annealing, LaB$_6$ is a
very promising candidate for a self-aligned gate material of GaAs IC.

The present researches are supported by the Agency of Industrial
Science and Technology, MITI.

1)   Yokotsuka T., Uchida Y., Kobayashi K.L.I., Narusawa T. and
      Nakashima H., Extended Abst. 17th Conf. on Solid State Devices
      and Materials, Tokyo 1985, p. 437.
2)   Svensson S.P. Kanski J., Andersson T.G., and Nilsson P.-O.,
      J. Vac. Sci. Technol. B2 235 (1984).
3)   Svensson S.P., Landgren G., and Andersson T.G., J. Appl. Phys.
      54, 4474 (1983)
4)   Wang W.I., J. Vac. Sci. Technol. B1 574 (1983).
5)   Narusawa T., Kobayashi K.L.I. and Nakashima H., Jpn. J. Appl.
      Phys. 24 L98 (1985).

CHARACTERIZATION OF Sb GROWTH ON GaAs(110)

W.Pletschen, N.Esser, J.Geurts, W.Richter[a]
A.Tulke, M.Mattern-Klosson, H.Lüth[b]

a) I.Physikalisches Institut b) II.Physikalisches Institut, RWTH Aachen, D-5100 Aachen, FRG

ABSTRACT

Overlayer structure and interface band bending in the system GaAs-Sb are studied by LEED, UPS, ARUPS and Raman spectroscopy. The first Sb layer grows epitaxially. Further growth is initially amorphous and stays amorphous for substrate temperatures of 80K even at 66 monolayers (ML). However at 300K it crystallizes around 12 ML. Strong band bending changes are observed for coverages below 0.1 ML, but also in a second stage beyond 1 ML. The detailed behaviour depends on the Sb structure.

1. EXPERIMENTAL

In situ measurements were performed for Sb coverages from sub-monolayer up to 66 ML. The Sb was grown by thermal evaporation in UHV at substrate temperatures of 80K or 300K. For ARUPS the angular resolution was achieved by tilting the sample while the angle between He-lamp and hemispherical analyzer was fixed. The resonant Raman measurements were carried out with a Kr ion laser, using the 531 nm line for the studies of the Sb whereas the band bending was investigated with the 407 nm line.

2. Sb STRUCTURE

After 1 ML of Sb the ARUPS results in Fig.1a show new structures in the valence band, labeled S_3 to S_6 (full lines), as compared to the clean surface (dashed lines). From these the band structure in Fig.1b (dashed lines) is derived. It agrees very well with calculations[1] (full lines in Fig.1b), based on the Sb chain model, in which each Sb atom is bound to two neighbouring Sb atoms (S_3, S_4) and to a Ga (S_6) or an As atom (S_5). The two remaining Sb-electrons form a lone pair.

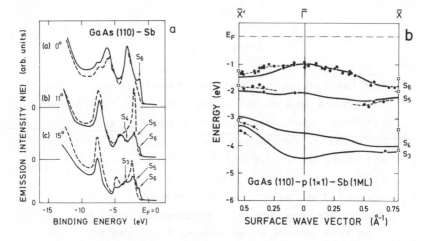

Fig.1: ARUPS spectra on GaAs-Sb (a) and resultant energy dispersion $E(k_{//})$ of electronic interface states (b)

This epitaxial structure for the first monolayer is also confirmed by dynamical LEED results[2]. For higher coverages, the LEED pattern disappears, characteristic for non-epitaxial growth. This can be understood in terms of the weak binding of additional Sb to the first ML via the lone pair electrons. Fig.2a shows Sb lattice vibrations, observed by Raman spectroscopy. The broad structure, around 150 cm^{-1}, is typical for amorphous Sb. At 80K the Sb film stays amorphous up to the highest investigated coverages (66 ML). However at 300K the E_g and A_{1g} modes of crystalline Sb appear around 12 ML.

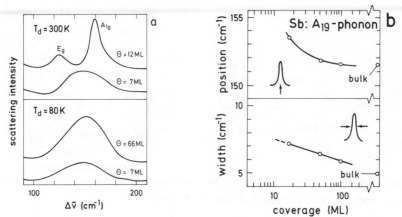

Fig.2: Sb phonon spectra (a), line width and frequency position of the A_{1g} mode (b) for different coverages

The Sb film is polycrystalline as shown by the broadening of the
phonon modes (Fig.2b). Moreover, the phonon modes shift to
higher frequencies (Fig.2b) due to stress in the crystallites.
From the broadening a crystallite size of about 10 nm is
estimated. Comparison to stress experiments on bulk Sb gives a
lattice dilation of about 1%. With increasing film thickness the
line width reduces and the shift decreases, indicating stress
relaxation in the now larger crystallites.

3. BAND BENDING

Results on band bending for n-type GaAs at submonolayer
coverage, determined by UPS[3], are shown in Fig.3a for a
deposition temperature of 300K. Band bending starts far below
0.1 ML and levels off below 1 ML at 650 meV. With Raman
scattering the band bending was determined from 0.01 ML up to 66
ML[4]. By this method the band bending is derived from the GaAs
LO phonon intensity, which at (110) surfaces depends mainly on
the surface electric field and thus on the band bending. The
results for 300K are shown in Fig.3b. At low coverages there is
excellent agreement to the UPS data (crosses, taken from Fig.
3a).

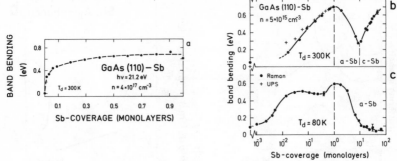

Fig.3: Band bending from UPS (a) and Raman scattering at deposition temperatures 300K (b) and 80K (c).

At high coverages, beyond 1 ML, in addition strong band bending variations occur. They are correlated to the Sb structure: amorphous Sb reduces the band bending (Fig. 3b,c) whereas the crystallization causes a reincrease (Fig.3b). A possible explanation is given by the generation of new interface states, caused by the relocation of the Sb atoms.

4. CONCLUSIONS

Our results confirm the Sb chain model for the first Sb layer. Higher coverages are amorphous for substrate temperature 80K, whereas crystallization occurs for 300K. The band bending has two stages: first a strong increase below 0.1 ML, then additional variations beyond 1 ML, which depend on whether the Sb structure is amorphous or crystalline.

5. REFERENCES

1) Mailhiot, C., Duke, C.B. and Chadi, D.J., Phys. Rev. B31, 2213 (1985)

2) Carelli, J. and Kahn, A., Surf. Sci. 116, 380 (1982)

3) Mattern-Klosson, M. and Lüth, H., Solid State Commun. 56, 1001 (1985)

4) Pletschen, W., Esser, N., Münder, H., Zahn, D., Geurts, J. and Richter, W., Surf. Sci. (1986,to be published)

CATHODOLUMINESCENCE SPECTROSCOPY OF METAL/III-V SEMICONDUCTOR INTERFACE STATES

R. E. Viturro, M. L. Slade, and L. J. Brillson,

Xerox Webster Research Center, Webster, NY 14580, USA.

We have employed optical emission techniques to characterize the formation and evolution of interface states with metal deposition on UHV cleaved (110) III-V semiconductor surfaces from submonolayer to multilayer coverages. We show that the evolution of the electron excited optical emission spectra of metal/InP and GaAs interfaces can be correlated to their E_F movements and macroscopic Schottky barrier heights.

Metal/semiconductor (M/SC) interfaces have attracted considerable attention over the past few decades for both scientific and technological reasons[1]. Still, the nature of the interface electronic states and basic mechanism of Schottky barrier formation are not yet well understood [2]. For clean, ordered InP or GaAs (110), intrinsic gap surface states are absent, and a few monolayers of deposited metal create new interface states which stabilize the Fermi level (E_F) in a limited range within the band gap[3]. Here we report the most direct observation of these metal/SC interface states thus far. We have detected luminescence from interface states by means of cathodoluminescence spectroscopy[4](CLS), using a chopped electron beam from a glancing incidence electron gun and a LN$_2$ cooled Ge detector (North Coast)[4,5].

Figure 1 shows CL spectra from p-type InP(110), for mirror-like areas and step cleaved areas, and for submonolayer coverage of different reactive and unreactive metals deposited on the mirror-like InP cleavage. The CL spectra of clean InP shows only one emission centered at 1.35 eV within the energy range 0.6-1.6 eV, corresponding to the near-band gap (NBG) of InP at room temperature. On the other hand, the CL spectrum of step-cleaved areas shows weak emission at sub-band gap energies. This spectrum is similar to those obtained from submonolayer deposits of metals on mirror-like areas. Thus, the initial metal deposition disrupts the (110) surface, causing the formation of broken bonds, such as those formed during a step-cleavage process. Figure 2 shows the CL spectra of clean InP(110) and GaAs(110) with a series of metal overlayer thicknesses. We observe new emission features which reflect the modification of the SC surface upon metal deposition and the consequent formation of new states.

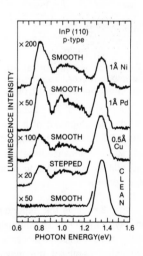

1. CL spectra of UHV mirror-like cleaved InP(110) surfaces before and after submonolayer Ni, Pd, or Cu deposition, and the step-cleaved surface

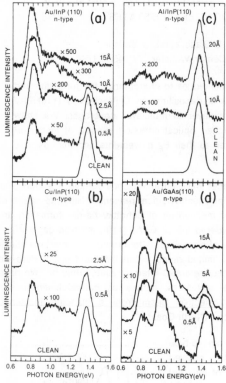

Fig. 2 demonstrates that the changes produced in the optical emission properties of InP upon metal deposition are strongly dependent on the particular metal. For example, the formation of the interface states for Au vs Cu deposited on InP has different dependences on metal thickness, evolving faster with Cu deposition.

On the other hand, the CL spectra of increasing Al thicknesses on InP show that the NBG transition dominates the spectral shape even after deposition of 20 Å of Al on InP, whereas the low energy emissions are similar to those of Fig. 1. The overall luminescence intensity is drastically reduced, but the spectral shape is not substantially changed by Al deposition. Fig. 2(d) shows CL spectra of the Au/GaAs(110) system. The bottom spectrum corresponds to a mirror-like cleaved surface. The intensity of the lower energy emissions depends on cleavage quality and doping level[6].

2.   CL spectra of (a) Au, (b) Cu, and (c) Al on clean mirror-like n-type InP (110) and (d) Au on clean mirror-like n-type GaAs (110).

Deposition of Au causes first a small shift of the 0.8 eV emission to lower energies, and the development of a peak centered at 0.75 eV that dominates the spectral shape after 15 Å of Au. The dependence of the CL transitions on excitation depth confirms the surface localized nature of the low energy transitions and demonstrates the capabilities of the CLS to probe a large range of depths with changes in the electron beam energy.

Fig. 3 depicts the dependence of peak intensity on excitation depth for 2.5 Å of Au on n-type InP(110). The intensities of the two lower energy transitions decrease and that of the NBG transition increases with increasing excitation depth. Fig. 3 also shows the calculated depth of the maximum energy loss to the plasmon vs. normal incidence electron energy, referred to the right-hand scale[7]. The surface localization of the low energy peaks is apparent. For glancing incidence we expect a still lower depth of the maximum energy loss[8].

An overall reduction in NBG luminescence intensity with metal deposition appears in all the studied systems. The formation of a "surface dead layer" due to band bending in the SC depletion region reduces the radiative recombination[9]. In our case, we modify the surface potential of the SC by metal deposition, thus changing the width of the depletion region. Photoemission experiments[10] show that the E_F shifts are both a large but relatively slow function of the Au thickness and fast function of the Cu thickness.

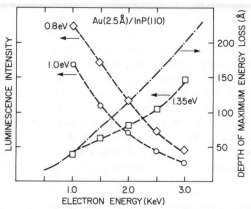

3. Dependence of the luminescence peak intensity on incident electron energy and excitation depth for 0.8 eV, 1.0 eV, and 1.35 eV emissions for 2.5 Å Au on InP(110).

These results correlate well with the observed evolution of the CL spectra upon Au and Cu deposition, and explains the corresponding disappearance of NBG emission. On the other hand, the deposition of Al on InP causes the formation of only a small band bending which builds up slowly with Al coverage[10], and the NBG transition is still detected after deposition of about 20 monolayers of Al.

Several possibilities exist for the origin and properties of the new detected transitions in the CL spectra of the metal/SC interface. Evidently, the perturbation caused by the initial metal deposition modifies the semiconductor surface structure and, consequently, the surface electronic structure. As more metal is deposited and the metal and SC interact, these states evolve into the interface states. These states cannot be assigned to metal induced gap states[11], because at such low coverage the overlayer is not yet metallic. At higher coverages, the spectral shape rules also out possibilities such as surface amorphization which would produce structureless optical emission spectrum or a broad NBG wing.

Diffusion of the metal in the SC, on the other hand, may cause the formation of a highly doped thin layer, which may account for the observed optical emission spectra. The high diffusion coefficient of Cu in InP, even at temperatures as low as 400 C[12], suggests that an in-diffusion process may form a thin highly doped layer, even near room temperature. However, we have not found clear correlation between the emission energies of the metal/InP interfaces and optical emission from the same metal-doped InP [12,13]. The results suggest that simple diffusion of the metal deep into the semiconductor is not the process which causes the formation of centers that give rise to the observed optical emission. More likely, interdiffusion of the different species and the formation of an interlayer with particular electronic properties will be responsible for the optically detected interface states.

Finally, the observed energies of the transitions appears to correlate well with the values of the SBH measured in the metal/InP and GaAs systems. The energy levels involved in the optical emission spectra can be established only with respect to either conduction or valence band edges. For a metal/n-type SC junction, because of the hole accumulation on the interface region, transitions that have as a final state the valence band maximum are more likely to occur. Under this assumption, for Au and Cu/InP systems, the interface states peaks at about 0.57 eV below the CBM, a value close to the SBH of those systems, ca. 0.43-0.5 eV [10,14], and confirmed by SPS[15]. The Al/InP system, on the other hand, shows a SBH of about 0.2 eV or smaller[10,14]. This correlated well with the persistence of the NBG transition and weak sub-band gap emission detected. For Au/GaAs, a similar analysis results in a SBH of about 0.7 eV, whereas electric measurements gives a value of 0.8-0.9 eV[14].

In conclusion, we have observed the formation and evolution of metal/SC interface states by optical emission techniques. We distinguish between interface states promoted by metal deposition from those of step-cleaved areas. The CL spectra show qualitative differences between metals, especially with different chemical reactivity. The experiments demonstrate that these new states are distributed over a wide range of energies, that they are localized on the metal/SC interface, and that they evolve with multilayer coverage. Dominant features of the CL spectra of metal/SC show interface levels at energies which can account for Schottky barrier heights.

Partial support by the Office of Naval Research (ONR N00014-80-C-0778) and fruitful discussions with Christian Mailhiot are gratefully acknowledged.

1) Sze, S.M., **Physics of Semiconductor Devices**, 2nd ed. (Wiley-Interscience, New York, 1981) ch.5.
2) See for example the Proceedings of the 13th Conference on Semiconductor Interfaces, J. Vac. Sci. Technol. A4 (1986).
3) Brillson, L.J., Surf. Sci. Rept. 2, 123 (1982), and references therein.
4) Viturro, R.E. Slade, M.L. and Brillson, L.J., Phys. Rev. Lett. 57, 487 (1986).
5) Brillson, L.J., Richter, H.W., Slade, M.L., Weinstein, B.A. and Shapira, Y., J. Vac. Sci. Technol. A3, 1011 (1985).
6) Viturro, R.E. and Brillson, L.J., unpublished results.
7) Wittry, D.B., **Electron Beam Interactions With Solids**, 99 (1984), and references therein.
8) Murata, K., Phys. Status Solidi A36, 527 (1976).
9) Wittry, D.B. and Kyser, D.F., J. Appl. Phys. 38, 375 (1967).
10) Brillson, L.J., Brucker, C.F., Katnani, A.D., Stoffel, N.G., Daniels, R. and Margaritondo, G., J. Vac. Sci. Technol. 21, 564 (1982).
11) Tersoff, J., Phys. Rev. Lett. 52, 465 (1984).
12) Skolnick, M.S., Foulkes, E.J. and Tuck, B., J. Appl. Phys. 55, 2951 (1984).
13) Temkin, H., Duff, B.V., Bonner, W.A. and Keramidas, V.G., J. Appl.Phys. 53, 7526 (1982).
14) Spicer, W.E., Lindau, I., Skeath, P. and Su, C.Y., J. Vac. Sci. Technol. 17, 1019 (1980); Wieder, H.H., ibid. 15, 1498 (1978); Williams, R.H., ibid.18, 929 (1981).
15) Shapira, Y., Brillson, L.J. and Heller, A., Phys. Rev. B29, 6824 (1984).

INFLUENCE OF EXCHANGE REACTIONS ON THE BUILD-UP OF THE METALLIC CHARACTER DURING THE GROWTH OF METAL/InP CONTACTS

F. Houzay, M. Bensoussan and J.M. Moison

Centre National d'Etudes des Télécommunications
Laboratoire de Bagneux
196 Avenue Henri Ravera - 92220 Bagneux - FRANCE

ABSTRACT

The quantitative UPS study of Al and Ga depositions on (100) InP is reported. In the monolayer coverage range, the incoming Al or Ga atoms exchange with substrate In atoms in a covalent bonding and the released In species are responsible for the metallic character of the surface. When the growth proceeds, due to non miscibility effects, a fraction of monolayer of In floats on the metallic Al or Ga layer. In these conditions, the metal-semiconductor contact is shown to tend towards the ideal Schottky barrier.

1. INTRODUCTION

In the growth of column III metals on III-V semiconductors, it is now well-established that an exchange reaction (ER) between incoming and substrate column III atoms can take place from the early stages of the growth, which are claimed to be central to the formation of the Schottky barrier. Such a reaction has been reported for Al and Ga deposited on (100) InP[1]. We present here a detailed ultra-violet photoemission spectroscopy (UPS) study of these systems showing the progress of the ER and the build-up of the metallic character in the monolayer range. The experimental procedure has been described elsewhere[2] and will not be detailed here. We just mention that in the following the number of metal atoms deposited is expressed in monolayer units, a monolayer (ML) being 5.8×10^{14} cm^{-2}, i.e. the atom density in a InP (100) plane.

2. RESULTS AND DISCUSSION

The UPS spectra of the In 4d core levels indicate that a strong ER between adsorbate and substrate columm III atoms occurs from the very initial stages of interface formation[1]. After metal deposition, the spectra show a superposition of two doublets. One of them originates from the covalently-bounded In of the substrate, slightly shifted (in the range of 0.1 eV) towards higher binding energies referred to the bare substrate. The other doublet is located at a lower binding energy close to the metallic In position and clearly arises from the In released in the ER. This new In seems to have a "metallic" bonding environment.

The extent of the ER is estimated from the coverage dependence of the intensities of the covalent In 4d 3/2 and "metallic" In 4d 5/2 peaks. The peak intensities, after substraction of background intensity, are referred to the clean InP signal. A separate experiment with bulk metallic In has shown that the intensity of the UPS peaks depends only on the number of In atoms contributing to the UPS signal and not on their bonding environment. Then, the results are given on figure 1 in terms of N_C and N_M which are respectively the number of covalent and "metallic" In atoms detected by UPS assuming an escape depth of 4 A. To get N_C, the ratio between the intensities before and after deposition are considered disregarding the attenuation by the released In atoms and N_M is obtained from a metallic In monolayer referred to a dense metallic In (111) plane[1].

After a 1-ML deposit of Al, N_C and N_M show a plateau at about 0.8 ML and 0.6 ML respectively. This is in agreement with a nearly completed exchange of the first substrate monolayer. At high deposits, the final structure is then likely to be formed by a thin AlP interface between bulk InP and the Al film with a fraction of "metallic" In monolayer atop, since the Al-In solubility is small[3]. This is also supported by AES measurements[4] which indicate that Al atoms are involved in covalent bonds in the monolayer coverage range. In the Ga case, as Ga 3d and covalent In 4d core levels are superimposed[5], only N_M can be accurately measured. The behaviour of Ga is similar to the one of Al but the ER is less important than in Al case : about only one third of In monolayer is involved in the ER with Ga atoms. Corre-latively, the build-up of the metallic character for the two systems is estimated from the density of states at the Fermi level. This den-sity appears clearly at 1 ML and 0.9 ML of Al and Ga respectively. Within the coverage range considered here, the metallic character of the surface is likely due to the released In and not to the deposited atoms.

In the monolayer range, the incoming element III atoms have a strong tendency to form covalent bonding. A good saturation of the dangling bonds as well as a good minimization of the defects are ex-pected : the metal deposited is then in close contact with the subs-trate and the nature of the interface is clearly related to the depo-sited metal. This process of interface formation has to lead at least to a minimization of the interface state density within the band gap and may be to a softening of the Fermi level pinning. This statement is supported by the observed shift of the covalent In 4d core levels towards higher binding energies which is more important for Al than for Ga showing a lower Schottky barrier height for Al-InP than for Ga-InP in agreement with previous data[6].

3.   CONCLUSION

The study of the Al (Ga) depositions on (100) InP shows that Al/In (Ga/In) exchanges take place during the deposition of the first monolayer and that two thirds of a monolayer (one third of a monola-yer) of released In floats on the Al (Ga) overlayer. The exchange creates an intermediate AlP (GaP) layer between the InP substrate and the Al (Ga) overlayer. The metallic character is then due to the re-leased In and not to the deposited atoms. The existence of such an interface leads to a close contact between the metal overlayer and the semiconductor. This kind of contact should minimize defects and dan-gling bonds, often responsible for the Fermi level pinning, and then tends towards the ideal Schottky barrier.

[1] Houzay, F., Bensoussan, M. and Barthe, F., Surf. Sci. 168, 347 (1986)

[2] Houzay, F., Hénoc, P., Bensoussan, M. and Barthe, F., J. Vac. Sci. Technol. B3, 1212 (1985)

[3] Hansen, M. and Anderko, K., "Constitution of Binary Alloys", Mc Graw Hill, New York (1958)

[4] Houzay, F., Moison, J.M. and Bensoussan, M., J. Vac. Sci. Technol. B3, 756 (1985)

[5] Houzay, F., Bensoussan, M., Guille, C. and Barthe, F., Surf. Sci. 162, 617 (1985)

[6] Williams, R.H., Dharmodasa, I.M., Patterson, M.H., Moani, C. and Forsyth, N.M., Surf. Sci. 168, 323 (1986)

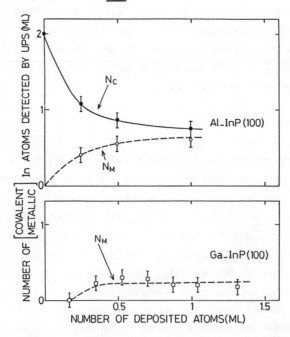

Figure 1 : Number of covalent and "metallic" In atoms detected by UPS versus the number of deposited metal atoms (expressed in monolayer units) obtained from the UPS spectra of the In 4d core levels taken at $h\nu$ = 40.8 eV.

MICROSCOPIC AND MACROSCOPIC STUDIES OF METAL ON SEMICONDUCTORS:
SCHOTTKY BARRIERS AT METAL-CdTe JUNCTIONS

I M DHARMADASA, W G HERRENDEN-HARKER and R H WILLIAMS
Department of Physics, University College, PO Box 78, Cardiff CF1 1XL, UK.

INTRODUCTION

The formation of stable and reliable contacts to semiconductors is of
the utmost importance in microelectronics technology, yet the physics of
such contacts is not well understood. Recent theories of Schottky barrier
formation involve mixed phases and defects at the metal-semiconductor
interface, and also the generation of metal induced gap states (MIGS) in the
semiconductor. In these models, as well as in the conventional Bardeen
model, interface state densities of $\sim 10^{14} cm^{-2}$ yield strong pinning of the
Fermi levels and, for metals on compound semiconductors in particular, these
different mechanisms may compete with the relative importance of each
depending on the details of the contact fabrication. In this paper we
describe measurements of Schottky barrier heights and microscopic
interactions at metal-CdTe interfaces. CdTe has a similar band gap to GaAs
and on the basis of Fermi level pinning by MIGS one might anticipate similar
Schottky barrier behaviour for the two materials. A wide range of metals on
n-GaAs yield values of ϕ_b between 0.6 eV and 0.9 eV and it has been reported
that these values are not significantly influenced by oxide layers at the
interface, ie. similar ϕ_b's are obtained whether the metal is deposited on
clean or oxidised surfaces.[1] In this paper we show that the situation for
CdTe is very different and is not consistent with MIGS theories in their
most elementary forms.

RESULTS

Figure 1 shows measured Schottky barrier heights for metals on clean,
air cleaved and chemically etched (using Br in methanol) n-CdTe crystals.
Data for some of these metals have been published previously.[2,3] Several
points may be made:- (a) The low barrier heights are subject to large
inaccuracies since they were too low to be measured by I-V and C-V methods
and, where possible, were determined by photoelectron spectroscopy.
However, this is often difficult due to 'chemical shifts' associated with
the core level photoemission. (b) All values of ϕ_b greater than 0.4 eV in
Figure 1 were obtained by the I-V method. Great care was taken to correct
for generation-recombination currents[4] and values of ϕ_b determined by C-V
were in reasonable agreement to those derived from I-V method, in most
cases.[5] (c) Au consistently appears to yield values of ϕ_b close to 0.9 eV

or close to 0.6 eV on clean surfaces. This two valued result is also consistently found for Au and Sb on chemically etched CdTe. (d) Metals such as Cr, V and Mn yield low values of ϕ_b for clean and oxidised surfaces. Au forms large barriers in both cases. These metals appear little influenced by the existence of an oxide layer a few Å thick. (e) For another group of metals the measured Schottky barrier height is significantly influenced by the nature of the surface. One group (Ag, Hg, Fe,

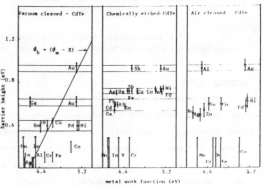

Figure 1. Plot of Schottky barrier heights against metal work functions for metals on clean, chemically etched and air-cleaved n-CdTe crystals.

Sb, Co, Cu, Ni and Pd) appear to give reproducible values of ϕ_b of around 0.7 eV on chemically etched surfaces. They are less reproducible on air cleaved surfaces, as indeed are the values for another group (Cd, Pb, Ga, Al, Sn) on chemically etched surfaces.

In order to examine these effects further, the detailed microscopic nature of the metal-CdTe interfaces have been investigated by photoelectron spectroscopy, using the synchrotron radiation source at Daresbury. Detailed studies of the interactions of a range of metals will be presented elsewhere.[3,5] Here we illustrate the studies by reporting data for Cr on air cleaved CdTe surfaces. These data are presented in Figure 2, where the emission from the Te 4d core level (hν = 100 eV) is shown both for the oxidised surface and following deposition of varying thicknesses of Cr. The tellurium $4d^{3/2}$ and $4d^{5/2}$ components show very pronounced satellite

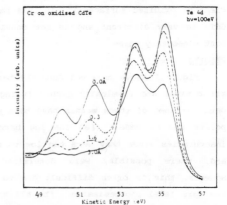

Figure 2. Flux normalised photoemission spectra of the Te 4d core level with increasing coverages of Cr, Taken at photon energy of 100 eV.

structure for photoemission from the oxidised surface. The satellite structure is shifted to larger binding energies by around 3.4 eV and arises from tellurium atoms in a TeO_2 environment.[3] Deposition of very thin layers of Cr on the oxidised CdTe surface leads to an attenuation of the Cd 4d emission as expected. The behaviour of the Te 4d emission, however, is entirely different. Now, the component associated with TeO_2 is rapidly reduced whereas the other component first increases and subsequently decreases in intensity. The data is entirely consistent with the possibility that the TeO_2 is reduced to tellurium initially, and subsequently the tellurium is incorporated in, or on top of the metal contact.

Detailed photoemission studies of the interaction of Ag with oxidised CdTe indicated that the overlayer reduced the oxide layer to a significantly lesser extent.[3] In the same studies, however, it has been found that Au does reduce the oxide in a manner rather analogous to the behaviour of Cr, Mn and V.

DISCUSSION

Photoelectron spectroscopy illustrates very clearly that the contacts formed between CdTe single crystal and a range of metals are not abrupt and simple. The metals illustrated in Figure 1 give rise to a range of values of ϕ_b, varying from close to zero to 0.9 eV for n-type crystals. Clearly, the data reported in Figure 1 is not consistent with the elementary model of Schottky, where a linear dependence of ϕ_b on ϕ_m is expected. The data is also quite inconsistent with models involving pinning of the Fermi level by metal induced gap states, or MIGS.[6] In its most elementary form the application of this model predicts a constant value of ϕ_b of ~0.7 eV on n-type CdTe.

Photoemission studies of metal interaction with clean CdTe surfaces demonstrate that the interfaces formed are rarely abrupt. The formation of mixed phases is very common and for that reason the data of Figure 1 is probably best accounted for on the basis of defect[7,8] and mixed phase[9] models. However, the data cannot be explained on the basis of a single defect energy and requires the existence of multiple defect levels.

Clearly the interaction of many metals with oxidised CdTe surfaces lead to mixed phase contacts. Several metals lead to a drastic reduction of the oxide layer and may make intimate contact with the CdTe surface. It is thought that this is the reason why Mn, V and Cr yield similar values of ϕ_b for both clean and oxidised surfaces. Au is believed to behave in a similar manner. Ag is representative of the group of metals where the barrier heights may be drastically influenced by the presence of the oxide layer at the interface. The surface oxide layer on chemically etched CdTe is believed to be Te rich and to lead to strong pinning of the Fermi level at approximately mid–gap. It appears that Ag may not be able to penetrate through this oxide layer so that the pinning energy remains that appropriate for the Te rich oxide rather than that for the metal overlayer. Partial reduction of the oxide may lead to a mixed phase contact where the phases present are kinetically limited. In this case non reproducible values of Schottky barrier heights are anticipated and are indeed observed.

In conclusion, metal contacts on clean cleaved CdTe surfaces lead to a wide range of Schottky barrier heights which are inconsistent with models involving metal induced gap states in their most elementary form. The range of barriers observed, and the way they are drastically influenced by oxide layers, suggest that models involving interfacial defects and mixed phases are equally, if not more important. In this complex system as well as at metal–semiconductor interfaces in general it is likely that many processes contribute to Schottky barrier formation and in certain situations one mechanism may dominate above all others. For metals on CdTe mixed phases and defects fall in this latter category.

REFERENCES

1. Spicer, W E, Newman, N, Kendeliwicz, T, Petro, W G, Williams, M D, McCants, C E and Lindau, I, J. Vac. Sci. Technol. B3(4), 1178 (1985).
2. Williams, R H and Patterson, M H, Appl. Phys. Lett. 40(6), 484 (1982).
3. Dharmadasa, I M, Herrenden–Harker, W G and Williams, R H, Appl. Phys. Lett. (to appear in July 1986).
4. McLean, A B, Dharmadasa, I M and Williams, R H, Semiconductor Science and technology (to appear in 1986).
5. Williams, R H, McLean, A B and Dharmadasa, I M (to be published 1986).
6. Tersoff, J, Proceedings of PCSI, 1986; J. Vac. Sci. Technol. 1986.
7. Spicer, W E, Chye, P W, Skeath, P R, Su, C Y and Lindau, I, J. Vac. Sci. Technol., 16(5), 1422 (1979).
8. Williams, R H, Varma, R R and Montgomery, V, J. Vac. Sci. Technol. 16, 1418 (1979).
9. Freeouf, J L and Woodall, J M, Appl. Phys. Lett. 39(9), 727 (1981).

IV. QUANTUM WELLS

QUANTIZED HALL EFFECT AND ZERO RESISTANCE STATE
IN A THREE-DIMENSIONAL ELECTRON SYSTEM

H. L. Störmer, J. P. Eisenstein, A. C. Gossard,
K. W. Baldwin, and J. H. English

AT&T Bell Laboratories
Murray Hill, New Jersey 07974
UNITED STATES

Quantization of the Hall effect and vanishing diagonal resistivity are observed in a GaAs-(AlGa)As superlattice which, in the absence of a magnetic field, conducts in all three spatial dimensions. In the quantized state, the conductivity parallel to the magnetic field tends toward zero. These findings suggest that rather than two-dimensionality of the electronic system, it is the absence of conductivity along the magnetic field which is a necessary condition for the observation of the quantized Hall effect.

Since the discovery of the integral quantized Hall effect (IQHE) in a Si-MOSFET[1], its confirmation in a variety of III-V heterojunctions[2] and its interpretation in terms of the singular density-of-states of a two-dimensional system in a high magnetic field[3], there has developed a prejudice that two-dimensionality of the electronic system is a prerequisite for its existence. Indeed, by now the IQHE has been observed in a variety of structures[4,5] that deviate from the simple one-layer, one-subband systems of the early work. Nevertheless, all those structures consist of electronically strictly two-dimensional systems lacking any dispersion and, therefore, conduction in the direction normal to the layers. Hence, they must be regarded as a stack of independent quantized Hall resistors connected in parallel. The question arises whether strict two-dimensionality is indeed a prerequisite or whether a generalization has to be adopted. In fact, Azbel has considered this question for highly anisotropic materials.[6] While he argues for vanishing diagonal resistivity, he does not predict quantization of the Hall effect.

In order to test this case we studied a system which is electrically three-dimensional by virtue of being a good, although anisotropic, conductor in all spatial dimensions. In spite of its three-dimensionality, this system exhibits the IQHE, $\rho_{xy} = h/ie^2$ and a pronounced zero-resistance state, $\rho_{xx} \rightarrow 0$.[7] While the conductivity, σ_{zz}, along the magnetic field (B) direction in general is non-zero, it vanishes ($\sigma_{zz} \rightarrow 0$) at values of B at which the system assumes the quantized state. This result suggests that rather than strict two-dimensionality, it is the absence of conductivity along the magnetic field direction which is a necessary condition for the occurrence of a IQHE and zero-resistance state.

The structure used in the experiment is a GaAs/AlGaAs superlattice (SL) with highly penetrable barriers. Two identical samples were grown via MBE on a semi-insulating substrate for in-plane transport and on a n^+-substrate for normal transport. The dimensions of the SL are illustrated in Fig. 1(a). Measurements of the Hall density and Hall mobility at 4.2K yield $n_H = 2.1 \times 10^{17}$ cm^{-3} and $\mu_H = 6400$ cm^2/V-sec, assuming a thickness of $30 \times (188\text{Å} + 38\text{Å})$.

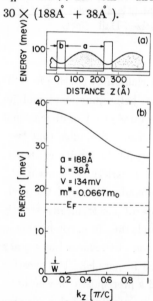

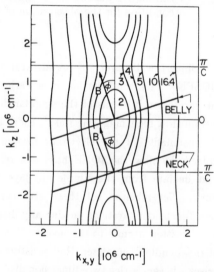

Fig. 1 (a) Kronig-Penney model and wave function for $k_z = 0$ for the GaAs/(AlGa)As superlattice employed. (b) Dispersion relation in the z direction.

Fig. 2 Contours of constant energy (E = 1,2,3,..., 16.4 meV) in the k_{xy}-k_z plane for the band structure of Fig. 1(b). Belly and neck represent the two extremal orbits of the Fermi surface.

Heterostructures fabricated from GaAs/(AlGa)As are well understood and extremely well represented by a simple square-well potential.[8] We calculate the miniband structure and wave function of the electronic system using a Kronig-Penney with the parameters listed in Fig. 1(b). Nonparabolicity of the conduction band, a small difference in mass between GaAs and (AlGa)As, and a slight distortion of the potential well are neglected. From the calculations we obtain the dispersion relation shown in Fig. 1(b). Fig. 1(a) shows the z dependence of the wave function for $k_z = 0$. The variation from maximum in the well to minimum in the barrier is less than a factor of 4, demonstrating the high degree of transparency of the barriers.

Fig. 2 shows contours of constant energy in the k_z, k_{xy} plane for various energies up to $E_F = 16.4$meV. The three-dimensionality of the electronic system is evident. In comparison, a strictly two-dimensional system is represented in such a plot as a set of straight lines parallel to the k_z axis, indicating the lack of dispersion in the z direction. For lateral transport the samples were fabricated into Hall bars, while the specimen for normal transport were etched into mesas with 100μm diameter, see insert Fig. 5. The in-plane conduction at zero-field is $\sigma_{xx} = 210(\Omega\text{cm})^{-1}$. The perpendicular resistance of the mesa amounts to R = 1.2Ω which translates into $\sigma_{zz} = 0.12(\Omega\text{cm})^{-1}$, suggesting a rather large anisotropy of $\alpha \sim 10^3$. This value of α represents an upper limit of the anisotropy since a major fraction of the perpendicular resistance is due to contact and substrate resistance.

Experiments on the in-plane transport allows to determine the actual shape of the Fermi surface. The magneto-oscillations shown in Fig. 3(a) indicate a beating pattern between two different oscillations close in frequency. A node is clearly visible around B $\sim$ 2.3 which establishes the difference between belly and neck orbit to be $\Delta A = 1.1 \times 10^{11}$ cm^{-2}. The calculated value is $\Delta A = (9.0 - 7.6) \times 10^{12}$ cm$^{-2} = 1.4 \times 10^{11}$ cm^{-2}. Tilting the magnetic field away from $\Phi = 0$ deg shifts the node to lower fields (see Fig. 4) i.e. smaller differences between belly and neck orbits as expected from the Fig. 2. The non-vanishing conductivity in the z-direction at zero-field and the observation of separated belly and neck orbits establish convincingly the three-dimensionality of the electronic system under study.

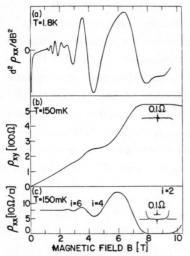

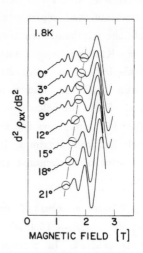

Fig. 3 (a) Second derivative of magnetoresistance ρ_{xx} vs magnetic field for $\Phi = 0$. (b) Hall resistance ρ_{xy} (c) Magneto resistance ρ_{xx} vs magnetic field.

Fig. 4 Angular dependence of $d^2\rho_{xx}/dB^2$. The node of the beating pattern shifts to lower fields as expected from Fig. 2

Fig. 3(b) and 3(c) show our lowest-temperature data on ρ_{xy} and ρ_{xx}. At $B \sim 9$ T, ρ_{xx} develops a clear zero-resistance state with $\rho_{xx} < 0.01$ $\Omega/\square$. For 0.15 K $< T < 4.2$ K ρ_{xx} is activated over more than two decades with an activation energy of $\Delta/2 = 0.13$ meV (Fig. 6). Concomitant with the minimum in ρ_{xx}, a plateau appears in ρ_{xy} which is defined to about 0.01 Ω over a range of 1.5 T. The normal conductivity σ_{zz} (Fig. 5) oscillates in phase with ρ_{xx} and tends towards $\sigma_{zz} -> 0$ at a field position where ρ_{xy} approaches a plateau and ρ_{xx} vanishes. With decreasing temperature, the Hall resistance converges upon a constant value $\rho_{xy} = 537.73 \pm 0.03$ Ω or $\rho_{xy} = h/48e^2$ to 5 parts in 10^5, well within the advertised accuracy of 1 part in 10^4 of the decade resistor. These findings suggest that in the quantized state the resistivity and the conductivity tensor assume the form

$$
\rho = \begin{pmatrix} 0 & \dfrac{h}{ie^2} & 0 \\ -\dfrac{h}{ie^2} & 0 & 0 \\ 0 & 0 & \rho_{zz} \to \infty \end{pmatrix} \qquad \sigma = \begin{pmatrix} 0 & -\dfrac{ie^2}{h} & 0 \\ \dfrac{ie^2}{h} & 0 & 0 \\ 0 & 0 & \sigma_{zz} \to 0 \end{pmatrix}
$$

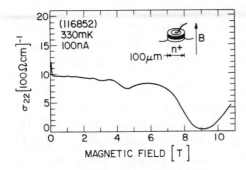

Fig. 5 Conductivity σ_{zz} normal to the layers of the superlattice as a function of magnetic field. This sample was grown on a conducting substrate.

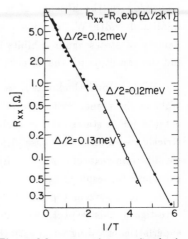

Fig. 6 Magnetoresistance in the i = 2 minimum of ρ_{zz} versus inverse temperature. Data are taken on 3 different specimen in 2 different systems.

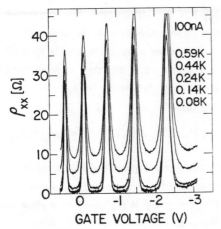

Fig. 7 Variation of ρ_{zz} at i = 2 minimum as a function of gate voltage.

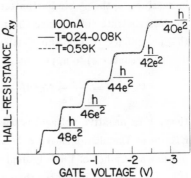

Fig. 8 Variation of ρ_{xy} at i = 2 plateau as a function of gate voltage.

One can develop an intuitive picture which is able to explain some of the experimental facts. In an ideal two-dimensional system in a high magnetic field along z, the Landau levels consist of δ functions separated by the cyclotron energy $\hbar\omega_c$. In a three-dimensional system, each quantized state in the plane is associated with a range of k_z, each having a slightly different energy. Therefore, each Landau level develops into a band. Since the z motion is not affected by the magnetic field, the shape of each band is field independent and reflects the one-dimensional density of states of the k_z

dispersion in the absence of a field. Its width reproduces the zero-field miniband width W. In high magnetic fields B, when $\hbar\omega_c$ exceeds W, the density of states again exhibits gaps, as in the ideal two-dimensional case, and the condition for the observation of a IQHE seems to be fulfilled.

With a total thickness L = jc there exist j different k_z states for each state in the plane. With a level degeneracy of d = eB/h in the x-y plane, a total of jeB/h states exist in each Landau band, and arguments used for the strictly two-dimensional case yield $\rho_{xy} = h/jie^2$. Therefore, with i = 2 and j = 30, one expects $\rho_{xy} = h/60e^2$ which, however, differs from the experimental result of $\rho_{xy} = h/48e^2$. This discrepancy is associated with the depletion of several top and bottom layers of the superlattice. In order to investigate this aspect, we fabricated a gate electrode on top of the superlattice allowing us to vary the depletion depth. Figs. 7 and 8 show the variation of ρ_{xx} and ρ_{xy} in the i = 2 state versus the gate voltage. Large oscillations in ρ_{xx} and concomitant step-like transitions from one quantized value to the next are being observed when the sample is depleted to increasing depth. A similar pattern results when a backside bias is applied to the substrate side. Since the repetitive pattern is assumed to arise from the periodicity of the superlattice, we expect the peaks in ρ_{xy} (the best defined structures) to appear at gate voltages.

$$V_k + \overline{V} = \frac{e}{2\epsilon\epsilon_0} Nc^2 [k - k_o + \phi]^2, \quad k,\ k_o = 0,1,2,..;\ 0 \leq \phi \leq 1$$

Here, $\overline{V}$ (a negative voltage) is the built-in surface potential measured from E_F, N is the 3D density, c is the period of the superlattice, k_o is the unknown number of initially depleted layers and ϕ is a phase factor close to 1/2. The voltage differences then are linear in k.

$$V_{k+1} - V_k = \frac{e}{2\epsilon\epsilon_0} Nc^2 [2(k - k_o + \phi) + 1]$$

Fig. 9 shows a plot of this quantity from which c = 207Å and $k_o = 4$ are deduced. The periodicity is in good agreement with the period determined by TEM. In the absence of a gate voltage there are $k_o = 4$ layers from the top and from $\rho_{xy} = h/48e^2$ we conclude that only two layers are depleted from the substrate side. It is striking that the transitions between subsequent states are so well defined and that a model assuming sequential depletion of individual layers seems to account for the observations. Yet, in a superlattice such a distinction between individual layers is no longer possible due to strong interlayer tunneling. Indeed, the transition between subsequent quantized states in this superlattice are not due to sequential depletion of layers but are transitions at which the whole bulk participates.

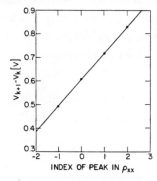

Fig. 9 Voltage difference between neighboring peaks of Fig. 7 versus peak index.

In a periodic one-dimensional system with finite boundaries, the variation of the uppermost state of each subband varies rather abruptly with decreasing distance between boundaries. These state which initially extend all across the superlattice are sequentially peeled off the continum and move rapidly to higher energy crossing the Fermi level whenever the confinement is reduced by one period. It is this periodicity which is responsible for Fig. 7 and 8.

A remaining puzzle is the small activation energy observed. At B $\sim$ 9T and with W $\sim$ 2.5meV, we arrive at a gap energy of $\Delta = \hbar\omega_c$ - W $\sim$ 12meV, while experimentally, we find only $\Delta \sim$ 0.26meV. In general, broadening of the Landau level (here Landau band) is assumed to be responsible for such a reduction. With a mobility of $\mu = 6400$ cm^2/Vsec one deduces a lifetime broadening of $\Delta E \sim \hbar/\tau \sim$ 2.8meV, which accounts in part for the small activation energy. A better understanding of the magnitude of Δ will require a better knowledge of the occurrence and position of a mobility edge and its relation to the IQHE in such an anisotropic three-dimensional system. This problem should be at least as challenging as the much simpler, strictly two-dimensional case where a conclusion has still not been reached.

In summary, our findings suggest a generalization of the conditions under which the IQHE can be observed. Two-dimensionality of the electronic system is not essential. Rather the absence of conductivity along the magnetic field, either pre-existing (as in the two-dimensional case) or field-induced (as in the above case) seems to be the required criterion for the existence of a IQHE.

At present the high scattering rate of the heterojunction superlattice structures prevent observations of the fractional quantized Hall effect (FQHE).[10] Yet other material combinations might ultimately provide such a possibility. Since the FQHE results from strong correlations among the carriers, we expect dramatic consequences for this phenomenon from such a strong interlayer coupling.

We would like to thank R. C. Dynes, F. D. Haldane, B. I. Halperin, M. G. Lamont, V. Narayanamurti, A. Pinczuk, M. A. Schlüter, D. C. Tsui, and D. J. Werder.

[1] von Klitzing, K., Dorda, G. and Pepper, M., Phys. Rev. Lett. *45*, 494 (1980).

[2] Tsui, D. C. and Gossard, A.C., Appl. Phys. Lett. *37*, 550 (1981); Guldner, Y., Girtz, J. P., Vieren, J. P., Voisin, P., Voos, M., and Razeghi, M., J. Phys. Lett. *43*, L613 (1982); Störmer, H. L., Schlesinger, Z., Chang, A. M., Tsui, D. C., Gossard, A. C., and Wiegmann, W., Phys. Rev. Lett. *51*, 126 (1983).

[3] Laughlin, R. B., Phys. Rev. *B23*, 5632 (1981).

[4] Haavasoja, T., Störmer, H. L., Bishop, D. J., Narayanamurti, V., Gossard, A. C., and Wiegmann, W., Surf. Sci. *142*, 294 (1984).

[5] Razeghi, M., Duchemin, J. P., Portal, J. C., Dmowski, L., Remeni, G., Nicholas, R. J., and Briggs, A., Appl. Phys. Lett. *48*, 721 (1986).

[6] Ya. Azbel, M., Phys. Rev. *B26*, 3430 (1982).

[7] Störmer, H. L., Eisenstein, J. P., Gossard, A. C., Wiegmann, W., and Baldwin, K., Phys. Rev. Lett. *56*, 85 (1986).

[8] For example, Esaki, L., in Proceedings of the 17th Intl. Conf. on the Physics of Semiconductors, San Francisco, 1984, edited by J. D. Chadi and W. A. Harrison (Springer, New York, 1984), p.473.

[9] Chang, L. L., Mendez, E. E., Kawai, N. J., and Esaki, L. Surf. Sci. *113*, 306 (1982).

[10] Tsui, D. C., Störmer, H. L., and Gossard, A. C., Phys. Rev. Lett. *48*, 1559 (1982).

THE FRACTIONAL QUANTUM HALL EFFECT IN GaAs-GaAlAs HETEROJUNCTIONS

R.G. Clark, R.J. Nicholas, J.R. Mallett, A.M. Suckling and A. Usher

Clarendon Laboratory, Parks Road, Oxford OX1 3PU, U.K.

J.J. Harris and C.T. Foxon

Philips Research Laboratories, Redhill, Surrey, U.K.

Important differences in the nature of the FQHE in the $N = 0$ and $N = 1$ Landau levels are identified in several ultra high mobility samples, which appear to be consistent with recent theoretical predictions. The previously unresolved questions surrounding preliminary indications of even denominator states are discussed.

In a previous report[1], we have identified significant differences in the nature of fractional quantisation in the $N = 0$ and $N = 1$ Landau levels in an ultra high mobility sample G63 at 30mK. For the $N = 0$ level, a well-defined hierarchical set of fractional states p/q was found for the upper spin split region $1 < \nu < 2$ with odd denominator values up to $q = 9$, as found for $\nu < 1$, whereas for the $N = 1$ level ($2 < \nu < 4$) resistivity minima were observed at even denominator occupancies $\nu = 2^1/_4$, $^1/_2$, $^3/_4$ and $3^1/_4$, $^1/_2$, $^3/_4$ which showed surprisingly little temperature dependence in the range 20mK to 1K, with no sign of exponential activation. G63 resistivity data taken at 60mK is shown in fig. 1(a) for the region $2 < \nu < 4$. Quantisation of the Hall resistance was not resolved in the $N = 1$ region. Emergent plateau-like features did occur at field values corresponding to the even fractional assignments, but were displaced above the straight line defined by the unquantised Hall coefficient. The G63 ρ_{xx}, ρ_{xy} v. B data have been digitised to produce σ_{xx} v. B traces and the even fractional assignments also accurately describe σ_{xx} minima for the $N = 1$ level. A point of concern was that the spin splitting where the even fractions are observed is only $\sim$ 1K, which might lead to a mixing of collective states in the two spin levels. Against this was the observation of accurately located $1^1/_3$, $1^2/_3$ and $1^2/_5$ minima in the $N = 0$ level at the same magnetic fields when the electron concentration was much lower (before illumination).

This work has now been extended by tilting the plane normal to

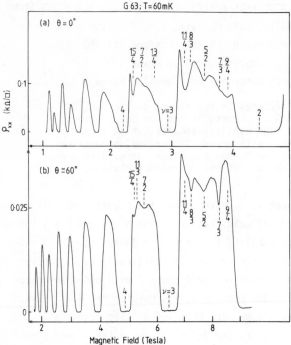

Fig. 1. Resistivity traces for sample G63 with i = 100 nA and the magnetic field at an angle (a) θ = 0° and (b) θ = 60° to the plane normal; μ = 2.15x10⁶cm²/Vs, n = 2.1x10¹¹/cm². Landau level occupancy ν = nh/eB.

sample G63 at θ = 60° to the field and examining the region 2 < ν < 4 in (untilted) samples of similar or higher mobility but at increased electron concentration and hence magnetic field. The tilted FQHE results are shown in fig. 1(b) and results for samples G148 and G71 are shown in fig. 2 and 3 respectively. From fig. 1 it can be seen that for 2 < ν < 3 the tilted field has significantly strengthened the previously weak feature at $2^1/_3$ and a relatively strong minimum at $2^2/_3$ has emerged. Although the minimum at $2^1/_2$ remains essentially unaltered, the deep minimum at $2^3/_4$ with θ = 0° is reduced to an inflection and the shallower minimum at $2^1/_4$ is lost in the now steeply falling resistivity edge. This strengthening of odd denominator resistivity minima is also observed for 1 < ν < 2 and is a significant aspect of the tilted field work. In the region 3 < ν < 4 however, no odd denominator minima appear when θ = 60°, but once again while the $3^1/_2$ minimum is slightly strengthened, the minimum at $3^3/_4$ is reduced in strength and the structure at $3^1/_4$ is lost in the steep resistivity edge.

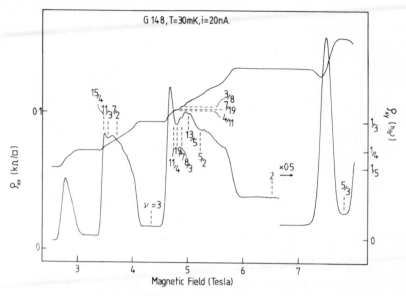

Fig. 2. ρ_{xx} and ρ_{xy} traces for sample G148 with μ = 2.74x10^{6}cm^2/Vs and n = 3.15x10^{11}/cm^2.

Besides increasing the small spin splitting in this region, the parallel component of the tilted field may act to compress the z component of the electron wavefunction and enhance the two-dimensionality of the system, thereby providing increased resolution in this low field regime.

The situation is further clarified by the G148 results,which unfortunately show the onset of bulk conduction at the high electron concentration and mobility required to resolve effects in the region $2 < \nu < 4$. Despite this, reproduceable structure in ρ_{xx} at $2^1/_2$, $2^2/_3$, $2^3/_5$ and most significantly at $2^5/_7$ is resolved, as are features at $3^1/_2$ and $3^2/_3$. Additionally, for the first time quantised Hall plateaus lying on the straight line defined by the unquantised Hall coefficient are seen at values very close to $1/(2 + 5/7)$ and $1/(2 + 2/3)h/e^2$. The Hall calibration in this region is extrapolated from the ν = 3, 4 and 5 plateaus,which show only small offsets from the bulk conduction. Significant effects do set in around ν = 2, the plateau of which is shifted downwards from its correct value. The strongest feature, at $2^5/_7$, is tantalizingly close to $2^3/_4$ but the high order odd denominator assignment is also found to be consistent

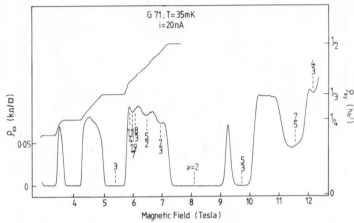

Fig. 3. ρ_{xx} and ρ_{xy} traces for sample G71 with $\mu = 1.4 \times 10^6 \text{cm}^2/\text{Vs}$ and $n = 3.9 \times 10^{11}/\text{cm}^2$.

with results for sample G71, as shown in fig. 3. Relatively well-resolved $2^1/_2$ and $2^1/_3$ minima are observed in G71 with a very weak feature at $2^2/_3$. Hall plateaus in this sample are unresolved.

In a recent work, MacDonald and Girvin[2] predict that while the fractional states in the region $^1/_5 < \nu < ^1/_3$ are quite weak for $N = 0$, for $N = 1$ they should be nearly as strong in this range of ν as in the range $\nu > ^1/_3$ and consequently that high-order odd denominator fractions may well emerge as the strongest features in weakly resolved structure[3]. We conclude that this appears to be the case for the hole analogue fractions, which is an important difference from the FQHE in the $N = 0$ level. Within this model, it can be anticipated that mixing of collective states where the spin splitting is small would lead to different results for the $N = 0$ and $N = 1$ levels. Nevertheless the appearance of a regular series of ρ_{xx} minima at even fractional occupancies in fig. 1(a) is puzzling. The constant and pronounced features at $2^1/_2$ and $3^1/_2$ and the very weak temperature dependence of fractional states in the $N = 1$ level are still unexplained.

REFERENCES

1)   Clark R.G., Nicholas R.J., Usher A., Foxon C.T. and Harris J.J.,
     Surf. Sci. 170 141 (1986); also Sol. Stat. Comm. to be published.
2)   MacDonald A.H. and Girvin S.M., Phys. Rev. B 33 4414 (1986)
3)   MacDonald A.H., private communication.

*Samples grown by MBE at Philips Research Laboratories, U.K.

MICROWAVE AND FAR-INFRARED INVESTIGATION OF THE HALL CONDUCTIVITY OF THE TWO-DIMENSIONAL ELECTRON GAS IN GaAs/AlGaAs IN THE QUANTUM HALL EFFECT REGIME

F.Kuchar[+], R.Meisels[+], K.Y.Lim[+], P.Pichler[§], G.Weimann[%], and W.Schlapp[%]

+ Institut für Festkörperphysik,Universität and L.Boltzmann Institut, Kopernikusgasse 15, A-1060 Wien, AUSTRIA

§ Institut für Physik, Montanuniversität, A-8700 Leoben, AUSTRIA

% Forschungsinstitut der Deutschen Bundespost, D-5100 Darmstadt, FRG

Integer Quantum Hall Effect (IQHE) behavior of σ_{xy} is observed at microwave frequencies in multiple and single heterostrcutres with 2D electron and hole gas layers. In the far-infrared (984 GHz) the IQHE structure is not observed in a multiple heterostructure. This implies that a high-frequency breakdown of the IQHE occurs at frequencies between about 1/30 and 1/3 of the cyclotron resonance frequency.

1. INTRODUCTION

Most of the Quantum Hall Effect (QHE) experiments have been performed with d.c.[1] For a complete understanding of the effect also the high-frequency behavior should be known. Experiments in the MHz range are discussed in Refs.2 and 3. Some theoretical treatments[4-6] based on a model of semiclassical orbits predicted a breakdown of the integer QHE at low frequencies in the MHz range. This is thought to be due to a delocalization of electrons on large, closed semiclassical orbits (equipotential lines). At microwave frequencies ($\approx$ 30 GHz), however, the integer plateaus in GaAs/Al$_x$Ga$_{1-x}$As multiple heterostructures were still observed.[2,3]

In this paper we report new experiments on GaAs/AlGaAs heterostructures with 2-dimensional electron and hole gas (2DEG, 2DHG) layers in the microwave as well as in the far-infrared (submillimeter) range at frequencies close to the cyclotron resonance.

2.   EXPERIMENTAL

The contactless technique applied to measure directly the Hall conductivity σ_{xy} of the two-dimensional electron gas of GaAs/Al$_x$Ga$_{1-x}$As heterostructures uses a crossed polarizer-analyser arrangement in Faraday geometry. As shown in Ref.3 the transmitted intensity in this arrangement is proportional to σ_{xy}^2 . In the microwave range the polarizer-analyser system consisted of crossed waveguides, in the submillimeter range (optically pumped CH$_3$OD laser) linear-grid polarizers were used outside the optical cryostat. The samples were grown by molecular beam epitaxy[7] (Table 1).

Table 1. Parameters of the samples at T=4.2K (details see Ref.7). N_s is the total carrier density.

Sample No.	Number of 2D layers	Type of the 2D carriers	N_s (cm^{-2})	μ (cm^2/Vs)
1408b	14	n	8.0×10^{12}	2.0×10^4
1320	1	n	5.2×10^{11}	3.0×10^5
1456	1	p	4.0×10^{11}	5.0×10^4

3.   EXPERIMENTAL RESULTS AND DISCUSSION

3.1   Microwave Experiments

In experiments on the multiple heterostructures 1408 and 1325[2,3,8] quantized behavior of the Hall conductivity σ_{xy} was clearly observed. The data of sample 1320 (Fig.1a) were analyzed by plotting the square root of the bolometer signal ($\sim \sigma_{xy}$) versus the reciprocal magnetic field (1/B) in order to obtain information about the periodicity of the Hallplateau structure (Fig.1b). Three interesting features can be observed: (i) The centers of the plateaus are equidistant in 1/B. (ii) The plateau values of σ_{xy} are multiples of a fundamental value. (iii) Also the i=3 plateau is clearly resolved as in the d.c. data. A comparison with d.c. data is given in Ref.8. In Fig.1a the two enlarged curves for T=2.1 and 4.2K show the temperature dependence of σ_{xy}^2. As in d.c. data the plateaus tilt and narrow with increasing temperature.

Results regarding the microwave Quantum Hall effect of the two-

dimensional hole gas (2DHG) are shown in Fig.2. The i=2 plateaus is
well developed at T=2.1K. Since the mobility of 2D holes still increases at lower temperatures[10] plateaus at higher filling factors are expected to appear there.

3.2   Far-infrared Experiments

Fig.3 shows the Germanium-bolometer signal at 305 μm (984 GHz)
for the multiple heterostructure 1408b. The peak corresponds to the
cyclotron resonance. For comparison, the microwave data (33.4 GHz) of
the same sample are also shown. In the far-infrared data there is no
clear evidence of the plateau behavior.

4.   CONCLUSIONS

1) Plateaus of the Hall conductivity σ_{xy} appear at microwave frequencies in the 2DEG and 2DHG of single and multiple heterostructures.
2) As in the d.c. data on the 2DEG-SHS, the i=3 plateau and hence the
spin splitting is clearly resolved. This is a strong indication that
the g factor enhancement is independent of the measuring frequency,
even if it is very close to the free-carrier spin resonance.
3) Edge effects seem to play no important role for the observation of
the IQHE at microwave frequencies: Corresponding results were obtained
with sample dimensions smaller and larger than the common cross section
of the two crossed waveguides.
4) The σ_{xy} plateaus occur equidistant on a 1/B scale, their absolute
values are multiples of a fundamental value.
5) At submillimeter frequencies, the plateaus are not observed when ω_c
is up to a factor of 3 higher than ω. This means that a high-frequency
breakdown of the integer quantization occurs at frequencies somewhere
between 1/30 and 1/3 of the cyclotron resonance frequency. An interpretation within the model of semiclassical orbits[4-6] would mean that
the closed orbits of localized states cannot be much longer than several cyclotron orbits, which limits the validity of such a model. Certainly they are much small than usual sample dimensions.

Work supported by "Jubliläumsfonds der Österreichischen Nationalbank",
project no. 2715.

5. REFERENCES

1) See e.g. Surface Sci. 131, Nos. 1-3 (1982; 142, Nos. 1-3 (1984)

2) F.Kuchar, R.Meisels, G.Weimann, W.Schlapp, Phys.Rev B 33, 1965 (1986)

3) R.Meisels, K.Y.Lim, F.Kuchar, G.Weimann,W.Schlapp, In Springer Ser. Solid-State Sci. 67, 184 (1986)

4) R.F.Kazarinov and Serge Luryi, Phys.Rev B 25, 7626 (1982)

5) S.M.Apenko and Yu.E.Lozovik, J.Phys.C 18, 1197 (1985)

6) R.Joynt, J.Phys. C 18, L 331 (1985)

7) G. Weimann and W. Schlapp, Appl.Phys. A 37, 3057 (1985)

8) F.Kuchar, R.Meisels, K.Y.Lim, P.Pichler, G.Weimann, and W.Schlapp (to be published) in Springer Ser.Solid-State Sci.

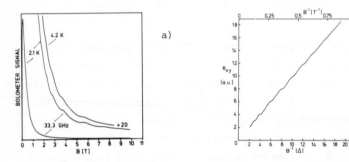

Fig.1: (a) Magnetic field dependence of the bolometer signal at 33,3 GHZ for sample 1320. 20x enlarged curves: 4.2K data not on same scale as 2.1K data. (b) σ_{xy} vs. B^{-1} at 2.1K. The lower B^{-1} scale in multiples of a fundamental value.

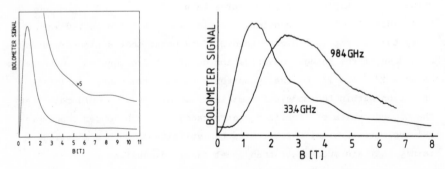

Fig.2:Bolometer signal for sample 1456 with a 2DHG. T=2.1K, ν=30.4 GHz

Fig.3:Submillimeter(λ=305 μm) and 33.4 GHz signal of sample 1408b.

THEORY OF FRACTIONAL AND INTEGRAL QUANTUM HALL EFFECT IN CORRELATED Q2D-ELECTRON SYSTEMS

R. Keiper and O. Ziep

Department of Physics, Humboldt University Berlin, GDR

The QHE gives strong indication for a new correlated electron state which seems to be connected with density structuring in the Q2D-system [1,2]. The hypothesis of a magnetic field induced electron crystal or liquid with pairs at the sites $\bar{R}_i$ is based on the balance of electron-magnetic field and electron-electron interactions whereas spin splitting is more than one order of magnitude lower (see table 1).

B	Si			GaAs		
	$\hbar\omega_c$	$e^2/4\pi\varepsilon_0\varepsilon l_0$	$\frac{1}{2}g\,\mu_B B$	$\hbar\omega_c$	$e^2/4\pi\varepsilon_0\varepsilon l_0$	$\frac{1}{2}g\,\mu_B B$
5	3	11.4	0.3	8.8	9.7	0.05
10	6.1	16.1	0.6	17.6	13.7	0.1
20	12.2	22.8	1.2	35	19.3	0.2

Table 1: Characteristic energies in meV, B in Tesla

$$\hbar\omega_c = \frac{eB}{M} \;,\; l_0 = (\frac{\hbar}{eB})^{1/2} \;,\; \mu_B = \frac{e\hbar}{2m_0}$$

Using localized one particle states $\Psi_\lambda(\bar{r})$ with $\bar{r}=r\,\varphi\,z$, $\lambda=$inm for one electric subband the radial wave functions

$$R_{nm}^{(i)}(\xi) = (\frac{n\,!}{(n+|m|)\,!})^{1/2} e^{-\frac{\xi}{2}}\varphi^{\frac{|m|}{2}} L_n^{|m|}(\xi), \quad \xi = \frac{r^2}{2l_0^2} \quad (1)$$

(L_n^m – Laguerre polynomials) correspond to the local gauge potential $\bar{A} = 1/2\,(\bar{r}-\bar{R}_i)\times\bar{B}$. In the following we (i) discuss the Coulomb energy in Hartree approximation, (ii) illustrate the local charge density for $l_0 \ll N^{-1/2}$ (N being the 2D-carrier concentration) and (iii) calculate the conductivity tensor for the quantum regime. The one-electron energy is characterized by the kinetic part for an electron state (nm) at the site i

$$E_{eB}^{\lambda} = \hbar \omega_c (n + \frac{|m| - m + 1}{2})$$ (2)

the Coulomb integral

$$E_{ee}^{\lambda \lambda'} = \frac{e^2}{4\pi\varepsilon_0 \varepsilon} \int d^3\bar{r} \int d^3\bar{r}\,' \; \frac{|\Psi_\lambda(\bar{r})|^2 \; |\Psi_{\lambda'}(\bar{r}\,')|^2}{|\bar{r} - \bar{r}\,'|}$$ (3)

which must be separated into an intra (i=i') - and an inter (i≠i') - site part, and the interaction with the homogeneous positive background E_{eb}^{λ}. Assuming the sites to be occupied by spin degenerate pairs we calculate the intra-site part of (3) (see table 2) which decreases with increasing n or m due to different shapes of $R_{nm}^{(i)}$ whereas E_{eB}^{λ} increases.

(nm)	E_{ee} in $e^2/4\pi\varepsilon_0\varepsilon l_0$	(nm)	E_{ee} in $e^2/4\pi\varepsilon_0\varepsilon l_0$
(00)	0.9	(01)	0.6
(10)	0.5	(02)	0.5
(20)	0.35	(03)	0.45

Table 2: Electronic structure dependent Coulomb integrals

But the interplaying electron-electron and electron-magnetic field interactions show an intriguing behaviour. Attractive forces between local magnetization currents lead in dependence on B to changes in the site number and occupation. In this model the Hall plateaus are explained as stability ranges described by definite electron configurations at the sites.

As represented in Fig. 1a-d for $l_0 \ll N^{-1/2}$ the spatial charge distribution depends on l_0, n, |m|. Therefore it is necessary to introduce a measure for the weighted area in the xy-plane covered by (nm)-states[3].

$$S_{nm} = \frac{\pi}{l_0^2} \int dr r^3 R_{nm}^2 (\mathcal{f}(r)) = 2\pi l_0^2 (2n + 1 + |\bar{m}|).$$ (4)

In Hall experiments the classical picture shows a carrier motion with a cyclotron radius r_c around $\bar{R}_i$ and a constant mean velocity $\bar{v} = \bar{E} \times \bar{B}/B^2 = \dot{\bar{R}}_i$. In the quantum regime the local charge distribution is characterized by the expectation value (4) that generalizes πr_c^2 The quantiza-

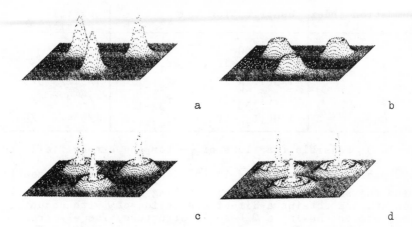

a   b

c   d

Fig. 1: Charge density in the xy-plane for a- $(nm)=(00)$.
b- $(nm)=(01)$, c- $(nm)=(10)$, d- $(nm)=(20)$

tion of S_{nm} influences both the local and the macroscopic
current density being a self-averaging quantity for a
fixed (nm)-configuration. Omitting in a tight binding concept the intersite contribution of the matrix elements of
the coordinates x, y with (nm)-states and changing over
from i-summation to the integral by means of S_{nm}^{-1} we obtain
in linear response theory the complex conductivity tensor

$$
\left.\begin{array}{l} \sigma_{yx}(B,T,\omega) \\ \sigma_{xx}(B,T,\omega) \end{array}\right\} = \frac{e^2}{h} \sum_{nmn'm'} \frac{\overline{\Delta_{nm,n'm'}}}{\Delta_{nm,n'm'}+\hbar\omega+i\Gamma_{nm,n'm'}}
$$

(5)

$$
\frac{(nm|\varphi^2/2|n'm')^2}{(nm|\varphi|nm)} \left\{ \begin{array}{l} [\delta_{m',m-1}-\delta_{m',m+1}] \\ i[\delta_{m',m-1}+\delta_{m',m+1}] \end{array} \right\} [f_{nm}-f_{n'm'}]
$$

Here $\Delta_{nm,n'm'}$ and $\Gamma_{nm,n'm'}$ denote excitation energies and
damping for the correlated quasiparticle system. The dimensionless matrix elements in (5) are taken with (1). In
case of extremely low temperature the occupation numbers
f_{nm} are 1 or 0. Expression (5) shows a resonance behaviour
for $\hbar\omega=\Delta_{nm,n'm'}$ leading to anomalies in the cyclotron
resonance spectra. In the limit $\Gamma_{nm,n'm'}\rightarrow 0$ the energy factor in $\text{Re}\,\sigma_{xx}$ approaches a δ-function. If there holds additionally $\omega\rightarrow 0$, $\text{Re}\,\sigma_{xx}$ vanishes and $\text{Re}\,\sigma_{yx}(B)$ has the form
of table 3. The coefficients of f_{nm} arise from the exactly

summation index n, (m = 0)	contribution to σ_{yx} in $\frac{e^2}{h}$	possible fractions ν
0	$1\ (1f_{00} - 1f_{0,-1})$	1
1	$1/3(1f_{10}+1f_{01} - 2f_{1,-1})$	1/3 1/3 2/3
2	$1/5(1f_{20}+2f_{11} - 3f_{2,-1})$	1/5 2/5 3/5
3	$1/7(1f_{30}+3f_{21} - 4f_{3,-1})$	1/7 3/7 4/7
4	$1/9(1f_{40}+4f_{31} - 5f_{4,-1})$	1/9 4/9 5/9

Table 3: Possible fractions of $\frac{e^2}{h}$ for the strong field case

evaluable matrix elements in (5). An excellent agreement with the sequence of measured filling factors $\nu \leq 1$ [4] is given and one obtains a direct connection of ν with occupied (nm)-states showing a Q2D-shell structure. The electron configurations of table 3 indicate a continuum in $\text{Re}\,\sigma_{yx}$ at $\nu=1/2$. Filling factors 4/5, 2/7, 5/7, 6/7, 2/9, 7/9, 8/9 cannot occur. The plateau at $\nu=1/3$ is realized if the state (1,0) is occupied, the 2/3-effect occurs if both the states (0,1) and (1,0) are occupied. For a shell $\{(n,0), (n-1,1), (n,-1)\}$ occupied with 6 electrons the total angular momentum and the contribution to the Hall conductivity are zero. The low field case $1 < \nu < \infty$ may be realized by an integral occupation of correlated (nm)-states where a priority seems to exist for integer values of ν . In the ranges of well developed Hall plateaus the condition $\Gamma_{nm,n'm'} \ll \Delta_{nm,n'm'}$ is fulfilled whereas in the transition regions it can be assumed that $\Gamma_{nm,n'm'} \approx \Delta_{nm,n'm'}$ and instabilities in the electronic structure lead to peaks in the longitudinal resistivity which may be approximated by

$$\rho_{xx}^{max}\ (B) = \frac{\text{Re}\,\sigma_{xx}}{\text{Re}\,\sigma_{xx}^2 + \text{Re}\,\sigma_{yx}^2} \approx \frac{h}{e^2\nu}\ . \tag{6}$$

The abrupt changes in the electron configuration at definite magnetic fields B or concentrations N may be understood as transitions between stability regions determined by correlated localized states (Hall plateaus).

1) von Klitzing,K., Dorda,G. and Pepper,M., Phys. Rev. Lett. 45, 494 (1980)

2) Tsui,D.C., Störmer,H.L. and Gossard,A.C., Phys. Rev. Lett. 48, 1559 (1982)

3) Keiper,R. and Ziep,O., phys. stat. sol. (b) 133, 769 (1986)

4) Chang,A.M., Berglund,P., Tsui,D.C., Störmer,H.L. and Hwang,J.C.M., Phys. Rev. Lett. 53, 997 (1984).

TRIPLET CORRELATIONS
IN THE FRACTIONAL QUANTUM HALL EFFECT

C.D. Chen and E. Tosatti

International School for Advanced Studies
Strada Costiera 11, I-34100 Trieste, Italy

ABSTRACT

We study three-electron short-range correlations in the exact
ground state of a system of N two-dimensional electrons
(mostly N=7)in the lowest Landau level, with a disc geometry.
Triangular "crystallite-like" maxima are found for filling
$\nu<0.55$. Their ν-dependence is studied in detail, and
discussed in the light of the existing theories of the
fractional quantum Hall effect.

When electrons fill only partly a highly degenerate 2-dimensional
Landau level, their ground state is entirely determined by correlations.
Two-electron correlations are very well known, since they determine the
energy[1]. Three-electron correlations, on the other hand, are important
for understanding the real-space structure, but remain so far unknown.

We have undertaken their calculation for a system of few spin-polarized electrons (mostly N=7). The symmetric gauge is chosen, with a
disc-shaped neutralizing background of radius $R=(2N/\nu)^{\frac{1}{2}}\ell_o$, with
ℓ_o^2=hc/$(2\pi eB)$. The hamiltonian matrix is set up among all Slater determinants obtained by distributing N electrons over $(N/\nu)=\ell_{max}+1$ orbitals,
for a given total angular momentum L (L is the only good quantum number
in this geometry). The background terms are purely diagonal. Only a

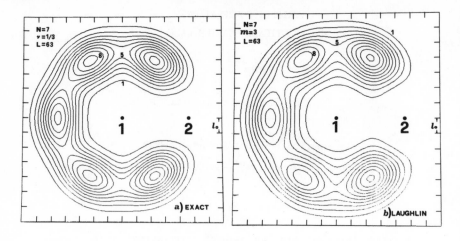

Fig.1 a)exact three–electron correlations for N=7, r_1=0,
$r_2=(4\pi/\nu\sqrt{3})^{\frac{1}{2}}$, in units of 0.0754 electrons/$(2\pi\ell_0^2)$.
b)same, for Laughlin's fluid wave function (Ref.4).

discrete set of fillings can be studied. At a given (ν,L), the ground
state $|0>$, the first excited state and the gap E_{gap} between them are
found with Davidson's algorithm. The value $L=L_0$ corresponding to the
minimal ground state energy identifies our best approximation to the true
ground state for that ν. Details of the calculations will be given else–
where.

The (unnormalized) three–body correlation function

$$G(\vec{r}_1,\vec{r}_2,\vec{r})=<0|\psi^+(\vec{r}_1)\psi^+(\vec{r}_2)\psi^+(\vec{r})\,\psi(\vec{r})\,\psi(\vec{r}_2)\,\psi(\vec{r}_1)|0>/(n(\vec{r}_1)n(\vec{r}_2))$$

is then calculated. Here, $n(\vec{r})=<0|\psi^+(\vec{r})\,\psi(\vec{r})|0>$ is the electron density
and $\psi(\vec{r})$ is the field operator. The most interesting results, with r_1=0,
$r_2=<r>=(4\pi/\nu\sqrt{3})^{\frac{1}{2}}$ (the typical inter–electron spacing), and ν=1/3, are
shown in Fig.1(a). The picture suggests a 7–electron "crystallite", with
typical triangular coordination (while of course the density and the
two–body correlations are rotationally invariant). This aspect is appeal–

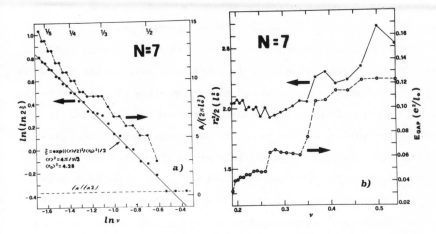

Fig.2 a)Corrugation ξ versus filling, compared with a superposition of gaussians from neighboring peaks and hexagon
area A (see text); b)Gaussian tangential peak width r_o^2 and
excitation gap.

ing, also in view of a recent revival[2] of interest in Wigner lattice
states[3],[4] . However, we have also calculated $G(0,\vec{r}_2,\vec{r})$ for the
Laughlin fluid state[5], and found it very similar (Fig.1(b)). Moreover,
we generally find a ring of N-2 peaks, even for N≠7, in which case the
triangular symmetry is lost. Hence, short-range correlations at ν =1/3
are definitely fluid-like. A similar result holds for ν =1/5. The
situation could in principle be very different for ν ≠1/m, where there is
no reference fluid state such as Laughlin's. We find a picture qualitatively similar to Fig.1(a) for all fillings up to ν ≈0.55, above which
the gaussian-shaped peaks of G merge into a ring, with no corrugation
left. For ν < 0.55, the peak-peak corrugation $\xi = G_{max}/G_{saddle}$, and a
tangential gaussian peak width r_o $(G \sim G_{max} \exp(-r^2/r_o^2))$ near each peak
along the ring), are shown on Fig.2. The increase of corrugation with

decreasing ν is roughly fitted by summing two equal gaussian contributions of constant width $<r_0^2>=4.28$ from neighboring peaks, whose distance is about $<r>$. In this respect, the behavior is reminiscent of a "warm" Wigner crystal. In detail, the corrugation exhibits plateaus,which reflect the incompressible nature of the system. This is confirmed by the plotted behavior of the hexagonal area A obtained by joining the five maxima and completing the hexagon by mirror symmetry. The widest plateau is found near $\nu=1/3$. This is the only signal of quantization detectable in this geometry. The other quantizations are not visible, because the disc boundary condition strongly perturbs states with long-range correlations. This aspect is made apparent by studying the lowest energy gap, identifiable with the exciton or magneto-roton gap. Our gap has a minimum instead of a maximum at $\nu=1/3$. Away from a strong quantization, the boundary conditions give rise to strongly pinned states with a large artificial gap. On the other hand, at the strong quantization the correlations become very short, i.e. the system is more "fluid", and the boundary is less important. This picture is in qualitative accord with ideas put forward in Ref.3 . Lastly, we note a gap drop, suggesting quantization, near $\nu=1/4$. More work is in progress on these points.

REFERENCES

1)Yoshioka, D., Phys. Rev. B29, 6833 (1984).

2)Tosatti, E. and Parrinello, M., Lett. Nuovo Cimento 36, 289 (1983).

3)Kivelson, S., Kallin, C., Arovas, D.P. and Schrieffer, J.R., Phys. Rev. Lett. 56, 873 (1986); Baskaran, G., Phys. Rev. Lett. 56, 2716 (1986).

4)Chui, S.T., Phys. Rev. B32, 8438 (1985).

5)Laughlin, R.B., Phys. Rev. Lett. 50, 1395 (1983).

ACTIVATION ENERGIES AND LOCALIZATION
IN THE FRACTIONAL QUANTUM HALL EFFECT

G. S. Boebinger
Department of Physics and Francis Bitter National Magnet Laboratory
Massachusetts Institute of Technology, Cambridge, MA 02139

D. C. Tsui*
Dept. of Electrical Engineering and Computer Science,
Princeton University, Princeton, N.J. 08544

H. L. Störmer*, J. C. M. Hwang** A. Y. Cho and C. W. Tu
AT&T Bell Laboratories, Murray Hill, N.J. 07974
UNITED STATES

We report the magnetic field dependence of the activated
behavior of the minima in the diagonal resistivity and
conductivity observed at fractional Landau level fillings,
ν=1/3, 2/3, 4/3 and 5/3. Results on the mobility
dependence are included.

In the fractional quantum Hall effect, deep minima are observed in the diagonal resistivity, ρ_{xx}, and conductivity, σ_{xx}, of a two-dimensional electron system at fractional Landau level filling factors, ν=nh/eB (n is electron density; B is magnetic field). These minima have been observed to be activated[1] and are ascribed to the existence of energy gaps above novel quantum fluid ground states at fractional ν.[2] Elementary excitations above these ground states include pair production of fractionally charged quasiparticles[2] and creation of uncharged magneto-rotons.[3,4] In an ideal system, these energy gaps scale as Δ=$Ce^2/\epsilon\ell_0$ (ℓ_0=$\hbar/eB^{1/2}$ is the magnetic length; ϵ is the dielectric constant; C is a constant). For an electron-hole symmetric system, Δ is the same for ν=1/3 and 2/3. For ν=4/3 and 5/3, Δ may be comparable since these states correspond to 1/3 and 2/3 filling of the upper spin-subband of the lowest Landau level.

Herein we report an extensive study of the activated behavior of the FQHE which affords a broad comparison between theory and experiment of

---

*   Guest scientist at Francis Bitter National Magnet Laboratory
** Gain Electronics, Somerville, N.J. 08876

the energy gaps involved. The experiments were performed in a dilution refrigerator and a 30T hybrid magnet. Our samples are modulation doped GaAs/AlGaAs heterojunctions with nominal electron densities $n_0 = (1.5 - 2.5) \times 10^{11} \text{cm}^{-2}$ and nominal mobilities $\mu_0 = (7-60) \times 10^4 \text{cm}^2/\text{V}$. The electron densities are linearly tunable by a backside gate bias which also alters the mobility, as $\mu \sim n^\alpha$ with $\alpha = 1.0 = 1.7$ for our samples. The magnitude of ρ_{xx} or σ_{xx} at the minimum was measured over a temperature range from 1.4K to 150mK typically.

In Figure 1, we plot ρ_{xx} and σ_{xx} at the $\nu = 1/3$ minimum versus $1/T$ as measured simultaneously on a single sample. The linear region of the data (dashed line) indicates activated conduction, $\rho_{xx} = \rho_0 \exp(-\Delta/2T)$ where Δ represents a quasiparticle pair production energy. Note that Δ is the same for ρ_{xx} and σ_{xx}. This is expected from $\sigma_{xx} = \rho_{xx}/(\rho_{xx}^2 + \rho_{xy}^2) \sim \rho_{xx}/\rho_{xy}^2$.

At magnetic fields below $\sim 10T$, purely activated conduction is observed down to $T \sim 100\text{mK}$. At higher magnetic fields, deviations from activated behavior are observed at the lowest temperatures. These deviations, which are attributed to hopping conduction of the localized quasiparticles, increase with magnetic field and are substantial at 18.6T(Fig. 1).

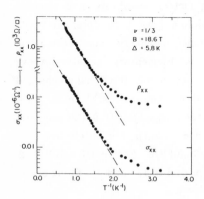

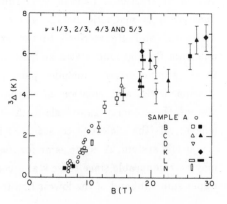

Fig. 1 Direct comparison of temperature dependence of ρ_{xx} and σ_{xx}.

Fig. 2 Pair creation energy for the thirds versus magnetic field. The open (filled) symbols correspond to $\nu = 2/3$ (1/3). The filled squares at 5.9T, 7.4T are on $\nu = 5/3$ and 4/3, respectively.

Fortunately, the pair creation energy is only slightly dependent on the formula chosen to fit the lower-T data and is consistent with the value determined from a straight line drawn through the high-T data. The use of a backside gate bias allows a quantitative study of the magnetic field dependence of the energy gaps. Fig. 2 contains all data for $\nu=1/3$, 2/3, 4/3 and 5/3 from samples of very high mobility, $\mu \gtrsim 4 \times 10^5 cm^2/Vs$. We observe a single magnetic-field-dependent activation energy for 1/3, 2/3, 4/3 and 5/3 filling. The magnitude of the associated energy gap is roughly a factor of five smaller than calculations for creation of either quasiparticle pairs or magneto-rotons.[3,4,6] The observed activation energy does not scale as $B^{1/2}$. Rather, there is a finite magnetic field threshold at $\sim 5.5T$ below which activation energies are vanishingly small. A simple model for inclusion of disorder suggests $\Delta = Ce^2/\epsilon \ell_0 - \Gamma$ where Γ is the line width of the excited state. We have fit the data of Fig. 2 to this model (dashed line in Fig. 3) with C=0.05 and $\Gamma=6.0$ K. A more physical determination of the effects of disorder yields a more satisfactory fit.[7,8]

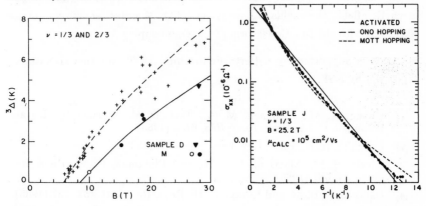

Fig. 3 Pair creation energies of the thirds minima for high- (intermediate-) mobility samples, represented by crosses (circles).

Fig. 4 Temperature dependence of σ_{xx} from a low-mobility sample. The three lines show least squares fits to the three conduction formulae.

By studying a wide range of samples, the mobility dependence of the energy gaps can be qualitatively determined. Samples with mobilities of $(1-3)\times 10^5 cm^2/Vs$ show reduced activation energies and increased magnetic field threshold. (circles in Fig. 3) (Previously, the datum at 4.7K/28T was erroneously reported to have a much higher mobility.)[5] The solid line is a fit with C=0.04 and Γ=5.9K.

For mobilities below $\sim 10^5 \text{cm}^2/\text{Vs}$, the data are not consistent with activated conduction. Fig. 4 shows data with fits to activated conduction, the 2D Mott hopping formula,[9] and 2D Ono hopping formula.[10] This set, like all eight sets for $B \gtrsim 23T$, exhibits conduction consistent with the functional form of the Ono hopping formula over our entire experimental temperature range. However, our data on the FQHE reveal the same quantitative discrepancy found in experiments on the IQHE.[11] The magnitude of the density of states at the Fermi level, derived from the Ono formula, is orders of magnitude too high.

In conclusion, a single activation energy is observed for ν=1/3, 2/3, 4/3 and 5/3 in the FQHE which can be reconciled with theoretical calculations once the effects of disorder and the finite thickness are taken into account. Lower mobility samples exhibit lower activation energies and higher magnetic field thresholds. Lowest mobility samples show conduction consistent with the functional form of the Ono hopping formula.

GSB is supported by a Hertz Fellowship. DCT and the FBNML are supported by NSF.

[1]   Chang, A. M., Paalanen, M. A., Tsui, D. C., Stormer, H. L., and Hwang, J. C. M., Phys. Rev. *B28*, 6133 (1983).

[2]   Laughlin, R. B., Phys. Rev. Lett. *50*, 1395 (1983).

[3]   Girvin, S. M., MacDonald, A. H., and Platzman, P. M., Phys. Rev. Lett. *54*, 581 (1985).

[4]   Haldane, F. D. M. and Rezayi, E. H., Phys. Rev. Lett. *54*, 237 (1985).

[5]   Boebinger, G. S., Chang, A. M., Stormer, H. L., and Tsui, D. C., Phys. Rev. Lett. *55*, 1606 (1985).

[6]   Morf, R. and Halperin, B. I. Phys. Rev. *B33*, 2221 (1986).

[7]   MacDonald, A. H., Lin, K. L., Girvin, S. M., Platzman, P. M. Phys. Rev. *B33*, 4014 (1986).

[8]   Gold, A., Europhys. Lett. *1*, 241 (1986).

[9]   Mott, N. F., and Davis, E. A. *Electronic Properties in Non-Crystalline Materials* (Clarendon, Oxford), 2nd ed. 1979.

[10]   Ono, Y., J. Phys. Soc. Jpn. *51*, 237 (1982).

[11]   Ebert, G., von Klitzing, K., Probst, C., Schubert, E., Ploog, K., and Weimann, G., Solid State Commun. *45*, 625 (1983).

DENSITY OF STATES MEASUREMENTS IN THE EXTREME QUANTUM LIMIT

T. P. Smith, III and W. I. Wang

IBM T. J. Watson Research Center, Yorktown Heights, NY 10598 U.S.A.

P. J. Stiles

Department of Physics, Brown University, Providence, RI 02912 U.S.A.

ABSTRACT

We have measured the magnetocapacitance of a two-dimensional electron gas from the weak-field limit to the extreme quantum limit and obtained quantitative information about the density of states at fractional filling factors. The minima in the density of states have a parabolic temperature dependence and the strength of the minima saturates at high (B > 25 Tesla) magnetic fields.

The fractional quantum Hall effect (FQHE)[1] is one of the most interesting results of many-body interactions in two-dimensional electronic systems (2DES). Experimental studies of the FQHE have primarily focused on transport properties associated with this effect. Recently, however, structure has been observed at fractional filling factors in capacitance measurements[2,3] . Capacitance measurements are attractive since the density of states (DOS) of a two-dimensional system is directly related to the capacitance[4]. However, in past experiments[2,3] conductivity effects influenced the measured capacitance at high magnetic fields, and the DOS was not extracted in the extreme quantum limit.

In order to make capacitance measurements and also avoid resistance effects, we fabricated GaAs heterostructure capacitors on conducting substrates. A 2DES is located at the interface between an undoped GaAs layer and a modulation doped (AlGa)As layer and separated from n^+ GaAs by 50 nm of undoped GaAs. The overlap of the wave functions of the electrons in the 2DES with the wave functions of the electrons from the conducting substrate allows electrons to flow in and out of the 2DES with negligible resistive loss, even in the extreme quantum limit. To form a capacitor an aluminum electrode was deposited on top of the structure. The DOS at fractional filling factors was studied in two samples. From analysis of the weak-field capacitance oscillations the mobility of sample A may be as high as 400,000 cm^2/V-s and the mobility of sample B as high as 200,000 cm^2/V-s.

Both samples have a zero-bias carrier concentration of $1.9 \times 10^{11}/cm^2$. The area of the capacitors on sample A is 7.5 times smaller than the area of the capacitors fabricated on sample B. The carrier concentration of sample A can be varied from 0.6 to 3.5×10^{11} /cm² and that of sample B from 1.4 to $2.4 \times 10^{11}/cm^2$.

Fig. 1 shows the measured magnetocapacitance of sample B at low magnetic fields and zero bias. The decrease in the capacitance between Landau levels is relatively small (~ 4% at $\nu = 2$) indicating that the DOS in the gap between Landau levels is large (~ 40% of the zero field DOS at $\nu = 2$). We modelled our capacitance data using a DOS of the form given by Gerhardts[5] with $\Gamma = \Gamma_0 B^\alpha$. To take into account the effects of spin splitting (observed at B $\simeq$ 8 Tesla) we used both a monotonically increasing g-factor and a self-consistent oscillatory g-factor used previously by Englert et al.[6]. The model using a monotonically increasing g-factor gave the best fit and is shown in Fig. 1. The fit is very good up to about 5 Tesla. The magnetic field dependence of the Landau level broadening is not too far from the $\sqrt{B}$ dependence predicted by Ando and Uemura[7] but the level width for our fit is almost an order of magnitude larger than predicted (3.2 meV vs 0.6 meV at 5 Tesla). However, our results agree fairly well with magnetization[8,9] ($\Gamma = 4.4$ meV at 5 Tesla) and specific heat [10] ($\Gamma = 2.1$ meV with a 20% background at 5 Tesla) experiments. Because there is negligible change in the conductance of the samples as the magnetic field is varied and because our results agree fairly well with other DOS measurements, we are confident that resistivity effects are not influencing our results.

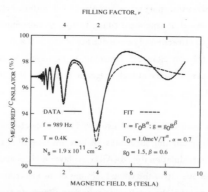

1. (a) The measured and calculated magnetocapacitance below the extreme quantum limit. The dashed line is the calculated capacitance using a monotonically increasing g-factor.

The high field magnetocapacitance and DOS for samples A and B is shown in Figure 2. Consistent with the results at low magnetic fields, the structure at fractional filling factors is much stronger in sample A which has a higher mobility. Nonetheless, the minima in the capacitance at filling factors of 2/3 and 1/3 are relatively weak for both samples. Although these small changes in the capacitance reflect large changes in the density of states, the DOS remains larger than the zero-field DOS, Do for $\nu < 1$.

The large DOS in the fractional gaps is consistent with the results at integer filling factors. However, the origin of the large number of states between Landau levels and at fractional filling factors is not yet well understood. Disorder, scattering, inhomogeneity, are all probably involved but this has not been confirmed.

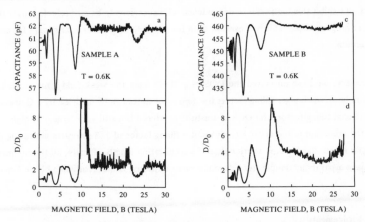

2.   (a) The measured magnetocapacitance of sample A and (b) the measured density of states. (c) The measured magnetocapacitance of sample B and (d) the corresponding density of states. The insulator capacitance is assumed to be 62.8 ± 0.6 pF for sample A and 463 ± 1 pF for sample B.

As the magnetic field strength at which $\nu = 1/3$ occurs becomes larger, the reduction in the DOS at $\nu = 1/3$ saturates. (See Fig. 3a.) This may indicate that the magnetic length is shorter than the localization length above 25 Tesla. The minima in the DOS cannot be further reduced since all the carriers can complete cyclotron orbits within localized areas. Both samples exhibit this behavior with the saturation occurring at higher magnetic fields for sample B which has the lower mobility.

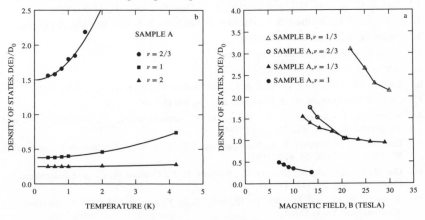

3.   The measured magnetocapacitance under different biases and temperatures.

Fig. 3b shows the temperature dependence of the measured capacitance for sample A. From a Sommerfeld expansion [11] of the density of states a T^2 temperature dependence is expected at a minimum. This temperature dependence is due to the spreading of the Fermi function about the Fermi energy. The DOS between Landau levels, between spin-split levels and at fractional filling factors all exhibit this temperature dependence with successively larger prefactors over the temperature range studied. This is consistent with the successively smaller energy gaps for each type of minimum.

In summary, we have measured the DOS of a 2DES from the weak-field limit to the extreme quantum limit. At integer filling factors the density of states can be modeled by a Gaussian DOS. At fractional filling factors the DOS is drastically reduced but still very large. At high magnetic fields the minimum in the density of states at a filling factor of 1/3 saturates indicating that the magnetic length has become shorter than the localization length. Finally, all the minima in the DOS have a parabolic temperature dependence as expected from broadening of the Fermi function.

Acknowledgements

The authors are grateful to L. Rubin, B. Brandt and the staff at the Francis Bitter National Magnet Laboratory for their support. We would also like to thank M. Christie and L. Alexander for fabricating the capacitors and D. A. Syphers, E. E. Mendez, D. H. Lee, L. L. Chang, L. Esaki, F. F.Fang, and T. Jackson. for their help and many useful discussions. This work was supported in part by the Army Research Office.

References

1. Tsui, D. C., Stormer, H. L. and Gossard, A. C., Phys. Rev. Lett. 48, 1559 (1982).
2. Smith, III, T. P., Goldberg, B. B., Heiblum, M. and Stiles, P. J., Surf. Sci. 170, 304 (1986).
3. Hickmott, T., private communication.
4. Stern, F., unpublished IBM internal report.
5. Gerhardts, R. R., Surf. Sci. 58, 227 (1976).
6. Englert, Th., Tsui, D. C., Gossard, A. C., and Uihlein, Ch., Surf. Sci. 113, 295 (1982).
7. Ando, T. and Uemura, Y., J. Phys. Soc. Jap. 36, 959 (1974).
8. Haavasoja, T., Stormer, H. L., Bishop, D. J., Narayanamurti, V., Gossard, A. C. and Wiegmann, W., Surf. Sci. 142, 294 (1984).
9. Eisenstein, J. P., Stormer, H. L., Narayanamurti, V., Cho, A. Y., Gossard, A. C., and Tu, C. W., Phys. Rev. Lett. 55, 875 (1985).
10. Gornik, E., Lassnig, R., Strasser, G., Stormer, H. L., Gossard, A. C. and Wiegmann, W., Phys. Rev. Lett. 54, 1820 (1985).
11. Ashcroft, N. W. and Mermin, N. D., Solid State Physics pp. 44-46(Holt, Rinehart and Winston, New York, 1976).

FRACTIONAL QUANTUM HALL EFFECT OF P-TYPE GaAs-(GaAl)As HETEROJUNCTIONS AT MILLIKELVIN TEMPERATURES

G. Reményi, G. Landwehr, W. Heuring, E. Bangert,
Physikalisches Institut der Universität Würzburg
D-8700 Würzburg, Fed. Rep. of Germany
and
G. Weimann, W. Schlapp,
Forschungsinstitut der Deutschen Bundespost
D-6100 Darmstadt, Fed. Rep. of Germany

Measurements of magnetoresistance and Hall resistance of
p-type GaAs-(GaAl)As heterojunctions at mK temperatures
revealed structures at fractional occupation of the
Landau-level N = 1 which were not reported before.

1. INTRODUCTION

P-type semiconductor inversion layers have been investigated in
much less detail than their n-type counterparts. The reason for this
is the rather complicated subband structure in conjunction with large
effective masses resulting in relatively low mobilities. However, p-
type GaAs-(GaAl)As heterostructures are an exception. In high quality
samples mobilities of more than 200.000 cm^2/Vs at He-temperatures have
been observed[1]. These properties are connected with the lifting of
the Kramers degeneracy at the top of the valence band for finite wave
vectors due to the lacking inversion symmetry in single heterojunctions. The scattering times in the best p-type devices are comparable
with those found in n-type samples so that it has been possible to
measure both the normal and the fractional Quantum Hall Effect. It
turns out that the magnetoresistance at moderate magnetic fields shows
interesting quantum effects, which can be explained by detailed theoretical calculations[2] of the subband structure. These properties have
been discussed recently elsewhere[3].

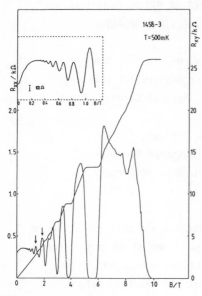
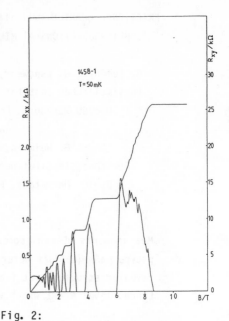

Fig. 1:

Magnetoresistance R_{xx} and
Hall resistance R_{xy} for a
p-type GaAs-(GaAl)As heterostructure at T = 0.5 K.
Insert: enlarged scale at low
magnetic fields B

Fig. 2:

Magnetoresistance R_{xx} and Hall
resistance R_{xy} for sample 1458-1
as a function of magnetic field B
at T = 50 mK. The structure
in R_{xy} around 8 T is due to the
fractional quantum Hall effect.

2. EXPERIMENTAL

Magnetoresistance R_{xx} and Hall resistance R_{xy} of modulation doped
p-type GaAs-(GaAl)As heterojunctions were investigated between 50 mK
and 4.2 K in magnetic fields B up to 12.5 T. The Hall mobility at 0.5
K of four specimens studied in detail was between 60.000 cm^2/Vs and
185.000 cm^2/Vs, and the hole concentration varied between 2 x 10^{11}/cm^2
and 2.5 x 10^{11}/cm^2. Here we shall discuss results for two samples
which were prepared from the same chip. The data obtained for sample
1458-3 at 500 mK are shown in Fig. 1. One can recognize well developed
Hall plateaus and pronounced Shubnikov-de Haas oscillations. The maxima and minima at magnetic fields below 1 T are periodic in 1/B (see

insert). At the B-values marked by arrows deviations from the periodicity show up which can be explained theoretically[2]. We shall not
discuss here the large non-oscillatory magnetoresistance component
below 0.3 T. The structure in R_{xx} between 6 T and 8 T belonging to the
Landau-level with the quantum number N = 1 can be attributed to the
fractional Quantum Hall Effect. The dip around 8 T can be assigned to
the occupation number η = 4/3. Obviously there is unresolved structure
in R_{xy} between 6 T and 10 T. In Fig. 2 data are shown which were
obtained at 50 mK for the sample with the highest mobility (220.000
cm^2/Vs at this temperature). One can recognize structures in the Hall
resistance which look like plateaus which have not yet fully developed. The resistance values in the middle of the three best developed
plateaus, expressed in units of h/e^2 are: 7/5, 5/4 and 7/6. This looks
like an observation of the fractional Quantum Hall Effect at occupation numbers corresponding to even fractions. Such a result would be
at variance with the theoretical model proposed by Laughlin[4], predicting quantization for the Hall conductivity

$$\sigma_{xy} = - p/q \ (e^2/h)$$

only for fractions p/q with odd denominator. However, caution is
necessary in the interpretation of the data. The Hall plateaus were
stable when the magnetic field was varied at constant temperature but
somewhat different resistance readings were obtained when the temperature was varied, leading to different fractions. One is confronted
with the problem whether it is correct to assign the quantum numbers
from the quantized R_{xy} values or from the filling factors as derived
from the magnetic field values which can be determined with rather
high accuracy from the integral Quantum Hall Effect. Because the
fractional Hall plateaus are not located on the classical Hall line
the correlation between R_{xy} and the filling factor which is valid for
the integral Quantum Hall Effect is lost.

The examination of ten registration curves, obtained at temperatures between 50 mK and 500 mK had the following result: Hall plateaus
and accompanying minima in R_{xx} showed up in nearly all measurements at
filling factors η = 10/7, 13/9, 4/3 and 1.37 ($\approx$ 11/8). However, the
accompanying R_{xy} values varied with temperature. On the other hand,

lead the analysis of the resistance values, at which the fractional
Hall plateaus showed up at the lowest temperatures, to the following
quantized p/q-values: 1.25 $\pm$ 0.03 (5/4), 1.20 $\pm$ 0.03 (6/5) and 1.17 $\pm$
0.03 (7/6). Within the given error limits the corresponding fractionals
were observed in at least five measurements. The corresponding magnetic fields were in the range 7.3 - 7.5 T (η = 1.45 - 1.40); 7.45 -
7.65 T (η = 1.43 - 1.39) and 7.5 - 7.9 T (η = 1.37 - 1.34). The Hall
resistance and the magnetic field were measured with a precision of
better than 1 %. Warming up the samples between two runs to 77 K did
not change the results.

The preceeding discussion has shown that we are not in a position
to be able to claim to have reliably identified fractional quantization of the Hall resistivity. Our samples were very homogeneous and
had a very high mobility and did not need illumination in order to
obtain the data presented here. But obviously there is still a superposition of structures caused by fractional quantization which cannot
be sufficiently resolved. One has to realize, that microinhomogenities
in samples can cause features which can simulate structures related to
the Quantum Hall Effect. Before one can decide whether the FQHE with
even fractions exists, further careful experiments on high quality
samples are required.

ACKNOWLEDGEMENT

Part of the experiments were performed at the Hochfeldmagnetlabor
Grenoble, Max-Planck-Institut für Festkörperforschung.

REFERENCES

1) E.E. Mendez, Surface Science 170, 564 (1986)

2) E. Bangert and G. Landwehr, Surface Science 170, 593 (1986)

3) G. Landwehr in: Application of High Magnetic Fields in Semiconductor Physics, Springer Series in Solid State Sciences (SSS), in
   print; G. Reményi, G. Landwehr, W. Heuring, G. Weimann and W.
   Schlapp, ibidem

4) R.B. Laughlin, Phys. Rev. Lett. 50, 1395 (1983).

THE FRACTIONAL QUANTUM HALL EFFECT IN Si MOSFETS: NEW RESULTS.

J. E. Furneaux[*]

Naval Research Laboratory, Washington, DC 20375-5000, USA

D. A. Syphers[*]

University of Oregon, Eugene OR 97403, USA

J. S. Brooks[*]

Boston University, Boston, MA 02215, USA

G. M. Schmiedeshoff

Francis Bitter National Magnet Laboratory, MIT, Cambridge, MA 02139,

USA

R. G. Wheeler

Yale University, New Haven, CN 06520, USA

G. Dorda

Siemens AG, D-8000 Munchen 83, FRG

R. P. Smith[*] and P. J. Stiles[*]

Brown University, Providence, RI 02912, USA

Since its discovery[1] the fractional quantum Hall effect (FQHE) has been one of the most interesting subjects in solid state physics. The FQHE is similar to the integer quantum Hall effect (IQHE) in which the Hall conductance is quantized in units of e^2/h when the Fermi level of the two-dimensional electron gas (2DEG) is pinned between Landau bands. This occurs at integral multiples of the degeneracy of a Landau level, eB/h, or equivalently at integral filling factor $\nu = nh/eB$ where n is the density of electrons in the 2DEG. This gives rise to plateaus in the Hall resistivity, ρ_{xy}, and a vanishing diagonal resistivity, ρ_{xx}. In the FQHE similar structures are observed at fractional ν. Since the first report[2] of the FQHE in Si metal-oxide-semiconductor field-effect transistors (MOSFETs), there have been a number of studies in this interesting system[3]-[7]. Silicon MOSFETs are particularly useful for the study of the FQHE because they provide a 2DEG system with a very different effective mass and band structure from GaAs and because it is very straightfoward to study the

[*] Visiting Scientist Francis Bitter National Magnet Laboratory, MIT, Cambridge, MA 02139

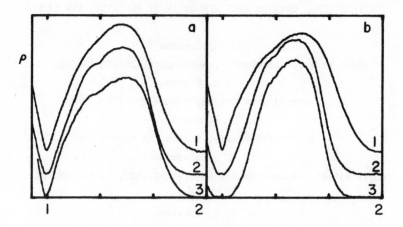

Figure 1. Diagonal magnetoresistance as a function
of filling factor with the curves offset for clarity. a) T = 1.3 K, 1) H = 29.3 T, 2) H = 21.3 T,
3) H = 16.7 T. b) H = 23.6 T, 1) T = 2.0 K, 2) T =
1.04 K, 3) T = 0.64 K.

density and magnetic field dependence of the FQHE. Here we present a
summary of our recent studies of the FQHE in Si MOSFETS. For a recent
review of the IQHE and FQHE see Prange and Girvin[8].

Our samples are Hall bars produced at Yale University and Siemens
AG. They have maximum mobilities of about 2.8 m^2/Vs at n = 7 X 10^{15}
m^{-2} and T = 4.2 K which increase to about 3.3 m^2/Vs at n = 1.8 X 10^{15}
m^{-2} and T = 0.5 K. For a more complete study of the temperature
dependence of the mobility see Smith and Stiles[9]. We have presented
our initial observations of the FQHE previously[7]. Here we summarize
the results of more extensive studies of the FQHE with emphasis on the
unexpected magnetic field and temperature dependence of the FQHE.

Figure 1 shows how the FQHE develops as a function of magnetic
field and temperature in these samples. Figure 1a shows that as the
magnetic field is increased the FQHE first increases in strength at
fields below about 22 T. Then it unexpectedly decreases in strength
at fields above about 25 T at 1.3K. Figure 1b shows an even more
surprising behavior; as the temperature is decreased at a constant
field of 23.6 T the strength of the ν = 4/3 FQHE structure first
increases then decreases. In fact at all fields above about 22 T the

FQHE shows evidence for a similar maximum strength as a function of temperature. This maximum moves to higher temperatures with increasing field. Below 20 T, although there is evidence for the saturation of the strength of the FQHE, this strength monotonically increases with temperature down to 40 mK.

While we do not have a quantitative model for this unexpected development of the FQHE with temperature and magnetic field, we have some tentative ideas which may explain this unique behavior. First, since this behavior has not been observed in $GaAs-Al_xGa_{1-x}As$ hetero-junctions, it may be characteristic of a multivalleyed band structure. Rasolt et al.[10] have predicted that the FQHE could be much weaker in Si due to extra dissipation associated with the multivalleyed nature of the 2DEG in that system. This does not seem to be the case for Si in general since the largest FQHE gaps reported have been in that system[6]. The valley splitting still may be an important factor in the low temperature behavior. The development of the IQHE at $\nu = 1$ is evidence for the valley splitting being important in this system. The peak in the strength of the 4/3 FQHE corresponds closely to the field or temperature where the $\nu = 1$ IQHE becomes well developed; ie. ρ_{xx} begins to develop a broad area where it is zero. A second factor which is probably relevant for the case reported here is the relatively large number of localized states in the system. This can be seen by the broad plateau at $\nu = 2$ and the relatively high (compared to $GaAs-Al_xGa_{1-x}As$ heterostructures) density of strong localization (up to about $= 0.9$) seen here. We therefore feel that screening these localized states is very important to the existence of the FQHE in these samples, and that the observed behavior may be due to changes in the screening. Finally, it should be noted that the mobility is different for the different magnetic fields. The peak mobility occurs at $\nu = 4/3$ at a field of 9.5 T. The mobility is fairly constant (within 5%) between 8 T and 16 T and then decreases approximately linearly to a total decrease of about 20% at 29 T at this ν. Structure in ρ_{xx} and ρ_{xy} at $\nu = 2/3$ are not observed even at 20 T when the maximum mobility occurs at that ν. The maximum strength of the FQHE is therefore only loosely related to the maximum mobility.

In conclusion we have made the first studies of the FQHE in a Si

MOSFET at fields above 20 T and at temperatures below 0.4 K. We have observed an unexpected peak in the strength of the FQHE as a function of magnetic field and as a function of temperature at fixed magnetic field. This temperature dependence is inconsistent with the single parameter scaling model of Laughlin et al.[11] and seems to be associated with screening and the valley splitting in these samples.

We would like to acknowledge useful discussions with S. Teitler. We would like to acknowledge support from the Office of Naval Research, from the National Science Foundation under contracts DMR 81-19550, DMR 81-13456, and DMR 83-14397, from the staff of the Francis Bitter National Magnet Laboratory and especially from Paul Emery for technical assistance.

1) Tsui, D.C., Störmer, H.L., and Gossard, A.C., Phys. Rev. Letters 48, 1562 (1982).

2) Pudalov, V.M. and Semenchinskii, S.G., Pis'ma Zh. Exsp. Teor. Fiz. 39, 143 [JETP Letters 39, 170] (1984).

3) Pudalov, V.M. and Semenchinskii, S.G., Solid State Commun. 52, 567 (1984).

4) Gavrilov, M.G., Kukushkin, V.I., Kvon, Z.D., and Timofeev, V.B., Proceedings of the 17th International Conference on the Physics of Semiconductors ed. J.D. Chadi and W.A. Harrison, 287 (New York: Springer, 1985).

5) Kukushkin, V.I., Timofeev, V.B., and Cheremnykh, P.A., Zh. Eksp. Teor. Fiz. 87, 2223 [Sov. Phys. JETP 60, 1285] (1984).

6) Kukushkin, V.I. and Timofeev, V.B., Surface Science, 170, 148 (1986).

7) Furneaux, J.E., Syphers, D.A., Brooks, J.S., Schmiedeshoff, G.M., Wheeler, R.G., and Stiles, P.J., Surface Science, 170, 154 (1986).

8) Prange, R.E. and Girvin, S.E. eds., The Quantum Hall Effect (Berlin: Springer, 1986).

9) Smith, R.P. and Stiles, P.J., Solid State Commun. 58, 511 (1986).

10) Rasolt, Mark, Perrot, F., and MacDonald, A.H., Phys. Rev. Letters 55, 433 (1985).

11) Laughlin, R.B., Cohen, Marvin L., Kosterlitz, J.M., Levine, H., Libby, S.B., and Pruisken, A.M.M., Phys. Rev. B 32 1311 (1985).

ACTIVATION ENERGIES OF THE 2/3 FRACTIONAL QUANTUM HALL EFFECT
IN A GaAs/AlGaAs HETEROSTRUCTURE WITH A BACKSIDE GATE

J. Wakabayashi, S. Sudou, S. Kawaji, K. Hirakawa[+] and H. Sakaki[+]

Department of Physics, Gakushuin University, Mejiro, Toshima-ku,
Tokyo 171, JAPAN.
+Institute of Industrical Science, University of Tokyo, Roppongi,
Minato-ku, Tokyo 106, JAPAN

The activated temperature dependence of the resistivity ρ_{xx} in the fractional quantum Hall effect is supposed to originate from the excitation of fractionaly charged quasiparticles which carry the current. Several theoretical calculations have been done for the excitation energy of quasiparticles in various approximations.[1]

Recently, Boebinger et al.[2] reported the magnetic field dependence of the activation energies for $\nu = 1/3$ and 2/3. They claimed that the activation energies for both $\nu = 1/3$ and 2/3 lied on a single line with a finite threshold field at $B \sim 5.5$ T. However, their results obtained in the quasi-Corbino samples are not reliable because σ_{xx} can not be measured by "quasi"-Corbino sample in a strong magnetic field. On the other hand, Wakabayashi et al.[3,4] have reported that both activation enegies at $\nu = 1/3$ and 2/3 showed no systematic dependence on the magnetic field B but showed systematic dependence on the μB, where μ is the electron mobility at zero field.

In the present paper, we report results on the activation energies at $\nu = 2/3$ as a function of the backside gate bias. The sample used is a modulation doped GaAs/$Al_{0.3}Ga_{0.7}As$ heterostructure whose backside gate electrode was made by gold deposition after polishing the GaAs substrate to the thickness about 100 μm. The electron density and the mobility with no gate bias were 1.9×10^{11} cm^{-2} and 4.9×10^5 cm^{-2}/Vs, respectively. The electron density was varied in the range between 1.5 $\times 10^{11}$ cm^{-2} and 2.5×10^{11} cm^{-2} by applying a gate bias between -58 V and +86 V. Concurrently the electron mobility was varied between 2.7 $\times 10^5$ cm^2/V.s and 8.8×10^5 cm^2/V.s. The electron mobility have shown the power law dependence on the electron density as $\mu \propto n^{2.4}$. The

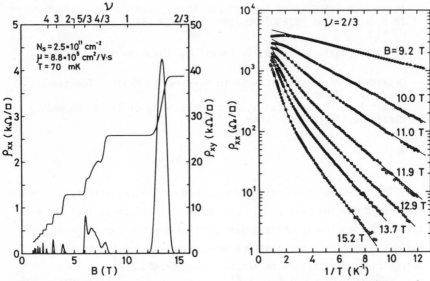

Fig. 1 Magnetic field dependence of resistivity ρ_{xx} and ρ_{xy} at the gate bias +86 V. The current used is 10 nA (2×10^{-6} A/m). The upper scale shows the filling factor ν.

Fig. 2 Temperature dependence of ρ_{xx} minima at $\nu = 2/3$ for several backside gate bias. The current used is 0.2 nA for the field between 9.2 T and 11.9 T, and 2.0 nA between 12.9 T and 15.2 T.

length and the width of the sample are 600 μm and 50 μm, respectively. The usual lock-in technique was used to measure ρ_{xx} and ρ_{xy}.

In Fig. 1, we illustrate the magnetic field dependence of ρ_{xx} and ρ_{xy} in order to show the characteristics of the present sample. The wide plateau of ρ_{xy} and concurrent vanishing ρ_{xx} are observed around the integral filling factor.

Figure 2 shows the temperature dependence of ρ_{xx} at $\nu = 2/3$ for each gate bias between -58 V and 86 V. Quite recently, Wakabayashi et al.[5] made a carefull experiments on the temperature dependence of ρ_{xx} at $\nu = 1/3$ and 2/3. Their results have confirmed that the temperature dependence at low temperature was also activated[6] and the temperature dependence was a sum of two activated conduction.[3,4] The present results are also expressed by a sum of two activated conduction as,

$$\rho_{xx}(T) = \rho_{01}\exp(-W_1/T) + \rho_{02}\exp(-W_2/T). \qquad (1)$$

Each solid line except top two curves in Fig. 2 represents a fitted curve obtained by the eq.(1). We denote the activation energy by W_1

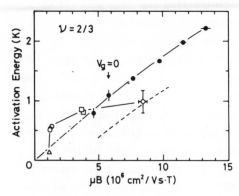

Fig. 3 Activation energies at $\nu = 2/3$ against μB. Closed symbols are data from Fig. 2. Open symbols are data from ref. 6. The broken line shows calculated result from the data of ref. 7.

for the high temperature region and W_2 for the low temperature region. When a large negative bias is applied, the temperature dependence of ρ_{xx} minima shows a single activated conduction as shown by the top two curves in Fig. 2. It is important that the activation energy in this case is not W_1 but W_2. The activation energy W_1 is corresponding to the theoretical excitation energy as discussed elsewhere.[5] Chang et al.[7] also observed a single activated conduction at $\nu = 2/3$ of a sample with a backside gate. Their results, however, may be due to very small value of ρ_{02} of their high mobility sample.[4]

The magnetic field dependence of both activation energy W_1 and W_2 determined from the data in Fig. 2 show a finite threshold field at about 8.5 T. On the other hand, when the activation energies W_1 are plotted against μB, they show zero threshold field as shown in Fig. 3. In fact W_1 is roughly proportional to $\mu\sqrt{B}$. This fact also supports that the activation energy is a function of the mobility as well as the magnetic field.[3,4] One of the possible function of the mobility in the activation energy is broadening of the energy levels due to disorder.[8] However the electron mobility at zero field does not represent all the effect of disorder as discussed below.

The activation energy W_1 shows stronger μB dependence than the results by Wakabayashi et al.[4] as shown in Fig. 3. Similar μB dependence of the activation energy can been seen in Chang et al.'s[7] data which is shown by a broken line in Fig. 3. These results suggest that there are another parameters other than the magnetic field and the mobility. A possible parameter to give such effect is the surface field in the inversion layer normal to the interface. One of the functions of the surface field is controlling the disorder. It is well known in silicon inversion layers[9] that a strong surface field increases disorder originating from the interface roughness. It is supposed that the interface is smooth in the present system owing to the lattice matched epitaxial growth of GaAs and AlGaAs crystal. However, it is considered that there are residual interface roughness at GaAs-AlGaAs heterojunction.[10] Further, it is also plausible to expect a residual impurities at the interface, which is considered to produce the localization of quasiparticles and the finite width of the fractional Hall plateau as shown in Fig. 1. If we assume these disorder at the interface, the effect of the backside gate bias on the activation energy W_1 is considered to be qualitatively reasonable because the positive bias pulls away the electrons from the interface and decrease the disorder which electrons feel, and the negative bias vice versa. The surface field, that is, introduces disorder which affect the activation energy more than the electron mobility. However, a quantitative discussions and the other functions of the effect of the surface field is left for the future study.

ACKNOWLEDGEMENTS: This work is supported by a Grant-in-Aid for Special Distinguished Research from the Ministry of Education, Culture and Science. One of the authors (J.W.) thanks the Yazaki Memorial Foundation for Science and Technology for the financial support.

REFERENCES
1) R. Morf and B.I. Halperin: Phys. Rev. B33, 2221 (1986), and references are cited there in.
2) G.S. Boebinger, A.M. Chang, H.L. Stormer and D.C. Tsui: Phys. Rev. Lett. 55, 1606 (1985).
3) J. Wakabayashi, S. Kawaji, J. Yoshino and H. Sakaki: Surf. Sci. 170, 136 (1986).
4) J. Wakabayashi, S. Kawaji, J. Yoshino and H. Sakaki: J. Phys. Soc. Jpn. 55, 1319 (1986).
5) J. Wakabayashi, S. Sudou, S. Kawaji, K. Hirakawa, J. Yoshino and H. Sakaki: Submitted to J. Phys. Soc. Jpn.
6) S. Kawaji, J. Wakabayashi, J. Yoshino and H. Sakaki: J. Phys. Soc. Jpn. 53, 1915 (1984).
7) A.M. Chang, M. A. Paalanen, D.C. Tsui, H.L. Stormer and J.C.M. Hwang: Phys. Rev. B28, 6133 (1983).
8) A.H. MacDonald, K. L. Liu, S.M. Girvin and P.M. Platzman: Phys. Rev. 33, 4014 (1986).
9) T. Ando, A.B. Fowler, and F. Stern: Rev. Mod. Phys. 54, p.502(1982).
10)B.A. Joyce: Proc. Int. Conf. on Modulated Semicond. Structures, Kyoto, (1985), p.1 (to be published in Surf. Sci.)

SCALING FUNCTION OF CONDUCTIVITIES IN QUANTIZED HALL EFFECT

Tsuneya ANDO

Institute for Solid State Physics, University of Tokyo
7-22-1 Roppongi, Minato-ku, Tokyo 106, Japan

It is demonstrated by a numerical study that there exists a scaling relation between diagonal and off-diagonal conductivities. The scaling relation is nearly independent of system parameters in strong magnetic fields but is drastically modified by mixings between different Ladau levels when the field becomes weaker.

1. INTRODUCTION

The quantized Hall effect requires the presence of both localized and extended states, and it cannot be discussed separately from the localization problem in strong magnetic fields.[1] The purpose of the present paper is to present results of extensive numerical study on the scaling relation between the diagonal σ_{xx} and Hall σ_{xy} conductivities.

There have been various theoretical investigations on the localization problem in strong magnetic fields. The most direct and reliable method is numerical study. One of the criteria which have turned out to be quite successful is the Thouless number g(L). Numerical study based on g(L) has shown that states are exponentially localized except those just at the center of Landau levels and the inverse localization length becomes continuously zero at the center.[2] The finite-size scaling has also been applied and given the inverse localization length in excellent agreement with the Thouless-number method.[3]

In this paper we calculate σ_{xx} and σ_{xy} as a function of the Fermi level for different system sizes. The conductivities are the functions of the Fermi level E_F and the system size L. With increasing L, σ_{xx} vanishes exponentially for sufficiently large L and σ_{xy} approaches the integer multiple of $-e^2/h$, except when the Fermi level lies just at the center of the Landau level or at extended states. Such system-size dependence of σ_{xx} and σ_{xy} for each E_F gives a flow line in the σ_{xy}-σ_{xx} space. This flow line is expected to depend on E_F. However, it is demonstrated that there exists only a single flow line, i.e., that there exists a scaling relation between σ_{xx} and σ_{xy}. This is in clear contradiction to the conclusion obtained in the nonlinear σ model.[4] A preliminary account of a part of the work has been reported elsewhere.[5]

2. METHODS

We consider a two-dimensional system with a finite size LxL in a strong magnetic field H. Scatterers are distributed at random. We shall use periodic boundary conditions in both x and y directions. The diago-

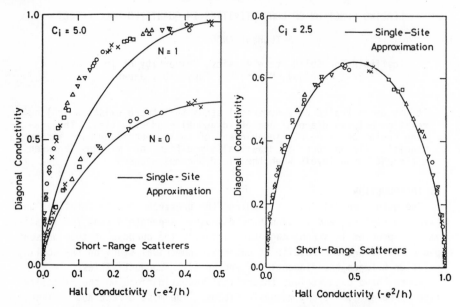

Fig. 1 Scaling relation between σ_{xx} and σ_{xy} for the lowest two Landau levels N=0 and 1. The diagonal conductivity proportional to the Thouless number and the Hall conductivity of different energies and four different system sizes are plotted. The four points with a same symbol correspond to the conductivities of a given Fermi energy for the four system sizes L/l = 15.0, 25.1, 35.1 and 45.1 for N=0 and L/l= 20.1, 30.1, 40.1 and 50.1 for N=1. The solid lines represent the relation calculated in the single-site approximation. Only the low energy side is shown because of the symmetry about the center of the Landau level and the origin of σ_{xy} is shifted by e^2/h for N=1.

Fig. 2 Scaling relation between σ_{xx} and σ_{xy} for the lowest Landau level N=0 in the case that the density of states is strongly asymmetric. The diagonal conductivity proportional to the Thouless number g(L) and the Hall conductivity of different energies and four different system sizes are plotted. The four points with a same symbol correspond to the conductivities of a given Fermi energy for the four system sizes L/l = 15.0, 25.1, 35.1 and 45.1. The solid lines represent the relation calculated in the single-site approximation. Note that the relation between σ_{xx} and σ_{xy} does not depend on signature of scatterers in the single-site approximation.

nal conductivity is determined by the Thouless number and the Hall conductivity is calculated directly using the Kubo formula represented by a correlation function of the velocities of the center coordinates.

The conductivities are calculated for systems having L ranging from 15.0 to 50.1 in units of l, the radius of the cyclotron orbit of the lowest Landau level defined by $l^2=c\hbar/eH$, and containing scatterers with

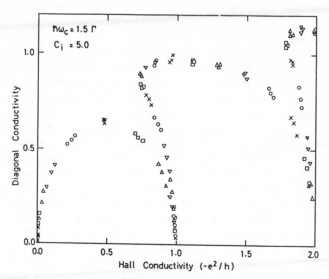

Fig. 3 Scaling relation between σ_{xx} and σ_{xy} when mixings among different Landau levels are taken into account, i.e., when $\hbar\omega_c=1.5\Gamma$), where $\hbar\omega_c$ is the energy separation between adjacent Landau levels and Γ is the broadening of the Landau level calculated in the self-consistent Born approximation. The diagonal conductivity proportional to the Thouless number $g(L)$ and the Hall conductivity of different energies and three different system sizes are plotted. The three points with a same symbol correspond to the conductivities of a given Fermi energy for the three system sizes $L/l=15.0$, 20.1 and 30.1.

short-range potentials.

3. RESULTS AND DISCUSSION

In Fig. 1 the diagonal and off-diagonal conductivities for different E_F's and L's are plotted. We assume the strong-field limit in which mixings among different Landau levels play a minor role. This figure demonstrates that σ_{xx} and σ_{xy} are mutually correlated and cannot be considered independent in the present system. A similar plot in logarithmic scales shows the relation $\Delta\sigma_{xy}\propto\sigma_{xx}^q$ for small σ_{xx} and $\Delta\sigma_{xy}$ with q being close to 2, where $\Delta\sigma_{xy}$ is the deviation from the quantized value.

The present calculation shows that starting from a given point the conductivities flow along a line and approach the stable fixed point corresponding to vanishing σ_{xx} and the quantized σ_{xy} with increasing system size. Only exception is the points corresponding to $\sigma_{xy} = -(N+1/2)e^2/h$, which are unstable fixed points corresponding to extended states associated with each Landau level. The flow lines are dependent on the Landau level N and qualitatively the same as that calculated in the single-site approximation[6,7] represented by the solid lines.

432   T. Ando

To see whether the flow lines are universal and independent of system parameters, we also calculate the flow line for the lowest Landau level in systems having low concentrations of scatterers with attractive δ-potential. In this system the density of states is strongly asymmetric. Figure 2 shows that the scaling relation does not depend on impurity concentrations or the asymmetry of the density of states. It is easy to show that in the single-site approximation[6,7] the relation is universal as long as impurity potentials are of short range.

Figure 3 gives the relation between σ_{xx} and σ_{xy} when the magnetic field becomes weaker. We immediately notice the drastic effect of level mixings' on the relation. The flow lines for higher Landau levels are deformed mainly due to the decrease of the Hall conductivity in proportion to the diagonal conductivity.[8] These scaling functions are qualitatively the same as that calculated in the single-site approximation.

This scaling relation can be observed experimentally by studying temperature "flow diagram". This is possible since inelastic scatterings present at nonzero temperatures introduce an effective system size. A model calculation in which effects of excitations of electrons into delocalized states are also considered shows that the envelope of flow lines for different Fermi levels becomes the scaling function of the conductivities.[9] Kawaji has obtained the temperature flow lines in inversion layers on the Si surface and shown that they have an envelope line qualitatively in agreement with the theoretical result.[10] Wei et al. reported similar results at InGaAs/InP heterostructures.[11]

This work is supported in part by Grant-in-Aid for Special Distinguished Research from Japanese Ministry of Education.

References

1 see, for example, Ando, T., Prog. Theor. Phys. Suppl. **84**, 69 (1985).
2 Ando, T., J. Phys. Soc. Jpn. **52** (1983) 1893; **53** (1983) 3101; **53** (1983) 3126.
3 Ando, T. and Aoki, H., J. Phys. Soc. Jpn. **54** (1985) 2238.
4 Levine, H., Libby, S.B. and Pruisken, A.M.M., Phys. Rev. Lett. **51** (1983) 1915; Nucl. Phys. B **240** (1984) 30; 49; 71.
5 Ando, T, Surf. Sci. **170** (1986) 243.
6 Ando, T. and Uemura, Y., J. Phys. Soc. Jpn. **36** (1974) 959.
7 Ando, T., J. Phys. Soc. Jpn. **36** (1974) 1167; **37** (1975) 622; **37** (1974) 1233.
8 Ando, T., Matsumoto, Y. and Uemura, Y., J. Phys. Soc. Jpn. **39** (1975) 279.
9 Aoki, H. and Ando, T., Surf. Sci. **170** (1986) 249.
10 Kawaji, S., Prog. Theor. Phys. Suppl. **84** (1985) 178.
11 Wei, H.P., Chang, A.M., Tsui, D.C., Pruisken, A.M.M. and Razeghi, M., Surf. Sci. **170** (1986) 238.

CURRENT INSTABILITY IN THE QUANTUM HALL EFFECT

T.Takamasu, S.Komiyama, S.Hiyamizu[†] and S.Sasa[†]

Department of Pure and Applied Sciences, University of Tokyo,
Komaba, Tokyo 153, Japan
† Semiconductor Materials Laboratory, Fujitsu Laboratories, Ltd.,
Ono, Atsugi 243-01, Japan

Current instability emerges in a two dimensional electron system in GaAs-AlGaAs heterostructure devices at the i=2 Hall plateau when current density in the channel increases beyond critical levels in the temperature range T≲35K. The origin of the instability is identified as the occurrence of a theoretically predicted NDC peculiar to the currents with a Hall angle $\Psi \approx \pi/2$ in strong magnetic fields.

One of most striking features of the Quantum Hall Effect (QHE) is the presence of macroscopic currents perpendicular to the electric field. On the other hand, there is a theoretical prediction that, as the Hall angle Ψ of the current in a system approaches $\pi/2$, the system more and more easily falls into a regime of negative differential conductivity (NDC) to cause a current instability.[1] Since the prediction is derived independently of microscopic mechanisms of relevant transport phenomena, the QHE can not be free from the prediction and the macroscopic stability should be examined. We report here the first observation of this type of NDC and resulting current instability occurring in a quasi-two dimensional electron system in GaAs-AlGaAs heterostructure devices at the i=2 Hall plateau.

Samples used are GaAs-AlGaAs heterostructure devices with channel width d=50μm and the distance between Hall electrodes l=100μm as shown in Fig.1. The density of electrons is $n_s \approx 3.8 \times 10^{11} cm^{-2}$ with the Hall mobility $\mu_H \approx 4 \times 10^5 cm^2$/Vsec.

It is well established experimentally that, when the current density J (or the Hall electric field E_y) increases at low lattice temperatures, energy dissipation of the electron system in the regime of QHE increases drastically at a critical electric field to break the QHE.[2,3] In our earlier studies, we have explained the breakdown as a discontinuous rise in the electron temperature by analyzing the electron heating effect.[4,5] In the course of the studies, we have found new kind of current anomalies to emerge in addition to the breakdown at the i=2 Hall plateau and pointed out that these current anomalies must have a physical origin different from that of the breakdown phenomena.[5]

To give an outline of the new anomalies in the conductivity, we re-
illustrate in Fig.1 a part of the data reported in Ref.5. The rapid increase of σ_{xx} at E_y=110V/cm(J=8.5mA/cm) for low T_L's≲4.2K is the
breakdown of the QHE. In addition, several less remarkable structures
are also noted in the σ_{xx} vs. E_y curves: At elevated temperatures T_L≳14K,
the σ_{xx} curves have kinks at E_y≈70V/cm(J≈5.4mA/cm), above which the
σ_{xx} curves have characteristic structures though not discerned in Fig.1:
Also at lower lattice temperatures T_L≲1K, σ_{xx} has zigzag-like dependence on E_y in a higher E_y-range above about 120V/cm(J=9.3mA/cm), where
the breakdown has been accomplished. All these anomalous structures,
though not very remarkable in Fig.1, are well recognized as striking anomalies in direct recorder traces made in linear scale. To elucidate the
anomalies at elevated temperatures, Fig.2(a) displays recorder traces of
the dissipative electric field E_x=(V_1-V_3)/21 against the current density
J=I_{DS}/d (I_{DS}; the drain-source current),including also the data at higher
T_L's up to 44K. (since tanΨ≫1, E_x and J are approximately proportional
to $\sigma_{xx}J$ and E_y, respectively.) At the highest temperature T_L=44K, E_x increases sublinearly with J and the smooth increase continues up to the
highest J studied (12mA/cm) without exhibiting any anomaly, (though the
curve in the higher J-range can not be included in the figure). In the

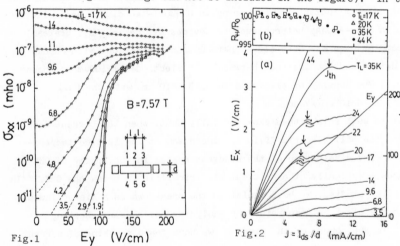

Fig.1 E_y (V/cm)   Fig.2 J = I_{ds} /d (mA/cm)

Fig. 1 σ_{xx} vs. E_y for the i=2 Hall plateau at different lattice
temperatures T_L's.
Fig. 2 (a) E_x and E_y against J. (b) The ratio of R_H=E_y/J and the
quantized value R_0=h/2e^2 vs. J.

temperature range, $35K{\geq}T_L{\geq}$
14K, on the other hand, E_x
curves exhibit anomalies at
critical values of J $(=J_{th})$
causing definite kinks in the

Table

T_L	(K)	17	20	22	24	35
J_{th}	(mA/cm)	5.4	5.8	6.0	6.3	9.0
E_{th}	(V/cm)	70	75	78	82	117
E_{NDC}	(V/cm)	70	75	76	80	116

curves, above which oscillatory structures follow. The values of J_{th} for
different T_L's are summarized in the table together with the corresponding E values; $E_{th}=J_{th}/\sigma_{xy}$. In the higher current region above J_{th}, measurements suffer large noises so that the oscillatory structures are over-
lapped with fine irregular structures not describable in Fig.2(a). The
shape of the oscillatory structures is therefore not reproduced in detail
for different runnings of J sweep. Furthermore the overall shape of the
structures systematically changes when the sweep direction of J is reversed as examples being shown for T_L=17K and 24K in Fig.2(a). (The other
data in Fig.2(a) are for the sweep direction of the J increase.) In addition to the fluctuations in time, spatial fluctuations of the potential in
the electron system were indicated in the range $J>J_{th}$ by additional voltage measurements at different Hall electrodes. All the anomalies in the
conductivity observed in $J>J_{th}$ for elevated temperatures $T_L{\geq}14K$ have been
equally observed also for lower temperatures $T_L{\lesssim}11K$ in the region of J
(or E_y) where similar structures appear ($J{\geq}8.5mA/cm$).

The Hall electric field $E_y=(V_2-V_5)/d$ is also shown with a straight
line in Fig.2(a), where the values of E_y for all the T_L's in the range
1.9K~44K agree with each other within the line width. Accurate measurements, however, reveal small but finite nonlinearities as elucidated in
Fig.2(b), where the ratio of the Hall resistance $R_H=E_y/J$ to the quantized
value $R_0=h/2e^2$ is shown with the common abscissa with Fig.2(a).

We wish to examine the origin of
the here observed instabilities. According to Ref.1, NDC and resulting instability emerge when a differential conductivity tensor σ_d of an electron
system has a negative determinant;
$\det\sigma_d$ 0, which, in case of $\tan\Psi{\gg}1$,
can be translated into the inequality

$$1-\alpha'-\beta'^2/4 < 0 \qquad (1)$$

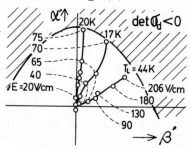

Fig. 3 Dependence of α' and
β' on E.

with $\alpha' \equiv \alpha \tan \Psi$ and $\beta' \equiv \beta \tan \Psi$. Here $\alpha \equiv \partial \Psi / \partial (\log E)$ and $\beta \equiv 1 - \partial (\log J) / \partial (\log E)$ are two parameters characterizing the nonlinearities of the system with the Hall angle Ψ, the magnitude of the current density J, and $E = (E_x^2 + E_y^2)^{1/2}$. (We can note from the relation (1) that, given certain nonlinearities α and β, the condition for NDC gets less stringent as $\tan \Psi$ increases.)

To examine the condition (1), α' and β' must be derived from the experimental data. Hence we have transformed the data of $E_x(J)$ and $E_y(J)$ as shown in Fig.2(a) and (b) into J(E) and $\Psi(E)$, and calculated α' and β' at several values of E. Since the measured E represents "the real E" only when the system is spatially homogenious, we limit the calculations to the lower E-range (or $J < J_{th}$), where the system remains stable. So derived values of α' and β' for different levels of E are marked with open circles on the $\alpha' - \beta'$ plane for several different T_L's in Fig.3. The parabola in the figure represents $1 - \alpha' - \beta'^2 / 4 = 0$, the region above which is the area where NDC is predicted. At every level of T_L, the point (α', β') departs from near the origin with increasing E, indicating an increasing nonlinearity with E. The points (α', β') for all the T_L's except 44K almost reach the boundary $1 - \alpha' - \beta'^2 / 4 = 0$ at certain electric fields (as will be designated as E_{NDC}), while the point (α', β') for 44K remains well within the stable region. In the table, values of E_{NDC} for respective T_L's are compared with E above which instabilities have been observed. The values of E_{TH} and E_{NDC} excellently agree with each other at all the temperatures listed. Furthermore, α' and β' remain within the stable region for $T_L = 44K$ (Fig.3), where the system exhibited no instabilities (Fig.2(a)). All these observations make us to conclude that the current instabilities appearing in the temperature range 14K~35K originate in the NDC predicted in Ref.1.

Although we can not definitely identify the origin of the similar instabilities at lower T_L's because of the difficulty in calculating α' and β', we tentatively suggest that the breakdown of the QHE, which drastically raises the electron temperature, drives the system into the regime of NDC described here.

(Reference)
1) Kurosawa,T.,Maeda,H. and Sugimoto,H.,J.Phys.Soc.Japan,36,491(1974).
2) Ebert,G.,Klitzing,K.v.,Ploog,K.,Weimann,G.,J.Phys.C16,5441(1983).
3) Cage,M.E.,Dziuba,R.F.,Field,B.F.,Williams,G.,Girvin,S.M.,Gossard,A.C. Tsui,D.C. and Wagner,R.J.,Phys.Rev.Lett.51,1374(1983).
4) Komiyama,S.,Takamasu,T.,Hiyamizu,S.,Sasa,S.,Solid State Com.54,479 (1985).
5) Takamasu,T.,Komiyama,S.,Hiyamizu,S.,Sasa,S.,Surf.Sci.170,202(1986).

FIR PHOTOCONDUCTIVITY IN THE INTEGRAL QUANTUM HALL REGIME IN GaAs/AlGaAs

M.J. Chou[†], D.C. Tsui[†] and A.Y. Cho[*]

†Department of Electrical Engineering, Princeton University, Princeton, NJ 08544
*AT&T Bell Laboratories, Murry Hill, NJ 07974

We report results on the far infrared photoconductivity of high mobility (μ=3 to 8×10^5cm^2/Vsec) 2D electrons in GaAs-AlGaAs. The cyclotron resonance (CR), when it is observed in the integral quantum Hall regime, is a sharp peak corresponding to an increase in ρ_{xx}. At lower B, in the classical magneto-transport regime, it manifests as an enhancement of quantum oscillations, riding on a CR background.

The magnetic field (B) dependence of the far infrared (FIR) photo-response of two-dimensional (2D) electrons has been the subject of several recent investigations[1-5]. In the early work on Si, only quantum oscillations[1] and an enhancement of their amplitude at the cyclotron resonance (CR)[2] were observed. In GaAs/AlGaAs, a sharp CR was observed by Maan et al[3]. It was demonstrated very recently by Horstman et al[5] that a well defined CR can also be extracted from their data, which are dominated by quantum oscillations due to electron heating. They attributed the CR to heating by the resonantly absorbed FIR.

In this paper, we report our results from GaAs-AlGaAs heterostructures with a considerably higher 2D mobility (μ=3 to 8×10^5cm^2/Vsec) in the 2D density range n_s=2 to 5×10^{11}/cm^2. We were able to vary the n_s of each sample continuously by the use of a semitransparent metal gate and

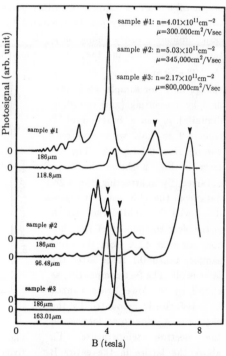

Fig.1: The photoresponse as a function of magnetic field for three samples and several laser wavelengths λ. The arrows indicate the cyclotron resonance (CR).

the photo-response was measured at a large number of fixed laser wavelengths (λ).

We find that for short λ, where CR is in the integral quantum Hall (IQH) regime, sharp CR are observed. The CR signal is an increase in the diagonal resistivity tensor (ρ_{xx}) regardless of the position of the Fermi level (E_F). This result demonstrates that electron heating is not the dominant cause for the CR. Moreover, structures not attributable to quantum oscillations are observed. Such structures, as well as the strength of the CR, show strong dependences on the bias current, the position of E_F, and the sample geometry. For long λ, when the CR occurs at lower B in the classical magneto-transport regime, the CR manifests as an enhancement in the quantum oscillations riding on a CR background.

Fig. 1 shows the data, taken with sufficiently short λ lasers to place the CR in the IQH regime (B≳3T), from three different samples. In the low n_s sample (#3), the CR is a sharp resonance on a flat background, but its width is ∼6 times broader than that from the absorption measurement. In the high n_s samples (#1 and #2), the photoresponse is more complicated, but the CR peaks are still discernible. The signal in general has a fast component, with a decay time $\tau < 10$ μsec, and a slow component with $\tau > 1$ msec. They can be separated by pulse measurements. We attribute the slow component to the change in ρ_{xx} due to increases in the lattice temperature under the laser irradiation and the fast component to some electronic processes.

The change in ρ_{xx} due to a change in T for sample #2 is studied by measuring (and recording digitally) ρ_{xx} as a function of T from 4.2 to 7.5K with a current varying from 0.1 to 5μA. The top trace in Fig. 2 shows the $\Delta\rho_{xx}$, obtained by subtracting the 4.2K data from the 6.0K data. It reproduces well the middle trace, which is the slow component of the photoresponse at 4.2K from the same sample, taken 0.1 msec after the laser is off. The fast component, as shown by the lower trace, cannot be satisfactorily accounted by the change in ρ_{xx} due to an increase in the electron temperature, T_e, above the lattice in the entire B range. For B≲2T, in the classical magneto-transport regime, the data can be explained by assuming $T_e - T = 0.5$K (at an FIR intensity of ∼10mW/cm²). In the IQH regime, the data appear more

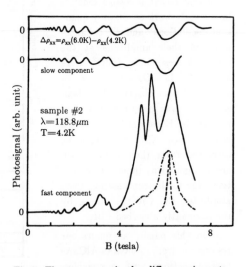

Fig.2: The top trace is the difference in resistivity, $\Delta\rho_{xx}$, between 6.0 and 4.2K. The middle and lower traces are the slow and fast components of photoresponse, respectively. The dashed curve is the CR, measured in transmission, and the normalized photoresponse is shown as the dash-dot curve.

than the calculated $\Delta\rho_{xx}$. Since the CR, measured by absorption (dashed curve), is close to a ρ_{xx} maximum of the N=2 Landau level in this sample, simple electron heating by resonant absorption of FIR is expected to produce a sharp dip at ∼6T. Instead, a peak is observed. If we assume that the data above 4T is a CR peak superimposed on a bolometric background with peaks and dips, similar to the quantum oscillations in the top trace, and normalize the data by ($\Delta\rho_{xx}$+constant), we obtain the dash-dot curve in Fig. 2. The resulting CR peak is at the expected B position, but ∼4 times broader than the CR measured in absorption, and not a dip, as expected from simple electron heating. Moreover, additional structure, not due to CR or quantum oscillations are observed in some samples. An example is the data shown in Fig. 3.

Fig. 3 illustrates the complicated bias current (I_b) dependences of the photoresponse in the IQH regime. For example, a huge resistance peak is observed at a Landau level filling factor,

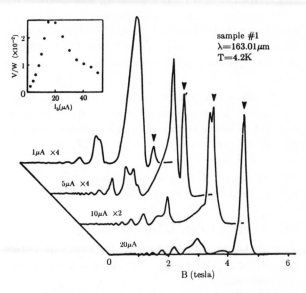

Fig.3: The I_b dependences of photoresponse showing additional structures at ∼3T. The inset shows the photoresponse at the CR as a function of I_b.

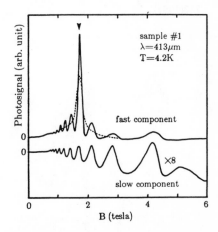

Fig.4: Photoresponse for long λ. The upper and lower traces are fast and slow components, respectively. The dashed curve is the background, drawn through the half way points of the oscillations.

$\nu \equiv n_s h/eB = 4.32$ (h is Plank's constant), with $I_b = 1\mu A$. With increased I_b, additional structures appear and this peak becomes sharper and moves closer to the CR. It merges with the CR peak at $I_b \simeq 20\mu A$. The I_b dependence of the CR peak is shown in the inset. It reaches a maximum of $2.4 \times 10^{-2} V/W$ at $I_b \simeq 15\mu A$ and starts to decrease at higher I_b when parallel conduction is observable in the quantum Hall measurements.

Finally, Fig. 4 shows the photoresponse for long λ, when the CR is at lower B in the classical magneto-transport regime. While the slow component can be well described by lattice heating, the fast component shows an enhancement of the heating induced quantum oscillations at the CR, riding on a background (the dashed curve), which is the CR.

We thank C.H. Yang for helpful suggestions. The work at Princeton University is supported by the Air Force Office of Scientific Research.

(1)   Y. Shiraki, J. Phys. C **10**, 4549(1977).

(2)   C.F. Lavine, N.J. Wagner, and D.C. Tsui, Surf. Science **113**, 112 (1982).

(3)   J.C. Maan, Th. Englert, D.C. Tsui, and A.C. Gossard, Appl. Phys. Lett. **40**, 609(1982).

(4)   D. Stein, G. Ebert, K. von Klitzing, and G. Weimann, Surf. Science **142**, 406(1984).

(5)   R.E. Horstman, E.J. v.d. Broek, J. Wolter, R.W. van der Heijden, G.L.J.A. Ridden, H. Sigg, P.M. Friflink, J. Maluenda, and J. Hallais, Solid State Comm. **50**, 753(1984).

MAGNETOCONDUCTIVITY OF INVERSION ELECTRONS IN PERIODIC MOS-MICROSTRUCTURES ON SI

M. Wassermeier, H. Pohlmann, and J. P. Kotthaus

Institut für Angewandte Physik, Universität Hamburg
Jungiusstraße 11, 2000 Hamburg 36
F. R. GERMANY

ABSTRACT

In lateral MOS-microstructures on Si(100) in which a periodic wire grid of many ($\leq 10^4$) narrow NiCr stripes with periodicity a$\geq$200 nm serves as gate we study the two-terminal conductance of inversion electrons at low temperatures and high magnetic fields. Giant maxima in the conductance are observed whenever the Fermi energy lies between two Landau levels.

MOS-microstructures on Si with periodical lateral modulation of the inversion electron density n_s exhibit novel electronic phenomena if the lateral length scale reaches submicron dimensions[1,2]. Here we report two-terminal conductance studies of large area MOS-structures fabricated on oxidized p-Si(100) ($d_{ox} \approx 20$ nm) in which the gate electrode consists of a periodic sequence of narrow NiCr-stripes of width $t \approx a/2$ and periodicity a down to 200 nm. In such microstructures many ($\leq 10^4$) parallel and very narrow inversion channels are formed at low gate voltages and low temperatures. We measure the quasistatic two-terminal conductance along the inversion channels at radio frequencies ($\nu \approx 20$ MHz) using large area capacitive source-drain contacts[3] on both sides of the wire grid gate which has a total width of 2 mm and length of 0.5 mm.

Typical conductance I/V vs. gate voltage curves at low temperatures and magnetic fields of up to 12 Tesla are shown in Fig.1 for a sample with a=800 nm consisting of $N \approx 2500$ channels. The B=0 conductance behaves as in unstructured samples and yields a peak mobility of about 3000 cm^2/Vsec. At high fields we observe giant Shubnikov-de Haas oscillations with conductance maxima whenever the Fermi level lies in the center between two Landau levels, i.e. whenever the filling factor $\nu = n_s \cdot h/(e \cdot B) = 4n$ (n=1,2,3,... Landau level index). The absolute value of I/V at $\nu = 4n$ depends exponentially on B as illustrated in Fig.2 for n=1 and two samples with different a. It also increases with decreasing source-drain voltage V. It exceeds by far the quantized Hall conductance $\sigma_H = (e^2 \cdot \nu)/h$ that is found as the limiting conductance of two-terminal measurements in unstructured MOS-samples[4].

The surprising maxima in I/V can be described phenomenologically as a consequence of the fact that the Hall conductance σ_H does not depend on sample dimensions whereas the dissipative conductance does. We start by assuming that in single narrow inversion channels of length L and width W the two-terminal voltage across the sample is $V = V_x + V_H$ as found in Ref.4 with $V_H = \rho_{xy} \cdot I$ and $V_x = (L/W) \cdot \rho_{xx} \cdot I$. For N parallel inversion channels we then calculate the conductance as $I/V = N(\sigma_{xx}^2 + \sigma_H^2)/((L/W)\sigma_{xx} + \sigma_H)$. At low temperatures and filled Landau levels, i.e. in the quantum Hall regime, σ_{xx} approaches zero approximately as $\sigma_{xxmin} \propto \exp(-(\hbar\omega_c - 2\Gamma)/2kT)$ where $\omega_c = eB/m^*$ and 2Γ is the width of the region of extended states within a given Landau level. If one has $\sigma_{xx} \ll \sigma_H$ and $(L/W) \cdot \sigma_{xx} \ll \sigma_H$ one may expect $I/V = N \cdot \sigma_H$ and recovers the result of Fang and Stiles[4] for N=1. If however $\sigma_{xx} \ll \sigma_H$ and $(L/W) \cdot \sigma_{xx} \gg \sigma_H$ one expects $I/V = N \cdot (W/L) \cdot \sigma_H^2/\sigma_{xx} \propto 1/\sigma_{xxmin}$, i.e. conductance exponentially rising with magnetic field in qualitative agreement with our observations.

This simplified phenomenological model that assumes homogeneous field distributions in the sample leaves many questions unanswered. If, e.g., we study a sample with a=20 μm ($N \approx 100$), we do not observe the $\nu = 4n$ conductance to rise above its B=0 value. This may be interpreted as capacitive shortening of the Hall voltage via the gate. But then it is unclear why this does not occur at values of a below 1 μm. Also for N·W/L$^\alpha$const. as in our samples one would not expect I/V to increase with increasing N or decreasing W. Finally, the model cannot explain the dependence of I/V on V. In Fig.3 we display the ν=4n conductance normalized to the B=0 conductance at same n_s vs. V. Within our range of measurement the conductance at ν=4n rises in proportion to 1/V, meaning the current is I=b·V+c (b,c=const) and has a dominant voltage independent part. This suggests that at ν=4n there is a large dissipationless current in parallel to a relatively small ohmic current. From this and the fact that the absolute current rises about in proportion to the number of wires per unit length, one may speculate that the dissipationless current is an edge current, quite in contrast to most of the present models on the quantum Hall effect.

In conclusion we observe that in microstructured MOS-samples a large dissipationless current contribution occurs in two-terminal measurements whenever the Fermi level is located between   two Landau levels, i.e. in the region of "localized states". It increases with the number of inversion channels per unit crossectional length. We hope that our results may provide a clue as to whether the dissipationless current in the quantum Hall regime is carried by bulk states or edge states. However, a deeper theoretical understanding is necessary to unambiguously explain our surprising observations.

We thank J. Hajdu and B. Kramer for stimulating discussions and acknowledge financial support of the Stiftung Volkswagenwerk.

1. Mackens, U., Heitmann, D., Prager, L., Kotthaus, J. P., and Beinvogl, W., Phys. Rev. Lett. 53, 1485 (1985).
2. Warren, A. C., Antoniadis, D. A., and Smith, H. I., Phys. Rev. Lett. 56, 1858 (1986).
3. Dolgopolov, V., Mazuré, C., Zrenner, A., and Koch, F., J. Appl. Phys. 55, 4280 (1984).
4. Fang, F. F., and Stiles, P. J., Phys. Rev. B27, 6487 (1983).

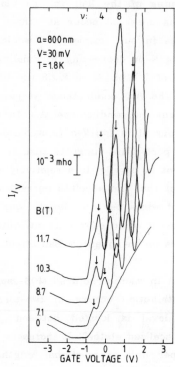

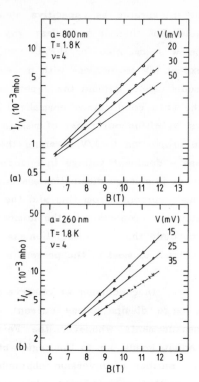

Fig. 1: Two-terminal conductance I/V vs. gate voltage in a periodic MOS-microstructure.

Fig. 2: Conductance I/V at filling factor $\nu=4$ vs. magnetic field B. (a) for the sample of Fig.1, (b) for a sample with a=260 nm but same total dimensions as in (a).

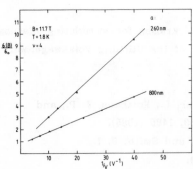

Fig. 3: Normalized conductance $\sigma(B)/\sigma_0$ vs. reciprocal source-drain voltage V for two samples.

A NEW APPROACH TO THE QUANTUM HALL EFFECT

R. Woltjer, R. Eppenga and M.F.H. Schuurmans

Philips Research Laboratories
5600 JA Eindhoven, The Netherlands

ABSTRACT

Both the Quantum Hall effect and the measured distribution of the Hall voltage across a Hall bar can be explained by a moderate inhomogeneity in the electron density across the sample and by the introduction of a local resistivity tensor which does <u>not</u> account for the effects of localization but <u>does</u> depend on the local electron density.

A new model for the explanation of the Quantum Hall effect is presented. We show that a moderate spatial inhomogeneity in the electron density n across the sample already leads to the Quantum Hall effect <u>without invoking the localization</u> corresponding to such an inhomogeneity. The existence of such an inhomogeneity has been demonstrated experimentally and a systematic inhomogeneity has been created by using a tilted gate on a sample[1].

In our model the inhomogeneous 2D electron gas, $n=n(x,y)$, is described by a local resistivity tensor $\hat{r}(x,y)$, with diagonal magnetoresistance components r_{xx} and $r_{yy}(=r_{xx})$ and off-diagonal Hall resistance components r_{xy} and $r_{yx}(=-r_{xy})$; the coordinates x and y will refer to the long and small axis of the rectangular Hall bar. The distribution of the current density $\vec{J}=(j_x,j_y)$ across the sample is determined by Kirchhoff's laws. Measured voltages are determined by integrals involving $\hat{r}(x,y)$. For example the Hall resistance is given by

$$R_{xy}(x) = \int (\hat{r}(x,y)\vec{j}(x,y))_y \, dy \; / \int j_x(x,y) \, dy \; .$$ (1)

Expressions for the local resistivity tensor of the inhomogeneous 2D-electron gas are borrowed from the corresponding expressions from the homogeneous gas. Ignoring the effects of scattering on the Hall resistivity we obtain

$$r_{xy} = h/ie^2; \quad i = hn(x,y)/eB$$ (2)

with i the filling factor of the Landau levels. Elastic scattering in the Born approximation gives for the magnetoresistivity the expression[2]

$$r_{xx} \propto \sum_k (k+1/2)(eB/h)^2 \; f(E_k)(1-f(E_k)),$$ (3)

where the sum runs over the Landau levels at E_k and the chemical potential μ in the Fermi-Dirac distribution function $f(E)$ is determined by the requirement that the sum of $f(E_k)$ over k equals the filling factor i; i.e. $\mu=\mu(n(x,y)/B)$. Note that in the derivation of (3) the spin-splitting and the broadening of the Landau levels is disregarded.

The Hall resistivity as described by (2) gives a straight line as a function of $i^{-1}=eB/hn$ and the magnetoresistivity as described by (3) has thermally activated minima as a function of i^{-1}; see Fig. 1. The spatial inhomogeneity is completely described by the dependence on $n(x,y)/B$. The resistivity tensor chosen does not contain the effect of localization of electron states, i.e. the Hall resistivity has no staircase-like structure and the magnetoresistivity does not show broad minima around the integral filling factors. Nevertheless, our model leads to the phenomena observed in the Quantum Hall regime. If $r_{xx}(B/n(x,y))$ (almost) goes to zero for integral filling factors i_0, Kirchhoff's laws tell us that (almost) all current will flow along paths with $hn(x,y)/eB=i_0$. In view of the inhomogeneity in the 2D-gas there will be a range of magnetic field values B at which such current carrying paths exist between the current contacts. The Hall resistance as described by (3) is then dominated by contributions from these

paths and R_{xy} becomes equal to $r_{xy}(i=i_0)$. We obtain plateaus in the
Hall resistance and the magnetoresistance is almost zero over the same
range of B-values!

RESULTS

We have studied the consequences of the inhomogeneity in a simple
1D approach[2], where $n(x,y)=n(y)$ and therefore $j_y=0$. Instead of
considering the variation of n as a function of y we specify the
distribution of n values as they may appear across the sample. The
distribution is taken to be Gaussian, with mean electron density n_0
and spread dn. Results for a relative spread $a=dn/n_0$ of 3% are given
in Fig. 2. The Hall resistance R_{xy} indeed shows broad plateaus and the
magnetoresistance R_{xx} shows the corresponding broad minima. At B=5
Tesla and T=6 Kelvin the plateau is flat to within 0.3% over 4% change
in the magnetic field; At T=2 Kelvin to within 10^{-6}%. Consistent with
experiment the minima in R_{xx} tend to zero exponentially with $T \rightarrow 0$.
Calculations at different temperatures show that the minima in R_{xx} and
dR_{xy}/dB at the B values where the minima occur, are linearly
interrelated as shown in Fig.3. Such behavior is consistent with
experimental results of Tausendfreund and von Klitzing[3].

Systematic inhomogeneity across the width of the sample can be
dealt with by appropriately choosing a y-dependence of the mean n_0,

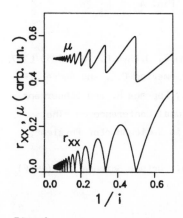

Fig. 1

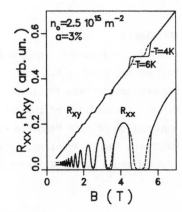

Fig. 2

keeping the relative spread $a=dn/n_0$ constant. Fig.4 gives the results
for a linear gradient of 6% in $n_0(y)$ and $a=2\%$. The results compare
very well with the experimental results of Ebert et al.[1]. For further
details on our model and the results we refer to ref. 2.

In conclusion we have shown that it is possible to explain (i)
the Quantum Hall effect and (ii) the distribution of the Hall voltage
across a sample in terms of a local resistivity tensor which does not
exhibit the effects of localization. In a particular experimental
situation it therefore remains to be decided to what extent the
various effects observed are due to localization or spatial
inhomogeneity in the electron density.

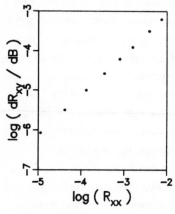   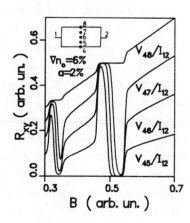

Fig. 3   Fig. 4

REFERENCES

1.   Ebert G., Klitzing K. von and Weimann G, J.Phys.C. 18 (1985) L257
2.   Woltjer R., Eppenga R., Mooren J., Timmering C.E. and André J.P.,
      Europhysics Letters 1986; Woltjer R., Eppenga R. and Schuurmans
      M.F.H., Proceedings of the International Conference on "The
      Application of High Magnetic Fields in Semiconductor Physics",
      Wurzburg, August 1986; ed. Springer.
3.   Tausendfreund B. and Klitzing K. von, Surface Science 142 (1984)
      220.

NOVEL PROPERTIES OF A NEW PHASE OF A TWO DIMENSIONAL ELECTRON GAS AT A SEMICONDUCTOR HETEROJUNCTION

E.J. Pakulis

IBM Thomas J. Watson Research Center
P.O. Box 218, Yorktown Heights, NY 10598
USA

ABSTRACT

New results on the condensed phase of a two dimensional electron gas at a GaAs/Ga$_{0.77}$Al$_{0.23}$As heterojunction are presented. An energy gap at the Fermi surface is directly observed in a microwave absorption experiment. Also observed are a high conductance condensed phase and switching behavior which is strongly influenced by light and by an applied magnetic field.

A phase transition in the two dimensional electron gas (2DEG) at a GaAs/Ga$_{0.77}$Al$_{0.23}$As heterojunction, exhibiting both hysteresis and critical magnetic field (H) behavior, was recently reported[1] and attributed to mixed charge/spin density wave[2] formation. The transition was observed only in samples in which the 2DEG forms in the semi-insulating GaAs substrate material, as opposed to an undoped epitaxial GaAs buffer layer. The transition to the condensed phase at a critical temperature T_c is accompanied by a sharp drop in electrical conductance parallel to the plane of the 2DEG. T_c is sample dependent, and values in the range 2-10K have been observed. For $T < T_c$ an activated component of conductance was observed, consistent with an energy gap at the Fermi surface. A gap of 9K (in the limit $T = 0$ and $H = 0$) was inferred for a sample with $T_c = 3.2$K.

The energy gap at the Fermi surface for $T < T_c$ can be directly observed in a microwave absorption experiment. The presence of a sharp microwave absorption threshold when the microwave energy becomes equal to the gap energy is experimental evidence for the existence of a gap. In this experiment, microwaves of constant energy equal to 1.1K were used, and the gap was tuned to the microwave energy by sweeping the magnetic field. This was possible because both T_c and the energy gap decrease[1,3] with increasing H. The microwave absorption was detected by measuring the microwave photoconductivity signal, since microwave excitation of electrons across the gap is accompanied by a change in the dc conductance of the 2DEG. Fig. 1 shows the microwave photoconductivity signal as a function of H for T = 1.6K. Curve (a) is for H perpendicular to the junction. The microwave related signal increases steeply to a maximum near 110 Gauss, 20 Gauss below the critical field for destruction of the condensed phase, H_c = 130 Gauss. For H>H_c, there is no longer a gap, and the microwave photoconductivity drops back to a low background level. Fig. 1(b) is the microwave photoconductivity signal as a function of H parallel to the junction. In the parallel configuration, H_c was 150 Gauss, and

no microwave absorption threshold was observed. This means that for parallel fields the energy gap is not tuned smoothly through the microwave energy as H is increased, but rather, drops discontinuously to zero from a value larger than the microwave energy. Fig. 2 is a schematic representation of the dependence of the energy gap on H for (a) a first order transition and (b) a second order transition. The horizontal dashed line shows the position of the microwave energy. It is clear that for case (b), the gap can be tuned smoothly through the microwave

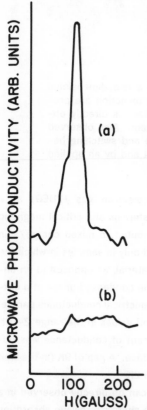

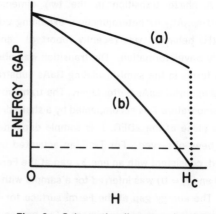

Fig. 1 The microwave induced change in dc conductance as a function of H for T = 1.6K. (a) H normal and (b) H parallel to the plane of the junction. The curves have been shifted vertically, but the scale is the same.

Fig. 2 Schematic diagram showing variation of energy gap with H for (a) a first order transition and (b) a second order transition. The horizontal dashed line represents the microwave energy relative to the gap energy. H_c is the critical field.

energy, whereas for case (a) it cannot. The microwave results are thus consistent with a transition first order in parallel H and second order in perpendicular H.

As reported in Ref. 1, in most samples the conductance decreases below the transition temperature. This is the behavior normally observed in charge/spin density wave transitions, due to pinning of the waves by impurities or by the crystal lattice itself. In some of the GaAs/GaAlAs samples, however, abrupt increases in conductance are observed on entering the condensed phase. Fig. 3 shows an example of such an increase, where the step size is approximately 1%. The increase in conductance suggests the absence of pinning and can be considered similar in some respects to a superconducting transition. The analogy to superconductivity will be discussed in detail elsewhere.[4]

The same sample exhibiting the conductance step in Fig. 3 also showed switching behavior. Switching phenomena, discontinuous jumps in conductance as a function of voltage, are often seen in charge/spin density wave systems above some threshold voltage. Fig. 4(a) shows

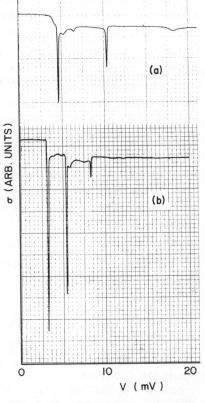

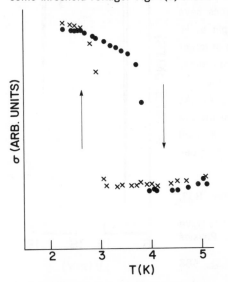

Fig. 3 Conductance as a function of T. The size of the step is approximately 1%. The arrows indicate the direction in which the data points were taken. H = 0.

Fig. 4 The differential conductance as a function of dc voltage in the condensed phase (a) before and (b) after illumination with white light. Both curves have the same scale, but have been shifted vertically.

the differential conductance as a function of applied dc voltage. The multiple switches are attributed to multiple domains of condensate, each switching at a different threshold voltage. The mechanism for the switches is not known, but may involve the abrupt destruction of the condensed phase above some threshold voltage. Fig. 4(b) is data taken after illumination of the sample with white light. Illumination clearly has a pronounced effect on the observed domain structure, as the number, sharpness, and positions of the switches change. This may be due to the change in carrier density. The domain structure also changes after heating the sample above T_c and recooling.

Fig. 5 shows the peculiar effect of an applied H on one switch. For H = 0, the switch occurred at a voltage V = 18.8 mV. As H is increased from zero, the threshold voltage gradually decreases. In curve 5(a), the threshold voltage has been reduced to V = 17.5 mV by a field H = 120 Gauss. A spontaneous splitting of the domain occurs at a field H = 240 Gauss, and the two resultant domains can be seen in curve 5(b). The mechanism for the splitting is not understood, but the following characteristics have been observed: not all domains are split by H, domains which split sometimes coalesce at larger H, and the splitting is reversible and reproducible.

In summary, several new results on the condensed phase of a 2DEG at a GaAs/Ga$_{0.77}$Al$_{0.23}$As heterojunction have been obtained. These include the observation of an energy gap, a high conductance condensed phase, and switching behavior.

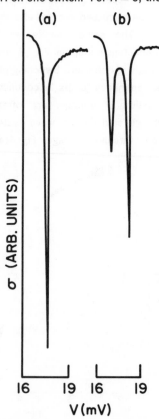

Fig. 5 The differential conductance as a function of dc voltage of a single domain (a) before and (b) after splitting of the domain at H = 240 Gauss.

1] Pakulis, E.J. and Fang, F.F., Phys. Rev. B33, 2916 (1986).
2] For a recent review of the charge density wave phenomenon see Gruner, G. and Zettl, A., Physics Reports 119, 117 (1985).
3] Pakulis, E.J., Bull. Am. Phys. Soc. 31, 588 (1986).
4] Pakulis, E.J., to be published.

ACCURATE CALCULATION OF [001] QUANTUM WELL CONFINEMENT STATES AND THE CORRESPONDING GAIN/ABSORPTION SPECTRUM

R. Eppenga , M.F.H. Schuurmans and S. Colak

Philips Research Laboratories
5600 JA Eindhoven
The Netherlands

ABSTRACT

We present calculated electron and hole confinement states and energies of a GaAs/AlGaAs [001] quantum well as well as the calculated optical matrix elements between the hole and the electron states as a function of $\vec{k}_\parallel$. Excellent agreement with recent tight binding results of Chang and Schulman[1] is found. We also present the calculated absorption and gain spectra.

MODEL

A $\vec{k}.\vec{p}$ envelope function approach for the calculation of electron and hole confinement states in a GaAs/AlGaAs [001] quantum well has been developed. In the Γ-point basis set we treat the electron (el), light hole (lh), heavy hole (hh) and the spin—orbit split-off (so) bands exactly and all other bands perturbatively. Apart from the gap and the spin orbit splitting the resulting model contains 6 parameters for each material. They may be taken to be the el, lh, hh and so effective mass m_{el}, m_{lh}, m_{hh} and m_{so} at the center Γ of the Brillouin zone along [001] , the hh effective mass m'_{hh} at Γ along [111] and a parameter B which describes the inversion asymmetry of each material.

The virtues of the model lie in its simplicity, the unified description of the valence and the conduction bands and the correspondingly unique choice of adjustable parameter values. These virtues allow for a sensible comparison with experimental excitation spectra.

Our $\vec{k}.\vec{p}$ model extends on that of Altarelli[2] in that we do take into account the non-parabolic behavior of the so band and the mixing of the lh and so band, even at $\vec{k}_{\shortparallel} = \vec{0}$. The vector $\vec{k}_{\shortparallel}$ describes the translational symmetry in the parallel plane of the quantum well.

RESULTS

Results for confinement energies at $\vec{k}_{\shortparallel} = \vec{0}$ have been published[3]. We present the calculated confinement energies and optical matrix elements between hole and electron confinement states as a function of $\vec{k}_{\shortparallel}$ for a GaAs/Al$_{.25}$Ga$_{.75}$As quantum well of width 192 Å. The parameters used are listed in Table 1 and are chosen to accurately reproduce the bulk band structures of GaAs and Al$_{.25}$Ga$_{.75}$As as obtained by Chang and Schulman[1]. For a complete correspondence to their work we also take the ratio of conduction to valence band discontinuity to be 85/15.

Our results, as shown in Figs. 1 and 2, are in excellent agreement with the results of the more complicated tight binding calculations by Chang and Schulman[1]. This is particularly gratifying since their model employs exact boundary conditions. In a $\vec{k}.\vec{p}$ envelope

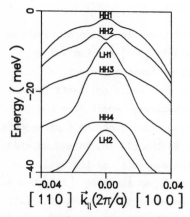

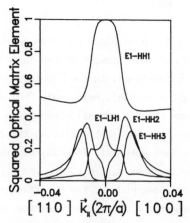

Fig. 1. Hole confinement energy spectrum for a GaAs/Al$_{.25}$Ga$_{.75}$As quantum well of width 192Å; a is the lattice constant.

Fig. 2. Optical matrix elements for light polarization in the quantum well plane.

function approach the alledged continuity of the rapidly varying u part of the true wavefunction is an approximation.

Fig. 2 shows that the so-called n=1 electron (E1) to n=2 and n=3 heavy hole (HH2, HH3) transitions may occur in the excitation spectrum: for $\vec{k}_{\parallel} \neq 0$ oscillator strength is borrowed from the n=1 electron to n=1 light hole (LH1) transition because of light hole-heavy hole admixture. Our $\vec{k} \cdot \vec{p}$ model is thus able to explain the recently observed n=1 to n=2 and n=3 transitions[4].

We have applied our model to the calculation of absorption (Fig. 3), gain (Fig. 4) and spontaneous emission spectra of a quantum well. These spectra exhibit the net effect of all such important aspects as the strongly non-parabolic behavior with $\vec{k}_{\parallel}$ of the hole confinement energies, upward curvature of various hole bands leading even to a singularity in the joint density of states and the $\vec{k}_{\parallel}$ dependence of the optical matrix elements. Note that the narrow peaks in Fig. 3 result from these singularities and should not be confused with exciton peaks. Two-particle interactions have not yet been added to the calculations, but this can be done easily. It is interesting to note that although both the energy spectrum and the optical matrix elements vary strongly with $\vec{k}_{\parallel}$, the absorption spectrum and the gain

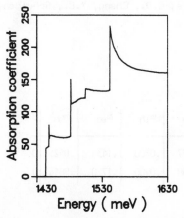

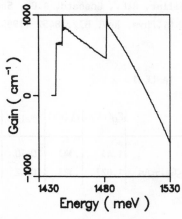

Fig. 3. Absorption spectrum for a GaAs/Al$_{.25}$Ga$_{.75}$As quantum well of width 192Å at 0 K.

Fig. 4. Gain spectrum. The pumped carrier density is taken to be 2.5×10^{18}cm^{-3} and the temperature is 300 K.

spectrum show more or less the step-like structure as expected from simple models which assume constant density of states (equivalent to bands parabolic in $\vec{k}_\parallel$) and constant optical matrix elements. However, the confinement energies and the step heights are different. Moreover, the plateaus in between the steps are not flat and narrow peaks occur near the edges of the steps. Modelling of our spectra by step-like ones from simple models requires at least adaption of the confinement energies and step heights. For example, the density of states of the first heavy hole subband requires the use of a light hole like mass because of subband mixing. We have also used the calculated spectra to obtain a gain/current relation for a quantum well laser. Such results may be pertinent for the description of lasers that operate on band to band transitions.

REFERENCES

1. Chang, Y.C. and Schulman, J.N., Phys. Rev. B31, 2069 (1985).
2. Altarelli, M., Physica 117B/118B, 747 (1983) and J. Lumin. 30, 472 (1985).
3. Schuurmans, M.F.H. and 't Hooft, G.W., Phys. Rev. B31, 8041 (1985)
4. Miller, R.C., Gossard, A.C., Sanders, G.D., Chang, Y.C., Schulman, J.N.,Phys. Rev. B32, 8452 (1985).

TABLE 1

	E_g(eV)	Δ (eV)	m_{el}	m_{hh}	m_{lh}	m_{so}	m_{hh}	B
GaAs	1.43	.343	.0670	.4537	.0700	.1434	.8525	0
Al.$_{25}$Ga.$_{75}$As	1.823	.327	.0942	.5100	.0900	.1720	.9814	0

EVIDENCE FOR A TEMPERATURE DEPENDENCE OF TWO-DIMENSIONAL ELECTRON GAS CONCENTRATION IN GaAs-GaAlAs HETEROSTRUCTURES

J. Beerens[+o], G.Grégoris[+], J.C. Portal[+], F. Alexandre[*] and M.J.Aubin[o]

[+]INSA-CNRS Dép. de Physique, Av. de Rangueil, 31077 Toulouse, France.
and SNCI-CNRS, B.P. 166X, 38042 Grenoble Cedex, France.

[*]CNET, 196 avenue Henri Ravera, 92220 Bagneux, France.

[o]Dép. de Physique, Université de Sherbrooke, Sherbrooke, Québec J1K2R1.

ABSTRACT

We report transport measurements under hydrostatic pressure on GaAs-GaAlAs heterojunctions in the temperature range $150<T<300K$. Our results show that the 2DEG concentration is temperature dependent; we interpret this as being due to a variation of the Si donor activation energy with T in GaAlAs.

1. INTRODUCTION

Measurements at 4.2K have recently demonstrated that hydrostatic pressure is an interesting tool for the study of GaAs-GaAlAs hetero-structures[1], since it provides a way to modify the two-dimensional electron gas (2DEG) concentration N_s without changing sample. This behavior is related to the pressure-induced deepening of the donor level in GaAlAs, which causes a reduction of the charge transfer to the quantum well. We report here an extension of this study to higher temperatures, where parallel conduction is most often non-negligible and complicates seriously the results interpretation. We restricted our measurements to the temperature range $150<T<300K$ where there is no persistent photoconductivity effects, and therefore the usual thermal equilibrium statistics can be used to describe the occupancy of the donor.

2. EXPERIMENTAL

We have performed Hall measurements under hydrostatic pressure up to 1.5GPa using the liquid pressure cell technique. Samples used were Hall-bridge shaped, high quality GaAs-$Ga_{0.72}Al_{0.28}As$ modulation doped heterostructures MBE grown on semi-insulating GaAs substrates and con-

sisting of a 1μm unintentionally doped GaAs buffer layer covered with
an undoped GaAlAs spacer layer (100 to 400Å) and a Si-doped (7 $10^{17}cm^{-3}$)
GaAlAs layer (500Å). Mobility at 4.2K was of the order of 3 to 4 10^5
cm^2/Vs. We have also studied bulk GaAlAs layers grown in the same conditions as the heterostructures. The Si-doped layer was 3μm thick, and
a large undoped spacer (0.3μm) was grown between the GaAs buffer and
the doped GaAlAs in order to avoid charge transfer to the interface.

3.   RESULTS AND DISCUSSION

Typical results obtained with an heterostructure are shown in
Fig.1 as a function of pressure and for different temperatures. At low
pressure, the decrease rate of n_H depends strongly on T. We attribute
this behavior to the contribution of the GaAlAs layer to the conduction
which increases with T because of thermal excitation of the donors.
Pressure causes a deepening of the donor level which brings a fast
decrease in the GaAlAs free electron concentration. The 2DEG concentration decreases less rapidly and, when P is sufficiently high, we can
therefore obtain a situation where the parallel conduction becomes
negligible. This is what happens for higher values of P in Fig.1: the
decrease rate is reduced and much less sensitive to temperature. The
value of the mobility (lying between 6000 to 27000cm^2/Vs, depending on
T) indicates that we are dealing with a 2DEG, and the temperature
dependence of the concentration of this gas is clearly observed.

In order to examine quantitatively these results, we have used a
model[2] in which the quantum well is treated in the triangular well approximation, taking into account the two first quantized levels and the
depletion charge in GaAs, and the GaAlAs conduction band is obtained
through the numerical integration of Poisson eq., taking into account
the finite width of this layer, the partial ionization of the donors,
the presence of compensating acceptors and Γ, L and X bands. All bands
were assumed to be parabolic; the pressure and temperature dependences
of each of them and of the effective masses were taken from ref.3 and
refs. cited therein. The conduction band discontinuity ΔE_c at the interface was taken as 60% of the difference between GaAs and GaAlAs band
gaps, and was considered to be independent of P and T.

The behavior of the donor level with respect to T and P has been
determined from the study of bulk GaAlAs samples. For the activation

energy ε_d (relative to the Γ conduction band-edge) at 0K, we used:

$$\varepsilon_d=14+110P \text{ meV} \quad \text{(P in GPa)} \quad (1)$$

and the temperature dependence was taken as

$$E_d(T)=E_d(T=0K)+0.75(E_\chi(T)-E_\chi(T=0K)) \quad (2)$$

where E_d and E_χ are the energies of the donor level and of the X minimum relative to the top of the valence band at Γ ($\varepsilon_d=E_\Gamma-E_d$). Considering only a single donor level, and using such T and P dependences, we have been able to obtain a good agreement with the experimental results for both the heterostructures (Fig.1) and the bulk GaAlAs samples (Fig.2). A variation of ε_d with T has also been invoked in another recent study of GaAs-GaAlAs heterostructures[4], and is not an unexpected behavior for a donor presenting strong lattice relaxation[5]. Care must be taken, however, before using our results as a definite argument for the association of the Si-donor state with the X minimum, since a lower symmetry state could well have the same type of behavior.

4. CONCLUSION

Using hydrostatic pressure, we have been able to reduce the parallel conduction to negligible values in GaAs-GaAlAs heterostructures and thereby to study the temperature dependence of the 2DEG concentration. Plausible assumptions concerning the variation of the Si-donor activation energy with T lead to changes in the calculated 2DEG concentrations which are in good agreement with the observed ones. A lot of work still has to be done to get a precise description of this T dependence, but these first results lay interesting grounds for its study.

5. ACKNOWLEDGEMENTS

We gratefully acknowledge financial support from the CNET and the Conseil Régional Midi-Pyrénées.

6. REFERENCES

(1) Mercy, J.M., Bousquet, C., Robert, J.L., Raymond, A., Grégoris, G., Beerens, J., Portal, J.C., Frijlink, P.M., Delescluse, P., Chevrier, J., Linh, N.T., Surf. Sci. 142, 298 (1984).

(2) Beerens,J.,Grégoris,G.,Portal,J.C.,Alexandre,F., to be published.

(3) Lee, H.J., Juravel, L.Y., Woolley, J.C., Springthorpe, A.J., Phys. Rev. B21, 659 (1980).

(4) Svensson, S.P., Swanson, A.W., J. Appl. Phys. 59, 2870 (1986).

(5) Schubert, E.F., Ploog, K., Phys. Rev. B30, 7021 (1984).

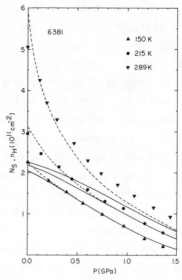

FIG.1 Symbols give the measured Hall concentration n_H at
      different pressures and temperatures for an hetero-
      structure having a spacer of 300Å. Full and dashed
      lines give respectively the values of N_S and n_H cal-
      culated with our model.

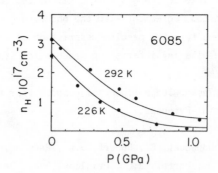

FIG.2 Circles give the Hall concentration measured on bulk
      GaAlAs samples. Full lines have been calculated using
      the description of the donor level given int the text
      and taking the Γ, L and X bands into account.

ENERGY RELAXATION OF TWO-DIMENSIONAL ELECTRONS AND DEFORMATION POTENTIAL CONSTANT IN SELECTIVELY-DOPED AlGaAs/GaAs HETEROJUNCTIONS

K. Hirakawa, H. Sakaki, and J. Yoshino

Institute of Industrial Science, University of Tokyo

7-22-1 Roppongi, Minato-ku, Tokyo 106, Japan

The electron heating process in selectively-doped AlGaAs/GaAs heterojunctions is investigated. It is shown that the dominant energy-relaxation mechanism of the degenerate two-dimensional(2D) electrons for the electron temperature below 40K is the emission of acoustic phonons via deformation potential coupling, and that the energy-loss rate is almost independent of 2D electron density. From a detailed analysis, it is derived that the deformation potential constant of the quantized 2D conduction band in GaAs is $11\pm1eV$.

The two-dimensional (2D) electrons accumulating in selectively-doped AlGaAs/GaAs heterojunctions are easily heated up by an external applied electric field E_x due to their high-mobility nature,[1,2] leading to a rapid rise in electron temperature T_e [3,4] and a reduction of electron mobility μ.[5] To clarify such hot electron phenomena is of great importance not only for electronic device applications but also for extracting informations about electron-phonon interactions.[6]

All the samples used in the present study were molecular-beam-epitaxially grown selectively-doped n-type $Al_{0.3}Ga_{0.7}As$/GaAs single heterojunctions. All the samples exhibit well developed zero resistance regions in diagonal resistivities and plateaus in Hall resistivities in the quantized Hall regime, which ensures the absence of parallel conduction in the doped AlGaAs layers. The samples cover a wide range of mobility μ and electron density N_s, as summarized in Table I.

Electron temperatures T_e are determined as functions of input power per electron P_e ($\equiv J_x E_x/N_s$), which is equal to the energy-loss rate $-\langle dE/dt \rangle$ in a steady state, by analyzing the amplitudes of Shubnikov-de Haas oscillations measured at lattice temperature T_L=4.2K under low magnetic fields (B<2T) at different current densities J_x.[3] Figure 1 shows the T_e versus $-\langle dE/dt \rangle$ (or P_e) characteristics. When

T_e<30K, measured T_e increases rapidly with P_e, reaching ~30K when
$P_e=10^{-13}$W. Furthermore, T_e is found to be almost uniquely determined
by P_e and fall on a single curve despite the wide variety of μ and N_s.

To understand such an electron heating process, we have performed
theoretical calculations, following the approach of Price.[7-9] The T_e
dependence of the energy-loss rate calculated for N_s = 5.0x10^{11}cm^{-2}, as
a typical case, is also plotted in Fig. 1. In the calculation, the
deformation potential constant D is set to be 11eV, as discussed below.
The calculated results are in excellent agreement with the experimental
data. It is clarified that the dominant energy-relaxation mechanism of
degenerate 2D electrons when T_e<40K is the emission of acoustic phonons
via deformation potential coupling, and that the effect of the piezoelectric coupling is only about 10% of that of the deformation potential coupling. The emission of polar-optical phonons, which is a very efficient energy-relaxation mechanism, becomes effective only when T_e>40K.

Table I. Electron densities N_s and mobilities μ at 4.2K of the samples studied here.

sample		N_s $(10^{11}$cm$^{-2})$	μ $(10^4$cm^2/Vs)
1	R-327	2.1	22.6
2	U-333	3.5	7.0
3	R-6	4.6	20.5
4	R-98	7.1	7.8
5	U-319	8.1	3.8

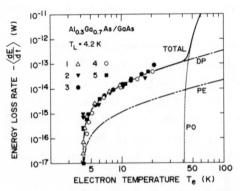

Fig. 1   Energy-loss rates -<dE/dt> measured at lattice
temperature of 4.2K are plotted as functions of electron
temperature T_e. Calculated results of energy-loss rates due to
individual scattering mechanisms are also plotted in the figure;
acoustic deformation potential scattering (DP), acoustic
piezoelectric scattering (PE), and polar-optical phonon
scattering (PO). The total energy-loss rate (TOTAL) is shown by
a solid line.

Figure 2 shows the N_s dependence of the energy-relaxation time τ_e determined when T_L=4.2K and T_e=10.0K, as a typical case, since τ_e does not show any significant changes when 4.2K<T_e<30K. τ_e varies from 1.3nsec to 0.45nsec when N_s is raised from $2.1\times10^{11}cm^{-2}$ to 8.1×10^{11} cm^{-2}, and is approximately proportional to $N_s^{-0.8}$. Such a large value of τ_e when T_e<30K indicates that the energy relaxation via emission of acoustic phonons is not efficient, and leads to a rapid rise in T_e.

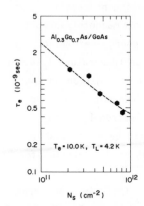

Fig. 2 The energy-relaxation time determined at T_e=10K and T_L=4.2K vs. electron density N_s. The caluculated result is also shown by a broken line.

Figure 3 shows the N_s dependence of the energy-loss rate when T_e=10.0K and T_L=4.2K. Measured energy-loss rates are almost independent of N_s and are $(1.4\pm0.2)\times10^{-14}W$. In the figure the calculated energy-loss rates are also plotted as functions of N_s by broken lines with the deformation potential constant D as a parameter, and are almost independent of N_s when N_s is in the range from $1\times10^{11}cm^{-2}$ to $1\times10^{12}cm^{-2}$. This fact is consistent with the experimental results.

From a careful comparison between the experimental data and the theoretical results we determine the deformation potential constant D. This method is expected to allow more accurate determination of D than the measurement of the temperature dependence of mobility,[10,11] since the electron heating processes are purely

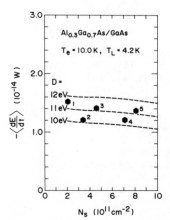

Fig. 3 The energy-loss rates measured when T_e=10K and T_L=4.2K are plotted as a function of electron density N_s. The results of theoretical calculations are also plotted by broken lines, with the deformation potential constant D as a parameter.

phonon-related phenomena and are not affected by Coulomb scattering. The determined value of D is 11±1eV, which is fairly larger than the commonly accepted value 7eV,[12] and is rather closer to the results of recent works.[10,11] The present value of D is, however, somewhat smaller than that (13.5eV) determined by Mendez, et al.[10] and Lin, et al.[11] by measuring the temperature dependence of the 2D electron mobility. The origin of this discrepancy is not clear at present.

In summary, we investigate the electron heating process in selectively-doped AlGaAs/GaAs heterojunctions from magnetotransport measurements at low temperatures. The dominant energy-relaxation mechanism of degenerate 2D electrons for electron temperature $T_e<40K$ is shown to be the emission of acoustic phonons via deformation potential coupling. The energy-loss rates are almost independent of N_s, and T_e is uniquely determined by the input power per electron P_e. The measurement of P_e dependence of T_e allows a sensitive and reliable determination of the deformation potential constant D. D=11±1eV for the conduction band in GaAs is probable.

We wish to express our sincere thanks to Prof. T. Ando and Prof. S. Kawaji for stimulating discussions. This work is mostly supported by a Grant-in-Aid from the Ministry of Education, Science, and Culture, Japan and also by the Toray Science Foundation.

References
1. Stormer,H.L., Dingle,R., Gossard,A.C., Wiegmann,W., and Sturge,M.D., Solid State Commun. 29, 705 (1979).
2. Hiyamizu,S., Saito,J., Nanbu,K., and Ishikawa,T., Jpn. J. Appl. Phys. 22, L609 (1983).
3. Sakaki,H., Hirakawa,K., Yoshino,J., Svensson,S.P., Sekiguchi,Y., Hotta,T., Nishi,S., and Miura,N., Surf. Sci. 142, 306 (1984).
4. Hopfel,R.A. and Weimann,G., Appl. Phys. Lett. 46, 291 (1985).
5. Shah,J., Pinczuk,A., Stormer,H.L., Gossard,A.C., and Wiegmann,W., Appl. Phys. Lett. 44, 322 (1984).
6. Hirakawa,K. and Sakaki,H., submitted to Applied Physics Letters.
7. Price,P.J., Ann. Phys. (NY) 133, 217 (1981).
8. Price,P.J., J. Appl. Phys. 53, 6863 (1982).
9. Price,P.J., Phys. Rev. B30, 2236 (1984).
10. Mendez,E.E., Price,P.J., and Heiblum,M., Appl. Phys. Lett. 45, 294 (1984).
11. Lin,B.J.F., Tsui,D.C., and Weimann,G., Solid State Commun. 56, 287 (1985).
12. Rode,D.L., Phys. Rev. B2, 1012 (1970).

ENERGY RELAXATION AND RECOMBINATION OF HOT CARRIERS IN GaInAs/InP QUANTUM WELLS USING PICOSECOND TIME- RESOLVED LUMINESCENCE SPECTROSCOPY

D.J.WESTLAND, D.MIHAILOVIČ and J.F.RYAN
Clarendon Laboratory, University of Oxford, Oxford, U.K.

M.SCOTT
Plessey Research Centre, Caswell, U.K.

P.A.CLAXTON
SERC Central Facility for III- V Semiconductors, Univ. of Sheffield, Sheffield, U.K.

We have measured the radiative and Auger recombination coefficients in GaInAs/InP quantum wells using time- resolved photoluminescence with a 3 ps resolution. Time-integrated luminescence measurements   confirm the values obtained. We find that carrier "heating" via Auger recombination cannot of itself account for the slow carrier cooling that we observe in these structures.

Nonradiative recombination in narrow- gap semiconductors is the subject of much current experimental and theoretical investigation because of its relevance to the performance of semiconductor lasers, Auger recombination being   the dominant nonradiative process which limits the quantum efficiency and threshold current. GaInAs lasers, especially those with multiple quantum well structures (MQWS), are important because their photoemission wavelengths lie in the spectral region of low loss and low dispersion for optical fibres. In this paper we report measurements of the radiative and Auger recombination coefficients for GaInAs MQWs, obtained using time- resolved spectroscopy. We have also investigated the energy relaxation of hot carriers in these structures and find that the energy loss rate is slower than in the bulk, which could be due in part to Auger recombination.

Two different $Ga_{.47}In_{.53}As/InP$ MQW samples were used in our experiments: (A) a ten period   MQW with 62Å wells and 100Å barriers, and (B) a stack of wells of widths 10,20,40,60Å with 0.1 μm barriers. All measurements were made with the samples at 4K. The time- resolved luminescence measurements were made using a frequency upconversion technique: A frequency- doubled, mode- locked Nd:YAG laser is used to synchronously pump an R6G dye laser producing 3ps pulses at 607nm. Part of the laser beam is used to excite luminescence from the sample, and part is taken via a variable time delayline and focussed colinearly with the luminescence into a $LiIO_3$ crystal. The sum frequency of the laser and the luminescence is generated only when the delayed laser pulse arrives at the crystal. By varying the delay time the intensity of luminescence as a function of time can be recorded.

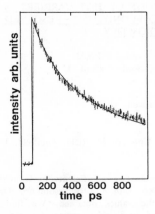

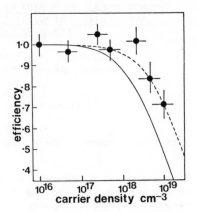

Figure 1. Total luminescence intensity
from Sample A as a function of time.
The solid curve shows a fit to eq (1).

Figure 2. Luminescence efficiency
(time−integrated)   of Sample A
for different carrier densities.

The luminescence from sample A, obtained at very high photoexcitation density
$\approx 10^{19} cm^{-3}$, was measured as a function of time at different energies throughout
the entire spectrum, and   by numerical integration over energy we obtained the *total*
luminescence as a function of time (Figure 1). The profile is not exponential, but
shows a rapid decay at early times which is the behaviour expected for Auger
recombination. The carrier kinetics may be described using the simple rate equation:

$$\frac{dn}{dt} = -\alpha n^3 - \beta n^2 \tag{1}$$

where n is the electron or hole density, α is the Auger recombination coefficient and
β is the band−to−band radiative recombination coefficient. Within the 1ns time
window of the measurement the carrier density is high compared to both the Mott
limit and the impurity concentration, and therefore we do not include a term linear
in n which might arise from excitonic or trapping effects. The time−dependent
luminescence intensity is given by :

$$I(t) = s \beta n^2 \tag{2}$$

where s is a constant. Using (1) and (2) we obtained a fit to the   data using α and
β as variable parameters. The best fit, obtained with $\alpha = 9\pm3 \times 10^{-30} cm^6 s^{-1}$
and $\beta = 4\pm1 \times 10^{-11} cm^3 s^{-1}$, is shown by the solid curve in Figure 1. The
principle source of uncertainty in these values is due to the uncertainty in the initial
carrier density.

Time−integrated luminescence has been measured for the 62Å MQWS as a function of laser excitation density, using a cooled Ge detector. Under our experimental conditions, ie. ultra short pulse excitation, the time— integrated luminescence efficiency η depends on the ratio of the radiative and Auger rates:

$$\eta \propto \frac{1}{P_1} \cdot \frac{\beta}{\alpha} \ln \left\{ 1 + \frac{\alpha}{\beta} n_0 \right\} \qquad (3)$$

where n_0 is the initial (ie $t=0$) photoexcited carrier density and P_1 is the laser power. For low values of n_0, η is independent of n_0 (this assumes that there are no other nonradiative decay channels, such as trapping, which may be carrier density dependent). At large n_0, η decreases rapidly due to Auger effects. Thus by fitting the experimental data over a wide range of n_0, a value for α/β can be obtained.

The experimental data for the 62Å MQW sample are shown in Figure 2 and display the general behaviour predicted by (3). The solid curve is the theoretical curve obtained for $\alpha/\beta = 2.25 \times 10^{-19}$ cm^3, the value obtained from the time—resolved measurements. For comparison we include a curve (broken line) obtained for $\alpha/\beta = 1.0 \times 10^{-19}$ cm^3. The conclusion from this is that the measurement of α/β is accurate to within a factor of 2. It must be stressed however that the carrier densities are not known with high precision, again a factor of 2 uncertainty is reasonable. Taking this into account gives $\alpha/\beta \simeq 1-5 \times 10^{-18}$ cm^3, so that values obtained by both methods agree within experimental error.

The value of α measured by us is a factor of 5—10 smaller than that reported by Sermage et al[2] who obtained a value of 6×10^{-29} cm^{6}s^{-1} for the Auger coefficient of GaInAs/AlInAs MQWs. The discrepancy may well be due to the different experimental conditions — in our experiment carriers are excited in both the barriers and the wells so that the actual carrier density in the wells may be higher than we estimate due to carrier transfer from the barriers to the wells. This effect would, however, cause a reduction of our value of α. On the other hand the Auger rate coefficient is temperature dependent; in our experiment the sample is maintained at 4K but the carrier temperature varies from 600K — 100K during the measurement (see below), whereas in the experiment of Sermage et al the sample is maintained at 300K and the average carrier temperature is therefore much higher. This effect is a more likely source for the dicrepancy

The energy relaxation rate of photoexcited hot carriers confined in QW structures has been shown by recent time resolved luminescence measurements to be substantially slower than expected[3,4]. One of the mechanisms proposed to explain this is the reabsorbtion of non— equilibrium phonons by the hot carriers. In narrow gap semiconductors like GaInAs, a further possibility exists, viz nonradiative recombination "heating", in which the recombination energy is retained by the carriers that remain. From the measurements presented above, we can estimate the mean effective input power per carrier from Auger recombination:

$$P \simeq \frac{1}{n} \left[\frac{dn}{dt} \right]_{Auger} E_g = \alpha \, n^2 \, E_g \qquad (4)$$

Using the value of α obtained above we estimate $P \approx 1 \mathrm{meV} \, \mathrm{ps}^{-1}$ at early times for our experimental conditions.

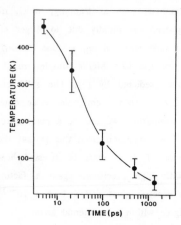

Figure 3. The carrier cooling curve for the 10Å QW of Sample B.

The cooling curve obtained from our time–resolved luminescence measurements on the 10Å QW of sample B is shown in Figure 3. At early times, when the Auger effect is strongest, the mean energy loss rate estimated from the slope of the cooling curve is $\approx 0.5 \mathrm{meV} \, \mathrm{ps}^{-1} \, \mathrm{carrier}^{-1}$. This value is about 3 times smaller than that observed for bulk GaInAs[5]. The magnitude of the energy loss rate by LO phonon emission is estimated to be $\geqslant 30 \mathrm{meV} \, \mathrm{ps}^{-1} \, \mathrm{carrier}^{-1}$; this value is substantially greater than the nonradiative "heating" rate estimated above, and so this mechanism is unable to explain the reduced cooling rate.

References

1. N.K.Dutta, J. Appl. Phys. **54**, 1236 (1982)
2. B.Sermage, D.S.Chemla, D.Sivco and A.Y.Cho, IEEE J.Quantum Electronics **QE–22**, 774 (1986)
3. J.F.Ryan, R.A.Taylor, A.J.Turberfield, A.Maciel, J.M.Worlock, A.C.Gossard and W.Wiegmann, Phys. Rev. Lett **54**, 1841 (1984)
4. K.Kash, J.Shah, D.Block, A.C.Gossard and W.Wiegmann, Physica **134 B&C**, 189 (1985)
5. K.Kash and J.Shah, Appl. Phys. Lett **45**, 401 (1984)

RESONANT INTERACTION OF PLASMONS AND INTERSUBBAND RESONANCES IN TWO-DIMENSIONAL ELECTRONIC SYSTEMS

D. Heitmann[a,b] and S. Oelting[a]

(a) Institut für Angewandte Physik, Jungiusstr. 11, D-2000 Hamburg 36
(b) Max-Planck-Institut f. Festkörperforschung, Heisenbergstr. 11, D-7000 Stuttgart 80, West-Germany

ABSTRACT

The influence of uniaxial stress on the occupation of subbands, on intersubband- (ISR) and plasmon resonances in metal-oxide-semiconductor-systems (MOS) has been studied experimentally. We will in particular report a situation where we can tune via stress the ISR and plasmon resonance and find a resonant interaction resulting in an enhanced amplitude and an antilevel crossing.

In MOS or heterostructure systems electrons can be confined in narrow potential wells. The energy spectrum of these quasi two-dimensional (2D) electronic system consists of separated subbands $E_i(k_x, k_y) = E_i + \hbar^2 k_x^2/2m_x + \hbar^2 k_y^2/2m_y$, i= 1,2,... resulting from the free motion parallel to the interface (x-y-plane) and the quantized motion perpendicular to the interface[1]. The subband structure arises from the projection of the volume energy ellipsoids onto the surface. Thus, for Si (100), which will be considered here, two subband systems exist. The first system ($E_0, E_1...$) with a valley degeneracy $g_v = 2$ and isotropic masses $m_x = m_y = 0.2 \, m_0$ is normally (stress p=0, T<10 K) occupied at low carrier densities N_S. A second subband system ($E_{0'}, E_{1'},...$), denoted by primes, with $g_v = 4$ and anisotropic masses ($m_x = 0.92 \, m_0$, $m_y = 0.2 \, m_0$) is occupied at high N_S. Here we will investigate the influence of uniaxial stress on 2D-plasmon resonances[2] and intersubband resonances[1] (ISR). This allows us to study the stress dependent occupation of subbands, and, as we will see, resonant plasmon-ISR interaction.

The experiments are performed on Si MOS capacitors with 50 nm oxides and size 5x6mm^2. In a cryostat a well defined uniaxial stress p can be applied to the samples at 12 K. Quasi accumulation conditions are established for the electrons in the p-type samples by band gap illumination. Plasmon and ISR excitation is observed in the transmission of normally incident farinfrared (FIR) laser radiation. The gate of the MOS capacitor consists of periodic stripes of alternating high and low conductivity with periodicity a= 514 nm. This gate couples the FIR radiation to plasmons of wave vector q= $2\pi/a$. (q is here in [001] direction). The periodic stripes also induce in the near field e_z-components of the FIR electric field and can excite ISR on Si(100) [3]. Spectra measured at a fixed FIR laser energy of 12.9 meV are shown in Fig. 1a. The pronounced resonances at $N_S \approx 4\cdot10^{12}$ cm^{-2} are 2D-plasmon resonances which shift with increasing p to higher N_S. The 2D-plasmon frequency[2] is $\omega_p^2 = N_S \cdot e^2 \cdot q/m_p \cdot 2\bar{\varepsilon}$. Here $\bar{\varepsilon}$ is an effective dielectric function and $m_p^{-1} = (N_{SO}/m_{p0} + N_{SO'}/m_{p0'})/N_S$ is the inverse plasmon mass in the two subband system. N_{SO}/N_S and $N_{SO'}/N_S$ are the relative occupations of the subbands. Stress in [001] direction lowers two of the originally fourfold valleys and the shift of the resonance positions in Fig. 1a reflects the increase of the plasmon mass due to the increasing occupation of E_0, with its larger mass ($m_{p0'} = 0.92\ m_0$) at increasing p. Since m_{p0} and $m_{p0'}$ are known we can evaluate the relative occupation N_{SO}/N_S in its dependence on p and can construct - using results from different N_S and frequencies - a stress diagram. We find that our experimentally evaluated stress diagram agrees quite well with calculated diagrams[4]. Since for these calculations the bulk shear modulus was assumed we can indirectly deduce from this agreement, that the bulk shear modulus can also be applied to the surface band structure. This has also been obtained from tunneling electron spectroscopy on MOS systems[5].

At low N_S in Fig. 1a additional resonances occur. The behavior in this density regime is in general rather complex. Here we can identify the resonances at p<1.5 kbar with 0→1 ISR transitions. With increasing p the resonance smears out and at p> 3.5 kbar a new resonance is observed. This resonance is the 0'→1' ISR. We know from stress diagrams that at low N_S with relatively low stress a nearly complete occupation of E_0, can be achieved. In this situation we find in Fig. 1a that the ISR resonance does not depend on p, reflecting the fact, that stress

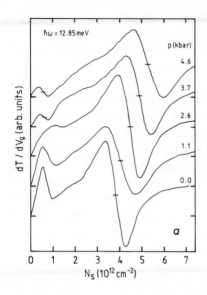

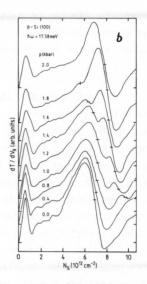

Fig. 1: Derivative of the transmission T with respect to the gate voltage V_g vs N_S for different values of the uniaxial stress at laser frequency 12.85 meV (a) and 17.58 meV (b)

shifts the whole subband ladder, but not the separation between sub-bands. For measurements at laser frequency 17.6 meV in Fig. 1b we have a very different situation. The plasmon resonance occurs for p=0 at $N_S = 7 \cdot 10^{12}$ cm^{-2}. It shifts with increasing p to higher N_S, similarly as in Fig. 1a. For certain values of p we observe a second resonance – which we will identify below as a 0'→1' ISR – appears and shifts to

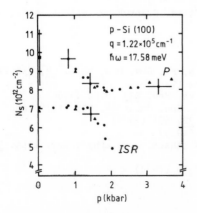

Fig. 2: Experimental resonance positions from Fig. 1b. The interaction of the plasmon (P) and ISR leads to a resonant anti-level crossing of the dispersions

lower N_S. In the crossing regime the two resonances show all features of a resonant interaction, an anti-level crossing of the dispersions, shown also in Fig. 2 and a resonant enhancement of the ISR excitation[6].

We have indicated the 0'→1' ISR by extrapolating our experimental positions in Fig. 2 to p=0 and comparison with experiments in Refs. 3 and 5. The strong stress dependence of the ISR that we observe here in contrast to Fig. 1a arises from the fact that at high N_S two different kinds of subbands, E_0 and $E_{0'}$, are occupied. Thus with increasing p redistribution of carriers from E_0 to $E_{0'}$ occurs. The wavefunctions of the 0'-subband extend deeply into the Si. They strongly influence the potential at large distances from the interface and thus the wavefunctions, energies and separation of higher subbands. The fact, that both the ISR and the plasmon resonance can be tuned with stress gives us the unique possibility to study the resonant interactions of the two characteristic dynamic excitations of the 2D system. We find that the interaction strength is, if we express the splitting in N_S via the plasmon dispersion in terms of an energetical splitting, $\Delta E = 2$ meV. This agrees with the resonant interaction strength estimated from calculations in Ref. 7. Another characteristic feature of the interaction observed here is the resonantly enhanced amplitude of the ISR excitation (Fig. 1b). The ISR can only be observed in the regime of the plasmon resonance. This enhanced excitation arises from the resonantly enhanced transverse electric field components, which accompany – in contrast to bulk plasmons – the 2D plasmon oscillation. Similar plasmon resonance enhanced processes are well-known for surface plasmons at semi-infinite boundaries, e.g., plasmon resonance enhanced excitations in Langmuir-Blodgett films, Giant Raman effect, etc.

REFERENCES

1 Ando, T., Fowler, A.B., and Stern, F., Rev. Mod. Phys. 54, 437(1982)
2 Heitmann, D., Surf. Sci. 170, 332 (1986)
3 Heitmann, D. and Mackens, U., Phys. Rev. B33, 8269 (1986)
4 Takada, Y. and Ando, T., J. Phys. Soc. Japan 44, 905 (1976) and Vinter, B., Solid State Commun. 32, 651 (1979)
5 Kunze, U., Phys. Rev. B32, 5328 (1985)
6 Oelting, S., Heitmann, D., and Kotthaus, J.P., Phys. Rev. Lett. 56, 1846 (1986)
7 Das Sarma, S., Phys. Rev. B29, 2334 (1984)

SURFACE ROUGHNESS SCATTERING THEORY FOR 2D ELECTRONS IN SI-MOSFET's

R. Lassnig, E. Gornik

Institut für Experimentalphysik, Universität Innsbruck

A-6020 Innsbruck, AUSTRIA

Several theoretical models for the description of surface roughness (SR) scattering in Si-MOSFETs have been derived within the past ten years /1-6/. However, two important aspects of the problem have not been resolved until now: First, all theoretical models rely essentially on an expansion of the surface roughness potential with respect to the deviation from the ideal interface. This expansion of a sharp step potential is questionable; already Ando et al. /7/ pointed out that the validity of the previous theories "has not been fully elucidated yet". Second, the experimentally observed dependence of the SR scattering strength on the depletion charge density (N_d) cannot be explained by the previous models. N_d can be well controlled through a bias voltage, and the resulting mobility variation represents an ideal test for the validity of the theoretical model. This is most essential since SR scattering theory generally relies on two sample-dependent fitting parameters, which makes a single fit of mobility curves easy.

Here we present a new approach in estimating the effective SR potential, which circumvents the expansion problem and yields good agreement with the experimental data. The effective scattering potential is calculated by solving the microscopic inhomogeneous Poisson Equation for the many-particle ground state. The scattering process is then determined by the mapping of the potential variation with the unperturbed wave functions.

Poisson's Equation is written in an integral form:

$$\phi(\vec{r}) = - \int d^3\vec{r}' \, \frac{\rho(\vec{r}') - \vec{\nabla}\kappa(\vec{r}')\vec{\nabla}\phi(\vec{r}')}{\kappa(\vec{r}')|\vec{r}-\vec{r}'|} \tag{1}$$

which determines the potential $\phi(\vec{r})$ created by a charge density $\rho(\vec{r})$

embedded in a medium of static dielectric function $\kappa(\vec{r})$. The continuum description corresponds to mean values over atomic distances. For the unperturbed system, charge neutrality requires the the second term under the integral is given by:

$$-\frac{d\kappa(z)}{dz}\,\frac{d\phi(z)}{dz}\,\frac{1}{\kappa(z)} = e(N_d + N_{el})\,(\frac{1}{\kappa_{ox}} - \frac{1}{\kappa_s})\delta(z) \tag{2}$$

N_{el} is the electron density, $\kappa_{ox} = 3.8$ and $\kappa_s = 11.8$ are the oxide and silicon dielectric constants, respectively. The effect of the dielectric discontinuity is equivalent to that of a local space charge.

Now we calculate the effective SR potential by calculating $\phi(\vec{x}=0,\ z_0 > 0)$ for the radial symmetric model structure shown in Fig. (1). The structure represents the penetration on the silicon through the "unperturbed" z=0 plane, and the electronic charge density is indicated by the shaded area. The smooth variation is approximated by a disk of radius 1 and thick-

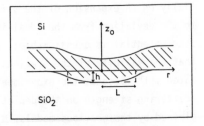

Fig.1: Schematic view of the model structure used for the description of the surface roughness potential.

ness h. The first iteration of Poisson's Equation is then:

$$\frac{\vec{\nabla}\kappa\vec{\nabla}\phi}{\kappa} \simeq e(N_d + N_{el})\,(\frac{1}{\kappa_{ox}} - \frac{1}{\kappa_s}) \cdot \begin{cases} \delta(z+h) & \dots \quad |\vec{x}| < 1 \\ \\ \delta(z) & \dots \quad |\vec{x}| > 1 \end{cases} \tag{3}$$

Equ. (1) is integrated for the shifted charge distribution, and the resulting local potential variation is mapped with the unperturbed ("flat") wave functions. Using the Fang-Howard functions /8/ with variational parameter b, the scattering strength is obtained as:

$$|<\delta V_d(\vec{x},z)+\delta V_{el}(\vec{x},z)>| = eF^*h(\vec{x}) = 2\pi e^2 h(\vec{x}) \{ \frac{1}{\kappa_{ox}}(N_d+N_{el}) -$$

$$- \frac{1}{\kappa_s}(N_d + \frac{1}{2} N_{el}) - (N_d + N_{el})\,(\frac{1}{\kappa_{ox}} - \frac{1}{\kappa_s})\frac{1}{2} \int_0^\infty dz\ \frac{z^3 e^{-z}}{\sqrt{z^2+1^2 b^2}} \} \tag{4}$$

The factor F* can be interpreted as an effective field and $h(\vec{x})$ describes the variation of the irregularities over the (x,y)-plane. Finally, the scattering time is calculated within the traditional correlation function approach /7/, identifying the correlation length for the roughness distribution with our lateral length scale 1. We found that an exponential decay law for the correlation function:

$$<h(\vec{x})h(0)> = h^2 e^{-|\vec{x}|/1}, \quad K(q) = 2\pi h^2 1^2 / (1+q^2 1^2)^{3/2} \quad (5)$$

describes the experimental results much better than a Gaussian one. The result for the mobility is:

$$\mu_{SR}^{-1} = \frac{m^2 F^{*2}}{eh^3\pi} \int_0^\pi d\theta (1-\cos\theta) K(2k_F\sin(\theta/2))/\epsilon^2(2k_F\sin(\theta/2)) \quad (6)$$

where the static screening $\epsilon(q)$ includes local field corrections. The influence of higher subbands is neglected, but these become effective only at the higher electron concentrations /5/.

From our theory we find that the lowest order scattering due to a variation of the depletion potential depends essentially on the difference of the dielectric constants of the two materials. For $\kappa_{ox} = \kappa_s$ no local variation of the depletion potential would be observable. A different behaviour is found for the variation of the electronic Hartree potential, since the many-particle charge density relaxes into the lowest energy state, meaning that the local potential variation must be calculated for the shifted charge density.

In Fig. (2) the inverse surface roughness limited

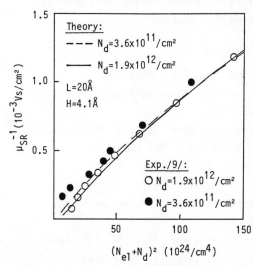

Fig.2: Inverse SR limited mobility versus the squared gate charge. The theory and the experimental points are shown for two depletion charge densities.

mobility is plotted as a function of the squared gate charge. The experimental data of Hartstein et al. /9/ for two different depletion charge densities are compared to our theoretical results for a correlation length of $l = 20$ A and $h = 4.1$ A. At low electron concentrations the dashed curve ($N_d = 3.6 \times 10^{11}/cm^2$) lies considerably higher than the full curve ($N_d = 1.9 \times 10^{12}/cm^2$), which is quantitatively in good agreement with the experimental results. At high electron densities, the theoretical as well as the experimental value are practically independent of the depletion charge. A comparable consistent description of the experiments was not possible with the previous theories.

In conclusion, we have derived a new theoretical model for the SR scattering which relies on a consistent derivation of the scattering potential, avoiding the expansion of the sharp interface potential. The experimentally observed depletion charge dependence of the electron mobility is successfully described.

Acknowledgements:
We thank E. Vass and A. Gold for valuable discussions. This work was partially sponsored by the Stiftung Volkswagenwerk.

References:
/1/ Prange R.E., Nee T.; Phys. Rev. 168, 779 (1968)
/2/ Cheng Y.C.; Proc. 3rd Conf. on Solid State Devices, Tokio 1971, Suppl. to ovo Buturi 41, 173 (1972)
/3/ Matsumoto Y., Uemura Y.; Proc. Int. Conf. Solid Surf., Kyoto, Jap. J. Appl. Phys., Suppl. 2, Pt. 2, p. 367 (1974)
/4/ Ando T.; J. Phys. Soc. Japan 43, 1616 (1977)
/5/ Mori S., Ando T.; Phys. Rev. B 19, 6432 (1979)
/6/ Gold A.; Proc. 6th Int.Conf. on Electronic Properties of 2D Systems, Kyoto 1985, p. 381
/7/ Ando T., Fowler A.B., Stern F.; Rev. Mod. Phys. 54, 505 (1982)
/8/ Fang F.F., Howard W.E.; Phys. Rev. Lett. 16, 797 (1966)
/9/ Hartstein A., Ning T.H., Fowler A.B.; Surf. Science 58, 179 (1976)

MANY-BODY EFFECTS IN THE MAGNETO-TRANSPORT PROPERTIES OF P-TYPE SILICON INVERSION LAYERS

G. Landwehr and R. Baunach*

Physikalisches Institut der Universität Würzburg, D-8700 Würzburg
FEDERAL REPUBLIC OF GERMANY

*present address:Siemens Forschungslaboratorien, Otto-Hahn-Ring 6,
D-8000 München 83, Fed. Rep. of Germany

Low temperature magneto-transport properties of p-type
silicon inversion layers can be explained on the basis
of the theory of many-body interactions in a disordered
two-dimensional system.

1. INTRODUCTION

In the last few years a considerable amount of work was devoted towards the elucidation of the low-temperature magneto-transport properties of disordered two-dimensional systems. A logarithmic increase of the resistance with decreasing temperature as well as the low-temperature magnetoresistance have been explained by weak localization and many-body interaction effects[1]. Contrary to n-type silicon inversion layers p-type layers show a positive magnetoresistance. This was found already many years ago[2]. At that time a proper interpretation was not possible because no adequate theory was available. Since the situation has changed, the studies were resumed. In the meantime some work on magneto-transport in p-type silicon inversion layers has been published by other authors[3] which, however, seems incomplete and not fully consistent.

2. EXPERIMENTAL

Magneto-resistance (MR), Hall-effect and resistivity of p-channel MOSFETs were studied in magnetic fields up to 0.65 T in the tempera-

ture range from T = 1.8 K to 4.2 K. The magnetic field was always
perpendicular to the current and could be rotated from the position
perpendicular to the inversion layer to the parallel configuration.
Devices of (111) and (110) orientation were investigated at various
gate voltages. Because of the anisotropy of (110) samples specimens
with {001} and {110} current orientation were studied.

3. RESULTS AND DISCUSSION

The channel resistance of the specimens investigated showed a
small logarithmic increase of the order of 1 % when the temperature
was lowered from 4 to 2 K. However, during the time consuming experimental runs it frequently happened that spurious induced charges
caused significant resistance changes which relaxed only very slowly.
As a consequence, the data were unsuitable for quantitative evaluation. No conclusions could be drawn from the missing temperature
dependence of the Hall constant because the experimental error of $\pm$ 2%
was of the same order as the expected variations due to interaction
effects.

In Fig. 1 the magnetoresistance at 4.2 K of a (111) MOSFET at a
gate voltage U_G corresponding to a hole-concentration of $6.9 \times 10^{12} cm^{-2}$
has been plotted as a function of the angle between magnetic field B
and the surface normal. The MR is enhanced, if the temperature is
lowered. If B is oriented parallel to the inversion layer, the magnetoresistance vanishes, indicating that spin effects can be neglected.
It is evident that the measured magnetoresistance is always positive.
This indicates that weak localization effects which lead to a negative
magnetoresistance must be of minor importance. The suppression of weak
localization effects can be attributed to strong spin-orbit interaction. A semilogarithmic plot of the magneto-conductivity gives for
0.1 T< B< 0.65 T straight lines for 1.8 K< T < 4.2 K. This is shown in
Fig. 2. The (110) samples showed a similar behavior, except that the
characteristic logarithmic dependence was shifted towards smaller
magnetic fields and that a linear magnetoresistance component - which
cannot be explained theoretically at present - was superimposed at
higher B values.

The analysis of the data was performed in terms of the interac-

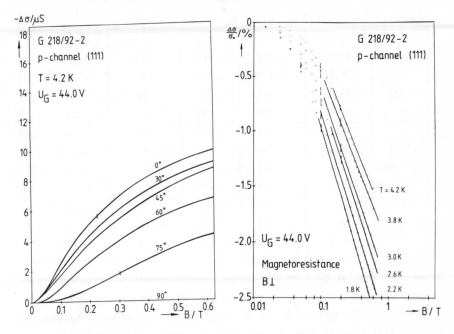

FIG. 1 Magnetoresistance at
a gate voltage of 44 V
at various orientations of B

FIG. 2 Magneto-conductivity
at U_G = 44 V on a
logarithmic scale

tion constants g_i as introduced by Fukuyama[4]. Neglecting weak localization and spin effects the theory predicts for the magneto-conductivity:

$$\Delta\sigma (B) = (e^2/2\pi^2\hbar)\ \alpha_{B\perp}\ \ln B \qquad \text{with}\ \alpha_{B\perp} = g_2 - 2g_4.$$

The constants g_1 - g_4 are functions of the interaction constant F,
which depends on the Fermi-wavevector and the screening constant. The
following approximations were compared with the experiment: **a)** dynamically screened Coulomb interaction **b)** higher order corrections
c) Fermi-liquid model. The results are shown in table 1.

One can recognize that the experimental values for $\alpha_{B\perp}$ for the
(111) sample are roughly two times larger than predicted for model a).
The same holds for the (110) sample with j $\parallel$ {001}, except for U_G= 14 V
where good agreement exists. Because it is known from previous experi-

TABLE 1: Theoretical and experimental values for $\alpha_{B\perp}$

Sample	U_G/V	a)	b)	c)	experimental
	32.0	-.402	-.141	-.124	-(.84 $\pm$ .22)
	38.0	-.400	-.139	-.122	-(.85 $\pm$ .10)
(111)	44.0	-.394	-.135	-.119	-(.84 $\pm$ .07)
	50.0	-.389	-.132	-.117	-(.95 $\pm$ .12)
	14.0	-.401	-.150	-.130	-(.40 $\pm$ .04)
(110)	28.0	-.386	-.136	-.119	-(.75 $\pm$ .15)
j$\parallel$ {001}	42.0	-.380	-.129	-.115	-(.75 $\pm$ .08)
	56.0	-.369	-.124	-.110	-(.77 $\pm$ .08)
(110)	14.0	-.402	-.153	-.132	-(.63 $\pm$ .16)
j$\parallel$ {110}	35.0	-.382	-.133	-.118	-(1.17 $\pm$ .30)
	56.0	-.369	-.125	-.111	-(1.20 $\pm$ .34)

ments[2] that at U_G = 28 V a second subband should be occupied, it seems that a second band, which is not taken into acccount in the theory, is the origin of the discrepancy. The differences between the two (110) samples indicate that the anisotropy on the subband structure is significant. The (111) data suggest that in this sample also two subbands are occupied. The obtained results, especially the failure of models b) and c) will be discussed in detail elsewhere.

ACKNOWLEDGMENT:

The samples were kindly provided by Dr. G. Dorda, Siemens Forschungslaboratorien, München

REFERENCES

1) See e.g. : Anderson Localization, Nagaoka Y. and Fukuyama H., eds., Springer Verlag 1982

2) von Klitzing K., Landwehr G. and Dorda G., Solid State Comm. 15, 489 (1974)

3) Gusev G.M., Kvon Z.D., Neizvestnyi I.G. and Ovsyuk V.N., Surface Science 142, 73 (1984)

4) Fukuyama H., J. Phys. Soc. Japan, 50, 3562 (1981).

DENSITY OF STATES IN SILICON INVERSION LAYERS
IN HIGH MAGNETIC FIELDS FROM RESONANT TUNNELING MEASUREMENTS

A. Hartstein and R. H. Koch

IBM Thomas J. Watson Research Center
P.O. Box 218, Yorktown Heights, NY 10598
USA

ABSTRACT

The density of states in a silicon inversion layer can be determined from resonant tunneling measurements. The electron mobility derived from Landau level broadening is smaller than that derived from conductance measurements. At high fields a spin splitting of the $n = 0$ Landau level is observed.

A number of experimental attempts have recently been made to determine the density of states in a 2D electron gas. Capacitance[1] and tunneling[2] measurements have been performed on single layer samples such as silicon inversion layers. Specific heat[3] and magnetization[4] measurements have only been performed on multi-layer samples. Resonant tunneling in MOS systems has been demonstrated[5-7] utilizing thin oxide (3.5 nm) silicon FET's which were deliberately doped with Na^+. Large peaks are observed in the tunneling conductance as a function of gate voltage which can be attributed to resonant tunneling of electrons from the silicon inversion layer into the aluminum gate via Na^+ induced localized states in the oxide. The magnetic field behavior of the resonant tunneling has also been investigated[8]. The peaks were observed to shift with field and the amplitude of the tunneling current showed Subnikov de-Haas (SdH) oscillations. In this paper we focus on the determination of the density of states in the silicon inversion from the observed SdH oscillations. We also report a spin splitting of the oscillations at high magnetic field.

MOSFET structures were fabricated with both wet and dry thermal SiO_2 gate oxides, 3 - 5 nm thick, on (100) p-type silicon substrates. NaCl, with a nominal density of 10^{12} to 10^{13} cm^{-2}, was evaporated onto the oxide immediately before the deposition of a 15 nm thick Al gate. The device area ranged from 1 - 4 mm^2 and the source and drain contacts were formed by n^+ diffusions. Na^+ ions were drifted toward the Si/SiO$_2$ interface at 320K under the influence of an electric field. The devices were subsequently cooled to prevent further drifting.

The source, drain and substrate were connected together; and the tunneling current between them and the gate electrode was measured using a lock-in technique at 1 Hz. For the low field (H ≤ 8T) measurements the temperature was held at 0.45K using a He$_3$ cryostat. For the high field (H ≤ 24T) measurements the temperature was held at 1.3K by pumping on a He$_4$ bath.

Fig. 1 shows the dynamic conductance at 0.45K as a function of gate voltage for several values of the applied magnetic field normal to the surface. Two resonant tunneling peaks are clearly evident. The effect of the magnetic field is to both shift the peak positions and to modulate the amplitudes of the peaks.

The variation of the tunneling amplitude with magnetic field is shown in Fig. 2 for one value of the gate voltage. It is clear that the amplitude shows a small modulation periodic in 1/H. The period is dependent on the value of the applied gate voltage.

It is possible to construct a Fan diagram[9] from SdH data shown in Fig. 2 along with additional data. This gives us a threshold voltage of 0.2V, which along with the oxide thickness, allows us to

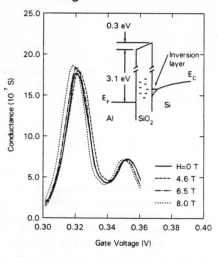

Fig. 1 Conductance as a function of gate voltage at 0.45K for a typical sample with d_{ox} = 4.4nm for several values of the applied magnetic field. The inset depicts the energy band structure of the samples.

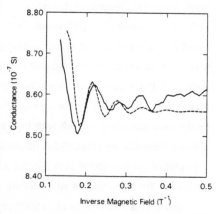

Fig. 2 Peak amplitude vs inverse magnetic field for V_g = 0.434V and T = 0.48K. The dashed curve is an estimate of the size of the SdH oscillations given τ_0 = 2.2 × 10^{-13} sec.

determine the inversion carrier density. By additionally measuring the channel conductance, a mobility of 4500 cm²/V-sec at $V_g = 0.4V$ was determined.

The inset to Fig. 1 illustrates the MOS band structure. Electrons in the silicon inversion layer can tunnel through the thin oxide into the conduction band of the Al electrode via localized states in the oxide barrier. In this process the tunneling current is directly proportional to the joint density of states. Since the density of states in the Al conduction band is smooth, structure in the inversion layer density of states will dominate any structure in both the joint density of states and the tunneling conductance as in Fig. 2.

From the positions of the maxima in Fig. 2, one can construct a diagram of $1/H$ vs. Landau index. Maxima in the density of states occur when the Fermi energy satisfies $E_F = (n + 1/2)\hbar\omega_c$, where n is the Landau level index, $\omega_c = \dfrac{eH}{m^*c}$ is the cyclotron frequency and m^* is the effective mass. One then obtains

$$\frac{1}{H} = (n + 1/2)\frac{e\hbar}{m^*cE_F}, \qquad (1)$$

for the maxima. This relationship is shown in Fig. 3 for the data of Fig. 2 and data obtained at other gate voltages. Note that points lie on straight lines for each value of the gate voltage. These lines intersect at an index of -1/2, and the slopes decrease for increasing gate voltage. All of these observations are in

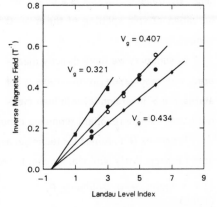

Fig. 3 Inverse magnetic field vs Landau index for the gate voltages indicated.

agreement with the assignment of the modulation of the resonant tunneling as due to a magnetic field modulation of the density of states in the inversion layer.

In a weak magnetic field the density of states in the inversion layer is[10]

$$D(E) = \frac{m^*}{2\pi\hbar^2}[1 - 2\cos(\frac{2\pi E_F}{\hbar\omega_c})\exp(-\frac{\pi}{\omega_c\tau_0})], \qquad (2)$$

where τ_0 is a scattering time. This density of states can be evaluated since the Fermi energy is determined by the gate and threshold voltages, and τ_0 can be calculated from the mobility. The dashed line in Fig. 2 is the predicted shape of the density of states in the inversion layer from Eq. 2 and the parameters indicated. It fits the tunneling conductance shape reasonably well. However, the value of $\tau_o = 2.2 \times 10^{-13}$ sec, used in the fit, implies a mobility of 2050 cm²/V-sec. This is smaller than the experimentally determined conductance mobility of 4500 cm²/V-sec. Perhaps this implies a locally inhomogeneous mobility which is lower in the vicinity of the Na⁺ ions which induce the tunneling.

In Fig. 4 data are shown for magnetic field up to 24T and T = 1.3K. The predicted positions of the 1/2 filled Landau levels are indicated. The SdH oscillations are clearly seen with the highest field oscillation showing a splitting. We believe that this is a direct observation of spin splitting of the n = 0 Landau level.

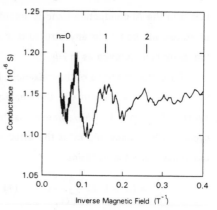

In summary we have directly measured the modulation in the density of states in a Si inversion layer in high magnetic field using a resonant tunneling experiment.

Fig. 4 Amplitude vs inverse magnetic field for $V_g = 0.427V$ and T = 1.3K.

We thank N. F. Albert for technical assistance. Part of this work was performed at the Francis Bitter National Magnet Laboratory, MIT, Cambridge, MA.

1] T. P Smith, B. B. Goldberg, P. J. Stiles and M. Heiblum, Phys. Rev. B 32, 2696 (1985).
2] E. Gornik, R. Lassnig, G. Strasser, H. L. Stormer, A. C. Gossard and W. Wiegmann, Phys. Rev. Lett. 54, 1820 (1985).
3] U. Kunze and G. Lautz, Surface Science 142, 314 (1984).
4] J. P. Eisenstein, H. L. Stormer, V. Narayanamurti, A. Y. Cho, A. C. Gossard and C. W. Tu, Phys. Rev. Lett. 55, 875 (1985).
5] R. H. Koch and A. Hartstein, Phys. Rev. Lett. 54, 16 (1985).
6] R. H. Koch and A. Hartstein, Proc. Int. Conf. Tunneling at Low Temp., Leuven, Belgium, p.17 (1985).
7] A. Hartstein and R. H. Koch, Surface Science 170, 391 (1986).
8] A. Hartstein and R. H. Koch, 'Magnetic Field Dependence of Resonant Tunneling in Thin MOS Structures', to be published.
9] A. B. Fowler, F. F. Fang, W. E. Howard and P. J. Stiles, J. Phys. Soc. Japan 21, Suppl., 331 (1966).
10] T. Ando, A. B. Fowler and F. Stern, Rev. Mod. Phys. 54, 437 (1982).

TEMPERATURE DEPENDENT RELAXATION RATES IN Si (100) MOSFET ELECTRON INVERSION LAYERS AT LOW TEMPERATURES

J. Nørregaard, J. Hanberg and P.E. Lindelof

Physics Laboratory, H.C. Ørsted Institute
University of Copenhagen, Universitetsparken 5
DK-2100 Copenhagen Ø
DENMARK

Logarithmic quantum corrections to the resistivity of silicon MOSFET inversion layers are well documented in the literature [1], and can be studied in the temperature dependence of the resistivity and the negative magnetoresistance. We have studied several silicon MOSFET structures for varying gate voltage, temperature and magnetic field. The mobilities of our samples were around 5000 cm^2/Vs. Analyzing the low field magnetoresistance, we find good agreement with the theory of Hikami et.al. [2], using only one adjustable parameter, the phase relaxation time τ_φ . The crosses in Fig. 1 show the slopes of the transport relaxation rate versus temperature, plotted as a function of the inverse electron sheet density n^{-1} . The residual transport scattering rate was always much larger than the temperature dependent part at temperatures below 4.2 K. We assume that the scattering rates add according to Mathiesens rule. In order to get extra information about the contributions to the temperature dependent scattering we measured the weak localization magnetoresistance. The phase relaxation time, which is thereby determined could be separated into two terms with different temperature dependence[3,4]

$$\frac{1}{\tau_\varphi} = AT + BT^2 \tag{1}$$

The temperature dependence of the phase relaxation rate was steeper than the temperature dependence of the transport relaxation rate. It turns out, however, that a large part of the phase relaxation rate is due to electron-electron scattering, which indeed do not relax the total momentum of the electrons and therefore do not contribute to the resistance. Fig. 2 shows B in eq.(1) plotted as a function of n^{-1}. Theory predicts a contribution of the form[5]

$$B = \frac{\pi}{2} F^2 \frac{k_B^2}{\hbar E_F} \ln\frac{E_F}{k_B T} \qquad (2)$$

Since $E_F/k_B T$ has only slight variations in the temperature and density region considered, we find that B' is inversely proportional to n. F is the interaction constant, which for silicon (100) electron inversion layers is around

Fig.1. Slope of the transport scattering versus temperature (crosses), Coefficient in eq.(1) (dots) and A calculated from eq.(3) (solid line) all plotted versus the inverse two-dimensional carrier density. Other solid curves are guide to the eye only.

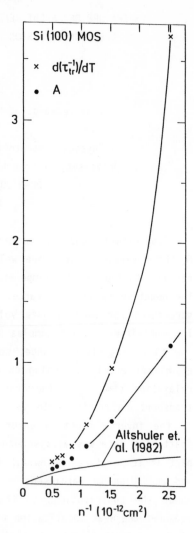

Fig. 2. Coefficient, B, to the quadratic temperature dependent term in the phase relaxation rate plotted versus the inverse carrier density. The straight line is eq. (2). Bended curve is just guide to the eye.

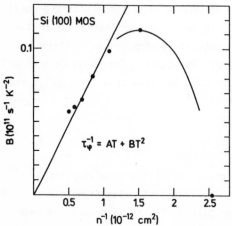

0.95. The expression, Eq.(2), is shown as the straight line in Fig. 2 and we see a perfect agreement at high electron density supporting our argument. Fig. 3 shows the coefficient for the linear temperature dependence of phase relaxation rate, A, as a function of inversion

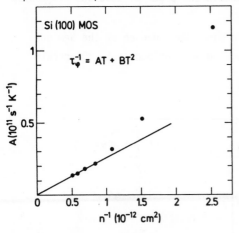

Fig.3. Coefficient to the linear temperature dependent term in the phase relaxation rate plotted versus the inverse carrier density. Straight line is a linear extrapolation of the theory by Altshuler et al.[6]

charge density. Altshuler et al.[6] predicted an electron-electron scattering contribution to the phase relaxation which was mediated by the impurity (Nyquist relaxation) giving a contribution to A' of the form

$$A' = \frac{k_B}{4E_F \tau_{tr}} \ln \frac{2E_F \tau_{tr}}{\hbar} \qquad (3)$$

If the logarithmic term is fixed at its maximum value (highest n in the experiment was $1.9 \cdot 10^{12} cm^{-2}$) this term gives the straight line shown in Fig. 3. As seen this term only explains the observed A at the highest electron concentration. Since this contribution should add to the transport relaxation rate, we show the data in Fig. 1 for comparison. As seen the impurity mediated electron-electron scattering rate almost accounts for the transport relaxation rate at high carrier densities. However, at low densities both the transport relaxation rate and the phase relaxation rate overshoots the calculated value. We believe that this is due to the temperature dependence of the impurity screening, first introduced by Stern[7] and later calculated in detail by Gold and Dolgopolov.[8] These last authors found a contribution to the transport relaxation rate of the form

$$\frac{d\tau_{tr}^{-1}}{dT} = \frac{1}{\tau_{tr}} C'(\alpha,n) \frac{k_B}{E_F} \qquad (4)$$

J. Nørregaard et al.

where α is a parameter which is -1 for ionized impurities in the oxide and $\alpha=0$ for surface roughing at the Si/SiO_2 interface. Table I gives the values of $C(\alpha,n)$ versus n we find from our experiment together with values $C'(\alpha,n)$ given in the paper by Gold and Dolgopolov[8].

Acknowledgement: We are indepted to Ejner Mose Hansen at the Semiconductor Technology Laboratory for advice and for access to MOS growth facilities there.

n ($10"cm^{-2}$)	τ_{tr} (10^{-3}s)	C (α,n)	$C(\alpha,n)$ $(\alpha=-1)$
3.9	3.41	3.61	2.2
6.6	5.12	2.37	1.9
9.3	5.61	1.90	1.8
11.9	5.69	1.58	-
14.6	5.69	1.46	-
17.2	5.51	1.52	-
19.9	5.34	1.48	-

TABLE 1

REFERENCES.

1) Ando, T., Fowler, A.B. and Stern, F., Rev.Mod.Phys. 54, 437 (1982).

2) Hikami, S., Larkin, A.I. and Nagaoka, Y., Prog.Theor.Phys. 63, 707 (1980)

3) Uren, M.J., Davies, R.A., Kaveh, M. and Pepper M., J.Phys.C: Solid State Phys. 14, 5737 (1981)

4) Choi, K.K., Phys.Rev. B28, 5774 (1983)

5) Fukuyama, H. and Abrahams, E., Phys.Rev. B27, 5976 (1983)

6) Altshuler, B.L., Aronov, A.G. and Khmelnitsky, D.E., J.Phys.C: Solid State Phys. 15, 7367 (1982)

7) Stern, F., Phys.Rev.Lett. 44, 1469 (1980)

8) Gold, A. and Dolgopolov, V.T., Phys.Rev. B33, 1076 (1986)

Temperature Dependence of Subband Spectra and DC Transport in Silicon MOSFETs: Impurity Bands and Band Tails

E. Glaser and B.D. McCombe

SUNY at Buffalo

Buffalo, New York 14260

USA

Far infrared subband spectroscopy and electrical transport measurements have been made at temperatures between 4.2K and 70K on n-inversion layers in (100) Si MOSFETs with oxides purposely contaminated with mobile positive ions as well as in "clean", high mobility samples. These studies have been carried out in order to provide further information about the nature of localized states in these structures.[1] The results provide evidence for the existence of long band tails and impurity bands associated with both sets of inequivalent conduction band valleys at low inversion layer electron densities (n_s).

The devices investigated were large (2.5mm x 2.5mm) thick gate, n-channel Metal-Oxide-Semiconductor Field Effect Transistors (MOSFETs) fabricated on 20 ohm-cm p-type (100) Si substrates with mobile positive ions deposited in the oxide (sample 1) and with "clean" oxides (sample 2). The positive oxide charge density was varied by drifting controlled amounts of positive ions (ΔN_{ox}) to the SiO_2-Si interface under gate bias-stress.

The FIR measurements were carried out in a light pipe system with a Fourier Transform Interferometric Spectrometer with the IR electric field polarized perpendicular to the SiO_2-Si interface. The sample mounting arrangement, light coupling scheme, and gate modulation technique have been described elsewhere.[2] Field effect mobility studies as a function of temperature were carried out on the same samples by the usual low frequency ac gate modulation

approach. As shown in the lower panel of Fig. 1, intersubband
resonances (0→1 and 0→2) for sample 1 (ΔN_{ox}=0, μ_{eff}(peak)≅4400cm^2/V-
sec) are observed at 4.2K due to transitions originating in the
ground subband states of the two-fold degenerate conduction band
valleys. Near 20K, an additional peak becomes observable at
energies above the 0→1 transition (but below the 0→2 transition);
this transition becomes <u>dominant</u> for temperatures greater than 30K
as shown in the figure. This peak is attributed to transitions
(0'→1') originating in the ground subband states of the
energetically higher-lying four-fold degenerate conduction band
valleys, which become populated at elevated temperatures. This
observation at such low temperatures (k_BT≅2meV) is unexpected within
an extended state model, since the energy of the E_0' subband has
been calculated[3] for this range of n_s to be very close to the E_1
subband energy (15-20meV above the E_0 subband).

     For comparisn purposes, temperature dependent studies were
also carried out on a "clean" oxide MOSFET (sample 2, μ_{eff}(peak)
≅15,000cm^2/V-sec) as shown in the upper panel of Fig. 1. The
absorption spectra of these two devices for similar values of n_s and
elevated temperature exhibit significant differences in the relative
intensities of the 0→1 and 0'→1' intersubband transitions.
Comparison of the relative occupation factors determined from
fitting the line intensities with Lorentzian oscillators (solid
circles in Fig. 1) with a three subband (E_0,E_0',E_1), extended state
model and Fermi-Dirac statistics yield <u>very</u> <u>small</u> apparent 0-0'
energy separations (≅$\frac{1}{3}$ of those predicted[3]) for sample 1 with
ΔN_{ox}=0. Similar analyses for the "clean" samples gives 0-0' energy
separations in good agreement with the theoretical calculations
(≅14meV). It is concluded that the small <u>apparent</u> 0-0' separations
in poorer quality samples are a result of long band tails associated
with the E_0 and E_0' subbands, which are ignored in the extended
state model.

     Representative differential absorption spectra for sample 1
for several temperatures with ΔN_{ox}=4.9x10^{11}cm^{-2} and n_s=4.2x10^{11}cm^{-2}
are shown in Fig. 2. The previously reported[4] "impurity-shifted"
intersubband transition resonance is observed at 4.2K (arrow). This

line is attributed to transitions originating in an impurity band associated with the ground subband (E_0) and terminating in impurity band states associated with the first excited subband (E_1). For temperatures between 15K and 25K the line broadens and the peak shifts rapidly to higher frequencies. The rapid shift of the peak is ascribed to unresolved contributions to the overall line profile from transitions due to populating bound states (an impurity band) associated with the ground subband of the four-fold degenerate primed valleys (0'). The observed shift is consistent with the results of a recent calculation[5] of the transition energies for electric dipole transitions between impurity states of the two lowest-lying primed subbands. The overall behavior of the resonance profile appears to be due to unresolved contributions from broad intersubband resonances derived from the population of impurity-bound states associated with both the primed and un-primed subbands, and a transfer of relative population from the 0 to the 0'- associated states with increasing temperatures.

This interpretation is consistent with field effect mobility data taken on the same sample (Fig. 3). As shown in a), in the presence of substantial interfacial charge, a second peaked structure (in addition to that due to the impurity band below the E_0 subband) rapidly develops at higher gate voltage with increasing temperature above 10K. Similar structure was reported some time ago on a nominally uncontaminated sample.[6] However as shown in panel b) this second peaked structure clearly develops only above a minimum value of ΔN_{ox}; and its position in gate voltage shifts down with increasing ΔN_{ox}. This peak is attributed to the population of bound states associated with the ground subband of the higher-lying valleys and the corresponding modifications of the conductivity and its derivative. It should be noted that such structure was not observed in measurements on high mobility samples. These results, together with the subband measurements, are consistent with a model of localized states in long band tails extending well below the mobility edges, and the formation of impurity bands for both sets of valleys above a minimum value of added positive oxide charge density $(\Delta N_{ox} \cong 3\text{-}4\times10^{11} cm^{-2})$.

492   E. Glaser & B.D. McCombe

The authors are grateful to G. Kramer and R.F. Wallis for helpful discussions and communication of unpublished results. This work was supported by the Office of Naval Research under contract #N00014-83-K-0219.

REFERENCES

1)   Hartstein, A. and Fowler, A.B., Phys. Rev. Lett. 34, 1435 (1975).

2)   McCombe, B.D., Holm, R.T. and Schafer, D.E., Solid State Commun. 32, 603 (1979).

3)   Vinter, B., Phys. Rev. B15, 3947 (1977).

4)   Glaser, E., Czaputa, R., McCombe, B.D., Kramer, G.M and Wallis, R.F., Surf. Sci. 170, 386 (1986).

5)   Kramer, G.M. and Wallis, R.F., Bull. Am. Phys. Soc. 31, 396 (1986) and to be published.

6)   Fang, F.F. and Fowler, A.B., Phys. Rev. 169, 619 (1968).

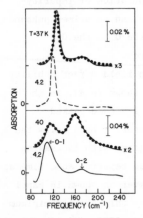

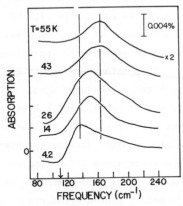

Fig. 1 Lower: differential absorption spectra, sample 1; $\Delta N_{ox}=0$; $n_s=3.7 \times 10^{11} cm^{-2}$. Upper: sample 2; $n_s=4.5 \times 10^{11} cm^{-2}$. $N_{depl.} \cong 1 \times 10^{11} cm^{-2}$ for both. Solid circles: Lorentzian oscillator fits.

Fig. 2 Subband spectra for sample 1; $\Delta N_{ox}=4.9 \times 10^{11} cm^{-2}$; $n_s=4.2 \times 10^{11} cm^{-2}$.

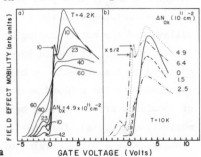

Fig. 3 Field effect mobility data for sample 1.

SUBBAND LANDAU LEVELS IN n-INVERSION LAYERS ON InSb

F. Malcher, G. Lommer, and U. Rössler

Institut für Theoretische Physik, Universität Regensburg,
Postfach 397, D-8400 Regensburg, FED. REP. GERMANY

ABSTRACT

Subband Landau levels in n-inversion layers on p-InSb are
calculated on the basis of self-consistent solutions of the
subband problem for zero magnetic field. We use a 2×2 conduction band Hamiltonian and consider the band effects of
nonparabolicity by higher order terms in the electron momentum $\hbar\underline{k}$. This method is equivalent to multiband-calcula-
tions for zero magnetic field and allows easily inclusion
of the magnetic field.

1. INTRODUCTION

Subband states in inversion layers on narrow-gap semiconductors
are strongly influenced by the coupling between conduction and valence
band. The standard way to consider this coupling in a subband calculation is a multiband formulation of the kinetic energy operator by means
of KANE's matrix Hamiltonian or its extensions.[1] Accordingly the subband problem becomes a set of coupled differential equations for conduction and valence band envelope functions, which form the multicompo-
nent spinor of the subband wave function. TAKADA et al.[2] diagonalized
KANE's Hamiltonian by a $\underline{k}$-dependent unitary transformation but neglected the band-band coupling introduced by the same transformation applied
to the interface potential. They solved the resulting coupled equations
for the conduction band envelope functions self-consistently for InSb.
MARQUES and SHAM[3] were the first to perform self-consistent calculations of the 6 coupled differential equations resulting from the Kane-
Hamiltonian (neglecting the split-off band) in application to InSb.
Finally ANDO[4] demonstrated the validity of the semiclassical approxi-

mation for subband calculations in space-charge layers on narrow gap
semiconductors.

The complexity of the problem caused by the occupation of many
subbands and by the band-band coupling has prohibited so far a self-
consistent calculation of subband Landau levels in n-inversion layers
on InSb.[5] Our approach to this open problem is to consider the nonparabolicity effects of band-band coupling in a 2×2 conduction band Hamiltonian by higher order terms in the electron momentum $\hbar\underline{k}$. Thus, we
avoid the difficulties of a multiband calculation and gain the possibility of considering a magnetic field and the effects of coupling to
remote bands in addition to a self-consistent solution of the subband
problem.

2.   SINGLE BAND FORMULATION OF THE SUBBAND PROBLEM

The subband Hamiltonian for a n-inversion layer including nonparabolicity, magnetic field, and spin-splitting effects arising from
the electric field in the surface layer can be formulated by means of
an invariant expansion as for AlGaAs/GaAs heterostructures.[6,7] We separate from this Hamiltonian, $H=H_0+H_1$, the leading term

$$H_0 = - \frac{\hbar^2}{2m^*} \frac{d^2}{dz^2} + V(z) \tag{1}$$

for which we find a self-consistent solution by numerical integration.
The Hartree potential, which besides the interface barrier, image and
exchange-correlation potential is the dominant part of $V(z)$, is calculated from the charge distribution

$$\rho(z) = N_{Depl}(z) + \sum_{i,occ.} N_i |\xi_i(z)|^2. \tag{2}$$

$N_{Depl}=|N_A-N_D|$ is the depletion charge density and N_i the electron concentration in the ith subband with wave function $\xi_i(z)$. We consider the
effect of nonparabolicity on the subband density of states in the self-
consistent calculation by assuming a subband dispersion

$$E_i(k_\parallel) = E_i - \frac{E_g}{2} + \{(\frac{E_g}{2})^2 + \frac{2}{3}P^2(k^2+\langle k_z^2\rangle)\}^{\frac{1}{2}} - \frac{2P^2}{3E_g}\langle k_z^2\rangle \tag{3}$$

where E_g is the gap energy, P Kane's matrixelement and $\langle k_z^2 \rangle$ the expectation value of $k_z^2=-d^2/dz^2$ for $\xi_i(z)$.

The corrections, H_1, due to nonparabolicity, magnetic and electric
field, are considered in a variational calculation with the expansion

$$\psi = \sum_{iNs} c_{iNs} \xi_i(z) |Ns> \qquad (4)$$

of the complete subband function in products of eigenfunctions of H_0 and Landau states $|N,s>$. This allows to consider the coupling between different Landau levels and subbands due to magnetic and surface electric field induced nonparabolicity corrections. The influence of the magnetic field on the self-consistent potential is not considered.

3.   RESULTS AND DISCUSSION

The validity of our approach has to be checked by comparison with the multiband calculation of MARQUES and SHAM[3] for zero magnetic field. Fig. 1 demonstrates that the subband energies, and the Fermi energy of Ref. 3 are reproduced with high accuracy. The same is true for subband occupations which also agree well with experimental data.[5] Apparently contributions from the valence band envelope functions to the charge density, which have been considered in Ref. 3 but not in our calculation, are of minor importance in InSb, as has been seen already in Ref. 2.

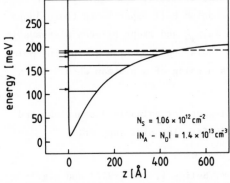

Fig. 1

Self-consistent results for interface potential, occupied subband energies, and Fermi energy of a n-inversion layer on p-InSb. The arrows indicate the results of the multiband calculation of Ref. 3.

In Fig. 2 we show Landau levels arising from the lowest two subbands. These results include contributions from the coupling between Γ_{6c} and Γ_{8v} in all orders of k by proper use of the square root formula for narrow-gap materials. In addition we consider the coupling to the split-off band Γ_{7v} up to fourth order in k. Within these limitations the coupling between different subbands is diagonal in the Landau quantum numbers Ns except for the spin-orbit coupling[7] $\sim \underline{\sigma}(\underline{k} \times \underline{E})$, which couples neighbouring Landau levels of different spin in different subbands.

In Fig. 2 this happens only between $|01{\downarrow}>$ and $|10{\uparrow}>$ around 7 Tesla.

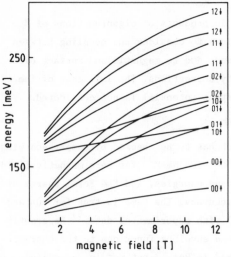

Fig. 2

Subband Landau energies arising from the two lowest subbands for the same parameters
as in Fig. 1. The dominant
component of the eigenvector
of each state is indicated by
the set of quantum numbers
i,N, and s.

The Fermi energy jumps at the magnetic field $\hbar N_s/ez$ between
Landau level z and z+1, irrespective of the quantum numbers i,N,s. In
Fig. 2 this happens e.g. at about 11, 8.8, 7.3, and 6.3 Tesla for
z=3,4,5, and 6, respectively. In quantum Hall experiments these jumps
are usually connected with maxima in ρ_{xx} and steps between plateaus in
ρ_{xy}. If, as e.g. in Fig. 2 at 7.3 Tesla the jump hits a crossing point
between different Landau levels, a missing of a plateau should be expected.

We would like to thank A. Ziegler for valuable discussions and
the Deutsche Forschungsgemeinschaft for financial support.

4.   REFERENCES

1. Kane, E.O., J. Phys. Chem. Solids $\underline{1}$, 249 (1957) and e.g.
   Trebin, H.R., Rössler, U., and Ranvaud, R., Phys. Rev. $\underline{B20}$,
   686 (1979).
2. Takada, Y., Arai, K., Uchimura, N., and Uemura, Y.,
   J. Phys. Soc. Japan $\underline{49}$, 1851 (1980).
3. Marques, G.E., Sham, L.J., Surf. Sci. $\underline{113}$, 131 (1982).
4. Ando, T., J. Phys. Soc. Japan $\underline{54}$, 2676 (1985).
5. Merkt, U., Horst, M., Evelbauer, T., Kotthaus, J.P.,
   Phys. Rev. B (1986 in press).
6. Lommer, G., Malcher, F., and Rössler, U., Phys. Rev. $\underline{B32}$,
   6965 (1985), Superlattices and Microstructures $\underline{2}$, 267 (1986).
7. Bychkov, Y.A., Rashba, E.I., J. Phys. $\underline{C17}$, 6039 (1984).

INVESTIGATION OF THE 2-DEG IN A HgCdTe INVERSION LAYER IN THE QUANTUM LIMIT

W. P. Kirk and P. S. Kobiela

Physics Department, Texas A&M University,
COLLEGE STATION, TEXAS 77843

R. A. Schiebel and M. A. Reed

Central Research Laboratories, Texas Instruments, Inc.
DALLAS, TEXAS 75265

ABSTRACT

Measurements of the quantum Hall effect and the diagonal resistivity in perpendicular and parallel magnetic fields have been used to study the two-dimensional electron gas in a II-VI narrow band-gap compound, $Hg_{1-x}Cd_xTe$.

1. INTRODUCTION

The observation of the quantum Hall effect has been used to study properties of the two-dimensional electron gas (2-DEG) in $Hg_{1-x}Cd_xTe$ in the quantum limit. Due to a high effective g-factor ($g^*=94$) and low effective mass ($m^*=.006\ m_o$), HgCdTe offers an interesting new system to compare with Si (MOSFETs) [1] and III-V (heterostructure) compounds. [2] A percolation threshold of the carriers into extended state behavior is noted at high magnetic fields. The diagonal resistivity displays a positive magnetoresistance at low fields with a logarithmic field dependence.

2. EXPERIMENTAL TECHNIQUE

The 2-DEG was formed in the inversion layer of a HgCdTe MISFET (metal-insulator semiconductor field effect transistor) device. Additional technical detail regarding the MISFET has been described elsewhere. [3] Two samples with significantly different interface preparations were studied. Sample 111 was prepared with a sizeable fixed positive charge at the interface with a cutoff wavelength

λ_c=9.7 μm and corresponding composition x=0.22. Sample 183 was prepared with a negative fixed interface charge with λ_c=10.7 μm and
x=0.21. The carrier density of either sample could be changed by adjusting the gate voltage. In sample 111, for example, the carrier
density was varied from 2.42×10^{11} cm^{-2} to 5.32×10^{11} cm^{-2}, corresponding to mobilities from 6.28×10^{4} cm^2/V·s to 4.89×10^{4} cm^2/V·s.

3.   RESULTS AND DISCUSSION

Figure 1 shows the Hall resistivity ρ_{xy} and the diagonal resistivity ρ_{xx} (at fixed carrier density) for sample 111 at 20 mK. Unexpected absence of plateaus in ρ_{xy} and corresponding ρ_{xx}=0 minima appear at lower fields. A systematic study of ρ_{xy} and ρ_{xx} for a series
of carrier densities indicates that a distinctive quantum Hall effect
(QHE) starts abruptly around 2.0 T. This field is much higher than
expected for observing pronounced Shubnikov-de Haas (SdH) minima, viz.
B ≈0.2 T, when $\omega_c \tau_0$=1, in HgCdTe. A corresponding plot of ρ_{xy} and ρ_{xx}
for sample 183 is shown in Fig. 2. Again there is an absence of well

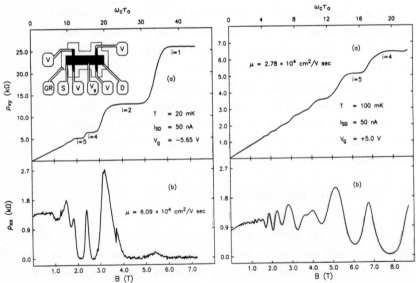

FIG. 1. ρ_{xy} and ρ_{xx} vs B for
sample 111. Inset shows top
view of MISFET.

FIG. 2. ρ_{xy} and ρ_{xx} vs B for
sample 183.
[$\omega_c \tau_0$ on top axis]

developed quantum Hall plateaus at low fields. However, more SdH
structure appears in this sample because its mobility is about a factor two smaller than sample 111; consequently $\omega_c\tau_o=1$ is established at
higher fields. In Fig. 3 we illustrate the abrupt turn on of the QHE
in sample 111 by plotting the reciprocal Hall resistivity ρ_{xy}^{-1} versus
the ratio of the carrier density to magnetic field, n/B. The absence
of quantized Landau levels for i>5, even at higher densities, shows a
percolation threshold of the extended states has been reached with a
sudden onset of the QHE as predicted by Kazarinov and Luryi. [4]

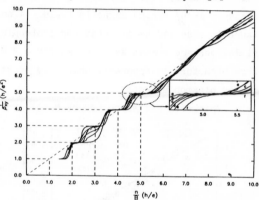

FIG. 3. Inverse of ρ_{xy}
versus n/B for sample
111. Six carrier den-
sities (a-f) are shown
ranging from n=2.48x10^{11}
cm^{-2} to n=4.04x10^{11} cm^{-2}.
Inset magnified 3x hor-
izontally by 2x verti-
cally.

The critical magnetic field B_c, at which the threshold occurs,
can be approximated by $B_c = (h/1.4e)\ n_{sc}$, where n_{sc} is the effective
density of scattering centers. [5] We find $n_{sc} = 4.3x10^{11}$ cm^{-2}, which
corresponds to an isotropic impurity spacing of 147 Å, as compared to
1000 Å determined previously in GaAs. [6] The density n_{sc} can be
compared with the fixed charge density n_{cv} at the interface obtained
by C-V measurements. We find $n_{cv} = 1.6x10^{12}$ cm^{-2}, about a factor four
larger than n_{sc}. This implies clustering or possibly an appreciable
3-D extent of the fixed charge at the interface.

At low fields the positive magnetoresistivity behavior can be
analyzed using the interaction theory of Lee and Ramakrishnan. [7]
The following equation for the magnetoresistance change, $\delta R=R-R_o$, is
used in this analysis, viz.,

$$\left(\frac{\delta R}{R_o^2}\right)\left(\frac{2\pi h}{e^2}\right)\left(\frac{\ell}{w}\right)\left(\frac{1}{(\omega_c\tau_o)^2-1}\right) = (2-F)\ell n(T/T_o)-\tilde{F}_\sigma\ell n(B/B_o).$$

Here $\omega_c \tau_o = \mu B$, $R = \ell \rho_{xx}/w$, with $\ell = 279$ μm, $w = 107$ μm, and $R_o = \ell \rho_{xx}^o/w$ is the sheet resistance per square at B=0. Positive magnetoresistance is associated with the $\ln(B/B_o)$ term. A plot of $\delta R/R_o$ (in reduced form as given by the above equation) versus $\ln B$ and $\ln T$ is shown in Fig. 4(a) and 4(b) respectively. The linear dependence in $\ln B$ yields a characteristic cut-off field of $B_o = 0.530$ T, and a Hartree factor F=0.951, which is about eight times larger than the calculated value, 0.115. There is no significant $\ln T$ dependence, which disagrees with prediction. Finally, Fig. 4(c) shows a plot of the longitudinal magnetoresistance $\delta R/R_o$ versus magnetic field. A small positive magnetoresistance is observed as expected for isotropic Zeeman splitting. SdH oscillations are absent as expected for a 2-D system. At large fields some structure is apparent, which may be caused by spin-orbit scattering or plasmon effects.

This work was supported by Nat'l. Science Foundation DMR8405197.

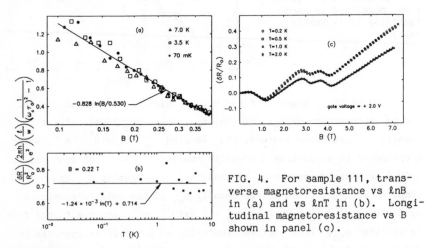

FIG. 4. For sample 111, transverse magnetoresistance vs $\ln B$ in (a) and vs $\ln T$ in (b). Longitudinal magnetoresistance vs B shown in panel (c).

REFERENCES

1. K. von Klitzing, G. Dorda, and M. Pepper, Phys. Rev. Lett. 45, 494 (1980).
2. D. C. Tsui and A. C. Gossard, Appl. Phys. Lett. 38, 550 (1981).
3. W. P. Kirk, P. S. Kobiela, R. A. Schiebel, and M. A. Reed, J. Vac. Sci. Technol. A4, 2132 (1986).
4. R. F. Kazarinov and S. Luryi, Phys. Rev. B 25, 7626 (1982).
5. R. Joynt and R. E. Prange, Phys. Rev. B 29, 3303 (1984).
6. M. A. Paalanen, D. C. Tsui, and J. C. M. Hwang, Phys. Rev. Lett. 51, 2226 (1983).
7. P. A. Lee and T. V. Ramakrishnan, Rev. Mod. Phys. 57, 287 (1985).

MAGNETIC DEPOPULATION OF 1D AND 2D SUBBANDS IN SEMICONDUCTOR STRUCTURES

D. J. Newson [a] and K-F. Berggren [a,b]

(a) Cavendish Laboratory, Madingley Road, Cambridge, CB3 OHE, U.K.

(b) Department of Physics and Measurement Technology, Linkoping University, Sweden.

ABSTRACT

A magnetic field parallel to the plane of a 2DEG (two dimensional electron gas) containing several 2D occupied subbands induces depopulation of these subbands, giving a highly anistropic magnetoresistance[1,2,3]. Similar depopulation occurs in a a narrow channel 2DEG in a perpendicular field, when only several 1D (one dimensional) subbands are occupied. We outline calculations of the effects in semiconductor systems where the confining electrostatic potential is that of the MESFET. We mention the behaviour of the Hall effect in quasi-one dimensional systems.

In a MESFET, electrons are confined by the potential [4],

$$V_E(x) = m^* \omega_0^2 (|x| - t/2)^2 \qquad |x| > t/2 \qquad (1)$$

and zero otherwise. Here $\omega_0^2 = e^2 N_D/(m^* \epsilon)$, N_D is the donor concentration, m^* is the electronic effective mass, ϵ the static dielectric constant and t is the conducting channel thickness. Solving for the subband energies (E_n) for $V_E(x)$ in eq. (1) by the WKB approximation gives an expression with the correct asymptotic limits [4]. A magnetic field in the z- direction (along the channel) gives an effective magnetic potential, $V_B(x) = m^*\omega_c^2(x-x_o)^2/2$ where $x_o = \hbar K_y/eB$. K_y is the electron wave vector in the y-direction and $\omega_c = eB/m^*$. The electrons then move in an effective potential

$$V(x) = V_E(x) + V_B(x) \qquad (2)$$

If $x_o > t/2$ then electrons are confined by a sharp potential well giving large energies at high K_y. E_n/K_y curves were calculated within the WKB approximation for the 3D geometry (depopulation of 2D subbands in $B_{\parallel}$) and the 2D geometry (depopulation of 1D subbands in $B_{\perp}$). Depopulation of 1D subbands is considered first. The parabolic free electron dispersions break up giving flat regions at low K_y (and hence x_o) as the state has strong Landau-like features. $B_{\perp}$ has two effects; increasing the subband energy levels and increasing the DOS allowing more electrons to occupy lower levels. The total electron density in the channel is constant and electrons redistribute to successively depopulate higher levels, giving maxima in ρ_{xx}/B data.

Figure (1) shows subband occupations obtained for a split gate heterojunction FET[5]. The potential across the channel is such that the device was modelled by eq. (1); this is not strictly accurate but is a good approximation to the self-consistent potential[6]. Figure (1) also shows the values of field at which maxima in $1/\rho_{xx}$ are observed. Also shown are the values of B at which 2D Landau level emptying occurs, the form of the data is substantially different to the true 2D case.

Magnetic depopulation of 2D subbands in MESFETs or JFETs can be obtained by extending the analysis to include the extra coordinate. Figure (2) shows depopulation of 2D subbands in $B_{\parallel}$ for a MESFET of doping $7 \times 10^{17} cm^{-3}$. Solid curves show the areal electron concentration, N_e (and the channel width t) at which the Fermi energy coincides the lowest state $E_n(K=0)$ of subband n. Vertical dashed lines show the field at which a Landau level (n_L) in an ideal 2DEG depopulates. Thus, in thin channels the magnetoresistance becomes strongly anisotropic. This model gives good agreement with experimental data[7].

A magnetic field perpendicular to a confined 2DEG induces no voltage between opposite, ideal Hall probes in the absence of current flow along the channel. This is because charge redistribution of state K_y is cancelled by charge redistribution of $-K_y$. Current along the device shifts the 'Fermi line', removing the symmetry between K_y and $-K_y$ at large K_y and a '1D Hall voltage' results because of magnetic field induced electron redistribution. Such curves will show features as subbands are depopulated; in high mobility material plateaux may be seen. However, in high magnetic fields ($h/eB \ll t^2$) the hybrid bands transform in to 2D Landau levels and the ordinary QHE will be seen.

DJN acknoweldges an SERC research studentship. KFB acknowledges partial financial support from the Swedish Natural Science Research Council. We have enjoyed discussions with many colleagues, in particular, M Pepper and T J Thornton.

REFERENCES

1. Nicholas R.J., Brummell M.A., Portal J.C., Cheng K.Y., Cho A. Y., Pearsall, T.P. Solid State Commun. 45, 911 (1983).

2. Ando T., Fowler A.B., Stern F. Rev. Mod. Phys. 54, 437 (1982).

3. Newson D.J., Berggren K-F., Pepper M., Myron H., Scott E.G., Davies G.J. J.Phys.C 19, L403 (1986).

4. Poole D.A., Pepper M., Berggren K-F., Hill G., Myron H.W. J.Phys.C 15, L21 (1982).

5. Thornton, T.J., Pepper M., Ahmed H., Andrews D., Davies G.J. Phys. Rev. Letters 56, 1198 (1986).

6. Laux S.E., Stern F. Applied Physics Letts. 49, 91 (1986)

7. Maan J.C. in 'Two Dimensional Systems, Heterostructures and Superlattices', Ed. G. Bauer et al (series in Solid State Physics 53, Springer Verlag, Berlin 1984).

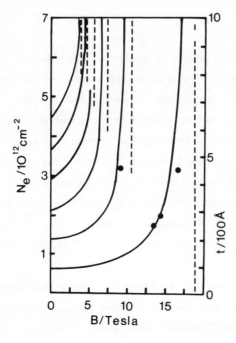

Fig.(1) shows the B dependence of the subband occupations for a Q1DEG. $m^* = 0.067$, $t = 1500A$, $N_e = 1.5 \times 10^{11} cm^{-2}$. Dashed arrows refer to the depopulation of Landau levels in a 2DEG. The full arrows correspond to exptl. peaks in the magnetoconductance of a split gate GaAs/AlGaAs heterojunction FET.

Fig.(2) shows depopulation of 2D subbands in a GaAs JFET doped $7 \times 10^{17} cm^{-2}$ by B . The solid curves give the electron concentration and width at which E_f coincides with the subband bottom. Vertical dashed lines show the field at which a 2D Landau level depopulates. Points are experimental data after Maan (1984). Subbands and Landau levels depopulate at different B, giving an anisotropic magnetoresistance

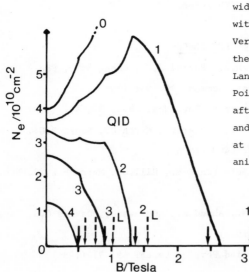

ENERGY LEVELS IN SOME QUASI UNI-DIMENSIONAL
SEMICONDUCTOR HETEROSTRUCTURES

J.A. Brum and G. Bastard

Groupe de Physique des Solides de l'Ecole Normale Supérieure

24 rue Lhomond, F-75005 Paris (France)

L.L. Chang and L. Esaki

IBM Thomas J. Watson Research Center

PO Box 218 Yorktown Heights, NY 10598 (U.S.A.)

ABSTRACT

We report energy level calculations in quasi one-dimensional semiconductor heterostructures. The emphasis is placed on a self consistent solution of the charge transfer in one-side modulation-doped GaAs-Ga(Al)As quantum wire.

Alongside the general efforts towards a reduction of the dimensions of microstructures, there are now some progresses in the achievement of quasi uni-dimensional semiconductor structures[1-3]. They are obtained via sophisticated growth, etching and cutting techniques. In such structures and in the absence of disorder, the carrier motion along two dimensions (z and x) is bound while it is free along the third one (y). It has been theoretically predicted that the carrier mobility in such a quantum wire is considerably enhanced over the bulk and two-dimensional values when the Electric Quantum Limit is reached[4]. Recently, Chang et al.[5] have proposed a new class of non-rigid quasi uni-dimensional semiconductor structures. Also, Laux and Stern[6] have succeeded in performing a self-consistent calculation for a narrow gate Si-Si O_2 MOSFET. Here, we propose approximate solutions of the three-dimensional Schrödinger equation which are well adapted to the present device configurations. The latter are such that the characteristic confinement energy for one direction (z) is much larger than for the other one (x). We then apply our model to a self consistent Hartree calculation of the energy levels and charge transfer in a one-side, spike-doped GaAs-Ga(Al)As semiconductor wire.

The Schrödinger equation for the envelope function is

$$\{ \frac{P_x^2}{2m^*} + \frac{P_y^2}{2m^*} + \frac{P_z^2}{2m^*} + V(x,z) \} \, \psi(\vec{r}) = \varepsilon \, \psi(\vec{r}) \tag{1}$$

where m^* is the carrier effective mass. The y motion is free, admits a parabolic dispersion and will no longer be considered here, except when evaluating densities of states. We assume that $V(x,z)$ can be splitted in the form $V_b(z) + w(x,z)$ where $w(x,z)$ is small compared to $V_b(z)$ in the (x,z) regions where V_b is large (see fig. (1) for rectangular wires). Let $\chi_n(z)$ (ε_n) denotes the nth eigenstate (n^{th} the eigenvalue) of the one dimensional Schrödinger equation for the z motion. $\psi(\vec{r})$ is expanded in the form :

$$\psi(\vec{r}) = \frac{e^{ik_y y}}{\sqrt{L_y}} \sum_n \chi_n(z) \, \alpha_n(x) \tag{2}$$

where $\alpha_n(x)$ is the solution of the set of coupled Schrödinger equations :

$$\sum \{ [\frac{P_x^2}{2m^*} + \int \chi_n^2(z) \, w(x,z) + \varepsilon_n] \delta_{nm} + \int \chi_n^*(z) \, w(x,z) \, \chi_m(z) \, dz \} \, \alpha_m(x) =$$

$$= [\varepsilon - E_n - \frac{\hbar^2 k_y^2}{2m^*}] \, \alpha_n(x) \tag{3}$$

At the zeroth order in the $w(x,z)$ coupling between the E'_ns, the wavefunction is separable in x and z. This approximation will be all the more valid when the size quantization along x is the smaller compared with that along z. We have numerically checked the validity of the decoupling procedure by considerieng the case of rectangular quantum wires. For these structures there is:

$$V(x,z) = V_b Y(z^2 - \frac{L_z^2}{4}) Y(x^2 - \frac{L_x^2}{4}) \ ; \ V_b(z) = V_b Y(z^2 - \frac{L_z^2}{4}) \ ; \ w(x,z) = V_b Y(x^2 - \frac{L_x^2}{4}) Y(\frac{L_z^2}{4} - z^2) \tag{4}$$

where V_b is the barrier height. We found that the zeroth order approximation is excellent for actual quantum well parameters (e.g. $L_z = 100 \,\text{Å}$, $V_b = 150$ meV and $m^* = 0.07 \, m_0$). Only the crossings (zeroth order term) between L_x-related subbands attached to different L_z-related subbands are replaced by anticrossings (full calculations) if the parity allows non-vanishing off-diagonal terms in eq. (3). In summary, for actual parameters and for conduction states the Schrödinger equation is effectively separable.

Consider now the case of a one-side modulation-doped GaAs-Ga(Al)As wire. The donors are assumed to be placed on a sheet separated by a distance d from the $z = -(L_z/2)$ interface (fig. (2)). To ensure thermodynamic equilibrium, some donors are ionized. If the doping is uniform along the x and y directions, the donors will be depleted over a distance l_x. Apart from the contribution due to the undoped structure (eq. 4) the Schrödinger equation should now include a term $-e\varphi_{sc}(x,z)$ which arises from the ionized donors and the carriers. We denote by k_{Fn} the Fermi wavevector for the nth x-related subband and by ε_d the binding energy of a donor in the barrier. We assume that only the ground z-related subband E is occupied. Then :

$$\varphi_{sc}(x,z) = \frac{4e}{\pi\kappa} \sum_{n} k_{Fn} \int_0^\infty \frac{dq_x}{q_x} \cos q_x x \; \{\frac{2}{q_x l_x} \sin(\frac{q_x l_x}{2}) \exp[-q_x |z+d+\frac{L_z}{2}|] - $$

$$- \int_{-\infty}^{+\infty} \chi_1^2(z') \exp[-q_x |z-z'|] \, dz' \int_{-\infty}^{+\infty} \alpha_n^2(x') \cos q_x x' \, dx' \tag{5}$$

where κ is the dielectric permittivity of the heterostructure and $\alpha_n(x)$ is the solution of :

$$\{\frac{p_x^2}{2m} + \int_{-\infty}^{+\infty} dz \, \chi_1^2(z) [V_b \, Y(x^2 - \frac{L_x^2}{4}) Y(\frac{L_z^2}{4} - z^2) - e\varphi_{sc}(x,z)]\} \alpha_n(x) =$$

$$= (\varepsilon - E_1 - \frac{\hbar^2 k_y^2}{2m^*}) \alpha_n(x) \tag{6}$$

In practice the $\alpha_n(x)$'s have been obtained by diagonalizing the effectively x-dependent self-consistent potential of eq.(6) within the basis spanned by the eigensolutions $\alpha_n^{(0)}(x)$ of the flat band x-dependent quantum well. The electrical neutrality has been used in eq.(5). The Fermi level ε_F being constant throughout the heterostructure it should take a value ε_F^{well} in the well which coincides with $V_b - \varepsilon_d - e\varphi_{sc}(-d-L_z/2, \pm l_x/2)$. The calculations have been performed with $V_b = 0.1$ eV, $m^* = 0.07 \, m_0$; $\varepsilon_d = 10$ meV, $L_z = d = 100$ Å and a sheet donor concentration $n_d = 10^{12}$ cm^{-2}. The fig. (3) shows the calculated self consistent potential for $L_x = 250$ Å, where 2 subbands for the x motion are occupied. Generally $-e\varphi_{sc}(x,z)$ displays an asymmetric dipole shape. The dipole is asymmetric inasmatch as the positive charges have a delta distribution along the z axis while the minus charges are spread out. The fig. (4) presents the calculated depletion length l_x versus L_x. One notices that the transfer is never complete. The equality between l_x and L_x is probably recovered at very large L_x, when many x-related subbands are occupied. The curve l_x versus L_x exhibits kinks any time a new x-related subband is occupied. These weak singularities reflect the square root-like behaviour of the occupancy of a given one-dimensional subband versus the Fermi energy.

If ever realized, one-side modulation-doped quantum wires will likely display fluctuations in the doping densities. Such fluctuations will affect the energy levels, the charge transfer and the carrier transport. Similarly, we have assumed that the quantum wire was translationally invariant along the y axis. In practice constrictions should be considered.

This work was supported by A.R.O. contracts and by the "Groupement de Recherches Coordonnées # 70 "Expérimentations Numériques". One of us (J.A.B.) would like to express his gratitude to CNPq (Brazil) for financial support.

REFERENCES

1.   A.B. Fowler, A. Hartstein and R.A. Webb, Phys. Rev. Lett. **48**, 196 (1982).

2.   T.J. Thornton, M. Pepper, H. Ahmed, D. Andrews and G.J. Davies, Phys. Rev. Lett. **56**, 1198 (1986).

3.   W.J. Skocpol, L.D. Jackel, E.L. Hu, R.E. Howard and L.A. Fetter, Phys. Rev. Lett. **49**, 951 (1982).

4.   H. Sakaki, Jpn. J. Appl. Phys. **19**, L735 (1980).

5.   Yia-Chung Chang, L.L. Chang and L. Esaki, Appl. Phys. Lett. **47**, 1324 (1985).

6.   S.E. Laux and F. Stern, Appl. Phys. Lett. **49**, 91 (1986).

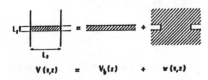

$$V(x,z) = V_b(z) + w(x,z)$$

Fig. 1– Splitting of the (x,y) dependent potential energy $V(x,y)$ of a rectangular quantum wire in terms of a z dependent term $V_b(z)$ and a correcting term $w(x,z)$. The dashed areas correspond to zero potential regions.

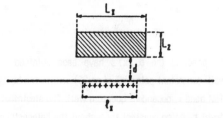

Fig. 2– One–side modulation–doped GaAs–Ga(Al)As semiconductor wire profile. The donors are placed on a sheet separated by a distance d from the $z = -L_z/2$ interface. l_x is the depletion length.

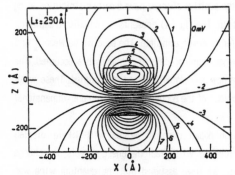

Fig. 3– Electronic potential energy contours for a GaAs–Ga$_{0.8}$Al$_{0.2}$As wire with $L_z = 100$ Å and $L_x = 250$ Å. $V_b = 100$ meV, $m^* = 0.07\, m_0$ and $\varepsilon_d = 10$ meV.

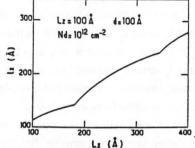

Fig. 4– Calculated depletion length l_x versus L_x. All the others parameters are the same as fig. 3.

ELECTRONIC PROPERTIES OF
QUASI-ONE DIMENSIONAL SEMICONDUCTOR QUANTUM WIRES

W. Y. Lai, S. Das Sarma, X. C. Xie, and Akiko Kobayashi

Department of Physics and Astronomy
University of Maryland
College Park, Maryland 20742, U.S.A.

ABSTRACT

Calculated results for the ground state variational
wavefunction, the density of states, and the
conductivity in a quasi-one dimensional system, as well
as the collective excitation spectra in a multiwire
superlattice are presented.

1. INTRODUCTION

Due to recent advances in technology, it has become possible to
fabricate[1] very narrow conducting channels ("quantum wires") where
electronic motion is quantized in two of the three spatial dimensions.
(y and z directions are chosen throughout in this paper.) We present
our results on the electronic structure specifically for Si- and InSb
based MOS systems, density of states and transport calculations in the
quantum wire, and collective excitation spectra in a multiwire
superlattice.

2. GROUND-STATE VARIATIONAL WAVEFUNCTION

Electronic properties pertaining to the ground state of a
quasi-one dimensional system are investigated. We use a variational
method and solve a two-dimensional Poisson's equation and Schrödinger
equation self-consistently. The following analytic variational
wavefunction is used:

$$\phi(y,z) = Az \, \exp(-b_0 z/2 - b_0 b_1 |y| z/2) F(y), \qquad (1)$$

where A is a normalization constant and F(y) is given by

$$F(y) = \begin{cases} \cos(\frac{\pi y}{w}) + \frac{\pi}{w\gamma} & |y| \leq w/2 \\ \\ \frac{\pi}{w\gamma} \exp(-\gamma(|y|-w/2)). & |y| \geq w/2 \end{cases} \qquad (2)$$

The variational parameters are b_0, b_1, and γ, whereas w is the width
of the strip. Our variational wavefunction defined by Eqs.(1) and (2)
has the character of a particle-in-a-box type wavefunction in the y
direction except that it has tails outside the strip because the
confining potential is not infinite. The wavefunction in z direction
has the Fang-Howard-Stern variational form[2] which has been quite

successful for a regular (i.e. non-narrow) two-dimensional silicon
inversion layer. In Fig.1(a) we show three dimensional plots of the
ground state probability density $|\phi(y,z)|^2$. In Fig.1(b), the ground
state energy E_0 and the width of wavefunction $z_{00}-<z>$ are shown as a
function of $N=N_s+N_d$, where N_s is the inversion layer electronic
density and N_d is the fixed depletion electronic density. For the
sake of comparison, our results for two-dimensional situation are
shown in Fig.1(b) also. From these results, we conclude that the
bending effect (i.e., the intermixing of the y and z parts) is small,
and that the tailing effect is important for narrow strips.

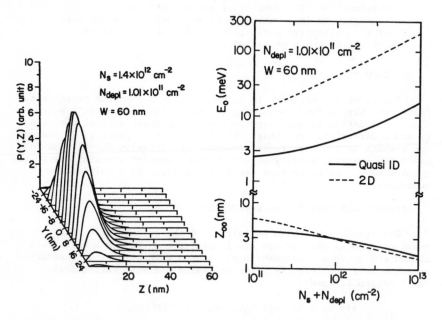

Fig.1(a): Ground state envelope
wavefunction for a quasi-one dimensional
system (along x direction).

Fig.1(b): Ground state energy E_0 and
z-spatial extent z_{00} of the ground state
wavefunction as a function of $N=N_s+N_d$.
Dashed lines refer to the corresponding
results for the two dimensional limit
($w=\infty$).

3. DENSITY OF STATES AND CONDUCTIVITY

   In this section, we present our results on the electronic
density of states and conductivity (σ_{xx}) in the presence of impurities
for a quasi-one dimensional system. We employ the Born approximation

to calculate the level broadening (Γ) due to charged impurities. Interband scattering and screening effect are also taken into consideration. For the sake of simplicity, the 2D dielectric constant[2,3] is used in the screening calculation. The resulting density of states is shown in Fig.2(a). It is seen that the peak intensity corresponding to each energy level becomes finite because of the impurity scattering. In Fig.2(b), σ_{xx} is shown as a function of the electron density $N_s (\alpha/E_F)$. Note that our calculated conductivity exhibits a sudden drop, instead of a jump, each time the Fermi energy passes through quantum level of the system. This is because, although more states are available at each energy level to contribute to the conductivity, we also have a sudden increase in Γ due to the interband scattering channel. The combined mechanisms thus give rise to the decrease in σ_{xx} as one passes through each level.

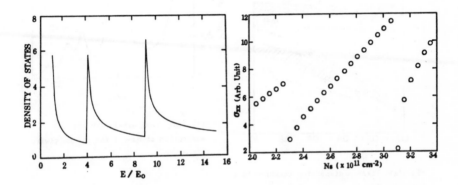

Fig.2(a): Density of states for a quasi-one dimensional system with impurities.

Fig.2(b): Conductivity σ_{xx} as a function of N_s for a quasi-one dimensional wire.

4. PLASMON BAND IN A LATERAL MULTIWIRE SUPERLATTICE

We consider a periodic lateral superlattice consisting of finite two dimensional strips of alternating electron density (multiwire superlattice). Such modulated two dimensional electron system where the electron density changes periodically along one direction has been fabricated recently[4] in silicon inversion layer structures. Using a semiclassical hydrodynamic approach,[5] we obtain the plasmon band structure and far-infrared absorption spectra. In Fig.3(a), the lowest three bands are shown for the following parameters: $b/a=2$; $n_A/n_B=0.3$; $\beta_A^2-\beta_B^2=0.1(\omega_0 a)^2$. a and b are the widths of the strip A and B; n_A and n_B are the equilibrium electron densities; β_A and β_B are the electronic compressibilities. $\omega_0=(2\pi e^2 n_A/\kappa m^* a)^{1/2}$ is taken as the

natural unit of plasma energy in the problem, where κ is the
dielectric constant and m^* is the effective mass. In Fig.3(a), the
band edges are at kd=0 or π where d=a+b and k is the wavevector along
the superlattice direction. The bands are classified by the integer
$m(=m_A=m_B)$ which defines the "local" plasmon modes in the individual
wires. Our result for the long wavelength optical absorption
spectrum[6] is shown in Fig.3(b). We note that the spectrum clearly
shows structures due to higher plasma modes with very weak intensity
for m>2.

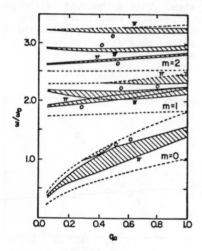

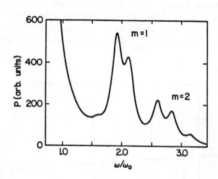

Fig.3(b): Long wavelength optical
absorption at qa=0.1 for the multiwire
structure.

Fig.3(a): Plasmon dispersion relation in a
lateral multiwire superlattice as a
function of wavevector q along the strip.

ACKNOWLEDGEMENT: This work is supported by the Office of Naval
Research and the Army Research Office.

REFERENCES

1. Fowler A. B. et al., Phys. Rev. Lett. 48, 196(1982); Wheeler R. G.
   et al., ibid 49, 1674(1982); Skocpol W. J. et al., Phys. Rev.
   Lett. 49, 951(1982); Petroff P. M. et al., Appl. Phys. Lett. 41,
   635(1985); Warren A. C. et al., IEEE Elec. Dev. Lett. 6, 294(1985).
2. Ando T., Fowler A. B., and Stern F., Rev. Mod. Phys. 54, 437(1982).
3. Das Sarma S. and Lai W. Y., Phys. Rev. B32, 1401(1985).
4. Mackens U., Heitmann D., Prager L., Kotthaus J. P., Beinvogl W.,
   Phys. Rev. Lett. 53, 1485(1985).
5. Eliasson G., Giuliani G. F., and Quinn J. J., Phys. Rev. B33,
   1405(1986).
6. Allen S. J., Tsui D. C., and Logan R. A., Phys. Rev. Lett. 38,
   980(1977).

EXCITONIC NONLINEARITIES IN SEMICONDUCTOR QUANTUM WELLS

by

D. S. Chemla

AT&T Bell Laboratories, Holmdel, NJ

1. INTRODUCTION

In semiconductor quantum wells (QW) excited near the absorption edge unusual transient nonlinear optical processes are observed. Some of these processes are qualitatively and quantitatively different from those seen in bulk material. They originate from the reduced dimensionality of bound and free electron-hole (e-h) pairs induced by the confinement in the ultrathin layers. In this article we review our investigations of the nonlinear optical processes in GaAs quantum well structures (QWS) excited below, at and above the lowest energy exciton resonances.

2. EXCITONIC RESONANCES IN QWS

The linear optical properties of III-V QWS have been reviewed recently[1,2]. For layer thickness Lz smaller or of the order of the bulk exciton Bohr diameter (e.g. ~300 $\overset{o}{A}$ for GaAs) the exciton structure and hence of the absorption resonances are strongly modified. A spectacular result of these changes in the exciton structure which has important consequences in nonlinear optics is the observation of well resolved excitonic resonances at room temperature in III-V QWS [3-5].

3. NONLINEAR OPTICAL EFFECTS

Large changes in the optical constants can be produced in semiconductor quantum wells by selective generation of excitons using resonant excitation or by production above the band gap of unbound electron-hole pairs [6,7,8]. Field-induced nonlinearities originating from virtual transitions can be observed using below- gap excitation [9]. However, even in this condition real population changes can also occur through phonon-assisted transitions, these in turn induce variations in the optical properties similar to those produced for excitation above the absorption edge [10,11].

When excitations long compared to the ionization time are used to probe the nonlinear optical response of QW at RT close to the exciton resonance, it is found to be quite insensitive to the wavelength and to the duration of the excitation [1,6]. It depends only on the density of free e-h pairs created directly or indirectly by thermal phonon ionization of selectively generated bound pairs. Even the very weak absorption in the band tail, many linewidths below the hh-exciton peak, produces the same effects, once phonon-assisted transfer promote e-h pairs into the bands [10]. The optical nonlinearity is huge ($X^{(3)}\sim$ 0.06 esu) it corresponds to very large changes in refractive index and in absorption coefficient [1,6]. These effects remain as long as the e-h plasma is present.

The extreme sensitivity of the exciton to the presence of carriers was utilized to study QW absorption phonon-sidebands. Residual absorption as low as 0.01cm^{-1} can be measured in 1 μm thick samples using a very weak probe centered at the excitons to monitor the transmission as a tunable pump excites the sample well below the gap [10]. On Fig.1 the changes in probe transmission as a function of pump detuning are shown for RT and 70K. On the low temperature curve, a shoulder is clearly resolved one optical phonon energy below the resonance [11].

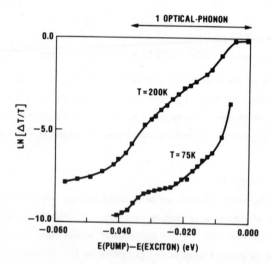

Figure 1 - Phonon assisted below gap absorption in GaAs QW showing the phonon-sidebands of the excitonic resonance.

At low temperature and enough below the resonance so that very few or no real transition occur, a strong transient change in the transmission of the probe is seen. Its temporal as well as spectral behavior are very distinct from that of the plasma induced effect as shown in Fig. 2. The cause of this transient has been identified as the optical Stark shift of the exciton resonances [9]. The transmission change is proportional to the shift of the resonance and hence it is inversely proportional to the the pump detuning and proportional to its intensity. A recent model treating simultaneously the Coulomb interaction responsible for exciton formation and the field induced stimulated absorption-emission of photon involved in virtual transitions, is in excellent agreement with the experiments [12].

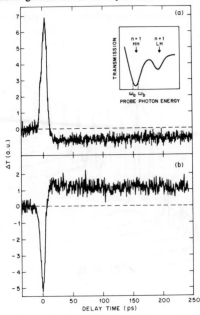

Figure 2 - Time resolved change of the probe beam transmission 1 meV below and 1 meV above the hh-exciton peak for an excitation 25 meV below the resonance. The fast response is due to the exciton AC Stark shift, the slow response to the effects of the e-h plasma.

When excitons are selectively generated by resonant excitation, they induce certain changes in the absorption spectrum. Subsequently as the excitons transform into free pairs, these changes evolve toward those seen using long duration excitations. Using ultrashort optical pulses to excite and probe the sample in times short compared to the exciton ionization time, this process has been resolved [13,14]. This is seen on Fig.3 where presented are the differential absorptions spectra measured with a wide band 80 fs continuum on a sample under excitation resonant with the hh-exciton by a 80 fs pump pulse [14]. A fast and extremely efficient transient bleaching of hh-exciton resonance is seen for the first 0.5 ps. Then the resonance recovers and after about 1 ps the changes in the absorption spectra stabilize to the magnitude and shape observed when e-h plasmas are generated and remain constant for tenths of ps. The ionization time of the hh-excitons at RT deduced from these measurements is t_i~300 fs. The most surprising results of these experiments is that the selective generation of excitons produce a much larger reduction of the excitonic absorption than the generation of free carriers. This behavior is in contradiction with theory [15] and experimental results [16] in bulk semiconductors.

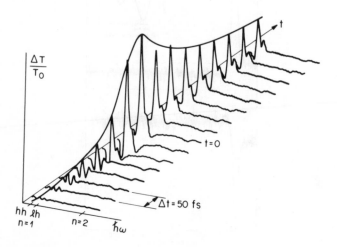

Figure 3 - Absorption differential spectra measured with a broad band (1.45-1.6 eV) 80fs continuum at various delays after the excitation with a 80fs pump pulse resonant with the hh-exciton. First the absorption at the peak of the exciton is very efficiently reduced then after the excitons are ionized by collisions with thermal phonons the spectrum evolve toward that measured with long duration excitations.

The generation of e-h pairs (bound or unbound) modify the optical transitions through several mechanisms [phase-space filling (PSF), the long range Direct Screening (DS) and the short range Exchange Interaction (EI)], whose effectiveness is different if the pairs are free (plasma) or bounded (exciton gas)[15]. In three-dimensions (3D) the dominant effect of e-h pair photogeneration is the DS of free e-h pairs, EI is only important in very dense plasmas [15]. Screening by the neutral excitons is much weaker [16]. The explanation of the transient response of excitons in QW resides in the very special conditions in which excitons are in QW and at RT. First the DS is strongly reduced in two dimensions [17]. In addition the excitons generated with ultrashort optical pulses and observed before they can interact with the thermal reservoir are not in thermodynamic equilibrium with the lattice. They have not yet a "temperature" and occupy the lowest energy state of the crystal. Then their first interaction with the thermal LO-phonons destroys them and releases a warm plasma, which then thermalizes in less than a ps. The PSF and EI are governed by the occupancy of the states that enter in the exciton wavefunction. For the warm plasma Boltzmann distributions, this occupancy is small. As for excitons it can be shown that N_x bound pairs are equivalent to special distributions of electrons and holes: $f_e(k) = f_h(k) = (N_x/2)|U(k)|^2$ where $U(k)$ is the exciton relative motion wavefunction in the phase space [15]. A theory of two-dimensional exciton saturation by e-h plasma and exciton, neglecting the DS, has been developed using the many body formalism [18]. The results are shown in Fig. 4. They explain the most salient features experimentally seen i.e. that at low temperature the plasma is more efficient than the excitons, but that as the temperature increases the plasma efficiency decreases as E_{1S}/kT. The model also gives an absolute value of the nonlinear cross section for exciton generation in very good agreement with experiment [14,18].

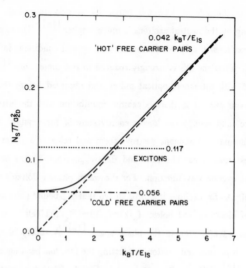

Figure 4 - Plot of the theoretical saturation density (normalized to the exciton area) for exciton gas (dotted line) and for a plasma (solid line) versus the temperature (normalized to the exciton binding energy). This show that the a cold plasma is more efficient than an exciton gas which is more efficient than a warm plasma.

The most important assumption of the theory is that DS is negligible compared to EI and PSF in QWS [18]. This was verified directly in the following manner. Since PSF and EI are effective only when the carrier wavefunction overlap with that of the excitons, a nonthermal carrier population generated in the continuum will first interact with the n = 1 resonance through DS. Then as the carrier thermalize to the bottom of the bands, the effects of the Pauli principle turn on. This switching of the PSF and EI has indeed been observed [14]. On Fig. 5 are presented the differential absorption spectra seen at 50fs interval during and after the generation of a non-thermal carrier distribution in the n = 1 continuum less than one LO-phonon energy above the gap at a density $N_{eh} \approx 2 \times 10^{10} cm^{-2}$. The turning on of the absorption at the n = 1 resonance as the carrier thermalize is clearly seen. Conversely the effects at the n = 2 excitons, which are only sensitive to the DS, do not vary in time, showing that DS remains essentially constant as the carriers thermalize. These results put an upper limit on the magnitude of DS which is at least six times smaller than the sum EI+PSF. In addition they provide direct information on the carrier scattering times these important heterostructures. The thermalization time is 200 fs and the carriers take to leave the states in which they are created in a time less than (but of the order of) 80 fs.

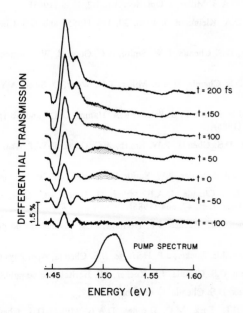

Figure 5 - Differential absorption spectra for a non-thermal carrier distribution generated by a 80fs pump centered at 1.509eV and observed at 50fs interval with a broad 80fs continuum. This resolves the effects of the exclusion principle from that of the direct Coulomb screening.

4. CONCLUSION

The nonlinear optical processes seen in semiconductor QWS present specific features not seen in bulk material. The magnitude of the mechanisms responsible for exciton resonance bleaching are reversed by the quantum confinement and at RT the optical response exhibit unusual transient due to exciton instability to collision with the large density of thermal phonons.

Acknowledgments: This work was performed in close collaboration with D.A.B. Miller and has benefited from extremely valuable contributions from R.L. Fork, A.C. Gossard, J.P. Heritage, C. Hirlimann, W.H. Knox, J. Shah, C.V. Shank, S. Schmitt-Rink, A. Von Lehmen, J.S. Weiner, J.E. Zucker.

REFERENCES

1. D.S. Chemla, D.A.B. Miller, J. Opt. Soc. Am. B2, 1155 (1985).

2. R.C. Miller, D.A. Kleinman, J. Lum. 30, 144 (1985) and D.S. Chemla, J. Lumin. 30, 502 (1985).

3. D.A.B. Miller, D.S. Chemla, P.W. Smith, A.C. Gossard, W. Wiegmann, Appl. Phys. B28, 96 (1982).

4. J.S. Weiner, D.S. Chemla, D.A.B. Miller, T.H. Wood, D. Sivco, A.Y. Cho, Appl. Phys. Lett. 46, 619 (1985).

5. H. Temkin, M.B. Panish, P.M. Petroff, R.A. Hamm, J.M. Vandenberg, S. Sunski, Appl. Phys. Lett. 47, 394 (1985).

6. D.A.B. Miller, D.S. Chemla, P.W. Smith, A.C. Gossard, W.T. Tsang, Appl. Phys. Lett. 41, 679 (1982).

7. D.S. Chemla, D.A.B. Miller, P.W. Smith, IEEE, JQE, QE-20, 265 (1984).

8. J.S. Weiner, D.S. Chemla, D.A.B. Miller, H. Haus, A.C. Gossard, W. Weigmann, C.A. Burrus Appl. Phys. Lett. 47, 664 (1985).

9. A. Von Lehmen, J.E. Zucker, J.P. Heritage, D.S. Chemla, to be published in Optics Lett. Sept (1986).

10. A. Von Lehmen, J.E. Zucker, J.P. Heritage, D.S. Chemla, Appl.Phys.Lett., 48, 1479 (1986).

11. A. Von Lehmen, D.S. Chemla, J.E. Zucker, J.P. Heritage, to be published.

12. S. Schmitt-Rink, D.S. Chemla, to be published.

13. W.H. Knox, R.L. Fork, M.C. Downer, D.A.B. Miller, D.S. Chemla, C.V. Shank, A.C. Gossard, W. Wiegmann, Phys. Rev. Lett. 54, 1306 (1985).

14. W.H. Knox, C. Hirlimann, D.A.B. Miller, J. Shah, D.S. Chemla, C.V. Shank, Phys. Rev. Lett. 56, 1191 (1986).

15. H. Haug and S. Schmitt-Rink, Prog. Quant. Electron. 9, 3, (1984).

16. G.W. Fehrenbach, W. Schafer, J. Treusch, R.G. Ulbrich, Phys. Rev. Lett. 49, 1281 (1982) and G.W. Fehrenbach, W. Schafer, R.G. Ulbrich, J. Lumin. 30, 154 (1985)

17. T. Ando, A.B. Fowler, F. Stern, Rev. Mod. Phys. 54, 437 (1982).

18. S. Schmitt-Rink, D.S. Chemla, D.A.B. Miller, Phys. Rev. B32, 6601 (1985).

OBSERVATIONS OF FORBIDDEN EXCITONIC TRANSITIONS IN GaAs/Al$_x$Ga$_{1-x}$As QUANTUM WELLS IN AN ELECTRIC FIELD

R. T. Collins, L. Viña, W. I. Wang, L. L. Chang, and L. Esaki, IBM Thomas J. Watson Research Center, P.O. Box 218, Yorktown Heights, NY 10598, USA

K. v. Klitzing and K. Ploog, Max-Planck-Institut für Festkörperforschung, D-7000 Stuttgart 80, Federal Republic of Germany

Abstract

Photocurrent spectroscopy has been used to study the unallowed exciton associated with the n=1 conduction subband and the n=2 heavy hole valence subband (h$_{12}$) in a quantum well in an electric field. For well widths greater than 120 Å two excitons are observed in the energy range where only h$_{12}$ should be seen. Based on the electric field, polarization and uniaxial stress dependence of the spectra we conclude that the extra peak arises from mixing between the first light hole and second heavy hole valence subbands.

The excitonic transitions with the largest oscillator strengths in square GaAs/Al$_x$Ga$_{1-x}$As quantum wells are those between conduction and valence subbands which have envelope functions with the same quantum number ($\Delta n=0$). When an electric field is applied perpendicular to the quantum wells, the symmetry is broken and optical matrix elements for "forbidden" transitions ($\Delta n \neq 0$) increase.[1] In particular the h$_{12}$ (n=1 conduction subband and n=2 heavy hole valence subband) exciton should become visible in absorption and related optical measurements. The energy of this exciton will be slightly greater than or less than that of the first light hole exciton (l$_1$) depending on the well width. We have made photocurrent measurements on quantum wells with widths between 80 and 160 Å. In samples with 80 Å wide wells we observe an electric field enhancement of the h$_{12}$ exciton, but, for samples with well widths larger than approximately 120 Å, two peaks become visible in the energy range where only the h$_{12}$ exciton is predicted to occur. Polarization dependent measurements show that the oscillator strength of the additional peak arises from the heavy hole valence band, while under uniaxial stress the energy of this peak shifts as if it is associated with the light hole valence band. We conclude that the ex-

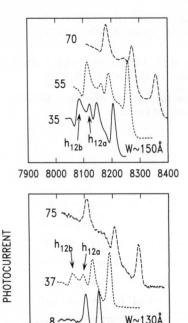

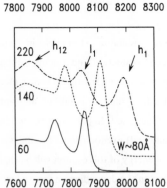

7900 8000 8100 8200 8300 8400

7800 7900 8000 8100 8200 8300

7600 7700 7800 7900 8000 8100
WAVELENGTH (Å)

Fig. 1: Electric field dependent photocurrent spectra showing an extra exciton peak in the energy range of h_{12} in wider quantum well (130 and 150 Å) samples. The approximate field is given to the left of each spectrum in kV/cm. Some spectra at lower fields have been multiplied by a constant to allow the spectra to be seen on the same scale. Measurements were made at 10K.

tra peak arises from a mixing between the first light hole and second heavy hole subbands, and speculate that it is due to mixing between an excited state of the l_1 exciton and the ground state of the h_{12} exciton.

The samples used in this study were GaAs/Al$_x$Ga$_{1-x}$As p-i-n and Schottky barrier photodiodes with two to ten quantum wells imbedded in the depletion region of the diodes. They were grown by molecular beam epitaxy. The top p$^+$ (p-i-n) or metal (Schottky) layer were sufficiently thin to allow light to be transmitted to the quantum wells. The widths of the Al$_x$Ga$_{1-x}$As intrinsic regions were between 0.2 and 1.5 μm.

Figure 1 shows photocurrent spectra for 80, 130 and 150 Å wide quantum wells as a function of electric field. The light is polarized perpendicular to the [100] growth axis. In the 80 Å sample, the h_{12} exciton becomes visible as the field is increased. It has been identified by comparing its energy as a function of field to the results of envelope function calculations. In the 130 and 150 Å wells, two peaks become visible with increasing electric field (h_{12a} and h_{12b}). In the 150 Å sample for fields of 35 and 55 kV/cm, steps are also visible in the spectrum at energies slightly greater than the h_1 (lowest energy heavy hole exciton) and h_{12b} excitons. Such steps have previously been observed and were identified as a superposition of the excited states of an exciton and the continuum edge of the subband to subband transition associated with the exciton.[2] It should be noted that, as

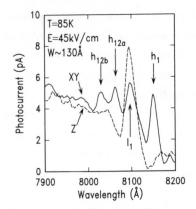

Fig. 2: Polarization dependent photocurrent spectra for 130 Å wide quantum wells.

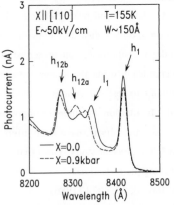

Fig. 3: Photocurrent spectra for 150 Å quantum wells under uniaxial stress.

the field increases, the h_{12a} exciton amplitude decreases until it becomes a step on the high energy side of l_1 and only h_1, l_1, and h_{12b} are seen.

The dependence of the photocurrent spectra on the polarization of the incident light and on uniaxial stress were also investigated. Figure 2 gives spectra for a sample with 130 Å quantum wells imbedded in an $Al_{0.3}Ga_{0.7}As$ waveguide.[3] The sample was illuminated from a cleaved [110] edge with light polarized parallel (Z) or perpendicular (XY) to the [100] growth axis. In the XY polarization h_1, l_1, h_{12a}, and h_{12b} are observed as in Fig. 1., but in the Z polarization only l_1 and the continuum step associated with l_1 are seen. Since Z polarized light should selectively excite optical transitions associated with the light hole valence band,[3] the oscillator strengths of h_{12a} and h_{12b} are predominantly heavy hole in character. Figure 3 gives the dependence of the exciton energies upon uniaxial stress (X) for 150 Å wide quantum wells. The stress was applied along a [110] crystal axis. Since only a small change in the energy of the h_1 exciton was observed, the spectra were shifted to align the h_1 peak. We, therefore, expect the heavy hole excitons to show small shifts, while excitons associated with the light hole valence band exhibit larger shifts.[4] The l_1 and h_{12a} peaks show a shift of approximately 2.0 meV, while the energy of h_{12b} shifts 0.6 meV. From this we conclude that the energy of h_{12a} is predominantly determined by the light hole valence band.

The above results, which show that h_{12a} has heavy hole oscillator strength, but is energetically related to the light hole valence band (for the electric field of Fig. 3) strongly suggest that the peak arises from mixing between the first light hole and second heavy hole subbands. This would explain the absence of extra peaks in the 80 Å wells of Fig. 1, since, as the well width is decreased, the subbands separate and the mixing is reduced. Also, as the electric field is increased, the first heavy and light hole subbands exhibit a larger Stark shift than the second heavy hole subband[1], and the mixing should again be reduced. This explains the decrease in h_{12a} as electric field is increased in Fig. 1 and also serves to identify h_{12b} as the true h_{12} exciton in the limit of large field. Since h_{12a} appears to merge with the continuum of the l_1 exciton, we speculate that it arises from a mixing between the ground state of the h_{12} exciton and an excited state of l_1. That such a mixing may occur has been predicted by Chan[5]. Another possible model for h_{12b} is that it is associated with points of low dispersion away from the center of the Brillouin zone in the first light hole or second heavy hole subbands as predicted in calculations of superlattice band structure which include the effects of valence band mixing.[6]

The observation of this mixed state exciton is significant for a number of reasons. The energies of the excitonic levels in quantum wells can generally be fit quite accurately with simple envelope function calculations which neglect valence band mixing effects. The strongest evidence for mixing has been the observation in optical measurements on symmetric quantum wells of peaks associated with forbidden transitions.[7] In the present study, we observe a peak which cannot be identified energetically using the simple envelope function model, thereby providing a better opportunity to test the various methods of calculating the electronic properties of quantum wells. In addition, a number of important physical parameters in heterojunction systems, such as band offsets, have been inferred from the energies of forbidden excitonic transitions.[8] The present study shows that care must be taken in interpreting the energies and identifying the transitions.

1.) R. T. Collins, K. v. Klitzing, and K. Ploog, Phys. Rev. B, **33**, 4378 (1986).

2.) R. C. Miller, D. A. Kleinman, W. T. Tsang, and A. C. Gossard, Phys. Rev. B, **24**, 1134 (1981).

3.) J. S. Weiner, D. S. Chemla, D. A. B. Miller, H. A. Haus, A. C. Gossard, W. Wiegmann, and C. A. Burrus, Appl. Phys. Lett., **47**, 664 (1985).

4.) F. H. Pollak and M. Cardona, Phys. Rev., **172**, 816 (1968).

5.) K. S. Chan, J. Phys. C, **19**, L125 (1986).

6.) Y. C. Chang and J. N. Schulman, Phys. Rev. B, **31**, 2069 (1985).

7.) R. C. Miller, A. C. Gossard, G. D. Sanders, Y. C. Chang, and J. N. Schulman, Phys. Rev. B, **32**, 8452 (1985).

8.) R. C. Miller, D. A. Kleinman, and A. C. Gossard, Phys. Rev. B, **29**, 7085 (1984).

REDUCED DIMENSIONALITY INDUCED DOUBLET SPLITTING OF HEAVY HOLE EXCITONS IN GaAs QUANTUM WELLS

R. BAUER[a], D. BIMBERG[a], J. CHRISTEN[a], D. OERTEL[a],
D. MARS[b], J.N. MILLER[b], T. FUKUNAGA[c], H. NAKASHIMA[c]

a) Institut für Festkörperphysik der Technischen Universität,
1000 Berlin 12, Germany: b) Hewlett-Packard Laboratories,
Palo Alto, Ca. 94304, USA; c) Optoelectronics Joint Research
Laboratory, Nakahara-Ku, Kawasaki 211, Japan.

ABSTRACT

A localization induced doublet splitting of the lowest energy two-dimensional exciton state $X(e,hh)_{n=1}$ in narrow quantum wells (QW's) is reported. The magnitude of the splitting increases from 0.4meV for 15nm GaAs QW's to 6.2 meV for 1 nm QW's. Excitation intensity and temperature dependent experiments show conclusively and in agreement with time resolved spectra that the splitting is caused by interaction of the $j = 1/2$ electron and the $j = 3/2$ heavy hole state which form $j_t = 1,2$ para - and ortho -exciton states, respectively.

GaAs quantum wells (QW's) present ideal and exceptional model systems for the study of effects of structurally induced localisation on various optical and electronic properties due to the unparalleled chemical and crystallographic quality of the wells and the interfaces to the AlGaAs barriers. Detailed investigations of the structure of the luminescence for wide GaAs QW's ($L_z > 8$ - 10 nm), the observation of a large variety of free and bound exciton states, as well as excited states of an excitonic molecule /3/ were reported in the past by several authors /e.g. 1,2/. The range of well widths covered in our work was not yet investigated hitherto in any detail.

In this contribution the results of a comprehensive investigation of the energetic structure of the radiative recombination of the lowest energy exciton state $X(e, hh)_{n=1}$ in 1 -15 nm wide GaAs QW's is presented. This state consists of an n=1 electron and heavy hole. The luminescence at any temperature /4/ from wells having a width $L_z < 10$ nm is dominated by the recombination of $X(e,hh)_{n=1}$ excitions and is purely intrinsic of character /5/. Excitonic properties equally dominate absorption and nonlinear optical effects of such wells /6/. We present the first unambigous evidence for a doublet character of the $X(e,hh)_{n=1}$ recombination, caused by a L_z-dependent localisation enhanced hyperexchange splitting of the $j = 1/2$ electron and $j = 3/2$ hole states. Our qualitative results and conclusions are found not to depend on the actual chemical composition of the QW or on the method by which it is grown.

Samples grown at two different laboratories under different growth conditions are investigated. A substrate temperature of 620° C and a growth rate r=2.8 Å/s is employed for the Hewlett-Packard samples. The samples from Optoelectronics Joint Research Laboratory are grown at 600° C at a rate of 1.3 Å/s. The growth is interrupted for times varying between 0 s and 120 s at both QW surfaces under As-stabilized conditions. More details on the growth process

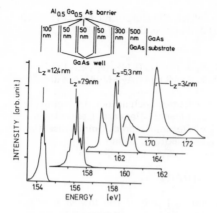

Fig.1: 1.5 K photoluminescence spectra of a sample containing 4 GaAs QW's of different widths. The widths are given on top of the spectra. The sample is grown with interruption of the growth of 120 s at both interfaces of the QW. The inset shows the structure of the sample.

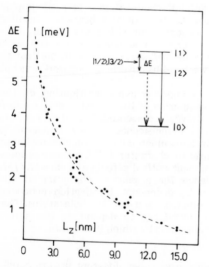

Fig.2: Quantum well width dependence of doublet splitting. Data from more than 20 different QW's are compiled in this figure.

are found in Refs. 7 and 8.

Photoluminescence (PL) spectra are taken at variable temperatures between 1.5 K and 300 K. Lowest excitation densities are 0.1 W/cm^2. Time resolved and time delayed cathodoluminescence (CL) spectra at variable temperatures and excitations are additionally taken in a way described in detail elsewhere /9/.

Fig.1 shows 1.5 K PL spectra taken at I_{exc}= 0.9 W/cm^2 for a sample (T_s = 600° C, r = 1.3 A/s) containing 4 single QW's surrounded by Al$_{0.5}$Ga$_{0.5}$As barriers. The width of the wells is calculated from the growth rate. Excellent agreement is achieved with an independent calculation which is based on a comparison of the spectral position of the luminescence with theoretically predicted one. Three primary peaks are observed for each of the 4 SQW's. The extremely narrow line width of the PL is ascribed to a dissapearence of the microroughness at the QW interfaces upon growth interruption /7, 8/. The interfaces consist of large growth islands differing by just one monomolecular GaAs layer in height. Thus each QW consists of areas m· a/2, (m - 1) · a/2, and (m + 1) ·a/2 wide, where m is an integer and a/2 is half the lattice constant. The center peaks of the 4 spectra of Fig. 1 are assigned with their appropriate QW widths. Closer examination of each of the triplets reveals a doublet substructure. (The doublets are not resolved in this Figure for the narrowest and widest QW's. A lineshape evaluation /7/ makes the splitting clearly visible also in these cases). The size of the doublet splitting ΔE is obviously a function of L_z and increases dramatically with decreasing L_z (Fig.2). The obser-

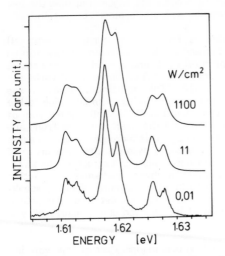

Fig.3: Dependence of the PL spectrum of the L_z = 4.8 nm SQW on excitation intensity. The intensity ratio of the two components of the doublet does not vary.

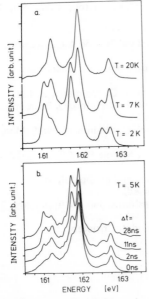

Fig.4: a) Temperature dependence of PL spectrum of a L_z = 4.8 nm SQW b) CL spectra taken at increasing delay time after the end of the exciting pulse.

vation of this splitting and its L_z dependence represents the key result of this paper.

We shall focus now on a few additional results which will help us to distinguish between various interpretations of our observations. One possible explanation, uniaxial stress generated by the small lattice mismatch between AlGaAs and GaAs, can be discarded already at this point as cause of the splitting or of its size, since we observe the strong L_z-dependence also in samples like the one of Fig.1 containing several SQW's of varying width. We also observe the same L_z-dependance of the splitting independent of the number of QW's in the specimen.

Fig.3 shows 1.8 K PL spectra of the same QW as in Fig. 4 for excitation intensities varying over 4 orders of magnitude(which is expected to influence the probability of forming an excitonic molecule). No intensity change of the low energy component with respect to the high energy component is observed. Thus a model assigning the low energy component to an excitonic molecule is definitely ruled out in agreement with results of magnetoluminescence investigations presented elsewhere.

Fig.4a shows the L_z = 5 nm PL spectrum of Fig. 1 at 3 different temperatures. At temperatures larger than 10 K the low energy component rapidly disappears. Fig.4b shows CL spectra of the same QW taken at 5 K and at increasing delay time after the end of a long excitation pulse. Even after a short delay the low energy component has rapidly increased in intensity. Detailed investigations of the transients prove the existence of two life-

times differing by more than one order of magnitude, suggesting that the low
energy component of the doublet is caused by a dipole forbidden transition.

Exchange splitting of the exciton state is found to be consistent with all
experimental results. The $j = 1/2$ electron and $j = 3/2$ hole states combine in an
exciton to yield (in a pseudospherical picture) $j = 1$ and 2   states. These states
are split due to the analytical exchange interaction. In 3-dimensional GaAs the
splitting is extremely small : 0.02 meV /10/. This splitting , however, was
observed to increase dramatically in large magnetic fields up to 20 T due to the
increased compression of the exciton wavefunction /10/. Similarly the size of
the exciton wavefunction decreases strongly and the overlap of electrons and
holes increases /11/ in a QW with decreasing L_z. This increased overlap causes
the increase of radiative recombination probability by more than one order of
magnitude as observed earlier /4, 5/. Apparently increased structural localization (and/or reduced dimensionality) leads to a "hyper-exchange" interaction
more than 2 orders of magnitude larger than in 3 D and, more generally speak-
ing, to largely enhanced many particle effects. A detailed theoretical
calculation of the exchange as a function of localisation and dimensionality is
desirable.

We are grateful to W. Ekardt for illuminating discussions. Part of this work is -
funded by DFG in the framework of SFB 6.

REFERENCES

/1/   J. Singh, K.U. Bajaj, D.C. Reynolds, C.W. Litton, P.W. Yu, W.T.
        Masselink, R. Fischer and H. Morkoc,
        J. Vac. Sci. Technol. B 3, 1061 (1985).
/2/   R.C. Miller, D.A. Kleinmann, W.A. Nordland Jr. and A.C. Gossard,
        Phys. Rev. B 22, 863 (1980).
/3/   R.C. Miller, D.A. Kleinmann, A.C. Gossard and E.O. Muntenau,
        Phys. Rev. B 25, 6545 (1982).
/4/   D. Bimberg, J. Christen, A. Werner, M. Kunst, G. Weimann, and W.
        Schlapp, Appl. Phys. Lett.,49,76 (1986), and J. Christen, D. Bimberg, A.
        Steckenborn, G. Weimann, ibid 44, 84 (1984).
/5/   D. Bimberg, J. Christen, A. Steckenborn, G. Weimann, W. Schlapp, J.
        Luminescence 30, 562 (1985).
/6/   S. Schmitt-Rink, D.S. Chemla, and D.A.B. Miller; Phys. Rev. B 32, 6601
        (1985).
/7/   D. Bimberg, D. Mars, J.N. Miller, R. Bauer, and D. Oertel,
        J. Vac. Sci. Technol.   (1986) ,(to be published).
/8/   T. Fukunaga, K.L. J. Kobayashi, H. Nakashima; Jap. J. App. Phys. 24, L
        510 (1985)/
/9/   D. Bimberg, H. Münzel, A. Steckenborn, and J. Christen,
        Phys. Rev. B 31, 7788 (1985).
/10/   W. Ekardt, K. Lösch, and D. Bimberg,
        Phys. Rev. B 20, 3303 (1979).
/11/   G. Bastard, E.E. Mendez, L.L. Chang, and L. Esaki,
        Phys. Rev. B 26, 1974 (1982).

ELECTRONIC AND OPTICAL PROPERTIES OF GaAs-AlGaAs QUANTUM WELLS IN APPLIED ELECTRIC FIELDS

G. D. Sanders[*]
Universal Energy Systems, Inc.
4401 Dayton-Xenia Road
Dayton, Ohio 45432, U.S.A.

K. K. Bajaj
U. S. Air Force Wright Aeronautical Laboratories
AFWAL/AADR
Wright-Patterson AFB, Ohio 45433, U.S.A.

ABSTRACT

A study of the electronic and optical properties of GaAs-AlGaAs quantum wells in external electric fields is presented using a theory which incorporates valence subband mixing effects. Electric field induced changes in the exciton binding energies and the total absorption spectra are calculated.

Calculations of electronic and optical absorption properties of $GaAs-Al_xGa_{1-x}As$ quantum wells in the presence of electric fields directed along the [001] growth direction are reported. The interest in electro-absorptive effects in quantum wells is strongly motivated by their potential applications in electro-optic devices such as high speed modulators.[1]

In our model, the effective-mass Hamiltonian for the spin 1/2 conduction band electron is given by $H_e = p^2/2M_e^* + V_e(z) - V_F(z)$, where M_e^* is the effective electron mass, $V_e(z)$ is a finite square well potential for the electron, and $V_F(z)$ is the potential due to the electric field of strength F. The effective-mass Hamiltonian for the spin 3/2 hole is given by $(H_h)_{m,m'} = T_{m,m'} + [V_h(z) + V_F(z)] \delta_{m,m'}$, where $V_h(z)$ is a finite square well potential for the hole and m labels the hole spin. The kinetic energy matrix $T_{m,m}'$ is given in the limit of infinite spin orbit splitting by the expression of Luttinger and Kohn.[2] Therefore, effects due to mixing of heavy and light holes are included.

The subband structure and envelope wavefunctions for the quasi-bound states are obtained variationally by expanding the wavefunctions in a basis of 15 even and odd parity Gaussian type orbitals. To stabalize the variational calculation the electric field potential $V_F(z)$ is cut off at the edges of the quantum well. After obtaining the envelope functions and subband energies of the quasi-bound states, the exciton wavefunctions and binding energies in the adiabatic approximation including valence subband nonparabolicity effects as well as the exciton and band-to-band optical absorption can be derived.[3] The results for the exciton binding energy and total absorption spectrum will be presented. A more complete discussion will be presented elsewhere.[4]

Figure 1 shows the variation with electric field of the binding energies of the five lowest lying excitons in a 200Å GaAs-Al$_{0.25}$Ga$_{0.75}$As quantum well. The kinetic energy of an exciton is determined by the joint density of states between the valence and conduction subbands and the potential energy is determined by the degree of overlap between the z-dependent probability densities of the electron and hole.[3,4] For strong fields such that the electric potential drop across the quantum well is much larger than the subband energies ($F > 10kV/cm$ in this example) the zone center hole effective masses are slowly varying while the potential energy varies rapidly due to strong perturbations in the envelope functions. For the HH1-CB1 and LH1-CB1 excitons, the application of a strong electric field results in a separation of charge density along z and a decrease in binding energy. The situation for the HH3-CB1 exciton is more complicated. The HH3 wavefunction along the z direction has three maxima and thus the exciton binding energy has a secondary maximum at a field strength where the CB1 wavefunction maximum overlaps strongly with a secondary maximum in the HH3 wavefunction. In the opposite extreme of small fields the exciton potential is slowly varying and the variation in the binding energy is due to changes in kinetic energy caused by rapid variations in the zone center effective masses of the holes. The very rapid decrease in the LH1-CB1 exciton binding energy as a function of F for F < 10kV/cm is due primarily to a rapid decrease in the exciton reduced mass.

Figure 2 shows the computed absorption spectrum for unpolarized light incident along the [001] growth direction of a 200Å GaAs-Al$_{0.25}$Ga$_{0.75}$As quantum well in the presence of applied field strengths of F = 0kV/cm and F = 30kV/cm. Lorentzian line broadening was used to approximate the experimental situation. As seen in Figure 2, there is a shift toward lower energies of the absorption edge as the field is applied. In addition, the oscillator strength of the $\Delta n = 0$ allowed excitonic transitions is transferred to the $\Delta n \neq 0$ transitions and at F = 30kV/cm the most pronounced excitonic transitions are seen to be those involving states with principal quantum numbers n = 1 and n = 2, i.e., HH2-CB1, HH1-CB2 and LH1-CB2. This is due primarily to changes in the overlap of the electron and hole zone center wavefunctions.

REFERENCES

1. Wood, T. H., Burrus, C. A., Miller, D. A. B., Chemla, D. S., Damen, T. C., Gossard, A. C., and Weigman, W., IEEE J. Quantum Electronics QE-21, 117 (1985).

2. Luttinger, J. M. and Kohn, W., Phys. Rev. 97, 869 (1956).

3. Sanders, G. D. and Chang, Y. C., Phys. Rev. B31, 6892 (1985); Phys. Rev. B32, 5517 (1985).

4. Sanders, G. D. and Bajaj, K. K., to be published.

* Work funded under AFOSR contract F33615-82-C-1716.

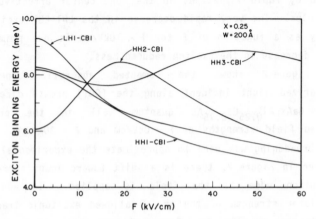

Figure 1. Exciton binding energies of the four lowest lying excitons as functions of the applied field strength F for a 200Å GaAs-Al$_{0.25}$Ga$_{0.75}$As quantum well.

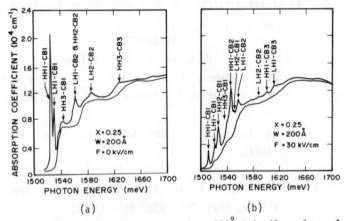

(a)   (b)

Figure 2. Computed absorption spectra in a 200Å GaAs-Al$_{0.25}$Ga$_{0.75}$As quantum well for unpolarized light incident along the [001] growth direction. Applied electric field strengths are (a) F = 0kV/cm and (b) F = 30kV/cm.

FIELD IONIZATON OF EXCITONS IN GaAs/AlGaAs QUANTUM WELLS

K. Köhler, H.-J. Polland, and L. Schultheis

Max-Planck-Institut für Festkoerperforschung, 7000 Stuttgart 80
Federal Republic of Germany

C.W. Tu

AT&T Bell Laboratories, Murray Hill, N.J. 07974, USA

1. INTRODUCTION

The photoluminescence properties of quantum wells (QWs) exposed to electric fields perpendicular to the layers significantly differ from the corresponding behavior of bulk GaAs.[1-3] While excitons in bulk GaAs ionize at fields of ~500V/cm as observed by the strong quenching of the photoluminescence, luminescence of QWs is even observable at electric fields up to 100kV/cm.[1,2] This observation is explained by the strong confinement of the electron hole pairs within the wells (Quantum Confined Stark Effect-QCSE).[4,5] However, ionization of excitons is also expected in QWs for still higher field strengths, limiting the regime of the QCSE.[5,6]

2. EXPERIMENTAL

Measurements for the investigation of the field ionization are performed with two GaAs/Al$_{0.3}$Ga$_{0.7}$As samples grown by Molecular Beam Epitaxy on a Si-doped substrate and containing single QWs with thicknesses of 5nm, 10.7nm, and 21.4nm in sample No.1 and 13.5nm and 27.7nm in sample No.2. Barrier thicknesses between two adjacent wells are 24nm and 14nm for the respective samples. The accurate thicknesses have been evaluated from transmission electron microscopy measurements. The electric field has been applied via a semitransparent Schottky contact formed by a 20nm thick Au film. The samples are kept at low temperatures (5K) and are excited by light pulses with a photon

energy at 1.66eV, i.e. below the bandgap of the AlGaAs. An excitation
intensity between 10^{11} and 5×10^{11} photons/cm^2 per pulse was used which
correspond to peak carrier densities of 10^{15} and 5×10^{15}cm^{-3}, respectively.

3. RESULTS AND DISCUSSION

Time-integrated photoluminescence spectra have been measured for
different voltages. For field strengths up to ~100kV/cm we find a low
energy shift of the emission line of the QW luminescence. Results of
the time- resolved measurements are summarized in Fig.1. At zero electric field (V_{ext}=1V) the lifetimes are shorter for the thinner wells
as a result of the carrier confinement.[7] For voltages down to -3V the
photoluminescence lifetime significantly increases for the thicker

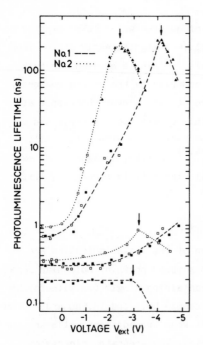

wells, whereas no changes are observed for the thinnest well. These
features are explained by the model
of a finite QW, which leads to an
increase of the lifetime, as shown
by the dotted lines in Fig.2a and
b.[1] However, the strong decrease of
lifetime for the 5nm-, 13.5nm- and
27.7nm- thick QWs as found for larger fields and indicated by arrows
in Fig.1 are not explained in the
theoretical model of the QCSE. We
attribute these features to the
tunneling of carriers through the
barriers which becomes dominant in
the high field regime and competes
with the carrier recombination in
the well. This interpretation is
supported by the strong concomitant
increase of the photocurrent and
sharp decrease in photoluminescence
intensity which we have e.g. observed for 27.7nm-thick QW.

Fig.1. Photoluminescence lifetime versus external voltage for
the QWs in sample No.1 (dashed
line) and No.2 (dotted line)
with barrier thicknesses w_B=24nm
and 14nm, respectively.

To quantitatively understand the carrier tunneling we have evaluated the tunneling time for the electrons. In the WKB approximation[8]
the tunneling time τ_t is given by

$$\tau_t = B \times \exp \left(2 \times \int \sqrt{2m_e/h^2 \times (V(z)-E_e)} \, dz\right)$$

m_e: effective electron mass in the barrier ($0.0916 \times m_0$); $V(z)$: potential for the electrons in the z-direction of the barriers (perpendicular to the layers) including the potential of the electric field; E_e:
electron energy in the well; h: Planck constant and B: cycle time of
the electron of the electron in the well which is assumed constant for
all QWs. B is evaluated by applying the simple WKB model to the
10nm-QW as used for the tunneling resonance calculations of energy
levels by D.A.B. Miller et al.[5] For B=2.3×10^{-14}s we optain accurate
agreement of our theoretical curve with that in Fig.9 of Ref.5. Theoretical curves for the tunneling time are shown by the dashed lines in
Fig.2a and 2b for the QWs with barrier thickness w_B of 24nm and 14nm,

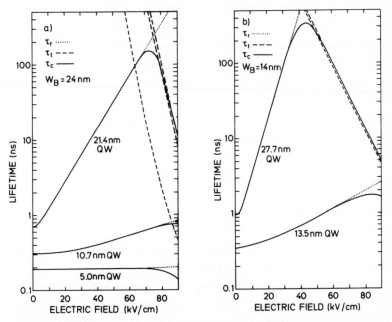

Fig.2a. Calculated radiative lifetime τ_r (dotted line), tunneling
time τ_t (dashed line) and carrier lifetime τ_c (solid line) as a
function of the electric field for sample No.1 (w_B=24nm)
Fig.2b. Corresponding calculated lifetimes for sample No.2 (w_B=14nm)

respectively. The curves for the carrier lifetime, which is given bei $1/\tau_e = 1/\tau_t + 1/\tau_r$ is represented by the solid lines in Fig.2a and 2b. The field-strength where tunneling is experimentally observed is well described by our simple WKB model: Especially the influence of barrier width on the tunneling time is well demonstrated. Extraction of carriers out of the well in sample No.2 (13.7nm- and 27.7nm-thick QW) is observed at weaker fields than in sample No.1 due to the smaller barriers with a thickness $w_B = 14$nm.

It should be noted that a simple estimation of the field strengths n Fig.1 as calculated from the thickness of the depletion layer and the corresponding voltage drop (sum of V_{ext} and the built-in voltage of 1V) leads to values, which must be reduced by a factor 2, to get best fit of the data of Fig.1 with the theory. The discrepancy is not completely understood but might be attributed to screening of the field by charged carriers. However, all experimental data are excellently described by theory, if it is assumed, that 1V corresponds to 15kV/cm.

In conclusion we have demonstrated, that for high electric fields (100kV/cm), field ionization limits the regime of the QCSE and is the dominant mechanism reducing the lifetime of the electron-hole pairs in the well.

REFERENCES

1) Polland, H.-J., Schultheis, L., Kuhl, J.,Göbel, E.O. and Tu,C.W.,
   Phys. Rev. Lett.55, 2610 (1985).

2) Vina, L., Collins, R.T., Mendez, E.E., and Wang, W.I.,
   Phys.Rev.B33,5939 (1986)

3) Yamanishi,M., Usami,M., Kan,Y. and Suemune,I.
   Jpn.J.Appl.Phys.24, L586 (1985).

4) Miller,D.A.B., Chemla,D.S., Damen,T.C., Gossard,A.C., Wiegmann,W.
   Wood,T.H. and Burrus,C.A., Phys.Rev.Lett.53, 2173(1984).

5) Miller,D.A.B., Chemla, D.S., Damen,T.C., Gossard,A.C., Wiegmann,W.
   Wood, T.H. and Burrus, C.A.,Phys.Rev.B32, 1043 (1985).

6. Austin, E.J., and Jaros, M., Appl.Phys.Lett.47, 274 (1985).

7. Göbel,E.O., Jung, H., Kuhl, J., and Ploog, K., Phys.Rev.Lett.51,
   1588 (1983)

8. Schiff, L.I.,"Quantum Mechanics" (McGraw Hill, Tokyo 1968).

THEORY OF QUANTUM WELL EXCITONS IN EFFECTIVE MASS APPROXIMATION

Gerrit E.W. Bauer and Tsuneya Ando

Institute for Solid State Physics, University of Tokyo
Roppongi,Minato-ku,Tokyo 106
JAPAN

Variational calculations of excitons in quantum wells in the effective mass approximation are presented which fully incorporate valence band degeneracies and finite potential barriers. The effect of a magnetic field is also investigated.

Since the pioneering work of Dingle[1] excitons in quantum wells have been subject to numerous studies motivated by the pronounced effects of two-dimensionality in the excitonic features of optical spectra. Reliable theoretical studies of the effect of the complicated valence band structure especially in the presence of external fields have been lacking so far, however. Here we present some results of our calculations of the exciton in GaAs/AlGaAs quantum wells in the many-band effective mass approximation.

The full Luttinger hamiltonian of the hole and the penetration of the wave function into the AlGaAs layer are taken into account. Energies and wave functions are calculated variationally with a basis consisting of the square well eigenfunctions normal to the quantum well and two-dimensional hydrogenic wave functions in the plane of the well with angular momentum components up to l=3 and non-orthogonal radial components of different spatial extension. Except for very narrow wells reasonably converged results are obtained by including 5 subband wave functions for electron, heavy hole and light hole, respectively, and 5 radial wave functions for each of the in-plane 1s, 2s, 2p and 3d angular momentum components. The effects of the cubic anisotropy in the plane of the well ("warping") and the 4f components have been studied and found to be small. Luttinger parameters from Ref. 2 and a valence band offset of 0.4 of the total band gap difference of the well materials have been used.

Results for the binding energy of the (1s) and (2s) heavy hole excitons are displayed in Fig. 1 together with representative experimental data. The agreement between the binding energy difference and the experiment of Miller et al.[3] is very satisfactory, but the magneto-optical experiments[4,5] yield (1s) binding energies which are significantly larger than the present theory which casts some doubt on the accuracy of the high magnetic field extrapolation method. For the heavy-hole oscillator strengths (Fig. 2) the agreement of experiment[6] and theory is reasonable in view of the large experimental error bar. The well size dependence of the oscillator strengths seems to be weaker than found by experiment, however.[6,7] In Figs. 1 and 2 the results for small Al-concentration (x=0.25) start to deviate clearly from those for higher potential barriers at small well widths which could indicate a transition from two to three dimensional behavior.

A difficulty occurs in the calculation of the light-hole binding energy due to the crossing of the heavy-hole continuum with the light-hole (1s) state, which becomes an excited state at well widths smaller than about 150 Å (Fig. 3). With our present basis set the effect of the continuum on the ground state is taken into account by a few discrete states at higher energies (labeled 5d and 6d in Fig. 3). The strong mixing of the light hole resonance with these states at small well widths must be therefore considered to be an artifact of the limited basis set. Still the resonance positions can be traced as indicated by the dotted lines in Fig. 3, although with decreased numerical accuracy. Nevertheless the agreement with the experimental results of Miller et al.[3] is good.

In Fig. 4 we compare our results for the spin-up and spin-down heavy-hole magneto-exciton with the experiment of Ossau et al.[7] and find a rather gratifying agreement. The theoretical spin splitting appears to be too large but will be reduced when the small and negative g-factor of the conduction band electrons is considered. Also indicated in Fig. 4 are the results of 1st and 2nd order perturbation theory which turn out to be accurate for up to only about 1 Tesla (see inset) and experimentally unaccessible, very small energy shifts.

A more detailed discussion of our results, also in the presence of electric fields, will be published in due course.

REFERENCES

1) Dingle, R., in Festkorperprobleme XV, 1975.
2) Hess, K., Bimberg, D., Lipari, N.O., Fischbach, J.U., Altarelli, M., Proc. of the 13th Int. Conf. on the Phys. of Semicond., 1976.
3) Miller, A.C., Kleinman, D.A., Tsang, W.T., and Gossard, A.C., Pys. Rev. B24, 1134 (1981).
4) Maan, J.C., Belle, G., Fasolino, A., Altarelli, M. and Ploog, K., Phys. Rev. B30, 2253 (1984).
5) Miura, N., Iwasa, Y., Tarucha, S., Okamoto, H., Proc. of the 17th Int. Conf. on the Phys. of Semicond., 1984.
6) Masselink, W.T., Pearah, P.J., Klem, J., Peng, C.K., Morkoc, H., Sanders, G.D., Chang, Y.C., Phys. Rev. B32, 8027 (1985).
7) Masumoto, Y., Matsuura, M., Tarucha, S., and Okamoto, H., Phys. Rev. B32, 4275 (1985).
8) Ossau W., Jackel, B., Bangert, E., Landwehr, G. and Weimann, G., Surf. Science, to be published.

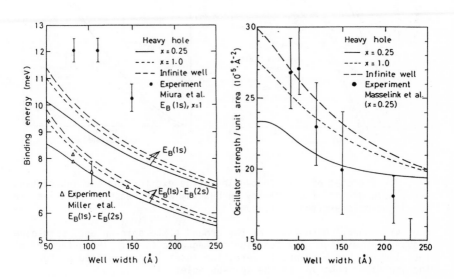

Fig. 1: Binding energy of the heavy-hole exciton ground state $E_B(1s)$ and the difference in binding energies of the ground and first excited state $E_B(2s)-E_B(1s)$ in GaAs/Al$_x$Ga$_{1-x}$As quantum wells.

Fig. 2: Oscillator strength per unit area of the heavy-hole exciton transition in GaAs/Al$_x$Ga$_{1-x}$As quantum wells.

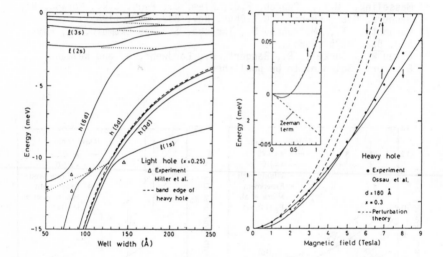

Fig. 3: Energies of the light-hole s-type exciton and the interacting heavy-hole states of d-symmetry in quantum wells measured relative to the light-hole exciton continuum.

Fig. 4: Energy dispersion of the ground state spin-up and spin-down heavy-hole exciton as a function of magnetic field. The g-factor of the electron has been put to zero. The inset is a magnified picture of the low field behavior of the spin-up hole.

NEAR BANDGAP PHOTOEMISSION IN GaAs ;
APPLICATION TO GaAs/GaAlAs 2D STRUCTURES

Romuald Houdré, Henri-Jean Drouhin, Claudine Hermann, Georges Lampel
and Arthur C. Gossard*

Laboratoire de Physique de la Matière Condensée,[†] Ecole Polytechnique
91128 Palaiseau (France)

*AT & T Bell Laboratories, Murray Hill, N.J. 07974 (USA)

Near bandgap photoemission experiments are presented
in bulk GaAs and 2D structures with surfaces activated to negative electron affinity (NEA).

1. INTRODUCTION

It is well known that Cs and O_2 coadsorption on a clean semiconductor surface in ultra high vacuum may lower the work function Φ to
about 1 eV above the Fermi level.[1] In p-type crystals, with bandgap
E_G larger than this value, a situation of NEA is then created : any
electron promoted into the conduction band near the surface by absorption of a photon of energy $h\nu \geqslant E_G$ may be emitted into vacuum and detected by measuring the photoemission current. These systems have a
technological interest for infra-red phototubes, spin polarized[2]
and/or monoenergetic electron sources.[3]

In this paper we shall first show that the energy distributions of
the photoemitted electrons measured in bulk GaAs provide very detailed
information on the band structure near the absorption edge and on the
energy relaxation and transport of conduction electrons.[4,5] Secondly,
these near bandgap photoemission experiments will be extended to layered structures : superlattices (SL) or quantum wells (QW).[6] It will be
indicated how the analysis of energy distribution curves (EDC's) may
reveal the energy spectrum of the conduction states in 2D systems.

[†]Groupe de Recherche du Centre National de la Recherche Scientifique.

2. BULK GaAs

The principle of near bandgap photoemission in NEA GaAs is described in Fig.1. Figure 1a presents a schematic diagram of the valence and
conduction bands of a highly doped p-type sample near and activated surface for which the work function Φ is less than the bandgap E_G, so that
the electronic affinity X_A is negative. Therefore any photoelectron
promoted into the conduction band from one of the three upper valence
bands may be emitted. Since the diffusion length L and the absorption
depth α^{-1} (hν) are both much larger than the mean free path between
collisions with optical phonons, most of the electrons are thermalized
before reaching the band bending region (BBR) of width $\ell_D \sim$ 10 nm. There
they may lose more energy and be emitted into vacuum if their energy is
above the vacuum level. The following remarks may be drawn from this
figure : (i) Most of the photoemitted electrons appear in the Γ-peak
originating from electrons first thermalized to the bottom of the conduction band. Then the photoemission yield may be high if the electron

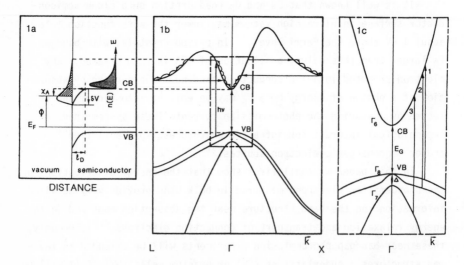

Fig. 1. Conduction (CB) and valence (VB) bands of bulk NEA GaAs ; 1a,
in real space showing the band bending δV on a distance $\ell_D \sim$10 nm ; 1b,
in k-space throughout the Brillouin zone ; 1c, in the vicinity of the
Γ-point with the vertical transitions induced by photons of energy hν,
from the heavy hole, the light hole and the spin-orbit-split valence
bands (respectively 1, 2 and 3).

affinity is low and if $L > \alpha^{-1}$ (hv) : this is the principle of high yield NEA GaAs commercial photocathodes. (ii) The experimental EDC schematized by the hachured area outside the sample reflects the electronic distribution in the bulk crystal (shadowed area inside the sample) modified by energy losses in the BBR and affinity cutoff. Therefore the low energy part of the EDC, rather independent of the excitation energy hv, gives the position of the vacuum level and may provide informations on the BBR. (iii) On the other hand the non-thermalized electrons appear on the high energy side of the EDC. Figure 1b, and the magnification Fig. 1c, evidence that the high energy threshold of the distribution corresponds to electrons excited from the heavy hole valence band (transition 1) and ballistically emitted into vacuum. Each final state of the various vertical transitions, induced by absorption of photons with energy hv, is a source for subsequent thermalization cascade or for photoemission of ballistic electrons. The same holds for the electrons accumulated in the side minima (L and X) before further thermalization in the lowest minimum Γ or emission into vacuum (Fig.1b). Each of these sources gives rise to a structure in the EDC as shown in Fig. 2. These structures are better revealed on the derivative curve, and appear more clearly at low temperature. In this situation the thermalized background is restricted to lower energies and the probability of energy gain by absorption of phonons is very low, leading to sharper thresholds. From a set of EDC's obtained for different hv's the dispersion of the valence and conduction bands in the vicinity of the Γ point, together with the locations of the L and X minima ($\Gamma - L = 0.30$ eV, $\Gamma - X = 0.46$ eV) was determined. It was shown that for kinetic energies in the Γ minimum as high as 1 eV, the Kane $\vec{k}.\vec{p}$ model perfectly fits the data.[5] This photoemission spectroscopy provides the same band information as hot luminescence[7] with a lower resolution but in a larger energy range.

It is known that absorption of circularly polarized light promotes spin polarized electrons in the conduction band of a zincblende semiconductor with a maximum polarization of 50% :[8] when photoemitted, these electrons are all the more polarized as they had less time to depolarize in the conduction band. Therefore the higher their energy, the higher their polarization. This is evident on the right part of Fig.2

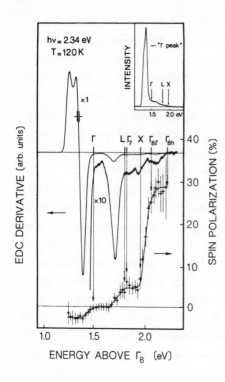

ENERGY ABOVE Γ_8 (eV)

Fig. 2. EDC derivative and PEDC at 120K of electrons emitted from NEA bulk GaAs ((100), p-type $\sim 10^{19}$ cm^{-3}) for 2.34 eV excitation energy. The inset visualizes the shape of the EDC and emphasizes that the main contribution to the photoemitted current originates from the lower energy peak, the so-called "Γ-peak". The experimental resolution (20 meV) is represented by the brackets. The location of the high energy threshold relative to the bottom of the conduction band is easily calculated using the $\vec{k}.\vec{p}$ Kane model (see Ref. 5). The creation energies corresponding to transitions 1, 2 and 3 on Fig. 1a are respectively labeled Γ_{8h}, $\Gamma_{8\ell}$ and Γ_7 ; the positions of the conduction band extrema in the bulk are noted Γ, L and X.

which presents a polarization versus energy distribution curve (PEDC). The various accidents on the PEDC reflect the final states of optical transitions or the effects of spin relaxation during the thermalization cascade. From the analysis versus hν of the maximum polarization, due to ballistic electrons, the mean free path between collisions is found to decrease from 150 nm for thermalized electrons to 75 nm for kinetic energies exceeding the L-valley energy.[9]

3.   2D GaAs/GaAlAs STRUCTURES

The main ideas underlying photoemission experiments from NEA surfaces may readily be extended to other systems and specially to layered structures. Figure 3 compares total photocurrent curves versus photon energy for a bulk GaAs sample and the GaAs/Ga$_{0.68}$Al$_{0.32}$As SL schematized in the upper part of the figure, both activated to NEA. The steps appearing in the SL curves are due to additional photon absorption

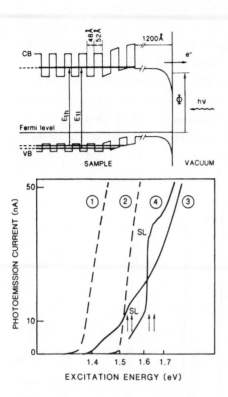

Fig. 3. Upper part : schematics of the GaAs/Ga$_{0.68}$Al$_{0.32}$As SL ; lower part : photocurrent in bulk GaAs 1,(2) and in the SL 3,(4) at 300K (30K).

between the valence and conduction minibands of the SL: they occur at the same photon energy (indicated by the arrows) as the structures observed in the excitation spectrum of the luminescence. This proves that electrons photoexcited in the SL are able to tunnel through the barriers and diffuse towards the GaAs overlayer before being emitted.[6,10] At 30K, the amplitude of the SL step is 1.7 times larger than that of the 120 nm GaAs overlayer. This implies that the photoemitted electrons originate from about 200 nm of GaAs in the SL layers so that at least 40 wells are concerned by the emission process.[11] This is consistent with the quantum mechanical description of SL's in contrast to incoherent tunneling from well to well. The increase of SL contribution when the temperature is lowered is probably due to the lowering of the barrier between the undoped SL and the heavily doped GaAs region.

It is thus very tempting to directly deduce the position of the various minibands from an EDC by observing the ballistic electrons emitted into vacuum from these states. In principle the conduction band offset could easily be measured from the energy separation between the structures related to the bottom of the conduction bands of GaAs and GaAlAs. Figure 4 compares EDC derivatives obtained in three samples for hν = 1.96 eV : (i) bulk GaAs ; (ii) the above described SL ;

(iii) a $Ga_{0.5}Al_{0.5}As$ sample terminated by a 40 Å wide GaAs QW, a 30 Å wide $Ga_{0.5}Al_{0.5}As$ wall and a 30 Å thick GaAs overlayer, all these layers located in the BBR. For the QW sample, the whole photoemission current is due to the GaAs layers since hν is smaller than the $Ga_{0.5}Al_{0.5}As$ bandgap (2.05 eV). The EDC structures are rather difficult to analyze because the energy levels in the BBR are not precisely calculated[12] and we shall pay more attention to the SL sample.

Figure 5 shows a set of EDC derivatives obtained with a He-Ne laser and some of the discrete lines of a Kr[+] laser. At first sight this

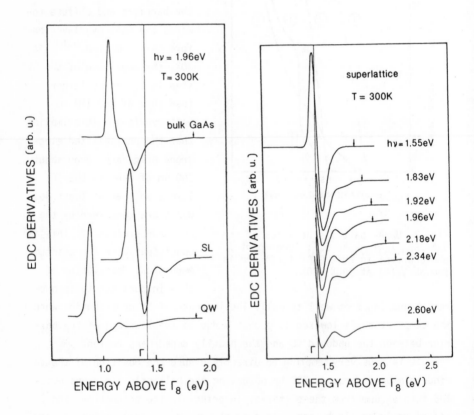

Fig. 4. EDC derivatives of the three samples described in the text for an excitation energy hν = 1.96 eV.

Fig. 5. Set of EDC derivatives of the SL for different excitation energies hν.

set is very similar to the one obtained for bulk GaAs[5] except that
the L feature is more pronounced in the SL. It is probably due to the
fact that once scattered into the L minima, electrons have a reduced
probability of reaching the bottom of the conduction band before emission than in a semi-infinite bulk GaAs sample. The high energy thresholds   are   consistent with high energy ballistic electrons all originating in the GaAs overlayer. No new structure is apparent, neither
at the position of the miniband nor for the SL continuum. We may then
conclude that no ballistic electrons from the SL reach the surface
through the 120 nm thick GaAs overlayer, even though the total current
curve (Fig. 3).evidences the contribution of the electrons excited in
the SL. The low temperature EDC's from the same SL with a diminished
GaAs overlayer should reveal the energy levels associated with the 2D
structures and experiments following these lines are under way.

4.   CONCLUSION

We have shown that near bandgap photoemission in layered structures located close to a NEA surface is potentially a powerful method to
detect the energy levels of 2D systems and to measure their positions
relatively to the bulk energy levels. These experiments have already
given a very direct proof of vertical transport by tunneling in these
systems and experimentally confirmed the quantum mechanical description
of superlattices.

5.   ACKNOWLEDGMENTS

An enlightening discussion with B. Vinter is gratefully acknowledged and we thank P.M. Frijlink for providing us the QW sample.

6.   REFERENCES

[1] Bell, R.L., "Negative Electron Affinity Devices" (Clarendon, Oxford,
1973).

[2] Pierce, D.T., Celotta, R.J., Wang, G.-C., Unertl, W.N., Galejs, A.,
Kuyatt, C.E., and Mielczarek, S.R., Rev. Sci. Instrum. 51, 478
(1980).

[3] Drouhin, H.-J., Hermann, C., and Lampel, G., Phys. Rev. B31, 3872 (1985).

[4] James, L.W. and Moll, J.L., Phys. Rev. 183, 740 (1969).

[5] Drouhin, H.-J., Hermann, C., and Lampel, G., Phys. Rev. B31, 3859 (1985).

[6] Houdré, R., Hermann, C., Lampel, G., Frijlink, P.M., and Gossard, A.C., Phys. Rev. Lett. 55, 734 (1985).

[7] Fasol, G. and Hughes, H.P., Phys. Rev. B33, 2953 (1986).

[8] "Optical Orientation", Vol. 8 of "Modern Problems in Solid State Sciences", edited by F. Meier and B.P. Zakharchenya (Elsevier, Amsterdam, 1984).

[9] Riechert, H., Drouhin, H.-J., and Hermann, C., to be published.

[10] Nozik, A.J., Thacker, B.R., Turner, J.A, and Olson, J.M., J. Am. Chem. Soc. 107, 7805 (1985). This paper deals with quantization effects in the photocurrent of superlattice electrodes in photo-electrochemical cells.

[11] Houdré, R., Hermann, C., Lampel, G., and Gossard, A.C., 6th General Conference of the Condensed Matter Division of the European Physical Society, Stockholm, 1986.

[12] Houdré, R., Hermann, C., Lampel, G., and Frijlink, P.M., Surf. Sci. 168, 538 (1986).

RADIATIVE RECOMBINATION OF TWO DIMENSIONAL CARRIERS IN Si-MOS
STRUCTURES: TIME DEPENDENCE.

F. Martelli

Fondazione "Ugo Bordoni", Viale Europa 190, 00144 Rome, Italy

R. Zachai

Physik Department E16, Technische Universität München,
8046 Garching, F.R.G.

H. Hillmer and G. Mayer

Physikalisches Institut Teil 4, Universität Stuttgart,

Pfaffenwaldring 57, 7000 Stuttgart 80, F.R.G.

Abstract

Time resolved photoluminescence measurements of the
surfacial radiative recombination associated with a two
dimensional space charge layer at the $Si-SiO_2$ interface
show a shortening of the carriers lifetime by increasing
the electric field applied at the interface. The results
are discussed in the frame of the existing models.

1. INTRODUCTION

It has been recently reported that a new radiative recombination
(S-band) occurs in photoexcited silicon in the presence of a two dimensional space charge layer at the $Si-SiO_2$ interface /1,2/. In particular it has been observed that this luminescence occurs only in the
presence of a two dimensional hole gas (2DHG) at the (100) surface /1/
and in the presence of a two dimensional electron gas (2DEG) at the
(110) and (111) surfaces /2/.

The origin of the S-band has been explained supposing that the
creation of e-h pairs by means of the incident light modifies the potential lineshape in the interface region in order to create in Si near
the usual space charge layer (I layer) a second (II), more internal
layer of opposite charge /1-3/. The radiative recombination then occurs
between carriers of the two different layers. Some attempt to support
this idea more rigorously has been done /4/.

The anomalous differences between the (100) and the other two
surfaces and the lack of the S-band in three of the six cases have
been tentatively explained on the basis of the different values of the

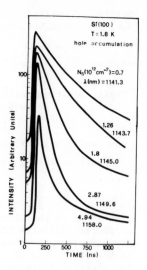

Fig. 1   Time resolved photoluminescence
of the S-band, for different electric
fields in the case of hole accumulation
at the (100) surface. The corresponding
2D carriers concentration is given. The
curves are vertically shifted for major
clarity.

effective electronic masses perpendicular to the interface, which depend on the surface orientation /5/. These differences may justify the
presence or the absence of bound states for the II layer. For a detailed discussion see ref. 2.

Recently it has been claimed that the S-band, for small electric
fields, represents the recombination of a 2D electron-hole liquid /6/.

In spite of the numerous experimental results /1-3,6,7/ the understanding of all the mechanisms underlying the existence of the S-
band are not yet complete. In the attempt to clarify moreover the origin of this luminescence and to check the proposed models we have performed the first time-resolved photoluminescence measurements of the
S-band. We will show that the results are consistent with the model
of ref. 2.

2.   EXPERIMENTAL AND RESULTS.

The samples used in this work were typical Si-MOS capacitors
described elsewhere /2,3/. The (111) and (100) surfaces were investigated. The samples were located in a temperature controlled dewar
(1.8-300 K) and were excited by Ar^+ laser pulses (514 nm, 12 ns pulse
linewidth). The luminescence signal was detected by a fast photomulti-

Fig. 2 Carriers lifetime for the surface luminescence in (111) Si (electron accumulation) and (100) Si (hole accumulation) for different electric fields. The corresponding 2D carriers concentration in the I layer is given.

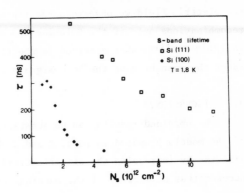

plier and processed by a single photon counting system. The excitation intensity was taken as low as the EHD recombination does not occur.

Fig. 1 shows the temporal evolution of the luminescence intensity at the energy of the maximum for different electric fields at the (100) surface. We remember that the energy position of the S-band shifts toward lower energies as the external electric field increases, slightly depending on the surface orientation /2/. We can see in the figure that the decay of the intensity is not purely exponential, expecially for the highest electric fields. In any case the reduction of the e-h lifetime by increasing the electric field is evident. If we take into account only the first part of the logarithmic decay which can be fitted by a straight line we see that τ varies from $\tau=310$ ns for an electric field corresponding to a 2D hole concentration at the equilibrium $N_s=0.7 \ 10^{12} \ cm^{-2}$ to $\tau \sim 50$ ns for $N_s=5 \ 10^{12} \ cm^{-2}$. Similar results are obtained for the (111) surface in the presence of a 2DEG. In this case τ varies from ~ 530 ns for $N_s=10^{12} cm^{-2}$ to ~ 190 ns for $N_s=10^{13} \ cm^{-2}$.

These results contradict what claimed in ref. 6 where the authors suggested, on the basis of steady state measurements, the e-h lifetime to be indipendent of N_s.

Fig. 2 summarizes our results, which show that for the same electric field the carrier lifetime is shorter in the case of the (100) surface (hole accumulation) than in the case of the (111) surface (electron accumulation).

Some interesting features in the effects of the external electric field on the bulk luminescences (EHD, BE) are also observed. Applying the electric field we observe a shortening of the lifetime of the bulk recombinations only if the S-luminescence is contemporarely observed /8/. This confirms what observed in the time-integrated luminescence about the quenching of the bulk signals when the S-luminescence occurs.

3.   DISCUSSION.

The obtained results can be substantially explained in the frame of the models proposed in refs. 1 and 2. The shortening of the carriers lifetime by increasing the electric field can be explained taking into account that an increse of the external field determivs a removal of the two charge layers from each other. This produces a reduction of the well which binds the carriers in the II layer, which arises from the extracarriers in the I layer /2,6/. This reduction permits to the carriers in the II layer to escape into the bulk with a consequential reduction of the lifetime. This picture is supported by the observation that the carriers lifetime in the states which give rise to the S-band is shorter for the (100) surface in which the II layer is composed of electrons. The electrons are lighter than the holes that compose the second layer for the (111) surface so that they easier migrate into the bulk. This may explain why the carriers lifetime in the II layer is shorter in the case of the (100) surface than in the case of the (111) surface.

REFERENCES

1) P.D. Altukhov, A.V. Ivanov, Yu.N. Lomosov and A.A. Rogachev, Pis'ma Zh. Exsp. Teor. Fiz. 38, 5 (1983) (JETP Lett. 38, 4 (1983));

2) F. Martelli, Solid State Communications 55, 905 (1985);

3) F. Martelli, PhD thesis, Munich 1985, unpublished;

4) P.D. Altukhov, A.M. Monakhov, A.A. Rogachev and V.E. Khartsiev, Fiz. Tverd. Tela 27, 576 (1985) (Sov. Phys. Solid State 27, 359 (1985);

5) T. Ando, A.B. Fowler and F. Stern, Rev. Mod. Phys. 54, 437 (1982);

6) P.D. Altukhov, A.V. Ivanov, Yu. N. Lomosov and A.A. Rogachev, Fiz. Tverd. Tela 27, 1690 (1985) (Sov. Phys. solid State 27, 1016 (1985);

7) Faustino Martelli, Surface Science 170, 676 (1986);

8) More results and details will be published elsewhere.

CARRIER CAPTURE OF QUANTUM WELLS STUDIED BY EXCITATION SPECTROSCOPY

A. Titkov[+], K. Kelting, K. Köhler
MPI für Festkörperforschung, Heisenbergstr. 1, 7000 Stuttgart 80, FRG
A. Shik; A. Ioffe Physico-Technical Inst. AS USSR, Leningrad

+) A. v. Humboldt fellow; permanent adress: A. Ioffe Physico-Technical Inst. AS USSR, Leningrad

Abstract

We observe resonant states in the quantum well continuum of heavy holes. The theory agrees well with the experiment. The various capture probabilities of quantum well regions with different well widths cause either a shift of the luminescence peak energy to other energies or cause a narrowing of the luminescence line.

Efficient carrier capture is of considerible importance for semiconductor devices. Because quantum wells (QW) are known as efficient carrier collectors, theories about the carrier capture of QW have been developed. Early theories [1,2] about carrier capture of QW are based on the classical calculation of LO-phonon-scattering-limited mean free path in bulk GaAs. Recent calculations performed by Kozyrev et al[3] and Blum et al [4] take the quantum aspects of the problem into account. They find that the capture probability of a QW is an oscillating function of the well width.

The samples studied are grown on (100)-orientated n^+-GaAs substrates. Sample 1 consists of three QW with thicknesses of 6.5 nm, 13.5 nm, and 27.7 nm which are separated by 14 nm thick $Ga_{.7}Al_{.3}As$-layers. The QW of sample 3 have thicknesses of 5 nm, 10.7 nm, and 21.4 nm. They are separated by claddinglayers of 23.3 nm $Ga_{.7}Al_{.3}As$. The luminescence is excited with a dyelaser at T = 2 K. A double grating monochromator (f = 0.75 m) together

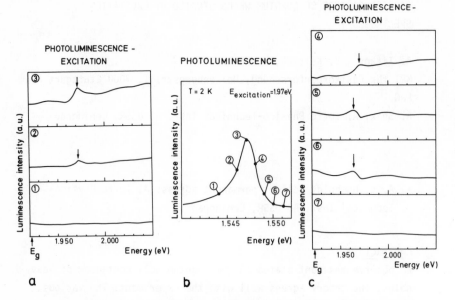

Fig. 1:PLE spectra taken at different detection energies are
shown in a, c. A PL spectrum of a 10.7 nm QW is prsented in b.
The detection energies are indicated by dots. The arrows in the
PLE spectra mark the resonant state.

with a cooled S20-photomultiplier are used for detection. The
excitation spot on the sample is not changed during the measurements and we keep the excitation power constant.

We present in figure 1 one photoluminescence (PL) spectrum
and several photoluminescence excitation (PLE) spectra of that PL
which are excited with photon energies larger than the bandgap
energy (E_g) of 1.922 eV. The PLE spectra exhibit a peak at 1.974 eV
The peak energy decreases to 1.962 eV when the detection energy
increases from 1.546 eV to 1.550 eV. We ascribe this peak to a
resonant state in the QW-continuum of heavy holes. We calculate
the thickness of the GaAs-layer of the QW using the equation[5] :

$$E_p = \frac{h^2 \pi^2}{2L_z^2 m_j} n^2 - V_{o,j} + E_g > E_g$$

j = electron, light hole, or heavy hole,

$V_{o,j}$ = depth of the potential,

$$m_j = \left\{ \begin{array}{l} m_{1h} \\ m_e(1 + E_p/E_g) \\ m_{hh} \end{array} \right. \quad \text{effective masses, data are taken from} \\ \text{reference [6]}$$

The calculated thicknesses are 10.1 nm (1.974 eV) associated with 1.546 eV detection energy and 10.4 nm (1.962 eV) which corresponds to 1.550 eV detection energy. We explain this inversion of luminescence energies with the recombination of bound excitons which determine the luminescence from the thinner region of the QW. A decrease in the intensity of the PLE spectra 5 and 6 at 1.97 eV indicates competion between regions of different well widths.

Furthermore we observe that either the linewidth or the peak energy of the PL-signal changes when the excitation energy is increased above E_g. This effect depends on the QW-thickness. Blum et al calculated the capture time τ of electrons and heavy holes. We take the capture probability P as proportional to $1/\tau$. We calculate the joint product of these single capture probabilities of capturing electrons and heavy holes. The joint product is a function of the well width L_z with sharp local maxima.

The local thickness of real QW can deviate from the nominal thickness. The various regions of different thicknesses within one QW have consequently different capture probabilities and they cause different recombination energies according to their thicknesses. These regions contribute to the PL-signal corresponding to their capture probability. As a result the PL peak energy shifts to lower energies when the thickest regions of the QW have the largest capture probability. The peak energy will be shifted to higher energies when the thinnes regions contribute most to the signal. The PL-signal will become narrower if the regions of intermediate thickness have the largest capture probability.

As a result of our experiments on carrier capture of QW with different well widths we find resonant states in the QW-continuum

of heavy holes. The peak energy of the resonant states agrees well
with the theory. We find also that - depending on the QW-thickness
- and increase of the excitation energy above the bandgap of the
claddinglayer results either in a shift of the PL peak energy
towards higher or lower energies or the PL line becomes narrower.

Acknowledgment
We gratefully acknowledge the TEM micrographs performed by
Dr. H. Oppholzer, Siemens AG, München.

References
[1] H. Shichijo, R. M. Kolbas, N. Holonyak, Jr., R. D. Dupuis,
    P. D. Dapkus; Solid State Commun. 27, 1027 (1978)
[2] J. Y. Tang, K. Hess, N. Holonyak, Jr., J. J. Coleman,
    P. D. Dapkus; J. Appl. Phys. 53, 6043 (1982)
[3] S. V. Kozyrev, A. Ya. Shik; Sov. Phys. Semicond. 19(3),
    1024 (1986)
[4] J. A. Blum, G. Bastard; Phys. Rev. B 33, 1420 (1986)
[5] G. Bastard; Phys. Rev. B 30, 3547 (1984)
[6] S. Adachi; J. Appl. Phys. 58, R1 (1985)

NEW TRANSITIONS IN LIGHT SCATTERING BY THE TWO DIMENSIONAL ELECTRON GAS IN HIGH MAGNETIC FIELDS

A. Pinczuk

AT&T Bell Laboratories, Murray Hill, N.J. 07974, U.S.A.

D. Heiman

MIT Francis Bitter National Magnet Lab., Cambridge,
Massachusetts 02139, U.S.A.

A. C. Gossard and J. H. English

AT&T Bell Laboratories, Murray Hill, N.J. 07974, U.S.A.

We report resonant inelastic light scattering by 2D electron gases in multiple GaAs quantum wells. In high magnetic fields normal to the plane we observe combined Landau level-intersubband excitations, Landau level transitions and quasi-3D magnetoplasmons.

In this communication we report the observation of new transitions in inelastic light scattering by the two dimensional electron gas in high magnetic fields. The electrons are confined to the conduction subband states of GaAs quantum wells in modulation doped GaAs-(AlGa)As heterostructures. The photon energies are close to resonance with the optical transitions from the upper valence subbands (derived from the Γ_8 heavy and light valence states) to the lower conduction subbands. We discovered that this approach carries new selection rules. They allow scattering by elementary excitations that were not observed in spectra obtained with photon energies in resonance with the spin-orbit split-off optical gap.[1] The most unexpected are those in which the electrons undergo a combined change of Landau level and subband state. Combined transitions have been observed in far infrared optical absorption experiments in which *tilted* magnetic fields provide the coupling between in-plane and normal to the plane motions required for a simultaneous change in subband index and Landau level number.[2,3] In light scattering the combined transitions are seen with magnetic fields *normal* to the plane. We attribute the new light scattering selection rules to the mixing between subband and in-plane motions in the heavy and light valence states of the quantum wells.[4-10] These states enter in the two *virtual* optical transitions of the light scattering process.

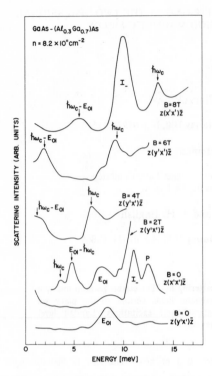

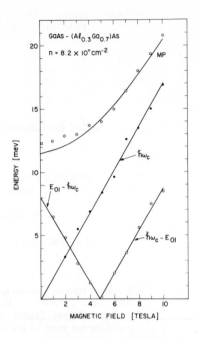

Figure 1: resonant inelastic light scattering spectra for different values of field B.

Figure 2: energies of combined Landau level intersubband transitions ($E_{01} - \hbar\omega_c$ and $\hbar\omega_c - E_{01}$) and Landau level transitions ($\hbar\omega_c$) as function of magnetic field. MP is the magnetoplasma mode of the multilayers.

Figure 1 shows spectra measured in a sample that consists of 15 periods of d = 980Å and the GaAs quantum wells have thickness d_1 = 270Å. The electron mobility is μ = 1.3 × 10⁵ cm²/Vsec. The sample was immersed in superfluid He and fields B ⩽ 10 Tesla were applied along the z ≡ (001) direction normal to the layers. The spectra were excited with a tunable IR dye laser (LD-700) operating cw between 7850−8050Å. The photon energies (1.54−1.58 eV) are 30-50 meV above the fundamental gap. A backscattering geometry was used in the Faraday configuration with F:4 collection optics. The in-plane component of the scattering wave-vector is k ⩽ 10⁴ cm⁻¹, and the normal component is k_z ≃ 5.9 × 10⁵ cm⁻¹. Polarized and depolarized spectra are labeled z (x'x') z̄ and z (y'x')z̄. The z (y'x') z̄ spectrum for B − 0 shows the peak at E_{01} = 8.0 meV, the spacing between the two lowest conduction subbands.[11] The z (x'x') z̄ spectrum shows the two additional peaks labeled I₋ and P. I₋ is assigned to the lowest collective intersubband excitation[11,12] and P to the plasmon[13,14] of the multilayers. The spectra for B ≠ 0 show two new bands. For B ⩽ 4 Tesla we have the bands

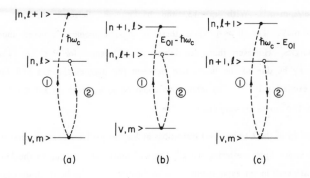

Figure 3: Optical transitions in resonant inelastic light scattering.
The numbers indicate the order of the transitions.

at $E_{01} - \hbar\omega_c$ and $\hbar\omega_c$; and for $B > 4$ Tesla the new bands are at $\hbar\omega_c - E_{01}$ and $\hbar\omega_c$ ($\omega_c = eB/m^{*c}$ is the cyclotron frequency of electrons). Figure 2 shows the positions of the bands at $E_{01} - \hbar\omega_c$, $\hbar\omega_c$ and $\hbar\omega_c - E_{01}$ for $B \leqslant 10$ Tesla. The lines are the calculated energies. Good agreement is found with $E_{01} = 8.0$ meV and $m^* = 0.068$ m_o.

Figure 3 shows the two optical transitions in light scattering at energies $\hbar\omega_c$, $E_{01} - \hbar\omega_c$ and $\hbar\omega_c - E_{01}$. $|n,\ell>$ are conduction states with subband index n and Landau level quantum number ℓ. $|v,m>$ are the valence states derived from the fourfold degenerate Γ_8 states of GaAs. v is the subband index and m the Landau level quantum number. For $B = 0$ Figs. 3(b) and 3(c) describe a new mechanism for stokes and antistokes scattering by intersubband excitations. *Valence band mixing* [4-10] allows the change in subband index within the *dipole approximation*. The observation of combined transitions for $B \neq 0$ can also be interpreted in terms of valence band mixing. It is well known that the states $|v,m>$ consist of admixtures of heavy and light states with harmonic oscillator wavefunctions of indices in the range $m - 2 \leqslant \ell' \leqslant m + 1$.[4,5,7] The additional mixing introduced by the magnetic field allows scattering by new transitions, including the excitations with energies $\hbar\omega_c$, $E_{01} - \hbar\omega_c$ and $\hbar\omega_c - E_{01}$.

In Fig. 2 MP are the magnetoplasma modes of the multilayer structure. Since $k_z d \approx 6 \sim 2\pi$ we should observe the mode in which all the planes oscillate in phase.[14] The line is the magnetoplasma energy calculated with $\omega_p(B) = \left[\omega_p^2(0) + \omega_c^2\right]^{1/2}$ and

$\omega_p^2(0) = 4\pi n e^2/\epsilon_s m^* d.$[15] With $\epsilon_s = 13$, the background dielectric constant, we obtain $\hbar\omega_p(0) = 11.4$ meV. For $B > 4$ Tesla there is good agreement between measured and calculated energies. However, there are substantial differences for $B \leqslant 4T$. The low field discrepancy may be due to the interaction between the magnetoplasmon and the collective intersubband excitation I_-.[16,17] The interaction could be strong even for $kd < 0.1$ because the energies of the two modes are very close for $B = 0$.

The full widths of the Landau level transitions at $\hbar\omega_c$ are ~ 1 meV, much smaller than those reported in previous light scattering work.[1] The difference may be due to the lower electron temperatures achieved in our experiments. Carrier heating is caused by the laser beam, which is less severe for the lower photon energies used in the present experiments. We hope that by using smaller incident powers it would be possible to carry out light scattering investigations of the two dimensional electron gas at low temperatures.

Acknowledgement: The Francis Bitter National Magnet Lab. is supported by the National Science Foundation through its Division of Materials Research.

[1] Worlock, J. M., et al., Solid State Commun. 40, 867 (1981); Tien, Z. J., et al., Surf. Sci. 113, 89 (1982).

[2] Beinvogl, W. and Koch, J. F., Phys. Rev. Lett. 40, 1736 (1978).

[3] Ando, T., Phys. Rev. B19, 2106 (1979).

[4] Altarelli, M., Ekenberg, U. and Fasolino, A., Phys. Rev. 32, 5138 (1985); Altarelli, M., Festkoerperprobleme 25, 381 (1985).

[5] Broido, D. A. and Sham, L. J., Phys. Rev. B31, 888 (1985); Yang, S. E., Broido, D. A. and Sham, L. J., Phys. Rev. B32, 6630 (1985).

[6] Sooryakumar, R., et al., Solid State Commun. 54, 859 (1985).

[7] Bangert, E. and Landwehr, G., Superlatt. Microstruc. 1, 363 (1985).

[8] Chang, Y. C. and Schulman, J. N., Superlatt. Microstruc. 1, 357 (1985).

[9] Ando, T., J. Phys. Soc. Japan, 54, 1528 (1985).

[10] Miller, R. C., Gossard, A. C., Sanders, G. D., Chang, Y. C. and Schulman, J. N., Phys. Rev. B32, 8452 (1985).

[11] Pinczuk, A. and Worlock, J. M. Surf. Sci. 113, 69 (1982).

[12] Burstein, E., Pinczuk, A. and Mills, D. L., Surf. Sci. 98, 451 (1980).

[13] Olego, D., Pinczuk, A., Gossard, A. C. and Wiegmann, W., Phys. Rev. B25, 7867 (1982).

[14] Pinczuk, A., Lamont, M. G. and Gossard, A. C., Phys. Rev. Lett. 56, 2092 (1986).

[15] Tsellis, A. C. and Quinn, J. J., Phys. Rev. B29, 3318 (1984).

[16] Das Sarma, S., Phys. Rev. B29, 2334 (1984).

[17] Jain, J. K. and Das Sarma, S., work in progress (private communication).

OBSERVATION OF SYMMETRY FORBIDDEN TRANSITIONS IN THE PHOTOREFLECTANCE
SPECTRA OF A GaAs/GaAlAs MULTIPLE QUANTUM WELL

H. Shen, P. Parayanthal and Fred H. Pollak

Physics Department, Brooklyn College, Brooklyn, N.Y. 11210, USA
and
Arthur L. Smirl, J.N. Schulman, R.A. McFarlane and Irnee D'Haenens

Hughes Research Laboratory, Malibu, CA 90265, USA

We have measured the photoreflectance spectra from a
$100\overset{\circ}{A}/150\overset{\circ}{A}$ GaAs/GaAlAs ($x \approx 0.17$) multiple quantum well (MQW)
at 300K and 77K. In addition to all the allowed quantum
transitions we have clearly observed features from several
symmetry forbidden transitions even at 300K. There also
is evidence for unconfined transitions. There is very
good agreement between experiment and a theoretical
calculation.

Electromodulation (electroreflectance and photoreflectance) is
rapidly becoming an extremely powerful tool to study microstructural
geometries (superlattices, quantum wells and heterojunctions) in semiconductors.[1,2] We have measured the photoreflectance (PR) spectra at
300K[3] and 77K from a 100A/150A GaAs/GaAlAs ($x \approx 0.17$) multiple quantum
well (MQW) grown by molecular beam epitaxy at Hughes Research Laboratory. The PR technique has been described in the literature.[1,2,4] We
used the 6328$\overset{\circ}{A}$ line of a He-Ne laser as the pump beam. The entire
spectra from the MQW have been fit by a third-derivative functional
form lineshape expression.[2,4] In addition to all the allowed quantum
transitions we have clearly observed features from several symmetry
forbidden transitions (SFT). There also is evidence for unconfined
transitions, i.e. energies above the band gap of the GaAlAs barrier
layer.[5] There is very good agreement between the experimentally determined energies of the various features and a theoretical calculation.[6]

Shown by the dotted lines in Figs. 1 and 2 are the experimental

PR spectra at 300K and 77K, respectively. The spectra were taken in the "low field" limit, i.e., the lineshapes were independent of pump beam intensity. The solid lines in these figures are a least-squares fit of the experimental data from the MQW to the Aspnes third-derivative functional form:[2,4]

$$\frac{\Delta R}{R} = Re\left[\sum_{j=1}^{P} C_j e^{i\theta_j}\left(E - E_{g,j} + i\Gamma_j\right)^{-m_j}\right] \qquad (1)$$

where E is the photon energy, p is the number of spectral features to be fit, C_j, θ_j, $E_{g,j}$, and Γ_j are the amplitudes, phases, energies and broadening parameters of the jth structure and m_j denotes critical point type. The energies of the various features are given by arrows at the top of the figure.

The E_o structure at 1.415 eV for 300K and 1.503 eV for 77K corresponds to the lowest direct band gap of GaAs and originates from the GaAs substrate. At 77K we have been able to fit this transition by Eq. (1) (see Fig. 2) but at 300K this feature appears to exhibit Franz-Keldysh oscillations and hence cannot be accounted for by Eq. (1).

In Figs. 1 and 2 the notation nmH(L) for the features from the MQW represents transitions between the nth conduction subband and the mth valence subband of heavy hole (H) or light hole (L) character. Allowed transitions have m=n while for SFT m≠n. The experimental energies and broadening parameters of the various features from the MQW at 300K and 77K are listed in Table I. Typical error bars for the energies are 20-30% of the corresponding Γ. The structure A is related to the direct band gap of the GaAlAs barrier layers. However, because of quantum effects it occurs at a somewhat higher energy, the energy difference being a function of the well and barrier parameters. We shall demonstrate below that this peak is caused by the 35H SFT. The observation of this feature in electromodulation is important since it is an indirect measure of the barrier height.

Comparison of Figs. 1 and 2 shows that while the SFT 12H, 13H and 35H are individual peaks at 300K other SFT (21L, 24H and 43H) be-

Table I. Experimental energies and broadening parameters (Γ) for the various features from a fit of Eq. (1) to the experimental data. Also listed are the theoretical values of the various transitions.

Spectral Features	300K			77K		
	Experiment		Theory	Experiment		Theory
	Energy (eV)	Γ (meV)	Energy (eV)	Energy (eV)	Γ (meV)	Energy (eV)
11H	1.448	8±1	1.449	1.538	6±1	1.537
11L	1.456	7±2	1.461	1.552	5±1	1.549
12H	1.470	25±5	1.471	1.561	6±1	1.559
13H	1.498	19±5	1.503	1.594	6±1	1.591
21L	1.529	10±2	1.531	1.615	7±1	1.618
22H	1.545	10±2	1.541	1.624	10±2	1.629
22L	1.580	13±2	1.581	1.669	13±3	1.669
24H	1.614[a]	30±5	1.602	1.694	10±3	1.690
33H	1.614[a]	30±5	1.618-1.632	1.716	22±5	1.703-1.723
A(35H)	1.653	13±2	1.650-1.673	1.731	18±5	1.738-1.761
33L	1.682[a]	11±2	1.654-1.680	1.743	20±5	1.742-1.768
43H	1.682[a]	11±2	1.678-1.640	1.802	24±5	1.766-1.825

a) Unresolved at 300K

come clearly resolved only at 77K. At 300K the fit to the amplitude was considerably improved in the region of the allowed feature 22H by including 21L.[3]

We have performed a calculation based on the two-band tight-binding model.[6] The energy of E_o was used for the energy gap of the quantum well layers at 300K and 77K. The mass parameters[7] were $m_e^* = 0.665$, $m_{\ell h}^* = 0.094$ and $m_{hh}^* = 0.34$, the band offset[7] Q = 0.60 and the well width (W) employed was 99.1Å (35 layers of thickness 2.83Å per layer). The various theoretical energies and corresponding transitions also are listed in Table I. For several of the transitions we give a range of energies due to the dispersion along the direction of the minizone. A good fit is achieved for the various transitions (including A) using a barrier height of 1.634 eV ($x \approx 0.17$)[8] at 300K and 1.722 eV at 77K. As can be seen from Table I, there is very good agreement between experiment and the theoretical values, thus verifying the origins of the different allowed and forbidden features. The

564   H. Shen et al.

value of W is in good agreement with the growth conditions.

The structure A cannot be interpreted as the energy gap of the
barrier layers. Calculations show that the barrier gap is not directly observable as an optical transition. The quantum effect due to the
finite layer widths raises the energy of the lowest allowed transition
above the barrier transition to slightly higher than the barrier band
gap. We find that A corresponds to the 35H transition, the first unconfined to unconfined feature. The tight-binding model indicates it
should be strong and easily observable.

REFERENCES
1. See, for example, Glembocki, O.J., Shanabrook, B.V., Bottka, N.,
   Beard, W.T., and Comas, J., Appl. Phys. Letts. 46, 970 (1985).
2. Parayanthal, P., Shen, H., Pollak, F.H., Glembocki, O.J.,
   Shanabrook, B.V., and Beard, W.T., Appl. Phys. Lett. 48, 1261
   (1986).
3. Shen, H., Parayanthal, P., Pollak, F.H., Smirl, A.L., Schulman,
   J.N., and D'Haenens, I., to be published in Solid State Comm.
4. See, for example, Aspnes, D.E. in Handbook on Semiconductors,
   Vol. 2, ed. by T.S. Moss (North Holland, N.Y., 1980) p. 109 and
   references therein.
5. Song, J.J., Yoon, Y.S., Fedotowsky, A., Yim, Y.B., Schulman, J.N.,
   Tu, C.W., Huang, D. and Morkoc, H., submitted to Phys. Rev. and
   references therein.
6. Schulman, J.N., and Chang, Y.C., Appl. Phys. Letts. 46, 571 (1985).
7. Miller, R.C. and Kleinman, D.A., J. Luminescence 30, 520 (1985).
8. Aubel, J.L., Reddy, U.K., Sundaram, S., Beard, W.T., and Comas, J.,
   J. Appl. Phys. 58, 495 (1985).

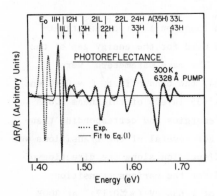

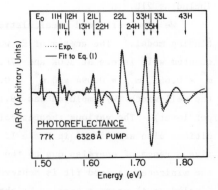

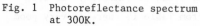

Fig. 1 Photoreflectance spectrum   Fig. 2 Photoreflectance spectrum
        at 300K.   at 77K.

Photoreflectance Study of Interband Transitions in
Multiple Quantum Wells: A Comparison With
Excitation Spectroscopy

B.V. Shanabrook and O.J. Glembocki
U.S. Naval Research Laboratory
Washington, DC 20375, USA

ABSTRACT

Photoreflectance (PR) and photoluminescence excitation (PLE) spectroscopy have been performed on GaAs/
AlGaAs quantum wells in the temperature range of 6K to
250K. The PLE spectra exhibit sharp peaks characteristic of bound excitonic transitions. Because at all temperatures, the PR and PLE lines agree in peak position
to within experimental error, we conclude that the
transitions observed in PR are also excitonic in
nature.

Photoreflectance (PR) measurements have been performed in a variety
of multilayer systems, which include multiple quantum wells (MQW)[1-3] and
modulation doped heterojunctions[1]. The PR technique has also been used
to study the hydrostatic pressure dependence of the optical properties
of quantum wells.[4] In spite of this flurry of activity in modulation
spectroscopy, several fundamental questions remain unanswered in the
simplest multilayer system, the multiple quantum well. Paramount among
these questions is whether the transitions observed in PR are due to
excitons or just band to band absorption. In this paper, we address
this question by performing simultaneous PR and PLE measurements over an
extended temperature range. Photoluminescence excitation spectroscopy
(PLE) measures the efficiency with which photoluminescence (PL) can be
excited by light of different frequencies. In MQW, it has been shown to
be roughly proportional to the absorption coefficient or the imaginary
part of the dielectric function, ε_2.[5,6] In contrast, near the fundamental gap of the systems of interest, reflectance or PR measurements are
predominantly sensitive to the real part of the dielectric function,
ε_1.[7] Since the real and imaginary parts of the dielectric function are
related through a Kramers-Kronig inversion, a comparison of the PLE and
PR spectra taken simultaneously should allow us to better understand the
nature of QW transitions observed in PR.
    Our measurements were performed on GaAs/Al$_{0.3}$Ga$_{0.7}$As multiple
quantum wells which had well and barrier widths of 200A and 150A,
respectively and a total thickness of 2 µ. The sample was capped with
3000A of AlGaAs in order to reduce the magnitude of the built-in electric field at the quantum wells that arises from the partial depletion
of surface states. The PLE and PR measurements were performed with a CW
dye laser pumped by an Ar ion laser. The PR modulation or pump light
was obtained from the 5145A and 4880A lines of the Ar laser and was
mechanically chopped at 1700 Hz. The radiation from the dye laser was
employed as the PR probe beam and to excite the luminescence for the PLE

experiment.   In this manner, the two spectra could be obtained simultaneously.   To ensure invariant PR lines,   neutral   density   filters   were
employed to reduce the intensities of both the probe and pump beams.

Shown in Fig. 1, is a series of
spectra obtained at 6K [Fig. 1(a)],
150K [Fig. 1(b)] and 250K [Fig.
1(c)]. The PLE detection window was
set in the region of the photoluminescence   spectrum   where   donor
related transitions occur.   While
the lowest energy peak in the PLE
spectrum of   Fig.   1(a) arises from
donor related absorption processes,
the three remaining higher energy
peaks result from bound excitonic
transitions.   The first number in
the notation for these peaks represents the conduction subband index,
while the second number denotes the
valence subband index. The letter
specifies whether the valence subband band is of heavy or light mass
character.   The PR spectrum is shown
below the PLE spectrum. We note the
correspondence in Fig. 1(a) between
the PLE peaks and the PR lines for
both the donor related features and
the   excitonic   transitions.   The
transition energies for the PR lines
have been estimated from the Aspnes
three point fit technique, assuming
an   excitonic   transition.[8]   Using
other critical point types such as
two dimensional or even three dimensional,   does not significantly
change   the   PR   peak   energies.[8]
Because the value of the transition

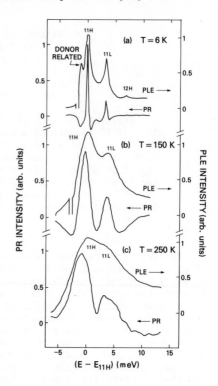

Figure 1. A comparison of PR and
PLE at (a) 6K for which the
energy of the 11H excitonic transition is E_{11H} = 1.5235eV, (b) at
150K for which E_{11H} = 1.4955eV
and (c) at 250K, for which E_{11H} =
1.4537eV.

energies obtained in PR agree with
the excitonic peak positions measured in PLE to within 0.2meV and
because the exciton binding energies
are   approximately   8meV in 200A wide
quantum wells, we conclude that the
PR features are due to an optical
modulation of bound excitonic transitions. Displayed in Figs. 1(b) and
1(c) are PR and PLE spectra taken at 150K and 250K, respectively. At
these temperatures, the positions of the PR and PLE lines agree to
within experimental error. In fact, even at 250K, where the broadening
of the excitonic transitions is so large that there is a significant
overlap between the absorption profiles of the 11H and 11L transitions,
the agreement between excitonic absorptions, probed in the PLE measurement, and PR transition energies is within ~1meV. This value is considerably less than the binding energy (~8meV) of the excitons and leads us
to conclude that even at 250K, excitons are observed in the PR spectrum.

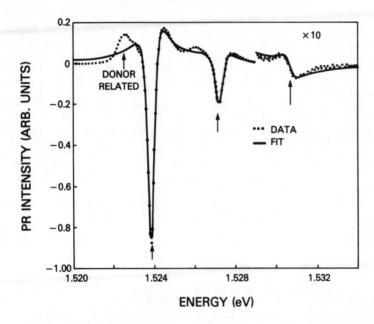

Figure 2. A comparison between the experimental data (dotted) and the fit (line) to derivatives of Eq. (1), as described in the text. The arrows indicate the positions of the gap energies obtained from the fit.

The dielectric function appropriate for excitons in cases of weak exciton-phonon coupling and at low temperatures is given by:[9]

$$\varepsilon - 1 \sim \frac{-I}{(E - E_g + i\Gamma)} \qquad (1)$$

where E is the photon energy, E_g is the exciton gap, Γ is the exciton broadening parameter and I is the integrated intensity of the transition. The optical modulation that is responsible for the PR effect produces a periodic variation in either the exciton energy gap, its lifetime $(1/\Gamma)$ or its integrated absorption strength. A modulation of either the exciton energy gap or its lifetime will result in PR lineshapes, described by a first derivative of Eq.(1), i.e., $\partial\varepsilon/\partial E_g$ or $\partial\varepsilon/\partial\Gamma$. Furthermore, in MQW, the total integrated intensity, I, can also be modulated.[10] Shown in Fig. 2 is a fit to 6K PR data using the dielectric function described by Eq. (1). We find that the allowed transitions, 11H and 11L, are dominated by a modulation of the exciton energy gap. In contrast, the forbidden transition, 12H, is best described by an intensity modulation.

We find that at higher temperatures, the fit using Eq. (1) does not properly describe the data. This result is anticipated, because Eq. (1) is valid only for weak exciton-phonon coupling and at low temperatures.[11] At higher temperatures or in the presence of inhomogeneous perturbations, theoretical studies of Toyozawa have indicated that the absorption profile changes from one described by a Lorentzian [Eq. (1)] to one described by a Gaussian.[11] This evolution of the dielectric function has been observed in a study of the temperature dependence of the PR line shape.[12] In Ref. (12), the 150K PR data of Fig. (1b) is shown to be well described by a dielectric function with a Gaussian absorption profile.

In summary, we have measured simultaneously, the PR and PLE spectra of GaAs/Al$_{0.3}$Ga$_{0.7}$As multiple quantum wells. A comparison of the results from these two spectroscopies indicates that the transitions observed in photoreflectance are excitonic in nature. The form of the PR lines have been discussed in terms of the optical modulation of the integrated oscillator strength, the exciton energy gap and the lifetime. At low temperatures, the PR line shapes exhibit forms that are well described by first derivatives of a dielectric function with a Lorentzian absorption profile. At high temperatures, the imaginary part of the dielectric function is better described by a Gaussian.

We are grateful to D.A. Broido and N. Bottka for useful discussions. This work is supported in part by the Office of Naval Research.

REFERENCES

1. O.J. Glembocki, B.V. Shanabrook, N. Bottka, W.T. Beard and J. Comas, Appl. Phys. Lett. 46, 970 (1985).
2. H. Shen, P. Parayanthal, Fred H. Pollak, Micha Tomkiewicz, T.J. Drummond and J.N. Schulman, Appl. Phys. Lett. 48, 653 (1986).
3. O.J. Glembocki, B.V. Shanabrook and W.T. Beard, in the Proc. of the 2nd International Conf. on Modulated Semiconductor Compounds, 1985 (Kyoto), to be published in Surf. Sci. (1986).
4. A. Kangarlu, H.R. Chandrasekhar, M. Chandrasekhar, F.A. Chambers, B.A. Vojak and J.M. Meese, this conference.
5. R.C. Miller, A.C. Gossard, D.A. Kleinman and O. Munteanu, Phys. Rev. B29, 3740 (1984) and R.C. Miller, D.A. Kleinman, O. Munteanu and W.T. Tsang, Appl. Phys. Lett. 39, 1 (1981).
6. W.T. Masselink, P.J. Pearah, J. Klem, C.K. Peng, H. Morkoc, G.D. Sanders and Y.C. Chang, Phys. Rev. B32 8027 (1985).
7. B.O. Seraphin and N. Bottka, Phys. Rev. 145, 628 (1966).
8. D.E. Aspnes, Surf. Science 37, 418 (1973).
9. D.E. Aspnes and A. Frova, Phys. Rev B2, 1037 (1970).
10. J.A. Brum and G. Bastard, Phys. Rev. B31, 3893 (1985).
11. Y. Toyozawa, Progess of Theoretical Physics 20, 53 (1958).
12. O.J. Glembocki and B.V. Shanabrook, Proc. 2nd Internat. Conf. on Superlattices, Microstructures and Microdevices, Göteborg, Sweden, August, 1986.

PHOTOREFLECTANCE SPECTROSCOPY OF GaAs - $Al_xGa_{1-x}As$ QUANTUM WELLS UNDER HYDROSTATIC PRESSURE

A. Kangarlu, H.R. Chandrasekhar, M. Chandrasekhar[†]

Department of Physics and Astronomy

University of Missouri, Columbia, MO 65211, U.S.A.

F.A. Chambers, B.A. Vojak and J.M. Meese

Amoco Research Center, Naperville, IL 60566, U.S.A.

We present the first study of the effect of hydrostatic pressure on the photoreflectance (PR) spectra of GaAs - $Al_xGa_{1-x}As$ multiple quantum wells (MQW). As the levels in the Γ CB move up in energy with pressure, they interfere with the L band signalled by a loss of intensity as they cross the latter. At pressures beyond 35 kbar the Γ CB of GaAs crosses the L band of GaAs and the X band of the $Al_xGa_{1-x}As$ barrier. Staggered transitions across the interface from the X CB of $Al_xGa_{1-x}As$ to the valence band of GaAs are seen signalling the formation of a type II heterostructure beyond these pressures. The valence band offsets are deduced from this data and are in agreement with our published data of photoluminescence under pressure.[1,2] The pressure dependence of the L band is also deduced from our results.

The MQW samples were grown by molecular beam epitaxy. The well and barrier widths are estimated from growth parameters. The Al mole fraction of barrier was accurately deduced from its PR signal. The experimental PR set up was similar to that described in the literature.[3] PR obtained using different wavelengths of an Ar^+ ion laser (5145Å to 4500Å) or He-Ne laser (6328Å) was the same. A variable temperature gasketted diamond anvil cell[4] was used. A large number of transitions (upto n = 10) were observed at room temperature and are studied in detail. The lineshapes are fit to the Aspnes third derivative functional form[5] (TDFF), and accurate values of widths, amplitudes and phases of the PR features are obtained. The oscillator strengths were deduced by the integrated area of the TDFF over the energy range. The notation used to identify the peaks are $CnHm$ or $CnLm$ for transitions between the nth conduction sub-band to the mth heavy (H) or light (L) hole valence sub-band. Both allowed (n = m) and forbidden (n ≠ m) transitions are observed.

Figs. 1 and 2 show the PR spectra at different pressures of a MQW sample with 400Å well widths and x = 0.3. The solid and dotted lines in each spectrum are due to experiment and a fit to TDFF function, respectively. Note that the C1H1 peaks (shown by an arrow) are aligned vertically. Fig. 3 shows the pressure dependence of various transitions.

The following features are worth noting. (1) The spectrum at 1 bar (Fig. 1 top) contains transitions derived from the Γ CB and VB sub-bands. The energies are in good agreement with a calculation using a Krönig – Penney model with appropriate band parameters[6] and a 70:30 band offset ratio[1] $(Q_c:Q_v)$. (2) The entire spectrum moves up in energy with increasing pressure at a rate of 10.4 meV/kbar which is the pressure coefficient (α) of C1H1. (3) The phases and amplitudes of a given transition do not remain constant with pressure. We do not understand the mechanisms responsible for this effect at present. (4) As the PR peaks cross an energy range of approximately 1.75 to 1.8 eV they lose their intensities and recover them. The minima occuring in the spectra as a function of pressure are shown by circles in Fig. 3. We interpret this feature as due to a mixing of the L band with the CB sub-bands. The line denoted E^L passes thru this data. The slope of this line yields an α of L band 3.2 meV/kbar with an intercept of 1.71 eV. in good agreement with accepted values[5]. (5) Spectra upto ~ 35 kbar (Figs. 1 and 2) show C1H1 to be the most intense peak. Beyond that pressure, it loses intensity and several new peaks appear both above and below the energy of C1H1. The specturm for 35.3 kbar shows a peak (denoted E^X) below C1H1. This peak moves down in energy with increasing pressure (Figs. 2 and 3). We have shown[1] that for GaAs – $Al_xGa_{1-x}As$ heterostructures, the energy of the X band of AlGaAs is lower in energy than that of GaAs as measured from the top of the VB of GaAs due to the band–offset between the well and barrier layers. We identify E^X as the staggered transition across the well-barrier interface from the X CB of AlGaAs

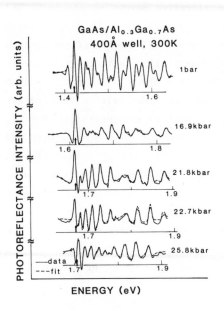

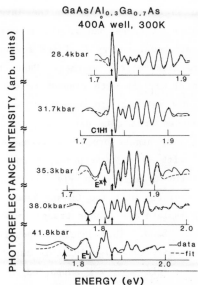

Fig. 1. The Photoreflectance spectra of a GaAs/Al$_{0.3}$Ga$_{0.7}$As MQW with a well width of 400Å at 300K at different pressures. The solid and dotted curves are due to the experiment and a fit to the TDFF functional form, respectively. Note the C1H1 transitions are aligned vertically.

Fig. 2. Same as Fig. 1 at higher pressures.

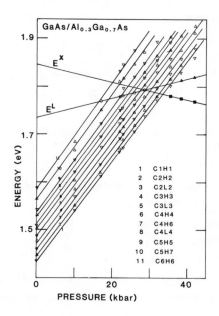

Fig. 3. Pressure dependence of the various MQW transitions associated with the ΓCB (open triangles), L CB (solid triangles) and X CB of the Al$_{0.3}$Ga$_{0.7}$As barrier (solid squares). The open circles correspond to the minima in intensity of the ΓCB transitions. The lines through the data are due to the least squares fits.

to VB of GaAs. Due to the large effective masses of electrons and
holes in these bands, the n = m = 1 transition will have a small
confinement energy ($<$ 1 meV for our sample) and the higher transitions
are close to one another. Hence E^X essentially tracks the X band edge
in $Al_xGa_{1-x}As$ with respect to the GaAs VB. The α of X band is - 1.50
± 0.2 meV/kbar in agreement with Ref. 1. The valence band offset
deduced from this data is Q_v = 0.30, also in agreement with Ref. 1.
(6) At higher pressures an additional peak was seen between ClH1 and
E^X with a positive pressure coefficient. It fell on L band line E^L
(solid triangles in Fig. 3) indicating that these transitions were
associated with the CB sub-bands of L. At this stage it is difficult
to isolate the higher transitions associated with L band from those of
the Γ band. This should be possible at much higher pressures. We are
currently investigating this aspect.

This work was supported by the U.S. Department of Energy under
contract NO. DE-AC02 84ER 45048.

REFERENCES:

†   Alfred P. Sloan Foundation Fellow.

1.   U. Venkateswaran, M. Chandrasekhar, H.R. Chandrasekhar, B.A.
      Vojak, F.A. Chambers and J.M. Meese, Phys. Rev. B <u>33</u>, 8416
      (1986).

2.   U. Venkateswaran, M, Chandrasekhar, H.R. Chandrasekhar, T.
      Wolfram, Phys. Rev. B <u>31</u>, 4106 (1985).

3.   J.L. Shay, Phys. Rev. B2, 803 (1970).

4.   For experimental details see U. Venkateswaran and M.
      Chandrasekhar, Phys. Rev. B31, 1219 (1985).

5.   D.E. Aspnes, in Handbook on Semiconductors, ed. by T.S. Moss
      (North-Holland, New York, 1980), p. 109.

6.   S. Adachi, J. Appl. Phys. <u>58</u>, R1 (1985).

7.   M. Chandrasekhar, et.al., <u>Proceedings</u> of the 18th International
      Conference on the Physics of Semiconductors, Stockholm, Sweden,
      Aug. 11-15 (1986).

Optical Spectroscopy of $2E_g$-transitions in GaSb/AlSb quantum wells

A. Forchel, U.Cebulla, G. Tränkle

4. Phys. Inst., Universität Stuttgart, D-7000 Stuttgart 80, FRG

T.L. Reinecke

Naval Research Laboratory, Washington D.C., 20375, USA

H. Kroemer, S. Subbanna, G. Griffith[+]

Dept. of Elec. and Comp. Eng., UCAL, Santa Barbara, CA 93106, USA

We have observed emissions due to the simultaneous recombination of two electrons and two holes in GaSb/AlSb MQWs. The well width-intensity relation and the excitation power dependence are explained by a model based on quantization induced symmetry changes in the subband edge wavefunctions.

We have investigated radiative transitions occuring at twice the difference in energy between the first electron and hole subbands ($2E_g$-transitions) in GaSb/AlSb quantum wells with well widths between 120 Å and 20 Å. These transitions are due to the simultaneous recombination of two electrons and two holes. Because the transition intensity depends strongly on the symmetry of the subband wavefunctions involved, $2E_g$-transitions in quantum wells can be used to study quantization induced symmetry changes. Previously similar transitions have been observed only at the indirect gap of bulk silicon.[1]

For our experiments the samples [2] were placed in an immersion cryostate (T_{Bath}=2 K). Most samples were excited close to the band gap ($1E_g$) by a NdYAG-laser operating in cw mode (P ≤ 100 mW) or with a Q-switch ($P_{max} \cong 100$ W, repetition rate 3 kHz) focussed to a 100 μm diameter spot. The 20 Å sample was excited by a cw IR dye laser. We used $1E_g$-excitation of the quantum wells at energies much lower than the observed transitions in order to avoid any emission from the cladding layer. The luminescence was dispersed by a double monochromator and detected by a sensitive low noise GaAs photomultiplier ($2E_g$-transitions) or a Ge detector ($1E_g$-transitions). Since the emission intensities of the $2E_g$-transitions are rather weak (typically 1 count/s) great care was used to suppress any background light level.

Fig. 1 displays for a 25 Å sample the luminescence spectrum in the $1E_g$ energy range (Top) in comparison with the emission observed at twice the band edge energy (Bottom). The $1E_g$ spectrum includes two clear structures on a broad background at about 1.095 eV and 1.02 eV, which previously have been identified as transitions between the first subbands at the Γ-point and at the L-point, respectively. Both transitions are observed in this spectrum be-

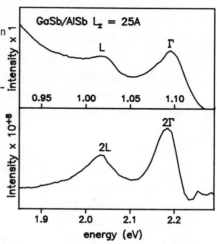

the GaSb which is direct in the bulk undergoes a size induced direct to indirect band gap transition in the QWs at L_z=40 Å.[3] The broad background band (peaked at 0.9 eV) observed in the $1E_g$ spectrum is due to defects in the multi quantum well structure, situated probably at the interfaces. The $2E_g$ spectrum (Fig. 1, Bottom) is dominated by two emission lines labeled 2 L and 2 Γ situated at twice the $1E_g$ transition energies. Qualitatively similar $1E_g$ and $2E_g$ spectra which show a characteristic variation with the well width are observed for the other indirect gap samples also. [4] For direct gap samples only the emission related to the Γ point is observed in the $1E_g$ and $2E_g$ data. Due to the short life time at the Γ point ($\tau_\Gamma \sim 500$ ps compared to $\tau_L \sim 100$ ns [3]) in the direct gap samples no significant population of the L-minimum occurs in the direct gap samples, whereas for indirect gap samples both minima are populated due to intervalley scattering.

We have investigated the excitation power dependence of the $2E_g$-transitions compared to the $1E_g$-transition. Fig. 2 displays the $2E_g^\Gamma$ (left) and $2E_g^L$ (right) transition intensities versus the respective $1E_g$-transition intensity. We observe in both cases approximately a qua-

dratic increase
of I_{2E_g} with I_{1E_g}.
This behaviour
is expected
since the $2E_g$-
transition in-
volves four
particles where-
as only two
are necessary
for the $1E_g$ re-

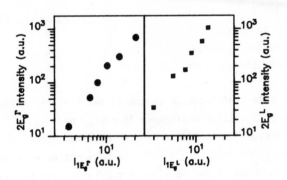

combination. The $1E_g$ intensity shows a square root increase with
respect to the pumping power whereas the $2E_g$ intensities depend
linearly on the excitation power. This may be due to nonlinear recombination processes which determine the population of the bands
under the comparatively high excitation intensities used here
(10^4 W/cm^2 $\leqslant I_{Laser} \leqslant 10^6$ W/cm^2).

We propose the following model to explain the $2E_g$-transitions.
These transitions are described by a dipole matrix element of the
form

$$< c_1 c_2 | \; d \; | v_1 v_2 >$$

where the c_i, v_i denote the wavefunctions of electrons and holes
and d is the dipole operator. This matrix element is negligible at
the Γ-point of 3D zincblende semiconductors since the related wavefunctions have almost pure s-type (conduction band) and p-type symmetry (valence band). [5] In the quantum wells, however, the transitions
occur between subband levels i.e. between energies above and below the
3D conduction and valence band extrema, respectively. The corresponding wavefunctions are composed from all bulk wavefunctions having
similar energy. From k·p theory [6] it is well known that the wavefunctions include increased admixtures of bands with different symmetry if one moves away from the band extrema. Hence we expect that
the intensity of the $2E_g$ transitions increases strongly for higher
quantization energies. Therefore we expect a strong well width dependence of the $2 E_g$ intensities. As shown in Table 1 this is indeed.

observed. In going from
L_z=42Å to L_z=20Å the
$2E_g^{\Gamma}$ intensity
increases by three
orders of magnitude.

Table 1

L_z (Å)	$2E_g^{\Gamma}$ intensity
42	4
30	20
25	700
20	4000

Note that the GaSb system is particularly suited for the present study due to the large difference in energy gaps $\Delta(E_{\Gamma}^{AlSb}-E_{\Gamma}^{GaSb})$ = 1.5 eV) of the quantum well and barrier material which occurs mainly in the conduction band ($\Delta E_c \sim 80\%$). In conjunction with the small electron mass at the Γ point one obtains particularly high subband energies in the conduction band (< 200 meV). According to our modell they lead to substantial modifications of the symmetry of the wavefunctions and hence are responsible for the $2E_g$-transitions.

We acknowledge stimulating discussions with M.H. Pilkuhn and H. Schweizer. The financial support of this work by the Deutsche Forschungsgemeinschaft, the Stiftung Volkswagenwerk and the Office of Naval Research is gratefully acknowledged.

References

+ now at C.S.I.R.O., Div. of Radio Physics, Epping, NSW 2121 Australia

1. K. Betzler, T. Weller, Phys. Rev. Lett. 26, 640 (1971);
   K. Betzler, T. Weller, R. Conradt, Phys. Rev. B 6, 1394 (1972).
2. G. Griffiths, K. Mohammed, S. Subbanna, H. Kroemer, J.L. Merz, Appl. Phys. Lett. 43, 1059 (1983).
3. A. Forchel, U. Cebulla, G. Tränkle, H. Kroemer, S. Subbanna, and G. Griffiths, Proc. 2nd Intl. Conf. on Modulated Semiconductor Structures, Kyoto, Japan 1985, to be published in Surface Science.
4. A. Forchel, U. Cebulla, G. Tränkle, E. Lach, T.L. Reinecke, H. Kroemer, S. Subbanna, and G. Griffiths, to be published.
5. The Matrix element at the Γ point is exactly zero in systems with inversion symmetry, e.g. in silicon. See Ref. 1.
6. E.J. Johnson, in Willardson and Beer, Semiconductors and Semimetals, Vol. 3, p. 153, and references therein;
   E.O. Kane, J. Phys. Chem. Solids 1, 83 (1956).

BAND FILLING OF PHOTOEXCITED
GaAs/Al$_{0.3}$Ga$_{0.7}$As SINGLE QUANTUM WELLS

S. Borenstain, D. Fekete, Arza Ron and E. Cohen
Solid State Institute
Technion-Israel Institute of Technology
Haifa 32 000, Israel

ABSTRACT

GaAs/Al$_{0.3}$Ga$_{0.7}$As single quantum wells (SQW) with widths of 20A and 40A were irradiated with intense dye laser pulses of 2 nsec duration. The broad band due to the radiative recombination of the electron-hole plasma (EHP) is studied as a function of ambient temperature and excitation intensity. It is found that the EHP reaches very high effective temperatures and that for T>100K stimulated LO-phonon emission affects the EHP carrier density.

1. INTRODUCTION

The luminescence spectrum of intensely photoexcited single and multiple quantum wells of GaAs/Al$_x$Ga$_{1-x}$As was reported by several authors[1,2]. The spectrum consists of a broad band due to the spontaneous radiative recombination of carriers in the EHP which fills the QW energy states[3]. In this work we study the dependence of the EHP parameters:the carrier density n and effective carrier temperature T_{eff}, on both excitation intensity and ambient temperature.

2. EXPERIMENT

The samples used in this study were MOCVD grown
$Al_xGa_{1-x}As/GaAs$ single QW structures with x=0.25 and
0.30 and well widths of approximately 20Å and 40Å,
respectively. The cladding layers thickness was 1μ.
The samples were either immersed in liquid He or in
a stream of cold He gas. A cw dye laser was used as
an excitation source for the determination of the
levels. A pulsed dye laser pumped with a N_2-laser
was used for high intensity photo-excitation. The

Fig. 1: Radiative recombination spectra of the EHP.
I_p denotes the pump beam intensity and T_a-the
ambient temperature. Solid line: experimental
spectra. Dashed lines: calculated with Eq. (1).

Fig. 2: a. The total emission intensity . b. The
EHP effective temperature. c. The carrier density
in the EHP.

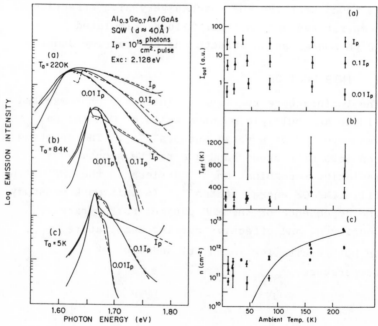

pulse duration was about 2 nsec. The beam was tightly focussed with photon flux of 5×10^{14}-5×10^{16} photons/cm^2 per pulse. Since the measured recombination lifetime is less than 0.5 nsec[4] and the intraband particle relaxation is less than 0.1 nsec, our experiments are done under essentially steady state conditions.

3. DISCUSSION

Fig. 1 shows a series of luminescence spectra from the SQW structure with d$\sim$40Å. They were observed in a backscattering configuration with a tightly focussed laser beam impinging perpendicular to the QW layer. The high energy part of the various spectra was fitted to the experimental data by the following expression for the EHP spontaneous emission:

$$I(\hbar\omega) \propto \sum_j \int_{E_e}^{\infty} \int_{-E_j}^{-\infty} f_c(E_e(\underline{k})) f_v(-E_j(-\underline{k})) \cdot$$

$$\cdot m_e^* m_j^* \delta(E_e(\underline{k}) - E_j(-\underline{k}) - \hbar\omega) dE_e dE_j \quad (1)$$

E_e and E_j (j stands for either the heavy- or the light-hole band) denote the lowest energies of the electron and hole bands, respectively. In the fitting procedure we assume a single effective temperature, T_{eff}, for both the electrons and the holes. We found that the fit is insensitive to a change in the effective hole temperature relative to that of the electrons. Such changes have been reported recently[5]. The calculated spectra are shown (in dashed curves) in Fig. 1. The parameters n and T_{eff} obtained from the fit are shown in Fig. 2, as a function of ambient temperature. Also shown is the temperature dependence of the total EHP luminescence, I_{out} (Fig. 2A). Similar results are obtained for the 20Å width SQW.

Several conclusions can be drawn: n , T_{eff} and I_{out}
are more strongly dependent on the excitation
intensity than on the ambient temperature T_a. n
increases with T_a (for $T_a > $ 100K) roughly as

$$n \propto \left(exp \left\{ \frac{\hbar \omega_{LO}}{k T_a} \right\} - 1 \right)^{-1}$$

which reflects the increase in population of
LO-phonons with T_a (the solid line in Fig. 2c). A
possible explanation for this is the increase in
capture rate of carriers from the cladding layer
into the SQW as a results of stimulated LO-phonon
emission. I_{out} decreases with T_a for the lowest
excitation intensity. This may be due to increased
effectiveness of deep traps to capture carriers. As
the excitation intensity increases, the deep traps
are saturated and I_{out} remains constant.

Acknowledgements: This work was supported by the
Fund for Basic Research administered by the Israel
Academy of Sciences and Humanities.Two of the
authors (E.C. and A.R.) thank the Fund for the
support of research at the Technion.

References

1. Ku Z.Y., Keismanis V.G. and Tang C.L., Appl.
    Phys. Lett. 44, 136(1984) .
2. Camras M.D., Holonyak N., Nixon M.A., Buruham
    R.d., Streifer W., Scifres D.K., Paoli T.L. and
    Lindstrom C., Appl. Phys. Lett. 42, 761(1983).
3. Fekete D., Borenstain S., Ron A. and Cohen E.,
    Superlattices and Microstructures 1, 245(1985).
4. Goebel E.O., Jung. H., Kuhl J. and Ploog K.,
    Phys. Rev. Lett. 51, 1588(1983).
5. Hopfl R.A., Shah Jagdeep and Gossard Arthur C.,
    Phys. Rev. Lett. 56, 765(1986).

ELECTRONIC PROPERTIES OF PbTe DOPING SUPERLATTICES UNDER ILLUMINATION

P. Pichler, G. Bauer and H. Clemens

Institut f. Physik, Montanuniversität Leoben, A-8700 Leoben, AUSTRIA

W. Jantsch, K. Lischka, A. Eisenbeiss and H. Heinrich

Institut f. Experimentalphysik, Johannes Kepler Universität Linz,
A-4040 Linz, AUSTRIA

ABSTRACT

PbTe doping superlattices exhibit excellent photoconductive response with the detectivity close to the
theoretical limit at the peak wavelength of 6 μm at
80 K. A detailed analysis of magnetotransmission and
lifetime measurements allows to attribute these properties to a lifetime enhancement by nearly two orders of
magnitude within the periodic nipi potential.

In the past few years a number of outstanding properties of
doping superlattices ("*nipi*"-structures) has been predicted[1] and some
of them have been verified experimentally for essentially two types of
semiconductors, GaAs[2] and PbTe[3]. Among the properties which are
caused by the spatial separation of photoexcited electrons and holes, a
lifetime enhancement and thus excellent photoresponse are expected. In
this paper we investigate optical properties of PbTe doping superlattices in this respect.

Fig. 1 shows the spectral detectivity of a PbTe–nipi structure,
used as a photoconductor at 80 K. Sample preparation is described
elsewhere[3,4]. For comparison, the spectral response of a single
junction photovoltaic PbTe detector[5] and the theoretical limit for the
maximum performance of photoconductors are also given.

In order to clarify the origin of the excellent photoresponse of
the nipi's, we present in the following results from FIR magneto-

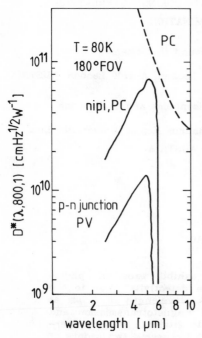

Fig. 1: Detectivity of a PbTe nipi at 800 Hz modulation frequency, 1 Hz bandwidth. For comparison, data for a single p-n junction are also given.

transmission investigations and lifetime measurements performed both with pulsed light and electron beams. The magnetooptical investigations show that free carriers are predominantly generated within the nipi structure but not in the buffer. This behavior is expected because of the lifetime enhancement which is anticipated for nipi's due to the periodic potential. Finally, we present experimental results for the lifetime enhancement.

Fig. 2a shows magnetotransmission data obtained at a wavelength of 118.8 μm in Faraday geometry using linear polarized light (Fig.2a and b) and circular polarization in the electron inactive mode (Fig.2c). The curves show distinct resonant structure which can be interpreted by fitting the spectra with a model including contributions from the polar optical lattice mode and a magneto-plasma term in the dielectric function (dashed lines in Fig. 2). Sample parameters are given in the insert of Fig. 2b. Arrows in Fig. 2a indicate the position of the free carrier resonances. The resonances marked 1 and 1' correspond to the transverse effective mass of electrons in the n-layers. The resonance of the holes in the ⟨111⟩ valley of the buffer oriented perpendicular to the sample surface causes resonance 2. Resonances 3 and 4 arise from electrons in the remaining three oblique valleys. Their splitting occurs, like the 1 - 1' splitting, due to resonances in the pseudo-Landau ladders of the i = 1 and i = 2 electric subbands. In Fig. 2b, the magnetotransmission with additional illumination with visible radiation is given. A clear enhancement of resonance 1 shows additional light induced free electrons in the n-layers. Resonance 2 does not change remarkably, whereas 3 and

4 merge together. The latter effect can be explained by the reduction of the periodic nipi space charge potential due to the additional photoinduced electrons in the n-layers and holes in the p-layers. This effect in turn reduces the subband spacings which are responsible for

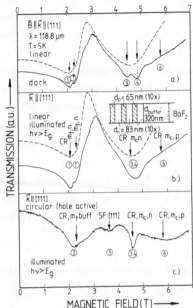

the splitting. From the electron inactive sense of circular polarization (Fig. 2c), resonance 2 can be attributed to a hole resonance in the perpendicular ⟨111⟩ valley. The unsplit resonances 3 and 4 of electrons in the oblique valleys show up for this polarization due to elliptic cyclotron orbits. Resonance 5 is attributed to a spin flip transition of holes in the perpendicular valley[6,7]. This transition is weakly allowed in bulk material due to nonparabolicity. Here we expect a strong enhancement due to the internal electric field which is present in the nipi's. The fitting parameters clearly show, that in steady state illumination produces electrons and holes in the n- and p-layers in equal numbers. A change in the buffer

Fig. 2: Magneto transmission (λ = 118.8 μm) of a PbTe nipi for (a) and (b) linear polarization, (c) electron- inactive circular polarization. (b): With additional visible illumination. Dashed curves: model calculations.

concentration could not be detected. This result clearly indicates, that the nipi structure is responsible for the excellent photoresponse, whereas the buffer contribution is negligible. This behavior can be explained by the spatial separation of e^- - h^+ pairs within the periodic potential of the nipi which causes a significant enhancement of the carrier lifetime as compared to the homogeneously doped buffer and hence a higher steady state concentration.

In order to demonstrate the lifetime enhancement directly, we investigate the time-resolved conductivity response both with respect to light pulses, generated by a GaAs laser, and a chopped electron beam in an Electron Beam Induced Current (EBIC) experiment. Both experiments yield nearly exponential transients with equal rise and fall times. At 80 K, the resulting time constant is 50 μsec both for photoconductivity at low current densities and EBIC. The value of 50 μsec obtained at 80 K exceeds the bulk lifetime[6] of 0.6 μsec for a carrier concentration of 8.5 x 10^{16} cm^{-3} by nearly two orders of magnitude. The latter value equals the average nipi carrier concentration as obtained by fitting the magnetooptical data (Fig. 2).

In conclusion, we have shown, that the investigation of magneto-optical properties allows a detailed analysis of the photo-induced free carrier effects in narrow gap nipi structures. The enhancement of the steady state photoconductivity as compared to homogeneous bulk material is caused by a dramatic increase of the carrier lifetime due to the nipi potential which also leads to a diffusion length in excess of the sample dimensions. This effect is evident from further EBIC investigations and will be presented elsewhere. As a result of the lifetime enhancement, the theoretical limit for photoconductive infrared detectors is approached.

Support by the "Fonds zur Förderung der wissenschaftlichen Forschung", Austria, is gratefully acknowledged.

REFERENCES

1. Döhler, G.H., in *Springer Series in Solid State Sciences,* **67**, 270 ed. by Bauer, G., Kuchar, F. and Heinrich, H. (Springer, Berlin 1986).

2. Ploog, K. and Döhler, G.H., Adv. Phys. **32**, 285 (1983).

3. Jantsch W., Bauer, G, Pichler, P. and Clemens, H., Appl. Phys. Lett. **47**, 738 (1985), and: Surf.Science, in print.

4. Clemens, H., Fantner, E.J. and Bauer, G., Rev. Sci. Instrum. **54**, 685, (1983).

5. Lopez-Otero, A., Haas, L.D., Jantsch, W. and Lischka, K., Appl. Phys. Lett. **28**, 546 (1976).

6. McCombe, B.D., Phys.Rev. **181**, 1206 (1969)

7. Schaber H. and Doezema R.E., Phys.Rev. **B20**, 5257 (1979)

8. Lischka, K. and Huber, W., J. Appl. Phys. **48**, 2632 (1977).

TEMPERATURE DEPENDENT CONDUCTION MECHANISM
FOR THE PPC EFFECT IN
MODULATION DOPED GaAs/AlGaAs HETEROJUNCTIONS*

Ulf Gennser and Peter Zurcher
Department of Physics and Energy Science
University of Colorado, Colorado Springs, CO 80933-7150

Marcel Py, Hausjorg Buhlmann and Marc Ilegems
Institute de Microelectronique, EPF Lausanne, Switzerland

ABSTRACT

The magnetic field dependence of the Hall coefficient in GaAs/AlGaAs 2-dimensional electron gas (2-DEG) heterostructures exhibiting persistent photoconductivity (PPC) shows that at 17K in the PPC-state, conduction takes place both in the 2-DEG channel and in the AlGaAs-layer, as opposed to 88K, where all of the PPC is sustained by the 2-DEG alone. Above $\sim$ 110K the decay of the PPC is governed by thermal excitation of electrons from the 2-DEG across the heterojunction barrier into the AlGaAs layer.

Persistent photoconductivity (PPC) as it is generally observed in III-V (2-DEG) heterojunction samples is a complicated phenomena involving photoexcitation and recombination properties of s-like deep donors and/or defects in a bulk semiconductor, e-tunneling through a heterojunction barrier, thermal excitation over such a barrier and properties of the 2-DEG at the heterointerface[1]. Over the last two years considerable experimental evidence has accumulated showing that the conduction in the PPC-state** at 4.2 K takes place in two parallel channels, the 2-DEG on the GaAs side of the heterojunction and in the bulk of the n-type $Al_{.3}Ga_{.7}As$ overlayer[2]. We have used the magnetic field dependence of the Hall coefficient parallel to the interface in such heterostructures to distinguish between conduction in one channel and conduction in two parallel channels of different individual mobilities and sheet carrier concentrations

* Research supported by the Swiss National Science Foundation.
**Here PPC refers to the case where the sample is left in a state of increased persistent conductivity after the illumination is switched off.

for the sample kept in different excitation states (dark, illuminated, PPC) at different temperatures. The decay of the photoconductivity, after switching off the light is measured and found to be strongly temperature dependent, i.e., between 70 and 160K it shows a double character consisting of a T-independent fast decay ($\tau \sim 1$ second) due to direct bulk recombination of ionized deep donors and a T- dependent slow decay with time constants ranging from months (at 70K) to seconds (at 150K) due to thermal excitation of electrons from the 2-DEG over the heterojunction barrier into the AlGaAs layer. These PPC-decay results nicely support the conclusions drawn from the magnetic field dependence of the Hall coefficient concerning the PPC conduction mechanism in GaAs/AlGaAs 2-DEG heterostructures. All experiments were made using van der Pauw structures and a standard Hall configuration with the magnetic field pointing perpendicular to the layers of the sample. Monochromatic light of energy $h\vartheta = 2.2eV$ was used to illuminate the sample.

At 17K the photogenerated increase in conductivity is fully persistent for periods of months or longer while at 88K the PPC decays within a few days, and at 188K there is no PPC observed. Fig. 1 shows the measured Hall coefficient for the three temperatures mentioned above. The measured sheet Hall coefficient, $R_H^{\square,meas}$, resulting from the parallel combination of two layers of mobilities μ_1 and μ_2, and sheet carrier concentrations n_1 and n_2, respectively is given as follows:

(a) In the extreme low magnetic field limit, $R_H^{\square,meas}$ has the assymptotic, field independent, value of $R_H^{\square,meas} = (1/e)(n_1\mu_1^2 + n_2\mu_2^2)/(n_1\mu_1 + n_2\mu_2)^2$.

(b) In the high field limit, $R_H^{\square,meas}$ approaches the field independent value of $R_H^{\square,meas} = (1/e)/(n_1 + n_2)$ and

(c) For an intermediate region, where $\mu_1 B \ll 1$ and simultaneously $\mu_2 B \gg 1$ one finds $R_H^{\square,meas} = (1/e)/(n_2 + \mu_1\mu_2 n_1^2 B^2/n_2^2)$.

One immediately sees that $R_H^{\square,meas}$ for a single conducting channel is independent of the applied magnetic field (if scattering factor effects are neglected), but in a two channel conduction system for $\mu^{meas}B \geq 1$, $R_H^{\square,meas}$ decreases with increasing $\vec{B}$-field, finally reaching a $\vec{B}$-independent value. The solid lines in Fig. 1 show the best fit for the intermediate $\vec{B}$-field region using above eq. Details on this fitting procedure and the unexplained logarithmic low field behaviour are given elsewhere.[3] The magnetic field dependence of $R_H^{\square,meas}$ shows that, at 17K, PPC is sustained in two channels clearly demonstrating the importance of bulk models for PPC such as the deep donor/defect model suggested by Lang et. al.[4] in 1979 or the recent deep donor-only model suggested by Hjalmarson and Drummond[5]. For temperatures between 80K and 160K,

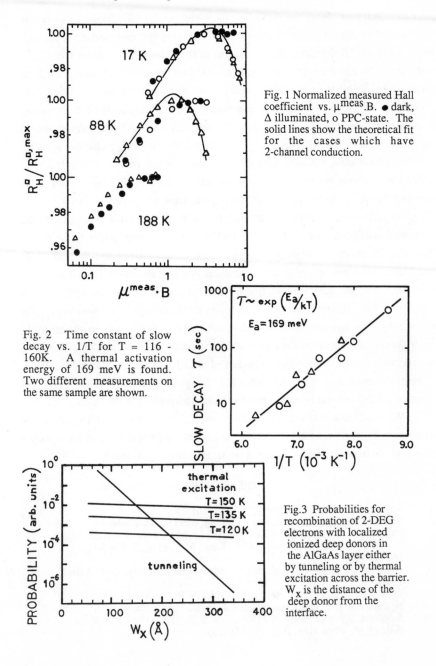

Fig. 1 Normalized measured Hall coefficient vs. $\mu^{meas}.B$. ● dark, Δ illuminated, o PPC-state. The solid lines show the theoretical fit for the cases which have 2-channel conduction.

Fig. 2 Time constant of slow decay vs. 1/T for T = 116 - 160K. A thermal activation energy of 169 meV is found. Two different measurements on the same sample are shown.

Fig.3 Probabilities for recombination of 2-DEG electrons with localized ionized deep donors in the AlGaAs layer either by tunneling or by thermal excitation across the barrier. W_x is the distance of the deep donor from the interface.

after an initial fast decay the remaining increased conductivity in the PPC-state is sustained by electrons transferred into the 2-DEG leaving only one conducting channel. For T = 116 - 160K the temperature behaviour of the following slow decay (see Fig. 2) is indicative for electrons being thermally excited over the potential barrier at the band discontinuity in order to recombine with ionized donors in the AlGaAs-layer. The probability of above decay path is compared with the probability of a competing T-independent decay path where electrons tunnel from the 2-DEG to localized ionized deep donor states in the AlGaAs-layer. Calculation of the tunneling probability using the WKB-approximation has been done similar to the approach described by Schubert et. al.[6]. Fig. 3 shows the results of these calculations for a 2-DEG sample having an undoped $Al_{.31}Ga_{.69}As$ buffer layer of 55Å thickness and a Si-doped (~ 1 x 10^{-18} cm^{-3}) $Al_{.31}Ga_{.69}As$ layer of 600Å thickness. In our model we have assumed that the measured activation energy E_a = 169 meV (see Fig. 2) corresponds to the energy difference between the highest occupied states in the 2-DEG and the top of the conduction band discontinuity at the interface. For ionized donors located within ~ 150-200Å from the interface, we find that recombination through tunneling is much more probable. However, since the tunneling electrons' wave functions are rapidly decreasing with distance, the donors that will be filled first are those closest to the heterojunction thus reducing the region where recombination can take place. Therefore, we conclude that for temperatures between ~110 and 160K, after an initial fast decay through electron tunneling, the remaining PPC decays by thermal excitation of electrons from the 2-DEG over the heterojunction barrier into the n-type AlGaAs layer.

We would like to thank H.P. Hjalmarson, R.E. Camley and T.J. Drummond for stimulating discussions on theoretical and experimental aspects of PPC. A big thank you also to G. Weimann who gave us one of the samples investigated here.

REFERENCES

1)   Nathan, M.I., Sol. St. Electr. 29, 167 (1986)
2)   Nicholas, R.J., et. al., Appl. Phys. Lett. 44, 629 (1984); Kastalsky, A. and Hwang, J.C.M., Appl. Phys. Lett. 44, 333 (1984); Luryi, S. and Kastalsky, A., Appl. Phys. Lett. 45, 164 (1984); Kane, M.J., et. al., J. Phys. C, 18, 5629 (1985)
3)   Gennser, U., et. al., Proc. of 1st Colorado Microelectronics Conf. March, 1986.
4)   Lang, D.V., et. al., Phys. Rev. B19, 1015 (1979).
5)   Hjalmarson, H.P. and Drummond, T.J., Appl. Phys. Lett. 48, 656 (1986).
6)   Schubert, E.F. and Ploog, K., Phys. Rev. B29, 4562 (1984); Schubert, E.F., et. al., Phys. Rev. B31, 7937 (1985).

THEORY OF THE CYCLOTRON RESONANCE SPECTRUM
OF MANY POLARONS IN QUASI-TWO-DIMENSIONS*

Wu Xiaoguang^a, F. M. Peeters$^\sharp$ and J. T. Devreese^b

*University of Antwerp (UIA), Department of Physics,
Universiteitsplein 1, B-2610 Wilrijk-Antwerpen, Belgium*

The magneto-optical absorption spectrum of the two-dimensional polaron gas is calculated by using a memory function approach. The cyclotron resonance frequency and the cyclotron resonance mass of the polaron are obtained for weak electron-phonon coupling. The influence of the many-body character of the system is studied by investigating the importance of the Fermi-Dirac statistics and of the electron screening. Good agreement is found between theory and experiment in the relevant magnetic field region.

1. Introduction

There have been several studies[1-6] on the polaron cyclotron resonance in two-dimensional (2D) systems both theoretically and experimentally (see also references given in Ref.1). In most of the theoretical studies the many-particle character of the 2D electron system was neglected.

In Ref.1 we calculated, within a single-electron approximation, the magneto-optical absorption of a polaron from which information on the cyclotron resonance frequency and the polaron effective mass could be extracted. The theoretical results were compared to the experimental data for GaAs-AlGaAs heterostructures[4,5] and InSb inversion layers[6]. It was found that the experimental data for InSb are well explained by the one-polaron theory of Ref.1. However, the experimental data for the GaAs-AlGaAs heterostructures could be fitted only if we assumed that the effective electron-phonon coupling strength is reduced by 25%. In Ref.1 we suggested that the many-particle effects have to be included in order to explain the experimental data for the GaAs-AlGaAs heterostructures. This is the purpose of the present paper.

2. Theory

In the present study we start from the Fröhlich polaron model (see for instance Ref.7). The non-zero width of the 2D electron layer is taken into account by considering the lowest subband for which we use the well-known variational wave function[8] $\psi_0(z) = (b^3/2)^{1/2} z e^{-bz/2}$ (the 2D electron gas is in the xy plane).

Using the memory function approach[9] the dynamic conductivity of the system (in the presence of a perpendicular magnetic field) can be expressed as

$$\sigma(\omega) = \frac{in_e e^2/m_b}{\omega - \omega_c - \Sigma(\omega)} \quad , \tag{1}$$

where $\Sigma(\omega)$ is the memory function, n_e the electron density and $\omega_c = eH/m_bc$ the unperturbed cyclotron frequency. To first order in the electron-phonon coupling constant the memory function can be written as

$$\Sigma(\omega) = \frac{1}{\omega} \int_0^\infty dt (1 - e^{i\omega t}) \mathrm{Im} F(t) \quad , \tag{2}$$

with (at zero temperature)

$$F(t) = -i \sum_{\vec{k}} \frac{k_\parallel^2}{n_e m_b \hbar} |V_{\vec{k}}|^2 D(k_\parallel, t) e^{-i\omega_{LO} t} \quad , \tag{3}$$

where $k_\parallel^2 = k_x^2 + k_y^2$ and $D(k,t)$ is the electron density-density correlation function[10] (without electron-phonon interaction) which contains the many-particle effects.

The occupation effect is included by approximating $D(k,\omega)$ by $D^0(k,\omega)$ the density-density correlation function of the non-interacting electron gas. We will refer to this approximation as the Hartree-Fock (HF) approximation. In order to fully incorporate screening a dynamical calculation is necessary which numerically is very involved. Therefore we have limited ourselves in the present work to a static screening approximation where $|V_{\vec{k}}|^2$ in Eq.(3) is substituted by $|V_{\vec{k}}| / \epsilon^2(k)$. In the present study a RPA approximation is made for the static dielectric function $\epsilon(k)$.

The cyclotron resonance frequency, which now is affected by the electron-phonon interaction, is obtained from the position of the cyclotron resonance peak in the magneto-optical absorption spectrum. The latter one is proportional to the imaginary part of the conductivity[8]. At zero temperature the position of the cyclotron resonance peak ω^* is given by the solution of the non-linear equation (cfr. Eq.(1))

$$\omega - \omega_c - \mathrm{Re}\Sigma(\omega; b, \omega_c) = 0 \quad . \tag{4}$$

In order to make a comparison between our theoretical results and the experimental data, it is necessary to incorporate the nonparabolicity of the electron energy band. The cyclotron resonance frequency $(\omega_c)_{np}$ due to nonparabolicity, without polaron effects, is obtained from the difference in energy of the two lowest Landau levels as calculated from Ref.11. Then, $(\omega_c)_{np}$ is inserted into Eq.(4) to replace the unrenormalized ω_c. This amounts to a *local parabolic energy band approximation* [1]. The solution of Eq.(4) ω^* then gives the cyclotron resonance frequency where both nonparabolicity and polaron effects are included. The cyclotron resonance mass is given by $m^*/m_b = \omega_c/\omega^*$.

3. Results

For the numerical calculation we use physical parameters corresponding to the GaAs-AlGaAs heterostructures.

In Fig.1 the correction to the cyclotron resonance mass and the shift of the cyclotron resonance frequency due to polaron effects (not including nonparabolicity) is shown as a function of the magnetic field strength. The calculation is performed for an ideal 2D system i.e. zero width of the 2D layer. The electron density is taken to be $4 \times 10^{11} \mathrm{cm}^{-2}$ which results in a filling factor[8] $\nu = 0.4 \times \omega_{LO}/\omega_c$.

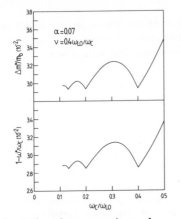

Fig.1 The polaron correction to the cyclotron resonance mass and the shift of the cyclotron resonance frequency as a function of the magnetic field strength for an ideal 2D polaron gas without nonparabolicity.

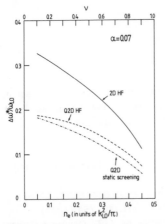

Fig.2 The splitting of the cyclotron resonance peak at the optical phonon frequency as a function of the electron density within different approximations.

The polaron mass correction shows an oscillation. This is a consequence of the occupation effect which at integer filling factors prohibits the electron of performing an intralevel transition. Naively this may be viewed as a reduction of the effective electron-phonon coupling.

The splitting of the cyclotron resonance peak at the optical phonon frequency is shown in Fig.2 as a function of the electron density. Only the polaron effect is included within the different approximations. The density is in units of k_{LO}^2/π which is about $2 \times 10^{12} cm^{-2}$ for the GaAs-AlGaAs heterostructures. The solid curve in Fig.2 gives the result for the splitting of the cyclotron resonance line in the HF approximation for an ideal 2D system. Inclusion of the non-zero width of the 2D electron layer reduces the splitting and results in the dashed curve. Incorporation of the static screening further reduces the splitting and the result is given by the dash-dotted curve.

For a typical electron density $n_e = 4 \times 10^{11} cm^{-2}$ the occupation effect reduces the splitting of an ideal 2D system with a factor of 1.3 (not shown in Fig.2, see Ref.1). The finite width of the 2D electron layer leads to a further reduction of the splitting with a factor of about 1.6. Static screening has only a small effect on the splitting and leads to a further reduction with 10%. Therefore to compare the theoretical results with the experimental data we will neglect the screening effect.

In Fig.3 the present theoretical calculation is compared to the experimental data of Horst et al.[4]. Good agreement is found for magnetic fields above 15 T.

It has been pointed out in Ref.1 that the experimental data of Ref.4 and the data of Ref.5 behave very differently for magnetic fields below 12 T, although the two experimental samples have almost the same electron densities: $n_e = 4.07 \times 10^{11} cm^{-2}$ in Ref.4 and $4 \times 10^{11} cm^{-2}$ in Ref.5. Also, the cyclotron resonance mass of Ref.5 shows a peak structure around 5 T which is not present in the results of Ref.4 and which does not seem to be due to polaron effects. More experimental data is needed to resolve this difference and to

592   X.-G. Wu et al.

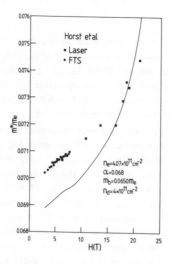

get a deeper understanding of the role played by the electron-phonon interaction in the cyclotron resonance experiments of Refs. 4 and 5.

Fig.3 Comparison between the theoretical (full line) and experimental results for the polaron cyclotron resonance mass in GaAs-AlGaAs heterostructures[4].

References
* Partially sponsored by F.K.F.O. project No.2.0072.80.
° Supported by the C.G.I.C.S. (Belgium).
‡ Senior research assistant of the N.F.W.O. (Belgium).
b Also at RUCA (Belgium) and THE (The Netherlands).

1 F. M. Peeters, Wu Xiaoguang, and J. T. Devreese, Physica Scripta (proceedings of the 6th CMD conference of the EPS, Stockholm, 1986); Wu Xiaoguang, F. M. Peeters, and J. T. Devreese, Phys. Rev. B (accepted for publication).

2 D. Larsen, Phys. Rev. B30, 4595 (1984).

3 R. Lassnig, Surf. Sci. 170, 549 (1986).

4 M. Horst, U. Merkt, W. Zawadzki, J. C. Maan, and K. Ploog, Solid State Commun. 53, 403 (1985).

5 H. Sigg, P. Wyder, and J. A. A. J. Perenboom, Phys. Rev. B31, 5253 (1985).

6 M. Horst, U. Merkt, and J. P. Kotthaus, Phys. Rev. Lett. 50, 754 (1983).

7 Polarons in Ionic Crystals and Polar Semiconductors, edited by J. T. Devreese (North-Holland, Amsterdam, 1972).

8 T. Ando, A. B. Fowler, and F. Stern, Rev. Mod. Phys. 54, 437 (1982).

9 R. P. Feynman, R. W. Hellwarth, C. K. Iddings, and P. M. Platzman, Phys. Rev. 127, 1004 (1962); J. T. Devreese, in Ref.7, p.83; F. M. Peeters and J. T. Devreese, Physica 127B, 408 (1984); W. Götze and P. Wölfe, Phys. Rev. B6, 1226 (1972).

10 A. L. Fetter and J. D. Walecka, Quantum Theory of Many-Particle Systems (McGraw-Hill, New York, 1971), p.190.

11 R. Lassnig and W. Zawadzki, Surf. Sci. 142, 388 (1984).

LANDAU LEVEL WIDTH AND CYCLOTRON RESONANCE IN 2D SYSTEMS

R. Lassnig, W. Seidenbusch, E. Gornik
Institut für Experimentalphysik, Universität Innsbruck
A-6020 Innsbruck, AUSTRIA

G.Weimann
Forschungsinstitut der Deutschen Bundespost
D-6100 Darmstadt, GERMANY

The density of states of quasi 2D electrons at high magnetic fields splits into Landau levels, which are separated by the cyclotron energy $\hbar\omega_c$. These levels (with Landau index n) are broadened due to material inhomogenities, impurities and phonons. In GaAs-GaAlAs heterostructures at low temperatures, the most important scattering mechanism is ionized impurity scattering, which is strongly influenced by screening.

In the theoretical part of this paper the temperature dependence of the screening and the Landau level (LL) width are calculated for the first time and a new theory of cyclotron resonance (CR) is presented. In addition, we have performed a detailed experimental analysis of the CR in GaAs-GaAlAs heterostructures, which will be compared with the theoretical results. Evidence is found for the influence of the fractional Quantum Hall states on the linewidth.

Due to resonant screening of ionized impurities [1,2], the level width Γ_n as well as the CR linewidth Γ_{CR} oscillate as a function of the filling factor ($\nu = 2\pi l^2 N_{el}$, where N_{el} is the electron density and l is the Landau radius). The reason for this effect is that for partially filled LLs the polarisability and hence the screening strength are strongly enhanced compared to a situation with filled LLs. Maxima of Γ are therefore obtained whenever a LL is fully occupied. The broadening parameter Γ_n of the LLs is calculated within the SCBA:

$$\Gamma_n^2 = 4 \sum_q U^2(q) |\langle n|e^{iqx}|n\rangle|^2 / \varepsilon(q)^2 \qquad (1)$$

where $U(q)$ denotes the unscreened electron-impurity interaction (including the form factors [1] and treating the long-range impurity interference in a Yukawa-type approximation). For partially filled levels the dielectric function $\varepsilon(q)$ depends essentially on Γ_n, which in turn depends on $\varepsilon(q)$; so the problem is self-consistent [1]. The fact that

this can lead to very small level widths is denoted as resonant screening, which is most pronounced at zero temperature.

Finite temperatures, however, represent a limit to this resonance
effect: For $k_B T/\Gamma > 1$ the electrons are distributed quite uniformly over
the LL, which results in a strong reduction of screening. This leads to
a hot electron effect, even at 1°K due to the small level widths.

Fig.(1a) shows the calculated temperature dependence of the level
width: Γ_0 is plotted versus the filling factor (neglecting spin splitting). N_{el} is kept constant, so the drawing corresponds to a magnetic
field sweep. N_i and d_{sp} are the bulk impurity concentration and the
spacer thickness. The curves correspond to temperatures of 0.1, 1 and
10°K. The difference is most pronounced if the LLs are half filled and
especially for $\nu < 2$ large variations due to temperature are found.

In Fig.(1b) the CR linewidth is plotted for the same experimental
situation. The dynamic conductivity is calculated by extending the
self-consistent current relaxation theory of Götze /3/ and Gold /4/
to the high magnetic field case, using the dynamic polarisability of
the unscattered electron gas. The oscillation of the screening strength
is much less pronounced in Γ_{CR} than in Γ_n. This behaviour is due to the
fact that CR represents a local excitation and thus probes rather short
range than long range potential fluctuations (only the latter are sensitive to screening).

This discussion is intended only to show the dominant trends that
are found in a LL system. For specific experiments (samples), the res-

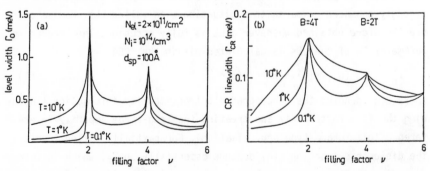

Fig.1: Landau level width (a) and CR linewidth (b) for a GaAs-GaAlAs heterostructure vs. the filling factor, for three different temperatures.

ults will depend strongly on the system parameters. For ν>2 overlapping
of different transitions due to nonparabolicity will be important and
for high densities overabsorption will become dominant.

Recently, several groups /5-8/ have observed oscillations of the
CR linewidth with the filling factor, performing CR transmission experiments. Here we report new careful investigations of the CR transmission for a number of GaAs-GaALAs heterostructure samples with concentrations varying from 1.2 to 3.7 x 10¹¹/cm² and mobilities between 10⁵ and
1.2 x 10⁶cm²/Vs. Different wavelengths from an optically pumped far infrared laser between 90 and 240 μm were applied. The carrier concentration was changed by LED illumination and by thin back gates.

The spectra were
analysed by a fit of
the transmission curve.
Fig.(2) shows a plot of
the determined linewidth
as a function of the fil-
ling factor for the low
density sample 1395 at
4.2°K. It shows a pro-
nounced maximum at ν=2.
The shape of Γ_{CR} over ν
is very similar to that
predicted by the theory
(see Fig. 1b).

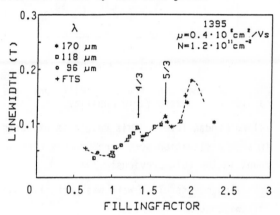

Fig.2: CR linewidth versus filling factor
for the sample with the lowest electron density. Fractional fillings are indicated.

Besides the quite rapid decrease of Γ_{CR} at low filling factors and
the maximum at ν=2 small maxima appear at filling factor value of 4/3
and 5/3. The maxima in Γ_{CR} at fractional filling factors are observed
somewhat stronger at 2°K. They can be interpreted as a local condensation into the fractional ground state. A resulting density of states
structure at fractional fillings could influence the CR due to reduced
screening. The maxima of Γ_{CR} are rather narrow and depend critically
on the correct determination of the filling factor from parallel measured Shubnikov - de Haas oscillations. We found that the incident far
infrared background radiation can change locally the carrier density.

596   R. Lassnig et al.

No clear structure is found at $\nu=1$. However, we also find only weak
spin effects in the static conductivity, leading to the conclusion
that the spin splitting is not very pronounced.

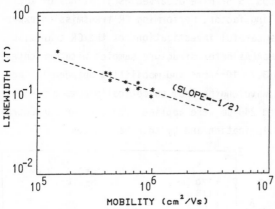

In addition, we have
plotted the CR linewidth
at $\nu=2$ as a function of
the zero field mobility
for several samples with
similar density in Fig.3.
A square root dependence
of Γ_{CR} on the inverse
mobility is found. If
one correlates the in-
verse mobility with the
number of impurities N_i,
one can state that Γ_{CR}
is proportional to $\sqrt{N_i}$ at

Fig.3: CR linewidth at $\nu=2$ for different
samples versus zero field mobility.

filled Landau levels. This result is expected from the theory, since
in such a situation screening is weak. The same behaviour has been
found in the bulk previously /9/.

Acknowledgements: This work was partially supported by the Stiftung
Volkswagenwerk.

References:

/1/ Lassnig R., Gornik E.; Solid State Comm. 47, 959 (1983)

/2/ Ando T., Murayama Y.; J.Phys.Soc.Japan 54, 1519 (1985)

/3/ Götze W.; Solid State Comm. 27, 1393 (1978)

/4/ Gold A.; Phys.Rev. B32, 4014 (1985)

/5/ Englert Th., Maan J.C., Uihlein Ch., Tsui D.C., Gossard A.C.;
    Physica 117B&118B, 631 (1983)

/6/ Seidenbusch W., Lindemann G., Lassnig R., Edlinger J., Gornik E.;
    Surf.Science 142, 375 (1984)

/7/ Seidenbusch W., Lassnig R., Gornik E., Weimann G.; Physica 134B,
    314 (1985)

/8/ Rikken G., Myron H.W., Wyder P., Weimann G., Schlapp W., Horstman
    R., Wolter J.; Surf. Science 170 (1986)

/9/ Gornik E.; Physica 127B, 95 (1984

THE DENSITY OF STATES IN THE GAPS OF TWO-DIMENSIONAL ELECTRONS ENERGY SPECTRUM IN THE PERPENDICULAR MAGNETIC FIELD

I.V.Kukushkin, M.G.Gavrilov

Institute of Solid State Physics, Academy of
Sciences of the USSR, 142432 Chernogolovka,
USSR

ABSTRACT

The density of states of 2D-electrons between Landau
levels (D_X) is defined by use of two different methods: from the analysis of thermoactivated magneto-
conductivity and from the radiative recombination of
2D-electrons with nonequilibrium holes. It is shown
that D_X decreases with the growth of the magnetic
field or the mobility of 2D-electrons: $D_X \sim (\mu H)^{-1/2}$

1. Thermoactivated magnetoconductivity

The idea of the thermoactivated magnetoconductivity
method is contained in works/1,2/, but as it will be shown
below it is necessary to take into account simultaneously
an electron and a hole inputs to the conductivity for obtaining the correct value D_X. At full filling of a Landau
sublevel($n_S = n_S^0 = \nu(eH/h)$, where n_S-the density of electrons,
ν- filling factor, integer) the magnetoconductivity σ_{xx}
(and magnetoresistivity ρ_{xx}) achieves minimum σ_{xx}^{min}, which
depends on temperature(T):

$$\sigma_{xx}^{min} \sim \rho_{xx}^{min} \sim \exp(-W/kT) , \qquad (1)$$

where the activation energy W equals a half of the energy
gap. When the density n_S differes from n_S^0 on a small value
Δn, the Fermi level moves from the middle of the gap to
the Landau level on the value ΔE_F(see insert in fig.1). In
this case one can get/3/;

$$\sigma_{xx}(\Delta n) \sim \rho_{xx}(\Delta n) = A \exp(-(W + E_F)/kT) + B \exp(-(W - E_F)/kT) \qquad (2)$$

and from (1) and (2):

$$\rho_{xx}(\Delta n) = \rho_{xx}^{min} \cdot ch(E_F/kT + C) , \qquad (3)$$

where A,B,C- are the constants depending on Δn (or $\Delta \nu$). In
fig.1 the dependences of $\text{Arch}(\rho_{xx}(\Delta n)/\rho_{xx}^{min})$ on T^{-1} are shown
for positive and negative Δn at $\nu=8$; $H=6,7T$; $\mu=1,4m^2/V \cdot s$. It
is seen that in the mentioned coordinates the experimental

points are located on direct lines and the slope of the
line defines ΔE_F for indicated Δn. It also follows from the
experiment that $C(\Delta\nu)=7\Delta\nu$ for $\Delta\nu\leqslant 0$, and $C=-4\Delta\nu$ for $\Delta\nu\geqslant 0$ at
the mentioned conditions. The dependence of ΔE_F on Δn ob-
tained by such method is shown in fig.2a for positive and
negative Δn. Fig.2b represents the energetic dependence of
the density of states between Landau levels obtained by
differentiation of the function $\Delta E_F(\Delta n)$. One can see that
the value D_x composes a considerable part of the density
of states of 2D-electrons at H=0: $D_0=2m/\pi\hbar^2=1,6\cdot10^{11}$ cm^{-2}meV^{-1}.
Analogous dependences of D on E were obtained in different
MOSFETs for various values H and μ. According to the works
/1,2/ the value D_x depends on μ but does not depend on H.
In our experiments at H=const , $D_x\sim\sqrt{\mu}$, and at μ=const ,
$D_x\sim\sqrt{H}$, so that D_x depends on product μH. This dependence
is shown in dimensionless coordinates in insert of fig.2b
for Si(001) MOSFETs by dark circles and squares and for
GaAs-AlGaAs heterostructures by dark triangles. The result
of the works/1,2/ are also shown by open triangles. It is
seen that the results obtained in Si MOSFETs and in heterostructures are in a good agreement and a slope of the de-
pendence of $\lg(D_0/D_x)$ on $\lg(\mu H)$ is equal to$\approx 0,5$: $D_x\sim D_0\sqrt{\mu H}$.

2.   The radiative recombination of 2D-electrons

The method of optical spectroscopy is founded on the
analysis of radiative recombination of 2D-electrons with
nonequilibrium holes bound to boron atoms, when a photoexcitation of MOSFETs occures (see insert of fig.3a).Under
condition of photoexcitation the depletion layer disappears
and the radiative recombination arises due to wave functions overlapping of 2D-electrons and holes bound to boron
atoms just close to the 2D-layer/4,5,6/. The shape of the
emission line (2D$_e$) is a convolution of distribution functions of 2D-electrons ($F_e=D\cdot f$, where f- filling numbers)
and holes (F$_h$). As follows from the experiment/5,6/ energy
width of F_h is small ($\approx 0,8$ meV), so the 2D-line reflects
the density of states of 2D-electrons. Fig.3a represents
emission spectra measured at H=0 (spectrum 2) and H=7T (3)
for $n_s=2,7\cdot10^{12}$ cm^{-2} and T=1,6K. Under the conditions of
spectrum 3 four Landau levels are completly filled and minimum in σ_{xx} corresponding to ν=16 was observed simultaneously in magnetotransport. At H=0 the 2D$_e$-line has a step-
like shape which directly reflects the fact that D(E)=const
and the width of 2D$_e$-line is equal to E_F/5,6/. In magnetic
field the emission spectra demonstrate Landau level quantization and cyclotronic splitting is sensitive only to
the normal projection of H to 2D-plane/5/ (compare emission
spectra at H=7T for θ=0 and θ=60° in fig.3a and 3b,respecti-
vely). The method of optical spectroscopy is an way of direct determination of the function D(E). By using of this
method the oscillations of Landau level width under variation of ν was observed for the first time/5/. It is seen
from insert of fig.3b, that in the region of quantum Hall

effect (at integer ν) under the conditions of strong localization the Landau level width Γ achieves a maximum and in metallic region (at half-integer ν) Γ has a minimal value. It is important that in the last case in most perfect MOSFETs (with high μ) the $2D_e$-line reflects the hole distribution function F_h , because the width of F_e approaches to zero. To determine the absolute value of $D(E)$ in magnetic field from $2D_e$-line it is necessary to compare the emission spectra (with equal integral intensities) obtained for given n_s at H=0 and H≠0. In fig.4a the $2D_e$-lines with equal integral intensities measured for $n_s=1,36 \cdot 10^{12}$ cm^{-2} at H=0 and H=7T are shown. Fig.4b represents $D(E)$ at H=0, Landau levels and total density of states at H=7T obtained after deconvolution of emission spectrum with function F_h measured separately. The values D_x defining by such optical method for different μ and H are shown in insert of fig.2b by half-open circles. It is seen that the values D_x measured by use of two different methods are in a good agreement. It means that the method of thermoactivated magneto-conductivity gives correct value D_x, although its application in a wide energy range is doubtful because function $D(E)$ doesn't remain constant under variation of $\Delta\nu$/5/.

References:
1. Weiss D.,Stahl E.,Weimann G.,Ploog K.,Klitzing K.v.,
   Proc. of EP2DS-VI, Japan, 307,(1985).
2. Stahl E.,Weiss D., Weimann G.,Klitzing K.v.,Ploog K.,
   J.Phys. C18, L783 (1985).
3. Gavrilov M.G., Kukushkin I.V., Pis'ma Zh.Eksp.Teor.
   Fiz., 43, 79 (1986).
4. Kukushkin I.V., Timofeev V.B., Pis'ma Zh.Eksp.Teor.
   Fiz., 40, 413 (1984).
5. Kukushkin I.V., Timofeev V.B., Pis'ma Zh.Eksp.Teor.
   Fiz., 43, 387 (1986).
6. Kukushkin I.V.,Timofeev V.B., Zh.Eksp.Teor.Fiz.,(1986)

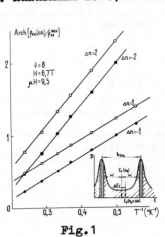

Fig.1

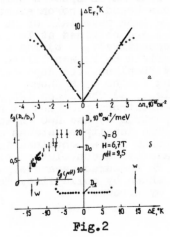

Fig.2

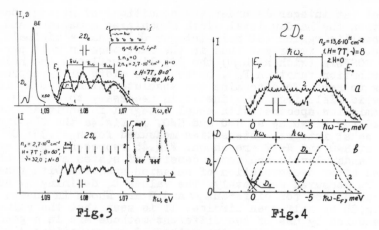

Fig.3   Fig.4

Figures captions:

Fig.1. The dependence of Arch($\varrho_{xx}(\Delta n)/\varrho_{xx}^{min}$) on T^{-1} at ν=8,H=6,7T, $\mathcal{M}(n_g^o)$=1,4m^2/V·s. The values Δn are represented in units 10^{10}cm^{-2}. The scheme of the density of states is shown in insert.

Fig.2. The dependences of ΔE_F on Δn (a) and D on E (b) obtained for ν=8,H=6,7T, μ=1,4m^2/V·s. The values D_o and D_x are indicated by arrows. The dependence of $lg(D_o/D_x)$ on $lg(\mu H)$ is represented in insert of fig.2b. Dark circles (ν=4) and squares(ν=8) correspond to Si-MOSFETs, dark triangles- to heterostructures, open triangles- results of works/1,2/. The results obtained from optical measurements are represented by half-open circles.

Fig.3. a) The insert represents the scheme of recombination of 2D-electrons with nonequilibrium holes. The intense line BE in spectra corresponds to luminescence of excitons bound to boron atoms from cristal volume. Spectrum 1 - long wave tail of BE-line obtained at n_g=0.Spectra 2 and 3 are obtained at T=1,6K, n_g=2,7·10^{12} cm^{-2}, H=0 and H=7T, respectively. b) The spectrum obtained when the magnetic field H=7T, was tilted on angle θ=60° from the normal to the 2D-plane. The dependence of Landau level width Γ on filling factor ν is shown in insert for H=7T and T=1,6K.

Fig.4. a) The lines (2D$_e$) of radiative recombination of 2D-electrons recorded at n_g=1,36·10^{12} cm^{-2}, T=1,6K, H=7T (spectrum 1) and H=0 (spectrum 2).

b)The density of states at H=0 (2) and Landau levels (1, broken lines) obtained from the emission spectra of fig.4a after deconvolution with the hole distribution funtion Fh measured separately. The density of states at H=7T is shown by solid line. The values D_o and D_x are indicated by arrows.

INTERSUBBAND SPECTROSCOPY IN GaAs-Ga$_{1-x}$Al$_x$As

HETEROJUNCTIONS VIA SUBBAND-LANDAU-LEVEL ANTICROSSING

A. D. Wieck and J. C. Maan

Max-Planck-Institut für Festkörperforschung, Hochfeld-Magnetlabor

Grenoble, 25, Avenue des Martyrs, 33042 Grenoble, FRANCE

and

U. Merkt and J. P.Kotthaus

Institut für Angewandte Physik, Universität Hamburg,

Jungiusstraße 11, 2000 Hamburg 36, F. R. GERMANY

and

K. Ploog

Max-Planck-Institut für Festkörperforschung, Heisenbergstraße 1

7000 Stuttgart 80, F. R. GERMANY

ABSTRACT

In magnetic fields tilted away from the surface normal we observe anticrossing of cyclotron resonance and combined intersubband-cyclotron resonance when their transition energies match. The capability of this method to systematically study subband spacings in GaAs-Ga$_{1-x}$Al$_x$As heterojunctions in modest magnetic fields is demonstrated.

The determination of subband spacings in GaAs-Ga$_{1-x}$Al$_x$As heterojunctions is of great interest to characterize the interface potential and its dependence on the two-dimensional electron density N_s and on the depletion charge N_{depl}. We detect the subband resonance $0\rightarrow1$ via a splitting of cyclotron resonance in tilted magnetic fields when the energy of the cyclotron resonance $\hbar\omega_c$ equals the one of the combined intersubband-cyclotron resonance $E_{01}-\hbar\omega_c$. This is possible since in tilted magnetic fields the electron motion parallel and perpendicular to the interface is coupled. Combined cyclotron-intersubband resonances have been excited with light polarized perpendicular to the interface in metal-oxide-semiconductor structures on

Si.[1] Previous cyclotron resonance experiments in GaAs-$Ga_{1-x}Al_xAs$ heterojunctions[2] with light polarized parallel to the interface have demonstrated coupling in quite high magnetic fields when the cyclotron transition is resonant with the intersubband transition itself ($E_{01}=\hbar\omega_c$). In our case ($E_{01}=2\hbar\omega_c$) we need only about half of the magnetic field intensity and still obtain sufficiently strong coupling provided the tilt angle is large enough.

In the experiments we measure the transmission of far-infrared light in a Fourier spectrometer in constant magnetic fields at liquid helium temperatures. The surface normal of the sample is tilted by an angle θ with respect to the magnetic field direction and the incident radiation propagates in the magnetic field direction. Typical spectra for a sample with electron density $N_s=6.1x10^{11}cm^{-2}$ are shown in Fig.1 for two tilt angles θ=33° and 12° in the upper and lower part of the figure, respectively. At the larger tilt angle θ=33° we observe clear splitting of the cyclotron resonance at frequency $\tilde{\nu}\approx90$ cm^{-1}. This splitting is explained by coupling of the n=2 Landau level of the ground electric subband E_0 to the n=0 Landau level of the excited subband E_1 as is depicted in the inset of Fig.2. Note, that the observation of two cyclotron modes (a,b) is only possible when more than one Landau level is occupied. At partial filling of the n=1 Landau level also cyclotron resonances n=0 → n=1 (c) are observed. At the highest magnetic field strengths B⊥>12T the cyclotron resonance broadens and decreases in amplitude as a result of the coupling at $\hbar\omega_c=E_{01}$. This coupling is studied in the lower part of Fig.1 at the smaller tilt angle θ=12°. At this tilt angle no coupling at $\hbar\omega_c=E_{01}/2$ is observed ($\tilde{\nu}\approx90$ cm^{-1}) but clear splitting is evident at $\hbar\omega_c=E_{01}$ ($\tilde{\nu}\approx180$ cm^{-1}). Figure 2 shows resonance positions extracted from spectra like those in the upper part of Fig.1. The transition c is not affected by coupling to intersubband resonance and lies slightly asymmetric in the gap between transitions a and b as a result of band nonparabolicity[3] that causes slightly higher transition energies for lower Landau transitions. Resonance positions extracted for various tilt angles in the two coupling regimes $E_{01}=\hbar\omega_c$ and $E_{01}=2\hbar\omega_c$ are depicted in Fig.3 for two electron densities $N_s=3.3$ and $6.1x10^{11}$ cm^{-2} obtained

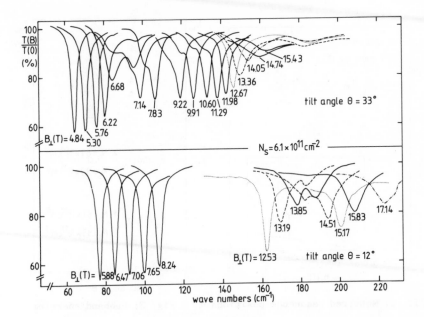

Fig. 1: Normalized transmission spectra of a GaAs-Ga$_{1-x}$Al$_x$As heterojunction in tilted magnetic fields.

in the same sample using bandgap illumination to generate the higher density. The subband energy E_{01} extracted at various tilt angles and consequently different total magnetic field strengths agree among each other within $\Delta E \le 0.5$ meV. In particular, this demonstrates that diamagnetic shifts arising from the magnetic field component $B_\parallel$ parallel to the interface are not important in the present experiments.

We have demonstrated that subband–Landau level anticrossing at cyclotron energy $\hbar\omega_c = E_{01} - \hbar\omega_c$ can serve to study subband spacings in GaAs-Ga$_{1-x}$Al$_x$As heterojunctions at moderate magnetic field strengths. The observed intersubband energies $E_{01} \approx 20.6$ and 22.7 meV for the two electron densities $N_s = 3.3$ and 6.1×10^{11} cm^{-2}, respectively, are in good agreement with theoretical values[4] calculated for a depletion charge $N_{depl} \approx 5 \times 10^{10}$ cm^{-2}. We currently employ this anticrossing mechanism to systematically investigate intersubband energies in

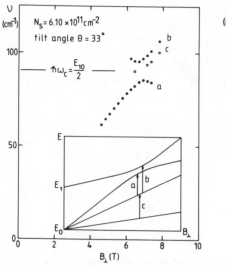

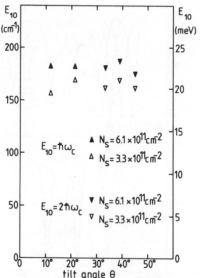

Fig. 2: Measured resonance positions near the coupling regime $\hbar\omega_c = E_{01}/2$. vs. magnetic field component $B\perp$ perpendicular to the interface. Three resonances a–c are observed as indicated in the inset.

Fig. 3: Subband energies E_{01} extracted from spectra taken at various tilt angles θ in both coupling regimes for two electron densities N_s.

samples with different electron density and different depletion potential. A detailed comparison of the results with theory will help to more quantitatively determine the potential at the heterojunction interface and its dependence on sample preparation.

We acknowledge financial support of the Deutsche Forschungsgemeinschaft.

1. Beinvogl, W. and Koch, J. F., Phys. Rev. Lett. __40__, 1736 (1978).
2. Schlesinger, Z., Hwang, J. C. and Allen, Jr., S. J., Phys. Rev. Lett. __50__, 2098 (1983).
3. Lassnig, R., Phys. Rev. B __31__, 8076 (1985).
4. Stern, F. and Das Sarma, S., Phys. Rev. B __30__, 840 (1984).

Optically Pumped Cyclotron Resonance and Resonant Electron-LO Phonon Interaction in GaAs/AlGaAs Multiple Quantum Wells

Y.H. Chang and B.D. McCombe

SUNY at Buffalo

Buffalo, NY, USA, 14260

and

J. Ralston and G. Wicks

NRRFSS Cornell University

Ithaca, NY, USA, 14853

Semiconductor systems of reduced dimensionality have attracted considerable attention in recent years. Electron-phonon interaction in such structures, particularly those fabricated from polar materials, is of interest from a fundamental point of view as well as because of its importance in device operation. In the latter case the AlGaAs-GaAs heterostructures and quantum well systems are of particular interest. Theoretical aspects of electron-LO phonon (polaron) interaction in two dimensional systems have been reported by several authors[1-4], and enhancement of polaron effects in systems of reduced dimensionality is predicted. Experiments carried out in these systems in the presence of a magnetic field have provided evidence for both enhanced and reduced polaron effects.[5-7] Screening, wavefunction extent[8] and occupation effects[2] have been invoked for the actual experimental systems in order to resolve apparent discrepancies, and qualitative agreement has been claimed. However, with such complications it is difficult to determine unambiguously the effects of reduced dimensionality on the polaron coupling.

We report experimental evidence for an enhanced polaron effect in GaAs-Al$_{.3}$Ga$_{.7}$As multiple quantum well (MQW) systems. These systems, with low electron concentration, are free of some of the above complications and provide the possibility of using well width

as a parameter to investigate systematically the transition of
polaron effects from 3D to 2D.

In the present experiments, a 3-level magnetopolaron
experiment was carried out by observing cyclotron resonance between
Landau levels (LLs) n=1 and n=2 over a range of magnetic field. As
the separation between LLs 0 and 2 approaches the LO phonon energy
the n=2 LL becomes degenerate with the n=0 LL plus one LO phonon.
In this situation, the electron-LO phonon interaction lifts the
degeneracy and gives rise to two distinct branches $(2^{\pm})$. The
difference in transition energy between the n=1 LL and the lower
branch (2^{-}), $\hbar\omega_c(1-2^{-})$, and that of $\hbar\omega_c(0-1)$ is a direct measure of
the interaction strength.

Measurements were carried out between 4.2°K and 60°K with a
far Infrared Fourier transform spectrometer in conjunction with a 9T
superconducting magnet system. Several samples have been
investigated. Results reported here are for a MBE-grown GaAs-
$Al_{.3}Ga_{.7}As$ MQW structure. The central one-third of each of the 210Å
wells was doped with silicon $(10^{16}cm^{-3})$. AlGaAs barriers were 150Å
wide, and the MQW structure was enclosed on both sides by ~2000Å
layers of $Al_{.3}Ga_{.7}As$. Free electrons in the wells were created
either by thermally ionizing the donors with a resistive heater, or
by optically pumping the sample with up to 10mW of He-Ne laser
light, or both.

Typical transmission spectra are shown in Fig. 1. The
dominant line in the upper trace of Fig. 1a) corresponds to the LL
0→1 cyclotron resonance. From the integrated intensity the free
carrier concentration is estimated to be 3×10^{9} electron/cm^2 per well
in approximate agreement with that expected from thermal ionization
of the QW donors. Effects of laser light pumping are shown in the
lower trace of Fig. 1a); the concentration of free electrons
generated in this manner is also ~ 3×10^{9} electrons/cm^2 per well.
The excess electrons observed in cyclotron resonance appear to
result from neutralization of compensating acceptors and an
increased electron temperature. The laser pumping thus provides a
unique way of varying electron density and temperature. It is
expected that with higher pump power it will be possible to produce

a substantial electron density in LL 1 and thus to provide the possibility of studying the electorn-LO phonon interaction while the lattice temperature is kept at 4.2°K and the electron density is varied.

In addition to the 0→1 cyclotron resonance transition depicted in Fig. 1a), weak satellite lines have also been observed for several magnetic fields as shown in Fig. 1b). The satellite lines are attributed to the transition between LLs n=1 and n=2⁻ as discussed above. Temperature dependent studies at a fixed magnetic field show that the optimum temperatures for observing these lines are between 35K and 45K. From the integrated intensities it is estimated that the ratio of the electron densities in the n=0 and n=1 LLs is about 20. However, for thermal equilibrium at 40°K and 8T the calculated ratio is 70. The discrepancy is presently not understood.

A comparison between a calculation[3] for the polaron effect in a strictly 2D system corrected for the nonparabolic GaAs conduction band and experimental data is shown in Fig. 2. Also included in this figure is a calculation for the 3D case[9,10], including both nonparabolicity and polaron effects. The experimental results indicate a stronger polaron effect than is predicted for 3D. The experiment data also show a smaller variation with field than is predicted by theory as resonance is approached. Considering the effect of wavefunction extent, the observed splittings are larger than expected for a 210Å well[8]. It is important to note that so far we have not considered the effects of a superlattice, in particular, the effects of zone folding of the LO phonons[11]. Since the major contribution to the polaron effect is the interaction between the electron and long wavelength LO phonons, the zone folding effect, which translates in the reduced zone the short wavelength LO phonons into long wavelength LO phonons is expected to enhance the polaron effect and produce a substantial splitting at the lower magnetic field than would otherwise occur. Such effects may provide a partial explanation for the present result; however, a detailed knowledge of the LO phonon dispersion relation along $\frac{2\pi}{a}[0,0,1]$

direction and a modified theoretical calculation are required to confirm this interpretation.

This work was supported in part by the Army Research Office and by NSF through a grant to NRRFSS.

REFERENCES

1.   S. Das Sarma, Phys. Rev. Lett. 52, 829 (1984).

2.   D.M. Larsen, Phys. Rev. B 30, 4595 (1984).

3.   F.M. Peeters and J.D. Devreese, Phys. Rev. B 31, 3689 (1985).

4.   R. Lassnig, Surf. Sci. 170, 549 (1986).

5.   M. Horst, U. Merkt and J.P. Kottaus, Phys. Rev. Lett. 50, 754 (1983).

6.   W. Seidenbusch, E. Gornik and G. Weimann, Physica 134, B+C (1-3), 314 (1985).

7.   W. Seidenbusch, G. Lindemann, R. Lassnig, J. Edlinger and E. Gornik, Surf. Sci. 142, 375 (1984).

8.   S. Das Sarma, Phys. Rev. B 27, 2590 (1983).

9.   G. Lindeman, R. Lassnig, W. Seidenbuch and E. Gornik, Phys. Rev. B 28, 4693 (1983).

10.   P. Pfeffer and W. Zawadzki, Solid State Comm. 57, 847 (1986).

11.   S. Das Sarma and A. Madhukar, Phys. Rev. B 22, 2823 (1980).

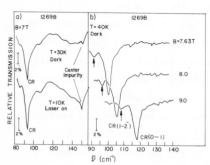

Fig. 1: a) Transmission spectra showing cyclotron resonances induced by thermal heating and laser pumping as well as impurity transitions. b) $0 \to 1$ cyclotron resonances and satellite lines attributed to LLs $1 \to 2^-$ transisions at different magnetic fields.

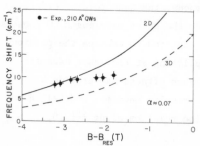

Fig. 2: Frequency shift between LLs $1 \to 2^-$ transitions and $0 \to 1$ transitions. B_{Res} is the magnetic field where the n=0 LL plus one phonon is degenerate with the n=2 LL. Calculations for 2-dimensions[3] and 3-dimensions[9] (both non-parabolicity and polaron effects included) are shown by the solid and dashed curves respectively. The parabolic 2-dimensional calculation has been corrected for non-parabolicity.

MAGNETOPHONON RESONANCE IN GaAs-GaAlAs HETEROJUNCTIONS

M.A. Brummell[*], M.A. Hopkins[*], R.J. Nicholas[*], J.J. Harris[+] and C.T. Foxon[+]

* Clarendon Laboratory, Parks Road, Oxford OX1 3PU, U.K.
+ Philips Research Laboratories, Redhill, Surrey, U.K.

We report cyclotron resonance studies of GaAs-GaAlAs heterojunctions which show that polaron effects are significantly reduced by screening at low temperatures, and magnetophonon resonance results which indicate that the electrons are interacting strongly with phonons lower in energy than the bulk GaAs LO phonon.

Polaron effects in two-dimensional (2D) systems are a topic of considerable current interest. Magnetophonon resonance[1] has shown that the interfaces can significantly modify the interactions, and cyclotron resonance experiments[2] have shown resonant polaron coupling near the TO phonon frequencies, in contrast to the expected interaction with LO modes. Simple theory[3,4] predicts enhanced polaron effects in two dimensions, but screening[5,6] and the finite wavefunctions in the third dimension[5,6] may reduce the coupling, and experiments on GaAs-GaAlAs heterojunctions[6-8] have suggested that the polaron effects are weaker than in bulk GaAs.

We have performed cyclotron resonance experiments as a function of temperature on several high quality GaAs-GaAlAs heterojunctions using 118.83 μm radiation from a far infra-red laser. At low temperatures, extremely narrow linewidths are seen ($\Delta B/B \approx 1/300$), but the resonances broaden rapidly with temperature, increasing as $T^{3/2}$ above $\sim$30 K. The nonparabolicity deduced from the frequency and electron concentration dependences of the cyclotron mass[9] was used to calculate the mass associated with the transition between the n = 0 and n = 1 Landau levels m^*_{01} assuming transitions weighted by the populations of the initial levels. The masses deduced are plotted in Fig. 1 and show an unexpected decrease as the temperature is reduced below $\sim$100 K, whereas similar measurements on bulk GaAs show no such change in m^*_{01}.

We believe that this behaviour is caused by temperature dependent screening of the effective mass enhancement due to polaron coupling. In zero magnetic field, screening reduces the strength of the polaron coupling by a factor of order 2-3[5,6]. In high fields, the effects should be much more extreme due to the quantisation of the density of states into

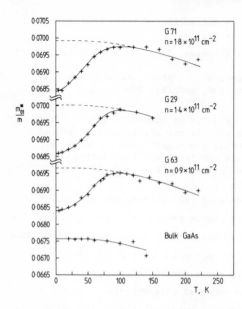

Figure 1: The temperature dependence of the cyclotron mass
due to transitions between the
two lowest Landau levels in
three heterojunction samples
and bulk GaAs. The decrease in
the heterojunctions below ∿100
K is attributed to screening
of the polaron coupling. The
dashed lines are extrapolated
using the bulk dependence.

Landau levels, as the screening depends upon the density of states within ∿kT of the Fermi energy[10]. At low temperatures the Landau levels
will be extremely sharp and will contain a very high density of states.
This leads to strong screening which suppresses the polaron mass enhancement and also leads to oscillatory cyclotron resonance linewidths[11]
due to variations in the screening with filling factor[12]. At higher
temperatures the increased level width and thermal energy will reduce
the screening and the mass enhancement reappears. Similar arguments[13]
were used to explain changes in the resonant polaron coupling in GaAs-
GaAlAs heterojunctions in high electric fields. For the conditions of our
experiments, simple theory[3,4] gives a polaron mass enhancement of 3.9%,
which is reduced to ∿1.5% when the finite wavefunctions in the third
dimension are taken into account[5,6]. Experimentally, the decrease in
m_{01}^{*} at low temperatures is ∿2%, suggesting that most or all of the enhancement is screened out. However, interactions with the LO phonons of
bulk GaAs have been assumed, whereas the magnetophonon resonance results
discussed below suggest that other phonons are involved.

Magnetophonon resonances occur when the cyclotron energy equals the
separation between two Landau levels, and may be observed as a series of
oscillations in the magnetoresistance, periodic in 1/B, at fields where

$$\hbar\omega_{LO} = N\hbar\omega_c = N\hbar eB/m^* \qquad\qquad N = 1,2,3\ldots \qquad (1)$$

The periodicity relates ω_{LO} and m^*, so either can be calculated if the other is known. We have used the measured cyclotron masses to deduce the energies of the phonons scattering the electrons, after corrections for the damping of the oscillations, nonparabolicity and resonant polaron effects[14]. Magnetophonon resonances in GaAs-GaAlAs heterojunctions were first observed by Tsui et al[15], and more detailed measurements have been reported[16,17]. However, these studies did not involve a comparison with cyclotron masses measured at the same temperatures, and the unexpected temperature dependence of the mass will cause errors.

The phonon frequencies deduced for two heterojunction samples are shown in Fig. 2, together with results for bulk GaAs[18] analysed in the same manner. The dashed lines are interpolations between Raman measurements on bulk GaAs[19], and the heterojunction results lie 12-15 cm^{-1} below the LO frequency whereas the bulk values are close to the LO frequency as expected. The heterojunction results lie closer to the bulk TO frequency, consistent with the resonant polaron coupling seen near the TO values in GaInAs-based heterojunctions[2]. It has been suggested that screening of the electric fields of the LO phonons will reduce their frequency towards the TO value[2,20], but this seems unable to explain the present results because the screening should be strongly temperature dependent, as discussed above. Although screening may be import-

Figure 2: The phonon frequencies deduced from the magnetophonon resonances in two heterojunction samples and in bulk GaAs. The dashed lines are interpolations between bulk Raman measurements, and the heterojunction results lie well below the LO frequency. In contrast, the bulk values are close to the LO frequency as expected.

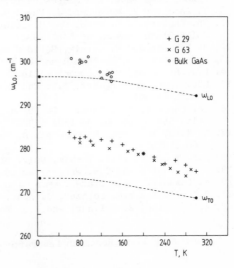

ant at very low temperatures, this suggests that other phonons are involved. The 'GaAs' LO phonon of the GaAlAs, with a frequency of 282 cm^{-1}
at $T = 0^{21)}$, is very near the measured values, but it seems unlikely
that this mode, involving coupling across the interface, should dominate
the scattering. The interface phonon frequencies satisfy the condition
$\varepsilon_1 + \varepsilon_2 = 0^{1)}$, which gives values of 290 and 270 cm^{-1} for a GaAs-GaAlAs
interface. These do not correspond to the measured frequency, but the
continuous media approximation used may be inadequate because of the
almost degenerate modes in both materials[22]. It is also possible that
the oscillations are due to two or more unresolved modes, but a phonon
lower in frequency than the GaAs LO mode must be involved.

1. J.C. Portal, G. Gregoris, M.A. Brummell, R.J. Nicholas, M. Razeghi,
    M.A. diForte-Poisson, K.Y. Cheng and A.Y. Cho, Surf. Sci. 142 368
    (1984)
2. R.J. Nicholas, L.C. Brunel, S. Huant, K. Karrai, J.C. Portal, M.A.
    Brummell, M. Razeghi, K.Y. Cheng and A.Y. Cho, Phys. Rev. Lett. 55
    883 (1985)
3. S. Das Sarma, Phys. Rev. Lett. 52 859 (1984)
4. F.M. Peeters and J.T. Devreese, Phys. Rev. B 31 3869 (1985)
5. S. Das Sarma, Phys. Rev. B 27 2590 (1983)
6. H. Sigg, P. Wyder and J.A.A.J. Perenboom, Phys. Rev. B 31 5253 (1985)
7. M. Horst, U. Merkt, W. Zawadzki, J.C. Maan and K. Ploog, Solid State
    Commun. 53 403 (1985)
8. W. Seidenbusch, G. Lindemann, R. Lassnig, J. Edlinger and E. Gornik,
    Surf. Sci. 142 375 (1984)
9. M.A. Hopkins, R.J. Nicholas, M.A. Brummell, J.J. Harris and C.T.
    Foxon, Superlatt. and Microstruct., Göteborg 1986, to be published
10. T. Ando and Y. Murayama, J. Phys. Soc. Japan 54 1519 (1985)
11. Th. Englert, J.C. Maan, Ch. Uihlein, D.C. Tsui and A.C. Gossard,
    Solid State Commun. 46 545 (1983)
12. R. Lassnig and E. Gornik, Solid State Commun. 47 959 (1983)
13. W. Seidenbusch, E. Gornik and G. Weimann, Physica 134B 314 (1985)
14. R.J. Nicholas, Prog. Quantum Electron. 10 1 (1985)
15. D.C. Tsui, Th. Englert, A.Y. Cho and A.C. Gossard, Phys. Rev. Lett.
    44 341 (1980)
16. Th. Englert, D.C. Tsui, J.C. Portal, J. Beerens and A.C. Gossard,
    Solid State Commun. 44 1301 (1982)
17. G. Kido, N. Miura, H. Ohno and H. Sakaki, J. Phys. Soc. Japan 51
    2168 (1983)
18. R.A. Wood, D. Phil. Thesis, Univ. of Oxford (1970)
19. A. Mooradian and G.B. Wright, Solid State Commun. 4 431 (1966)
20. R. Lassnig, Solid State Commun. 170 to be published
21. O.K. Kim and W.G. Spitzer, J. Appl. Phys. 50 4362 (1979)
22. A. Fasolino, E. Molinari and J.C. Maan, Phys. Rev. B 33 8889 (1986)

* grown at Philips Research Laboratories, Redhill.

DE HAAS-VAN ALPHEN AND DE HAAS-SHUBNIKOV OSCILLATIONS IN
QUANTUM WELLS WITH MULTIPLE SUBBAND OCCUPANCY

D G Cantrell and P N Butcher

Department of Physics
University of Warwick
Coventry, CV4 7AL, England

ABSTRACT

We discuss de Haas-van Alphen and de Haas-Shubnikov oscillations
in a quantum well with two electric subbands occupied at B=0
and examine the effect of inter-subband Landau level scattering
on the Landau level broadening.

1.   INTRODUCTION

The staircase-like density of states (DOS) for electrons confined
in one spatial dimension eg in MOSFETS and heterojunctions is well
known. Evidence for this DOS has been searched for in transport
properties. In the electrical conductivity for example discontinuities
are predicted to occur in the conductivity as new subbands are
occupied. Extracting information from such results is encumbered by
calculation of scattering rates and by lifetime broadening effects
which can mask the discontinuities[1] and are difficult to treat in
transport theory.

Shubnikov-de Haas (SdH) measurements have been made on two-
dimensional electron gases (2DEGs) when only one subband is occupied
and show periodicity in B^{-1}. When more than one subband is occupied
a beat phenomenon has been observed[2]. Recently it has been possible
to measure the oscillatory magnetisation of a 2DEG - the de-Haas van
Alphen (dHvA) effect[3]. To analyse the data no scattering rates or
transport theory need be evaluated. The results may be fitted to a DOS.
The calculation of the dHvA effect is much simpler than the SdH and
when several subbands are occupied we expect it to show a similar beat

structure. We calculate the dHvA effect for a system with two
occupied subbands as an indication of the behaviour of the SdH effect
which is discussed briefly. The effect of multiple subband occupation
on the broadening of the δ-function Landau levels (LLs) is examined as
is the effect of LL broadening on the magnetisation.

The DOS without broadening when a magnetic field (B) is applied
perpendicularly to the present system is two series of regularly
spaced δ-functions offset by an energy E_2 which is the energy
difference of the two subband origins. We take our energy zero to be
at the origin of the ground subband. Hence $\rho(\varepsilon) = D\left[\sum_{r_1} \delta(\varepsilon - \varepsilon_{r_1}) + \sum_{r_2} \delta(\varepsilon - \varepsilon_{r2})\right]$
where $\varepsilon_{r_1} = (r + \frac{1}{2})\beta B$, $\varepsilon_{r_2} = E_2 + (r_2 + \frac{1}{2})\beta B$ and $D = eB/\pi\hbar$ and $\beta = e\hbar/m^*$ where
D is the degeneracy of each LL.

2.   THE MAGNETISATION FOR FIXED FERMI ENERGY

We imagine a system in which all LLs up to the $n_1 - 1$ and $n_2 - 1$
levels are fully occupied and all other levels are empty. There is a
reservoir of electrons at the Fermi energy (ε_f) such that as B is
changed the original ε_f is retained. The thermodynamic potential at
T=0 for a field B is given by $\Omega = E - N\varepsilon_F$ where E is the energy and N the
number of electrons in the LLs. This quantity is simple to calculate,
involving only the sums of arithmetic progressions. The magnetisation
can be found from $M = -(\partial\Omega/\partial B)_{\varepsilon_F, T}$. The result is the addition of the
results for two separate bands with different ε_F[4] and may also be
written as

$$M = \frac{e}{\pi\hbar}(S_n \varepsilon_F - S_n^2 \beta B - n_2(E_2 - 2n_1\beta B)) \qquad (1)$$

where $S_n = n_1 + n_2$. The first two terms are the single band result, the
last term is the "interference" due to the two subbands. The results
will be discussed with the results of the next section.

3.   MAGNETISATION FOR A FIXED NUMBER OF ELECTRONS

We now imagine a system having N electrons which, at a field B,
completely fill $n_1 - 1$ subband 1 LLs and n_2 subband 2 LLs leaving the
n_1th subband 1 LL partially occupied. On raising B the $(n_1 - 1)$th level
depopulates gradually until either it is completely empty or $E_{n_1 - 1} = E_{n_2}$.
If the second equality holds before the $(n_1 - 1)$th level is completely
empty the electrons in this level transfer into the n_2th LL and it
starts to depopulate. To examine the magnetisation of the system when
a subband 2 LL is depopulating we take the situation where n_1 subband

1 LLs and n_2-1 subband 2 LLs are completely filled leaving the n_2th LL partially filled.

The energy of the system in the two situations can be evaluated and magnetisations M_1 and M_2 found as in section 2. We find

$$M_1 = \frac{e}{\pi\hbar} \left[(S_n-1)S_n \beta B - (S_n-\tfrac{1}{2})(2\varepsilon_F - E_2) + 2n_2 \left(\frac{\varepsilon_{F2}}{\beta B} - n_2 \right) \beta B \right] \quad (2)$$

and

$$M_2 = \frac{e}{\pi\hbar} \left[(S_n-1)S_n \beta B - (S_n-\tfrac{1}{2})(2\varepsilon_F - E_2) + 2n_1 \left(\frac{\varepsilon_F}{\beta B} - n_1 \right) \beta B \right] \quad (3)$$

Here ε_F and ε_{F_2} are the zero field Fermi energy relative to the origins of subband 1 and subband 2 respectively. (2) and (3) differ only in the last term, the first two terms are the single band result with n replaced by S_n[4]. This is more complicated than the fixed ε_F case as the bands are coupled by electrons transferring between them. However, in both cases the M depends on E_2 and ε_F. For many 2DEGs these depend on gate voltage and hence these systems have an electric field dependent M. Measurement of the dependence may yield information on the sensitivity of devices to changes in gate voltage.

4.   RESULTS

We present results for the situation where at B=0 ε_F=3E_2/2. The plot for fixed N has short flatter portions due to electron transfer between LLs in different subbands.

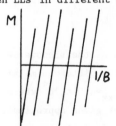

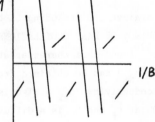

FIG. 1: M against 1/B for fixed ε_F FIG. 2 M against 1/B for fixed N

5.   LANDAU LEVEL BROADENING

We have extended the damping theoretical approximation[5] to include subbands as a first investigation of the effect of inter subband LL scattering on broadening. Here even at high fields LLs from different subbands can be close in energy. We examine the situation where pairs of LLs in different subbands are close in energy and neglect scattering to all other LLs. The imaginary parts of the Greens functions are calculated and hence the DOS. An example is shown below

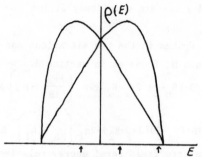

Fig.3 Example DOS. The arrows mark the origin and extremities of a semi ellipse centred at the same energy as one of the LLs.

The effects of LL broadening on M are shown in Fig. 4. The discontinuities are rounded which produces a "beat" like structure. The illustration is for the fixed N case with $\varepsilon_F = 3E_2/2$ At B=0.

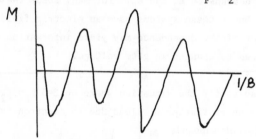

Fig. 4 M against 1/B including LL broadening.

6.   SHUBNIKOV-DE HAAS OSCILLATIONS

Ando[5] found that characteristic features of the transverse conductivity (σ_{xx}) can be obtained from a diffusion picture similar to hopping conductivity. σ_{xx} at T=0 is given by $\sigma_{xx} = e^2 D(\varepsilon_F) D^*$ where $D(\varepsilon_F)$ is the DOS at ε_F and D^* is a diffusion constant given approximately by $(\Delta x)^2/\tau$ where Δx is the mean hopping distance of an electron moving between cyclotron orbits and τ is a lifetime. To generalise to the subband case we sum over a subband index: $\sigma_{xx} = e^2 \sum_\alpha D_\alpha(\varepsilon_F) D_\alpha^*$. To calculate D_α^* we note there are two cyclotron orbits of different radii and that electrons have different lifetimes τ_α in each orbit.

7.   REFERENCES

1 Cantrell.D G and Butcher P N, J. Phys. C18 5111 (1985)
2. Landwehr G, Bangert E and Uchida S, Solid State Elec. 28 171 (1985)
3. Eisenstein J, Stormer HL, Narayanamurti V and Gossard AC, Superlatt. and Microstruct. 1 11 (1985)
4. Schoenberg D, J. Low T. Phys. 56 417 (1984)
5. Ando T and Uemura Y, J. Phys. Soc.Jpn 36 959 (1974)

MAGNETIC-FIELD TUNING OF VALENCE SUBBANDS
IN GaAs/(Al,Ga)As MULTIPLE QUANTUM WELLS

D. Heiman
Francis Bitter National Magnet Laboratory
Massachusetts Institute of Technology
Cambridge, MA, 02139 USA

A. Pinczuk and A.C. Gossard
AT&T Bell Laboratories
Murray Hill, NJ 07974 USA

A. Fasolino
SISSA, Strada Costiera 11
I-34100 Trieste, ITALY

M. Altarelli
Max Planck Institut fur Festkorperforschung
HML, 166X Grenoble FRANCE

The valence subband states in GaAs/(Al,Ga)As quantum wells
are probed as a function of applied magnetic field by resonant Raman scattering from high-mobility two-dimensional hole
gases. The complex splitting of interband transitions show
good agreement with effective-mass calculations.

Optical studies of semiconductor energy levels in GaAs/(Al,Ga)As
quantum-layered systems have revealed important features due to the
two-dimensional quantum confinement of carriers. In particular, the
complex character of the valence subbands has been exposed using
photoluminescence,[1,2] Raman scattering,[3,4] far-infrared absorption[5] and
cyclotron resonance[6,7] experiments. These complexities result from
confinement effects on the degenerate valence band states of zincblende
structure, which splits and mixes light- and heavy-hole valence band
states of the bulk. Effective-mass theories have been useful in
predicting these properties and providing quantitative results.[8]
Although there has been substantial progress toward understanding the
character of the valence subbands, few direct experimental
investigations have been made of the hole states.

This study examines the valence subband energy levels using
resonant Raman scattering from high-mobility, p-type, modulation-doped
GaAs/(Al,Ga)As multiple-quantum-well structures. These measurements
allow direct spectroscopic determinations of the energy spacings
between the occupied ground heavy-hole subband (h_0), and both ground
light-hole (l_0) and first-excited heavy-hole (h_1) valence subbands.

At zero magnetic field, the energies of the vertical intersubband hole
transitions $h_0 \rightarrow l_0$ and $h_0 \rightarrow h_1$ were measured previously and showed
good agreement with effective-mass, envelope-function calculations.[4]
Here, we find that an applied magnetic field B splits the simple
subbands into complex multiplets. The measured energies are correlated
with theory. Among several remarkable features, the light scattering
spectra show in the $h_0 \rightarrow l_0$ multiplet a transition energy that
decreases with increasing field. This result is interpreted within an
effective-mass calculation as arising from an electron-like light-hole
subband.[8] We also observe a disappearance of the $h_0 \rightarrow h_1$ transition at
finite field, arising from the quantized change in filling factor.

The sample is a modulation-doped GaAs-(AlGa)As multiple-quantum-
well heterostructure grown by molecular beam epitaxy. The centers of
the $Al_{0.51}Ga_{0.49}As$ barriers were Be-doped (4.9 nm thickness) and are
separated from the GaAs wells (11.2 nm width) by undoped $Al_{0.51}Ga_{0.49}As$
layers (29.4 nm thickness). The hole mobility is 54,000 cm^2/V-sec and
the density is $p = 1.5 \times 10^{11}$ cm^{-2}. The sample was immersed in liquid
helium at 2 K. Spectra were excited by a CW dye laser operating
between 730 and 780 nm wavelength. The sample was mounted in the
Faraday configuration (superlattice axis parallel to the field). A
near-backscattering configuration was employed with the incident and
scattered light turned 90 deg inside the dewar by a small mirror above
the sample to accommodate the radial-access 10 tesla Bitter solenoid
magnet. The laser was tuned in resonance with the first-excited
conduction subband state, which also minimizes the photoluminescence
background.[3]

Figure 1 shows envelope-function calculations of the first three
valence subbands, as a function of B, using parameters appropriate to
our sample. Details of the calculation are given in Ref. 8. The Fermi
level is displayed by the heavy line on the uppermost subbands. Raman-
allowed transitions originate from one of the filled levels and terminate on an empty level, with a change of Landau quantum number $\Delta n = 0$,
± 2. Some of these transitions are indicated by vertical lines. Figure
2 displays the energies of the strongest peaks observed in the Raman
spectra. The two features at B=0 were observed previously.[3,4] The

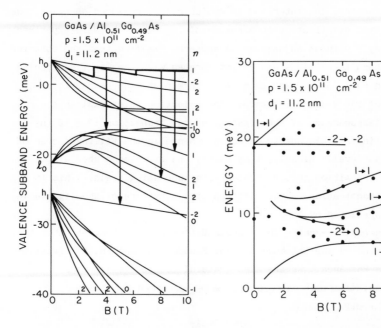

Fig. 1. Calculated valence subband energies versus magnetic field B. The Landau numbers are shown.

Fig. 2. Peak energies from Raman scattering versus magnetic field B. Points are experiment and lines are theory (see Fig. 1).

solid lines show the calculated transitions with no adjustable parameters - the only inputs were the sample parameters. The $1 \to 1$ ($h_0 \to l_0$) transitions show good agreement above B=5 T, but poor coincidence at low fields probably due to the large number of occupied levels. The $-2 \to -2$ and $-2 \to 0$ ($h_0 \to h_1$) transitions are expected to disappear abruptly at B=6.2 T when the $n = -2$ level is no longer occupied. We observe two peaks that disappear between 6 and 7 T. The upper peak at 18 meV is field-independent, which is characteristic of the calculated $-2 \to -2$ transition. The data is one meV above the theory. The peak at lower energy labeled $-2 \to 0$ agrees only qualitatively with theory. The data, however, display two features indicating an initial level on $n = -2$; disappearance above 6 T, and decreasing energy for increasing field. Such a decrease in energy is a verification of the negative-mass curvature introduced by band mixing.

In summary, the interband selection rules, measured energies, and Landau level quantization show reasonable agreement between experiment and theory - without adjustable parameters. We find that the observed intrasubband transitions can be interpreted with the light scattering selection rules $\Delta n = 0$ and $\Delta n = \pm 2$, where n are the Landau level indices of the valence subbands. The selection rules explain the sudden disappearance of the constant-energy transition $-2 \rightarrow -2$ ($h_0 \rightarrow h_1$) at finite field due to an abrupt change in the filling factor of the Landau level. Finally, both theory and experiment show a $h_0 \rightarrow l_0$ ($-2 \rightarrow 0$) transition that decreases with increasing field. This is a direct demonstration of the electron-like curvature of the light-hole subband predicted by theory.[8]

Acknowledgement - The Francis Bitter National Magnet Laboratory is supported by the National Science Foundation through its Division of Materials Research. We are also grateful to E.D. Isaacs for assistance. Numerical calculations were performed with support of CCVR, Palaiseau, France.

REFERENCES

1. R. Sooryakumar, D.S. Chemla, A. Pinczuk, A.C. Gossard, W. Wiegmann and L.J. Sham, Sol. State Commun. 54, 859 (1985).

2. J.C. Maan, A. Fasolino, G. Belle, M. Altarelli and K. Ploog, Phys. Rev. B30, 2253 (1984).

3. A. Pinczuk, H.L. Stormer, A.C. Gossard, and W. Wiegmann, Proc. 17th Int. Conf. Phys. Semicond., Eds. J.D. Chadi and W. Harrison (Springer-Verlag, New York, 1985), p.329.

4. A. Pinczuk, D. Heiman, R. Sooryakumar, A.C. Gossard and W. Wiegmann, Surf. Sci. 170, 573 (1986).

5. J.C. Maan, Y. Guldner, J.P. Vieren, P. Voisin, M. Voos, L.L. Chang and L. Esaki, Sol. State Commun. 39, 683 (1981).

6. H.L. Stormer, Z. Schlesinger, A. Chang, D.C. Tsui, A.C. Gossard and W. Wiegmann, Phys. Rev. Lett. 51, 926 (1983).

7. Y. Iwasa, N. Miura, S. Tarucha, H. Okamoto and T. Ando, Surf. Sci. 170, 587 (1986).

8. M. Altarelli, Festkoerprobleme 25, 381 (1985); and references cited therein.

HYDROSTATIC PRESSURE STUDY OF GaAs - $Al_xGa_{1-x}As$ QUANTUM WELLS AT LOW TEMPERATURES*

U. Venkateswaran, M. Chandrasekhar[†], H.R. Chandrasekhar

Department of Physics and Astronomy

University of Missouri, Columbia, MO 65211 U.S.A

B.A. Vojak, F.A. Chambers and J.M. Meese

Amoco Research Center, Naperville, IL 60566, U.S.A.

It has recently been demonstrated[1-3] that the valence band (VB) offset (ΔE_v) can be determined from accurate photoluminenscence (PL) measurements on type II superlattices (SL). By the application of hydrostatic pressure, the GaAs-$Al_xGa_{1-x}As$ quantum well system, which is normally a type I SL, can be transformed into a type II SL. In this paper, we focus on the determination of ΔE_v and how it changes with pressure. The well width dependence of the pressure coefficients (α) will also be discussed.

We have studied PL from a SL with 40 periods of 150Å GaAs wells separated by 100Å $Al_{0.25}Ga_{0.75}As$ barriers grown on a GaAs substrate by molecular beam epitaxy (MBE). Hydrostatic pressure measurements were done in a gasketed diamond anvil cell[4] with argon as the pressure medium and ruby R-line fluorescence as the manometer.

The lineup of the bands in GaAs - $Al_xGa_{1-x}As$ quantum wells is shown in Fig.1 for x$\gtrsim$0.15. At low pressure, we have a type I SL (electrons and holes both confined to the Γ bands in the GaAs layer). With increasing pressure, the energy of the Γ CB increases while that of the X CB decreases. After they cross at a pressure P_c, the electrons are confined in the X CB of the $Al_xGa_{1-x}As$ layer, while the holes are in GaAs, forming a type II SL. (Fig.1). The energies of the PL transitions illustrate this phenomenon. Below P_c (35.5 kbar in this sample), the heavy (E_{1h}^{Γ}) and light hole ($E_{1\ell}^{\Gamma}$) excitons and the e – A^o transitions[5] follow the Γ CB (Fig.2). Above P_c a new transition, E^X is seen (inset, Fig.2) in addition to E_{1h}^{Γ}. Its low

U. Venkateswaran et al.

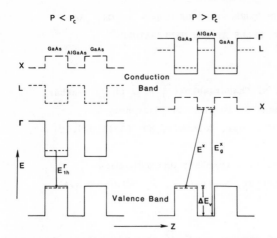

Fig. 1. Lineup of the bands in GaAs-Al$_x$Ga$_{1-x}$As heterostructures before and after the Γ-X crossover.

intensity and small negative α (-1.3 ± 0.1 meV/kbar) indicate that it is an X-band derived transition. Extrapolating E^X to P = 0, we find that its value lies about 40 meV <u>below</u> the X CB and ~25 meV <u>below</u> the neutral donors[6] (D_X^0) in bulk GaAs. We identify E^X as a staggered transition between the electrons in the X CB of AlGaAs and heavy holes in GaAs.

Referring to Fig. 1, one can write the valence offset as

$$\Delta E_v = E_g^X \text{ (AlGaAs)} - E^X + h \tag{1}$$

where h is the confinement energy in the wells. Due to the large effective masses of the electrons in the X CB and heavy holes, h is small (~4 meV for 150Å well). If one knows the X CB edge in AlGaAs and E^X accurately, ΔE_v can be determined from Eq. (1) to within the binding energy of the excitons. From our measurements on bulk AlGaAs under pressure[7] and other available data, we take E_g^X to be 2.024 ± 0.012 eV. E^X is known from Fig. 2 to be 1.936 eV. With these, we obtain ΔE_v = 90 ± 12 meV, i.e., a fractional valence offset Q_v = 0.30 ± 0.04.

In order to investigate whether the band offset $\Delta E_g = \Delta E_v + \Delta E_c$ changes with pressure, we have measured the PL in a bulk (~2.5 μm) Al$_{0.3}$Ga$_{0.7}$As sample grown by MBE on a GaAs substrate. The band gap at Γ CB in the GaAs substrate is found to have α = 10.7 ± 0.1 meV/kbar while that of Al$_{0.3}$Ga$_{0.7}$As has α = 9.9 ± 0.1 meV/kbar. We confirm this difference in α's by fitting the energy difference between the Γ-peaks of AlGaAs and GaAs as a function of pressure and obtain a slope of −0.78 ± 0.04 meV/kbar. This implies that there is a pressure-induced <u>decrease</u> of the total band discontinuity (ΔE_g) in the SL. As the α of AlGaAs decreases[8] with x ($0.25 < x < 0.5$), we expect the decrease in ΔE_g to be pronounced at high pressures for SLs with barriers of large x.

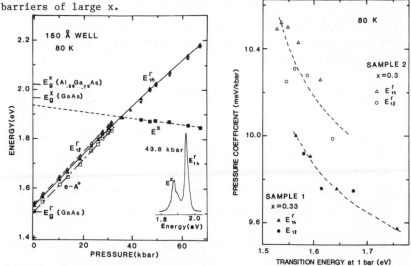

Fig. 2. Pressure dependence of the PL transitions.

Fig. 3. Pressure coefficients (α) versus transition energies for different wells.

To study the well width (L_z) dependence of α's accurately, several isolated GaAs quantum wells of different widths separated by 750Å wide Al$_x$Ga$_{1-x}$As barriers were grown on the same substrate by MBE. Sample 1 consisted of four wells (L_z = 26, 48, 70 and 96Å) and an Al mole fraction x = 0.33. Sample 2 consisted of five wells (L_z = 47, 70, 93, 117 and 140Å) and an x = 0.3. PL was measured at 80 and 150K as a function of pressure[1]. Fig. 3 is a plot of α versus the transition energy (E) rather than L_z since E can be measured accurate to 0.1 meV. There is a systematic decrease (~5%) in α with increasing E (or decreasing L_z) in both the samples. We confirm this trend by verifying that the energy <u>difference</u> between the various pairs of wells decreases at higher pressures. This difference analysis eliminates the uncertainties in the pressure calibration, thus

establishing the variation of α with transition energy to within 1%.
There is also a small decrease in α with increasing x (Fig.3).

In summary, our study shows that there is a small decrease in α
with decreasing L_z and increasing x.   The Γ-X subband crossover
leading to the formation of a type II SL occurs at lower pressures for
narrower wells.   We determine the valence band offset to be .30 ± .04
of the total band gap discontinuity.   Due to the difference in the α's
of GaAs and AlGaAs, the total band offset (ΔE_g) decreases with
increasing pressure.

REFERENCES

*   Supported by Amoco Corporation, Research Corporation and U.S.
        Army Grant number DAAL03-86-K-0083.

†   A.P. Sloan Foundation Fellow.

1.   U. Venkateswaran, M. Chandrasekhar, H.R. Chandrasekhar, B.A.
        Vojak, F.A. Chambers and J.M. Meese, Phys. Rev. B 33, 8416
        (1986).

2.   P. Dawson, B.A. Wilson, C.W. Tu and R.C. Miller, Appl. Phys.
        Lett. 48, 541(1986).

3.   D.J. Wolford, T.F. Kuech, J.A. Bradley, M.A. Gell, D. Ninno and
        M. Jaros in Physics and Chemistry of Semiconductor Interfaces,
        Pasadena, CA, Jan. 28-30 (1986).

4.   For further experimental details see, U. Venkateswaran and M.
        Chandrasekhar, Phys. Rev. B 31, 1219 (1985).

5.   For a detailed study, see U. Venkateswaran, M. Chandrasekhar,
        H.R. Chandrasekhar, T. Wolfram, R. Fischer, W.T. Masselink and
        H. Morkoc, Phys. Rev. B 31, 4106 (1985).

6.   D.J. Wolford and W.A. Bradley, Solid State Commun. 53, 1069
        (1985).

7.   M. Chandrasekhar, U. Venkateswaran, H.R. Chandrasekhar, B.A.
        Vojak, F.A. Chambers and J.M. Meese, this conference.

8.   N. Lifshitz, A. Jayaraman, R.A. Logan and R.G. Maines, Phys.
        Rev. B 20, 2398 (1979).

EFFECT OF CONFINEMENT ON UNIAXIAL STRESS DEPENDENCES
OF EXCITONS IN SEMICONDUCTOR QUANTUM WELLS

Emil S. Koteles, C. Jagannath, Johnson Lee,
Y. J. Chen, B. S. Elman and J. Y. Chi
GTE Laboratories Incorporated
40 Sylvan Road
Waltham, MA 02254 USA

ABSTRACT

The effect of spatial confinement on the uniaxial stress dependences of excitons in semiconductor quantum wells was investigated using low temperature photoluminescence excitation spectroscopy in order to reveal inter-subband transitions between various quantum levels in the conduction and valence bands. Strong repulsion between the ground state of the light-hole exciton, E_{11L}, and the parity allowed heavy-hole exciton transition, E_{13H}, gave direct evidence of valence band mixing. The effect of quantum well width on the stress behavior of the ground state excitons was interpreted using a model based on the overlap of the light- and heavy-hole wavefunctions.

INTRODUCTION

The band structure of two dimensional (2D) semiconductors (e.g. GaAs quantum wells) has recently become a topic of considerable interest. It is well established that, as a result of spatial confinement, electrons exist in discrete energy levels in the confining direction and that the 3D degeneracy of the light hole (LH) and heavy hole (HH) bands is lifted[1]. Thus excitons, when produced in such an environment, exhibit two series of discrete energy levels, $E_{ijL(H)}$, corresponding to transitions between the i-th conduction subband and the j-th light (heavy) hole subband. Besides allowed exciton transitions ($\Delta n=i-j=0$), various parity allowed (Δn even) and parity "forbidden" (Δn odd) transitions have also been reported in the literature[2]. The observation of these forbidden transitions has been recently explained by invoking wavefunction mixing between light and heavy valence bands[3]. In this connection the application of uniaxial stress is of particular interest in that the light-hole subband levels can be tuned through heavy-hole subband levels thereby precisely controlling the amount of mixing between these valence band states. A recent theoretical study[4] predicted dramatic changes in valence band structure due to variations in the coupling between various LH and HH subbands as they are shifted relative to each other in energy by uniaxial stress.

In this paper we present experimental results on the effect of uniaxial stress parallel to [100], on excitons in GaAs/AlGaAs quantum wells grown along [001] obtained using photoluminescence excitation spectroscopy (PLE) at low temperatures (5K). This external perturbation not only allows us to identify the various light- and heavy-hole transitions due to their different stress behaviors, but also reveals a level repulsion between LH and HH exciton levels. This yields direct experimental evidence of a strong interaction and therefore a strong mixing between the 1L and 3H valence subbands. Further, the stress dependences of E_{11L} and E_{11H} are observed to be functions of the quantum well width, L_Z. This is interpreted in terms of a model based on the overlap of the LH and HH wavefunctions and on the energy differences between the LH and HH excitons.

EXPERIMENT

The samples used were grown by molecular beam epitaxy on semi-insulating GaAs substrates. A typical sample consisted of approximately 1 micron of undoped GaAs buffer followed by several quantum wells. The experimental technique used has been described in detail in Ref. 5.

RESULTS AND DISCUSSION

Figure 1 shows the PLE spectra of a sample with $L_Z=22$nm for five different values of stress, X. The zero of the energy scale corresponds to the 11H transition. This figure illustrates the sensitivity of this technique to inter-subband transitions, even to those parity forbidden. The peaks labelled a-d could not be precisely identified. As X was increased it was observed: 1) the energies of E_{11L} and E_{22L} increased faster with X than the other features which we therefore associated with heavy hole (ijH) transitions, 2) the intensity of E_{22L} increased rapidly between 0 and 1 kbar and then remained constant relative to the intensity of E_{11H}, 3) as E_{11L} moved through E_{12H} its width broadened and it increased in intensity, a signature of interacting transitions and 4) an interaction between E_{11L} and E_{13H} was also evident from the transfer of intensities between the two transitions at about 2 kbar. Further evidence of a strong interaction between E_{11L} and E_{13H} is presented in Figure 2, a plot of the energy difference between the various transitions and E_{11H} as a function of stress. A pronounced anticrossing (level repulsion) is observed between these states at about 2.5 kbars. Such a strong interaction is evidence of these levels being comprised of mixtures of light- and heavy- hole valence band wavefunctions. A quantitative theoretical explanation of this interaction is beyond the scope of this report and will be discussed elsewhere. Further consequences of this interaction are small, but finite, shifts in the energies of E_{12H} and E_{13H} (to higher energy with respect to E_{11H}) after these transitions have been "crossed" by E_{11L}. In these same regions the halfwidth of E_{11L} increased significantly. Such broadening has recently been predicted due to the Fano interaction between E_{11L} and the heavy hole continuum states. A more detailed investigation of this phenomenon will be presented elsewhere[6].

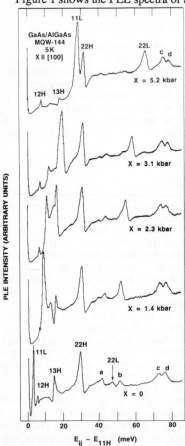

Fig.1 Low temperature PLE spectra of a GaAs/AlGaAs quantum well ($L_Z=22$nm) under five different values of uniaxial stress. The labels refer to excitonic transitions between conduction and valence subbands as explained in the text.

The effect of increasing spatial confinement on the uniaxial stress dependences of excitons in GaAs quantum wells is presented in Figure 3.

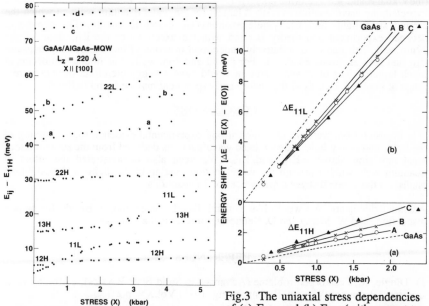

Fig.2 The uniaxial stress dependencies of higher energy excitonic transitions with respect to E_{11H}.

Fig.3 The uniaxial stress dependencies of (a) E_{11H} and (b) E_{11L} (with respect to their X=0 values) as a function of well width (A, B and C refer to L_Z = 22, 11 and 4nm respectively).

The increase in energy of E_{11H} and E_{11L} (relative to X=0 values) as a function of X are given for three different quantum well widths [open circles (L_Z=22nm), crosses (L_Z=11nm) and triangles (L_Z=4nm)]. Dashed lines represent the stress dependences of HH and LH excitons in the buffer GaAs layer which was measured simultaneously for calibration purposes. A systematic increase (decrease) in the slope of E_{11H} (E_{11L}) is observed for decreasing L_Z. These dependences can be explained by considering the total Hamiltonian of excitons in quantum wells. In our experimental configuration, where X is perpendicular to the growth axis, the change in energy of the $E_{11L(H)}$ transitions as a function of X is given by[5)]

$$\Delta E_{11L(H)}(X) = a(s_{11}+2s_{12})X$$

$$\pm [b(s_{11}-s_{12})X + Q^2X^2/2[(E_{11L}(0)-E_{11H}(0))+2b(s_{11}-s_{12})X]] \qquad (1)$$

where $[E_{11L}(0)-E_{11H}(0)]$ is the energy separation of the ground state LH and HH excitons at X=0, s_{11} and s_{12} are elastic compliance coefficients of GaAs, a and b are the hydrostatic and shear deformation potential constants respectively and Q is the overlap integral of the LH and HH wavefunctions. The theoretical fits to the data in Figure 3, solid lines labelled A, B and C for the L_Z=22nm , 11nm and 4nm wells respectively, were calculated using Eq. 1 with Q^2 as an adjustable parameter. $E_{11L}(0)-E_{11H}(0)$ was determined from the experiment and a =7.93 eV and b=0.98eV for all three curves. The theory predicts that b for quantum wells is a factor of 2 smaller than that for bulk GaAs.[5)] Best fits to the data were obtained when Q^2 =7.0 ± 1.0 (meV/kbar)[2], 7.5 ± 1.0

$(meV/kbar)^2$ and $4.0 \pm 2.0(meV/kbar)^2$ for $L_Z = 22$, 11 and 4nm respectively. Thus, as expected, the wavefunction overlap is largest for the wider quantum wells in which the ground state light- and heavy-hole excitons are closer in energy. The agreement between experiment and theory is good with the exception of the E_{11L} data for the 22nm well. This can be qualitatively understood in terms of the anticrossing between E_{11L} and E_{13H} as discussed above. For the 11 and 4nm wells, this interaction occurs at much larger values of X and therefore would have a negligible effect over the stress range of Figure 3. Details of this theory and experiment may be found in Ref.5.

CONCLUSIONS

In conclusion, we have presented strong experimental evidence for the mixing of the light- and heavy-hole valence band wavefunctions deduced from the stress induced level repulsion between E_{11L} and E_{13H}. We have also demonstrated the effect of quantum well width (through the wavefunction overlap of the light- and heavy-hole bands) on the uniaxial stress dependences of 2-D excitons.

Acknowledgements We would like to thank Drs. L. Sham and D. Broido for helpful discussions and W. Menzi and D. Kenneson for technical assistance.

REFERENCES

1) Dingle, R., Festkorperprobleme (Advances in Solid State Physics), edited by H. J. Queisser (Pergamon/Vieweg, Braunschweig, 1975) Vol.XV, p.21.

2) Miller, R. C. and Kleinman, D. A., J.of Lum. 30, 520 (1985) and references therein.

3) Miller, R. C., Gossard, A. C., Sanders, G. D., Chang, Yia-Chung and Schulman, J. N., Phys. Rev. B32, 8452 (1985).

4) Sanders, G. D., and Chang, Yia-Chung, Phys. Rev. B32, 4282 (1985).

5) Jagannath, C., Koteles, Emil S., Lee, Johnson , Chen, Y. J., Elman, B. S. and Chi, J. Y., to appear in Phys. Rev.B.

6) Broido, D., Sham, L., Koteles, Emil S. and Jagannath, C., to be published.

PRESSURE STUDIES OF GaInAs/InP HETEROJUNCTION

WITH THREE OCCUPIED ELECTRIC SUBBANDS

D. Gauthier, L. Dmowski[+], S. Ben Amor, R. Blondel and J.C. Portal
CNRS-INSA, F-31077 and CNRS-SNCI, 166X, F-38042 Grenoble (France)

[+]also High Pressure Research Centre, PAS Unipress, Warsaw (Poland)

M. Razeghi, P. Maurel and F. Omnes
Thomson CSF, BP 10, F-91401 Orsay (France)

ABSTRACT

The influence of hydrostatic pressure on electric subbands in GaInAs/InP heterojunction quantum well was studied by magnetotransport effects. A considerable decrease of the distances between the subbands has been revealed.

In recent years hydrostatic pressure has found a new field of application in the transport studies of two dimensional systems owing to its ability to decrease strongly their electron concentration. This effect was related to the deepening of donor levels in the doped layer of heterostructure with pressure [1]. However in the case when these levels are shallow or are placed high enough above the bottom of the conduction band they give no contribution to the change of the free electron concentration with pressure. The GaInAs/InP heterojunction is just such a case. We have studied it using magnetic field up to 18 T and hydrostatic pressure up to 15 Kbars (1.5 GPa). Previous studies on the same sample at atmospheric pressure [2] revealed three occupied electric subbands with two contributing to the quantum Hall effect. From Shubnikov-de-Haas and quantum Hall effect, in the way described in [2] we have deduced the population of each subband and the total electron concentration as a function of pressure. The results are shown in Fig. 1. The population of the third subband is fairly constant, while the more pronounced decrease takes place in the lowest subband. Depopulation studies of higher lying subbands in magnetic field parallel to the interface have confirmed the occupation of the third subband even at the highest pressure applied. Still at 15 kbars there is a fall in resistance at about 1.5 T due to the suppression of intersubband scattering indicating that the third subband is occupied in zero magnetic field (Fig. 2). Extrapolating the part of the curve where only one subband is populated to $B = 0$ we have estimated the relative contribution of intersubband scattering to the resistivity. It decreased from 0.57 at atmospheric pressure to 0.47 at 15 kbars.

We have calculated the Fermi energy of each E_i subband as $E_F - E_i = n_i \cdot \pi \hbar^2/m^*$ taking into account the increase of effective mass with pressure. Thus we could deduce the pressure dependence of the Fermi energy E_F and the positions of the subbands E_1 and E_2 relatively to the ground subband E_0 (Fig. 3). For atmospheric pressure we have taken the value $m^* = 0.0493\, m_0$ determined from cyclotron resonance. The pressure change of the effective mass was derived from magnetophonon resonance measured as a function of pressure. The detailed results will be published elsewhere. In Fig. 4, we present only the relative shift of the resonant magnetic fields $B_N(p)/B_N(O)$ for $N = 1, 2, 3$. From this shift we have determined the increase of mass of 1 ± 0.1 %/kbar. The observed fall in Hall mobility of about 30 % for 15 kbars indicates a strong dependence of mobility on effective mass as it is expected for GaInAs due to alloy scattering [3]. For intrasubband alloy scattering the increase of mass of 15 % should give a drop in mobility of 33 % what agrees well with our results. Relatively poor decrease of the total electron concentration with pressure (16 % for 15 kbars) corresponds well to the pressure change of the conduction band discontinuity if we assume constant band offset 40 % i.e. $\Delta E_c(p) = 0.4 \cdot \Delta E_g(p)$ and triangular well approximation. However, this approximation is not able to describe the pressure change of the population of each subband separately. The considerable decrease of the distances between the subbands with pressure confirms that triangular well approximation could not work for higher lying subbands. Such decrease can originate either in a shift of the subbands towards higher energies in strongly bent potential well or in an increase of the band bending with pressure.

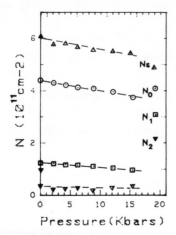

Fig1:Subbands populations NO,N1
N2 and total electron population
Ns versus pressure.

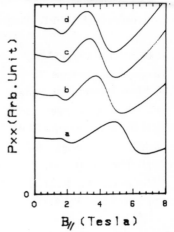

Fig2:Resistivity versus parallel
magnetic field (a=0Kbar,b=6Kbar
c=12Kbar,d=15Kbar).

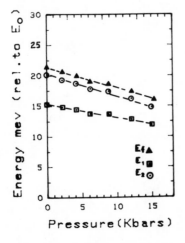

Fig 3:Energy position of the exci
ted subbands E1 and E2 and the Fer
mi level Ef versus pressure.The
origin is taken at the ground sub
band E0.

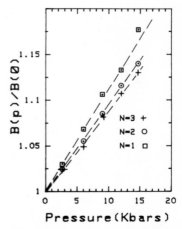

Fig4:Relative shift of the ma
gnetophon resonance position BN
versus pressure for different
resonance index N .

We   acknowledge   the   financial   support   from   Conseil   Régional Midi-Pyrénées and ESPRIT Programs.

1. Robert J.L., Mercy J.M., Bousquet C., Raymond A., Portal J.C.,
   Gregoris G., Beerens J.,
   in "Two Dimensional Systems Heterostructures and Superlattices",
   vol. 53, 253, Springer Verlag (1984)
2. Razeghi M., Duchemin J.P., Portal J.C., Dmowski L., Remenyi G.,
   Nicholas R.J., Briggs A.,
   Appl. Phys. Lett., 48, 712 (1986)
3. Walukiewicz W., Ruda H.E., Lagowski J., Gatos H.C.,
   Phys. Rev., B 30, 4571 (1984).

EFFECT OF UNIAXIAL STRESS ON ELECTRONIC STATES IN QUANTUM WELLS AND SUPERLATTICES

G. Platero and M. Altarelli

Max-Planck-Institut für Festkörperforschung, Hochfeld-Magnetlabor
166X, F-38042 Grenoble (France)

ABSTRACT

The envelope function approximation is used to describe electronic states in quantum wells and superlattices under uniaxial stress. The effect of internal strains on InAs-GaSb superlattices and that of external (110) stress on GaAs-AlGaAs p-doped quantum wells are discussed and compared to experiment.

1. INTRODUCTION

A theoretical description of the effects of uniaxial stress on quantum wells and superlattices is important for two reasons: (a) built-in strains, which, apart from a hydrostatic component, can be regarded as uniaxial [1], arise whenever the lattice mismatch is significant; (b) uniaxial stress can be applied externally as a tool of analysis of electronic states [2]. We perform calculations of electronic states under arbitrary uniaxial stress, within the framework of the envelope function approximation [3]. A six-band k.p description of the bulk band structure is adopted, neglecting the split-off valence band. The valence band part is modified according to the Pikus-Bir [4] strain Hamiltonian description already applied to heterostructures [5], while the modification of the valence-conduction k.p. coupling is obtained with similar methods [6]. The variation of the band offsets, when the gaps of the two materials change under the effect of stress, is an unknown quantity. In the case of internal strain in InAs-GaSb superlattices, the experimentally determined offset [7], which includes

of course this effect, is used in the calculation. In the analysis of
GaAs-AlGaAs quantum wells under external stress, however, some assumption has to be made. We chose, within the arbitrariness resulting
from the lack of experimental information, to keep the valence-band
offset unchanged under external stress. Any other reasonable choice,
however, would affect out results very slightly in the 0-3 kbar range
of interest.

2.   InAs-GaSb SUPERLATTICES: INTERNAL STRAIN EFFECTS

It is well known that this system is characterized by a 0.62 %
lattice mismatch. Since samples are generally grown on a thick GaSb
substrate or buffer layer, this is accommodated by a biaxial dilation
of the InAs layers, equivalent to a large uniaxial compression
($\sim$ 5 kbar) which produces a 46 meV splitting of light and heavy holes
plus a hydrostatic shrinking of the fundamental gap. As a result, the
conduction band is separated by 0.35 eV from the light hole band (to
be compared with the 0.41 eV gap at zero pressure), with a considerable reduction of the conduction band effective mass. In the superlattice with 12 nm InAs and 8 nm GaSb layers, the lowest InAs-like
subband is shifted upwards by $\sim$ 5 meV (with respect to a calculation
which does not include strain), while the next one is shifted downwards. Calculations with or without strain assume a 0.15 eV offset
between the GaSb valence band and the InAs conduction band. At the
anticrossing point [8], however, effects are even smaller, with changes
of $\sim$ 2 meV. Thus the conclusions of previous comparisons of theory
and experiment [9] are not much affected by the inclusion of strain.
This is so even when some strain in the GaSb layers is allowed for,
as long as the 0.15 eV offset occurs between GaSb heavy holes and InAs
electrons, because light holes do not play a significant role.

3.   GaAs-AlGaAs QUANTUM WELLS UNDER EXTERNAL STRESS

External stress can be used to obtain information on electronic
states in heterostructures. This is particularly interesting for the
valence subbands, because of their coupled character. For example,
stress parallel to the (001) growth direction changes the relative

position of light-hole and heavy hole subbands, and permits to tune
the non-parabolicities that result from their mixing at nonzero wave
vector. For this case we recover the results obtained by similar
methods in Ref. 5). The (110) case however, is more interesting, as
the shear components mix light and heavy character of the holes at all
wave vectors. Recent Raman experiments [2] indicate a very pronounced
dependence of inter-subband transitions on stress in the 0 - 1.5 kbar
region. The dispersion of valence subbands was computed as a function
of wave vector in the (001) plane for several values of the uniaxial
(110) stress, for the 9 nm GaAs wells investigated in Ref. 2). Results
in the (110) direction are shown in Fig. 1 for stress T = 0 and
T = 2.6 kbars. Notice the dramatic difference in the splitting of the
second and third subbands, which nearly touch at high stresses. The
difference in the joint density of states for transitions from the
first subband between 0 and k_F is shown in Fig. 2. The broadening of
the h_0-ℓ_0 transition which practically merges with the h_0-h_1 peak,
and its higher energy threshold are in good agreement with experiment,
although they take place at lower stress values than expected from the
calculation. The analysis of the stress results provides further
support to the envelope-function picture of valence subbands as revealed by Raman experiments [10].

ACKNOWLEDGEMENTS

Numerical calculations were performed with support of the Centre
de Calcul Vectoriel pour la Recherche, Palaiseau.

REFERENCES

1) Osbourn, G.C., J. Appl. Phys. 53, 1586 (1982)
2) Pinczuk, A., Heiman, D., Sooryakumar, R., Gossard, A.C. and Wiegmann, W., Surf. Sci 170, 573 (1986)
3) see e.g. Altarelli, M., in "Semiconductor Superlattices and Heterojunctions", edited by Allan, G., Bastard, G., Boccara , N., Lannoo,
   M. and Voos, M., (Springer, Berlin, 1986)
4) Pikus, G.E. and Bir, G.L., Soviet Phys., Solid State 1, 1502 (1959)
5) Sanders, G.D. and Chang, Y.C., Phys. Rev. B, 32, 4282 (1985)

6) Aspnes, D.E. and Cardona, M., Phys. Rev. B, <u>17</u>, 726 (1978)

7) Chang, L.L., in ref. 3

8) Altarelli, M., Phys. Rev. B <u>28</u>, 842 (1983)

9) Fasolino, A. and Altarelli, M., Surface Sci., <u>142</u>, 322 (1984)

10) Pollak, Fred H. and Cardona, M., Phys. Rev. <u>172</u>, 816 (1968)

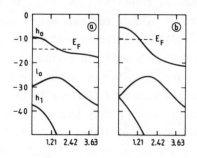

Fig. 1   Valence subband dispersion for k (in units of 10^6 cm^{-1}),
parallel to the 110 stress direction of a 9 nm
GaAs-Al$_{0.48}$Ga$_{0.52}$As quantum well with 3×10^{11} holes cm^{-2} for
(a) T = 0, (b) T = 2.6 kbar. Energies are in meV.

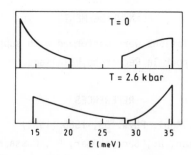

Fig. 2   Joint density of states (arbitrary units) for transitions
$h_0 \rightarrow \ell_0$, $h_0 \rightarrow h_1$, derived from the subband structure.
(a) T = 0, (b) T = 2.6 kbar.

V. SUPERLATTICES

THEORETICAL STUDY OF PHASE TRANSITIONS IN SUPERLATTICES UNDER PRESSURE

Richard M. Martin

Xerox Palo Alto Research Center,

3333 Coyote Hill Road,

Palo Alto, CA 94304, U.S.A.

ABSTRACT

It has recently been domonstrated by Weinstein, et al, that superlattices of AlAs/GaAs undergo phase transitions under pressure which are significantly modified compared to ordinary bulk crystals. In the present paper we discuss possible new superlattice phases that could occur in such systems. Preliminary results of density functional calculations indicate that AlAs/GaAs can transform to a pseudomorphic strained-layer structure, but most likely in a regime in which dislocations will partially relieve the strain.

Experiments at high pressures on superlattices of AlAs/GaAs by Weinstein, et. al., [1,2] have shown that the transformations to compressed phases are significantly modified from bulk transitions. This can be understood as a basic competition in the superlattice between the tendencies for the two constituent materials to break apart and undergo separate transitions to incoherent bulk-like phases and, on the other hand, to cooperate to form coherent crystalline superlattice structures in the high pressure phases. Thus the problems to be addressed are: Under what conditions will coherent structures be formed? Does the transition occur in a single step or in multiple layer-specific phase transitions? And what are the characteristic structures?

Under pressure each material, in bulk form, transforms to a structure with higher coordination (probably 6-fold) and reduced volume ($\sim$17%), at $\sim$ 123 Kbar in AlAs [1-2] and 174 Kbar in GaAs [3]. For thick layers there are, of course, two transitions to independent bulk-like phases. However, the pressure, P_T, for the AlAs transition may be modified because the pressure medium must supply the energy needed to create the interfaces between transformed AlAs and zincblende GaAs. As the thickness ℓ_A of the AlAs layers decreases, this energy becomes a larger fraction of the total energy; hence P_T increases and it may become favourable to create new structures. In Ref. 1-2, the increases in P_T were measured for ℓ_A $\sim$300A and 60A and it was concluded that ℓ_A $\sim$60A is near a crossover in which the thick-layer behaviour is modified. In this thin-layer regime, coherent structures are favoured, [4] but their nature is not known.

The crucial next question is whether there continue to be two phase transitions or they merge into a single transition. In the latter case both materials remain very similar, i.e. both 4-fold coordinated or both 6-fold. However, if two transitions occur, then in the intermediate pressure range, AlAs is 6-fold coordinated but GaAs 4-fold. Such a "4-6" superlattice involves pseudomorphic strained layers and interfaces with novel types of bonding. To compare these possibilities theoretically, we must determine the relative energies of these very different types of structures.

A convenient way to construct structures having the desired properties is by stacking hexagonal planes of atoms, which can describe directly 111 orientation superlattices. Let us denote the three positions for the planes by the conventional letters A, B and C with capital letters denoting As and small letters Al or Ga. Then the most common bulk structures are donoted:

NaCl	A–c–B–a–C–b–A–
NiAs	A–c–B–c–A–c–A–
ZnS	A––aB––bC––cA–

Here a sequence like A–c–B denotes c in an octohedral 6–fold coordinated site; c–B–c, a hexagonal 6–fold B site; and A––aB a tetrahedral 4–fold coordinated a site. Dashes denote spaces.

We have carried out calculations on bulk AlAs and GaAs in the ZnS and NaCl structures using density functional methods described in Ref. 5 and discussed elsewhere [1]. The results are in reasonable agreement with previous work by Froyen and Cohen (FC) [6] and predict phase transitions at about 90 and 170 Kbars for AlAs and GaAs. Also in agreement with FC we find NiAs slightly lower in energy than NaCl for AlAs.

Supercells can be constructed simply by appropriate sequences of Al and Ga. For example, the NiAs structure with alternating Al and Ga is found to have energy essentially the same as the sum of bulk AlAs and GaAs. From this we conclude that a transition to such a structure would occur at the weighted average pressure, e.g. ∿148 Kbar for equal thickness layers and ∿174 Kbar for thin AlAs sandwiched between thick layers of GaAs.

This is to be contrasted with a 4–6 type superlattice, a simple example of which is

$$A\text{--}a\ B\text{--}c\text{--}A\text{--}\quad (a = Ga,\ c = Al)$$

The energy of this structure is found to be 0.7eV/(2 atoms) higher than the average of GaAs (4–fold) and AlAs (6–fold). This extra energy makes such 4–6 superlattices higher in energy than 4–4 or 6–6; we speculate that only for layers ∿ many atoms thick could such structures be stable. In addition, there are strain energies for the layers discussed in Ref. 1. Our tentative conclusion is that there may be a limited range of layer thicknesses where such

4-6 superlattices could occur, but probably only for layers thick enough that dislocations will (partially) relieve the strain.

The occurrence of two transitions and pseudomorphic strained layers would be more likely to occur in systems where the difference in transition pressures is larger or with less strain in the intermediate state. For example, replacing GaAs by $GaAs_{1-x}P_x$ alloys would help satisfy both conditions. In fact, the reduced lattice constant of $GaAs_{1-x}P_x$ could cause AlAs to transform at lower pressures. The limit of such effects would be to move a transition to P = 0, i.e. stabilize a new phase. A likely candidate is InSb grown on substrates of smaller lattice constant, since InSb transforms to NaCl structure at a low pressure and has been previously formed metastably [7].

This work was supported by ONR Contract N00014-82-C-0244. Discussion with B. Weinstein, S. Hark, R. Burnham, C. Van de Walle, R. Needs and F. Ponce and the hospitality of the TCM Group, Cavendish Laboratory, Cambridge, are gratefully acknowledged.

REFERENCES

1. Weinstein, B. A., Hark, S. K., Burnham, R. D. and Martin, R. M., to be published.

2. Weinstein, B. A., Hark, S. K. and Burnham, R. D., present proceedings.

3. Yu, S. C., Spain, I. L. and Skelton, E. F. Solid State Comm. 25, 49 (1978); and Baublitz, M. A. and Ruoff, A. L. J. Appl. Phys. 47, 2821 (1976).

4. See, for example, Matthews, J. W. and Blakeslee, J. Cryst. Growth 27, 118 (1974); J. Vac. Sci. Technol. 14, 989 (1977).

5. See, e.g., Biswas, R., Martin, R. M., Needs, R. J. and Nielsen, O. H., Phys. Rev. B 30, 3210 (1984).

6. Froyen, S. and Cohen, M. L., Phys. Rev. 28, 3258 (1983).

7. Asoumi, K., Shimomara, O. and Minomura, S. J. Phys. Soc. Japan 41, 1630 (1976).

Self-Consistent Band Structure Calculations of $(GaAs)_n/(AlAs)_n$ Superlattices of Ultrathin Layers with n=1 to 10

Hiroshi Kamimura and Takashi Nakayama

Department of Physics, University of Tokyo,

Bunkyo-ku, Tokyo, Japan 113

ABSTRACT

The band structures and the dielectric function $\varepsilon_2(\omega)$ of ultrathin layered $(GaAs)_n/(AlAs)_n$ with n=1 to 10 are calculated by a selfconsistent pseudopotential method in a modified LDF formalism. Features of the absorption spectrum such as the change of three- to two-dimensional nature, the appearance of conspicuous dichroism, etc., are clarified.

Although the band structures of superlattices GaAs/AlAs with a thick layer width have been understood well by a Kronig-Penny model, the detailed investigation on the band structure of ultrathin layered superlattice systems has just recently started. In the present paper, following the method by Nakayama and Kamimura[1] , we extend the band structure calculations to the systems of $(GaAs)_n/(AlAs)_n$ which we denote SL(n,n) with n=5 to 10, and then calculate the dielectric function $\varepsilon_2(\omega)$ for n=1 to 10.

Since we have described the method of calculations in detail in ref.1, here we just present the calculated results in Fig.1. This figure shows the calculated band structures of valence and conduction bands along the layered direction ΓZ for $(GaAs)_n/(AlAs)_n$ with n=4,7,8 and 10 . As seen in Fig.1, the higher energy states of the valence band and the lower energy states of the conduction band become flat as n increases from 4 to 10. In particular, for SL(10,10) the valence band states in the energy region of the valence band offset which is 0.3eV in the present calculation and the conduction band states in the energy

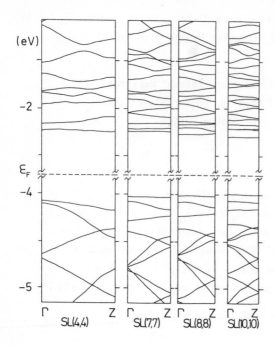

Figure 1. The calculated band structures of $(GaAs)_n/(AlAs)_n$ with n=4, 7, 8 and 10.

region of about 0.3eV which is much smaller than the value of the conduction band offset (1.46eV in the present calculation) are almost constant along ΓZ. Thus the present result indicates that the Kronig-Penny model is applicable for a valence band of SL(n,n) with n>10, while for a conduction band the interactions of quantum wells of GaAs through a barrier AlAs are still strong.

Then, using the energy dispersion and eigenfunction of each energy band, the dielectric function $\varepsilon_2(\omega)$ is calculated as functions of the photon energy $\hbar\omega$ for polarization of light parallel (z) and perpendicular (xy) to the superlattice direction. This can be compared with the observed absorption spectra. Here the excitonic effect is not included. In Fig.2 the calculated $\varepsilon_2(\omega)$ for xy and z polarizations of light for n=1, 4, 6, 7 and 10 are shown by solid line and dotted line, respectively. The following features emerge from this result:

(1) For n=1, the fundamental absorption edge shows the feature of a three-dimensional van Hove singularity M_0.

(2) With increasing n the feature of the absorption edge changes from

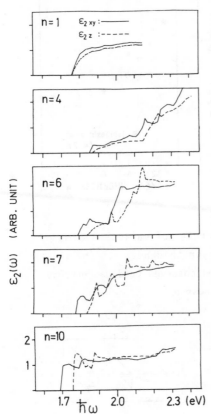

Figure 2. The dielectric function $\varepsilon_2(\omega)$ of $(GaAs)_n/(AlAs)_n$ with n=1, 4, 6, 7 and 10 as a function of photon energy $\hbar\omega$.

three-dimensional to two-dimensional nature.

(3) The weak intensity near the absorption edge in SL(4,4) is due to the decrease of GaAs character in the lowest conduction band. As seen in Fig.1, the lowest conduction band has the GaAs character only near Γ and the AlAs character in the most region of ΓZ, while in the second lowest conduction band the GaAs character predominates.

(4) There appears a large dichroism for n≥6. In the spectra of z polarization for n≥6 the low energy peaks correspond to transitions from the light holes to the low-lying conduction bands while in the spectra of xy polarization the low energy peaks correspond to transitions from the heavy hole states to the low-lying conduction bands.

(5) Since the electron-phonon interaction has not been taken into account in the present calculation, all the transitions are direct irrespective of the strength of transitions.

In Fig.3(a) we show the calculated band gap energies of the superlattices with n=1 to 10 at Γ, L and X_{xy} points, where the Brillouin zone and the Γ, L and X_{xy} points are shown in Fig.4.

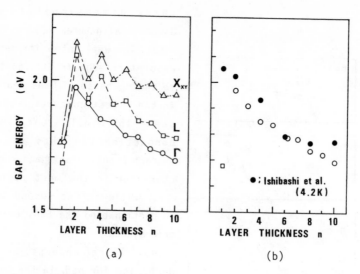

(a)   (b)

Figure 3. (a) Calculated energies of various minima of
SL(n,n) as a function of layer thickness n. (b)Calcu-
lated lowest gap energies and observed band gap energies
as a function of layer thickness n.

Figure 4. Brillouin zone
of SL(n,n).

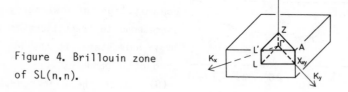

In Fig.3(b) the lowest band gap energies are compared with the
experimental results recently reported by Ishibashi et al.[2] A general
trend of decreasing the band gap energies with increasing a value of n
is consistent with the experimental results.

REFERENCES

1) T.Nakayama and H.Kamimura, J. Phys. Soc. Jpn. 54, 4726 (1985)
2) A.Ishibashi, Y.Mori, M.Itabashi and N.Watanabe, J. Appl. Phys. 58,
   2691 (1985).

QUANTITATIVE THEORY OF ZONE-FOLDING PHENOMENA IN
SEMICONDUCTOR SUPERLATTICES

M. Jaros, M.A. Gell, D. Ninno, I. Morrison, K.B. Wong and P.J. Hagon

Department of Theoretical Physics, The University,

Newcastle upon Tyne, United Kingdom.

We have performed full-scale pseudopotential calculations of the electronic structure and optical properties of $GaAs-Ga_{1-x}Al_xAs$, $InAs-GaSb$, $Si-Si_{1-x}Ge_x$ and $Si-Si_{1-x}Sn_x$ superlattices with well widths from 10 to 400 Å. Our predictions show that there is a wealth of largely unexplored and conceptually novel phenomena which is accessible to both quantitative theory and experiment.

1. INTRODUCTION

In an effort to provide a more accomplished theory of the electronic structure we have recently implemented our pseudopotential matrix method[1] and presented energy levels and charge densites in $GaAs-Ga_{1-x}Al_xAs$[2], $GaSb-InAs$[3], and $Si-Si_{1-x}Ge_x$[4] superlattices. We have demonstrated that our calculations are capable of predicting transition energies and oscillator strengths with a meV accuracy[5]. High pressure experiments have shown that the higher-lying states and their pressure dependence predicted in our calculations can be used to determine accurately the conduction and valence band offsets[6]. Deviations from simple models occur whenever the short-wave-length components of the superlattice potential are strong enough to mix waves of different momenta across a significant fraction of the bulk Brillouin zone. This so called zone folding drastically alters transition rates of the confined states in question since both radiative and nonradiative (e.g. Auger) cross sections are strongly wave vector dependent.

2. OPTICAL SPECTRA AND THE EFFECT OF ZONE FOLDING IN
$GaAs-Ga_{1-x}Al_xAs$ SUPERLATTICES

The most apparent examples of zone folding can be obtained when
the alloy composition is altered so that the confined states derived
from the conduction band minima at Γ and X cross. In that situation
one can observe the continuous change of the momentum wave function
and its manifestation in both the nodal and envelope character of the
charge density as a function of x[7]. An analogous effect occurs when
hydrostatic pressure is used to drive the lowest Γ -like level
confined in GaAs wells towards and above the states derived from the X
minima of $Ga_{1-x}Al_xAs$[8]. The coupling between Γ and X components at the
crossing is strong in spite of the fact that the Γ component is
spatially localised in GaAs whereas the X-like confined state is
localised in the barrier. Indeed, when the calculation is repeated
with the Dingle offset[9] the strength of the mixing effects is of the
same order of magnitude although in this case the spatial overlap[2] of
the Γ and X-like envelope functions is nearly 100%! Clearly, the
difference in the phase of the Bloch function across the interface in
the former case (staggered band offset) compensates for the lack of
spatial overlap and enhances the coupling as well as the corresponding
optical matrix element[8].

In narrow and ultrathin layers the zone folding effect reaches
giant proportions and, in addition to the familiar heavy-light hole
mixing, provides a fertile ground for testing full scale theory[10].

In the course of these studies, another remarkable feature has
been identified which lies outside the scope of the effective mass
model and which affects conduction band states of X character and
light-hole states in the valence band. It transpires that
localisation in these minibands may occur only in certain networks of
bonds and may lead to distinct polarisation properties of the
corresponding optical spectra. More specifically, the charge density
in these confined states is preferentially localised in those bonds
lying in the $\{101\}$ planes. A similar effect was reported by Pickett
et. al.[11] who studied ultrathin (110) superlattices. Our calculations
show that this effect is not peculiar to very thin layers. In a
microscopic picture the bonds lying along the zig-zag chain enveloping
the $\langle 001 \rangle$ axis are not in general equivalent because of the asymmetry

of the crystal potential in the bulk primitive cell. It is worth
pointing out that this effect has also been obtained in GaSb-InAs
(001) superlattices[3].

3.   ZONE FOLDING IN $Si-Si_{1-x}Ge_x$ AND $Si-Si_{1-x}Sn_x$
SUPERLATTICES

In this structure the substrate is chosen so as to match the
lattice constant of the superlattice period and the strain is taken up
by both constituent layers (e.g. compressive strain in $Si_{0.5}Ge_{0.5}$ and
tensile strain in Si). Two calculations were performed which will be
referred to as the strained and unstrained calculations. In the
unstrained calculation the positions of the Si and $Si_{0.5}Ge_{0.5}$ atoms
are at the same sites as the bulk $Si_{0.76}Ge_{0.24}$ atoms. This represents
a superlattice made from bulk silicon with a slightly larger lattice
constant and bulk $Si_{0.5}Ge_{0.5}$ with a slightly compressed lattice
constant. This enlargement and compression of the lattice constants
is the same in all directions. In the strained calculation the
positions of the silicon atoms and $Si_{0.5}Ge_{0.5}$ atoms were shifted from
the bulk lattice sites of $Si_{0.76}Ge_{0.24}$ in the ⟨001⟩ direction to
change the lattice spacing between adjacent atoms to that of the
actual lattice constants of the two respective materials, the
positions of the atoms on directions perpendicular to (001) being left
unchanged. In the unstrained case one obtains well confined electron
states in the alloy layers. The effect of strain is to shift the
energies of the confined states at the conduction band edge and to
localise the electron levels in the silicon layers. Since the hole
state wave function is highly localised at Γ (k = 0) the strength of
the optical transition across the superlattice gap is measured by the
magnitude of the Γ component in the state at the bottom of the
conduction band. It transpires that although some enhancement does
take place the optical matrix element is small compared to that
obtained in direct gap materials. This is so even in structures
consisting of ultrathin Si and SiGe layers or in structures with an
appreciable Sn component. However, the effect of strain does reflect
the geometrical properties of the system and it remains to be seen

whether the above conclusions are generally valid. We also find that the effective height of the confining barrier depends on the way strain is accommodated in the system. This offers interesting possibilities for transport applications.

ACKNOWLEDGEMENTS

We would like to thank Professor D. Herbert of RSRE (Malvern) and Dr D.J. Wolford of IMB (Yorktown Heights) for stimulating conversations. We thank the SERC (U.K.) and RSRE Malvern for financial support.

REFERENCES

1.   Jaros, M., Reports on Progress in Physics 48, 1091 (1985).

2.   Wong, K.B., Jaros, M., Gell, M.A. and Ninno, D., J. Phys. C19, 53 (1986).

3.   Gell, M.A., Wong, K.B., Ninno, D., and M Jaros, J. Phys. C (in press), (1986).

4.   Morrison, I., Jaros, M. and Wong, K.B., J. Phys. C19, 935 (1986).

5.   Ninno, D., Gell, A.M., and Jaros, M., J. Phys. C (in press), (1986).

6.   Wolford, D.J., Kuech, T.F., Bradley, J.A., Gell, M.A., Ninno, D. and Jaros, M., J. Vac. Sc. Technol. (in press), (1986).

7.   Ninno, D., Wong, K.B., Gell, M.A. and Jaros, M., Phys. Rev. B32, 1586, (1985).

8.   Gell, M.A. Ninno, D., Jaros, M., Wolford, D.J., Bradley, J.A. and Kuech T.F., Phys. Rev. B (in press), (1986).

9.   Dingle, R., Festkorperprobleme 15, 21 (1975).

10. Gell, M.A., Ninno, D., Jaros, M. and Herbert, D.C., Phys. Rev. B (in press), (1986).

11. Pickett, W.E., Louie, S.G. and Cohen, M.L., Phys. Rev. B17, 815, (1978).

ASPECTS OF ELECTRON-PHONON INTERACTION IN
SEMICONDUCTOR HETEROSTRUCTURES AND SUPERLATTICES

S. Das Sarma, W. Y. Lai, and Akiko Kobayashi

Department of Physics and Astronomy
University of Maryland
College Park, Maryland 20742, U.S.A.

ABSTRACT

We present results on the phonon emission rate by
energetic electrons in a theory that includes
dynamical screening effects in a two-dimensional
heterostructure. We also discuss polaronic transport
along the superlattice growth axis.

1. INTRODUCTION

Many of the electronic properties of semiconductor
heterojunctions, quantum wells, and multilayer superlattices are
modified by electron-LO phonon scattering via the Fröhlich
interaction. Examples of such properties are polaronic
renormalization of carrier effective mass,[1] carrier mobility, hot
electron energy loss,[2] and quasiparticle damping rate.[3] In this
paper, we are concerned with dynamical screening of electron-LO phonon
interaction in a two-dimensional semiconductor heterostructure, and
small-polaron like effects in the electronic transport along the
superlattice growth direction.

2. PHONON EMISSION RATE IN TWO-DIMENSIONAL STRUCTURE

The dynamical screening of the Fröhlich interaction by free
carriers in modulation-doped heterostructures is calculated using a
diagrammatic many-body approach. The quantity we calculate here is
the leading-order electronic self-energy due to the electron-LO phonon
interaction. This is given by[4]

$$\Sigma(k,i\omega_n) = -\frac{1}{\beta} \Sigma_{i\nu_m} \int \frac{d^2q}{(2\pi)^2} V(q,i\nu_m) G(k-q;i\omega_n-i\nu_m) \qquad (1)$$

with effective interaction

$$V(q,i\nu_m) = V(q) D_{LO}(q,i\nu_m) [\epsilon(q,i\nu_m)]^{-2} \qquad (2)$$

where $\beta=(k_BT)^{-1}$ is the inverse temperature; $\nu_m=2m\pi/\beta$ (with m=0, ±1, ±2
...) and $\omega_n=(2n+1)\pi/\beta$ (with n=0, ±1, ±2...) are the standard boson and

fermion Matsubara frequencies respectively; k and q are the 2D
wave-vectors in the plane of the electron layer; V(q) is the square of
the unscreened two dimensional Fröhlich interaction strength; G is the
one electron Green's function. It is important to note that the
dielectric function $\epsilon(q,i\nu_m)$ is both wavevector and frequecy

dependent, and that the LO-phonon propagator $D_{LO}(q,i\nu_m)$ includes the
renormalization due to the phonon self-energy correction arising from
the electron-phonon interaction. The poles of $D_{LO}(q,i\nu_m)$ therefore
correspond to the coupled plasmon-LO phonon modes. In Fig.1(a), we
show dispersion relations of these modes. For the sake of comparison,
the coupled modes for a three-dimensional case are shown in Fig.1(b).
It is seen that the simple LO phonon (ω_{LO}) and plasmon (ω_p) become new
hybridization modes $(\omega_+$ and $\omega_-)$ in the mode-coupling region due to the
dynamical screening effect.

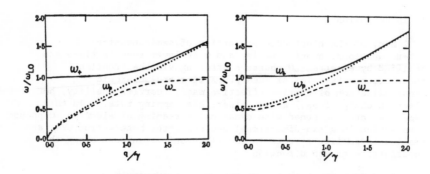

Fig.1(a): Dispersion relations of
plasmon mode (ω_p), and hybridization
modes $(\omega_+$ and $\omega_-)$ for a two-dimensional
system. $\gamma=(2m\omega_{LO})^{1/2}$ is chosen as the
unit for the wavevector q.

Fig.1(b): Dispersion relations of
plasmon mode (ω_p), and hybridization
modes $(\omega_+$ and $\omega_-)$ for a
three-dimensional system.

We take the hydrodynamic approximation for the dielectric
function, $\epsilon(q,\omega)$, in Eq.(2) to obtain the phonon emission rate
(imaginary part of the self-energy). In Figs.2(a) and (b), we display
our zero-temperature results for the case of non-degeneracy (no fermi
statistics effects) for a two-dimensional and a three-dimensional
system respectively. Our results including dynamical screening
effects are shown for different electronic densities N, and are
compared with the unscreening and the static screening results.[5] From
these results, we conclude that the phonon emission rate for the case
of dynamical screening lies in between the unscreening and the static
screening cases. We also note that, with increasing density, the rate
becomes close to that for the static screening case (especially at

high E). The sudden depletion in rate at low E is caused by the high
plasmon emission rate that exceeds the ω_+ and ω_- emission rates.
These structures seen at low E, however, may have been caused by the
specific dynamical screening model employed in this paper.

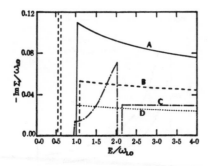

Fig.2(a): Emission rate for a
two-dimensional system as a function of
electronic energy E: (A) unscreening;
(B) dynamical screening with
$N=0.2\times10^{12}cm^{-2}$; (C) dynamical screening
with $N=10^{12}cm^{-2}$; (D) static screening.

Fig.2(b): Emission rate for a
three-dimensional system as a function of
electronic energy E: (A) unscreening;
(B) dynamical screening with
$N=0.2\times10^{18}cm^{-3}$; (C) static screening
with $N=0.2\times10^{18}cm^{-3}$; (D) dynamical
screening with $N=10^{18}cm^{-3}$; (E) static
screening with $N=10^{18}cm^{-3}$.

3. POLARONIC TRANSPORT IN A SUPERLATTICE

Most of the recent work on polaronic transport in semiconductor
superlattices concerns carrier current flow parallel to the layers.
In this section, we consider the drift velocity perpendicular to the
layer (i.e., superlattice growth direction). We use a model in which
a series of parallel two-dimensional electronic layers with small
overlap is embedded in the uniform relevant semiconductor material.
At low temperatures and without disorder, the motion of electron is
Bloch-like, due to the periodicity along the growth axis. As
temperature increases, the electron-LO phonon interaction will
increase the effective mass of the electron (polaron) decreasing the
bandwidth. As a result, motion of the electron will change to
hopping-like. By using a tight-binding scheme for this motion, we
note that the coupled electron-LO phonon field problem can be mapped
onto an equivalent one-dimensional Holstein polaron problem.[6] Figure
3(a) shows our results on the diffusion coefficient as a function of
temperature for different electron-LO phonon coupling strength. It
can be seen that the polaronic self-trapping transition occurs at
lower temperature for a material with stronger electron-phonon
coupling. To illustrate this point, in Fig.3(b), we plot the

S. Das Sarma et al.

transition temperature as a function of electron-LO phonon coupling
strength. Our results indicate that the effects arising from the
transition can be observed, for example, in CdTe multi-quantum-wells
at low temperatures, since it has low value of ω_{LO}($\cong$20 meV) and large
coupling strength.

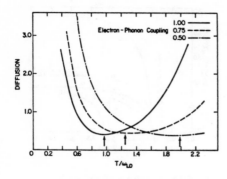

Fig.3(a): Diffusion coefficient as a
function of temperature for three
different coupling strengths.

Fig.3(b): Transition temperature as a
function of electron-LO phonon coupling
strength (A_Q) for two different LO
phonon energies.

ACKNOWLEDGEMENT: This work is supported by the Army Research Office.

REFERENCES

1. Nicholas R. J. et al., Phys. Rev. Lett. $\underline{55}$, 883(1985).
2. Shah J. et al., Phys. Rev. Lett. $\underline{54}$, 2045(1985).
3. Tsui D. C., Phys. Rev. B$\underline{10}$, 5088(1974).
4. Das Sarma S. and Mason B. A., Ann. Phys. $\underline{163}$, 78(1985).
5. Das Sarma S. and Mason B. A., Phys. Rev. B$\underline{15}$, 1418(1986).
6. Holstein T., Ann. Phys. $\underline{8}$, 343(1951); Munn R. M. and Silbey R., J.
   Chem. Phys. $\underline{83}$, 1843(1985).

POLAR INTERFACE MODES of SUPERLATTICES as SELF-
SUSTAINING OSCILLATIONS of INTERFACE CHARGES

R. Enderlein, F. Bechstedt

Humboldt-Universität zu Berlin, Sektion Physik
G.D.R.

ABSTRACT

A new approach to polar interface modes of superlattices and other microstructures is developed. The frequency spectrum and spatial dispersion of these modes is calculated from the
set of Newtons equations of the interacting
interface charges. The electron-interface mode
interaction Hamiltonians of Fröhlich and deformation potential type are derived.

1. INTRODUCTION

The occurence of polar interface modes /1,2/ is one of
the qualitatively new features of superlattices (SLs) and
other semiconductor microstructures. Polar modes, in general, are self-sustaining oscillations of ionic and electronic polarization charge densities of a solid. In SLs
these charge density oscillations show certain peculiari-
ties. They reach large amplitudes in the vicinity of interfaces while they are finite or even zero within SL
layers. Interface modes are the type of charge density
oscillations which locate entirely on interfaces forming
there a set of interface charges. The mechanical eigen-
oscillations of these charges can be understood as interface modes. In the present paper we will develop the interface charge approach to polar interface modes in more
detail. It will be shown that the same mode spectra and
wave-vector dispersions are obtained as in previous treatments, where interface modes follow as self-sustaining
solutions of the electrodynamics of layered media /1,2/.
Like in electrodynamic treatments, here we rely on the
continuum model, although the interface charge approach
is basically a microscopic one. It can be also applied to
a solid state model which takes the atomic structure ex-

plicitly into account. Within the continuum model the interface charges become area charges located on interfaces and interacting among each other via Coulomb forces. The corresponding set of Newtons equations is given in section 2. In section 3 we treat the Coulomb potential and the atomic displacement field of an interface mode. For the sake of simplicity we give explicit results only in the case of a double heterostructure which, however, exhibits already important features of a SL. The electron-interface mode interaction Hamiltonians of Fröhlich and deformation potential type are given in terms of creation and annihilation operators of interface phonons.

2. FREQUENCY SPECTRUM of INTERFACE MODES

The basic element of our approach is a single interface and, still more elementary, a free surface. Thus, to give an idea of our method, we start with a semi-infinite solid facing an empty half space. The polarization field $\bar{P}(\bar{x},t)$ of the semi-infinite solid is a step-like function $\bar{P}(\bar{x},t)= \theta(z)\bar{P}^{o}(\bar{x},t)$ with $\bar{P}^{o}(\bar{x},t)$ the polarization field of the corresponding infinite solid. The polarization charge density $\varrho(\bar{x},t)= -div\bar{P}(\bar{x},t)$ reads

$$\varrho(\bar{x},t) = -\theta(z) \, div\bar{P}^{o}(\bar{x},t) - \delta(z)\, \mathbb{G}\,(x,y,t) \tag{1}$$

with $\mathbb{G}\,(x,y,t)=P_{z}^{o}(x,y,z=0,t)$ an area charge density. Dealing with interface modes only, we suppose the bulk polarization charge density $-div\bar{P}$ to be zero. Then the Fourier transform $\phi(\bar{Q},z,t)$ of the Coulomb potential due to ϱ is given by

$$\phi(\bar{Q},z,t) = -\frac{2\pi}{Q}\, \mathbb{G}\,(\bar{Q},t)\, e^{-Q|z|}. \tag{2}$$

Here the electronic dielectric constant ε_{∞} has been set equal to 1. The polarization field $\bar{P}(\bar{x},t)$ may be expressed in terms of the relative displacement field $\bar{u}(\bar{x},t)$ of the polarizable continuum by means of the relation $\bar{P}(\bar{x},t) = Ne^{*}\bar{u}(\bar{x},t)$ with e^{*} the transverse effective charge per elementary cell and N the number of elementary cells per unit volume. Here and in what follows we refer to a polarization field due to phonons. Final results, however, apply also to plasmons if symbols are adequately reinterpreted.

The reduced mass density $Nm^{*}\theta(z)$ of the half-infinite solid performs a mechanical motion under the influence of a short range force density $-Nm^{*}\omega_{TO}^{2}\,\theta(z)\bar{u}(\bar{x},t)$ and a Coulomb force density $-Ne^{*}\theta(z)grad\,\phi(\bar{x},t)$. The corresponding Newtons equation transforms in an equation for the z-component of the surface displacement field $u_{z}(\bar{Q},z=+0,t)$ if the operator "div" is applied to it. For the surface area charge density $\mathbb{G}\,(\bar{Q},t)=Ne^{*}u_{z}(\bar{Q},z=+0,t)$ one gets an equation which defines the Fuchs-Kliewer surface phonon /1/. In the case of a single interface between two solids S and Ŝ (Ŝ for z<0 and S for z>0) one has two area charge distributions, one, $\hat{\mathbb{G}}\,(\bar{Q},t)$, on the left hand side of the interface, and one, $\mathfrak{S}(\bar{Q},t)$, on its right hand side. The two coupled Newtons equations for $\hat{\mathfrak{S}}$ and $\mathfrak{S}$ define the two interface modes of a single heterostructure. If the conditions $(\omega^2_{LO} - \omega^2_{TO})$, $(\hat{\omega}^2_{LO} - \hat{\omega}^2_{TO}) \ll |\omega^2_{TO} - \hat{\omega}^2_{TO}|$ are simultaneously fulfilled, the two modes decouple approximately. The displacement field are essentially different from zero either in solid S(S-like mode) or in solid $\hat{S}$($\hat{S}$-like mode). In the following part of the paper we assume this simplification to be valid. It applies, e.g., to the vibronic excitations of the GaAs-(Al,Ga)As-material system and to the plasmon excitations of a SL with alternatingly filled and unfilled layers of free carriers. For a double hetero (DH)structure (solid $\hat{S}$ for z<0 and z>d, solid S for 0<z<d) one has four different area charges, two ones, $\hat{\mathfrak{S}}_1$ and $\mathfrak{S}_1$, at the interface z=0, and two ones, $\mathfrak{S}_2$ and $\hat{\mathfrak{S}}_2$, at the interface z=d. With the simplification mentioned above one gets two sets of coupled equations for $\mathfrak{S}_1$, $\mathfrak{S}_2$, and two ones for $\hat{\mathfrak{S}}_1$, $\hat{\mathfrak{S}}_2$. They yield the Q-dependent eigenfrequencies of the interface modes of a DHstructure in accordance with previous results /1,2/. Finally, in the case of an infinite SL (solid S for layers nd<z<nd+d_1, $\hat{S}$ for layers nd+d_1<z<(n+1)d, d=d_1+ d_2), the S-like area charges $\mathfrak{S}_{1n}$,$\mathfrak{S}_{2n}$, of the n-th SL-cell obey the following set 2x2 matrix equations:

$$(\frac{d^2}{dt^2} + \omega^2_{TO} + \frac{1}{2}\omega^2_p) \begin{pmatrix} 1 & 0 \\ 0 & 1 \end{pmatrix} \begin{pmatrix} \mathfrak{S}_{1n} \\ \mathfrak{S}_{2n} \end{pmatrix} - \frac{1}{2}\omega^2_p \left\{ e^{\alpha_1}\begin{pmatrix} 0 & 1 \\ 1 & 0 \end{pmatrix} \begin{pmatrix} \mathfrak{S}_{1n} \\ \mathfrak{S}_{2n} \end{pmatrix} + \right.$$

$$\left. \left[\sum_{m=-\infty}^{n} e^{-(n-m)\alpha}\begin{pmatrix} 1 & -e^{\alpha_1} \\ e^{-\alpha_1} & -1 \end{pmatrix} + \sum_{m=n}^{\infty} e^{(n-m)\alpha}\begin{pmatrix} -1 & e^{-\alpha_1} \\ -e^{\alpha_1} & 1 \end{pmatrix} \right] \begin{pmatrix} \mathfrak{S}_{1m} \\ \mathfrak{S}_{2m} \end{pmatrix} \right\} = 0 \tag{3}$$

with α =Qd,α_1 =Qd_1 and ω^2_p=ω^2_{LO} - ω^2_{TO} . The assumption $e_\infty=\hat{e}_\infty$ has been made in (3). For Bloch-like propagating interface modes of wave number q∥z(-π/d<q<π/d) the two eigenfrequencies ω_+ of equation (3) follow from

$$\omega^2_\pm = \omega^2_{TO} + \frac{1}{2} \omega^2_p(1 \pm (\frac{\cosh Q(d_1 - d_2)- \cos qd}{\cosh Q(d_1 + d_2)- \cos qd})^{1/2}). \tag{4}$$

The corresponding eigenvectors ($\mathfrak{S}_{1q+}$, $\mathfrak{S}_{2q+}$) and ($\mathfrak{S}_{1q-}$, $\mathfrak{S}_{2q-}$) are, in general, elliptically polarized. In the limit q→0 the polarization becomes linear. For d_1<d_2,ω_- corresponds to a symmetric mode ($\mathfrak{S}_{1q-}$ =$\mathfrak{S}_{2q-}$), and ω_+ to an antisymmetric one ($\mathfrak{S}_{1q+}$ = $-\mathfrak{S}_{2q+}$)(reversed for d_1>d_2).

3. DISPLACEMENT-FIELDS AND INTERACTION HAMILTONIANS of INTERFACE MODES

The displacement field $\bar{u}(\bar{x},t)$ of an interface mode may be calculated as follows. First, one takes the Coulomb potential of interface charges connected with a given interface mode. This potential represents a superposition of single interface charge potentials as given by equation(2). Second, one calculates the dielectric displacement field $\bar{D}$ from the two equations $\text{div}\bar{D}=0$ and $\text{rot}\bar{D}=4\pi\text{rot}\bar{P}$. Since $\text{rot}\bar{P}$ does not vanish throughout, the D-field of an interface mode differs from zero. Third, one takes the $\bar{P}$-field from the relation $\bar{D}=\bar{E}+4\pi\bar{P}$. Finally, the displacement field $\bar{u}$ follows from the relation $\bar{P}=Ne^{*}\bar{u}$. We apply this procedure to the DHstructure of section 2. For the antisymmetric and symmetric S-like modes of this structure it follows $\bar{u}_{\pm}(\bar{Q},z,t) =$

$$\theta(z)\theta(d-z)\frac{u_{\pm}(\bar{Q},t)}{(1\pm e^{-Qd})}((-i\frac{\bar{Q}}{Q} + \bar{e}_{z})e^{-Qz}\pm(-i\frac{\bar{Q}}{Q} - \bar{e}_{z})e^{Q(z-d)}), \quad (5)$$

where $\bar{e}_{z}$ means the unit vector parallel to z and $u_{\pm}(\bar{Q},t)$ the z-component of the displacement field of the interface at z=0. The $u_{\pm}(\bar{Q},t)$-field represents the true field variable of an interface mode. It is object to quantization. According to the general quantization rules one gets

$$u_{\pm}(\bar{Q},t) = (\frac{\hbar}{2M_{\pm}(\bar{Q}) \omega_{\pm}(\bar{Q})})^{1/2} (b^{+}_{-\bar{Q}} + b_{\bar{Q}}) , \quad (6)$$

where $M_{+}=(G^{2}_{\perp} G_{\parallel}m^{*}) (2/Qd)\tanh(Qd/2)$ and $M_{-}=(G^{2}_{\perp} G_{\parallel}m^{*}) (2/Qd)\coth(Qd/2)$ mean reduced masses attributed to the interface displacement fields $u_{\pm}(\bar{Q},t)$ of the central slab of the DH-structure with $G_{\parallel}$ elementary cells parallel and $G^{2}_{\perp}$ perpendicular to z. Note that in the limit of small Qd which is met in light scattering, the symmetric interface mass M_{-} becomes large compared to, and the antisymmetric interface mass M_{+} equal to the bulk mass $G^{2}_{\perp} G_{\parallel}m^{*}$.

Knowing the displacement field $\bar{u}(\bar{Q},z,t)$, one can easely calculate the interaction Hamiltonian H_{F} and H_{D} due to, respectively, the Fröhlich and deformation potential coupling of an electron to an interface mode. For the DH-structure of section 2 and S-like modes it follows

$$H_{F} = \sum_{Q} (\frac{2\pi ee^{*}}{\epsilon_{\infty} Q})(\frac{\hbar}{2M_{\pm}\omega_{\pm}})^{1/2}(e^{-Qz} \mp e^{Q(z-d)})(b^{+}_{\bar{Q}} e^{-i\bar{Q}\cdot\bar{x}}+ c.c.)$$

$$H_{D} = \sum_{Q} D(\frac{\hbar}{2M_{\pm}\omega_{\pm}})^{1/2} \theta(z)\theta(d-z)\frac{e^{-Qz} \pm e^{Q(z-d)}}{1 \pm e^{-Qd}}(b^{+}_{\bar{Q}} e^{-i\bar{Q}\cdot\bar{x}} +$$

$$+ c.c.) \quad (7)$$

with D the optical deformation potential.

REFERENCES

/1/ R.Fuchs and K.L.Kliewer, Phys.Rev. 140, A2076 (1966)
/2/ E.P.Pokatilov and S.J.Beril,phys.stat.sol.(b)118,567
    (1983); R.E.Camley and D.L.Mills, Phys.Rev.B29,1695
    (1984); F.Bechstedt and R.Enderlein,phys.stat.sol.(b)
    131,53 (1985).

POLARITON STRUCTURE OF MULTILAYERED SEMICONDUCTING MATERIALS

J.P. Vigneron, A. Dereux, Ph. Lambin and A.A. Lucas

Facultés Notre-Dame de la Paix
61, rue de Bruxelles, B-5000 Namur, Belgium

ABSTRACT

With multilayered structures, the accumulation of interfaces gives rise to new excitation modes, that can be found both in the phonon and plasmon energy range. The detailed polariton structure of a general stratified structure, finite or infinite, and its effective dielectric response function can be assessed in terms of a new, Riccati-type initial value problem which can be used to quantitatively account for reflectance measurements, attenuated total internal reflection (ATR) and, using a non-retarded limit of the same formalism, high-resolution electron energy-loss spectroscopy (HREELS).

Semiconducting multilayer structures, including synthesized semiconductor superlattices[1] have been grown from a variety of semiconducting materials. Considerable efforts are devoted to the improvement of the growing techniques and sample quality of these heterogeneous structures, while, concurrently, their physical properties, mainly electronic and vibrational are beeing actively investigated. Recent works [2] reported experimental and theoretical results on the application of several new spectroscopies to such multilayer structures, demonstrating some of the consequences of the accumulation of interfaces on their response to electromagnetic probes.

The basic theory presented in this communication is applicable to any stratified materials with a graded dielectric or magnetic response, provided that the definition of the local response functions remains acceptable on the scale of the system inhomogeneities. The stratified dielectric material considered here consists of a semi-infinite medium characterized by a long-wavelength dielectric constant $\varepsilon(z,\omega)$ and a permittivity $\mu(z,\omega)$ which vary only in a direction perpendicular to its surface (z axis), that is, usually, the direction of the material growth. Due to stratification, the application of an electric field to such a material results in a macroscopic polarization involving induced charges associated with the dielectric function gradients. The polariton modes can be predicted from electrodynamics, which describes the wave propagation in terms of a one-dimensional impedance function Z_p. For the transverse magnetic polarization (the transverse electric modes will not be considered here) one writes:

$$Z_p(z,k_y,\omega) = i \frac{k_y c}{\omega} \sqrt{\frac{\mu_0}{\varepsilon_0}} \frac{1}{\xi_p(z,k_y,\omega)},$$

(1.a)

$$\xi_p(z,k_y,\omega) = \varepsilon(z,\omega) k_y [U_p(z) / \frac{dU_p(z)}{dz}]$$

(1.b)

related to the x component of the magnetic field through $H_x = e^{i(k_y y - \omega t)} U_p(z)$. The other fields components can be extracted from (1.b) through maxwell's equations. $\xi_p(z)$ is a continuous function of z even if the response functions $\varepsilon(z)$ or $\mu(z)$ assume finite discontinuities. It verifies a Riccati equation

$$\frac{1}{k_y} \frac{\partial \xi_p(z,k_y,\omega)}{\partial z} + \frac{1 - \left[\frac{\omega \varepsilon(z,\omega)\mu(z,\omega)}{k_y c} \right]^2}{\varepsilon(z,\omega)} \xi_p^2(z,k,\omega) = \varepsilon(z,\omega)$$

(2)

In these relations, k_y is the two-dimensional radiation wavevector measured along any direction chosen parallel to the sample surface, c is the speed of light in vacuum, and ω is the field frequency. In a homogeneous material such as a substrate, we easily note that the only acceptable solution of the Riccati equation is a constant. For a semi-infinite superlattice, made of repeated periods of two different layers with response functions ε_1, μ_1 and ε_2, μ_2, and thicknesses d_1 and d_2, one obtains a semi-periodic impedance ($\xi_p(z) = \xi_p(z-L)$, for z<0,) which at the surface (z=0) takes the value

$$\xi_p(0^-,k_y,\omega) = \frac{g_1^2 - g_2^2}{2(a_1 + a_2)} \pm \sqrt{\left(\frac{g_1^2 - g_2^2}{2(a_1 + a_2)} \right)^2 + \frac{a_1 g_2^2 - a_2 g_1^2}{a_1 + a_2}}$$

(3)

with

$$a_j = \frac{\varepsilon_j(\omega)}{\beta_j \tanh k_y \beta_j d_j}, \quad g_j = \frac{\varepsilon_j(\omega)}{\beta_j}, \quad \beta_j = \sqrt{1 - \left[\frac{\omega \varepsilon_j(\omega) \mu_j(\omega)}{k_y c} \right]^2}$$

(4)

Assuming real response functions $\varepsilon(\omega)$ and $\mu(\omega)$, the superlattice will transport electromagnetic energy at points in the (k_y,ω) diagram where ξ_p takes complex values (shaded

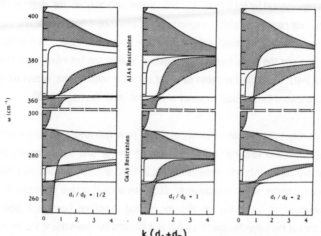

Fig. 1 Phonon–polariton structure of an ideal GaAs–AlAs superlattice. The shaded regions correspond to Bloch combination of surface/interface modes, while the solid lines are isolated, evanescent modes.

regions in Fig. 1). This occurs even within the restrahl regions of the constituent materials, where continua of polariton modes develop. These modes can be viewed as propagating excitations of interface Fuchs–kliewer[3] modes wich combine into Bloch waves. When ξ_p is real (non-shaded areas), it eventually matches the corresponding real value of the vacuum impedance ($\xi_p(0^+) = -1/[1-(\omega/k_y c)^2]^{1/2}$). These coincidences describe the isolated

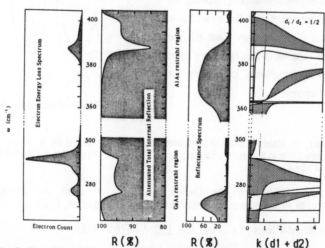

Fig. 2. Polariton spectrum of a GaAs–AlAs superlattice, with computer simulations of Infrared reflectance, ATR and HREELS spectra.

surface modes represented by solid lines in non-shaded regions, in the diagram of Fig.1. These modes correspond to electromagnetic fields evanescent both inside and outside the superlattice. For wavelengths much smaller than the layers thickness, the interface modes cease overlapping and the mode frequencies converge toward the interface mode frequencies $(\varepsilon_1(\omega_i)+\varepsilon_2(\omega_i)=0)$, inside the restrahl regions of the layer constituants, except for one mode which converges toward the surface frequency of the first layer $(\varepsilon_1(\omega_s)+1=0)$.

In order to determine the observability of theses polariton modes, computer simulations of reflectance, attenuated total reflection and electron energy loss measurements have been performed and typical results are presented in Fig.2. EELS and ATR are both sensitive to the isolated evanescent modes and to the continua. EELS, because it probes more specifically the modes with wavevectors satisfying the electron surfing condition $(\omega \approx k_y v_y)$ describes the non-retarded part of the polariton spectrum and is dominated by losses due to the excitation of the isolated mode which develops close to the surface. ATR, at incidence θ and using a prism of refractive index n_{prism}, probes the modes found along the line $\omega = k_y c/(n_{prism}\sin\theta)$, below the frequencies of the vacuum photons, while reflectance measurements probe the continuum areas in the radiative region, along $\omega = k_y c/\sin\theta$. Fig. 3 displays the result of recent EELS measurements[4], showing an isolated mode (peak A) and the losses due to the presence of a continuum (peak B). The calculations reported on this figure are based on the above description of the superlattice dielectric response.

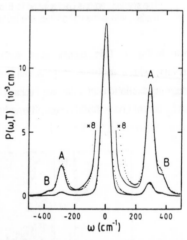

Fig. 3. EELS spectrum on a GaAs-GaAl$_{0.3}$As$_{0.7}$ superlattice.

1. L. Esaki, in *Recent Topics in Semiconductors Physics*, edited by H. Kamimura and Y. Toyozawa (World Scientific, Singapore, 1983), pp. 1-71; J. Phys. (Paris) Colloq. 45, C5-3 (1984).

2. A.A. Lucas, J.P.Vigneron, Ph. Lambin, P.A .Thiry, M. Liehr, J.J. Pireaux, and R. Caudano, in *Proceedings of he Sanibel Symposium*, St. Augustine, 1985.

3. Fuchs, R. and Kliewer, K.L., Phys. Rev. 140, A 2076 (1965)

4. Ph. Lambin, J.P. Vigneron, A.A. Lucas, P. A. Thiry, M. Liehr, J.J. Pireaux, and R.Caudano, and T. Kuech, Phys. Rev. Letters, 56,1842 (1986)

PHONONS IN GaAs/AlAs SUPERLATTICES

E. Molinari

CNR - Istituto di Acustica "O.M.Corbino", Via Cassia 1216, I-00189 Roma,Italy

A. Fasolino

ISAS, Strada Costiera 11, I-34100 Trieste, Italy

K. Kunc

CNRS, Tour 13, 4 pl. Jussieu, F-75230 Paris, France

ABSTRACT

We calculate the phonon spectra of GaAs/AlAs (001) superlattices along the growth direction by means of a new theoretical approach which provides realistic dispersions and displacement patterns for both acoustical and optical phonons at the same time. In particular we compare the confined superlattice optical modes with the corresponding bulk modes; for layers thicker than 3 monolayers we find very small deviations between the two. This suggests the possibility of getting information about bulk phonons from light scattering measurements in superlattices.

1. INTRODUCTION

Much work is recently being devoted to the understanding of the lattice dynamics of superlattices from the theoretical and experimental point of view[1-6]. The use of simple models has led to a good qualitative understanding of the main features of superlattice phonons, i.e. folding of acoustic branches and confinement of optical modes. Conversely more sophisticated models, aiming at a more comprehensive, quantitative description, are rather cumbersome when dealing with big cells and with many different thicknesses. The scheme we have developped describes superlattice vibrations along the growth direction by an exact mapping to one dimension (1D) of the full three-dimensional (3D) lattice dynamical problem; we show that the difference in lattice dynamics of the two components can be obtained by changing only masses and effective charges. We are then in a position to use a realistic description of bonding based on an ab-initio determination of the bulk force constants without losing the conceptual and computational simplicity of a 1D formalism (linear chain).

Our method lifts many of the approximations used sofar when comparing superlattice confined optical frequencies with the bulk dispersions (such as nearest neighbour approximation, perfect confinement, no difference in the dielectric properties of the two components), thus providing a test for their validity. We find that indeed the effect of the superlattice on the optical frequencies is very small for thicknesses down to 3 monolayers. This suggests a very intriguing possibility, that is to use the superlattices themselves to measure bulk phonons when a direct determination is lacking due to intrinsic problems either in growing bulk samples or in performing neutron scattering experiments, as e.g. in InAs and AlAs. These materials on the other hand can often be grown in high quality superlattices with different combinations of components. Light scattering measurements on confined modes can yield the full dispersion, that turns out to be equal to the bulk phonon dispersion in the growth direction with an error of the same order of magnitude or smaller than neutron scattering.

2. METHOD

We describe the lattice dynamics of bulk crystals and superlattices along (001) in terms of vibrating (001) planes of atoms connected by interplanar forces; two distinct sets of force constants are then needed to describe longitudinal (L) and transverse (T) modes. This 1D representation can be exactly derived from the conventional 3D description of zincblende materials by a simple transformation of cohordinates. In order to have the dynamics of bulk GaAs and AlAs easily transferable to superlattices, we want to regard the substitution of a Ga by an Al atom as a "perturbation" localized on the sites and not affecting the pair interactions.

The quantitative accuracy of this description relies on the quality of the force constants used at the onset. For GaAs their numerical values were determined ab-initio in Ref. 7 within the local density approximation and they extend up to the 3rd neighbor plane for the longitudinal and to the 5th for the transverse. It would then be rather illusory to obtain them from a fit to the experimental dispersions; this explains why earlier works restricted to 1st neighbor forces could only obtain either a rough description of $\omega(k)$ or a realistic description of just a part of the spectra. However, these force constants[7] suffer from imprecisions in their ab-initio determination, mainly due to the use of local pseudopotentials. The calculated spectrum (Fig.1) underestimates the frequency of TA(X) and the LO-TO splitting at Γ (7.2 cm^{-1} versus an experimental value of[8] 24.3 cm^{-1}).

The same force constants can be used to describe AlAs by simply replacing the mass of Ga with that of Al (in the spirit of Ref.9) and provide the dispersion shown in Fig.1, which proves that this "mass approximation" is rather successful except that the LO-TO splitting in AlAs is even more underestimated than in GaAs as a consequence of neglecting differences in the dielectric properties of the two materials.

In order to go beyond this approximation we correct the ab-initio dynamical matrix $C_T(k)$ (which, although formally "short range", includes both short range term and long range Coulomb interaction) with an extra Coulomb term evaluated within the Rigid Ion Model:

$$C = C_T + (q_{act} C_{coul} q_{act} - q_{ref} C_{coul} q_{ref}).$$

Here q_{act} is the actual effective charge leading to the observed LO-TO splitting ($|q_{act}| = 0.673$ |e| in the GaAs- and 0.763 |e| in the AlAs-part of superlattice), while the reference charge ($|q_{ref}| = 0.367$ |e|) reflects the LO-TO splitting in GaAs as predicted (imperfectly) by the ab-initio force constants; ideally there would be $q_{ref} = q_{act}$(GaAs) if perfect force constants were available for GaAs. The Coulomb coefficient matrix C_{coul} depends only on the crystal structure of the superlattice

studied. The resulting spectra shown in Fig.1
(solid lines) allow to judge the improvement
of this "mass and charge approximation" over
the "mass approximation"; the comparison
with experiments[8] is very favourable and
the superlattice dynamics is yet fairly easy
to deal with.

3. RESULTS AND DISCUSSION

In Fig. 2 we show the frequency spectrum
of a $(GaAs)_3$ $(AlAs)_3$ superlattice. From the
calculation of the corresponding displacement patterns we find that already at this
thickness the optical modes show some confinement in one or the other layer. Their small
dispersion is due to the weak interaction
between layers of the same compound. The
confinement is found to be more pronounced
for the AlAs-like modes than for the GaAs-
like ones and for the transverse more than
for the longitudinal. This is a general feature
in these systems, related in part to the
position of the modes with respect to the
bulk acoustical continua. It is interesting to
note that confined AlAs like modes also
appear in the acoustical range of frequencies
since the AlAs TA continuum extends higher
than that of GaAs.

Given the well defined assignment of
GaAs-like and AlAs-like modes, a natural
question that has been recently debated[11] is
the relation of the superlattice modes with
the corresponding bulk modes, i.e. how the
superlattice configuration affects the frequency and displacements of these modes
with respect to the bulk. Once a mode is
confined, the relevant length is no longer the
one of the superlattice unit cell, but only the
part of it where the mode extends. This
region is ill-defined whenever the modes substantially spill in the other layer. However,
for L modes Jusserand and Paquet[11] have
shown that, assuming 1st neighbor interactions and perfect confinement, the modes
should be considered as confined in a slab of
n+1 monolayers, where n is the number of
GaAs (or AlAs) monolayers in the layer (in
the present case n=3 for both GaAs and

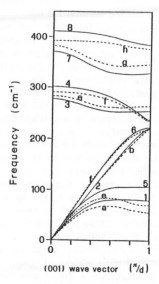

Fig.1   Bulk phonon dispersions
calculated in the "mass approximation" (dashed lines) and with the
inclusion of different Coulomb
fields (solid lines) for GaAs (curves
1-4 and a-d) and AlAs (curves 5-8
and e-h).

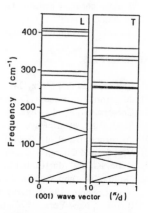

Fig.2   Longitudinal and transverse
phonon dispersion for a $(GaAs)_3$
$(AlAs)_3$ superlattice.

AlAs). The superlattice modes at the Γ point thus can be "unfolded" back into the bulk Brillouin Zone (BZ), by assigning the m-th confined mode of the superlattice to a q in the bulk BZ given by $q=m\pi/[(n+1)\,a/2]$ (here $1{\leqslant}m{\leqslant}n$ and a is the bulk lattice constant). Neglecting the 1 in the denominator would mean neglecting that both As interface atoms vibrate in any (GaAs-like and AlAs-like) confined L mode; this does not make much difference for thick layers, but already for n=7 it has brought to overestimate the superlattice effects both experimentally and in our previous theoretical work [11] . The "unfolding rule" is tested for two different thicknesses in Fig. 3. Although this rule was originally obtained within the approximations mentioned before, our calculation, based on a substantially more precise description, shows that it gives surprisingly small deviations of the superlattice frequencies from the bulk values even for relatively thin layers, where the confinement of superlattice modes is far from perfect. However we stress that the above arguments do not hold for T vibrations, due to the fact that the highly asymmetric

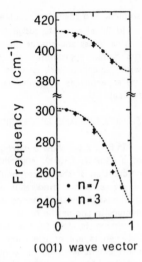

(001) wave vector

Fig.3   Test of the "n+1 rule" (see text): the confined modes of $(GaAs)_n$ $(AlAs)_n$ superlattices for n=7 and n=3 are "unfolded" onto the bulk LO phonons of GaAs and AlAs (solid curves). The wave vector is in units of $\pi/(a/2)$.

interactions make the two As interfacial atoms inequivalent[10] . Finally we notice that our results, even with the inclusion of the Coulomb term, cannot explain any deviation toward higher frequencies for the superlattice LO modes with respect to the corresponding bulk [4] without introducing some interface grading[5,10].

A more detailed presentation of the method and further results will be given in a forthcoming paper[10] .

We acknowledge partial support by GNSM-CISM (Italy) and CCVR (France).

REFERENCES

1) Barker, A.S. Jr., Merz, J.L., and Gossard, A.C., Phys. Rev. B17, 3181 (1978).
2) Jusserand, B., Paquet, D., and Regreny, A., Phys. Rev. B30, 6245 (1984).
3) Colvard, C., et al., Phys. Rev. B31, 2080 (1985).
4) Sood, A.K., et al., Phys. Rev. Lett. 54, 2111 (1985).
5) Jusserand, B., et al., Appl. Phys. Lett. 47, 301 (1985).
6) Yip, S., and Chang, Y.C., Phys.Rev. B30, 7037 (1984).
7) Kunc, K., and Martin, R.M., Phys. Rev. Lett. 48, 406 (1982).
8) Dolling, G., and Waugh, J.L.T., in "Lattice Dynamics", Wallis, R.F., ed. (Pergamon London, 1965), p.19; Onton, A., Proc. 10th Int. Conf. Phys. Semicond. (USAEC, New York, 1970), p.107.
9) Kunc, K., and Bilz, H., Solid State Commun. 19, 1927 (1976).
10) Molinari, E., Fasolino, A., and Kunc, K., to be published.
11) Molinari, E., Fasolino, A. and Kunc, K., Phys. Rev. Lett. 56, 1751(C) (1986); Jusserand, B., and Paquet, D., ibid., 1752(C); Sood, A.K., Menéndez, J., Cardona, M. and Ploog, K., ibid., 1753(C).

COLLECTIVE CHARGE DENSITY EXCITATIONS IN
POLYTYPE SEMICONDUCTOR SUPERLATTICES

P. Hawrylak, G. Eliasson, J. J. Quinn
Brown University
Department of Physics
Providence, RI   02912
USA

ABSTRACT

We present a theory of Raman scattering from collective excitations in finite polytype superlattices. The theory is applied to type I superlattices with two subband occupied, and the effects of intersubband scattering and many-body effects is studied.

1. INTRODUCTION

Resonant inelastic light scattering has been used to study single-particle and collective excitations of semiconductor superlattices [1, 2]. Theoretical calculations have been done on Raman scattering and electron energy loss spectra [3-5].

In this paper we will present a theory of Raman scattering of finite polytype superlattices, and apply the theory to type I and type II superlattices. Here we give the results for the type I systems, while the results for the type II systems will be published elsewhere [6].

2. THE MODEL

The system consists of N unit cells, each containing M quantum wells filled with different kind of carriers. The wells are embedded in a medium with dielectric constant ε_o. The carriers are localized in the wells (flat minibands), and calculations are done for zero temperature. The subband wave-functions have been calculated using self-consistent band structure calculation [5, 7], and coupling to phonons is included.

The Raman intensity is proportional [3-6] to the imaginary part
of the density-density correlation function Π, which can be expanded
in the single-particle subband wave-functions [4, 6]. The RPA equation
for the expansion coefficients is solved by the method used by Jain and
Allen [3] and by Hawrylak et al. [4]. Here we keep the coupling between different intersubband excitations. The Raman intensity is proportional to [6]

$$F(\omega,\vec{Q}) = \iint dz dz' \, \text{Im}(-\Pi(\omega,\vec{q},z,z')) e^{-2ki(z-z')} e^{-(z+z')/\lambda} \tag{1}$$

where λ is the photon decay length ($\lambda = 6000\overset{o}{A}$), and $\vec{Q} = (\vec{q}, 2k)$ is the
momentum transfer to the plasmon from a photon. (See ref. 6 for de-
tails.)

In this paper, we apply the formalism to a type I (GaAs - GaAℓAs)
superlattice with two subbands occupied. The Raman intensities for two
systems, one studied experimentally by Olego et al. [1], and the other
by Sooryakumar et al. [2] are evaluated. Figure 1 shows the calculated intrasubband peak of system 1. The parameters used are repeat
distance a = 890$\overset{o}{A}$, well width L = 250$\overset{o}{A}$, qa = 0.43, n_s = 0.73 $\cdot$ 10^{12} cm^{-2}
(two subbands filled), 2ka = 4.92, N = 20. Jain and Allen showed that
this peak can be fit well by a model neglecting subband structure and
finite size effects. For smaller values of k (2ka = 0.5), the discrete
plasmon structure in the intensity is shown in fig 2. The agreement
between a self-consisten calculation and a 2d layered electron gas
model (dashed line) is again striking. This is a pure coincidence as
this agreement is due to intersubband scattering, which becomes important when more than one subband is occupied. It is illustrated by the
dotted line in fig 2, where intersubband scattering is neglected. One
observes a difference in the positions of the peaks, and in the number
of peaks. This is due to the fact that in diagonal approximation, the
1-1 intrasubband peak appears, but is shifted down below the single-
particle continuum by non-diagonal elements in the full calculation.
The 0-0 peak is at the same time shifted up. The result of these
effects is now in good but superfluous agreement with a naive LEG
theory.

Figure 3 shows the 0-1 intersubband peak of system 2 [2]. The
parameters are N = 30, a = 818$\overset{o}{A}$, L = 221$\overset{o}{A}$, n_s = 0.6 $\cdot$ 10^{12} cm^{-2},

2ka = 4.6, qa = 0.47. The peak is centered around 21,8 meV, while the experimental result is 22.1meV, which is a good agreement. We find that many-body effects reduce the depolarization shift by 1meV. Introducing intersubband scattering reduces the energy further by 0.4meV. This softening of the intersubband excitation is also an effect of having two subband occupied, since it is of order 1. For one subband filled, Das Sarma [7] has shown that intersubband scattering gives a softening $O(q^2)$.

In conclusion, we have calculated Raman intensities for type I superlattices with two subbands occupied, studied the many-body effects, and noticed the importance of intersubband scattering. The authors wish to acknowledge the support of the U.S. Army Research Office, Durham, North Carolina.

REFERENCES

1. D. Olego, A. Pinczuk, A. C. Gossard, W. Wiegmann, Phys. Rev. B25, 7867 (1982).
2. R. Sooryakumar, A. Pinczuk, A. Gossard, W. Wiegmann, Phys. Rev. B31, 2578 (1985).
3. J. K. Jain and P. B. Allen, Phys. Rev. Lett. 54, 947 (1984).
4. P. Hawrylak, J.-W. Wu, J. J. Quinn, Phys. Rev. B32, 5169 (1985).
5. S. Katayama and T. Ando, J. Phys. Soc. Jpn. 54, 1615 (1985).
6. P. Hawrylak, G. Eliasson, J. J. Quinn, Phys. Rev. B, in press.
7. T. Ando and S. Mori, J. Phys. Soc. Jpn. 47, 1518 (1979).
8. S. Das Sarma, Phys. Rev. B29, 2334 (1984).

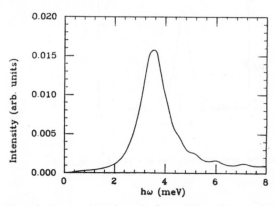

Fig. 1 Intrasubband peak for system 1. Parameter as in text.

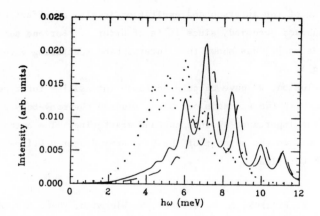

Fig. 2 Intrasubband peak for system 1. 2ka = 0.5. Solid
line is self-consistent calculation, dashed line layered
2DEG, dotted line diagonal approximation.

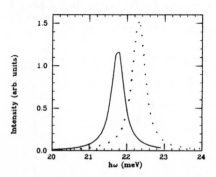

Fig. 3 Intersubband peak for system 2. Solid line self-
consistent calculation, dotted line diagonal
approximation.

A METHOD FOR CALCULATING ELECTRONIC STRUCTURES OF SEMI-CONDUCTOR SUPERLATTICES

H. M. Polatoglou, G. Kanellis, and G. Theodorou

Physics Department, Aristoteles University
Thessaloniki GR - 54006
GREECE

ABSTRACT

A similarity transformation of the superlattice Hamiltonian is presented. It is shown that the transformed Hamiltonian includes two parts. One part describes an average crystal, in terms of the primitive cells of the underlying simple structure, and the other the interaction between average crystal states. This interaction has the symmetry of the superlattice and is small, because it corresponds to differences of interactions of the original Hamiltonian. This method is applied to type-II superlattices.

1. INTRODUCTION

Considerable interest has been drawn on the electronic and other properties of the artificially grown superlattices (SL)[1], due to their physical and technological importance. The study of the electronic properties of SL's and other modulated structures with a large unit cell needs a great deal of computations [2-4]. In addition the interpretation of the results is not obvious. Recently Kanellis[5] has proposed a method for the calculation of the phonon spectra of modulated structures. The purpose of this work is to extend the method to the electronic properties problem of modulated structures. The method is applied to type-II superlattices. As a numerical example we treat the case of GaSb/InAs superlattices.

2. DESCTRIPTION OF THE METHOD

In a superlattice, which is build up from two binary compounds AC and BD the crystalline structure can be assumed to be the same as that of the constituent compounds. The structure of each compound is described in terms of identical primitive cells, each containing two atoms. With this structure a Brillouin Zone (BZ) is associated which will be refered to as 1st original BZ (1^{st}OBZ). The SL can be descri- bed in terms of a larger primitive cell, containing N_0 primitive cells of the constituent compounds. This larger primitive cell will be cal- led supercell. There are N_0 superlattice 1st BZ's (1st SBZ) which are contained in the 1st OBZ. For a wavevector k' lying in the 1st SBZ there are N_0 "equivalent" wavevectors k, lying in the 1st OBZ,

$$\bar{k} = \bar{g}_i + \bar{k}' \tag{1}$$

where $\bar{g}_i$ are reciprocal lattice vectors of the SL.

In the Linear Combination of Atomic Orbitals (LCAO) method, the basis for the SL wavefunctions (and the Hamiltonian) consists of the following Bloch sums,

$$\bar{y}_{k',n,j,\alpha} = \frac{1}{\sqrt{N_s}} \sum_m e^{i\bar{k}'\cdot(\bar{R}_m + \bar{r}_n + \bar{r}_j)} \varphi_\alpha(\bar{r} - \bar{R}_m - \bar{r}_n - \bar{r}_j), \tag{2}$$

where N_s is the number of supercells in the SL, $\bar{k}'$ is the   wavevector in the 1st SBZ, $\bar{R}_m$, $\bar{r}_n$, $\bar{r}_j$ are position vectors of supercells, pri- mitive cells inside the supercell and atoms inside the primitive cell, respectively, and $\varphi_\alpha(\bar{r} - \bar{R}_m - \bar{r}_n - \bar{r}_j)$ is the a symmetrically orthogona- lized atomic orbital centered at $\bar{R}_m + \bar{r}_n + \bar{r}_j$.

Instead of using the above ordinary basis, we choose the follow- ing one

$$\Psi_{k',g_i,j,\alpha} = \frac{1}{\sqrt{N_0 N_s}} \sum_{m,n} e^{i(\bar{k}' + \bar{g}_i)(\bar{R}_m + \bar{r}_n + \bar{r}_j)} \varphi_\alpha(\bar{r} - \bar{R}_m - \bar{r}_n - \bar{r}_j), \tag{3}$$

which will be shown to be more convinient. The above choice corresponds to the description of the SL in terms of the same primitive cells used for the constituent compounds and leads to the following picture: the subbands of the SL result from the interaction of the average crystal subbands. The symbol ε_{nj}^α denotes the on-site matrix element of an α orbital at position (n,j), and the symbol $t_{nj,s}^{\alpha\beta}$ denotes the Hamiltonian matrix element between an α orbital at site (n,j) and a β orbital at the s neighbor site. The Hamiltonian in the new basis can be considered as consisting of blocks labeled by pairs $(\bar{g}_j,\bar{g}_{j'})$. The elements of the diagonal block $(\bar{g}_j,\bar{g}_{j'})$ are given by,

$$<\bar{k}',\bar{g}_j,j,\alpha|H|\bar{k}',\bar{g}_j,j',\beta> = \delta_{\alpha\beta}\delta_{jj'}<\varepsilon_{nj}^\alpha>_n + \sum_s e^{-i(\bar{k}'+\bar{g}_j)\cdot\bar{r}_s}<t_{nj,s}>_n \qquad (4)$$

where the brackets $<>_n$ denote averaged values over the supercell.

It is evident from the above relation (4) that each diagonal block of the Hamiltonian matrix contains the averaged interactions inside the supercell and describes a subband of the average crystal, corresponding to the given SL.

For the elements of the off-diagonal blocks $(\bar{g}_j,\bar{g}_{j'})$ we get

$$<\bar{k}',\bar{g}_j,j,\alpha|H|\bar{k}',\bar{g}_{j'},j',\beta> = \delta_{\alpha\beta}\delta_{jj'}\frac{1}{N_0}\sum_n e^{-i(\bar{g}_j-\bar{g}_{j'})\cdot(\bar{r}_n+\bar{r}_j)}\delta\varepsilon_{nj}^\alpha +$$
$$+ \sum_s \frac{1}{N_0}\sum_n e^{-i(\bar{g}_j-\bar{g}_{j'})\cdot(\bar{r}_n+\bar{r}_j)} e^{-i(\bar{k}'+\bar{g}_j)\bar{r}_s}\delta t_{njs}^{\alpha\beta} \qquad (5)$$

where

$$\delta\varepsilon_{nj}^\alpha = <\varepsilon_{nj}^\alpha>_n - \varepsilon_{nj}^\alpha, \quad\text{and}\quad \delta t_{njs}^{\alpha\beta} = <t_{n,j,s}^{\alpha\beta}>_n - t_{n,j,s}^{\alpha\beta}$$

Relation (5) shows that the off-diagonal blocks depend only on the differences of the SL interactions from the average crystal values, which are usually small and can be treated by perturbation theory.

3.   ELECTRONIC ENERGY BANDS OF GaSb|InAs SL

In Fig. 1, the energy subbands of GaSb|InAs (100) SL, consisting of alternate layers each 8 atomic layers thick, are presented inside the 1st OBZ (solid lines), along with the average-crystal bands (dashed lines). A nearest neighbor approach is used within an sp^3s* basis[6]. Within the context of the theory presented above we can interpret the SL subbands as resulting, either from weak interactions between subbands showing strong dispersion (Δ_2' valence band and central part of Δ_5) or from strongly interacting subbands which appear in a small e-

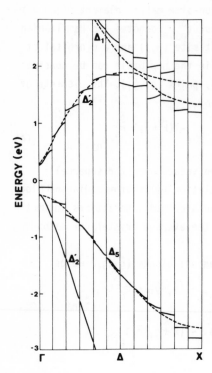

nergy range. These latter lead to
confined states or to the semicon-
ductor-semimetal transition. A more
detailed discussion will be given
elsewhere.

Fig. 1. Energy band structure of a
4 x 4 GaSb|InAs (100) SL
(solid lines), and average-
crystal (dashed lines).

4. REFERENCES

1] L. Esaki, in Proceeding of the 17th ICPS, edited by J.D. Chadi
and W.A. Harrison (Springer-Verlag, New York, 1985).

2] C. Tejedor, J.M. Calleja, F. Meseguer, E.E. Mendez, C.-A. Chang
and L. Esaki, Phys. Rev. B 32, 5303 (1985).

3] K.B. Wong, M. Jaros, M.A. Gell and D. Ninno, J. Phys. C 19, 53
(1986).

4] R.D. Graft, G.P. Parravicini and L. Resca, Solid State Commun.
57, 699 (1986).

5] G. Kanellis, Solid State Commun. 58, 93 (1986).

6] P. Vogl, H.P. Hjalmarson and J.D. Dow, J. Phys. Chem. Solids 44,
365 (1983).

QUASIPERIODIC SEMICONDUCTOR SUPERLATTICES

R. Merlin, K. Bajema, R. Clarke and J. Todd

Department of Physics, The University of Michigan
Ann Arbor, MI 48109-1120, U.S.A.

ABSTRACT

A brief overview of the unique structural, electronic and vibrational properties of quasiperiodic (Fibonacci) superlattices is presented, together with results of X-ray, Raman scattering and optical absorption experiments on the $GaAs-Al_xGa_{1-x}As$ system.

1. INTRODUCTION

It has long been recognized that quasiperiodic (incommensurate) superlattices could offer interesting possibilities for experimental studies of novel physical phenomena.[1] The motivation for this was, in a large part, theoretical work on quasiperiodic one-dimensional (1D) wave equations revealing spectra and eigenstates that are quite unlike those of periodic or random 1D systems.[2] The major problem in fabricating quasiperiodic structures has been the fact that simple incommensurate modulations require increasingly larger layer thicknesses to approach the irrational limit. Recently, it has been shown that layer deposition in sequences generated by special production rules provide a solution to this problem.[3] Superlattices grown according to these sequences show a degree of quasi-periodicity that is determined not by the width of individual layers (which is arbitrary), but by the thickness of the sample.[3] So far, most studies[3-9] have focussed on the class of superlattices derived from the Fibonacci sequence. Their properties and results of experiments on Fibonacci $GaAs-Al_xGa_{1-x}As$ heterostructures are discussed in this paper.

2. STRUCTURAL PROPERTIES AND X-RAY SCATTERING

The Fibonacci sequence $ABAABABAA...$ is defined by the production rules $A \rightarrow AB$ and $B \rightarrow A$, with A as the single element of the first generation. To obtain a Fibonacci superlattice, one simply replaces A and B by two arbitrary blocks of layers.[3,7] This arrangement leads to a structure which shows two basic reciprocal periods in a ratio given by the golden mean $\tau=(1+\sqrt{5})/2$, perpendicular to the layers.[3] The associated structure factor consists of a dense set of components such that diffraction peaks are expected at wavevectors given by $k=2\pi d^{-1}(m+n\tau)$, where m and n are integers and $d=\tau d_A+d_B$; d_A and d_B are the thicknesses of the building blocks A and B.[3]

X-ray diffraction patterns have been obtained[4] from a Fibonacci superlattice consisting of $A \equiv [17\text{Å } AlAs-42\text{Å } GaAs]$ and $B \equiv [17\text{Å } AlAs-20\text{Å } GaAs]$; the sample was grown by molecular beam epitaxy on a (001) $GaAs$ substrate. The patterns for $\vec{k}$ parallel to [001], in Fig. 1, demonstrate many

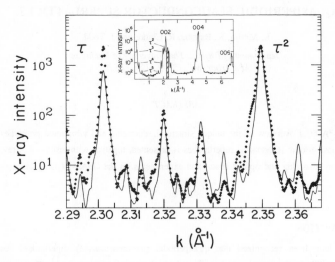

Fig. 1:   Synchrotron X-ray data (dots) and calculated diffraction profile for the ideal Fibonacci struc-
          ture (solid line); the sample is described in Section 2. The inset shows indexing of strong
          peaks in a low-resolution scan.

of the unusual properties of quasiperiodic ordering. The synchrotron data indicate that, at least up to
the instrumental resolution ($\approx$0.0015Å FWHM), the diffraction peaks do indeed form a dense set.
Moreover, the measurements agree remarkably well with the calculated profile for an ideal Fibonacci
structure (solid curve in Fig. 1).[4] This and also numerical simulations[4] show that quasiperiodic ordering is largely insensitive to the unavoidable random fluctuations in the growth parameters. A
further interesting effect is observed in the extended low-resolution scan of the inset; the dominant
superlattice reflections and satellites can be expressed as $k = 2\pi d^{-1}\tau^p$, with integer p. This behavior
results from d_A / d_B being close to τ in our sample.[4] It is not a general feature of Fibonacci superlattices.

3. ELECTRONIC AND OPTICAL PROPERTIES

Tight-binding models describing electrons in Fibonacci structures have been studied by many
authors.[6-9] The spectrum is a Cantor set characterized by clusters of eigenvalues that divide into
three subclusters.[6-9] The wavefunctions are critical (i.e., neither localized nor extended) exhibiting
either self-similar or chaotic behavior.[8,9] This applies only to bulk states; finite samples can further
show solutions localized at the surfaces.[8]

Investigations of the electronic structure of Fibonacci $GaAs-Al_x Ga_{1-x} As$ superlattices using
standard optical probes reveal mainly excitonic features. The example of a superlattice with $A \equiv [20Å$
$Al_{0.3}Ga_{0.7}As - 40Å\ GaAs]$ and $B \equiv [20Å\ Al_{0.3}Ga_{0.7}As - 20Å\ GaAs]$ is given in Fig. 2(a). Fig. 2(b)
shows results of effective-mass calculations of the electron, heavy-hole and light-hole spectra of this

sample, for motion normal to the layers. These results were used to further determine the Im(χ) vs photon-energy plots shown in Fig. 2(a). The comparison with the experimental data indicates a correlation between the positions of the exciton peaks and the largest calculated plateaus; the latter reflect major gaps in the 1D spectrum of Fig. 2(b). Since surface states may also occur at these gaps, it is not clear whether the excitons derive from bulk critical states or from quasi-2D states localized at the surface.

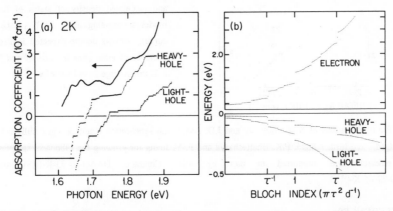

Fig. 2: (a) Optical absorption coefficient and calculated Im(χ) for transitions involving heavy- and light-hole states, in arbitrary units. (b) Energy vs Bloch index. Superlattice parameters are indicated in Section 3.

4. RAMAN SCATTERING BY ACOUSTIC PHONONS AND VIBRATIONAL PROPERTIES

The spectrum of phonons in 1D Fibonacci lattices shows a self-similar hierarchy of gaps which decrease in size with the phonon frequency.[6,8,9] As expected, the eigenfunctions are extended in the continuum limit. Their high-frequency behavior is not as yet well understood although there is some evidence favoring localization.[6] If this turns out to be correct, a transition between extended and localized states may take place at intermediate frequencies.

Raman scattering has been used to study longitudinal acoustic (LA) phonons propagating parallel to the growth axis in Fibonacci $GaAs-Al_xGa_{1-x}As$ superlattices.[5] The results reveal important differences between resonant and non-resonant spectra. For the latter, the scattering is largely determined by structural effects as in the case of periodic superlattices.[10] Off-resonance data from the sample described in Section 2 show doublets centered at frequencies that follow a τ^p-behavior,[5] consistent with the X-ray findings. The Raman spectrum of the same sample obtained under resonant conditions is shown in Fig. 3. The scattering reflects now a weighted density of states of LA modes, providing an experimental demonstration of the richness of the phonon spectrum. Resonances with electronic states localized at the surface can possibly account for this behavior.[5]

678   R. Merlin et al.

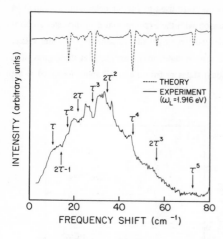

Fig. 3: Room temperature resonant Raman spectrum of the superlattice considered in
Section 4 corrected for thermal factors,
and calculated density of states of LA
modes propagating along [001] (dashed
curve). Arrows denote expected midfre-
quencies of main gaps in units of $\pi c d^{-1}$;
c is the average sound velocity.

The contributions of K.M. Mohanty and J.D. Axe to the synchrotron results are gratefully ack-
nowledged. We further thank P.K. Bhattacharya and F.-Y. Juang for growing the Fibonacci samples.
The research was supported in part by ARO Contracts DAAG-29-83-K-0131 and
DAAG-29-85-K-0175.

5. REFERENCES

1.   Sokoloff J.B., Phys. Rev. B **22**, 5823 (1980); DasSarma S., Kobayashi A. and Prange R.E.,
     Phys. Rev. Lett. **56**, 1280 (1986).

2.   See, e.g., Ostlund S. and Pandit R., Phys. Rev. B **29**, 1394 (1984).

3.   Merlin R., Bajema K., Clarke R., Juang F.-Y. and Bhattacharya P.K., Phys. Rev. Lett. **55**, 1768
     (1985).

4.   Todd, J., Merlin R., Clarke, R., Mohanty K.M. and Axe J.D., to be published.

5.   Bajema K. and Merlin R., to be published.

6.   Lu J.P., Odagaki T. and Birman J.L., Phys. Rev. B **33**, 4809 (1986).

7.   Odagaki T. and Friedman L., Solid State Commun. **57**, 915 (1986).

8.   Nori F. and Rodriguez J.P., to be published.

9.   Kohmoto M. and Banavar J.R., to be published; see also Kohmoto M., Phys. Rev. Lett. **51**,
     1198 (1983).

10. See, e.g., Colvard C., Gant T. A., Klein M. V., Merlin R., Fischer R., Morkoc H. and Gossard
     A. C., Phys. Rev. B **31**, 2080, (1985).

RAMAN MEASUREMENTS OF COUPLED LAYER PLASMONS, SINGLE PARTICLE EXCITATIONS AND SCATTERING TIMES IN THE LAYERED 2D ELECTRON GAS OF MODULATION DOPED GaAs/AlGaAs MULTIQUANTUM WELLS.

G. Fasol, N. Mestres, A. Fischer and K. Ploog.

Max-Planck-Institut, Heisenbergstr. 1, D-7000 Stuttgart 80

ABSTRACT

We present light scattering measurements of the excitations of the layered 2D electron gas parallel to the
layers. The single particle excitation spectra show
narrow bands, peaking just below $hk_\perp v_F$, typical for 2D
systems. They are described by the 2D Lindhard-Mermin dielectric function and we deduce the single particle relaxation time. We present the dispersion of the coupled
layer plasmon eigenmodes, which manifest the Coulomb-
interaction between the different layers.

We report in this paper Raman measurements of the excitations
associated with the degrees of freedom parallel to the layers of the
2D electron gas in GaAs/AlGaAs modulation doped quantum wells (MQW)
and signals which we attribute to coupling with the perpendicular
degree of freedom. The present work therefore links optical and transport measurements.

We have performed the measurements on a series of sample structures, grown by MBE specially for these measurements. The basic period of the structures is shown in an insert of Figure 1. The quantum
wells are intentionally asymmetric and show very high electron mobility (up to 4×10^5 cm^2/Vs) presumably due to the quality of the
binary AlAs/GaAs interface. From Shubnikov - de Haas and from Hall
effect measurements we deduce that only a single sheet of electrons
in each well of our MQWs contributes. SdH and Hall measurement yield
the same value for the density of carriers per well. The carrier
density is 6.8×10^{11} cm^{-2} per layer in case of sample 4849 presented
here. The insert of Figure 2b describes the geometry of the measurement.

The dielectric response of the 2D electron gas[1] and the layered 2D electron gas has been studied theoretically in a large number of papers, which mostly concentrated on plasmons[2]. Plasmons in the 2D and layered 2D electron gas have been previously studied experimentally by infrared[3] and Raman techniques[4].

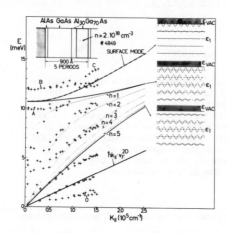

Fig. 1. Energy of Raman peaks as a function of $k_\parallel$ for five electron gas layers. The thin lines are plasmon layer eigenmodes (labelled 1 to 5), calculated without fitting parameter. Inserts show basic period of sample structure and a snapshot of the modulation of the carrier density for three of the eigenmodes.

The plasmon in a single 2D electron gas disperses as $k_\parallel^{\frac{1}{2}}$. In a layered electron gas the motion of electrons in the different layers becomes correlated due to the Coulomb-interaction between the electrons on different layers. Therefore the plasmon dispersion of a system of n electron layers fans out into n branches corresponding to n discrete coupled layer plasmon modes[5-6]. The thin lines in Figure 1 show a calculation of the $k_\parallel$ dispersion of these eigenmodes (labelled n = 1 to n = 5) for the case of a particular sample of five layers, in the approximation of infinitely thin electron layers. Figure 2a shows inelastic light scattering spectra, exhibiting peaks labelled 1 to 5, corresponding to these plasmon eigenmodes. As seen in Fig. 1, where the experimental peak positions are compared with theory, the theoretical curve deviates considerably from the experimental behaviour of the mode with lowest energy (n = 5), indicating the limits of the simplifying theory[5]. We found this deviation consistently in many different structures. In addition we observe peaks (A, B, C) attributed to coupling of the layer plasmons with <u>intersubband</u> excitations.

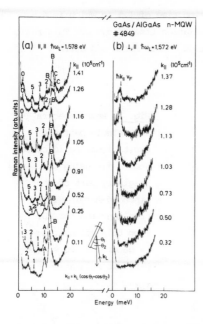

Fig. 2.

a) Raman spectra from layer plasmon modes

b) Raman spectra from single particle excitations.

In crossed polarisation we obtain Raman spectra (Fig 2b and Fig. 3), which are proportional to $\text{Im}(\chi(k_\parallel, \omega))$ and therefore to the single particle excitation spectra. The crosses in Fig. 1 mark the measured values for the peaks of the single-particle spectra. Their dispersion is well explained by $hk_\parallel v_F^{2D}$ (dashed line).

The single particle excitations are the elementary excitations of the 2D electron gas, where an electron is lifted from a state within the Fermi circle to a state outside, leaving a hole behind. Each single particle excitation is characterized by a wavevector $k_\parallel$ and an energy ω. Their spectra can be described approximately by the imaginary part of the Lindhard-Mermin[7] polarisability applied to the 2D electron gas, which we have calculated for the appropriate electron temperature using the relaxation time τ_s as a fitting parameter. It is crucial for this experiment to have accurate control of the temperature. We have stabilised the temperature, monitoring with a calibrated Ge resistor and measured the electron temperature with the exponential high energy tail of the luminescence. The laser power was reduced until the heating was minimal. For sample 4849 at 12 K we

682   G. Fasol et al.

determine a single particle relaxation time of 1.5 ps which is 4.5
times shorter than the mobility relaxation time of 6.4 ps. This ratio
confirms almost exactly the prediction of Das Sarma and Stern[8].

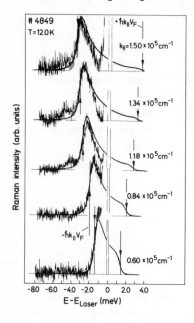

Fig. 3. Raman spectra
from single particle
excitations. Solid curves
show a fit using the
imaginary part of the
Lindhard-Mermin dielec-
tric function for 2D.
From this calculation
we determine the single
particle relaxation time
τ_s.

REFERENCES

1.   F. Stern, Phys.Rev.Lett. 18, 546 (1967)
2.   See for example:
     S. Das Sarma and J.J. Quinn, Phys.Rev. B25, 7603 (1982);
     J.K. Jain and P.B. Allen, Phys.Rev. B32, 997 (1985),
     and references therein.
3.   For a review see:
     D. Heitmann, Surface Science 170, 332 (1986)
4.   D. Olego, A. Pinczuk, A.C. Gossard and W. Wiegmann,
     Phys.Rev. B26, 7867 (1982);
     R. Sooryakumar, A. Pinczuk, A. Gossard and W. Wiegmann,
     Phys.Rev. B31, 2578 (1985);
     A. Pinczuk, M.G. Lamont and A.C. Gossard, Phys.Rev. Lett. 26,
     2092 (1986);
     G. Fasol, H.P. Hughes and K. Ploog, Surf.Science 170, 497 (1986)
5   J.K. Jain and P.B. Allen, Phys.Rev.Lett. 54, 2437 (1985)
6.   G. Fasol, N. Mestres, H.P. Hughes, A. Fischer and K. Ploog,
     Phys.Rev.Lett 56, 2517 (1986)
7.   N.D. Mermin, Phys.Rev. B1, 2362 (1970)
8.   S. Das Sarma and F. Stern, Phys.Rev. B32, 8442 (1985)

DETERMINATION OF CONFINED STATES AND BAND GAP OFFSET FROM
EXCITATION SPECTRA OF RESONANT RAMAN AND LUMINESCENCE
IN p-TYPE GaAs MULTI-QUANTUM WELLS

T. Suemoto[+], G. Fasol, and K. Ploog

Max-Planck-Institut für Festkörperforschung
Heisenbergstrasse 1, 7000 Stuttgart 80,
Federal Republic of Germany

ABSTRACT

In asymmetric p-type GaAs quantum wells, we have found
a series of sharp peaks in the resonance profiles of the
LO phonon and hole excitation bands. These structures
are assigned to transitions from the uppermost hole state
to the conduction band well states.

Resonant Raman spectroscopy in p-type multi-quantum wells (MQW)
has so far been directed toward the investigation of their valence band
structure.[1] In this report we show that p-type asymmetric MQW's are
ideally suited for detailed study of the conduction band well states by
resonant Raman spectroscopy.

The structure of one period of our MBE-grown MQW sample is shown
in the inset (a) of Fig 1. The structure was made deliberately asymmetric by putting an AlAs layer on one side of the well, and the
middle of the $Al_xGa_{1-x}As$ (x=0.33) layer was doped with acceptors (Be).
This sample structure has two main features; firstly, it has high hole
mobility (typically 20,000 cm^2/Vs) owing to the good binary/binary
AlAs/GaAs interface of the well. Secondly, the potential has an asymmetric shape. The carrier concentration for the sample discussed in
this paper has been determined to be 1.7 x 10^{11} cm^{-2}/layer by Hall
measurement at 77 K.

The Raman spectra were taken with parallel polarization in nearly
back scattering geometry with the sample held near 4 K in a He gas
atmosphere. We used an optical multichannel analyzer combined with a
triple grating monochromator to measure the Raman spectra, which
allowed us to take numerous spectra as a function of laser frequency in
a reasonable amount of time (typically 1-2 min/spectrum).

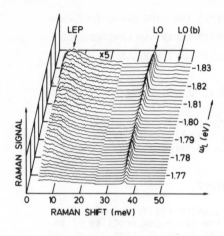

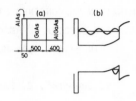

Fig. 1 Raman spectra at different laser frequencies (ω_L). The inserts show (a) the structure of one period of the MQW, and (b) the expected shape of the potential and wavefunctions in one period of the MQW.

Figure 1 shows a set of spectra taken with various laser frequencies ω_L. The peak at 36 meV corresponds to the LO phonon of the GaAs layer. The small peak at 45 meV which is most prominent for ω_L=1.825 eV corresponds to the AlAs-like LO phonon in the AlGaAs layer (LO(b)). The broad tail of the Rayleigh peak with a small bump at around 8 meV is characteristic of p-type samples; in n-type MQW's with a similar structure, we do not observe such a feature in the laser energy range investigated. We therefore assign this band to a hole transition in the valence subbands. Resonance profiles of these three features are shown in Fig. 2. The areas of the peaks above the background are plotted for the LO and the LO(b) phonons while an integration between 6-10 meV is plotted for the low energy peak (LEP). The resonance profiles of the LO phonon and the LEP consist of a series of peaks. Their positions coincide if we shift the LO curve toward lower energy by 33 meV, which is close to the LO phonon energy 36 meV. This suggests that the LO phonon resonance occurs when the energy of the scattered light is equal to that of the resonant transition. If we assume an ingoing resonance for the LEP, the difference in the positions would be 36 meV, while for an outgoing resonance it would be 28 meV. Therefore the observed value 33 meV can be understood by assuming an averaged position of ingoing and outgoing light for the LEP.

One of the important features of this LEP band is the independence of the spectrum on the incident energy, unlike the electron excitation bands in n-type MQW's.[2] This suggests that the Raman process causing the LEP corresponds always to the same pair of valence subband states regardless of the incident photon energy. The most probable resonance process is shown in the inset of Fig. 2(A) and (C) for the LO peak and the LEP, respectively. In terms of these processes, the positions of the ingoing resonance should correspond to those of the conduction subbands measured from the <u>uppermost hole state</u>. The vertical solid bars in Fig. 2(A) and (C) show the results of a calculation for energy levels in a square well potential; the dashed bars correspond to expected peak positions for outgoing resonance. Here we assumed 1.5177 eV for the AgAs band gap, and 268 meV for the conduction band offset between GaAs and AlGaAs layers.[3]

In the energy region where strong luminescence made measurement of the Raman scattering difficult, we used luminescence excitation spectoscopy to find the positions of the subband levels (Fig. 3).

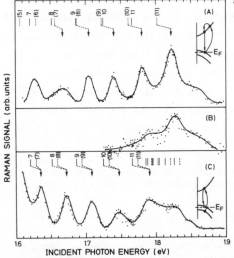

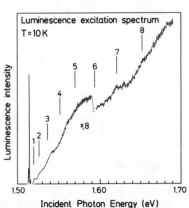

Fig. 2 Resonant Raman profiles for (A) the LO phonon of GaAs, (B) LO(b)-phonon of AlGaAs, and (C) the hole excitation. The inserts show the relevant resonance processes.

Fig. 3 Excitation spectrum of the luminescence that appears at 1.514 eV.

Combining the resonance profile and the luminescence excitation spectrum, we can number the levels up to n = 11. Agreement with the calculation is reasonably good, even without adjusting the well width in the calculation.

The fact that we observe resonances between the uppermost hole state and all conduction subbands in the relevant energy range shows a breakdown of the selection rule typical for symmetric wells, which allows only transitions combining the same n. Even the more general selection rule Δn = even which applies to any symmetric well is broken. This means that the system is really asymmetric as claimed. This situation can be understood, if one assumes that the hole is confined to the triangular part of the well as shown in the inset (b) of Fig.1.

The termination of the series of peaks in the LO resonance profile at n = 11 (Fig.2(A)) suggests that this is the uppermost state confined in the GaAs layer. For the LO(b) resonance, a peak appears only for n = 11 as seen in Fig.2(B). This means that the wavefunctions of the level n = 11 have appreciable amplitude in the AlGaAs layer, which can efficiently couple this electronic state to the phonons of AlGaAs. This supports our above interpretation that n = 11 is the uppermost bound state. The well depth measured from the n = 1 level should thus be in the range 264-312 meV, where the upper limit corresponds to the condition of having no 12th bound state. For zero band bending, this well depth range gives experimental bounds 65-76% for the conduction band offset, or it gives 57-69 % with an estimated upper limit of the band bending of 30meV.

[+] Alexander von Humboldt Research Fellow. On leave from Research Institute for Scientific Measurements Tohoku University, Katahira Sendai 980 Japan

REFERENCES

1) A. Pinczuk, H.L. Störmer, A.C. Gossard, and W. Wiegmann, Proc. of the 17th Int. Conf. on the Physics of Semiconductors, eds. J.D. Chadi and W. Harrison (Springer Verlag, New York. 1985), p.329.
2) A.Pinczuk, H.L.Störmer, R.Dingle, J.M.Worlock, W.Wiegmann, and A.C. Gossard, Solid State Commun. 32, 1001 (1979).
3) H.Kroemer, Proceedings of MSS-II (Kyoto 1985) to be published in Surface Science 1986.

PRESSURE-LUMINESCENCE BEHAVIOUR IN MODULATION-DOPED AlGaAs-GaAs MULTIPLE QUANTUM WELLS

C.H. Perry[*†]

Physik-Department E 16, Technische Universität München,
D-8046 Garching, FRG

B.A. Weinstein, S.K. Hark, and C. Mailhiot
Xerox Webster Research Labs., Rochester, NY 14644, U.S.A.

ABSTRACT

Photoluminescence spectroscopy has been used to monitor
and analyze the pressure-induced metal-insulator transition
in modulation doped MQW heterostructures. The pressure
shift is influenced by band-bending changes. Band crossings
at high pressures are evident from the intensity data.

1. INTRODUCTION

Photoluminescence (PL) is an established technique for the study
of 2D electron systems in semiconductors as it is a powerful means of
band structure determination. In the case of n-type modulation (Si)
doped multiple quantum well (MQW) heterostructures it has been applied
principally to the investigation of interband electron-hole (band-to-
band, BB) recombination processes and of transitions between conduction
band and neutral acceptor states (BA)[1]. We have extended the technique
by the application of hydrostatic pressures up to 50 kbars at low temperatures. From our studies several significant effects are observed:
Analysis of the main BB luminescence line-shape, shift and intensity
indicate a pressure-induced metal-insulator phase transition. This
occurs at $\sim$ 9 kbar et low temperatures in an $Al_{.21}Ga_{.79}As$-GaAs MQW
with an electron density $\sigma = 3 \times 10^{11}$ cm^{-2}.

---

*) Permanent address: Physics Department, Northeastern University,
Boston, Ma 02115, U.S.A.

†) Supported by NSF grant DMR 8121702

The effect is caused by a metastable deep level associated with
the Si impurities located at the L point[2] which can be driven out of
the gap by pressure[3].

The results also provide information on the change in the band-
bending E_b and the carrier concentration σ with pressure. For P > 9 kbar
in the intrinsic region we observe evidence in both the BB and BA intensity data for band-crossings associated with the AlGaAs and the
GaAs layers.

2.   EXPERIMENTAL

The sample was liquid argon loaded into a diamond anvil cell incorporated into the high-pressure-variable temperature cryostat[4]. High
pressures up to 50 kbars were sequentially applied at 180 $\pm$ 5 K,
where the PL spectrum was recorded prior to cooling to 13 $\pm$ 1 K. This
pressure-temperature cycle was repeated so that a hydrostatic regime
was obtained for the PL measurements at low temperatures.

3.   RESULTS AND DISCUSSION

Above $\sim$ 100 K, both $BB(E_0)$ and $BB(E_1)$ were observed. At 180 K
each shifts with pressure with $dE/dp = 9.7 \times 10^{-6}$ eV/bar; the linewidth remains constant. At $\sim$ 35 kbars $BB(E_1)$ disappears and $BB(E_0)$
diminishes in intensity above 40 kbars. This behaviour is the result
of the different signs of the pressure coefficients of E_G^Γ and $X_1^c - \Gamma_8^v$
gaps[2]. A cross-over occurs at $\sim$ 40 kbars causing the direct-gap
luminescence to rapidly decrease[6].

Cooling the sample to 13 $\pm$ 1 K after application of pressure
provided several new features. A dramatic decrease in the main $BB(E_0)$
line-shape in the p < 10 kbar range was observed. From the high energy
tail ($exp(-E/kT_e$ dependence) we deduced that the effective electron
temperature $T_e \overset{\sim}{=} 60$ K. E_F was determined from the PL spectrum as a
function of pressure using the analysis described in Refs. 5) and 6).
The results are shown in Figs. 1) and 2). It can be seen that the
slope dE_F/dp follows that observed from Hall studies[3]. The plot in
Fig. 2) was rescaled to obtain σ vs. p. These results are consistent
with the explanation put forward by Mercy et al.[3], namely that the
Si donors in the AlGaAs provide deep traps which can be activated by

the application of high pressure. From the measurements of Saxena[2]
we deduce that this impurity level is tied to the L point for x = 0.21.
At this concentration we estimate that the deep state lies $\sim$ 205 meV
below the L point. This gives the impurity binding energy E_I to be
$\sim$ 55 meV with respect to the Γ point. The pressure shifts are influenced by band-bending changes resulting in the non-linear behaviour in
the 0 - 9 kbar region.

To estimate the band-bending, E_b at p = 0 we used the results of
Batey et al.[7] to obtain ΔE_c for our sample; Hartree calculations incorporating the known sample parameters provided E_0, the lowest quantum
level transition. E_b is given by $\Delta E_c - (E_F + E_0 + E_I)$. Using $E_F \simeq 11$ meV;
$E_0 \simeq 7$ meV; $E_I \simeq 55$ meV and $\Delta E_c \simeq 147$ meV, we obtain $E_b \simeq 75$ meV at
1 bar. E_b goes to zero at $\sim$ 9 kbars.

The BA transition at $\sim$ 30 meV below BB mirrors the BB behaviour
for p < 9 kbars, but its intensity decreases and it cannot be observed
above $\sim$ 18 kbars (see Fig. 3). Its disappearance occurs in the region
of the $\Gamma \rightarrow X$ cross-over in $Al_{.21}Ga_{.79}As$. In contrast the BB transition
for p > 9 kbars shifts at 9.9 meV/kbar, its width remains constant and
its intensity increases to a maximum at $\sim$ 25 kbars. Thereafter it slowly decreases but remains observable to at least 5 kbars above the GaAs
$\Gamma \rightarrow X$ cross-over at $\sim$ 40 kbars.

ACKNOWLEDGEMENTS

We wish to thank A.C. Gossard and W. Wiegmann for the
MQW samples and their continued interest in this project.

REFERENCES
1) Perry, C.H., Petrou, A., Smith, M.C., Worlock, J.M., and Aggarwal,
   R.L., J.Luminescence 31 & 32, 491 (1984)
2) Saxena, A.K., J. Phys. C. 13, 4323 (1980)
3) Mercy, J.M., Bousquet, C., Robert, J.L., Raymond, A., Gregoris,
   G., Beerens, J., Portal, J.C., Frijlink, P.M., Delescluse, P.,
   Chevrier, J., and Linh, N.T., Surf. Sci. 142, 298 (1984)
4) Weinstein, B.A., Phil. Mag. B 50, 709 (1984)
5) Vilkotskii, V.A., Domanevskii, D.S., Kakanakov, R.D., Krasovskii,
   V.V., and Tkachev, V.D., phys. stat. sol.(b) 91, 71 (1979)
6) Olego, D., Cardona, M., and Muller, H., Phys. Rev. B 22, 894(1980)
7) Batey, J., Wright, S.L., and Dimaria, D.J., J. Appl.Phys. 57,
   484 (1985)

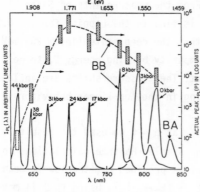

Fig. 1: PL spectrum of Al.21Ga.79As-GaAs MQW as a function of pressure (T_L = 13 K; $T_e \simeq$ 60 K) for the BB and BA transitions. Note the decreasing line-width for p < 10 kbars. The dashed curve shows the BB intensity (right-hand scale).

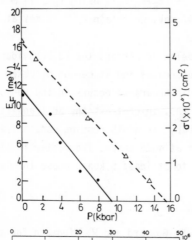

Fig. 2: E_F and σ show a linear decrease with increasing pressure. The slope of the PL measurements (circles) is similar to Hall data (triangles) taken from Ref. 3). The metal-insulator transition occurs at ∿ 9 and ∿ 15 kbars, respectively.

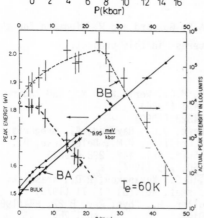

Fig. 3: The energy shift in the BB and BA transitions with pressure (solid curves). Note the non-linear behaviour for p < 9 kbars. The dashed curves show the relative intensities (right-hand scale). The Γ → X cross-over is ∿ 40 kbars in GaAs and ∿ 15 kbars in Al.21Ga.79As.

EXCITON LOCALIZATION IN GaAs/AlGaAs MULTIPLE QUANTUM WELL STRUCTURES

H. Stolz, D. Schwarze and W. von der Osten
Universität-GH-Paderborn
D-4790 Paderborn, W. Germany

G. Weimann
Forschungsinstitut der Deutschen Bundespost
D-6100 Darmstadt, W. Germany

ABSTRACT

We report stationary and picosecond time-resolved
luminescence measurements under resonant excitation
and excitation spectroscopy of bound excitons to
probe the exciton mobility in MQW structures. The
experiments show the coexistence of mobile and localized excitons across the exciton absorption band.
The results can be explained by assuming a local
mobility edge in regions of a few exciton radii and
efficient exciton migration between them.

1. INTRODUCTION

In multiple quantum well (MQW) structures the dynamics of excitons is strongly affected by their two-dimensional character and inhomogeneous broadening of states due to fluctuations in quantum well
width. Recently evidence was found [1,2] for a mobility edge (ME) i.e.
a transition from localized to mobile exciton states in going across
the exciton absorption band. In this contribution, we present measurements of resonantly excited luminescence of intrinsic (X) and impurity
bound excitons (BX) in MQW structures, which confirm the existence of
a ME, but provide detailed information on its nature. These measurements together with results from picosecond time-resolved luminescence
give a consistent picture of the exciton states and exciton energy
relaxation and localization.

2. RESONANT EXCITATION OF INTRINSIC EXCITON STATES

Our MBE grown samples vary in GaAs layer thickness from $L_z = 28$ Å
to 115 Å with barriers thick enough to prevent tunneling between
adjacent layers. The results are qualitatively the same for all samples, so we concentrate on the most intensely studied one characterized in fig. 1. The figure shows luminescence spectra excited at high
excitation energies and in resonance with the (n=1, e-hh) exciton
state. Due to the high sample quality the broadening of the exciton is

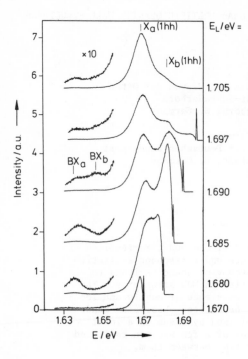

Fig. 1. Luminescence spectra of a
GaAs/Al$_x$Ga$_{1-x}$As MQW structure
(x = 0.45; L$_z$ = 45 Å; L$_{barrier}$ =
112 Å; 30 periods) excited with
different laser photon energies E$_L$
as indicated. Absorption peak at
1.687 eV is energetically shifted
compared to excitation spectrum
shown in Fig. 2, due to a different
excitation spot on the sample.
Excitation power density 50 mW/cm^2.
T = 1.8K.

sufficiently small for transitions corresponding to changes in well width of one atomic layer to be resolved. Besides the intrinsic exciton transitions (X$_a$, X$_b$) we observe luminescence due to impurity bound excitons (BX$_a$, BX$_b$, localization energy 40 meV). Under resonant excitation, the X$_a$, X$_b$ exciton states can be excited independently (excitation energy E$_L$= 1.697...1.670 eV) as seen by the strong increase in intensity of the corresponding luminescence line. In tuning through the exciton absorption, spectral narrowing of the luminescence lines occurs for each exciton, while at the same time the corresponding BX lines vanish (E$_L$ = 1.685 and 1.670 eV). This suggests the existence of a ME for each exciton near the centre of its absorption line. While excitons excited above are able to migrate and are trapped at the impurity states, this is not the case for excitation below the ME. Assuming exciton-phonon scattering (fast) and hopping transfer (slow) to be the dominant energy relaxation processes [3] above and below the ME, one would expect drastic differences in the time behaviour of luminescence as function of exciton energy.

3.   PICOSECOND TIME-RESOLVED LUMINESCENCE

For these measurements the sample was excited resonantly with optical pulses from a synchronously pumped dye laser (pulse duration 10 ps, pulse energy at the sample 10 pJ). The luminescence was analyzed by a high resolution double monochromator and a synchroscan streakcamera with combined spectral and time resolutions of 0.2 meV and 30 ps. The error in determining the various time constants by a least square fitting procedure is estimated to be ± 8.5 ps. The temporal dependence measured at various energies across the exciton luminescence band, was analyzed in terms of a 3-level model involving excitation followed by relaxation (time τ$_R$) and decay (time τ).

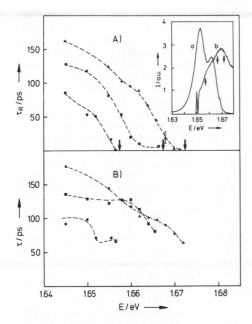

In fig. 2 results are shown for different E_L (arrows) selected above and below the ME for each exciton state. We observe only slight variations of τ with an average value of $\sim$130 ps. τ_R shows an almost linear increase to lower energies independent of E_L. The energy loss rate obtained is 0.09 meV/ps in good agreement with the theoretical expectation for LA-phonon scattering [3]. This implies that the dominant energy loss mechanism of excitons in MQW structures is phonon scattering, irrespective of excitation. From this it may be concluded that the ME, although existing, is not sharp with mobile and localized excitons coexisting across the absorption band.

Fig. 2. Energy dependence of relaxation and decay times (τ_R, τ) obtained from analyzing the time behaviour of luminescence. Inset shows excitation photon energies (arrows) relative to luminescence (a) and excitation (b) spectra.

4. BOUND EXCITON LUMINESCENCE AND EXCITON RELAXATION

Further evidence for this behaviour comes from the BX luminescence excited via the intrinsic exciton absorption. In these measurements only one exciton state dominates due to a selected excitation spot at the sample (absorption peak at 1.67 eV). As shown in fig. 3B, even for excitation below the absorption peak BX luminescence occurs, implying mobile excitons. The shift in peak position and variations in lineshape which occur by changing E_L reflect the relaxation of the excitons before trapping. To describe this process we have developed a simple model of relaxation and localization based on the following assumptions:

- For the potential fluctuations causing the broadening of the exciton state two different correlation lengths are important. One is smaller than the exciton radius giving rise to spatial fluctuations of exciton energy on an atomic scale, characterized by a distribution function [4] $g_X(E)$. The other is larger than the exciton radius providing islands in which the exciton is mobile. The combined effect results in a different ME in each of these islands (distribution function $f_{ME}(E)$)

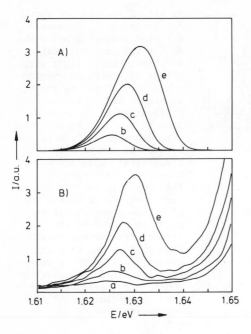

leading to the coexistence of mobile and localized excitons at the same energy.
- Due to their high mobility[1] for excitons above the local ME the diffusion length during the exciton lifetime is in the order of 1 μm. This allows diffusion into neighbouring islands and subsequent energy relaxation or trapping, characterized by a transfer probability $f_T(E)$.
- For excitons below the local ME we neglect hopping transfer during their lifetime.
The curves in fig. 3A are the result of a calculation of the BX lineshape, assuming Gaussian energy distributions for g_X and f_{ME} (spread 5 meV each) and constant f_T, an assumption justified by the statistical nature of the fluctuations. The good agreement of the fit proves the validity of the presented model to describe energy relaxation and localization of excitons in MQW structures. The model, presuming alternating stages of exciton diffusion and relaxation, in particular would explain in a straightforward way the slow energy relaxation and its dependence on well thickness reported previously [3].

Fig. 3. Bound exciton luminescence excited at various energies across intrinsic exciton absorption
(a: 1.664, b: 1.666, c: 1.668, d: 1.670, e: 1.675 eV).
A) Model calculation
B) Experiment.

5.   REFERENCES

1) Hegarty, J., Goldner, L. and Sturge, M.D., Phys. Rev. B30, 7346 (1984).

2) Schultheis, L. and Hegarty, J., J. Phys. Coll. C7, 167 (1985).

3) Masumoto, Y., Shionoya, S. and Okamoto, H., Proc. 17th. Int. Conf. Phys. Semiconductors, San Francisco, 1984, p. 349, Springer New York, 1985.

4) Petroff, P.M., Gossard, A.C., Wiegmann, W. and Savage, A., J. Cryst. Growth 4, 5 (1978).

OBSERVATION OF SUBBAND WIDTH IN THIN LAYER SUPERLATTICES
WITH EXCITATION SPECTROSCOPY IN A MAGNETIC FIELD

B.Deveaud*, A.Regreny*, M.Baudet*, A.Chomette*, J.C.Maan**, R.Romestain***

* CNET - LAB/ICM - 22301 LANNION - FRANCE

** Max Planck Institut for Festkoperforschung, 166 X, 38042 Grenoble Cedex

***Lab de Spectrométrie Physique, USMG, BP 87, 38402 St Martin d'Heres

Abstract

High quality GaAs/GaAlAs superlattices have been studied in excitation spectroscopy under a strong magnetic field (up to 10 T). In some samples (such as 50 $\overset{\circ}{A}$/50 $\overset{\circ}{A}$ or 60 $\overset{\circ}{A}$/60 $\overset{\circ}{A}$) a structure is already observed, between the two excitonic peaks, at zero magnetic field. The behaviour of this structure leads to an interpretation in terms of π/d exciton. In the same way, another series of Landau levels is obtained corresponding to the π/d maximum in the electronic density of states. In the samples with the thinest layers, reduction of the exciton binding energy is observed.

The study of GaAs/GaAlAs multiple quantum wells (MQW) by excitation spectroscopy in a magnetic field has been introduced by Maan et al [1]. Transitions between Landau levels are observed as resonances in the spectra and the analysis of the data leads to the effective masses of the carriers. In such systems, the excitonic transitions only show a weak diamagnetic shift and a double series of Landau levels is observed respectively corresponding to heavy hole → electron or light-hole → electron transitions. These two series extrapolate to their respective band edges, thus allowing to determine the binding energy of each kind of exciton [1,2]. Evidence for the occurrence of vertical transport (perpendicular to the layers) has also been given by excitation experiments with a magnetic field in the plane of the layers [3].

Our samples were grown by Molecular Beam Epitaxy (MBE) and their quality is demonstrated by their very small linewidth [4], in some of them a splitting of the excitonic peak is observed [5]. Their parameters L_Z (well width), L_B (barrier width) and x (Al content of the barriers) are obtained from x ray diffraction. The quality of our samples (referred to as L_Z/L_B in Å) is further demonstrated by the fact that we observe Landau resonances on most of the samples down to 2.5 T. The experimental set-up is the same as described in ref. 1.

Without application of any magnetic field, it is to be noted that the shape of the PLE spectra changes when one changes from MQW structures to superlattices (SL) with thin layers. One of the striking features is displayed in figure 1 where we evidence the apparition of a new transition between the heavy and light hole excitonic resonances. This structure is observed for samples between 50/50 and 60/60, i.e. when the miniband width is a few meV, which is too far from the heavy hole exciton to be due to a monolayer interface step [5].

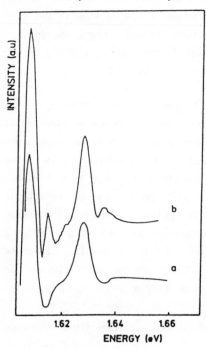

Fig. 1 : Luminescence excitation at zero magnetic field and 2K of :
a) a 60/150 MQW sample showing the usual shape with the onset of band to band transitions appearing as a step.
b) a 60/60 superlattice sample where an extra peak is evidenced between the heavy hole and the light hole excitons.

In the multi quantum well sample (60/150) the so–called onset of band to band transitions is observed as a step in the PLE spectrum without magnetic field. Except for the two lowest Landau levels, all extrapolate precisely towards that onset giving a binding energy for the exciton equal to 10 meV ± 1 meV.

The first characteristic of our intermediate samples : 50/50, 50/60, 60/60 is that, starting from the extra peak shown in fig. 1, one of the transitions shows a weak diamagnetic shift (see fig. 2) and we are tempted to interpret the extra peak as due to some excitonic transition.

In the same series of samples (see fig. 2 for the case of the 50/50 S.L) the plot of the Landau transition series evidences an extra series when compared to the MQW case. A series of Landau levels converges to the band edge for heavy hole-electron transitions, the second series extrapolates to the light-hole band edge. The last series extrapolates to some point in between, 8 meV above the heavy hole band edge.

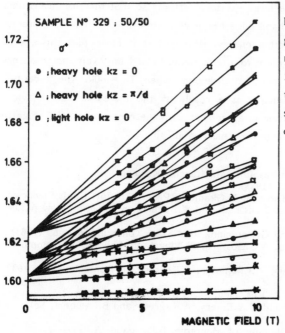

Fig. 2 : Transition energies, as a function of magnetic field, for a 50/50 sample, The excitonic states are represented by additional crosses.

Both experimental facts : the extra excitonic peak and the extra series of Landau levels are interpreted as due to the same physical origin. In a SL such as a 50/50, the quantum level for electrons begins to spread into a miniband in Z (allowing for vertical transport [6-7]). The density of states in Z of this miniband shows two strong maxima, as long as the miniband is not too large : one for $k_Z = 0$ and the other one for $k_Z = \pi/d$. As a consequence, excitonic transitions as well as band to band transitions peak at two different energies, the distance between these two energies being the miniband width. A value of 8 meV in a 50/50 SL is in reasonable agreement with theoretical predictions.

In SLs with shorter period (30/30, 20/20, 11/11) the experiments are more difficult to interpret : as a matter of fact, the Landau resonances are much weaker although the quality of the samples is comparable. The third structure, described in the preceeding section is no more observed, presumably due also to the diminution of the density of states in π/d because of the increase of the miniband width. The series of Landau level transitions evidence quite large excitonic effects so that the first transitions do not converge at the band edge but lower. This makes difficult a precise determination of the exciton energy. However, it is clear that this energy is reduced, and becomes close to a bulk value. Indeed the excitonic resonance seen in the excitation spectrum disappear when T > 60k. The reduction of the exciton binding energy is a result of the increased 3D behaviour of our samples (a conclusion in agreement with the experiments of Chomette et al [7], Belle et al [3], and Stievenard et al [8]).

REFERENCES :

[1] J.C.Maan, G.Belle, A.Fasolino, M.Altarelli, K.Ploog,
     Phys. Rev. B 30, p 41 (1984)
[2] N.Mura, Y.Iwasa, S.Tarucha, M.Okamoto, Proc. 17th Int. Conf. Physics
     of semiconductors Ed. J.D.Chadi and W.A.Harrison (NY, Springer) p 359
     (1984)
[3] G.Belle, J.C.Maan and G.Weimann, Solid State Commun, 56, p 65 (1985)
[4] C.Weisbuch, R.Dingle, A.C.Gossard, W.Wiegmann, Solid State Commun.
     39, p 709 (1981)
[5] R.Deveaud, J.Y.Emery, A.Chomette, A.Regreny, J. Appl. Phys. 59, 1633
     (1986)
[6] L.Esaki, R.Tsu, IBM J. Res. Dev. 14, p 61 (1970)
[7] A.Chomette, B.Deveaud, J.Y.Emery, A.Regreny and B.Lambert, Solid
     State Commun. 54, p 75 (1985)
[8] D.Stievenard, D.Guillaume, J.C.Bourgoin, B.Deveaud, A.Regreny,
     to appear in Europhys. Lett.

EXPERIMENTAL AND THEORETICAL STUDIES OF QUANTIZED ELECTRONIC STATES ABOVE THE ENERGY BARRIER OF GaAs/Al$_x$Ga$_{1-x}$As SUPERLATTICES

J.J. Song, Y.S. Yoon, P.S. Jung and A. Fedotowsky
University of Southern California, Los Angeles, CA 90089, USA

J.N. Schulman
Hughes Research Laboratory, Malibu, CA 90265, USA

C.W. Tu and R.F. Kopf
AT & T Bell Laboratory, Murray Hill, NJ 07974, USA

D. Huang and H. Morkoc
University of Illinois, Urbana, IL 61801, USA

Optical transitions between unconfined states, i.e., states above the conduction or below the valence band potential barriers of GaAs/Al$_x$Ga$_{1-x}$As superlattices have been observed by photoluminescence excitation spectroscopy (PLE) and interpreted with the aid of theoretically generated absorption spectra. The strengths of these transitions are found to vary strongly with the barrier layer width changes.

The confined electron and hole states in GaAs/Al$_x$Ga$_{1-x}$As superlattices have been investigated by a variety of experimental techniques and theoretical approaches, and are considered to be reasonably well understood[1,2]. Unconfined energy subbands which are formed above (below) the conduction (valence) band barriers of periodic potential, however, have not received much attention[3,4]. In this work, we report on experimental and theoretical studies of optical transitions involving unconfined states in GaAs/Al$_x$Ga$_{1-x}$As superlattices (SLs). Our calculations show that the strengths of the transitions from the first unconfined heavy-hole to the first unconfined conduction states depend strongly on the barrier layer widths. These calculations are consistent with the data taken from a series of molecular beam epitaxial (MBE) GaAs/Al$_x$Ga$_{1-x}$As (x ~ 0.2) with L_z ~ 150 Å, and L_b ~ 150 Å, 70 Å, and 30 Å, where L_z and L_b denote the well and the barrier thickness, respectively[5]. We have subsequently observed a variety of peaks above the barrier energy gap E_b in SLs with different sample parameters[6].

An example of PLE spectra in the vicinity of E_b is shown in Fig. 1. The sample is GaAs/Al$_x$Ga$_{1-x}$As with x ~ 0.22, L_z ~ 186 Å and L_b ~ 100 Å. The peak at 1.7 eV arises from confined excitonic transitions from n=4 heavy-hole states to n=4 conduction states (4HH). The broader peak at 1.77 eV is a superposition of several different kinds of transitions including n=4 light-hole excitonic transitions (4LH). The strikingly steep increase of the PLE signal above E_b (~ 1.79 eV) is attributed to excitonic transitions from the first unconfined hole states to the first unconfined conduction states.

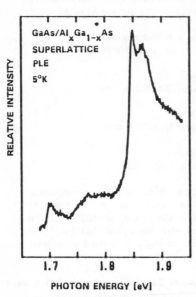

Fig. 1. PLE spectrum in the vicinity of
the barrier energy gap.

In contrast to confined transitions, which are
not so sensitive to L_b values (except for very thin
barrier layers), the strengths of unconfined
transitions are greatly affected by L_b. This can be
seen in Fig. 2 which depicts theoretically
generated absorption spectra of three
GaAs/Al$_x$Ga$_{1-x}$As SLs with fixed well widths
(L_z = 150 Å) and Al. compositions (x = 0.2) but
different barrier widths. The three spectra
illustrate the sensitivity of the absorption to the
value of L_b. In particular, the steep increase at
~ 1.8 eV in Fig. 2(a) is nearly absent in two other
samples. This is in excellent agreement with our
experimental results which will be discussed
below. The absorption curves shown in Fig. 2
were calculated using a two-band tight binding
model[2l,7l]. The energy subbands and the wave
function were calculated as a function of k_z, the
wave vector in the SL growth direction[8l]. A
simple parabolic subband dispersion was
assumed for the wave vectors parallel to the SL
planes. Excitonic effects were not included.

Figure 3 shows PLE spectra taken from a series of three samples with parameters
nearly the same as those in Fig. 2. The three samples exhibit essentially the same low
energy spectra as shown in the inset. Forbidden transitions are
clearly observed between 1LH and the 2HH. The high energy regions of the spectra,
Figs. 3(a), (b), and (c), however, are quite different from one another. The distinct peak at
~ 1.8 eV in the sample with L_b = 150 Å is hardly detectable in two other samples, as
expected from theoretical absorption spectra. This peak is attributed to excitonic
transitions between the first unconfined heavy-hole subband to the first unconfined
electronic subband,and observed essentially the same L_b dependence of the PLE spectra
in another series of SLs with nearly the same sample parameters as those in Fig. 3, but
fabricated under different MBE growth conditions.

The barrier-width-dependence of the optical transitions between the first unconfined
heavy-hole and conduction subbands can easily be understood by examining the wave
functions of these states. They are the sixth heavy-hole states (HH6) and the fourth
conduction states (C4) in samples studied in Figs. 2 and 3. The comparison of these
wave functions is illustrated in Fig. 4. The wave functions have two components: s-

orbital-like on the cations and p-orbital-like on the anions. The main difference between the wave functions of Fig. 4(a) and Figs. 4(b), (c) is in the number of the nodes in the barrier and the overlap of the wave functions for the valence and conduction states. The appearance of the distinct peak indicated by an arrow in Fig. 3(a) is associated with the large overlap of the wave functions shown in Fig. 4(a). Our calculations also reveal that there exists a critical barrier width L_c, below which the intensity of the transition between the first unconfined states is insignificant[5]. For GaAs/$Al_{0.2}Ga_{0.8}$As with L_z = 150 Å, L_c is between 85 Å and 100 Å.

We have also observed discrete peaks above the $E_o + \Delta$ gap of GaAs, which arise from transitions between confined hole states associated with the split-off valence band and confined conduction states. Details of this work will be discussed in Ref. 6.

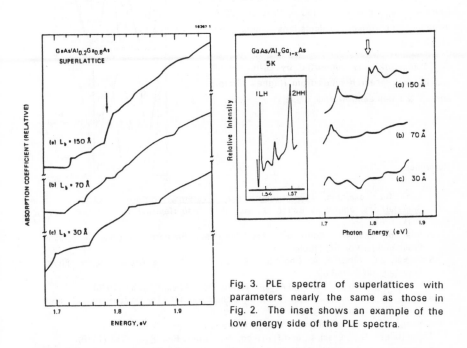

Fig. 3. PLE spectra of superlattices with parameters nearly the same as those in Fig. 2. The inset shows an example of the low energy side of the PLE spectra.

Fig. 2. Theoretically generated absorption spectra for GaAs/$Al_{0.2}Ga_{0.8}$As with L_z = 150 Å. The transition between the first unconfined states is marked by an arrow.

Fig. 4. Comparison of wave functions of the first unconfined valence (HH6) and conduction (C4) states at $k_z = 0$ of the samples in Fig. 2. The wave functions have two components which are drawn separately: s-orbital-like on the cations and p-orbital-like on the anions. The smaller amplitude dashed and solid lines for each L_b are for the HH6 s-component and p-component, respectively. N_b is the number of GaAs atomic layers, 2.83 Å per layer.

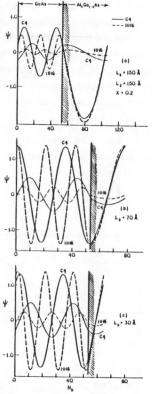

In conclusion, we have observed a variety of resonance peaks in the PLE spectra above the barrier gap of GaAs/Al$_x$Ga$_{1-x}$As superlattices. The importance of the barrier widths in observing unconfined states is pointed out with a series of GaAs/Al$_{0.2}$Ga$_{0.8}$As with L_z = 150 Å, 70 Å, and 30 Å, through theoretical calculations and PLE experiments.

This work was supported by ONR, AFOSR and internal research funds from Hughes Aircraft and AT & T Bell Laboratories.

References

1. See for example, Two-Dimensional Systems, Heterostructures, and Superlattices, edited by Bauer, G., Kuchar, F., and Heinrich, H. (Springer, New York, 1984).
2. See for example, Schulman, J.N., Materials Research Society Symposia Proceedings, Vol. 56, (1985).
3. Zucker, J.E., Pinczuk, A., Chemla, D.S., Gossard, A., and Wiegmann, W., Phys. Rev. B29, 7065 (1984).
4. Wong, K.B., Jaros, M., Gell, M.A., and Ninno, D., J. Phys. C19, 53 (1986).
5. Song, J.J., Yoon, Y.S., Fedotwosky, A., Schulman, J.N., Tu, C.W., Huang, D., and Morkoc, H., to be published.
6. Song, J.J., and Schulmann, J.N., in preparation.
7. Sai-Halasz, G.A., Esaki, L., and Harrison, W.A., Phys. Rev. B18, 2812 (1978).
8. Sample parameters used in our calculations are as follows; the energy gap difference $E_g(Al_xGa_{1-x}As)-E_g(GaAs)$ = 1.247 x eV, the valence band offset is 40% of the band gap difference, effective masses m_e = 0.0671 m_o, m_{hh} = 0.34 m_o, m_{1h} = 0.094 m_o for GaAs and m_e = 0.0837 m_o, m_{hh} = 0.378 m_o, m_{1h} = 0.105 m_o for Al$_{0.2}$Ga$_{0.8}$As.

CAPACITANCE VOLTAGE AND DEEP LEVEL TRANSIENT SPECTROSCOPY IN A GaAs/GaAlAs SUPERLATTICE

B. Deveaud, A. Regreny
CNET (LAB/ICM) 22301 LANNION FRANCE

D. Stievenard, D. Vuillaume
Laboratoire de Physique des Solides, ISEN
39 Boulevard Vauban 59800 LILLE FRANCE

J.C. Bourgoin, A. Mauger
Groupe de Physique des Solides de l'ENS
Université Paris VII, Tour 23 75221 PARIS FRANCE

ABSTRACT

Capacitance voltage (C(V)) and deep level transient spectroscopy (DLTS) measurements have been performed on a 20A/20A superlattice. They first show that the superlattice behaves, in the growth direction, as a conductive uniformly doped material. After irradiation with 1 MeV electrons, four main peaks are observed in DLTS at 0.140, 0.185, 0.340 and 0.500 eV from the conduction band. Inspection of the results shows that the conduction band offset with an Al concentration of 0.3 is 360meV. Emission of levels both from the GaAs and the GaAlAs layers are observed towards a common conduction miniband.

One of the questions still not completely solved in GaAs/GaAlAs superlattices is the quality of the vertical transport : i.e. transport along the growth direction, accross the GaAlAs barriers. This transport was prevented by the poor quality of the interfaces. As this quality has now been largely improved, indications of the occurence of such a kind of transport [1, 2] have been obtained. Other techniques can help to assess the quality of the structure in this direction : C(V) and DLTS.

These techniques also enable one to study deep levels in superlattice systems. The study of such deep levels is very promising as they may act as probes inside the very superlattice, being sensitive to energy changes, stress, changes in phonon energies,... Alternatively, superlattices or quantum wells can give informations about the behaviour of deep levels thanks to the very precise control of layer widthes than can be achieved.

We have studied by capacitance techniques a very small period superlattice. A 20A/20A superlattice with 30 % GaAlAs barriers has been grown by Molecular Beam Epitaxy with a doping level of $3 \cdot 10^{16} cm^{-3}$. The sample has first been characterized by photoluminescence and by X-ray diffraction (see [3] for experimental details). Gold Schottky contacts are then evaporated on the surface of the superlattice (1.6 μm thick). The structure readily shows a rectifying behaviour with a reasonable barrier height. Capacitance-voltage (C(V)) measurements show the usual linear behaviour of $1/C^2$ versus V. The slope of $1/C^2$ leads to a constant carrier concentration of $2.7 \times 10^{16} cm^{-3}$ [4].

Such a behaviour demonstrates the existence of a common conduction miniband in the system with very small residual disorder (C(V) measurements can be performed down to 4 K). This demonstrates in a very direct way the quality of the vertical transport in this structure. DLTS studies show a very small number of levels except for a large concentration of very deep centers seemingly corresponding to interface defects.

After irradiation with 1 MeV electrons to a dose of $3 \times 10^{15} cm^{-2}$, four main peaks are observed on the DLTS spectrum. The corresponding levels have energies of 140, 185, 340 and 500 meV below the conduction band. The same irradiation process in GaAs would create 3 levels at 45, 140 and 360 meV, and also 3 levels in GaAlAs (x = 0.3) at 200, 360 and 720 meV. The full interpretation of the spectra (see [5]) shows that :
- First, the levels created in GaAs and in GaAlAs are both observed with correct introduction rates. This again shows that the wave functions of electrons extend in the whole structure.
- Second, the relative positions of GaAs and GaAlAs conduction bands as well as the position of the conduction miniband extremum can be quite safety deduced. We find that the conduction band offsets amounts to 360 meV and that the conduction miniband begins 140 meV above the original band edge of GaAs. This leads to an unexpected consequence

as an offset such as 360 meV cannot occur with the usual variation assumed for the GaAlAs band gap.

Most teams working in this field (see [6] for example) use in their calculations a GaAlAs band gap amounting to : 1.52 eV + 1.247 x (x < 0.3, T=4K)

However, some workers [7] found quite different results and propose that the GaAlAs band gap would rather be : 1.52 eV + 1.6 x.

Using the first value, Chomette et al [8] did not succeed to fit the luminescence properties of a whole series of very carefully characterized samples using as a model the now well admitted Effective Mass Approximation (EMA) model [9]. We have first tried to calculate the electronic properties of our 20/20 superlattice using EMA and the 1.247 x band gap for GaAlAs. There is no way to fit the experiment. On the contrary, if we introduce as a value for GaAlAs gap variation 1.6 x, all results become coherent for an offset partition of 70 % is the conduction band and 30 % in the valence band (see table 1).

	Experiment	EMA 70,30 1.6 x	EMA 60,40 1.247 x
Exitonic Energy	1.698 eV	1.714	1.670
Heavy holeLight hole splitting	7 meV	8	7
Conduction band offset	360 meV	336	224
Conduction sub band minimum	140 meV	150	104

Let us note that a valence band offset of 30 % with an 1.6 x gap variation law gives a very good agreement with a large number of recently published papers [10, 11]. So we have calculated again the properties of the whole series of samples studied by Chomette et al [8]. Once agin, a very good agreement is obtained for the 70 %, 30 % offset partition [12]. The discreapencies are limited to errors smaller than what would be induced by one monolayer (2.8 A) variation of the well width (which can be considered as the minimum uncertainty of the parameter determination of our structures.

As a conclusion, our C(V) and DLTS measurements on a small period superlattice has allowed us first to check the real ability for this kind of superlattice to vertical transport. The existence of a comon miniband over the whole structure is shown by the observation of emission from the defect levels both from the GaAs and the GaAlAs layers. Interpretation of the results leads us to propose that, at least in small period superlattices, the band gap of GaAlAs varies as 1.6 x (eV) and the offset partition is 70 %, 30 %.

ACKNOLEDGMENTS : The authors wish to thank G. Dupas for his assistance in sample growth, G. Bastard, A. Chomette and B. Lambert for helpfull discussion and P. Auvray and M. Baudet for X-rays characterization of the samples.

REFERENCES

[1]   A. Chomette, B. Deveaud, J.Y. Emery, A. Regreny, B. Lambert
       Solid State Commun. 54, p. 75 (1985)

[2]   G. Belle, J.C. Maan, G. Weimann, Solid State Commun. 56, p.65 (1985)

[3]   B. Deveaud, A. Chomette, J.Y. Emery, A. Regreny
       J. Appl. Phys. 59, 1633 (1986)

[4]   J.C. Bourgoin, A. Mauger, D. Stievenard, D. Vuillaume, B. Deveaud,
       A. Regreny, Submitted to Phys. Rev. Lett.

[5]   D. Stievenard, D. Vuillaume, J.C. Bourgoin, A. Mauger, B. Deveaud,
       A. Regreny, to appear in Euro Phys. Lett.

[6]   M.H. Meynadier, C. Delalande, G. Bastard, V. Voos, F. Alexandre and
       J.L. Lievin, Phys. Rev. B 31, 5539 (1985)

[7]   B. Monemar, K.K. Shi, G.D. Petit
       J. Appl. Phys. 47, 2604 (1976)

[8]   A. Chomette, B. Deveaud, M. Baudet, P. Auvray, A. Regreny
       J. Appl. 59,   (1986)

[9]   G. Bastard, Phys. Rev. B24, p. 5693(1981) and Phys. Rev .B25 p.7584(1982)

[10]   J. Batey, S.L. Wright, D.J. di Maria
       J. Appl. Phys. 57 p. 484 (1985)

[11]   M.O. Watanabe, J. Yoshida, M. Mashita, T. Nakanisi, A. Hojo
       J. Appl. Phys. 57 p. 5340 (1985)

[12]   B. Deveaud, B. Plot, A. Chomette, A. regreny, B. Lambert, J.C.Bourgoin
       D. Stievenard
       E. MRS Strasbourg France 1986 to be published.

LAYER-SPECIFIC PHASE TRANSITIONS IN AlAs/GaAs
HETEROSTRUCTURES UNDER HIGH PRESSURE

B.A. Weinstein and S.K. Hark

Xerox Webster Research Center 114-41D, Rochester, New York, 14644 USA

R.D. Burnham

Xerox Palo Alto Research Center, Palo Alto, California, 94304 USA

ABSTRACT

We report transitions to opaque states in multilayer AlAs/GaAs heterostructures under hydrostatic pressure. Depending on layer width, reversible transitions of the AlAs layers, both individually and collectively, are observed at pressures well above the transition in bulk AlAs (measured here also for the first time.) The occurence of strained layers and/or interface defects in the transformed phase is considered. High pressure "growth" of novel metastable superlattices is proposed.

Pressure-induced phase transitions in covalent semiconductors are caused by the necessity of the relatively open tetrahedral structure to become closer packed under compression.[1] The resulting volume decrease is typically 15-20%. For group IV and weakly ionic III-V materials, the lowest pressure transition is generally to a 6-fold coordinated β-Sn structure[2,3] that exhibits metallic conductivity[4] and can be superconducting.[5] With increasing ionicity in III-V and II-VI compounds, the lowest pressure transition is to a semiconducting NaCl structure,[2,3] or, for intermediate ionicity cases such as GaAs,[6] to more complicated structures. After releasing pressure, metastable phases sometimes occur.[7] Considerable progress in understanding these transitions has resulted from recent density functional treatments.[8,9]

The present work discusses the first measurements of pressure-induced transitions in multilayer (ML) heterostructures, where the interface boundary conditions play an important role. We study the lattice-matched AlAs/GaAs system. Because the transitions in its bulk components occur at different pressures ($P_t^G = 172$ kbar for GaAs[6] vs $P_t^A = 123$ kbar for AlAs as established below), this system is a candidate for new effects not present in any bulk material. Here we deal primarily with the occurrence and consequences of separate phase changes, specific to either the AlAs or the

GaAs layers. Issues of crystal structure and energetics are explored elsewhere.[10]

Six different samples are studied -- bulk GaAs, a bulk-like film of AlAs, three superlattices, and a ML structure having different width AlAs layers. Their detailed characteristics are given in Table I. Note the sample nos. for future reference. All the samples, except bulk GaAs, were grown by metalorganic chemical vapor deposition (MOCVD) with (001) oriented epitaxial layers that were not intentionally doped. To simulate bulk AlAs, which decomposes in air, we studied a 1.1 μm thick AlAs layer sandwiched between protective GaAs layers (sample no. 2). The compositions (in particular the alloy nature of sample no. 6) and the layer widths of the superlattices were confirmed by Raman and luminescence measurements.[11] Initially, and at low pressure, all the superlattices emitted intense luminescence, easily seen by the eye when visible. After a transition had occurred, no emission could be detected over 0.5-1.0 μm at any pressure, including 1 atm.

Specimens were prepared by lapping and chemical polishing their GaAs substrates down to a 10-30 μm thickness. Samples nos. 3-6 were further processed with a GaAs-selective etch[12] to obtain freestanding epilayers. Our hydrostatic pressure, room temperature experiments used conventional diamond-anvil cell (DAC) techniques.[13] Details concerning the experiment are given elsewhere.[10,11]

Under pressure (P), all the samples became transparent prior to any phase transitions due to the increase of the GaAs bandgap.[14] Hence, the transitions of thin layers to metallic (or small bandgap) states could be detected quite unambiguously and photographed using visible microscopy. Furthermore, any macroscopic markings ($\gtrsim 2$ μm in size) could be noted for later comparisons. A collection of photographs showing the phase transitions in our samples appears in Fig. 1. (For a detailed explanation of this figure, see the caption provided.)

In agreement with previous work,[6] we find that bulk GaAs (Figs. 1a-1c) transforms at $P_t^G = 172 \pm 4$ kbar. A photograph was recorded just prior to completion of this sluggish transition (taking ~15 min), which shows one corner of the sample still untransformed (Fig. 1b). The sample did not revert to a transparent state when the pressure was lowered to $P = 0$ and then raised to what was formerly the transparent regime (Fig. 1c, at 77 kbar).

Recent x-ray studies indicate that bulk GaAs assumes a metastable structure different from zincblende upon pressure release.[15]

Table I. Summary of sample characteristics and transition pressures.

Sample No.	1	2	3	4	5	6
Geometry	Bulk GaAs	GaAs/AlAs/ GaAs	AlAs/GaAs	AlAs/GaAs	AlAs/GaAs *M.L.	$Al_{0.7}Ga_{0.3}As/$ $Ga_{0.7}Al_{0.3}As$
Thicknesses	~20 μm	0.3/1.1/0.3 μm	290/160 Å	60/25 Å		~19/19 Å
Cycles	1	1	60	200		320
P_t (kbar)	172 $\pm$ 4	123 $\pm$ 4	142 $\pm$ 4	166 $\pm$ 4	123-127 130-133	155 $\pm$ 4
R_t (kbar)	No reversal	60 $\pm$ 10	70 $\pm$ 10	115 $\pm$ 10	80 $\pm$ 10	No reversal

*A multilayer with the thickness sequence (in Å, AlAs underlined): bulk GaAs/1400/350/700/ 350/350/350/175/350/175/3500.

The transition in bulk AlAs has not been measured previously; however, a density functional calculation has predicted it to occur at 70 kbar.[8] Our bulk-like film (1.1 μm thick) transformed at $P_t^A = 123 \pm 4$ kbar (Fig. 1e). The transition was abrupt, taking <0.1 sec, and homogeneous across the sample. Upon lowering pressure, the sample regained transparency at 60 $\pm$ 10 kbar, undergoing an essentially backward replay of the increasing-P transition. The sample exhibited no new macroscopic markings after this sequence, but its color was darker than initially. (Compare Figs. 1d and 1f.) These transitions were recycled with similar qualitative and quantitative results. The discrepancy between the observed and predicted[8] transition thresholds may not be meaningful because of the large hysteresis.

Sample no. 3 transformed at $P_t^{(1)} = 142 \pm 4$ kbar (Fig. 1g). The transition proceeded as follows: For P slightly less than $P_t^{(1)}$, this superlattice was transparent (actually light yellow), similar to the no. 4 specimens in Fig. 1g. At $P_t^{(1)}$, there was a sudden change to a darker but still transparent color; again the process took <0.1 sec and was uniform across the sample. In quick succession, six similar changes occurred, separated by ~15 sec intervals, until sample no. 3 was completely opaque. From this sequence, we conclude that the transitions involved only the AlAs layers. If GaAs layers also transformed, there would be no isolation between layers, and the entire transformation would have occurred in one continuous

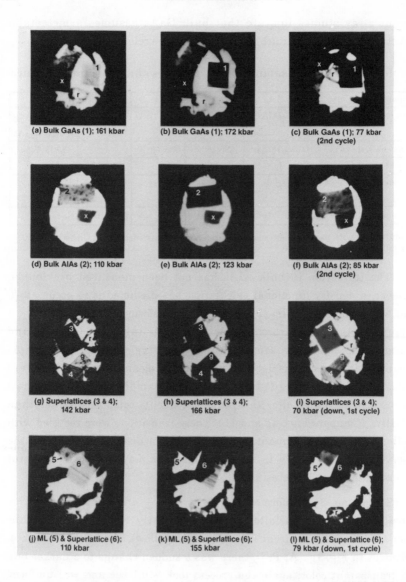

(a) Bulk GaAs (1); 161 kbar   (b) Bulk GaAs (1); 172 kbar   (c) Bulk GaAs (1); 77 kbar
                                                              (2nd cycle)

(d) Bulk AlAs (2); 110 kbar   (e) Bulk AlAs (2); 123 kbar   (f) Bulk AlAs (2); 85 kbar
                                                              (2nd cycle)

(g) Superlattices (3 & 4);   (h) Superlattices (3 & 4);   (i) Superlattices (3 & 4);
    142 kbar   166 kbar   70 kbar (down, 1st cycle)

(j) ML (5) & Superlattice (6); (k) ML (5) & Superlattice (6); (l) ML (5) & Superlattice (6);
    110 kbar   155 kbar   79 kbar (down, 1st cycle)

Fig. 1. Microphotographs of transitions. All pressures (P) are on upstroke and first cycle
(except as noted). Each row shows a different measurement series. Samples are labeled by
no. (ruby by r) following Table I. Those labeled X were nonuniformly or insufficiently
etched and their results are not discussed. The two no. 4 samples overlap and are labeled
with a single "4"; one is free floating, the other (labeled g) is attached with a dab ($\sim$10 μm) of
grease to an anvil-face. Strain gradients affect the latter, showing up in its sluggish
transition (156-160 kbar) and darkened retransformed color (1i). At high-P, gasket debris
sometimes appears (1a-c, 1j-l), floating with the samples.

process, instead of in discrete steps. Given that seven steps were needed to produce complete opacity, it is likely that transitions in individual AlAs layers were observed (since 7 x 290 Å ≈ 2000 Å, a typical metallic penetration depth).

The transition in sample no. 4 took place at a hydrostatic pressure of $P_t^{(1)} = 166 \pm 4$ kbar, substantially higher than for sample no. 3 (see Figs. 1g and 1h). Individual layer-transitions were not observed in sample no. 4, but instead several dark nucleation centers appeared and grew, in both size and opacity, until after ~15 min. the specimen was completely black (as shown in Fig. 1h). We attribute this behavior to collective transitions of many AlAs layers in unison.

The transitions in samples no. 3 and 4 were cycled twice. Similar results were obtained on increasing pressure, but the effect of reducing pressure depended drastically on the maximum pressure reached, 166 kbar in the first cycle and 275 kbar in the second. As shown in Fig. 1i, both superlattices regained transparency when P was decreased from equal or slightly above $P_t^{(1)}$ on the first cycle. The character and kinetics of these reversals were similar to the corresponding increasing-P transitions, but their thresholds R_t were 50-70 kbar below $P_t^{(1)}$ (see Table I). No new markings were observed after transparency reappeared, but samples no. 3 and 4(g) were darker in color. In contrast, when P was decreased from substantially above $P_t^{(1)}$ on the second cycle, transparency was not regained in either sample, even after decreasing to P = 0. This is significant; it implies that, in both superlattices, a second transition (unobserved because of opacity) occurred at a pressure $P_t^{(2)}$ above $P_t^{(1)}$. Most likely, this transition took place in the GaAs layers (expected to transform at $P_t^{(2)} \leq P_t^G$) since bulk GaAs also does not recover transparency when P is reduced, but remains in a metastable phase.[5] This further supports the deduction that only the AlAs layers transformed at $P_t^{(1)}$ in sample no. 4 and implies that both superlattices adopted metastable phases after releasing P on the second cycle.

The first transition in sample no. 5, to a brown, faintly transparent phase, occurred throughout the range 123-127 kbar. A second such transition, resulting in complete opacity, occurred throughout the range 130-133 kbar. We attribute these two changes, respectively, to the 1400 Å

and 700 Å AlAs layers in this sample (see Table I). Strain gradients due to
gasket contact probably account for the sluggishness of these transitions.
On reducing P from 155 kbar, sample no. 5 reverted to a slightly darkened
transparent state (see Fig. 1l) with no new markings. (This implies that any
strain gradients were small.)

A transition commenced at 155 kbar in sample no. 6, proceeding via a
sluggishly expanding nucleation center that stopped growing part way
through the sample, as shown in Fig. 1k. Surprisingly, sample no. 6 did not
retransform when P was decreased from 155 kbar (Fig. 1l). This leads us to
tentatively assign the observed change to a <u>monolithic</u> transition of the well-
and barrier-layers together.

From the above results, we conclude that, in ML AlAs/GaAs
heterostructures, the zincblende phase of AlAs can be superpressed, i.e.,
retained, to pressures far above its bulk stability regime. For finite GaAs
layers (25 Å or more seems sufficient from the sample no. 4 results), the
amount of superpressing increases for decreasing AlAs layer width ℓ and
approaches the limit set by the GaAs transition. This shows that the
boundary constraint of lattice-matching to GaAs provides an increasingly
effective barrier against the AlAs transition as ℓ becomes smaller. These
statements apply regardless of the state after transformation. Figure 2
summarizes our results for $P_t^{(1)}$ vs ℓ.

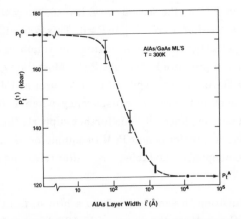

Fig. 2. Dependence of observed transition pressures on AlAs layer width. Point to left or
ordinate is for bulk GaAs. Rectangles give transition range for sample no. 5. A typical error
flag is shown. Horizontal lines at P_t^A and P_t^G bound the possible $P_t^{(1)}$ values.

We find that separate phase transitions can occur, specific to either the AlAs or the GaAs layers of a ML heterostructure. In fact, transitions occur within individual AlAs layers for ℓ exceeding a few hundred Å. A corollary result is that the interface coherence is destroyed during such transitions, presumably by extensive formation of misfit defects[16] to accommodate the ~15% volume decrease in the AlAs layers. In turn, for this range of ℓ, the AlAs layers should be free to adopt the high pressure phase of bulk AlAs, which is expected from calculations to have the NaCl structure.[8]

As the superlattice layers narrow further, our evidence from sample no. 4 is that individual AlAs-layer transitions become less likely than collective transitions of many AlAs layers. This is due to the growing competition between decoupling the layers by producing interface defects, or preserving the interface coherence by forming a strained-layer system. The latter would require a totally new AlAs phase that has ~15% less volume but still remains bonded to zincblende GaAs across the interface.[9]

We can estimate the critical layer thickness λ_c, below which strain wins over misfit defects from the criterion[16] $\lambda_c \approx (ln\ \lambda_c/b + 1)b/4f$ where b is the dislocation strength, typically ~4 Å,[16] and f is the misfit, which should be ~3-5%. This gives λ_c ~100 Å, in support of our interpretation for sample no.4. For this superlattice, it is realistic to expect intermediate behavior in which the transformed interface is characterized by misfit defects at some regions, and strain at others.

When the AlAs and GaAs layers both become atomically thin, one expects a single phase transition of the entire ML structure to a new lattice-matched state. We believe this occurred in sample no. 6 because its transition did not reverse when P was reduced from a maximum not exceeding $P_t^{(1)}$ = 155 kbar. Although the ~19 Å layers in sample no. 6 are not atomically thin, their ability to transform in unison is enhanced by their alloy composition (see Table I), which shifts the relevant bulk transition pressures closer together.

Lastly, consider several novel heterostructures made possible by the layer-specific nature of superlattice transitions. If, as for bulk Si,[5] the high-P phase of AlAs exhibits superconductivity, then a superconductor/semiconductor superlattice could be realized. In such a superlattice, it is likely that the pressure-tuning range for T_c could be extended compared to the bulk. The InAlSb and InSbAs systems are interesting in this regard,

714   B.A. Weinstein et al.

because the bulk thresholds, 22 kbar for InSb and 84 kbar for InAs or AlSb,[3]
are widely separated. Furthermore, we saw that the reverse transition in
AlAs depends critically on the presence of zincblende GaAs, which
effectively acts as a regrowth template. Raising P far above $P_t^{(1)}$ produces
different GaAs templates, possibly allowing one to retrieve metastable AlAs
phases not stable in the bulk. Thus, pressure might be exploited for
"growth" purposes.

We are grateful to R.M.Martin for much useful advice, and to D. Cobb
and S. Leo for skilled photographic and artistic assistance.

1. Klement, W. and Jayaraman, A., "Progress in Solid State Chemistry," 3,
    (Pergamon Press, London, 1966), p. 289.
2. Jamieson, J.C., Science 139, 762 (1963); Science 139, 845 (1963).
3. Yu, S.C. and Spain, I.L., Solid State Commun. 25, 49 (1978).
4. Minomura, S. and Drickamer, H.G., J. Phys. Chem. Solids 23, 451
    (1962).
5. Chang, K.J., Dacorogna, M.M., Cohen, M.L., Mignot, J.M., Chouteau,
    G., and Martinez, G., Phys. Rev. Lett. 54, 2375 (1985).
6. Baublitz, M., Jr. and Ruoff, A.L., J. Appl. Phys. 53, 6179 (1982).
7. Kasper, J.S. and Richards, S.M., Acta. Cryst. 17, 752 (1964).
8. Froyen, S. and Cohen, M.L., Phys. Rev. B28, 3258 (1983).
9. Biswas, R., Martin, R.M., Needs, R.J., and Nielsen, O.H., Phys. Rev.
    B30, 3210 (1984); also Martin, R.M. this conference.
10. Weinstein, B.A., Hark, S.K., Burnham, R.D., and Martin, R.M., to be
    published.
11. Weinstein, B.A., Hark, S.K., and Burnham, R.D., J. Appl. Phys. 58, 4662
    (1985); Hark, S.K., Weinstein, B.A., and Burnham, R.D., to be
    published.
12. Logan, R.A. and Reinhart, F.K., J. Appl. Phys. 44, 4172 (1973).
13. Piermarini, G.J. and Block, S., Rev. Sci. Instrum. 46, 973 (1975).
14. Zallen, R. and Paul, W., Phys. Rev. 155, 703 (1967).
15. Sites, J. and Spain, I.L., private communication.
16. Matthews, J.W. and Blakeslee, A.E., J. Vac. Sci. Technol. 14, 989 (1977);
    J. Crystal Growth 27, 118 (1974).

MAGNETO-ABSORPTION SPECTRA OF EXCITONS IN $Al_xGa_{1-x}As$ ALLOY MULTI-QUANTUM WELLS IN HIGH MAGNETIC FIELDS

N. Miura, S. Takeyama and Y. Iwasa

Institute for Solid State Physics, University of Tokyo
Ropppongi, Minato-ku, Tokyo 106, Japan

ABSTRACT
Magneto-optical absorption spectra of excitons were measured in $Al_xGa_{1-x}As$-AlAs multi-quantum wells in pulsed high magnetic fields up to 150T. The field dependence of the diamagnetic shift of heavy hole excitons was investigated.

1. INTRODUCTION

$Al_xGa_{1-x}As$ alloy quantum wells provide promising optical device applications in the visible wavelength range[1]. The direct band gap of the alloy at the Γ-point increases with increasing x, and the absorption edge enters the visible range above x > 0.2. The indirect band gaps for the X and L points also increase with x, but with a smaller rate. The cross-over of the direct gap with the indirect gap occurs at x = 0.43 for the X − minimum and at x = 0.48 for the L − minimum, thus the alloy becomes an indirect gap semiconductor for larger x. Moreover, in AlGaAs-AlAs quantum wells, the band offset makes the X-minimum in the AlAs layer the lowest electron state among various high symmetry points in the alloy and AlAs layers for x $\gtrsim$ 0.29. The alloy quantum wells exhibit distinct exciton absorption peaks. Tarucha et al. obtained the binding energy of heavy hole excitons from magneto-optical absorption spectra.[2] It was found that for quantum wells with width $L_z \simeq 90$ Å, the binding energy increases with increasing x. If the effective masses of elections and holes and the dielectric constant are assumed to vary linearly from GaAs (x=0) to AlAs (x=1), the x dependence of the binding energy cannot be explained by the theory of excitons confined in quantum wells[3], but it appears to show a three dimensional dependence[2]. In this paper, we investigate the magneto-absorption spectra in alloy quantum wells in very high magnetic fields up to 150T.

2. EXPERIMENTAL PROCEDURE

Samples of $Al_xGa_{1-x}As$–AlAs multi-quantum wells were grown on Cr-doped (100) - oriented GaAs substrates by the NTT group using MBE. The Al composition x was changed from 0 to 0.59. Nearly the same well width L_z = 90 - 100 Å, barrier width L_B = 30 Å and the number of quantum well layers N_1 = 100 were chosen for all the samples. For the absorption measurement, GaAs substrates were removed off by selective etching after mounting the sample on a glass plate.

Magneto-optical measurements were performed at liquid helium temperatures in two kinds of pulsed high magnetic fields. One is a long pulse field (duration time is 20 ms) up to 35T generated by a copper wire wound magnet and the other is a very high field (duration time 6 μs) up to 150T generated by the single-turn coil technique[4]. Magneto-optical spectra were obtained by means of OMA in case of the former field, and by means of a streak spectrometer using an image converter camera in the latter[5].

3. RESULTS AND DISCUSSION

Fig.1 shows the absorption spectra of the excitons in alloy quantum wells for various x in the presence and absence of magnetic field. At zero field, the exciton peaks were discernible up to x = 0.51. In high magnetic fields, oscillatory magneto-absorption was observed as well as the shift of the ground state exciton levels. A typical example of streak pictures of the magneto-absorption spectra in megagauss fields is shown in Fig.2 for a sample with x = 0.34. The ground state line of the heavy hole exciton as well as the first excited state are seen to shift to the high energy side.

The streak photographs were analyzed by a micro-photo-densitometer using a CCD camera. Fig.3 shows an example of the micro-photo-densitometer trace for a sample with x = 0.51. From such traces, the energy shift of the absorption lines were obtained

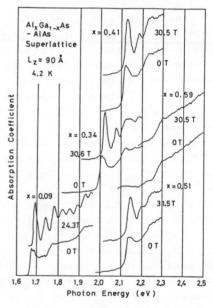

Fig.1 Magneto-absorption spectra of excitons in various samples of AlGsAs-AlAs multiquantum wells at H = 0 and H ≈ 31T (H = 24T for x = 0.09). L_z ≈ 90 Å. T = 4.2K.

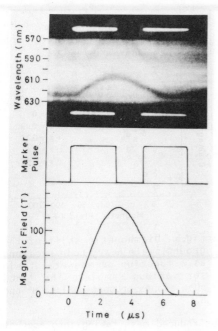

Fig.2　Streak picture of magneto-absorption spectra of excitons in megagauss fields for a sample with x = 0.34.　L_z = 92 Å.　T = 30K.

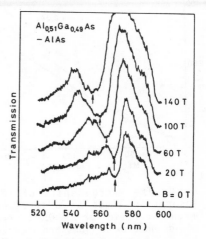

Fig.3 Micro-photo-densitometer trace of the magneto-absorption spectra for a sample with x = 0.51.　L_z = 90 Å.　T = 41K.

as shown in Fig.4 for x = 0.34 . The open circles and the closed circles indicate the ground state and the first excited state of the heavy hole exciton. The ground state shows an almost linear dependence against magnetic field in the high field range, while the first excited line bends downwards indicating the band non-parabolicity. In other samples, x = 0.41 and x = 0.51, the ground state line also showed a nearly linear dependence in megagauss range.

The overall magnetic field dependence of the ground state diagmagnetic shift is shown in Fig.5 for various samples including the low field data. It is quite evident that a quadratic dependence on field at low fields gradually tends to a linear dependence at higher fields, above the field where $\gamma = \hbar\omega_c/2R_y$ exceeds 1 for each sample (30-40T for x = 0.34 - 0.51).

The field dependence of the ground state diamagnetic shift was calculated by the variational method, assuming a reduced exciton mass obtained by a linear interpolation between GaAs and AlAs. As variational functions for the hydrogen atom-like state and the high field state, wavefunctions introduced by Greene and Bajaj(GB)[6] and Yafet et al.(YKA)[7] were employed, respectively. The result for x = 0.34 is shown in Fig.4. At low fields, the GB line is in agreement with the experimental data, but in higher fields, the experimental shift is

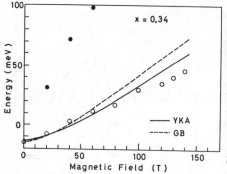

Fig.4 The energy shift of the
absorption lines as a function
magnetic field for x = 0.34. The
solid and broken lines are theoretical lines (see text).

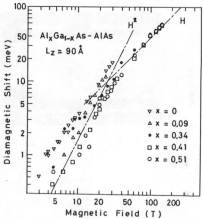

Fig.5 Diamagnetic shift of the
ground state heavy hole excitons
as a function of magnetic field
for various samples. The lines
show slopes proportional to H^2 and
H.

considerably smaller than the
theoretical lines. To obtain
better agreement, a calculation
should be carried out using a proper reduced mass taking into account
the valence band degeneracy and the band non-parabolicity.

As is seen in Fig.3, the ground state absorption line was found to
grow dramatically for x = 0.51 including an increase of the line width,
in high fields above 40T. A smaller but noticiable growth of the line
was also observed in a sample with x = 0.41. This phenomenon was
indiscernible for a sample with smaller x (x = 0.34). The reason of
the field-enhancement of the absorption is not clear at this moment,
but it may possibly be related to the effect of the indirect gap in
either AlGaAs or AlAs layers.

This work was done in collaboration with S. Tarucha and H. Okamoto
of NTT. The anthors are indebted to them for valuable discussion.

REFERENCES
1. Iwamura, H., Saku, H., Hirayama, Y., and Okamoto, H., Jpn. J. Appl.
    Phys., 24 L101 (1985).
2. Tarucha, S., Iwamura, H., Saku, T., Okamoto, H., Iwasa, Y. and
    Miura, N., To be published in Surface Science.
3. Shinozuka, Y., and Matsuura, M., Phys. Rev. B28, 4878 (1983).
4. Nakao, K., Herlach, F., Goto, T., Takeyama, S., Sakakibara, T., and
    Miura, N., J. Phys. E, Sci. Instrum. 18 1018 (1985).
5. Kido, G., Miura, N., Katayama, H. and Chikazumi, S., J. Phys. E,
    Sci. Instrum. 14 349 (1981).
6. Greene, R. L. and Bajaj, K. K., Solid State Commun. 45 831 (1983).
7. Yafet, Y., Keyes, R. W. and Adams, E. N., J. Phys. Chem. Solids. 1
    137 (1956).

DIRECT-INDIRECT BAND-GAP TRANSITION
IN GaAs/AlAs SHORT PERIOD SUPERLATTICES.

G.Danan*,A.M.Jean-Louis**,F.Alexandre**,B.Etienne*,B.Jusserand**,
G.Leroux**,J.Y.Marzin**,F.Mollot*,R.Planel* and B.Sermage**.

L2M-CNRS (*) and CNET-BAG (**)
196 Avenue H.-Ravera,F-92220 Bagneux,France

ABSTRACT

The "direct" or "indirect" character of GaAs/AlAs Short
Period Superlattices is investigated through optical
experiments. Due to spatial transfer of electrons in AlAs,
we get an accurate value of the offset parameter $\Delta=67\%$.

1. INTRODUCTION

GaAs/AlAs Short Period Superlattices (SPS) have been substituted
to the $Ga_{1-x}Al_xAs$ ternary alloys in several types of epitaxial
microstructures on GaAs[1-3]. However, little has been published on
fundamental electronic properties of these "new" materials since the
pioneer work of Gossard *et al.*[4,5]. In this paper, our purpose is to
present some of these properties, obtained from systematic optical
studies of a large number of samples.

2. ENVELOPE FUNCTION APPROACH

The envelope function approximation[6] has been applied by many
authors to the usual system GaAs/$Ga_{1-x}Al_xAs$ (x<0.4) SPS. If AlAs is
the barrier material, the maximum confinement of Γ states is increased
and may cross over the X states. We calculated the lower energy SPS
states from Γ,L and X minima in the envelope function approximation,
using a non-parabolic band structure in Γ (3-band Kane model), and a
parabolic model for X and L minima and the heavy hole valence band ,
for both GaAs and AlAs. At X and L points, the effective mass tensor
is anisotropic and the confinement energy must be calculated using

masses along the (001) direction (SPS axis). States built with X
minima along the (001) direction easily appear to be at lower energy.

In Fig.1, we plotted representative results of calculated[7] SPS
energy gaps E^Γ, E^X, and E^L, as a function of Al average concentration
x, for a SPS period P=4nm, using the offset parameter $\Delta = \Delta E_c / \Delta E_g = 67\%$,
as justified below. For this value, we point out that electrons and
holes are separated in real space since AlAs is a well for X
conduction states whereas GaAs becomes the barrier. As a consequence,
one may expect that the direct-indirect transition is also a type I -
type II transition. Such an effect has been suggested in other
structures[8,9].

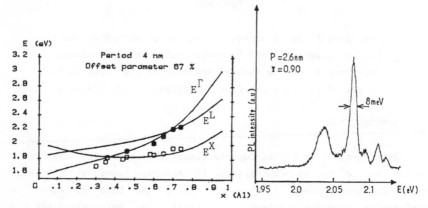

Figure 1: Solid lines: calculated
energy gaps of GaAs/AlAs SPS as a
function of x for P=4nm at T=1.7K.
o = experimental luminescence energy
● = experimental absorption edge
   obtained through PLE.

Figure 2: Photoluminescence
spectrum of a representative
indirect sample at T=1.7K.
Exciting power density
is 50 W/cm².

Optical transitions associated to Γ- or X-like states are direct
in the SPS Brillouin zone; this is not the case for L-like conduction
states since their transverse momentum never vanishes. However,
transitions involving X-like states should be weakly allowed. In this
paper, SPS with $E^\Gamma < E^X$ (resp. $E^X < E^\Gamma$) will be referred to as "direct"
(resp. "indirect").

We can make some comments on this theoretical approach.

i) for short-period samples, the crossover composition only slightly
depends on period.

ii) the structure is always direct for all x if $L_{GaAs} > 4$ nm.

iii) the envelope function approximation becomes questionable for
very short periods (P<2nm), and in the crossover region (i.e. $E^X \cong E^\Gamma$).

3.   EXPERIMENTS

All the samples (typically 50) were grown by MBE, two series at T_s=600°C and 650°C with 2nm<P<20nm and 0.15<x<0.90, and one serie keeping P=4nm and varying T_s and x. They have been characterized by X-ray simple and double diffraction and some of them by Raman scattering. Except a few poor quality samples, the 3 series exhibit the same behaviour with respect to the direct-indirect transition.

We performed low temperature photoluminescence (PL) measurements. In the structures which are expected to be direct, we most often get one main intense and narrow line. The line width ranges typically from 2 to 7 meV. The PL energy is in good agreement with the calculation presented in Part 2. As expected, it strongly depends on the structure parameters (see Fig.1).

In the indirect structures, the PL spectra are often less intense (typically 2 or 3 orders of magnitude) and exhibit several narrow lines with comparable intensities (see Fig.2). They correspond to transitions between Γ-like valence states and X-like conduction states confined in AlAs. An assignment of these lines is beyond the scope of this paper. We presume we observe zero-phonon lines coexisting with phonon replicas. At higher temperatures, we also often observe a broad band at higher energy corresponding to the Γ-Γ transition.

The excitation spectra of photoluminescence (PLE) appears to provide the same essential information as absorption. Two representative spectra are shown in Fig.3. They confirm the direct or indirect character of the energy gap. Thus, PLE is an easy and rather unambiguous criterion to classify our structures. Our results are summarized in Fig.4. Obviously, the assignment of the structures

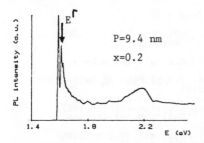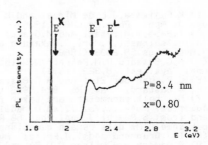

Figure 3: *Luminescence excitation spectra of representative direct (a) and indirect (b) samples at T=1.7K. The sharp low energy peak corresponds to luminescence. Calculated energy gaps are pointed out.*

722   G. Danan et al.

situated in the crossover region requires more sophisticated
experiments.

The direct or indirect character of the SPS band gap has been
also confirmed by PL decay time measurements. The lifetime ranges from
40 to 100 nsec in indirect type samples but is shorter than our
experimental resolution (20 nsec) in direct ones.

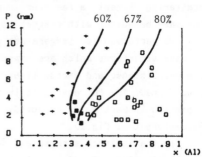

*Figure 4: All investigated samples
are displayed on this P versus
x graph. Following our experimental criteria, we distinguish
the direct (+) from indirect (o)
and ambiguous (■) ones. The
theoretical crossover curves for
several values of Δ are plotted.*

4.   CONCLUSION

We evidenced a direct to "indirect" transition using various
optical experiments on a large number of GaAs/AlAs SPS. It provides a
new determination of the offset parameter, Δ=67% which agrees with
recent results in GaAs/GaAlAs systems. Its sensitivity is due to the
type I – type II transition, and its precision is essentially limited
by the knowledge of AlAs indirect band gap. In addition, though the
agreement of our experimental results with the envelope function
approach is surprisingly good, theoretical works using more
sophisticated methods[10],[11] could be of interest with the Δ=67% value.

The authors especially thank Dr Paquet for helpful discussions.

(1) K.Fujiwara,J.L.de Miguel and K.Ploog
Jpn.J. Appl. Phys.24,L405 (1985)
(2) T.Baba,T.Mizutani and M.Ogawa; J.Appl.Phys.59,526 (1986)
(3) A.Ishibashi,Y.Mori,F.Nakamura and N.Watanabe
J.Appl.Phys.59,2503 (1986)
(4) A.C.Gossard,P.M.Petroff,W.Weigmann,R.Dingle and A.Savage
Appl.Phys.Lett.29,323 (1976)
(5) J.P. van der Ziel and A.C.Gossard; J.Appl.Phys.48,3018 (1977)
(6) G.Bastard; Phys.Rev.B24,5693 (1981)
(7) All bulk parameters are taken from: Landolt-Börnstein,
Numerical data...,Vol.17,ed.O.Madelung (Springer-Verlag,1982)
and from: S.Adachi,J.Appl.Phys.58,R1 (1985) and ref.therein;
but some AlAs data remain questionable.
(8) T.J.Drummond and I.J.Fritz; Appl.Phys.Lett.47,284 (1985)
(9) P.Dawson,B.A.Wilson,C.W.Tu and R.C.Miller
Appl.Phys.Lett.48,541 (1986)
(10) J.N.Schulman and T.C.McGill; Phys.Rev.B19,6341 (1979)
(11) W.Andreoni and R.Car; Phys.Rev.B21,3334 (1980)

DISORDER BROADENING AND ACTIVATION OF MODES FROM THE FOLDED ACOUSTIC BRANCHES IN GaAs-AlAs SUPERLATTICES.

J.SAPRIEL,J.HE,J.CHAVIGNON,G.LE ROUX,J.BURGEAT,F.ALEXANDRE,R.AZOULAY

Centre National d'Etudes des Telecommunications
Laboratoire de Bagneux
196 Avenue Henri Ravera 92220-BAGNEUX-FRANCE
and R.VACHER
USTL,F34060,MONTPELLIER CEDEX.

ABSTRACT

Different causes of broadening of the folded longitudinal acoustic branches are analyzed.The effect of the optical absorption is experimentally evidenced but actually the role of the defects is prevalent in the broadening. Folded transverse acoustic modes due to a polarization leakage are clearly observed on all samples and allow a complete acoustic characterization of the superlattices.New modes are found which correspond to the Brillouin-zone edge of the folded longitudinal and transverse acoustic branches which are called Disorder Activated Folded Longitudinal Acoustic modes (DAFLA) and Disorder Activated Folded Transverse Acoustic modes (DAFTA).The frequencies of these modes depend on the superlattice period D contrary to those of the previously studied DALA and DATA.

Several Raman studies have been devoted to the frequency dependence of the folded longitudinal acoustic modes (FLA) [1,2] in GaAs-AlAs superlattices but so far nothing ,except theoretical considerations, has been published on their linewidth [3].In good superlattices samples the Raman peaks corresponding to the first folded longitudinal acoustic modes are the narrowest ($\simeq 0.5$ cm-1) ever observed in III-V materials.Since the correction to be applied is much larger than the linewidth itself (the resolution of the Raman spectrometer is $\simeq 1$ cm-1) , the results of the measurements are quite uncertain.To overcome this problem,we have built up an experimental set up composed of a Raman double-grating spectrometer and a Fabry-Perot-interferometer.This tandem has a resolution of 0.1 cm-1 and a good optical transmission.The Raman spectrometer acts as a filter

whose frequency is adjusted at the frequency of the line under investigation and the Fabry-Perot is scanned with a variable gaz pressure,its free
spectral range being equal to 5 cm-1.

Besides the FLA we observed in every sample folded transverse acoustic
modes (FTA) corresponding to the same wavevector K as the FLA and modes
corresponding to the Brillouin-zone edge (K=Π/D) on the FLA and FTA.These
last modes are not Raman active.Only a certain kind of disorder can
induce them.Therefore we have called them DAFLA and DAFTA.

A-LINEWIDTH MEASUREMENTS OF THE FLA.

We have probed about 20 superlattice samples whose period D ranged
between 20 and 80 Å.The linewidth of the 1st FLA modes varied from a few
tenth of cm-1 (for the best samples) to a few cm-1 (for the poorest
quality superlattices).The following results have been obtained:

          1-For a given sample the width increases with the folding order
(Fig1)
          2-The broadening of the FLA with increasing values of the optical
absorption has been predicted in   Ref[3] .One will find experimental
confirmation of this effect in Fig2 for the 1st FLA in a good sample.
          3-A series of superlattices of approximately the same structure and
composition but grown at different substrate temperature [4] have been
investigated.A high Raman intensity combined with a small linewidth have
been taken as a criterion of compositional and thickness homogeneity of
the alternating layers.Samples A and B grown at 510 and 580°C respectively
are compared in Fig 3.Obviously the samples grown at low temparetures are
of better quality.There is a strong correlation between the intensity and
the width of the FLA and the corresponding intensity and width of the
satellites observed by X-Ray double diffraction [5].
          4-The linewidth decrease with decreasing values of the measurement
temperature is rather small. From room temperature to liquid nitrogen
temperature this decrease is about of 0.1 cm-1.Consequently the role of
the anharmonicity on the linewidth can be neglected with respect to that
of the inhomogeneities for all the investigated samples.

B-NEW PHONON MODES IN THE ACOUSTIC REGION.

The deviation from strict backscattering due to Brewster incidence and
to the aperture of the collecting lens,creates a small light polarization

component along z [2] which activates the FTA.Similarly to the FLA,the FTA appear as doublets.We sometimes observed both the first and the second doublets .The frequency distance between components of the doublets is proportional to the acoustic velocity and is therefore smaller for the FTA doublet than for the FLA doublet.If Ω_1 and Ω_2 are taken as the frequencies of the doublet component,BZE the center of the first gap at the Brillouin zone edge $(K=\Pi/D)$ is given by $(\Omega_1 + \Omega_2)/4$.One can consider BZE_L and BZE_T corresponding to the FLA and FTA respectively [5].The new modes DAFLA and DAFTA fall at frequencies very close to Ω_L and Ω_T respectively. The low frequency part of the phonon spectrum in a superlattice is given in Fig 4 and shows the 1st FTA doublet as well as the DAFLA and the DAFTA.Fig 5 corresponds to the inhomogeneous sample B: the DAFTA and the DAFLA become structureless (inhomogeneous broadening)in this case.More details are given elsewhere [5].

REFERENCES.

[1]-Colvard C. et al ,Phys.Rev.Lett.45,298,(1980).

[2]-Sapriel J. et al ,Phys.Rev.B,28,2007(1983).

[3]-Sapriel J.,Michel J.C.,Superlattices and Microstructures 353,(1985).

[4]-Jusserand B. et al,Appl.Phys Lett. 47,301 (1985).

[5]-Sapriel J. et al Proceedings of the 1985 Meeting of the MRS to be published in the Journal de Physique(France).

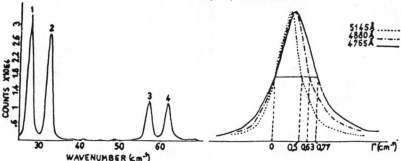

Fig 1-Raman spectrum of the FLA showing the folding orders 1-4. With the tandem we measured for the linewidths $\Gamma_1=0.56cm^{-1}$; $\Gamma_2=0.60cm^{-1}$;$\Gamma_3=0.84cm^{-1}$; $\Gamma_4=0.93cm^{-1}$(D=55 A ;mean Al concentration $\bar{x}$ =0.43).

Fig2-Dependence of the linewidth Γ on the wavelength of the incident light.The increase of the linewidth vs the energy of the photons is interpreted as a consequence of the optical absorption.(D=40 A,$\bar{x}$ =0.35).

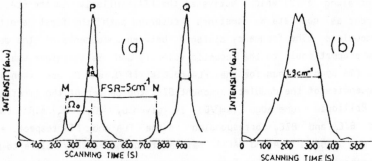

Fig3-a:Typical spectrum obtained with our tandem for the 1st FLA in
sample $A(d_{GaAs}=25$ A $,d'_{AlAs}=15$ A $,100$ periods).Growth temperature
510°C.M and N correspond to the reduced Rayleigh line.P and Q are the
1st FLA analyzed by the tandem;we measured $\Gamma_a=0.5$ cm^{-1};
FSR:free spectral range.

b:Linewidth of the 1st FLA obtained in a superlattice B of the
same structure with a growth temperature of 680°C.The quality of B is
poor with the respect to that of A and the consequence on the linewidth is obvious.

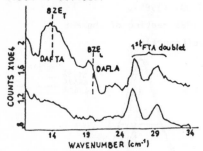

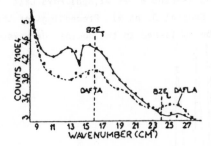

Fig4-Low frequency part of the
Raman spectrum in a homogeneous
superlattice.The Brillouin zone
edge BZE_L and BZE_T are calculated
from the components of the first
FLA and FTA doublets.The upper
curve is for symmetry A_1.The other
symmetry (polarizations of the incident and diffused light orthogonal)is given by the lower curve.

Fig 5-Raman spectra on an inhomogeneous superlattice for two different points on the surface(full
and dashed lines).

INDIRECT GAP EFFECTS IN GaAs-AlAs SUPERLATTICES

L.Brey[+], C.Tejedor[+], J.L.de Miguel[*], F.Briones[*] and K.Ploog[**].

[+]Departamento de Fisica del Estado Solido, Universidad Autonoma, 28049 Madrid, Spain.
[*]Centro Nacional de Microelectronica, Serrano 144, 28006 Madrid, Spain.
[**]Max-Planck-Institut FKF, Heisenbergstr.1 7000 Stuttgart 80, F.R.Germany.

We present the first observation of photoluminescence (PL) transitions from states having both Γ and X bulk character in some particular superlattice (SL) configuration. The actual structures consist of 30 period GaAs-AlAs SL with a few monolayer thick spikes of AlAs in the center of the wells (see inset in Fig.1). The samples under study were grown by molecular beam epitaxy (MBE)[1]. The well and barrier thicknesses were kept constant at a value of $L_z=90\text{Å}$ and $L_B=50\text{Å}$, respectively, for the whole set of samples. From sample to sample, only the width l_B of the AlAs spike at the center of each well was varied, with $l_B= 0, 2, 4, 6$ and 8 monolayers (ml).

In Fig. 1 we plot low temperature (2K) PL spectra obtained from a set of samples with varying thickness of the AlAs spike. All spectra were recorded by exciting at the 2.335 eV line of a cw Kr^+ laser with a power density of about 1 W/cm^2. The single peak associated to the sample without AlAs spike has a full width at half maximum (FWHM) of 6 meV. The important result of our investigation is that even for very narrow AlAs spikes (2ml) there is primarily a large shift of the PL peak to higher energy. In addition,

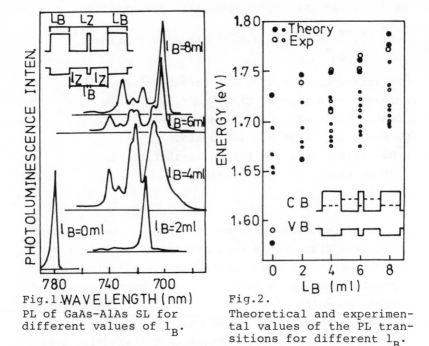

Fig.1. WAVE LENGTH (nm)
PL of GaAs-AlAs SL for
different values of l_B.

Fig.2.
Theoretical and experimental values of the PL transitions for different l_B.

an increase of the tickness of the spikes results in
the appearance of a set of peaks regularly distributed
at intervals of the order of 20 meV, which scarcely
evolve in energy when l_B is further increased up to 8
ml. Kronig-Penney calculations, simulating the effect
of the SL on the states of GaAs, do not supply the
number and energy of the observed transitions.

A complete understanding of the observed PL spectra
requires a careful description of the band structure
of the specific SL configurations. In particular, it
is necessary to take into account that AlAs is an indi-
rect-gap semiconductor with the minimun of the conduction band (CB) close to the X point. Due to the SL
potential along the (001) direction, we have for the
Γ point of the SL both a well spatially centered on
GaAs for Γ-like CB states and a well on AlAs for X-
like CB states. States at these wells will coexist at

the Γ point of the SL. In order to properly describe
all these features, we use a tight-binding Hamiltonian
including spin-orbit interaction. We follow a simple
perturbative approach[2] which gives us the SL eigenstates both spatially and in terms of the bulk states of
the constituents semiconductors. The valence band (VB)
offset ΔE_v is the only parameter we have adjusted in
order to get the best possible agreement between theory and experiment. Fig. 2 shows the experimental results and calculated transitions for ΔE_v=0.67 eV.
Large open circles indicate the energy of the PL peaks
that essentially dominate the spectrum at higher temperatures or at high excitations levels, while small
open circles represent transitions surviving only
under low temperature or low excitation. Large closed
circles correspond to calculated transitions which
involve CB states concentrated on GaAs. These states
give the largest overlap
between CB and VB, as can
be seen in Fig. 3, and are
responsible for the most
intense peaks in the experiment. As far as the
other PL peaks are concerned, they are clearly connected with CB states originated from X-like states
in AlAs (see Fig. 3). Now
a detailed explanation of
the experimental results
becomes possible. In the
absence of any AlAs spike,
the Γ-well is a few tenths
of an eV deeper than the
X-well, so that the lowest

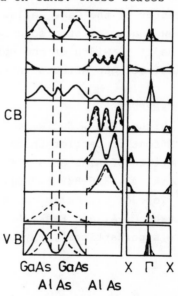

Fig.3. Eigenstates connected
by the transitions of the
figure 2 for
l_B=0(----); l_B=4(——)

CB state of the SL originates from a GaAs Γ-like state. This state is responsible for the only PL transition to the VB state concentrated on GaAs. The introduction of the AlAs spike implies for CB the appearance of a barrier for GaAs Γ-like states and a well for AlAs X-like states. So, for a few monolayers spike width, the CB GaAs Γ-like state shifts upwards in energy crossing a set of AlAs X-like states. The main PL peak remains that originating from transitions between states concentrated on GaAs. At the same time, electrons transferred to AlAs X-like states cannot thermalize to GaAs Γ-like states of the CB because the latter are higher in energy. Therefore, those electrons in AlAs recombine with holes in GaAs giving a set of smaller peaks in the PL spectra.

In summary, we have presented PL spectra of a set of novel GaAs-AlAs SL with AlAs spikes placed at the center of the GaAs wells. The appearance of a set of new characteristic spectral features as a function of the   width of the spike is explained in terms of indirect gap effects introduced by AlAs.

REFERENCES

[1] Miguel J.L., Fujiwara K., Tapfer L. and Ploog K., "Effect of barrier thicness on the luminescence properties of AlAs-GaAs MQW grown by MBE", Appl. Phys. Lett. 47, 836-838 (1985).

[2] Tejedor C., Calleja J.M., Meseguer F., Mendez E.E., Chang C.-A. and Esaki L., "Raman resonance on E_1 edges in superlattices", Phys.Rev.B 32, 5303-5311 (1985).

Structure of Hydrogenated Amorphous Silicon/Silicon Oxide Superlattices

B. Abeles, L. Yang, W. Eberhardt,
and C. B. Roxlo,

Exxon Research and Engineering Co.
Annandale, NJ 08801, U.S.

IR, optical and photoemission spectroscopies show that a-Si:H layers on a-SiO$_x$:H form atomically abrupt interfaces, with the first few monolayers strongly disordered, while the a-SiO$_x$:H on a-Si:H interface is ~ 5Å wide. The offset between the valence bands is 4.1± 0.15 eV. Electroabsorption measurements indicate that there is a 0.17 eV asymmetry in the off-sets of the two interfaces.

Compositionally modulated amorphous semicondutors based on hydrogenated amorphous silicon[1] (a-Si:H) provide a new tool for studying fundamental properties of amorphous materials. The ability to make interfaces which are nearly atomically abrupt[2,3] with a relatively low density of defects[4,5] makes it possible to observe two dimensional behavior in ultra thin amorphous films, such as quantum size effects[1,6] and enhanced photoluminescence[5]. In this paper we focus on a-Si:H/a-SiO$_x$:H superlattices and investigate how the structure and electronic properties of the interfaces depend on the sequence in which the interfaces are grown. This system is particularly attractive because the large contrast in the electronic properties in the two materials facilitates the determination of the bonding and the electronic structure at the interfaces. Moreover in view of the extensive knowledge of the crystalline Si/SiO$_2$ interface, a comparative study of the amorphous interfaces is of considerable interest.

GROWTH, COMPOSITION AND BONDING

The superlattices were made by plasma assisted chemical vapor deposition using SiH$_4$ for the a-Si:H layers and a mixture of 2% SiH$_4$ in N$_2$O for the a-SiO$_x$:H layers[7]. Growth of the superlattice films was

monitored by the reflectance of a He-Ne laser (6328Å) incident at 67°
from the normal to the film, with E-field in the plane of incidence.

The reflectance in Fig. 1 exhibits
interference fringes (inset in Fig.
1) modified by fine structure which
is due to the interfaces formed by
the sublayers. It provides information on the structure of the
interfaces as well as on the optical
constants and thicknesses of the
sublayers[8]. After the plasma is
turned on during the N_2O +2% SiH_4
cycle, the initial steep rise in the
reflectance (A in Fig. 1) is due to
growth of ~10 Å oxide by plasma
oxidation of the underlaying silicon
layer. The more gradual increase in
the reflectance (B) is due to

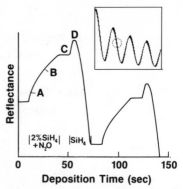

FIG. 1 Optical reflectance from
a growing a-Si:H/a-SiO$_x$:H (20
Å/30 Å) superlattice film.
Expanded section circled in
inset; gas flow periods
indicated by vertical lines.

deposition of the oxide by plasma CVD. The plasma is then turned off
for 10 sec. (C) to flush out the N_2O with SiH_4. The peak in the
reflectance immediately after turning on the SiH_4 plasma (D) is caused
by macroscopic roughness of the interfaces[8] of the order of 10-15A with
a lateral length scale of >100 A, as observed in TEM[9], and by extra
hydrogen at the interface.

The atomic composition of the films was determined by RBS and N^{15}
nuclear resonant reaction techniques[10]. The average composition of Si
and O in the superlattice films agreed with that calculated from the
measured composition of the bulk films ($SiH_{0.1}$ and $SiO_{1.9}H_{0.09}$) and the
sublayer thicknesses. However, the H concentration was substantially
higher in the superlattices and corresponded to ~$10^{15}cm^{-2}$ extra H per
repeat distance. More detailed information on the extra hydrogen at the
interfaces was obtained from the IR Si-H stretch bands shown in Fig.
2. In the bulk a-Si:H and in the superlattices with large a-Si:H layer
thickness, d_s, the 2000 cm^{-1} mode dominates indicating that H is bonded
primarily as monohydride. With decreasing d_s the interface Si-H bands
become more prominent. The 2080 cm^{-1} band is associated with dihy-
dride[10] or H bonded on the surfaces of microvoids[11] in an ~20Å a-Si:H

layer at the interface which is formed when a-Si:H is deposited on a-SiO$_x$:H (Si on SiO$_x$ interface). The 2100-2300 cm^{-1} band we attribute to Si-H bonds with Si back bonded to 1, 2 or 3 oxygen atoms at the interface formed when a-SiO$_x$:H is deposited on a-Si:H (SiO$_x$ on Si interface).

More detailed information on the atomic bonding is obtained from core level spectroscopy on single heterojunctions[12]. The Si on SiO$_x$ heterojunctions were made by depositing first a 40 Å a-SiO$_x$:H layer on top of a 400 Å a-Si:H buffer layer followed by a a-Si:H overlayer. The SiO$_x$ on Si heterojunctions were made by depositing the a-SiO$_x$:H overlayer on top of the a-Si:H buffer layer. The samples were transferred under UHV to the photoemission chamber.

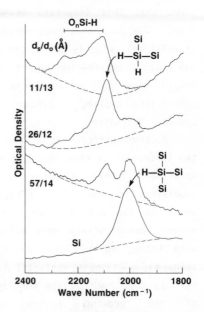

FIG. 2 Si-H stretching bands in a-Si:H/a-SiO$_x$:H superlattices and bulk a-Si:H. a-Si:H and a-SiO$_x$:H sublayer thickness d$_s$ and d$_o$ and vibrational modes are indicated.

The Si-2p core level spectra plotted for the two types of heterojunctions with different overlayer thicknesses are shown in Figs. 3 and 4 respectively. The growth of the Si overlayer in Fig. 3 manifests itself by an abrupt chemical shift of the overlayer photoemission peak to a kinetic energy 3.7 eV above that of the SiO$_x$:H underlayer. This chemical shift is due to the different Si coordination, four nearest Si neighbors in the a-Si:H overlayer, compared to four nearest O neighbors in the a-SiO$_x$ underlayer (0.9 eV chemical shift per oxygen bond). The near total absence of any suboxide in Fig. 3 suggests that the overlayer Si atoms bond only to Si in the underlayer and that oxygen takes no part in the bonding to the overlayer. On the other hand, the growth of the SiO$_x$ overlayer in Fig. 4, manifests itself by a considerably more gradual downward shift in the overlayer photoemission peak. A quantitative analysis of the data

shows that this interface is ~ 5Å
wide. The fact that this interface
is relatively wide is not surprising,
since the first few monolayers of a-
SiO_x:H grow by oxidation of the
underlaying a-Si:H, a process that
involves diffusion.

Another difference between the
two interfaces is that the spin orbit
splitting (0.6eV) in the a-Si:H
overlayers is smeared out (Fig. 3),
while in the a-Si:H underlayer it
becomes progressively more resolved
with increasing overlayer thickness
(Fig. 4). A simple explanation for
this is that the top few monolayers
of the growing a-Si:H layer
contribute a broad core line, as a
result of being strongly disordered
and hydrogenated. Plasma oxidation
consumes the defective Si layer and
the hydrogen in it, leaving behind a
less disordered a-Si:H layer with a
correspondingly narrower line
shape. We associate the hydrogen
which is incorporated in the oxide
with the 2100-2250 cm^{-1} band observed
in the IR (Fig. 2).

VALENCE BAND, CONDUCTION BAND AND
EXCITONS

The photoemission intensities of
the valence bands (V.B.) for the same
series of Si on SiO_x heterojunctions
as in Fig. 3, are shown in Fig. 5.
The peaks A and B correspond to the
0-2p bonding and non-bonding states[13]

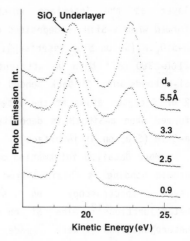

FIG. 3 Si-2p core levels for Si
on SiO_x. Thickness of a-Si:H
overlayers d_s indicated.

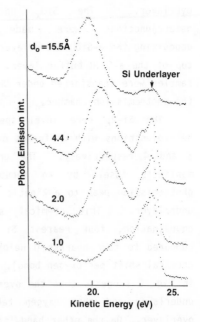

FIG. 4 Si-2p core levels for
SiO_x on Si. Thicknesses of SiO_x
overlayers d_o are indicated.

of the oxide V.B. The V.B. of the a-Si:H overlayer manifests itself by the shoulder C which becomes more prominent with increasing overlayer thickness. The off-set between the a-Si:H and a-SiO$_x$:H V.B. edges was 4.1 $\pm$ 0.15 eV for both types of heterojunctions. A small difference in the off-set energies for the two interfaces was in fact detected by the more sensitive electroabsorption spectroscopy (see below).

Figure 6 shows the absorption edges for a 15 Å a-SiO$_x$ overlayer on Si determined by partial yield spectroscopy. The absorption is due to transitions from the Si-2p core levels to conduction band states modified by core hole exciton interactions.[12,14] In the case of a-SiO$_x$:H the core hole exciton manifests itself as a peak, 2eV below the conducting band edge while in the a-Si:H layers, the exciton band is merged with the bottom of the conduction band and manifests itself as a steep edge. The energy band diagram determined from the photoemission measurements is shown in Fig. 7. Since the conduction band edge in the a-Si:H layers is not resolved in the UV absorption measurements we have placed it at 1.7eV above the valence band edge, where 1.7 eV is the optical band gap of bulk a-Si:H.

A technique which provides information on the asymmetry of the two

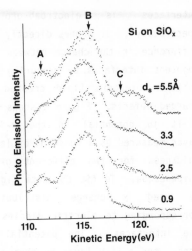

FIG 5. Valence bands for Si on SiO$_x$. Thicknesses of Si overlayers d$_s$ are indicated; features A, B, C discussed in text.

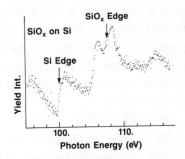

FIG. 6 Partial yield spectrum for 15Å a-SiO$_x$:H overlayer on a-Si:H. a-Si:H and a-SiO$_x$:H absorption edges are indicated.

interfaces is electroabsorption spectrocopy[7]. It probes directly the difference in the charge between two adjacent interfaces by measuring the built-in electric field created by charged defects. Figure 8 shows the built in potential in the a-Si:H layer measured as a function of layer thickness for two different oxide thicknesses. To fit the data Fig. 8 we used a charge distribution corresponding to electric dipoles at the interfaces. The potential across the a-Si:H layer in that case is , $\Phi_s = \psi d_s/(d_0\varepsilon_s + d_s\varepsilon_0)$, where ε is the dielectric constant and ψ is the potential step corresponding to the electrical dipole. The curves in Fig. 8 correspond to ψ = 170 mV. The potential step ψ gives rise to the asymmetry in the off-set in the band edges of the two interfaces and built in fields shown schematically in the inset of Fig. 8.

INTERFACE DEFECTS

The large difference in the structure and atomic spacing of the a-Si:H network compared to a-SiO$_x$:H or a-SiN$_x$:H leads to a defective interface region in superlattices made of these materials. The process of defect introduction at the interfaces is inherently asymmetrical. This is due to the fact that it is easier to introduce defects into material as it is

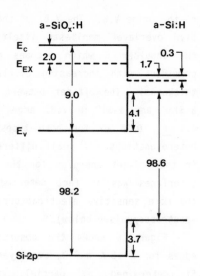

FIG. 7 Energy diagram of a-Si:H/a-SiO$_x$:H heterojunction showing the relative positions of Si-2p core levels, valence band edges, E_v, conduction band edges E_c and core hole exciton levels E_{ex}.

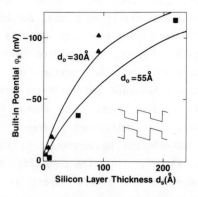

FIG. 8 Built-in potential in a-Si:H sublayer as a function of sublayer thickness d_s for different oxide sublayer thicknesses d_0. The solid lines are calculated; inset shows asymetry of V. B. offsets.

growing than into material which has already been deposited. Thus, at the Si on SiO_x interface one would expect that the first few monolayers of silicon would be defective. At the opposite interface most of the defects should appear in the oxide layer. Indeed such an asymmetry is consistent with the core level spectra. This is in contrast to the a-Si:H/a-Ge:H superlattices where the lattice mismatch is relatively small and the interfaces exhibit a very low density of defects[5].

Evidence for a disordered interface layer comes for instance from broadening of the Raman lines[15] and broadening of the optical absorption edge[1] in superlattices with a-Si:H layers thinner than ~40 Å. Figure 9 shows the absorption coefficient, measured by photothermal deflection spectroscopy and optical transmission, of bulk a-Si:H and two a-Si:H/SiO_x:H superlattices. The spectra exhibit an exponential tail and, outside the range of Fig. 9, a shoulder between 0.8 and 1.3 eV. The spectrum of the thinnest Si sublayer in Fig. 9 shows an increased band gap, which has been attributed to quantum size effects.[1]. The exponential tail is caused by shallow localized defects near the valence band maximum which have been attributed to topological disorder in the network.[16] This tail broadens as the a-Si:H layer spacing decreases, from a slope of 60 meV for bulk a-Si:H silicon to 150 meV for the 7Å a-Si:H sublayers. This slope provides a measure of the topological disorder present in the interface region.

The absorption in the 0.8 to 1.3 eV region (~2 cm^{-1} for bulk a-Si:H) is found to increase linearly with the density of interfaces. This absorption is attributed to deep defect states associated with dangling bonds at the Si on SiO_x interface. The extra absorption induced by the interfaces which corresponds to ~ 2 x 10^{11} cm^{-2} defects per interface is of the same magnitude as was observed in a-Si:H/a-SiN_x:H superlattices[4].

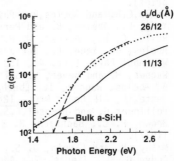

FIG. 9 Optical absorption α vs photon energy for bulk a-Si:H and a-Si:H/a-SiO_x:H superlattices. The a-Si:H and a-SiO_x:H sublayer thicknesses d_s and d_o are indicated.

SUMMARY

At the Si on SiO_x interface the top silicon layer bonds only to the Si atoms in the $a-SiO_x$:H underlayer resulting in an atomically abrupt interface. Because of the large lattice mismatch, the first few monolayers of a-Si:H are strongly disordered and hydrogenated. There is evidence that this highly disordered layer appreciably slows down the decay of picosecond induced optical absorption[17]. On the SiO_x on Si interface, the first ~10 Å of $a-SiO_x$:H grow by plasma oxidation of a-Si:H, forming a ~5 Å wide interface which is graded in oxygen and hydrogenated. The off-set of the valence bands determined by photoemission is 4.1± 0.15eV; electro-absorption measurements indicate that the off-set is 0.17 eV larger in the Si on SiO_x interface.

We thank W. Lanford, D. Sondericker, P. D. Persans and H. Stasiewski for their contributions. The photoemission work was done at the Brookhaven National Laboratory NSLS.

1. Abeles,B. and Tiedje,T., Phys Rev. Lett. 51 2003 (1983).
2. Roxlo, C.B., Abeles, B., and Tiedje, T., Phys Rev. Lett. 1994, 52 (1984).
3. Persans,P.D., Ruppert, A.F, Abeles,B, and Tiedje,T, Phys Rev. B31, 5558 (1985).
4. Tiedje, T. and Abeles, B., Appl. Phys Lett. 45, 179 (1984).
5. Tiedje,T., Abeles,B and Brooks,B., Phys Rev. Lett. 54, 254 (1985).
6. Wronski,C.R., Persans,P.D. and Abeles,B., Appl Phys Lett. (in print).
7. Roxlo,C.B. and Abeles,B., Phy Rev. B (in print).
8. Yang, L., Abeles, B. and Persans, P., Appl Phys Lett. (in print).
9. Roxlo, C. B., Deckman, W. H. and Abeles, B., (unpublished).
10. Abeles, B., Yang, L., Persans, P. D., Stasiewski, H. C. and Lanford, W., Appl Phys. Lett. 48, 158 (1986).
11. Wagner, H. and Beyer, W. Solid State Commun. 48, 585 (1983).
12. B. Abeles, Wagner, I., Eberhardt, W., Stohr, J., Stasiewski, H. and Sette, F, AIP Conf. Proc. 120, 394 (1984).
13. Hollinger,G. and Himpsel,F.S., J. Vac Sci. Technol. Al, 640 (1982).
14. Brown, F. C., Bachrach, R. Z. and Skibowski, M., Phys. Rev. B. 4781 (1977).
15. Maley, N. and Lannin, J. S., Phys. Rev. B31 5577 (1985).
16. Cody, G. D., Tiedje, T., Abeles, B., Brooks, B. and Goldstein, Y., Phys. Rev. Lett. 47, 1480 (1981).
17. Grahn, H. T., Stoddart, H. A., Zhou, T., Vardeny, Z., Tauc, J and Abeles, B., Prodceedings 18th Int. Conf. on Physics of Semicon.

RAMAN SCATTERING FOR STUDIES OF SEMICONDUCTOR-HETEROSTRUCTURES AND SUPERLATTICES

G. Abstreiter and H. Brugger

Physik-Department E 16, Technische Universität München
D-8046 Garching, Fed. Rep. of Germany

ABSTRACT

The application of inelastic light scattering for studies of semiconductor heterostructures and superlattices is discussed. Emphasis is put on phonon spectroscopy in strained overlayers and multilayer structures. Built-in strains, critical thickness, and Brillouin zone folding effects are studied using various aspects of phonon Raman scattering.

1. INTRODUCTION

Semiconductor heterostructures and superlattices exhibit a large number of new effects which are caused by the reduced dimensionality and by specially tailored band structures. The new properties led to a strongly increased interest in such systems, both from fundamental physics point of view as well as for possible near future device applications. Important for the achievement of the desired properties are the interfaces of the two semiconductors involved. Until recently, high quality interfaces with a negligible density of interface states were only realized in nearly lattice matched systems such as GaAs/AlAs. Non-lattice matched and consequently strained heterostructures and superlattices of good quality are only achieved when a certain critical thickness of the individual layers is not exceeded. The built-in strain can be used as an additional parameter for band structure engineering.

The large number of different systems require also a variety of different experimental techniques to characterize and analyze the fabricated semiconductor structures. Inelastic light scattering has been developed in the past years besides others, as a powerful probe of various properties of semiconductor heterostructures and superlattices. This involves mainly three types of excitations (see Fig. 1). Electronic light scattering can usually only be performed under resonance conditions. It measures the elementary excitations of two- and three-dimensional electron and hole gases, which involves single particle excitations, plasmons and coupled phonon-plasmon modes. In two-dimensional systems both inter- and intrasubband transitions of single particle and collective nature are observed. The extensive work on electronic excitations has been reviewed in the past already several times /1,2,3,4/. Therefore it is not repeated here. We concentrate on phonon Raman scattering in strained layer systems. Emphasis is put

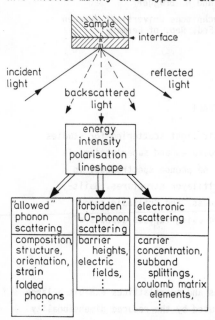

Fig. 1: Schematics of inelastic light scattering in semiconductor hetero-structures

on Si_xGe_{1-x} overlayers on GaAs and on Si/Si_xGe_{1-x} superlattices. Allowed phonon scattering is used to determine structure and composition, built-in strains, critical thickness and superlattice effects. Forbidden LO-phonon scattering in polar semiconductors like GaAs is used to learn about the formation of barriers, e.g. Ge on GaAs, and on the critical thickness of Si_xGe_{1-x} strained overlayers on GaAs. Selected examples will be discussed in the following.

2. DETERMINATION OF CRITICAL THICKNESS BY RAMAN SCATTERING

One of the most important quantities to know in order to achieve
high quality strained layer structures is the critical thickness of
the strained film up to which the lattice mismatch   is accomodated by
elastic deformation. At the critical thichness d_c the deformation energy
is partly reduced by formation of misfit dislocations. Knowledge of the
onset for this strain relaxation is necesary before one starts growing
superlattices. Both allowed and forbidden Raman scattering can be used
to achieve information on the critical thickness.

2.1 ALLOWED RAMAN SCATTERING

Phonon Raman scattering has been used to study "in situ" the nature
of growing Si_xGe_{1-x} films on GaAs (110) cleavage surfaces. The samples
are grown in an UHV chamber by evaporating Ge and Si onto GaAs substrates.
Pure Ge (x=0) is nearly lattice matched to GaAs. High quality two-dimensional growth is achieved for substrate temperatures of the order of
350^0 C to 400^0 C. The optical phonons of Ge can be detected for overlayers
as thin as 1 nm /5,6/. The intensity is increasing strongly with increasing
film thickness. At T=100K Ge grows amorphous. With increasing x the
difference in lattice constant can be changed continuously from 0 to
about 4% for pure Si. Thick Si-films on GaAs exhibit a Raman spectrum
which is typical for amorphons layers up to substrate temperatures of
500^0 C. The phonon spectrum of Si_xGe_{1-x} alloys shows three characteristic
peaks which are related to the Si-Si, Si-Ge, Si-Si vibrations in the
random alloy. The composition can be determined from the energy positions
and the intensity ratios of these quasi-optical modes. Fig. 2 shows a
series of Raman spectra obtained from a cleavage (110) surface with
different $Si_{0.5}Ge_{0.5}$ overlayers. The three $Si_{0.5}Ge_{0.5}$ modes appear in
the spectrum already at layer thickness of about 1 nm. Their peak energies,
however, are shifted quite strongly to smaller energies. This shift is
caused by the built-in strain due to the lattice mismatch of 2%. The
smaller average lattice constant of $Si_{0.5}Ge_{0.5}$ leads to a tetragonal
distortion of the unit cell which corresponds to a tensile in-plane
strain responsible for the downward shift of the phonons. With increasing
layer thickness the phonon lines grow in intensity.

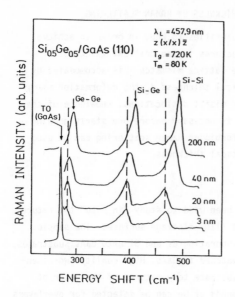

Fig. 2: Allowed phonon scattering
from GaAs (110) cleavage surfaces
with various overlayers of $Si_{0.5}Ge_{0.5}$

Above 20 nm they shift slowly to higher energies and approach the values for unstrained $Si_{0.5}Ge_{0.5}$. Formation of dislocation lines is responsible for this strain relaxation. Similar measurements have been performed for various x-values. The onset of the phonon shift is directly related to the critical thickness d_c. Approximate values are $d_n \simeq 200$ nm, 20 nm and 2 nm for differences in lattice constants of 1%, 2% and 3%, which corresponds to $x \simeq 0.25$, 0.5 and 0.75. It should be mentioned, however, that the phonon spectra measure an average strain distribution over the information volume. They are consequently not very sensitive to the onset of strain relaxation and tend to give to high values for d_c.

2.2 FORBIDDEN LO-PHONON SCATTERING

Symmetry forbidden LO-phonon scattering has been used as a sensitive technique to study the formation of semiconductor heterostructures /5,6/. Under resonance conditions and in the presence of a macroscopic electric field, the LO-phonon in GaAs, which is forbidden in back scattering from (110) surfaces, is comparable in intensity with the allowed TO-phonon line. The strengh is proportional to the square of the electric field. Close to the surface or interface therefore it measures directly the barrier height. Details of the formation of lattice matched Ge/GaAs interfaces are discussed in Refs. /5,6/. A freshly UHV-cleaved (110)

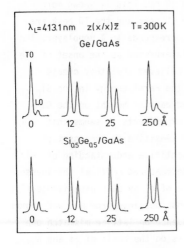

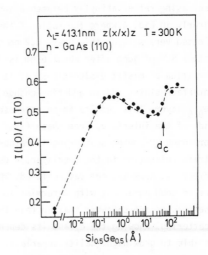

Fig. 3: Resonant excitation of forbidden LO-phonon scattering in GaAs for various overlayers of Ge and $Si_{0.5}Ge_{0.5}$

Fig. 4: Normalized intensity of forbidden LO-phonons versus coverage with $Si_{0.5}Ge_{0.5}$

GaAs surface exhibits a negligible small surface barrier. This gives rise to a small forbidden LO-phonon line which can be seen in the spectra of Fig. 3. The LO intensity is, however, strongly increasing already at submonolayer coverages, reaches a maximum at a layer thickness of about 1,2 nm, and decreases again for thicker overlayers on n-GaAs. The behaviour directly reflects the formation of the barrier at the interface starting from nearly flat band condition for the clean surfaces. In the sub-monolayer region the formation of surface states leads to an increase of the barrier height due to Fermi level pinning. In the case of high quality growth and thick overlayers the barrier height is reduced again due to the small conduction band offset in the GaAs/Ge system and the lack of Fermi level pinning. This is reflected directly by the intensity of the forbidden LO-phonon shown in the series of spectra in Fig. 3.

For $Si_{0.5}Ge_{0.5}$ strained overlayers the behaviour is very similar

in the region of small coverages as it is also shown in Fig. 3 (lower
part). With increasing layer thickness the LO intensity is, however,
increasing rather abruptly between 6 and 10 nm. This is shown more
clearly in Fig. 4 where the ratio of the LO peak and the TO peak are
plotted versus layer thickness. 0,2 nm corresponds to 1 monolayer. The
rather abrupt jump after about 6 nm is interpreted as the onset of the
formation of misfit dislocations. It is believed that they create
interface states and in addition change the Fermi energy in the $Si_{0.5}$
$Ge_{0.5}$ layer. This results in an increased barrier height on the GaAs
side of the interface. From these first experiments it seems that
forbidden LO-phonon scattering is rather sensitive to the onset of
strain relaxation in the overlayer. A quantitive understanding of this
effect is, however, not yet achieved. The measured critical thicknesses
are in good agreement with determinations of d_c by LEED experiments /7/.
The results are somewhat smaller than those obtained from allowed phonon
scattering. Knowing the d_c and its dependence on lattice mismatch one
is able to grow high quality superlattices on the basis of Si and Ge.

3.   SYMMETRICALLY AND ASYMMETRICALLY STRAINED LAYER SUPERLATTICES

In the work of Kasper et.al. /8/ it was shown that it is possible
to achieve a strain symmetrisation in $Si-Si_xGe_{1-x}$ superlattices by
choosing the right substrate. Superlattices grown directly on Si substrates are asymmetrically strained, the Si layers are unstrained, the
Si_xGe_{1-x} layers are biaxial compressed. A symmetrical strain distribution
is achieved by growing first a thick buffer layer of intermediate composition. In such a way the total strain energy can be minimized. It
also leads to a band offset such that the conduction band of Si is
lowered strongly below the conduction band edge of Si_xGe_{1-x} which leads
to mobility enhancements for electrons in selectively doped samples
/9,10/. The strain distribution can easily be measured by allowed phonon
Raman scattering as discussed already in section 2. Fig. 5 shows two
phonon Raman spectra of symmetrically and asymmetrically strained superlattices. The built-in strain is directly reflected by the shift of
the phonon lines. Contrary to Si_xGe_{1-x} overlayers on GaAs, the three
Si_xGe_{1-x} modes are shifted upwards which is caused by the compressive

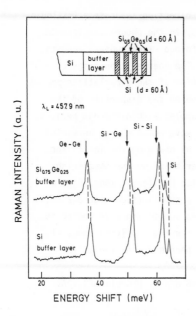

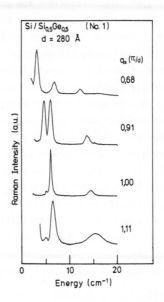

Fig. 5: Phonon spectra of symmetrically and asymmetrically strained Si/Si$_{0.5}$Ge$_{0.5}$ superlattices. The insert shows schematically the layer sequence

Fig. 6: Folded acoustic phonons for various scattering wave vectors in the range of the superlattice Brillouin zone boundary (π/d)

biaxial strain in the present situation. The shift is approximately twice as large for the sample grown directly on the Si substrate compared with the Si$_{0.75}$Ge$_{0.5}$ buffer layer. The Si optical phonon mode at around 65meV (520cm^{-1}) is not shifted at all in the asymmetrically strained superlattice, while it is shifted downwards in the symmetrical case. The positions of the phonons in unstrained semiconductors are marked by the arrows. The phonon energies reflect directly the strain distribution in the samples. The built-in strain has a strong influence on the band gaps and the band ordering. This has been discussed for example in Ref. /11/.

In the last part we concentrate on superlattice effects which show up in the Raman spectra via folded acoustic phonons. The periodicity

of multilayer structures along the growth direction leads to a Brillouin
zone folding and the appearence of new gaps in the phonon spectra. In
Si/Si_xGe_{1-x} strained layer superlattices folded longitudinal acoustic
phonons have been observed recently /12/. The energetic position of
these modes allowed an accurate determination of the superlattice period.
Very recently Brugger et al. /13/ have studied the dispersion of the
acoustic superlattice modes for various samples with different periods.
In backscattering geometry the scattering wavevector q_s can be varied
from about $0.7 \times 10^6 cm^{-1}$ to $1.5 \times 10^6 cm^{-1}$ by changing the laser line.
For superlattices with periods of the order of 30 nm the scattering
wavevector is comparable to the superlattice Brillouin zone. Fig. 6
shows some Raman spectra of a $Si/Si_{0.5}Ge_{0.5}$ superlattice with d=28nm.
The dispersion of the modes with increasing scattering wave vector is
clearly observed. The finite splitting of the two low lying modes at
$q_s = \pi/d$ shows directly the new gap in the phonon dispersion at the
zone boundary.

This work was supported by Siemens AG and by DFG via SFB 128.

REFERENCES

/1/   G. Abstreiter, M. Cardona, A. Pinczuk, in Topics in Applied Physics
       54, Springer-Verlag (1984) p. 5
/2/   A. Pinczuk and J.M. Worlock, Surface Science 113, 69 (1982)
/3/   G. Abstreiter, in Molecular Beam Epitaxy and Heterostructures, eds.
       L.L. Chang and K. Ploog, Martinus Nijhoff Publishers (1985), p.425
/4/   G. Abstreiter, R. Merlin, A. Pinczuk, IEEE special issue, in press
/5/   H. Brugger, F. Schäffler, G. Abstreiter, Phys.Rev.Lett 52, 141 (1984)
/6/   G. Abstreiter, in Festkörperprobleme XXIV, Ed. P. Grosse, Vieweg
       Braunschweig (1984), p. 291
/7/   K. Eberl, H. Brugger, G. Abstreiter, to be published
/8/   E. Kasper, H.J. Herzog, H. Dämbkes and Th. Ricker, Springer Series
       in Solid State Sciences 67, Springer Verlag (1986), p. 52 and ref.
       therein
/9/   H. Jorke and H.J. Herzog, in Proc. 1st Int. Symp. Si MBE ed. J.C.
       Bean, The Electrochem. Soc. Pennington (USA) (1985), p. 360
/10/   G. Abstreiter, H. Brugger, T. Wolf, H. Jorke, H.J. Herzog, Phys.
       Rev. Lett. 54, 2441 (1985)
/11/   G. Abstreiter, H. Brugger, T. Wolf, R. Zachai, Ch. Zeller in Ref.8
       p. 130
/12/   H. Brugger, G. Abstreiter, H. Jorke, H.J. Herzog, E. Kasper, Phys.
       Rev. B 33, 5928 (1986)
/13/   H. Brugger, H. Reiner, G. Abstreiter, H. Jorke, H.J. Herzog, E. Kasper
       Proc. of the 2nd Int. Conf. on Supperlattices, Göteborg (1986)
       to be published

GAP STATE OPTICAL ABSORPTION AND LUMINESCENCE STUDIES OF a-Si:H/a-SiN$_x$:H AND a-Si:H MODULATION DOPED MULTILAYERS

S KALEM, M HOPKINSON, T M SEARLE, I G AUSTIN, W E SPEAR* and

P G LeCOMBER*

Department of Physics, The University, Sheffield S3 7RH, England

*Department of Physics, The University, Dundee DD1 4HN, England

We have investigated gap state absorption and
luminescence properties of a-Si:H/a-SiN$_x$:H and a-Si:H
modulation doped multilayers. We suggest interface
states, rather than quantum confinement effects, may
determine the optical properties when layer thicknesses
are reduced below about 30Å.

Despite several investigations(1, 2, 3), quantum mechanical
confinement has not yet been unambiguously identified in amorphous
multilayer structures. Size effects are clearly seen, but they may be
due to chemical contamination or to interface defects. In this paper,
we present the results of photothermal deflection spectroscopy(PDS)
and photoluminescence(PL) measurements on a-Si:H/a-SiN$_x$:H and a-Si:H
modulation doped multilayers(ML) in an attempt to clarify the roles of
confinement and defects in these structures.

Multilayers consisting of a-Si and a-SiN$_x$ were prepared by the
glow discharge technique from SiH$_4$ and NH$_3$/SiH$_4$ mixtures (gas phase
ratio R) at 330°C. The a-SiN$_x$ (R ∼ 3.8) layer thickness was kept
constant at either 17Å or 30Å, but the a-Si layer thickness d$_{Si}$ was
varied from 12Å to 500Å. The a-Si:H modulation doped ML were
deposited at ∼ 275°C using B$_2$H$_6$ and PH$_3$ doping levels of about
100vppm. The PDS used a 450W Xe arc lamp, PL was excited with a Kr$^+$
laser.

Typical PDS spectra of some a-Si/SiN$_x$ ML are shown in Fig 1.
These samples have different d$_{Si}$ thicknesses, but each contains 40

layer pairs. The d_{Si} = 95Å ML is an exception, containing 20 pairs. Spectra of the individual sublayer materials are also shown. In addition the figure contains the absorption spectrum of a typical modulation doped ML with a sublayer thickness of ~ 150 A and containing 20 layer pairs.

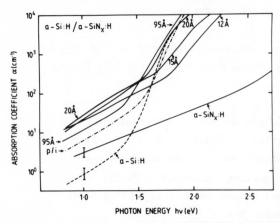

Fig 1   Thickness dependence of gap state
        absorption of a-Si:H/SiN$_x$:H ML

Fig 1 shows that the absorption edges are strongly blue shifted when d_{Si} is less than 20Å. Similar shifts have also been observed with transmission measurements (2). The Tauc gap E_T is plotted versus d_{Si} on Fig 2a. The most striking feature is the rapid change at low d_{Si}.

Fig 1 also shows that the edge broadens as d_{Si} decreases, and Fig 2b plots the Urbach energy E_0. For $d_{Si} > 20$Å E_0 is almost thickness independent for the 40 layer samples but increases abruptly in the thinnest ML. We note that E_0 is smaller in ML with fewer layers and also in the modulation doped ML.

In the region below about 1.5–1.6 eV, the absorption coefficient at a given energy is almost independent of the sublayer thickness. The absorption in this region reveals a wider tail due to the opening gap. Thus the integrated absorption I_{ex} from the lowest energies to the Urbach tail increases with decreasing d_{Si} (Fig 2c).

The PL band at 0.9 eV is also attributed to defects. At low temperature and with above gap excitation, the spectra of all low defect films are dominated by the 1.3 eV emission, and higher temperatures or lower excitation energies are required to see the other band. Our earlier work showed that the 0.9 eV band was not visible in the ML even at 300K; however, the new data of Fig 3 obtained with subgap excitation show the increasing 0.9 eV component in thinner layers. As in the absorption data of Fig 2, the PL spectra only alter significantly when $d_{Si} < 30\text{Å}$.

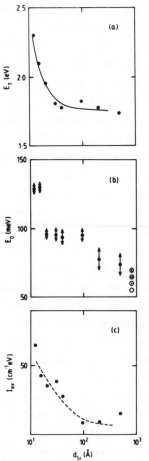

As others (2) have observed, the main evidence for quantum confinement effects is the blue shift of the gap. The full line on Fig 2a is the prediction of a Kronig-Penney model using plausible masses, gaps and band offsets, to be described in detail elsewhere (4), and the agreement with the data is encouraging. However, other features of the absorption, for example the broadening of the Urbach tail and its dependence on layer number seen in Fig 2b, are less readily explained by confinement and instead suggest increased disorder. Recently, several reports have emphasised interface effects, and in particular the presence of excess H and N in the first 10-20Å of a-Si grown onto a-SiN$_x$ (5). E_T and E_0 depend on H and more sensitively on N (4, 6) content, and so the results of Fig 2a and 2b are at least qualitatively consistent with the existence of such layers. The strong broadening, Fig 2b, in the ML suggests the involvement of N rather than H. We have pointed out the similarities of the PL behaviour of the ML and films containing either excess H or N (3)

Fig 2 Band gap (a), Urbach energy (b), excess absorption (c) vs thickness. Open circles show p(or n)/i (◖), i/p(or n) (◗), n/p (◉), and a-Si:H (○)

The PDS measurements in the defect region also suggest the presence of interface defect states which are similar in all Si/SiN$_x$ ML. Absorption due to interfaces depends on the number of layers not the ML thickness. This interpretation is confirmed by the observation that the absorption of the 95Å film, which contains 20 rather than 40 layers is reduced by a factor 2.0. In single films, we find that increased H enhances the defect absorption (4).

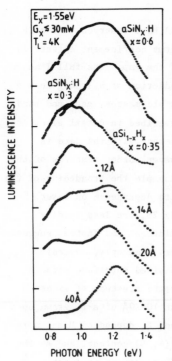

Fig 3 shows how the 0.9 eV band dominates the spectra for $d_{Si} < 20\text{Å}$. These observations probably reflect the decreasing volume of material unoccupied by the interface states. PL picks out the regions of lowest defect density material and so the absence of the 1.3 eV emission strongly suggests that the interface volume fills all the Si sublayer of the 12Å ML; its weakness in the 14Å and 20Å ML suggest that here too most of the volume is so occupied. The upper spectra in Fig 3 compare low excitation energy PL spectra of two SiN_x films, and of a low deposition temperature, high H content film. In contrast with the above gap excited PL spectra, the N alloys and the ML look quite different. The high H film shows the 0.9 eV band characteristic of low efficienty (∼ 4%) material. These comparisons suggest that H is more likely to be associated with the interface defects than N.

Fig 3 Sub gap excited PL spectra of a-SiN_x:H, a-Si:H and a-Si:H/SiN_x:H multilayers

In conclusion, our optical measurements support the suggestion that interface states, probably associated with H, dominate the properties when $d_{Si} < 30\text{Å}$. We cannot exclude quantum confinement effects, but believe that their existence will only be unambiguously demonstrated when N and H effects can be excluded.

REFERENCES

(1)   Ibaraki, N and Fritzche, H, Phys Rev, B30, 5791 (1984)
(2)   Abeles, B and Tiedje, T, Semicond & Semimet, 21, C, 407 (1984)
(3)   LeComber, P G et al, J Non-Cryst Solids, 77 & 78, 1081 (1985)
(4)   Kalem, S et al, to be published
(5)   Abeles, B et al, J Non-Cryst Solids, 77 & 78, 1065 (1985)
(6)   Jackson, W A et al, J Non-Cryst Solids, 77 & 78, 909 (1985)

ENERGY DISTRIBUTION OF EXCESS ELECTRONS TRAPPED IN THE TAIL STATES OF
AMORPHOUS SILICON DOPING SUPERLATTICES

F. Walloch, L. Ley, and P. Santos

Max-Planck-Institut für Festkörperforschung, Heisenbergstraße 1,
D 7000 Stuttgart 80, Federal Republic of Germany

The excitation spectra of the quasi-stationary distribution
of trapped electrons and holes in doping superlattices at 10 K
are measured using spectrally resolved ir photoconductivity.
The spectra are interpreted in terms of optical transitions
between carriers distributed in energy according to Monroe's
model and their respective band edges.

1. INTRODUCTION

In amorphous silicon photogenerated carriers thermalize in an
exponential distribution of localized band tail states before they
recombine. In the multiple trapping (MT) model for dispersive transport
the occupation of the tail states is determined through the dynamics of
trapping and thermal reemission[1,2]. This leads to a time (t) and temperature (T) dependent demarcation energy E_D that separates tail states
($E<E_D$; $E_C\equiv0$) that are in thermal equilibrium with extended states from
those ($E>E_D$) where thermal reexcitation is no longer possible during
the observation time: $E_D(t) = kT \ln(\nu_0 t)$; ν_0 is the attempt-to-escape
frequency. This model, deduced under the proviso that trapping is only
possible from extended states, has recently been modified by Monroe[3]
insofar as at low temperatures direct tunneling transitions of trapped
carriers to deeper states contribute to their energy distribution. This
leads to a different demarcation energy which now separates essentially
occupied from essentially empty tail states:
$E_D(t) = 3kT_0 \ln(\ln(\nu_0 t)) - kT_0 \ln(8\gamma^3/N_L)$. Here N_L is the total number
of localized states, $1/\gamma$ the decay length of wave function, and T_0
characterizes the width of the exponential tails. Using values for the
parameters appropriate for electrons in a-Si:H (kT_0=25meV, $1/\gamma$=10Å,
N_L=1x10²⁰cm⁻³) we calculate for an observation time of 100 sec and a
temperature of 10 K the demarcation energy in the MT model to be 27 meV
and in Monroe's model to be 150 meV. The demarcation energies are in
both cases independent of the number of carriers in the tail, i.e. E_D
does not have the character of a quasi-Fermi-level.
In doping superlattices based on amorphous silicon the concentration of

metastable carriers residing in band tails exceeds that in unstructured material by more than an order of magnitude: recombination at low temperatures (T≤50K) is strongly supressed due to the spatial separation of photogenerated carriers[4]. Moreover, the total concentration of trapped carriers can be adjusted in a controlled fashion by ir quenching. Under these favourable conditions it is possible to determine the excitation spectrum of the quasi-stationary distribution of excess carriers in the band tails of amorphous doping superlattices using spectrally resolved photoconductivity and to determine the value of the demarcation energy if it exists.

2. SAMPLE PREPARATION AND EXPERIMENTAL PROCEDURE

Measurements were performed on an amorphous silicon superlattice of the nipi type with a period of 400 Å and equal sublayer thicknesses[4]. The maximum concentration of trapped excess carriers at low temperatures in this sample is expected to be N_{sat}=2x10^{18}cm^{-3} assuming a modulation amplitude of 0.9 eV[4]. The sample is provided with interdigitized Cr surface contacts with an effective gap of 0.4x70mm^2 which are used to measure the photoconductivity with an applied voltage of 100 V. The sample is maintained at 10 K. Illumination with band gap light (hν=1.92eV, 0.5mW/cm^2) for 15 sec generates the maximum possible concentration of spatially separated carriers, N_{sat}. After a dark time of 60 sec the sample is illuminated with light from a Ge filtered tungsten halogen lamp (hν<0.7eV, 60mW/cm^2) which reduces the concentration of excess carriers through enhanced recombination. Under the assumption that the photocurrent measured during illumination with Ge filtered light is proportional to the instantaneous concentration of excess carriers N_{sep}, we adjust fill factors N_{sep}/N_{sat} between 100% and 5%. After a further dark time of 60 sec photoconductivity spectra are measured between 0.1 and 0.9 eV photon energy using digital lock-in techniques.

3. RESULTS AND DISCUSSION

The measured photoconductivity spectra normalized to the incoming photon flux are shown in Fig.1. Dark and infrared conductivities of the sample in state A, i.e. cooled down from room temperature in the dark, are below our detection limit. Consequently, the conductivities shown in Fig.1 are solely due to trapped carriers. The spectra are increasing functions of photon energy and their amplitude scales roughly with the fill factor N_{sep}/N_{sat}. The spectral shape of σ_{ph} exhibits two distinct regimes. Between 0.15 and 0.18 eV the spectra rises proportional to $h\nu^{-1}(h\nu-E_D)^{1/2}$ with a threshold of E_D=(0.15±0.01)eV.

Above 0.3 eV the power-law exponent changes to 3/2 with a threshold of (0.26±0.01)eV. The absorption band at 0.25eV is due to Si-H stretching modes in the film and the dips at 0.4eV and 0.46eV are due to bands of H_2O covering the windows of the cryostat. The fixed threshold energies are at odds with a simple picture of trapped carriers filling the band tails. In that case an increase in threshold is expected as the quasi--Fermi-levels move deeper into the band tails with decreasing fill factor. A fixed threshold is expected in both the MT model an its modification due to Monroe. The threshold energy of 150meV, however, clearly favours Monroe's description of electrons trapped in the conduction band tail.

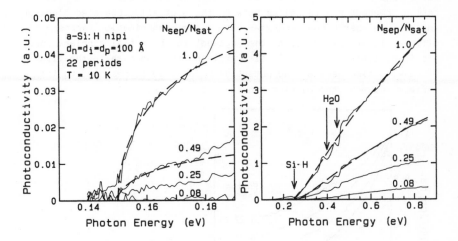

<u>Fig.1</u> Photoconductivity spectra normalized to the incoming photon flux. Parameter is the fill factor N_{sep}/N_{sat}. Dashed lines are power-law fits as described in the text.

At low temperatures the normalized photoconductivity is proportional to the absorption coefficient of the sample[5]. Applying the O'Connor-Tauc model for optical transitions between a sharp initial distribution at E_D and a free electron like conduction band ($N_C \propto (E-E_C)^{1/2}$) gives the observed power-law dependence of σ_{ph} on hν for the lower energy part of the spectra. By the same token (constant matrix element and $\eta\mu\tau$ product) the 3/2 power-law above 0.3eV requires a constant density of occupied states starting 0.26eV below E_C and extending into the gap provided we ascribe these transitions also to electrons. From our analysis we calculate for a fill factor of 100% the density of occupied states in the flat portion to be two orders of magnitude lower than in the peak at E_D. This ratio is reduced to a factor of ten for a fill factor of 10%. Alternatively we might ascribe this part of the spectrum to hole emission from the valence band tail providing $N_V \propto (E-E_V)^{3/2}$. The threshold energy, when interpreted as E_D in the framework of Monroe's model, yields for the valence tail slope a value of (45 ± 5)meV in reasonable agreement with optical[6] and photoemission[7] data. A problem with this interpretation is the larger integrated contribution from trapped holes compared to that from electrons.

The time decay of the photoconductivity during illumination with different photon energies is shown in Fig.2. We observe a faster decay for higher photon energies, i.e. the recombination rate increases with the excess energy of the carriers above the mobility edges. Since recombination proceeds via tunneling between the spatially separated electrons and holes higher excess energies enable the recombination partners to diffuse closer towards each other against the potential barriers inherent in nipi structures. This also explains the increase in the lifetimes with time because the potential modulation is restored to its maximum value as the carrier concentration decreases[4].

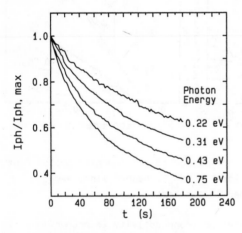

Fig.2
Time decay of photocurrent during illumination with different photon energies.

4. REFERENCES

1) Tiedje, T., Rose, A., Solid State Commun. **37**, 49(1980)
2) Orenstein, J., Kastner, M.A., Sol. State Commun. **40**, 85(1981)
3) Monroe, D., Phys. Rev. Lett. **54**, 146 (1985)
4) Hundhausen, M., Ley, L., Carius, R., Phys. Rev. Lett. **53**, 1598(1984)
5) Hoheisel, M., Carius, R., Fuhs, W., J. Non-Cryst. Solids **59+60**, 457(1983)
6) Cody, G.D., Semiconductors and Semimetals **21B**, 11(1984)
7) Miramontes, R., Winer, K., Ley, L., in these proceedings

A SPECTROSCOPIC ELLIPSOMETRY STUDY OF a-Ge:H/a-Si:H INTERFACES

A.M. ANTOINE and B. DREVILLON
Equipe Synthèse de Couches Minces pour l'Energétique (ER 258)
LPNHE, Ecole Polytechnique, 91128 Palaiseau, FRANCE.

The formation of a-Ge:H/a-Si:H interfaces is observed in real-time by in-situ spectroscopic ellipsometry. The structure of the a-Ge:H/a-Si:H and a-Si:H/a-Ge:H is found different. A density deficit is observed on the growth of the first 20 Å of a-Si:H while the early stage of the growth of a-Ge:H on a-Si:H is compatible with an uniform growth model.

A great deal of attention has been recently given on amorphous semiconductors superlattices generally consisting of thin layers of hydrogenated amorphous silicon (a-Si:H) interleaved periodically with a-Ge:H[1]. The knowledge of the structure, at the atomic scale, of the interface region between various layers is an important issue of the understanding of the electronic properties of amorphous or crystalline superlattices. High sensitivity and the capability of obtaining data in-situ make spectroscopic ellipsometry an useful technique for addressing such a problem. In this paper, we present the results of the first real-time SE study of the growth and microstructure of a-Ge:H/a-Si:H and a-Si:H/a-Ge:H heterojunctions.

These heterojunctions are prepared from a-Si:H and a-Ge:H films deposited on metallic substrates heated at 250°C by glow discharge decomposition (13.56 MHz) of SiH_4 and 50% (GeH_4, H_2) mixture respectively, the plasma being interrupted when alternating the gas composition. The film preparation conditions are described in detail elsewhere[2]. The growth processes of the heterojunctions are analyzed in-situ using a spectroscopic phase modulated ellipsometer[3]. During deposition, the conventional ellipsometric angles ψ and Δ are measured at 0.25 s intervals (every $\sim$ 0.4 Å of film growth), the energy of the incoming light being fixed. A shutter located immediately in front of the substrate is used during plasma stabilization. An energy corresponding to a very low penetration depth of the light ($\sim$ 100 Å) in the growing film is chosen: 2.7 eV and 3.5 eV are respectively used for a–Ge:H and a–Si:H in order to get a good contrast with the substrate.

    Experimental (ψ , Δ) trajectories corresponding to the growth of a–Ge:H and a–Si:H films on metallic substrates are displayed in Figs 1 and 2 (solid lines). The characteristic decrease of Δ at fixed ψ immediately after opening the shutter, observed in both cases, is an evidence of a nucleation mechanism [2,4]. The incomplete coalescence of the nuclei generally leads to the formation of a surface roughness [3, 4]. This behaviour is illustrated in Figs. 1 and 2. The dotted lines correspond to an homogeneous growth model (A) while the dashed curves correspond to the following inhomogeneous growth assumption (B): In the first step it is assumed that hemispherical nuclei are created with an average distance d ($\simeq$ 30–60 $\overset{\circ}{A}$) between them, the radius of the nuclei increasing with time. In the second step, the absence of coalescence of the hemispheres results in an d/2 thick overlayer above the growing film. In each case, the refractive index of the hemispheres corresponds to the bulk material value. Figs 1 and 2 show that a rather good fit is obtained with model B in both cases. In contrast model A gives a very bad agreement with the experimental trajectories.

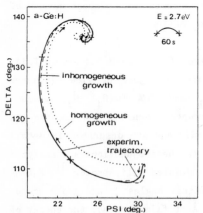

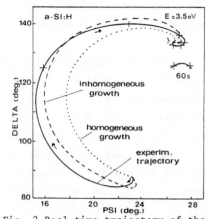

Fig.1 Real-time trajectory of the growth of a–Ge:H on a metallic substrate compared to the two theoretical models described in the text.

Fig. 2 Real-time trajectory of the growth of a–Si:H on a metallic substrate compared to the two theoretical models described in the text.

The real-time inspection of a-Ge:H/a-Si:H and a-Si:H/a-Ge:H heterojunctions are respectively displayed in Figs. 3 and 4. In order to obtain a detailed description of both interfaces different models are considered, in each case the substrate and its surface roughness are considered as an effective medium. The dotted and dashed-dotted lines correspond to the homogeneous growth of a-Ge:H (resp. a-Si:H) on a-Si:H (resp. a-Ge:H) substrates. The refractive index of the growing film is determined from the fit to the model B of the experimental (ψ , Δ) trajectories displayed in Figs 1 and 2. The dotted lines correspond to the homogeneous growth of these bulk materials. In contrast, the dashed-dotted curve corresponds to an introduction of a void volume fraction f_v = 0.25 in the bulk material. The dashed lines correspond to the nucleation stage of model B described above (with d = 30 Å). An interaction between the growing film and the rough substrate (a-Ge:H or a-Si:H) is considered with the physical interface

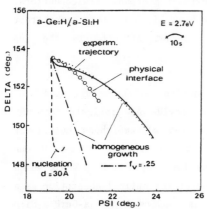

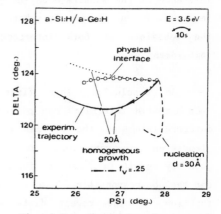

Fig. 3 Experimental and computed trajectories of the creation of a-Ge:H/a-Si:H heterojunction.

Fig. 4 Experimental and computed trajectories of the creation of a-Si:H/a-Ge:H heterojunction.

model (white circles). In this latter case, the filling of the surface roughness of the substrate by the material deposited above is assumed to be realized together with the growth of this material. A chemical interface model assuming that a thin a-Ge$_x$Si$_{1-x}$:H layer is produced at the interface between a-Ge:H and a-Si:H cannot be distinguished from the physical interface model because it has been checked that the

758 A.M. Antoine & B. Drevillon

dielectric function of a–Ge$_x$Si$_{1-x}$:H alloys can be deduced from a–Ge:H and a–Si:H reference material using the effective medium approximation.

The comparison between the results displayed in Figs. 3 and 4 shows that the structure of a–Ge:H/a–Si:H and a–Si:H/a–Ge:H interfaces is clearly different. A density deficit (of $\sim$ 25%) is observed on the growth of the first 20 Å of a–Si:H deposited on a–Ge:H (see Fig. 3). A qualitatively comparable behaviour is obtained when depositing a–Si:H on a metallic substrate (first step of model B). In contrast, the early stage of the growth of a–Ge:H on a–Si:H is compatible with an uniform growth model. However, Fig. 4 shows that the physical interface model cannot be definitely ruled out in this latter case. The comparison between the results displayed in Figs. 1 and 2 on one hand to those displayed in Figs 3 and 4 on an other hand evidences the influence of the substrate on the growth of amorphous semiconductors. In contrast, by applying a substrate bias, it has been checked that the behaviour of both interfaces is not affected by the ion bombardment.

In conclusion, this first in–situ ellipsometry study of a–Ge:H/a–Si:H interfaces can possibly provide new insights into the interpretation of the optical properties of amorphous multilayer structures. In particular, the initial nucleation of a–Si:H can possibly result in a density deficient layer at a $\sim$ 20 Å scale which can then be responsible for an increase of the optical absorption as a function of the repeat distance as observed in a–Ge:H/a–Si:H supperlattices [1].

REFERENCES

1. Tiedje, T. Wronski, C.R., Persans, P. and Abeles, B., J. Non–Cryst. Solids 77 and 78 1031 (1985).
2. Antoine, A.M., Drevillon, B. and Roca i Cabarrocas, P., Submitted to J. Appl. Phys.
3. Drevillon, B., Perrin, J., Marbot, R., Violet, A. and Dalby, J.L., Rev. Sci. Instr. 53 969 (1982).
4. Drevillon, B., Thin Solid Films 130 165 (1985).

SHORT LENGTH SCALE CONDUCTIVITY IN DEGENERATE SUPERLATTICES

M. Capizzi,* H. K. Ng, G.A. Thomas,**
R. N. Bhatt and A. C. Gossard

AT&T Bell Laboratories
Murray Hill, NJ 07974, USA

ABSTRACT

We observe conductivities in disordered metallic superlattices that are enhanced above Drude values at short length scales. Far infrared spectroscopy is used to probe lengths near the elastic mean free path in degenerately doped GaAs:Si.

1. INTRODUCTION

We report the finite frequency conductivity of disordered metals as a function of length scale. We formulate a theoretical description of the results based on localization and Coulomb interactions which inhibit the conduction at long lengths (low frequencies), as is well documented[1], but which also enhance it at lengths of order of the elastic mean free path, ℓ (intermediate frequencies), before the conductivity is attenuated in the characteristic Drude fashion at still higher frequencies.

The use of a nini superlattice[2] (alternating n-type and insulating layers) allowed us to tailor the density, mean free path, and sample thickness to values nearly optimum for far infrared absorption spectroscopy. An unusual feature was a scattering rate significantly larger than in bulk doped material[3], allowing us to probe a regime inaccessible in the careful study of MOSFETs[4] where agreement with Drude behavior was found in relatively high mobility devices.

2. SPECTROSCOPIC MEASUREMENTS

The data were obtained using fast Fourier transform far infrared spectroscopy which utilized a Michelson interferometer encorporating either a wire grid or a 6 μm mylar beam splitter and a bolometer operating at 0.3 K. Figure 1 plots the apparent absorption coefficient $\alpha(\omega)$ in μm^{-1} as a function of frequency for three samples of GaAs:Si, obtained from the transmitted intensity through the sample, I, normalized to that through a reference crystal of GaAs, I_0:

$$\alpha = \ln (I_0/I)/d \tag{1}$$

Sample A is uniformly doped, while the 11Å thick dopant layers are 53Å and 267Å apart in samples B and C. The (3D) dopant concentrations, n, for A, B and C are 16, 4 and 1.3×10^{17} cm^{-3}, while the total thicknesses of the superlattices, d, equal 0.212, 1.28 and 5.56 microns.

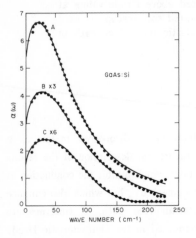

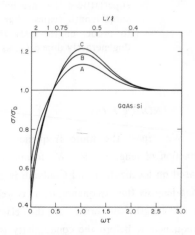

Fig. 1   The *apparent* absorption coefficient as a function of frequency, ω, for three samples of GaAs:Si, multiplied by the factors indicated. Solid dots are data and the curves are theoretical fits described.

Fig. 2   The conductivity (normalized to Drude value, Eq. 2) derived from the fit to the data in Fig. 1. Note the enhancement compared to the Drude prediction as a function of length (top scale).

Figure 2 shows the conductivity, obtained from a fit to the absorption data in Fig. 1 as described below, normalized to the Drude value:

$$\sigma_D = \sigma_B/(1 + \omega^2\tau^2) , \tag{2}$$

where the dc (Boltzmann) conductivity $\sigma_B = ne^2\tau/m^*$. The fitted σ_B values are 680, 157 and 50 $(\Omega\,cm)^{-1}$ for A, B and C, in close agreement with measured values. The scattering rate, τ, is obtained by fitting the absorption at high frequency. At the top is plotted the distance L (normalized to ℓ) that an electron diffuses in the time scale of order of the inverse frequency of the incident radiation[5], given by $L = (D/\omega)^{1/2}$. The diffusion coefficient $D = \ell^2/3\tau$, where $\ell = v_F\tau$ is calculated using the known density n and effective mass ($m^* = 0.0665 \, m_e$) in GaAs.

3. THEORY

At low frequencies $\omega\tau << 1$, a correction to the Drude conductivity (Eq. 2) arises in disordered systems due to localization and Coulomb interaction effects:[6]

$$\sigma' = \sigma_B (k_F\ell)^{-2} \{-b+c(\omega\tau)^{1/2}\} , \tag{3}$$

where the theoretical constant c is 2.24 (with 1.84 from localization and the rest from Coulomb interactions, assuming only Hartree contributions).[6] Since these effects must die out for $\omega\tau >> 1$, we include a phenomenological Gaussian cutoff with range $(1.1\tau)^{-1}$,

$$\sigma'' = \sigma' \exp(-(1.1 \, \omega\tau)^2) , \tag{4}$$

and we satisfy the conductivity sum rule by constraining the integral of σ'' over ω to 0. We calculate the dielectric constant consistently from $\sigma_D + \sigma''$ using the Kramers-Kronig relation and put both into the reflectivity and absorption[7] to obtain the apparent absorption coefficient α for the fitting in Fig. 1.

4. RESULTS

We find values of $c/(k_F\ell)^2 = 0.66$, 0.92 and 1.05 for samples A, B & C which are larger than the corresponding theoretical estimates, 0.004, 0.032 and 0.145 obtained using our $k_F\ell$. Note, however, that our fits are for $\omega\tau \sim 1$ whereas the theory has been developed for $\omega\tau << 1$. The values of $c/(k_F\ell)^2$ increase with decreasing n, qualitatively as predicted by the theory for small $\omega\tau$, but the variation is much slower than predicted.

Based on our analysis, the peak at σ arises from the increase at small ω (large L) due to localization and Coulomb interactions followed by the usual Drude decrease in σ as ω gets large. The dielectric constant reflects this behavior through the Kramers-Kronig relation.

In conclusion, we find an enhancement of the conductivity at short length scales in a relatively clean disordered metal. We expect this enhancement to be universal.

We wish to thank F. J. DiSalvo, P. A. Lee, and N.F. Mott for helpful discussions, and W. W. Wiegmann and J. B. Mock for technical assistance.

REFERENCES

* Permanent address: Istituto di Fisica G. Marconi, P. le A. Moro, Rome, Italy.

** Also at Dept. of Physics, Harvard University, Cambridge, MA 02138.

1. G. A. Thomas, Phil. Mag. **B52**, 479 (1985).

2. L. L. Chang and K. Plog, *Molecular Beam Epitaxy in Heterostructures*, (Martinus Nijhoff, Dordrecht, 1985).

3. R. N. Bhatt and T. V. Ramakrishnan, Phys. Rev. **B28**, 6091 (1983).

4. A. Gold, S. J. Allen, B. A. Wilson and D. C. Tsui, Phys. Rev. **B25**, 3519 (1982).

5. L. P. Gorkov, A.I. Larkin and D. Khmelnitskii, Pis'ma Zh. Eksp. Teor. Fiz. **30**, 248 (1979) [JETP Lett. **30**, 228 (1979)].

6. P. A. Lee and T. V. Ramakrishnan, Rev. Mod. Phys. **57**, 287 (1985).

7. O. S. Heavens, *Optical Properties of Thin Solid Films*, (Dover, New York, 1969).

TIME RESOLVED PHOTOLUMINESCENCE IN InP DOPING SUPERLATTICES

R. Ranganathan, M. Gal,* J.M. Viner and P.C. Taylor

Department of Physics, University of Utah
Salt Lake City, UT 84112

J.S. Yuan and G.B. Stringfellow

Department of Materials Science and Engineering
University of Utah
Salt Lake City, UT 84112

ABSTRACT

Spectrally-resolved studies of the time decay of the PL
in InP doping superlattices are presented. These
measurements permit one to follow the time evolution of
the internal electric fields after intense optical
excitation and yield information on recombination rates
for spatially-separated electrons and holes in different
subband levels.

I. INTRODUCTION

Doping superlattices, which consist of alternate layers of n-doped
and p-doped semiconductors, have many novel optical and electrical
properties.[1] The energy gap, which is indirect in real space, can be
tuned by varying the optical excitation. This tunability has been
demonstrated recently for InP doping superlattices using cw photoluminescence (PL)[2] and modulated reflectance (PR)[3] techniques.

In a doping superlattice the valence and conduction band edges are
modulated in real space by the space charge potentials generated by the
donors and acceptors. Optically excited electrons and holes partially
screen this periodic potential and result in an increase in the effective band gap with carrier concentration. Because of the spatial
separation of excited electrons and holes, the radiative lifetimes can
be very long.[2,4]

---

*Present address: Department of Physics, University of New South Wales,
Kensington, P.O.B. 1, Sydney, Australia 2033.

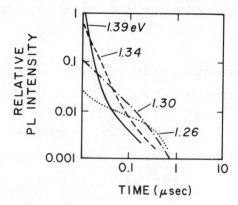

Fig. 1. Time decay of PL at various energies in an InP
         doping superlattice. Data are normalized such
         that the PL intensity is unity at 1.39 eV and
         0.01 μsec delay time.

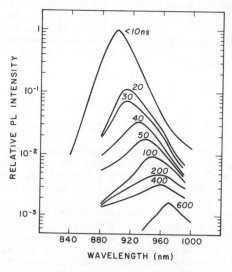

Fig. 2. Energy dependence of the PL spectra in an InP doping
         superlattice as a function of delay time after pulsed
         excitation. Data are normalized such that the PL
         peak right after optical excitation is unity.

    The samples used in the present study were grown by the organo-
metallic vapor phase epitaxial (OMVPE) technique. Details of the
preparation procedures have been described elsewhere.[2,5] The samples
consisted of six 200 Å layers of alternate n and p doping with

$n_n = 1 \times 10^{18}$ cm^{-3} and $n_p = 2 \times 10^{18}$ cm^{-3}. The PL measurements were made at 100 K using a Lambda Physik (model FL2001) excimer-pumped dye laser at 5400 Å as the excitation source. Excitation pulse widths were approximately 10 nsec. The power densities employed were < 100 kw/cm^2. PL signals were recorded with a Jobin-Yvon 1/3 m mono-chromator on a S1 photomultiplier detector and a LeCroy (Model 3500SA) signal averaging system.

2. EXPERIMENTAL RESULTS AND DISCUSSION

Figure 1 shows four typical decay curves of the PL in an InP doping superlattice at several PL energies. Background PL from the InP substrate has been subtracted from these curves. The spectra have been normalized to the short time PL intensity at 1.39 eV. It is apparent from this figure that a broad distribution of radiative recombination times exists at all observed PL energies. At lower energies the initial PL intensities are greatly reduced, but at longer times the PL at lower energies is seen to dominate.

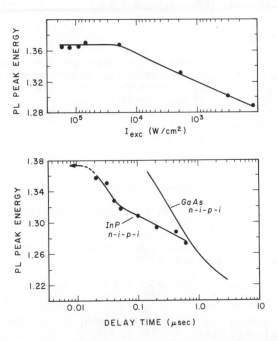

Fig. 3. Top: Position of the PL peak in an InP doping superlattice (n-i-p-i) as a function of cw pump power at 100 K. Bottom: Position of the PL peak energy as a function of delay time after pulsed optical excitation at 100 K. Data on a GaAs doping superlattice (after ref. 4) are presented for comparison.

A more transparent picture of the time evolution of PL in these doping superlattices can be obtained by plotting the PL spectra (energy dependence of the PL) at different delay times after the pulsed excitation as shown in Fig. 2. These data, which are effectively vertical slices through many curves such as those shown in Fig. 1, exhibit a well defined PL peak which shifts to lower energy with time. The peak PL intensity also decreases with time but the half-width remains relatively constant. These data are similar to those observed previously in GaAs doping superlattices grown by the molecular beam epitaxial (MBE) technique.[4]

Figure 3 shows the variation of the peak PL energy with decay time for the data of the InP doping superlattice in Fig. 2 and the data on GaAs doping superlattices of ref. 4. Although the absolute energies are different because of the different band gaps of InP and GaAs, the trends are very similar. Also plotted in Fig. 3 are cw PL data taken under identical conditions (T = 100 K in the same experimental set up) which show the variation of the peak PL energy position with cw excitation intensity. In principle, these two curves can be directly related because the peak PL energy is uniquely determined by the densities of photo-excited electrons and holes in both cases.[6]

4. ACKNOWLEDGMENTS

The research in the Department of Materials Science and Engineering was supported by the Army Research Office, under contract number DAAG20-83-K-0153 (JSY, GBS). Research in the Department of Physics was supported by the Office of Naval Research under contract number N00014-83-0535 (PCT, JMV) and the National Science Foundation under grant number DMR-83-04471 (MG).

REFERENCES

1. Döhler, G.H., Phys. Status Solidi B52, 79 (1972); 533 (1972).

2. Yuan, J.S., Gal, M., Taylor, P.C., and Stringfellow, G.B., Appl. Phys. Lett. 47, 405 (1985).

3. Gal, M., Yuan, J.S., Viner, J.M., Taylor, P.C., and Stringfellow, G.B., Phys. Rev. B33, 4410 (1986).

4. Rehm, W., Künzel, H., Döhler, G.H., Ploog, K., and Ruden, P., Physica 117B + 118B, 732 (1983).

5. Kuo, C.P., Yuan, J.S., Cohen, R.M., Dunn, J., and Stringfellow, G.B., Appl. Phys. Lett. 44, 550 (1984).

6. Ruden, R. and Döhler, G.H., Phys. Rev. 27, 3538 (1983).

Substrate Orientation Effects on Band Alignments
for Pseudomorphic (Ge,Si) Alloys on Silicon

R. People, J. C. Bean and D. V. Lang
AT&T Bell Laboratories
Murray Hill, N.J. 07974

ABSTRACT

We have investigated the effects of substrate orientation on the band gap and band alignment of coherently strained Ge_xSi_{1-x} alloys on Si. Uniaxial splittings of the six-fold degenerate valence bandedge are treated exactly for the (001), (111), and (110) orientations. Photocurrent measurements of the stained alloy bandgaps for growth on Si(001) substrates show excellent agreement between these deformation potential calculations and experiment. Further, the present strained bandgap estimates have been combined with existing pseudomorphic pseudopotential calculations of the Si/Ge valence band offset in order to obtain the anticipated band alignments.

Phenomenological deformation potential theory[1] has proved highly successful in estimating the bandgaps of coherently strained Ge_xSi_{1-x} bulk alloys on Si(001) substrates.[2,3] These coherently strained (001) bandgap estimates have been combined with existing pseudomorphic pseudopotential calculations of the Ge/Si valence band offset,[4] in order to provide estimates of the anticipated band alignments.[5] Indeed, these band alignment results provided the first quantitative resolution of the seeming paradox which arose from interpretations of various transport measurements on two-dimensional holes[6] and two-dimensional electrons[7,8] in the (Ge,Si)/Si system. In the present paper we extend the coherently strained bandgap calculations to substrate orientations other than (001). Results for (001), (111), and (110) orientations, which include an exact treatment of the uniaxial splitting of the six-fold degenerate valence bandedge, are given. These results are then combined with pseudomorphic pseudopotential estimates of the Si/Ge valence band offset, in order to predict the anticipated band alignments versus substrate orientation.

The *conventional* strain components may be defined in terms of the in-plane strain, $e_{\parallel}$, and the elastic constants.[2] Note that $e_{\parallel} < 0$, for growth of (Ge,Si) alloys on Si substrates. It is appropriate here to point out the difference between the strain tensor, S_{ij} and the *conventional* strain components, e_{ij}, namely:

$$S_{ij} = \frac{1}{2} e_{ij}(1 + \delta_{ij}) , \tag{1}$$

i.e. the off-diagonal components of S_{ij} differ by $1/2$ from the corresponding e_{ij}. This differentiation is important since the valence band deformation potential Hamiltonian is generally defined in terms of the *conventional* strain components,[9,10] whereas the conduction band uniaxial splittings are generally defined in terms of the strain tensor.[11,12] The splitting of the valence band states (at $\vec{k} = 0$), produced by lattice mismatch-induced coherency strain, is obtained from the strain Hamiltonian.[9,10] In general couplings between the states of the $J = 3/2$ and $J = 1/2$ manifolds is induced by the coherency strain. For the case $\vec{\ell} \parallel \{001\}$ or $\{111\}$, the 6×6 eigenvalue determinant breaks up into $3 - (2 \times 2)$ determinants. For the case $\vec{\ell} \parallel \{110\}$, one obtains two equivalent (3×3) determinants, whose solution is readily obtained numerically. We have used 3.35, 3.81 eV[13] and 2.04, 2.68 eV[9] for the valence band deformation potentials D_u, D_u' for Ge and Si, respectively.

Herring and Vogt[11] considered the uniaxial splitting of the *i-th* conduction band:

$$\Delta E_{c,u}^{(\gamma)} (\vec{k}_i; \ell_1, \ell_2, \ell_3) = \Xi_u^{(\gamma)} [\{\hat{a}_i \hat{a}_i\} - \frac{1}{3} \vec{1}] : \vec{S}(\ell_1, \ell_2, \ell_3) . \qquad (2)$$

Here γ denotes the symmetry of the given band edge (i.e. Δ, L, etc.), $\hat{a}_i$ is a unit vector parallel to the $\vec{k}$-vector of valley i, $\vec{1}$ is the unit tensor and ℓ_i are the direction cosines of the surface normal. We have considered the Δ and L minima, since these are the lowest conduction band minima in the (Ge,Si) system.[14] We have used $\Xi_u^{(\Delta)}$ (Si) = 9.2 eV,[12] $\Xi_u^{(L)}$ (Ge) = 16.2 eV.[12]

The hydrostatic terms describe the relative motion of the centers of gravity of a given conduction edge (γ) with respect to the valence bandedge. They are defined in terms of the dilation of the unit cell as follows:

$$\Delta E_{HYDRO}^{(\gamma)} (\ell_1 \ell_2 \ell_3) \equiv \{\Xi_d + \frac{1}{3} \Xi_u - a\}^{(\gamma)} \vec{1} : \vec{S}(\ell_1, \ell_2, \ell_3) . \qquad (3)$$

We have used 1.5 eV and -4.5 eV for the Δ and L hydrostatic deformation potentials, $\{\Xi_d + \frac{1}{3} \Xi_u - a\}^{(\gamma)}$, respectively.[15]

Given the alloy dependence of a given unstrained bandedge (i.e. Δ,L), the uniaxial conduction and valence band splittings, along with the hydrostatic contribution to the bandedge, we may determine the various strained bandgaps. We have combined the unstrained alloy bandgap data of Braunstein et al.[15] with the aforementioned strain contributions to obtain the two lowest conduction bandedges associated with the Δ-bandedge along the lowest strained L-bandedge of $Ge_x Si_{1-x}$ alloys. These results are shown in Figs. (1)-(3) for (001), (110), and (111) substrate orientations, respectively. The data in Fig. (1) are the photocurrent measurements of Lang et al.[3] It is apparent that phenomenological deformation potential theory provides an excellent description of the strained alloy bandgaps. Note further that the lowest bandedge of the strained alloy is derived from the Δ-bandedge over the entire range of composition for $\vec{\ell} \parallel \{001\}$ or $\{110\}$. For $\vec{\ell} \parallel \{111\}$ however, the present results indicate that the lowest bandedge of the strained alloy is derived from the L-bandedge for $x \geq 0.9$.

The conduction band offsets $\Delta E_c^{(\Delta)}(x)$ associated with the Δ-bandedge for $\vec{\ell}$ ||{001}, {110} and {111} are shown in Figs. (4)-(6), and include spin-orbit effects.[17] Note that $\Delta E_c > 0$, implies a type I alignment wherein the strained alloy bandedges lie within the bandgap of the cubic Si; $\Delta E_c < 0$, implies a type II (staggered) alignment wherein the alloy conduction bandedge lies higher in (electron) energy than the Si conduction bandedge.

ACKNOWLEDGEMENTS

We would like to thank V. Narayanamurti and A. Y. Cho for providing continued encouragement, during the course of these studies. We would also like to acknowledge helpful discussions with J. C. Hensel, C. G. Van de Walle, and M. Schluter.

REFERENCES

1)   Kleiner, W. H. and Roth, L. M., Phys. Rev. Lett. *2*, 334 (1959).

2)   People, R., Phys. Rev. *B32*, 1405 (1985).

3)   Lang, D. V., People, R., Bean, J. C. and Sergent, A. M., Appl. Phys. Lett. *47*, 1333 (1985).

4)   Van de Walle, C. G. and Martin, R. M., J. Vac. Sci. Technol. *B3*, 1256 (1985).

5)   People, R. and Bean, J. C., Appl. Phys. Lett. *48*, 538 (1986).

6)   People, R., Bean, J. C., Lang, D. V., Sergent, A. M. Störmer, H. L., Wecht, K. W., Lynch, R. T. and Baldwin, K., Appl. Phys. Lett. *45*, 1231 (1984).

7)   Jorke, H. and Herzog, H. J., in "Proceedings of 1st Int'l Symposium on Si Molecular Beam Epitaxy", ed. by J. C. Bean (Electrochemical Society Press; Pennington, N.J., 1985), p. 352.

8)   Abstreiter, G., Brugger, H. and Wolf, T., Phys. Rev. Lett. *54*, 2331 (1985).

9)   Hensel, J. C. and Feher, G., Phys. Rev. *129*, 1041 (1963).

10) Hasegawa, H., Phys. Rev. *129*, 1029 (1963).

11) Herring, C. and Vogt, E., Phys. Rev. *101*, 944 (1956).

12) Balslev, I., Phys. Rev. *143*, 636 (1965).

13) Hensel, J. C. and Suzuki, K., Phys. Rev. *B9*, 4129 (1974).

14) Braunstein, R., Moore, A. R. and Herman, F., Phys. Rev. *109*, 695 (1958).

15) Paul, W. and Warschauer, D. M., J. Phys. Chem. Solids, *6*, 6 (1958).

16) Van de Walle, C. G. and Martin, R. M., in *Proceedings of the M.R.S. 1985 Fall Meeting;* Boston, MA (to be published).

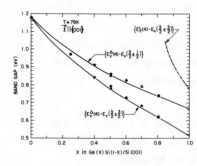

Fig. 1. Strained bandgaps derived from the Δ- and L-bandedges for growth on Si(001). Data are from Ref. 3.

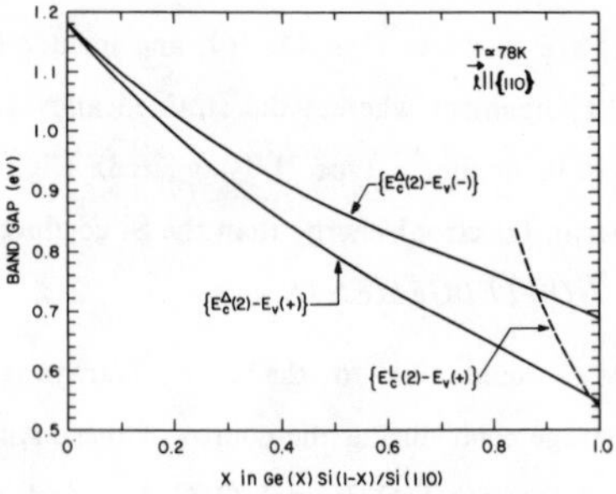

Fig. 2. Strained bandgaps derived from the Δ- and L-bandedges for growth on Si(110).

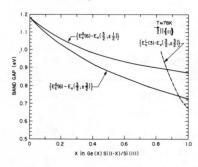

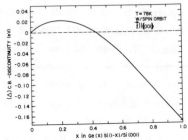

Fig. 3. Strained bandgaps derived from the Δ- and L-bandedges for growth on Si(111).

Fig. 4. Δ-C.B. discontinuity for coherently strained Ge_xSi_{1-x} on Si(001).

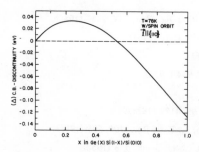

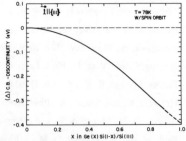

Fig. 5. Δ-C.B. discontinuity for coherently strained Ge_xSi_{1-x} on Si(110).

Fig. 6. Δ-C.B. discontinuity for coherently strained Ge_xSi_{1-x} on Si(111). Dashed curve indicates that the Δ-derived bandedge is no longer of lowest energy.

LIQUID PHASE EPITAXY OF SiGe ALLOYS

M.I. Alonso, E. Bauser, T. Suemoto and M. Garriga

Max-Planck-Institut für Festkörperforschung, Heisenbergstr. 1,
D-7000 Stuttgart 80, Federal Republic of Germany

ABSTRACT

We report on growth conditions for $Si_{1-x}Ge_x$ $(0 \leq x \leq 1)$ alloys with liquid phase epitaxy. The influence of different solvents on growth process is discussed. The samples are homogeneous and their surface morphology indicates a lateral microscopic growth mechanism.

Epitaxial layers of semiconductor alloys are interesting both for fundamental research and technical application since it is possible to vary continually their properties by changing their composition. In this paper we present studies on the growth of $Si_{1-x}Ge_x$ alloys from metallic solutions by liquid phase epitaxy, which should enable us to grow heterostructures of SiGe alloys.

The layers are grown in a tipping boat apparatus. The substrate material is usually Si(111), occasionally Ge(111). The maximum size of the substrates is 5 cm^2. They are cleaned by standard procedures, and the silicon ones are always given an HF-dip immediately before growth. We use mainly In as solvent because of its low solid solubility in the alloy, but experiments have also been done using Sn. The low In-concentrations in the epitaxial layers allow us to grow n-type samples by adding InAs or InP to the In solution. For growing highly doped p-type layers we use In:Ga solutions. Tin is also convenient as solvent because it forms an electrically inactive impurity in the alloy. Several growth parameters of our experiments are summarized:

SOLVENT	SATUR. TEMP.($^{\circ}$C)	COOLING RATE($^{\circ}$/h)	SUBSTRATE	$Si_{1-x}Ge_x$
In, In:As	950-900	50-75	Si	$0 \leq x \leq .1$
In:P	900	50-75	Si	$0 \leq x \leq .3$
Sn	930-900	20-10	Si	$.5 \leq x \leq .9$
In, Sn	800-550	20-10	Ge	$.95 \leq x \leq 1$

With the conditions above we obtain growth rates between 80 nm/min and
0.5 μm/min. We can reduce the growth rates by choosing smaller cooling
speeds and/or lower growth temperatures. We obtain layers which are
typically between 1 and 20 μm thick.

The different results obtained with In or In:As as compared to
In:P solutions are probably related to two favourable influences of
the phosphorus in the growth process. Firstly P reduces residual oxide
on the Si substrate, and thus facilitates the nucleation process of
the epitaxial layer. Secondly, high concentrations of P decrease the
lattice mismatch between substrate and epilayer. The reduction of the
lattice constant by the phosphorus dopant, however, is not sufficient
to explain the presently obtained 20% difference in the Ge concentration of our SiGe alloys. We suspect that alloys having higher Ge concentrations may be obtained on Si substrates by further improving starting growth conditions. This may be achieved using a buffer layer
which can be pure Si or some SiGe alloy. We succeeded for example in
growing graded layers from undoped In solutions, the upper one contained up to 50% Ge.

The surface morphology of the SiGe LPE-samples confirms lateral
microscopic growth [1]. We observe changes in the appearance of layers
with increasing Ge content, as shown in Fig. 1. The SiGe layers with
lower Ge content have generally smoother surfaces than layers with
higher Ge content when grown on Si substrates. The features we observe
are generally terrace growth surface on slightly misoriented substrates
and in addition stress induced slip lines. Figure 2 shows a part of a
layer where dislocation-controlled facet growth has taken place and
where the surface, consequently, is extremely flat. This type of
growth is most favorable for the growth of thin multilayer structures.

It requires the substrate surface to be almost exactly orientated in the (111) plane.

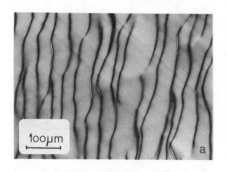
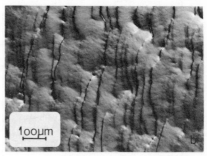

Fig. 1a,b. Layer surfaces of SiGe on Si. Nomarski differential interference contrast (NDIC) micrographs. a) 2% Ge, b) 9% Ge.

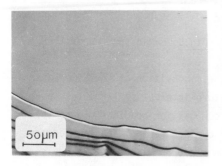
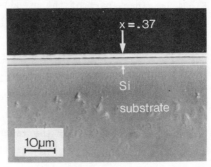

Fig. 2. Dislocation-controlled facet growth of $Si_{0.92}Ge_{0.08}$ on Si. Part of the terraced surface is also shown. Optical micrograph-NDIC.

Fig. 3. NDIC micrograph of a photo-etched cleavage face. Sample grown in a sliding boat.

The specimens were characterized with automatic spectroellipsometry (Fig. 4). Assuming linear variation with x for the interband transition E_1 [2] allows us a rapid composition determination. An increase of transition width due to alloy broadening is also observed. The composition determined with X-ray diffraction agrees well with that obtained by ellipsometry. For layers grown in a small temperature interval (i.e. $\lesssim 50$ K), the X-ray intensity peak originating in the layer is at most slightly broader than the peak originating in

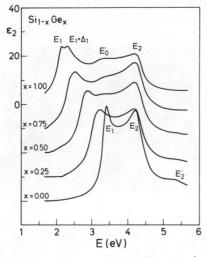

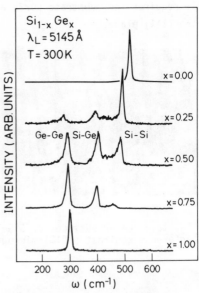

Fig.4.Imaginary part of the pseudodielectric function of some selected
samples obtained with ellipsometry.
The main structures are indicated.

Fig.5. Raman spectra of the
same samples of Fig.4.

the Si substrate, indicating that the composition of the SiGe layer is
homogeneous. If the growth temperature interval is large the layer
peak broadens, denoting a vertical composition gradient due to the
changes in composition of the layer during growth. The samples are
laterally homogeneous, as found with electron microprobe analysis.
The phonon Raman spectra of the samples (Fig. 5) show a good agreement with those of bulk materials indicating incommensurate layers.[3]

We acknowledge financial support from "La Caixa", Barcelona,
Spain (M.G.), and CIRIT, Catalunya, Spain (M.I.A.)

REFERENCES

1.   Bauser, E., in : Crystal Growth of Electronic Materials, ed.E.
     Kadis (Elsevier Science, 1985), pp. 41-55.
2.   Kline, J.S., Pollak, F.H. and Cardona, M., Helv. Phys. Acta 41,
     968-977 (1968)
3.   Renucci, M.A. Renucci, J.B. and Cardona, M., Light Scattering in
     Solids, ed. Balkanski (Paris, 1971), pp. 326 - 329.

A TWO-DIMENSIONAL MAGNETIC SEMICONDUCTOR SUPERLATTICE MnSe/ZnSe

A.V. Nurmikko, D. Lee, and Y. Hefetz
Division of Engineering and Department of Physics
Brown University, Providence RI 02912 USA

L.A. Kolodziejski and R.L. Gunshor
School of Electrical Engineering
Purdue University, West Lafayette IN 47907 USA

The new MnSe/ZnSe superlattice is studied with emphasis on ultrathin MnSe layers, including the monolayer limit. Unexpectedly large exciton Zeeman effects suggest the presence of significant paramagnetic contribution of the Mn-ion spin system near the 2D magnetic limit.

Among II-VI semiconductor superlattices (SL) a recent development has been the growth by MBE of versatile thin layer zincblende ZnSe/MnSe strained layer structures (1). This follows earlier work on (Zn,Mn)Se based SLs as reviewed elsewhere (2). The structures discussed below were of 'comb' type: thin layers of MnSe separated by ZnSe layers. Typically, 30 to 100 periods were grown on a ZnSe buffer layer. A useful aspect in preparation of such SLs was the use of RHEED intensity oscillations to 'calibrate' the thickness of the layers on the monolayer scale (3). The presence of such oscillations during epitaxy indicates high degree of two-dimensional growth and interface sharpness; the number of monolayers of MnSe per SL period was determined by switching growth from MnSe to ZnSe (and vice versa) at peaks of corresponding RHEED oscillations. Four samples were grown for the following nominal number of monolayers of MnSe per SL period: 1, 3, 4, and 10, respectively. An additional monolayer structure where growth was interrupted at the heterointerfaces was also prepared. (A single monolayer of MnSe is approximately 3.2 A in the zincblende phase.) The usual ZnSe 'spacer' layer thickness was 40-45 A, except for the 10 monolayer thick MnSe structure where, to avoid misfit dislocations, the ZnSe layers were reduced to approximately 24 A.

To examine electronic and magnetic properties of these structures photoluminescence (PL) methods were used because of their sensitivity. Low temperature PL from the MnSe/ZnSe SLs showed two distinct contributions (6): (a) exciton recombination across the SL bandgap and (b) emission from the d-electron excitation of the Mn-ions. To first order, the exciton recombination photon energies were in the range expected for the superlattices with very thin MnSe barriers (1, 3, and 4 monolayers). The 'thick' MnSe barrier structure (10 monolayers) had the significant strain in the ZnSe layers reducing the emission energies (of the order of 100 meV assuming a 'free standing' SL). Figure 1 shows the exciton recombination spectrum at T=1.8 K for the sample with 3 monolayers of MnSe per SL period at $B_z=0$ and $B_z=5$ Tesla. The exciton lifetime measured by time-resolved techniques yields values on the order of 150 psec; this contains a component due to energy transfer into the d-electron states in the MnSe 'barrier' layers (5). In a magnetic field the exciton PL is strongly circularly polarized. Figure 2 summarizes Zeeman shifts for several SL samples including both monolayer MnSe structures. Except for the thick MnSe layer structure, all other SLs showed a pronounced redshifts with significant field anisotropy (shown for one SL at top portion of the graph). With increasing temperatures, the Zeeman shifts decreased monotonously as shown in Fig. 3 (4).

Figure 4 shows an example of the Mn-ion d-electron emission for two MnSe/ZnSe SLs and a $Zn_{0.67}Mn_{0.33}Se$ single crystal zincblende thin film (6). Within the energy resolution for these spectra (<2 meV), no detectable changes were observed in the SL lineshapes or their positions as function of temperature (up to 100 K) or external field (up to 5 Tesla). The appearance of a distinct low energy peak in their emission spectrum is probably related to reduced coordination in the nearly two-dimensional MnSe layers and associated changes in the crystalline fields.

Since MnSe is expected to be an antiferromagnetic (AF) insulator at a low lattice temperature, a striking aspect of data in Fig. 2 are the large magnetizations which are implied in the ultrathin MnSe layer limit. When normalized against the MnSe layer thicknesses, the monolayer structures exhibit the largest magnetizations per unit area. In pronounced contrast, the 10 monolayer MnSe SL shows very little magnetization. These connections are derived in first order by using the standard formulation of 'semimagnetic' (or diluted magnetic) II-VI semiconductors where such Zeeman effects are due to giant spin splittings of conduction and (particularly) valence band edges by the exchange interaction of the bandedge states with Mn-ion d-electron spins. Under several drastic assumptions which include (i) zero valence band offset, (ii) considering p-d exchange only, (iii) ignoring complications from the elecron-hole Coulomb interaction, and (iv) neglecting magnetic polaron and other exciton localization processes, we estimate that the apparent magnetizations in monolayer SLs imply up to 20% of the Mn-ion d-electron spins being esentially free (fully paramagnetic). Stated differently, a pronounced reduction in the AF interactions appears to have occurred in going towards the magnetic 2D limit. (The exciton spectral shifts also appeared to saturate at fields B> 6 Tesla).

The question of AF ordering in a pure Heisenberg system (zero orbital angular momentum) in lower dimensions is of fundamental interest since in idealized case the critical behavior has been predicted to be absent (7). However, even a small degree of anisotropy in the d-d exchange (say, on the order of 10^{-4}) is expected to stabilize ordering. The data in Fig. 2 shows clear field anisotropy (which can have contributions from both p-d and d-d exchanges). Hence it would be surprising if the large magnetizations implied in our data were simply due to an ideal 2D spin isotropic system. Yet, temperature dependence of the Zeeman shifts for ultrathin MnSe layer SLs shows no evidence of a phase transition in our interpretation; rather, the shifts decrease in magnitude monotonously with increasing temperature in a modified paramagnetic way. (Some measurements of magnetization were made with a SQUID instrument; however, these were hampered by the small amounts of Mn in the samples.

The Mn-Mn spin exchange has been studied in semimagnetic semiconductors such as $Zn_{1-x}Mn_xTe$ where the short range, anion intermediated superexchange dominates (8). A rough mean field estimate yields an anticipated AF ordering temperature for 'hypothetical' 2D MnSe is on the order of 50K (reduced from 3D because of fewer nearest neighbor spins). The absence of temperature anomalies in Fig. 3 and in the d-electron spectra suggest frustration of AF order in quasi-2D magnetic limit in our MnSe. In MBE growth of heterostructures, the layer-by-layer growth kinetics are still incompletely understood; however, some interfacial 'roughness' on atomic scale (submonolayer) can generally be expected. This means that an interface is subject to some interdiffusion or, more likely at our relatively low growth temperatures, that segregated islands of MnSe and ZnSe constitute the microscopic interface fluctuations. We propose that such fluctuations at ZnSe/MnSe interfaces provide sufficient short range disorder to inhibit AF order even at low lattice temperatures. Small islands can lack critical behavior, particularly in the absence of an ordered 3D antiferromagnetic 'substrate' as disorder in a random 2D system inhibits ordering much more effectively than in 3D case. Estimates suggest that near our monolayer limit, a high density (0.3 to 0.5 fractional monolayer coverage) of rather small noninteracting islands contributes the magnetically 'active' part of the interface, with an average of roughly 5-7 spins per island randomly distributed on a plane. Qualitatively, the reduction in the 'susceptibility' with increasing MnSe layer thickness is controlled by the increasing AF interactions which couple the interface islands to an increasingly ordered MnSe underlayer. The anisotropy shown in Fig. 1 may imply a preferred ordering direction of the AF-coupled Mn-ion spins in the MnSe layer planes (xy antiferromagnet).

We thank S. Datta and N. Otsuka at Purdue and R. Pelcovits, S.K. Chang, and H. Nakata at Brown for contributions. Work at Brown was supported by Office of Naval Research; at Purdue by Office of Naval Research, AFOSR and a grant from Hughes Research Laboratories.

References:

(1) L.A. Kolodziejski, R.L. Gunshor, N. Otsuka, B.P. Gu, Y. Hefetz, and A.V. Nurmikko, Appl. Phys. Lett. 48, 1482 (1986)

(2) L.A. Kolodziejski, R.L. Gunshor, N. Otsuka, S. Datta, W.M. Becker, and A.V. Nurmikko, IEEE J. Quant. Electr. QE-22, September 1986

(3) L.A. Kolodziejski and R.L. Gunshor, N. Otsuka, S. Datta, and A.V. Nurmikko, Spring Meeting of the Materials Research Society, Palo Alto CA (1986)

(4) D. Lee, Y. Hefetz, A.V. Nurmikko, L.A. Kolodziejski, and R.L. Gunshor, to be published

(5) Y. Hefetz, W.C. Goltsos, A.V. Nurmikko, L.A. Kolodziejski, and R.L. Gunshor, Appl. Phys. Lett. 48, 372 (1986)

(6) Y. Hefetz et. al., Proc. 2nd Int. Conf. on Superlattices and Microstructures, Gothenburg (1986)

(7) N.D. Mermin and H. Wagner, Phys. Rev. Lett. 17, 1133 (1966)

(8) J. Spalek et. al. Phys. Rev. B33, 3407 (1986); B.E. Larson, K.C.
Hass, H. Ehrenreich, and A.E. Karlsson, Solid State Comm. 56, 347 (1985)

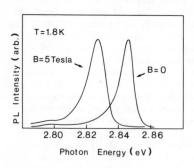

Figure 1: Exciton PL from a
MnSe/ZnSe superlattice (3 MnSe
monolayer per SL period) and the
Zeeman shift at B_z =5 Tesla
(perpendicular to layer plane).

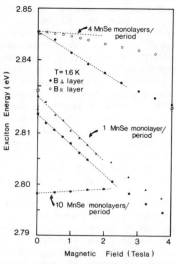

Figure 2: Position of exciton PL peak in a magnetic field for MnSe/ZnSe
SLs at T= 1.6 K. The thickness of MnSe layers is indicated as the
nominal number of monolayers (dark triangles refer to second monolayer
SL with growth interupted epitaxy). The dashed lines are a guide to eye
for low field slopes.

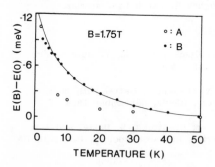

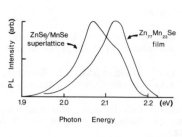

Figure 3: Zeeman shift (B_z = 1.75 Tesla) as a function of temperature for
samples A and B (one monolayer and three monolayer nominal MnSe
thickness, respectively). The solid line is a modified Brillouin
function applied to sample B with a contribution of AF interactions by
T_o = 9K.

Figure 4 : Comparison of the yellow Mn-ion d-electron luminescence at T=
1.6 K from a SL sample (3 monolayers of MnSe per SL period) and a thin
$Zn_{0.77}Mn_{0.23}Se$ film. The amplitudes are arbitrarily normalized.

Properties of P-type Hg1-xCdxTe-CdTe Superlattice

S. RAFOL, K. C. WOO AND J.P. FAURIE

DEPARTMENT OF PHYSICS

UNIVERSITY OF ILLINOIS AT CHICAGO

CHICAGO, IL 60680

We have studied several p-type $Hg_{1-x}Cd_xTe$-CdTe superlattices with composition that changes the superlattices from Type III to Type I superlattices using magneto-transport measurements.

There has been a great deal of interest in HgTe-CdTe superlattices since the fabrication of such superlattices by Faurie et.al using the Molecular Beam Epitaxial (MBE) technique[1]. These materials are not only excellent candidates for infrared detectors, but they also provide excellent opportunities to study semicondutor superlattices because of their unique band structures. HgTe is a semimetal of a negative band overlap of 0.3 eV and CdTe is a semiconductor of a 1.6 ev band gap. The conduction band in CdTe becomes the light hole band in HgTe. As a consequence of the matching up of bulk states belonging to the conduction band in HgTe with the light-hole valence band in CdTe, there exists a quasi-interface states[2]. Because of the peculiar character of this superlattice, it is expected that its properties are different from that of GaAs-AlAs (Type I) and GaSbInAs (Type II). Thus it constitutes a new type of superlattice system called a Type III superlattice.[3] In this paper we present magneto-transport measurements in p-type $Hg_{1-x}Cd_xTe$-CdTe superlattices with $0 < x < 0.22$ in magnetic fields as high as 23 tesla and temperatures as low as 0.5K. In this range of x, the superlattices change from Type III to Type I. Details of this experimental techniques will be published elsewhere.[4]

One of the most interesting unanswered questions of HgTe-CdTe superlattices is the mobility enhancement in the p-type structures. Hole mobilities have been reported as high as 30,000 cm^2/V.sec, but all are above 1,000 cm^2/V.sec.[5] The mobility of bulk p-type $Hg_{1-x}Cd_xTe$

is usually less than 500 cm^2/V.sec. Mixing of light and heavy holes
has been suggested for the enhancement of the hole mobilitities.[5]
Several theoretical investigations have been carried out to study
this problem. The band structure calculation has been refined using
a multi-band tight binding model[6] and the effect of the lattice mismatch
between the HgTe and CdTe has been investigated.[6, 7] These studies
conclude that the light holes should not contribute to the in-plane
transport properties. We have determined the effective mass of
$Hg_{1-x}Cd_xTe$-CdTe superlattices from the temperature dependence of the
amplitude of the Shubnikov-de-Hass oscillations and find that the
dominating carrier at low temperatures is the heavy hole.[4]

Unlike the GaAs-GaAlAs system, this mobility enhancement does
not result from the growth conditions nor is alloy scattering a
factor.[4]

Interestingly, the mobility enhancement ceases when the $Hg_{1-x}Cd_xTe$
in the HgCdTe-CdTe superlattices changes from a semimetal to semiconductor.[8] This strongly suggests that the mobility enhancement
only occurs for the Type III superlattices and not in Type I superlattices
in the HgCdTe-CdTe. The interfacial strain and the valance band
offset in all these samples should be the same. One of the differences
between the Type III and Type I superlattices is the existence of
interface states in the Type III superlattices but not in the Type I
superlattices. It is possible that the drastic difference in mobility
in these superlattices is related to these interfacial states.

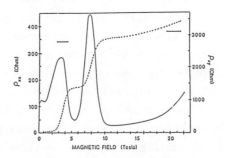

Fig. 1: Quantized Hall Effect of $Hg_{0.92}Cd_{0.08}Te$-CdTe Superlattice at
0.5K.

Fig.1 shows the ρ_{xx} and ρ_{xy} of a $Hg_{0.92}Cd_{0.08}Te$-CdTe superlattice. This sample has 100 periods of $Hg_{0.92}Cd_{0.08}Te$ (70 Å) and
CdTe (40 Å). The Quantized Hall Effect (QHE) is observed in ρ_{xy}.
Such an effect has been observed in n-type superlattices.[9],[10] This
represents the first p-type superlattice that shows QHE. Assuming
the minina in pxx at 5.5 T and 11 T to be the Landau level, index i =
2 and i = 1, the two-dimensional hole density is 2.65 x 10^{11} per
sq.cm. The value of the plateau (R (in ohm) = 25,812 / (i x n),
where n is the number of layers in contacted) indicates nine layers
of the superlattice is contacted. With the strong oscillations above
5 T, it is surprising that no oscillation is detected below 5 tesla.
Preliminary results of the n-type heterojunctions show the expected
gradual increase in the amplitude of the Shubnikov-de-Hass oscillations
and the sudden onset of the quantum oscillations in the p-type structures
is a peculiar property of the system.

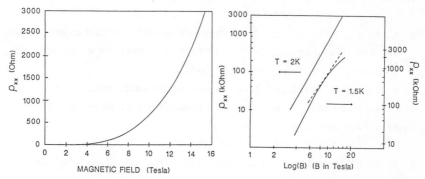

Fig. 2 (a): Magneto-resistance of   Fig. 2(b): Log-log plot of
$Hg_{0.84}Cd_{0.16}Te$-CdTe at 2K. $\rho(H)$ vs. H at 2K and 1.5K.

Fig. 2(a) shows the magneto-resistance of a Type I superlattice
($Hg_{0.84}Cd_{0.16}Te$-CdTe). No Shubnikov-de-Hass oscillation is observed
because of the low mobility of the sample. The resistance has a
strong magnetic field dependence above 8 T. Fig. 2(b) shows the
log-log plot of ρ vs B. Above 2.2K, the resistance is proportional
to B^{α} where α is about 5. This indicates the scattering mechanism is
scattering from impurities. At 1.5K the magneto-resistance can be

split into two regions, one with power a of 4.8 and the second region
with power a of 3.8 at higher magnetic fields.   Similar behavior
observed in bulk $Hg_{1-x}Cd_xTe$ has been presented as evidence for an
electronic phase transition at high magnetic fields.[11] Although this
interpretation is being disputed,[12] we speculate the origin of such a
"kink" in these two systems could have the same origin.

This work is supported in part by a grant of the Research Corporation
and the Defense Advanced Research Projects Agency under contract
No. MDA-903-85K-0030.   Part of this work was done when two of us
(K.C.W and S. R.) are Guest Scientists at the National Magnet Laboratory
at Cambridge, MA.

REFERENCES

1. J. P. Faurie, A. Million and J. Piaguet, Appl. Phys. Lett. 41, 713
   (1982).

2. Y.C. Chang, J.N. Schulman, B. Bastard, Y. Guldner and M. Voos,
   Phys. Rev. B 31, 2557 (1985).

3. L. Esaki, Proceedings of the 17th International Conference on the
   Physics of Semiconductors, edited by J. D. Chadi and W. A. Harrison,
   Springer-Verlag Inc., New York (1983), p. 473.

4. K.C. Woo, S. Rafol and J.P. Faurie, to be published.

5. J. P. Faurie, M. Boukerche, S. Sivananthan, J. Reno and C. Hsu,
   Superlattice and Microstructures 1, 237 (1985).

6. G.Y. Wu and T. C. McGill, Appl. Phys. Lett. 47, 634 (1983).

7. J. N. Schulman and Y. C. Chang, B33, 2594 (1986) and reference
   therein.

8. J. Reno, I.K. Sou, P.S. Wijewarnasuriya and J. P. Faurie, Appl. Phys.
   Lett., 48, 1069 (1986).

9. H.L. Stomer, J.P. Eisenstein, A.C. Gossard, W. Wiegmann and K. Baldwin,
   Phys. Rev. Lett., 56, 85 (1986).

10. J.T. Cheung, G. Nizawa, J. Moyle, N.P. Ong, T. Vreeland and B.
    Paine, J. Vac. Sci. Technol., July-August 1986.

11. G. Nimtz and Schlicht, Festkorperprobleme XX, 369 (1980).

12. V.J. Goldman, H.D. Drew, M. Shayegan and D.A. Nelson, Phys. Rev.
    Lett., 56, 968 (1986).